T0321901

ELECTRONIC STRUCTURE

The study of the electronic structure of materials is at a momentous stage, with new computational methods and advances in basic theory. Many properties of materials can be determined from the fundamental equations, and electronic structure theory is now an integral part of research in physics, chemistry, materials science, and other fields. This book provides a unified exposition of the theory and methods, with emphasis on understanding each essential component.

New in the second edition are recent advances in density functional theory, an introduction to Berry phases and topological insulators explained in terms of elementary band theory, and many new examples of applications.

Graduate students and research scientists will find careful explanations with references to original papers, pertinent reviews, and accessible books. Each chapter includes a short list of the most relevant works and exercises that reveal salient points and challenge the reader.

RICHARD M. MARTIN is Emeritus Professor of Physics at the University of Illinois Urbana–Champaign and Adjunct Professor of Applied Physics at Stanford University. He has made important contributions to many areas of modern electronic structure, including more than 200 papers, and is a coauthor of another major book in the field, *Interacting Electrons: Theory and Computational Approaches* (Cambridge University Press, 2016).

Electronic Structure
Basic Theory and Practical Methods

Richard M. Martin

University of Illinois Urbana–Champaign

CAMBRIDGE
UNIVERSITY PRESS

CAMBRIDGE
UNIVERSITY PRESS

University Printing House, Cambridge CB2 8BS, United Kingdom

One Liberty Plaza, 20th Floor, New York, NY 10006, USA

477 Williamstown Road, Port Melbourne, VIC 3207, Australia

314–321, 3rd Floor, Plot 3, Splendor Forum, Jasola District Centre, New Delhi – 110025, India

79 Anson Road, #06–04/06, Singapore 079906

Cambridge University Press is part of the University of Cambridge.

It furthers the University's mission by disseminating knowledge in the pursuit of education, learning, and research at the highest international levels of excellence.

www.cambridge.org
Information on this title: www.cambridge.org/9781108429900
DOI: 10.1017/9781108555586

First published 2020

A catalogue record for this publication is available from the British Library.

ISBN 978-1-108-42990-0 Hardback

To Beverly

Contents

Part II Density Functional Theory

Part V   From Electronic Structure to Properties of Matter

Part VI Electronic Structure and Topology

Preface

The preface to the first edition begins with "The field of electronic structure is at a momentous stage, with rapid advances . . .". Indeed, that is exactly what has happened in the 16 years since the book was first published – and it is even more true today. There have been so many developments that the field is very different now. More and more properties can now be determined accurately. Electronic structure is even more an essential part of research in condensed matter physics and materials science, and hence is even more important to understand the basic theory.

Even the definition of electronic structure has changed. Not only does it include bands, total energy, forces, and so forth, it also includes the global topology of the electronic system. The new Part VI in this edition is devoted to the topology of electronic bands and the resulting edge and surface bands. The role of spin and spin–orbit interaction is now a much more important aspect of electronic structure. In the first edition, simple models were used primarily as instructive examples. In this edition there is more extensive use of models. Even though quantitative calculations are needed to establish the topology of the electronic structure of any particular system, the essential topological properties can be formulated in terms of a two-band model and a 2×2 hamiltonian.

Density functional theory has changed drastically because of developments of new, improved functionals, and it is now standard to calculate many properties much more accurately than before. It is, thus, very important to understand the conceptual foundations of the new functionals. In this edition the functionals are described in greater depth. Chapter 8 treats the local density and generalized-gradient approximations, which are mainly unchanged. Chapter 9 is devoted to generalized Kohn–Sham theory that includes hybrid functionals and other wavefunction-dependent functionals, which are background for van der Waals functionals. Many of the basic ideas were already established and discussed in the first edition, but the functionals have been developed much more extensively and are essential in much of modern density functional theory.

The power of computation has increased dramatically. Nowadays it is a given that any of the methods should yield the same result if done carefully. It is still important to understand whichever method is being used and how to use the method carefully, but it is not so important to demonstrate that different methods give the same result. It is much more important to understand what can be calculated accurately using different functionals, no matter which method is used. There are ongoing efforts to create databases and use machine learning, etc. to create materials by design; for present purposes, the essential point is that the theoretical methods in this book are the foundation upon which those efforts stand.

The companion book for this volume [1] has now been published, and this book is enriched by references to that volume for extensive discussion of many-body effects. The central tenet of both volumes is that electronic structure is at its heart a many-body problem of interacting electrons, which is among the major problems in theoretical physics. Both volumes stress that independent-particle methods are very important and useful, and it is essential to understand why and what they capture of the essential physics.

It is an even more humbling experience than before to attempt to bring together the vast range of excellent new work, together with older work that is still relevant. Many ideas and examples are omitted (or given short shrift) due to lack of space, and others are not covered because of the speed of progress in the field. I am grateful that the preface affords the opportunity to apologize to the community for omissions of excellent work of which I am not aware or that I could not weave into a coherent picture of electronic structure.

Outline

Part I is introductory material in the first five chapters. Chapter 1 provides historical background and early developments that are the foundations for more recent developments. Chapter 2 is placed at this point in the book to emphasize properties of materials and modern understanding in terms of the electronic structure, with almost no equations. It is the longest chapter of the book, but it contains only a few examples that are chosen to illustrate the goals of electronic structure theory and some of the achievements of the past few decades. Chapters 3–5 present background theoretical material: Chapter 3 summarizes basic expressions in quantum mechanics needed later; Chapter 4 provides the formal basis for the properties of crystals and establishes the notation needed in the following chapters; and Chapter 5 is devoted to homogeneous electron gas, the idealized system that sets the stage for the electronic structure of condensed matter.

Part II, Chapters 6–9, is devoted to density functional theory, upon which is based much of the present-day work in the theory of electronic structure. Chapter 6 presents the existence proofs of Hohenberg and Kohn and the constrained search formulation of Levy and Lieb. Chapter 7 describes the Kohn–Sham approach for calculation of ground-state properties of the full many-body system using practical independent-particle equations. Chapters 8 and 9 are devoted to exchange–correlation functionals that are at the heart of the method; the purpose is not to catalog functionals but to provide the understanding that is needed to appreciate the underlying theory, limitations, and avenues for possible improvements.

Part III, Chapters 10 and 11, addresses the solution of mean-field Hartree–Fock and Kohn–Sham equations in the spherical geometry of an atom, and the generation of pseudopotentials. Atomic calculations illustrate the theory and are used directly as essential parts of the methods described later. Pseudopotentials are widely used in actual calculations on real materials, and, in addition, their derivation brings out beautiful theoretical issues.

Part IV is devoted to the core methods for solution of independent-particle equations in solids, which fall into categories: plane waves and grids (Chapters 12–13); localized atom-centered functions (Chapters 14–15); and augmented methods (Chapters 16–17). The goal is to describe the methods in enough detail to show key ideas, their relationships, and relative advantages in various cases. Many noteworthy aspects are placed in appendices. The last chapter, Chapter 18, is devoted to linear-scaling approaches that bring out theoretical issues and hold promise for large-scale simulations.

Part V, Chapters 19–24, describes important ways in which the basic theory and methods are used to address varied properties of materials. The "Car–Parrinello" method (Chapter 19) has revolutionized the field of electronic structure, making possible *ab initio* calculations on previously intractable problems such as solids at finite temperature, liquids, molecular reactions in solvents, etc. Chapter 20 describes developments in the understanding and use of response functions that provide methods for computing phonon dispersion curves and related phenomena. Time-dependent density functional theory in Chapter 21 is the basis for calculations of optical (and other) excitation spectra. Chapter 22 is new in this edition; it is devoted to surfaces, interfaces, and lower dimensional systems, which have added importance owing to the development of new material systems and the advent of topological insulators. New developments in the understanding and use of Wannier functions (Chapter 23) and the theory of polarization in solids (Chapter 24) have led to new understanding of issues resolved only in the last decades and that lead up to the final part.

Part VI, Chapters 25–28, is entirely new in this edition. The premise for the entire presentation is that topological insulators and related phenomena can be understood with only knowledge of bands at the level of an undergraduate textbook and the principle of superposition on quantum mechanics. Chapter 25 is an introduction to aspects of electronic structure that can be classified by very simple topological classification. Chapter 26 sets up the two-band formulation that is sufficient for topological topological insulators, and presents one-dimensional examples that progress from the Shockley transition to Thouless quantized transport in an insulator, which, along with the quantum Hall effect, are the first examples of topological classification in electronic systems. Chapters 27 and 28 are the culmination of this part of the book: topological insulators in two and three dimensions, the Z_2 classification, and examples with models and theory and experiments on real materials. Chapter 28 also contains a short introduction to Weyl semimetals and their fascinating properties.

The appendices are devoted to topics that are too detailed to include in the main text and to subjects from different fields that have an important role in electronic structure. New appendices not in the previous edition provide the background for the Dirac equation and spin–orbit interaction (Appendix O); Berry phases and Chern numbers (Appendix P); and the quantum Hall effect (Appendix Q), which is a precedent for topological insulators. Perhaps most important to emphasize is that Berry phases and the relevant aspects of Chern numbers can be understood in only a few pages starting from the superposition principle, as described in Appendix P.

Preface to First Edition

The field of electronic structure is at a momentous stage, with rapid advances in basic theory, new algorithms, and computational methods. It is now feasible to determine many properties of materials directly from the fundamental equations for the electrons and to provide new insights into vital problems in physics, chemistry, and materials science. Increasingly, electronic structure calculations are becoming tools used by both experimentalists and theorists to understand characteristic properties of matter and to make specific predictions for real materials and experimentally observable phenomena. There is a need for coherent, instructive material that provides an introduction to the field and a resource describing the conceptual structure, the capabilities of the methods, limitations of current approaches, and challenges for the future.

The purpose of this book and a second volume in progress is to provide a unified exposition of the basic theory and methods of electronic structure, together with instructive examples of practical computational methods and actual applications. The aim is to serve graduate students and scientists involved in research, to provide a text for courses on electronic structure, and to serve as supplementary material for courses on condensed matter physics and materials science. Many references are provided to original papers, pertinent reviews, and books that are widely available. Problems are included in each chapter to bring out salient points and to challenge the reader.

The printed material is complemented by expanded information available on-line at a site [ElectronicStructure.org maintained by the Electronic Structure Group at Ecole Polytechnique]. There one can find codes for widely used algorithms, more complete descriptions of many methods, and links to the increasing number of sites around the world providing codes and information. The on-line material is coordinated with descriptions in this book and will contain future updates, corrections, additions, and convenient feedback forms.

The content of this work is determined by the conviction that "electronic structure" should be placed in the context of fundamental issues in physics, while at the same time emphasizing its role in providing useful information and understanding of the properties of materials. At its heart, electronic structure is an interacting many-body problem that ranks among the most pervasive and important in physics. Furthermore, these are problems that must be solved with great accuracy in a vast array of situations to address issues relevant to materials. Indeed, many-body methods, such as quantum Monte Carlo and many-body perturbation theory, are an increasing part of electronic structure theory for realistic problems. These methods are the subject of the second volume.

The subjects of this volume are fundamental ideas and the most useful approaches at present are based upon independent-particle approximations. *These methods address directly and quantitatively the full many-body problem because of the ingenious formulation of density functional theory and the Kohn–Sham auxiliary system.* This approach provides a way to approach the many-body problem, whereby certain properties can be calculated, in principle exactly, and in practice very accurately for many materials using feasible approximations and independent-particle methods. This volume is devoted to independent-particle methods, with emphasis on their usefulness and their limitations when applied to real problems of electrons in materials. In addition, these methods provide the starting point for much of the work described in the planned second volume. Indeed, new ideas that build upon the construction of an auxiliary system and actual independent-particle calculations are critical aspects of modern many-body theory and computational methods that can provide quantitative description of important properties of condensed matter and molecular systems.

It is a humbling experience to attempt to bring together the vast range of excellent work in this field. Many relevant ideas and examples are omitted (or given short shrift) due to lack of space, and others not covered because of the speed of progress in the field. Feedback on omissions, corrections, suggestions, examples, and ideas are welcome in person, by e-mail, or on-line.

Acknowledgments

In addition to the institutions listed in the first edition, I want to express my appreciation to Stanford University and the Department of Applied Physics, where much of this was written, and to the Donostia International Physics Center, where I spent an extended visit. The spirit and enthusiasm of the participants and lecturers at the African School for Electronic Structure Methods and Applications (ASESMA) continue to be an inspiration to me and the electronic structure community.

The second edition benefited greatly by discussion, advice, and help from many people, in addition to those for the first edition. Special thanks are due to Lucia Reining and David Vanderbilt for inspiration and advice in many areas, and to David, Xiaoliang Qi, Raffaele Resta, Ivo Souza, and Shoucheng Zhang for patiently helping me understand issues associated with topological insulators. It is a pleasure to acknowledge the contribution of Kiyo and Ikuko Terakura, who made a Japanese version that is more than a translation; it is a critical analysis that found errors that I have tried to incorporate in the new edition. Heartfelt thanks to Beverly Martin, Hannah Martin, and Megan Martin, for reading the manuscript. And also great thanks for discussions, advice, and help to Janos Asboth, Jefferson Bates, David Bowler, David Ceperley, Yulin Chen, Alfredo Correa, Michael Crommie, Defang Duan, Claudia Felzer, Marc Gabay, Alex Gaiduk, Giulia Galli, Vikram Gavini, Don Hamman, Javier Junqera, Aditi Krishnapriyan, Steve Louie, Stephan Mohr, Joachim Paier, Dimitrios Papaconstantopoulos, John Pask, Das Pemmaraju, John Perdew, Eric Pop, Sivan Refaely-Abramson, Daniel Rizzo, Dario Rocca, Biswajit Santra, Matthias Scheffler, Chandra Shahi, Juan Shallcrass, Z. X. Shen, David Singh, Yuri Suzuki, Chris Van de Walle, Sam Vaziri, Johannes Voss, Maria Vozmediano, Renata Wentzcovitch, Binghai Yan, Rui Yu, and Haijun Zhang.

From the First Edition

Four people and four institutions have played the greatest role in shaping the author and this work: the University of Chicago and my advisor Morrel H. Cohen, who planted the ideas and set the level for aspirations; Bell Labs, where the theory group and interactions with experimentalists provided diversity and demanded excellence; Xerox Palo Alto Research Center (PARC), in particular, my stimulating collaborator J. W. (Jim) Allen and my second mentor W. Conyers Herring; and the University of Illinois at Urbana–Champaign, especially my close collaborator David M. Ceperley. I am indebted to the excellent colleagues and students in the Department of Physics, the Frederick Seitz Materials Research Laboratory, and the Beckman Institute.

The actual writing of this book started at the Max Planck Institut für Festkorperforschung in Stuttgart, partially funded by the Alexander von Humboldt Foundation, and continued at the University of Illinois, the Aspen Center for Physics, Lawrence Livermore National Laboratory and Stanford University. Their support is greatly appreciated.

Funding from the National Science Foundation, the Department of Energy, the Office of Naval Research, and the Army Research Office during the writing of this book is gratefully acknowledged.

Appreciation is due to countless people who cannot all be named. Many colleagues who provided figures are specifically acknowledged in the text. Special thanks are due to David Drabold, Beverly Martin, and Richard Needs for many comments and criticisms on the entire volume. Others who contributed directly in clarifying the arguments presented here, correcting errors, and critical reading of the manuscript are: V. Akkinseni, O. K. Andersen, V. P. Antropov, E. Artacho, S. Baroni, P. Blöchl, M. Boero, J. Chelikowsky, X. Cheng, T. Chiang, S. Chiesa, M. A. Crocker, D. Das, K. Delaney, C. Elliott, G. Galli, O. E. Gunnarsson, D. R. Hamann, V. Heine, L. Hoddeson, V. Hudson, D. D. Johnson, J. Junquera, J. Kim, Y.-H. Kim, E. Koch, J. Kübler, K. Kunc, B. Lee, X. Luo, T. Martinez, J. L. Martins, N. Marzari, W. D. Mattson, I. I. Mazin, A. K. McMahan, V. Natoli, O. H. Nielsen, J. E. Northrup, P. Ordejon, J. Perdew, W. E. Pickett, G. Qian, N. Romero, D. Sanchez-Portal, S. Satpathy, S. Savrosov, E. Schwegler, G. Scuseria, E. L. Shirley, L. Shulenburger, J. Soler, I. Souza, V. Tota, N. Trivedi, A. Tsolakidis, D. H. Vanderbilt, C. G. Van de Walle, M. van Schilfgaarde, I. Vasiliev, J. Vincent, T. J. Wilkens. For corrections in 2008 I am indebted to K. Belashchenko, E. K. U. Gross, I. Souza, A. Torralba, and J.-X. Zhu.

Notation

Abbreviations

BZ	first Brillouin zone
wrt	with respect to
+c.c.	denotes adding the complex conjugate of the preceding quantity

General Physical Quantities

E	energy
Ω	volume (to avoid confusion with V used for potential)
$P = -(dE/d\Omega)$	pressure
$B = \Omega(d^2E/d\Omega^2)$	bulk modulus (inverse of compressibility)
$H = E + P\Omega$	enthalpy
$u_{\alpha\beta}$	strain tensor (symmetrized form of $\epsilon_{\alpha\beta}$)
$\sigma_{\alpha\beta} = -(1/\Omega)(\partial E/\partial u_{\alpha\beta})$	stress tensor (note the sign convention)
$\mathbf{F}_I = -(dE/d\mathbf{R}_I)$	force on nucleus I
$C_{IJ} = d^2E/d\mathbf{R}_I d\mathbf{R}_J$	force constant matrix
$n(\mathbf{r})$	density of electrons
$t(\mathbf{r})$	kinetic energy density $t(\mathbf{r}) = n(\mathbf{r})\tau(\mathbf{r})$

Notation for Crystals

Ω_{cell}	volume of primitive cell
$\mathbf{a}_i$	primitive translation vectors
$\mathbf{T}$ or $\mathbf{T}(\mathbf{n})$	lattice translations
	$\mathbf{T}(\mathbf{n}) \equiv \mathbf{T}(n_1, n_2, n_3) = n_1\mathbf{a}_1 + n_2\mathbf{a}_2 + n_3\mathbf{a}_3$
$\tau_s, s = 1, \ldots, S$	positions of atoms in the basis
$\mathbf{b}_i$	primitive vectors of reciprocal lattice
$\mathbf{G}$ or $\mathbf{G}(\mathbf{m})$	reciprocal lattice vectors
	$\mathbf{G}(\mathbf{m}) \equiv \mathbf{G}(m_1, m_2, m_3) = m_1\mathbf{b}_1 + m_2\mathbf{b}_2 + m_3\mathbf{b}_3$
$\mathbf{k}$	wavevector in first Brillouin zone (BZ)
$\mathbf{q}$	general wavevector ($\mathbf{q} = \mathbf{k} + \mathbf{G}$)

Hamiltonian and Eigenstates

$\hat{H}$ — hamiltonian for either many particles or a single particle

$\Psi(\{\mathbf{r}_i\})$ — many-body wavefunction of a set of particle positions $\mathbf{r}_i$, $i = 1, N_{\text{particle}}$; spin is assumed to be included in the argument $\mathbf{r}_i$ unless otherwise specified

E_i — energy of many-body state

$\Phi(\{\mathbf{r}_i\})$ — single determinant uncorrelated wavefunction

$H_{m,m'}$ — matrix element of independent-particle hamiltonian

$S_{m,m'}$ — overlap matrix elements of states m and m'

$\psi_i(\mathbf{r})$ — independent-particle wavefunction or "orbital," $i = 1, \ldots, N_{\text{states}}$

ε_i — independent-particle eigenvalue, $i = 1, \ldots, N_{\text{states}}$

$f_i = f(\varepsilon_i)$ — occupation of state i where f is the Fermi function

$\psi_i^{\sigma}(\mathbf{r}), \varepsilon_i^{\sigma}$ — used when spin is explicitly indicated

$\psi_l(r)$ — single-body radial wavefunction $(\psi_{l,m}(\mathbf{r}) = \psi_l(r)Y_{lm}(\theta, \phi))$

$\phi_l(r)$ — single-body radial wavefunction $\phi_l(r) = r\psi_l(r)$

$\eta_l(\varepsilon)$ — phase shift

$\psi_{i,\mathbf{k}}(\mathbf{r}) = e^{i\mathbf{k}\cdot\mathbf{r}}u_{i,\mathbf{k}}(\mathbf{r})$ — Bloch function in crystal, with $u_{i,\mathbf{k}}(\mathbf{r})$ periodic

$\varepsilon_{i,\mathbf{k}}$ — eigenvalues that define bands as a function of $\mathbf{k}$

$\hat{H}(\mathbf{k})$ — "gauge-transformed" hamiltonian given by Eq. (4.37) with eigenvectors $u_{i,\mathbf{k}}(\mathbf{r})$

$w_{i\mathbf{n}}(\mathbf{r})$ — Wannier function for band i and cell $\mathbf{n}$

$w_{in,\mathbf{k}_\perp}(\mathbf{r})$ — hybrid Wannier function for band i and cell n in one direction and momentum $\mathbf{k}_\perp$ in the other directions

$\tilde{w}_{i\mathbf{n}}(\mathbf{r})$ — nonorthogonal transformation of Wannier functions

$\chi_\alpha(\mathbf{r})$ — single-body basis function, $\alpha = 1, \ldots, N_{\text{basis}}$. $\psi_i(\mathbf{r}) = \sum_\alpha c_{i\alpha}\chi_\alpha(\mathbf{r})$

$\chi_\alpha(\mathbf{r} - (\tau + \mathbf{T}))$ — ocalized orbital basis function on atom at position τ in cell labeled by translation vector $\mathbf{T}$

$\chi^{\text{OPW}}(\mathbf{r}), \chi^{\text{APW}}(\mathbf{r}), \chi^{\text{LMTO}}(\mathbf{r})$ — basis function for orthogonalized, augmented, or muffin-tin orbital basis functions

Spin and Spin–Orbit Interaction

σ_i or τ_i — Pauli matrices

H_{SO} — hamiltonian for spin–orbit interaction (Appendix O)

ζ — matrix element of H_{SO} for atomic-like orbitals

Density Functional Theory

$F[f]$	general notational for F a functional of the function f
$E_{xc}[n]$	exchange–correlation energy in Kohn–Sham theory
$\epsilon_{xc}(\mathbf{r})$	exchange–correlation energy per electron
$V_{xc}(\mathbf{r})$	exchange–correlation potential in Kohn–Sham theory
$V_{xc}^{\sigma}(\mathbf{r})$	exchange–correlation potential for spin σ
$f_{xc}(\mathbf{r},\mathbf{r}')$	response $\delta^2 E_{xc}/\delta n(\mathbf{r})\delta n(\mathbf{r}')$

Response Functions and Correlation Functions

$\chi(\omega)$	general response function
$\chi_0(\omega)$	general response function for independent particles
$K(\omega)$	kernel in self-consistent response function $\chi^{-1} = [\chi_0]^{-1} - K$
$\epsilon(\omega)$	frequency-dependent dielectric function
$n(\mathbf{r},\sigma;\mathbf{r}',\sigma')$	pair distribution
$g(\mathbf{r},\sigma;\mathbf{r}',\sigma')$	normalized pair distribution (often omitting the spin indices)
$G(z,\mathbf{r},\mathbf{r}')$ or $G_{m,m'}(z)$	Green's function of complex frequency z
$\rho(\mathbf{r},\sigma;\mathbf{r}',\sigma')$	density matrix
$\rho_\sigma(\mathbf{r},\mathbf{r}')$	density matrix diagonal in spin for independent particles

Berry Phases and Topological Insulators

ϕ	Berry phase
$\mathcal{A}_\alpha(\lambda)$	Berry connection as function of a parameter λ
$\Omega_{\alpha\beta}(\lambda)$	Berry curvature
$\Omega = \Omega_{xy} = -\Omega_{yx}$	Berry curvature in two dimensions
Φ	Berry flux
C	Chern number for closed surface
Z	set of integers, topology class of quantum Hall systems
Z_2	set of integers modulo 2, equivalent to even and odd, topology class of topological insulators with time-reversal symmetry

PART I

OVERVIEW AND BACKGROUND TOPICS

1

Introduction

Without physics there is no life.

Taxi driver in Minneapolis

To love practice without theory is like the sailor who boards ship without rudder and
compass and is forever uncertain where he may cast.

Leonardo da Vinci, notebook 1, *c.* 1490

Summary

Since the discovery of the electron in 1896–1897, the theory of electrons in matter
has ranked among the greatest challenges of theoretical physics. The fundamental
basis for understanding materials and phenomena ultimately rests upon understanding electronic structure, which means that we must deal with the interacting many-electron problem in diverse, realistic situations. This chapter provides
a brief outline with original references to early developments of electronic structure
and the pioneering quantitative works that foreshadowed many of the methods in
use today.

Electrons and nuclei are the fundamental particles that determine the nature of matter in
our everyday world: atoms, molecules, condensed matter, and man-made structures. Not
only do electrons form the "quantum glue" that holds together the nuclei in solid, liquid,
and molecular states, but also electron excitations determine the vast array of electrical,
optical, and magnetic properties of materials. The theory of electrons in matter ranks
among the great challenges of theoretical physics: to develop theoretical approaches and
computational methods that can accurately treat the interacting system of many electrons
and nuclei found in condensed matter and molecules.

Throughout this book there are references to a companion book [1], *Interacting Electrons*
by Richard M. Martin, Lucia Reining, and David M. Ceperley (Cambridge University
Press, 2016). Together these two books are meant to cover the field of electronic structure
theory and methods, each focusing on ways to address the difficult problem of many

interacting electrons. They are independent books and references to [1] indicate where more information can be found, especially about many-body theory and methods outside the scope of this book.

1.1 Quantum Theory and the Origins of Electronic Structure

Although *electric* phenomena have been known for centuries, the story of *electronic structure* begins in the 1890s with the discovery of the electron as a particle – a fundamental constituent of matter. Of particular note, Hendrik A. Lorentz[1] modified Maxwell's theory of electromagnetism to interpret the electric and magnetic properties of matter in terms of the motion of charged particles. In 1896, Pieter Zeeman, a student of Lorentz in Leiden, discovered [3] the splitting of spectral lines by a magnetic field, which Lorentz explained with his electron theory, concluding that radiation from atoms was due to negatively charged particles with a very small mass. The discovery of the electron in experiments on ionized gases by J. J. Thomson at the Cavendish Laboratory in Cambridge in 1897 [4, 5] also led to the conclusion that the electron is negatively charged, with a charge-to-mass ratio similar to that found by Lorentz and Zeeman. For this work, the Nobel Prize was awarded to Lorentz and Zeeman in 1902 and to Thomson in 1906.

The compensating positive charge is composed of small massive nuclei, as was demonstrated by experiments in the laboratory of Rutherford at Manchester in 1911 [6]. This presented a major problem for classical physics: How can matter be stable? What prevents electrons and nuclei from collapsing due to attraction? The defining moment occurred when Niels Bohr (at the Cavendish Laboratory for postdoctoral work after finishing his dissertation in 1911) met Rutherford and moved to Manchester to work on this problem. There he made the celebrated proposal that quantum mechanics could explain the stability and observed spectra of atoms in terms of a discrete set of allowed levels for electrons [7]. The work of Planck in 1900 and one of the major works of Einstein in 1905 had established the fact that the energy of light waves is quantized. Although Bohr's model was fundamentally incorrect, it set the stage for the discovery of the laws of quantum mechanics, which emerged in 1923–1925, most notably through the work of de Broglie, Schrödinger, and Heisenberg.[2] Electrons were also the testing ground for the new quantum theory. The famous Stern–Gerlach experiments [14, 15] in 1921 on the deflection of silver atoms in a magnetic field were formulated as tests of the applicability of quantum theory to particles in a magnetic field. Simultaneously, Compton [16] proposed that the electron possesses an intrinsic moment, a "magnetic doublet." The coupling of orbital angular momentum and an intrinsic spin of 1/2 was formulated by Goudschmidt and Uhlenbeck [17], who noted the earlier hypothesis of Compton.

---

[1] The work of Lorentz and many other references can be found in a reprint volume of lectures given in 1906 [2].
[2] The development of quantum mechanics is discussed, for example, in the books by Jammer [8] and Waerden [9]. Early references and a short history are given by Messiah [10], chapter 1. Historical development of the theory of metals is presented in the reviews by Hoddeson and Baym [11, 12] and the book *Out of the Crystal Maze* [13], especially the chapter "The Development of the Quantum Mechanical Electron Theory of Metals, 1926–1933" by Hoddeson, Baym, and Eckert.

One of the triumphs of the new quantum mechanics, in 1925, was the explanation of the periodic table of elements in terms of electrons obeying the exclusion principle proposed by Pauli [18] that no two electrons can be in the same quantum state.[3] In work published early in 1926, Fermi [20] extended the consequences of the exclusion principle to the general formula for the statistics of noninteracting particles (see Eq. (1.3)) and noted the correspondence to the analogous formula for Bose–Einstein statistics [21, 22].[4] The general principle that the wavefunction for many identical particles must be either symmetric or antisymmetric when two particles are exchanged was apparently first discussed by Heisenberg [23] and, independently, by Dirac [24] in 1926.[5]

By 1928 the laws of quantum mechanics that are the basis of all modern theories of electronic structure of matter were complete. Two papers by Dirac [26, 27] brought together the principles of quantum mechanics, special relativity, and Maxwell's equations in an awesome example of creativity that Ziman characterized as "almost by sheer cerebration," i.e., by sheer mental reasoning. From this work emerged the Dirac equation for fermions that have spin 1/2 with quantized magnetic moments and spin–orbit interaction. As described in more detail in Appendix O, this has profound consequences for electronic structure, where it leads to effects that are qualitatively different from anything that can be produced by a potential. In molecules and condensed matter it is usually sufficient to include spin–orbit interaction in the usual nonrelativistic Schrödinger equation with an added term $\hat{H}_{SO}$, as given in Eq. (O.10). Spin–orbit interaction has come to the fore in condensed matter theory with the discovery of topological insulators described in Chapters 25–28.

Further progress quickly led to improved understanding of electrons in molecules and solids. The most fundamental notions of chemical bonding in molecules (the rules for which had already been formulated by Lewis [28] and others before 1920) were placed upon a firm theoretical basis by quantum mechanics in terms of the way atomic wave functions are modified as molecules are formed (see, for example, Heitler and London in 1927 [29]). The rules for the number of bonds made by atoms were provided by quantum mechanics, which allows the electrons to be delocalized on more than one atom, lowering the kinetic energy and taking advantage of the attraction of electrons to each of the nuclei.

1.2 Why Is the Independent-Electron Picture So Successful?

A section with the same title is in the introduction to the companion book [1], and is devoted to methods to deal with the correlated, interacting many-body problem. The central tenet of

---

[3] This was a time of intense activity by many people [13] and Pauli referred to earlier related work of E. C. Stoner [19].

[4] There is a striking similarity of the titles of Fermi's 1926 paper, "Zur Quantelung des Idealen Einatomigen Gases" and Einstein's 1924 paper, "Quantentheorie des Idealen Einatomigen Gases."

[5] According to [13], Heisenberg learned of the ideas of statistics from Fermi in early 1926, but Dirac's work was apparently independent. In his 1926 paper, Dirac also explicitly pointed out that the wavefunction for noninteracting electrons of a given spin (up or down) can be written as a determinant of one-electron orbitals. However, it was only in 1929 that Slater showed that the wavefunction including spin can be written as a determinant of "spin orbitals" [25].

both volumes is that electronic structure is at its heart a many-body problem of interacting electrons, which is among the major problems in theoretical physics. The full hamiltonian is given in Eq. (3.1) and it is clear that the electron–electron interactions are not at all negligible. Nevertheless, theoretical concepts and methods involving independent electrons play a central role in essentially every aspect of the theory of materials. Both volumes stress that independent-particle methods are very important and useful, and it is essential to understand why and when they capture the essential physics.

The leading scientists of the field in the 1920s were fully aware of the difficulty of treating many interacting particles and it is edifying to examine the masterful ways that independent-particle theories were used to bring out fundamental physics. Such a system can be described by the eigenstates ψ_1 of an effective one-particle hamiltonian

$$H_{\text{eff}}\psi_i(\mathbf{r}) = \varepsilon_i \psi_i(\mathbf{r}), \tag{1.1}$$

with

$$H_{\text{eff}} = -\frac{\hbar^2}{2m_e}\nabla^2 + V_{\text{eff}}(\mathbf{r}), \tag{1.2}$$

where $V_{\text{eff}}(\mathbf{r})$ includes potentials acting on the electrons, e.g., due to the nuclei, and an effective field that takes into account at least some average aspects of interactions. The state of the system is specified by the occupation numbers f_i, which in thermal equilibrium are given by

$$f_i = \frac{1}{e^{\beta(\epsilon_i - \mu)} \pm 1}, \tag{1.3}$$

where the minus sign is for Bose–Einstein [21, 22] and the plus sign is for Fermi–Dirac statistics [20, 24]. Among the first accomplishments of the new quantum theory was the realization in 1926–1928 by Wolfgang Pauli and Arnold Sommerfeld [30, 31] that it resolved the major problems of the classical Drude–Lorentz theory.[6] The first step was the paper [30] of Pauli, submitted late in 1926, in which he showed that weak paramagnetism is explained by spin polarization of electrons obeying Fermi–Dirac statistics. At zero temperature and no magnetic field, the electrons are spin paired and fill the lowest energy states up to a Fermi energy, leaving empty the states above this energy. For temperature or magnetic field nonzero, but low compared to the characteristic electronic energies, only the electron states near the Fermi energy are able to participate in electrical conduction, heat capacity, paramagnetism, and other phenomena.[7] Pauli and Sommerfeld based their

---

[6] Simultaneous to Lorentz's development [2] of the theory of electric and magnetic properties of matter in terms of the motion of charged particles, Paul K. L. Drude developed a theory of optical properties of matter [32, 33] in a more phenomenological manner in terms of the motion of particles. Their work formed the basis of the purely classical theory that remains highly successful today, reinterpreted in the light of quantum mechanics.

[7] Sommerfeld learned of the ideas from Pauli in early 1927 and the development of the theory was the main subject of Sommerfeld's research seminars in Munich during 1927, which included participants such as Bethe, Eckhart, Houston, Pauling, and Peierls [11]. Both Pauli and Heisenberg were students of Sommerfeld, who went on to found the active centers of research in quantum theory, respectively in Zurich and Leipzig. The three centers were the hotbeds of activity in quantum theory, with visitors at Leipzig such as Slater, Peierls, and Wilson.

successful theory of metals on the model of a homogeneous free-electron gas, which resolved the major mysteries that beset the Drude–Lorentz theory. However, at the time it was not clear what would be the consequences of including the nuclei and crystal structure in the theory, both of which would be expected to perturb the electrons strongly.

Band Theory for Independent Electrons

The critical next step toward understanding electrons in crystals was the realization of the nature of independent noninteracting electrons in a periodic potential. This was elucidated most clearly[8] in the thesis of Felix Bloch, the first student of Heisenberg in Leipzig. Bloch [35] formulated the concept of electron bands in crystals and what has come to be known as the "Bloch theorem" (see Chapters 4 and 12), i.e., that the wavefunction in a perfect crystal is an eigenstate of the "crystal momentum." This resolved one of the key problems in the Pauli–Sommerfeld theory of conductivity of metals: Electrons can move freely through the perfect lattice, scattered only by imperfections and displacements of the atoms due to thermal vibrations.

It was only later, however, that the full consequences of band theory were formulated. Based on band theory and the Pauli exclusion principle, the allowed states for each spin can each hold one electron per unit cell of the crystal. Rudolf Peierls, in Heisenberg's group at Leipzig, recognized the importance of filled bands and "holes" (i.e., missing electrons in otherwise filled bands) in the explanation of the Hall effect and other properties of metals [36, 37]. However, it was only with the work of A. H. Wilson [38, 39], also at Leipzig in the 1930s, that the foundation was laid for the classification of all crystals into metals, semiconductors, and insulators.[9]

Development of the bands, as the atoms are brought together, is illustrated in Fig. 1.1, which is based on a well-known figure by G. E. Kimball in 1935 [41]. Kimball considered diamond-structure crystals, which were difficult to study at the time because the electron states change qualitatively from those in the atom. In his words:

> Although not much of a quantitative nature can be concluded from these results, the essential differences between diamond and the metals are apparent.
>
> G. E. Kimball in [41].

Figure 1.1 also shows a surface state that emerges at the transition between the two types of insulating gaps. This was derived in a prescient paper in 1939 by Shockley[10]

---

[8] Closely related work was done simultaneously in the thesis research of Hans Bethe [34] in 1928 (student of Sommerfeld in Munich), who studied the scattering of electrons from the periodic array of atoms in a crystal.

[9] In his classic book, Seitz [40] further divided insulators into ionic, valence, and molecular. It is impressive to realize that Seitz was only 29 years old when his 698-page book was published.

[10] This is the same Shockley who shared the Nobel prize, with Bardeen and Brattain, for development of the transistor. His work on surfaces was no idle exercise: the first transistors worked by controlling the conductivity in the surface region and the properties of surface states was crucial. Shockley was born in Palo Alto, California, and it is said that his mother is the one who put the silicon in Silicon Valley because Shockley moved back to Palo Alto from Bell Labs in New Jersey to take care of her and to start Shockley Semiconductor Laboratory. Sadly, his later years were sullied by his espousal of eugenics.

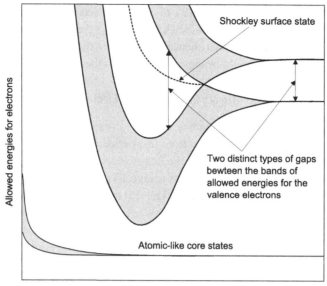

Figure 1.1. Schematic illustration taken from the work of G. E. Kimball in 1935 [41] of the evolution from discrete atomic energies to bands of allowed states separated by energy gaps as atoms are brought together in the diamond structure. One aspect is independent of the crystal structure: the division of solids into insulators, where the bands are filled with a gap to the empty states, and metals, where the bands are partially filled with no gap. However, there is additional information for specific cases: Kimball pointed out that there are two distinct types of energy gaps illustrated by the transition from atomic-like states to covalent bonding states and, in this example, the transition occurs at a point where the gap vanishes. There is a similar figure in the 1939 paper by Shockley [42], who realized that transition in the nature of the states in the crystal also leads to a surface state in the gap, which is a forerunner of topological insulators, one of the major recent discoveries in condensed matter physics this century, which is the topic of Chapters 25–28.

[42] who used a simple one-dimensional model to elucidate the nature of the transition in the bulk band structure and the conditions for which there is a surface state in the gap. Shockley had previously carried out one of the first realistic band structure calculations for NaCl [43] using a cellular method, but covalent semiconductors are very different. This is a story of choosing an elegantly simple model that captures the essence of the problem and an elegantly simple method, the same cellular method he had used before but now applied to a covalent crystal with surfaces. Shockley's analysis is explained in some detail in Chapters 22 and 26 because it is a forerunner of topological insulators.

Classification of Materials by Electron Counting

The classification of materials is based on the filling of the bands illustrated in Fig. 1.1, which depends on the number of electrons per cell:

- Each band consists of one state per cell for each spin.

- Insulators have filled bands with a large energy gap of forbidden energies separating the ground state from all excited states of the electrons.
- Semiconductors have only a small gap, so thermal energies are sufficient to excite the electrons to a degree that allows important conduction phenomena.
- Metals have partially filled bands with no excitation gaps, so electrons can conduct electricity at zero temperature.

Relation to the Full Interacting-Electron Problem

The independent-particle analysis is a vast simplification of the full many-body problem; however, certain properties carry over to the full problem. The fundamental guiding principle is *continuity* if one imagines continuously turning on interactions to go from independent particles to the actual system. It was only in the 1950s and 1960s that the principles were codified in the form used today by Landau and others.[11] A concise form now called the "Luttinger theorem" [46, 47] states that the volume enclosed by the Fermi surface does not change so long as there are no phase transitions. This is not a theorem in the mathematical sense. It is an argument formulated with careful reasoning, which makes it clear that an exception indicates a transition to a new state of matter, for example, superconductivity. So long as there is no transition to some other state, a metal with partially filled bands in the independent-particle approximation should remain a metal in the full problem. An insulator with filled bands must either remain an insulator with a gap or change into a semimetal with Fermi surfaces with equal numbers of electrons and holes. The issues are discussed much more extensively in [1] but arguments like this capture the basic concepts.

The emergence of topology as a property of the electronic structure provides a new angle on this old problem. Essentially all work on what are called topological insulators is for independent electrons, and the principle of continuity can be used in a different way to invoke the principle that the topology can change only if a gap goes to zero.

1.3 Emergence of Quantitative Calculations

The first quantitative calculations undertaken on multielectron systems were for atoms, most notably by D. R. Hartree[12] [50] and Hylleraas [51, 52]. Hartree's work pioneered the self-consistent field method, in which one solves the equation numerically for each electron moving in a central potential due to the nucleus and other electrons, and set the stage for many of the numerical methods still in use today. However, the approach was somewhat

---

[11] Reprints of many original papers can be found in the book by Pines [44], and theory is presented in the text known by the initials "AGD" [45], which refers to Luttinger for the original reference for the "theorem."

[12] D. R. Hartree was aided by his father, W. R. Hartree, a businessman with an interest in mathematics who carried out calculations on a desk calculator [48]. Together they published numerous calculations on atoms. D. R. went on to become one of the pioneers of computer science and the use of electronic computers, and he published a book on the calculation of electronic structure of atoms [49].

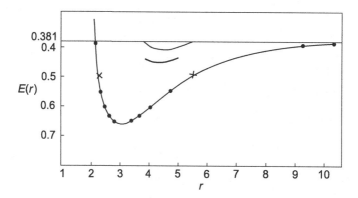

Figure 1.2. Energy versus radius of Wigner–Seitz sphere for Na calculated by Wigner and Seitz [54, 57]. Bottom curve: energy of lowest electron state calculated by the cellular method. Middle curve: total energy, including an estimate of the additional kinetic energy from the homogeneous electron gas as given in Tab. 5.3. From [54]. The upper curve is taken from the later paper [57] and includes an estimate of the correlation energy. The final result is in remarkable agreement with experiment.

heuristic, and it was in 1930 that Fock [53] published the first calculations using properly antisymmetrized determinant wave functions, the first example of what is now known as the Hartree–Fock method. Many of the approaches used today in perturbation theory and response functions (e.g., Section D.1 and Chapter 20) originated in the work of Hylleraas, which provided accurate solutions for the ground state of two-electron systems as early as 1930 [52].

The 1930s witnessed the initial formulations of most of the major theoretical methods for electronic structure of solids still in use today.[13] Among the first quantitative calculations of electronic states was the work on Na metal by Wigner and Seitz [54, 57] published in 1933 and 1934. They used the cellular method, a forerunner of the atomic sphere approximation, which allows the needed calculations to be done in atomic-like spherical geometry. Even with that simplification, the effort required at the time can be gleaned from their description:

> The calculation of a wavefunction took about two afternoons, and five wavefunctions were calculated on the whole, giving ten points of the figure.
>
> Wigner and Seitz [54].

The original figure from [54], reproduced in Fig. 1.2, shows the energy of the lowest electronic state (lower curve) and the total energy of the crystal including the kinetic energy (middle curve). The upper curve [57] includes an estimate of the correlation energy; the result is in remarkable agreement with experiment.

The electron energy bands in Na were calculated in 1934 by Slater [58] and Wigner and Seitz [57], each using the cellular method. The results of Slater are shown in Fig. 1.3;

---

[13] The status of band theory in the early 1930s can be found in the reviews by Sommerfeld and Bethe [55] and Slater [56] and in the book *Out of the Crystal Maze* [13], especially the chapter "The Development of the Band Theory of Solids, 1933–1960," by P. Hoch.

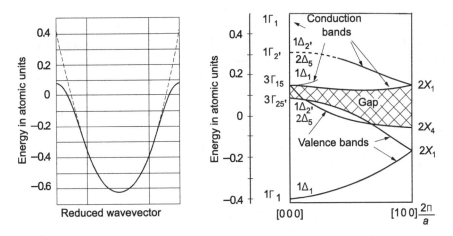

Figure 1.3. Left: Energy bands in Na calculated in 1934 by Slater [58] using the cellular method of Wigner and Seitz [54]. The bands clearly demonstrate the nearly-free-electron character, even though the wavefunction has atomic character near each nucleus. Right: Energy bands in Ge calculated by Herman and Callaway [59] in 1953, one of the first computer calculations of a band structure. The lowest gap in this direction is in reasonable agreement with experiment and more recent work shown in Fig. 2.23, but the order of the conduction bands at Γ is not correct.

very similar bands were found by Wigner and Seitz. Although the wavefunction has atomic character near each nucleus, nevertheless the bands are very free-electron-like, a result that has formed the basis of much of our understanding of sp-bonded metals. Many calculations were done in the 1930s and 1940s for high-symmetry metals (e.g., copper bands calculated by Krutter [60]) and ionic solids (e.g., NaCl studied by Shockley [43]) using the cellular method.

The difficulty in a general solid is to deal accurately with the electrons both near the nucleus and in the smoother bonding regions. Augmented plane waves (Chapter 16), pioneered by Slater [61] in 1937 and developed[14] in the 1950s [62, 63], accomplish this with different basis sets that are matched at the boundaries. Orthogonalized plane waves (Chapter 11) were originated by Herring [64] in 1940 to take into account effects of the cores on valence electrons. Effective potentials (forerunners of pseudopotentials; see Chapter 11) were introduced in diverse fields of physics, e.g., by Fermi [65] in 1934 to describe scattering of electrons from atoms and neutrons from nuclei (see Fig. 11.1). Perhaps the original application to solids was by H. Hellmann [66, 67] in 1935–1936, who developed a theory for valence electrons in metals remarkably like a modern pseudopotential calculation. Although quantitative calculations were not feasible for general classes of solids, the development of the concepts – together with experimental studies – led to many important developments, most notably the transistor.[15]

---

[14] Apparently, the first published use of the term "augmented plane waves" was in the 1953 paper by Slater [62].

[15] Shockley's work is noted on page 5. One of the other inventors of the transistor, John Bardeen, was a student of Wigner; his thesis work on electronic structure of metals is quoted in Chapter 5.

The first quantitatively accurate calculations of bands in difficult cases like semiconductors, where the electronic states are completely changed from atomic states, were done in the early 1950s.[16] For example, Fig. 1.3 shows the bands of Ge calculated by Herman and Callaway [59] using the orthogonalized plane wave (OPW) method (Section 11.2) with a potential assumed to be a sum of atomic potentials. They pointed out that their gap was larger than the experimental value. It turns out that this is correct: the gap in the direction studied is larger than the lowest gap, which is in a different direction in the Brillouin zone – harder to calculate at the time. Comparison with recent calculations, e.g., in Fig. 2.23, shows that the results for the valence band and the gap are basically correct but there are discrepancies in the conduction bands.

1.4 The Greatest Challenge: Electron Interaction and Correlation

Even though independent-particle theory was extremely successful in many ways, the great question was: What are the consequences of electron–electron interactions? One of the most important effects of this interaction was established early in the history of electronic structure: the underlying cause of magnetism was identified by Heisenberg [70] and Dirac [71] in terms of the exchange energy of interacting electrons, which depends on the spin state and the fact that the wavefunction must change sign when two electrons are exchanged.[17] In atomic physics and chemistry, it was quickly realized that accurate descriptions must go beyond the effective independent-electron approximations because of strong correlations in localized systems and characteristic bonds in molecules [75].

In condensed matter, the great issues associated with electron–electron interactions were posed succinctly in terms of metal–insulator transitions described by Eugene Wigner [76] and Sir Nevill Mott [77–79]. A watershed point was a conference in 1937 where J. H. de Boer and E. J. W Verwey presented a paper on "Semi-conductors with partially and with completely filled 3d lattice Bands," with NiO as a prominent example [80], followed by Mott and R. Peierls, in a paper entitled "Discussion of the paper by de Boer and Verwey" [77]. They posed the issues in the same form as used today: the competition between formation of bands and atomic-like behavior where the effects of interactions are larger than band widths. The competition is especially dramatic for localized partially filled 3d states, which leads to some of the most fascinating problems in physics today, such as the high temperature superconductors.[18] Methods to deal with such problems are a main topic

---

[16] See, for example, reviews by Herman [68, 69]. In his readable account in *Physics Today* [69], Herman recounts that many of the earlier hand calculations were done by his mother (cf. the role of D. R. Hartree's father).

[17] This was another milestone for quantum mechanics, since it follows from very general theorems [72–74] that in classical mechanics, the energy of a system of charges in equilibrium cannot depend on the magnetic field.

[18] The issues are expressed eloquently by P. W. Anderson in a paper [81] entitled "More is different," where it is emphasized that interactions may lead to phase transitions, broken symmetry, and other collective behavior that emerge in systems of many particles. The lasting character of these notions are brought out in the proceedings *More Is Different: Fifty Years of Condensed Matter Physics* [82].

of the companion book [1] and some of the characteristic examples are summarized in this book in Chapter 2, especially Section 2.17.

1.5 Density Functional Theory

Two papers by Pierre Hohenberg and Walter Kohn in 1964 [83] and Kohn and Lu Sham in 1965 [84] put the theory of electronic structure on a whole new level. Density functional theory (DFT) is now the basis of essentially all the qualitative work on the electronic structure of condensed matter and many other fields,[19] and it is a central topic of this book. The basic theory is the subject of Part II and applications of DFT are the primary topic of most of the rest of the book. *Perhaps the most important point to emphasize is that DFT is a theory of the interacting, correlated, many-body system of electrons.* The Kohn–Sham version of DFT uses independent-particle methods, but it is *not* an independent-particle approximation; it constructs an *auxiliary system* that in principle provides the exact density and total energy of the actual interacting system. The original Hohenberg–Kohn theorem is that all properties are functionals of only the density; however, as described in Chapter 6 this is really just a Legendre transformation with no prescription for doing anything else. The key to the phenomenal success of DFT is the stroke of genius of Kohn and Sham to formulate a theory that is supposed to give only certain properties – and not other properties – of the interacting system. By choosing only the ground-state density and total energy, it has proven to be possible to find approximations that are remarkably accurate and useful.

In Kohn–Sham DFT all the effects of exchange and correlation are incorporated in the exchange–correlation functional $E_{xc}[n]$, which depends on the density $n(\mathbf{r})$. The approximations for E_{xc} all involve some approach to utilize information derived by some many-body method. In the first edition of this book, the main examples were the local density and generalized-gradient approximations with almost all information coming from calculations on the electron gas. The success of these methods has led to much new work on improved functionals. In this edition there are two chapters on functionals (Chapters 8 and 9) with much more on the concepts and many-body ideas behind the new functionals.

1.6 Electronic Structure Is Now an Essential Part of Research

Since the advent of density functional theory many developments have set the stage for new understanding and applications in condensed matter physics, materials science, and other fields. The most influential development since the work of Kohn and Sham is due to Car and Parrinello in 1985 [85]. As explained in Chapter 19, their work has led to many new

---

[19] A tribute to the progress in crossing the boundaries between physics, chemistry, and other disciplines is the fact that the 1998 Nobel Prize in Chemistry was shared by Walter Kohn "for his development of the density-functional theory" – originally developed in the context of solids with slowly varying densities – and by John A. Pople "for his development of computational methods in quantum chemistry."

developments that have opened the door for an entire new world of calculations. Before their work almost all calculations were limited to simple crystals and heroic calculations on larger systems. After their work there was rapid development of many different techniques, and it is now routine to treat systems with hundreds of atoms, complex structures and surfaces, reactions, liquids as a function of temperature, and a host of other problems.

The theory and computational methods of electronic structure are now an integral part of research in physics and other fields. The basic methods for calculations are in Part IV, and the extensions to determine many properties of matter are the subject of Part V, and many examples are given in Chapter 2, which provides an overview of the types of problems that are the province of electronic structure today. The profound effect upon research is typified by modern-day searches for new materials. Of course, it is finally the role of experiment to create materials that can actually be made by feasible methods, but quantitative theory is almost always an essential part of research on new materials and systems: fullerenes, nanotubes, graphene, high-temperature superconductors, two-dimensional systems in layer materials and oxide interfaces, and materials for geophysics at pressures and temperatures in the earth, to cite just a few examples. In more and more cases discoveries of new materials have followed predictions by theory, such as the sulfur hydride superconductors at high pressure in Chapter 2 and topological insulators in Part VI that comprise an entire new field of research.

1.7 Materials by Design

The power of computers has made it possible to calculate the energies for enormous numbers of structures and to use a variety of approaches to search for new materials with desired properties, i.e., materials by design. This means much more than molecules or macroscopic solids. The term "materials" denotes clusters that may have many possible sizes, shapes and composition; surfaces and interfaces between different systems; defects that control the desired phenomena; liquids, solutions, and solid-liquid interfaces; layered systems that can be stacked in a controlled manner; and many other possibilities. Certainly the most important property is whether or not it can actually be made. A successful theory must be able to determine the stable compounds and their structure, and in many cases examine other cases that may be metastable or may be created by nonequilibrium conditions. It should be able to treat systems from molecules to clusters to solids. It should also be able to predict the desired properties accurately, e.g., strength, bandgaps, optical phenomena, magnetism, and superconductivity.

Such a theory does not exist. But there have been great steps toward the goal. At the center of the progress is density functional theory (DFT), which is essentially the only method capable of quantitative predictions for stable structures and other ground-state properties. In recent years there has been progress in new functionals that can treat a wider range of materials, e.g., van der Waals functionals and ones that incorporate other effects such as polarizability in the functional itself. There are famous deficiencies like the "bandgap problem," but great progress has been made in DFT methods to predict bandgaps. Characteristic examples of the theory and methods are the topic of this book. In addition, combination with many-body methods – the topic of the companion book [1] – is a powerful approach that can overcome many deficiencies in present-day DFT calculations.

There are now a number of large-scale collaborations to create the infrastructure for theorists and experimentalists to work together to discover and utilize new materials much more efficiently and effectively. Two examples are the Materials Genome Initiative (MGI), centered in the United States, and Novel Materials Discovery (NOMAD), centered in Europe, which are readily accessible on the Internet. The development of the tools and the means to utilize the vast amounts of data that are generated are exciting areas, but they are only touched on in this book. The topics of this and the companion [1] book are the theory and methods that are the foundation upon which these developments depend.

1.8 Topology of Electronic Structure

One of the greatest developments in the conceptual structure of condensed matter theory since the Bloch theorem is the recognition of the role of topology in electronic structure and the discovery of topological insulators in 2005–2006 in the seminal papers by Kane and Mele [86, 87] and Bernevig and Zhang [88]. The famous TKNN paper [89] by Thouless and coworkers in 1982 showed that the precise integer multiples in the quantum Hall effect (QHE) (see Appendix Q) can be explained as a topological invariant. However, the QHE occurs only in the presence of a strong magnetic field; it was the discovery of topological insulators, where spin–orbit interaction leads to related effects in the absence of a magnetic field, that has brought topology squarely into electronic structure. In hindsight we can see that the work of Shockley in 1939 [42] (see especially Fig. 1.1 and Chapters 22 and 26) was a forerunner of topological insulators, and it can be viewed as a crystalline topological insulator.

It is an inspiration that after so many years since the advent of quantum mechanics and the Bloch theorem that there are still new discoveries! Just as surprising is the fact that topological insulators can be understood with only knowledge of band structure and spin–orbit interaction at the level of an undergraduate solid state physics textbook. This is the topic of Chapters 25–28.

SELECT FURTHER READING

The companion book [1]:

Martin, R. M., Reining, L. and Ceperley, D. M., *Interacting Electrons: Theory and Computational Approaches* (Cambridge University Press, Cambridge, 2016). Together these two books are meant to cover the field of electronic structure theory and methods.

Examples of classic books that have much wisdom that is relevant today:

Dirac, P. A. M., *The Principles of Quantum Mechanics* (Oxford University Press, Oxford, 1930), reprinted in paperback. Insights from one of the penetrating minds of the era.

Mott, N. F. and Jones, H., *The Theory of the Properties of Metals and Alloys* (Clarendon Press, Oxford, 1936), reprinted in paperback by Dover Press, New York, 1955. A landmark for the early development of the quantum theory of solids.

Seitz, F., *The Modern Theory of Solids* (McGraw-Hill Book Company, New York, 1940), reprinted in paperback by Dover Press, New York, 1987. Another landmark for the development of the quantum theory of solids.

Slater, J. C., *Quantum Theory of Electronic Structure*, vols. 1–4 (McGraw Hill Book Company, New York, 1960–1972). A set of volumes containing a trove of theoretical information and many references to original works.

2

Overview

It is of great importance how atoms are put together. Of particular relevance are
arrangement, distances, bonding, collisions, and motion.
Titus Lucretius Carus, *On the Nature of Things*, bk. 2, 60 BCE

Summary

The theory of electronic structure of matter provides understanding and quantitative
methods that describe the great variety of phenomena observed. A list of these
phenomena reads like the contents of a textbook on condensed matter physics, which
naturally divides into ground-state and excited-state electronic properties. The aim of
this chapter is to provide an introduction to electronic structure without recourse to
mathematical formulas; the purpose is to lay out the role of electrons in determining
the properties of matter and to present an overview of the challenges for electronic
structure theory.

2.1 Electronic Structure and the Properties of Matter

The properties of matter naturally fall into two categories determined, respectively, by the
ground state and by *excited states* of the electrons. This distinction is evident in the physical
properties of materials and also determines the framework for theoretical understanding
and development of the entire field of electronic structure. In essence, the list of ground-
state and excited-state electronic properties is the same in most textbooks on condensed
matter physics:

- Ground state: cohesive energy, equilibrium crystal structure, phase transitions between
  structures, elastic constants, charge density, magnetic order, static dielectric and magnetic
  susceptibilities, nuclear vibrations and motion (in the adiabatic approximation), and
  many other properties.
- Excited states: low-energy excitations in metals involved in specific heat, Pauli spin
  susceptibility, transport, etc.; higher-energy excitations that determine insulating gaps
  in insulators, optical properties, spectra for adding or removing electrons, and many
  other properties.

The reason for this division is that materials are composed of nuclei bound together by electrons. Since typical energy scales for electrons are much greater than those associated with the degrees of freedom of the more massive nuclei, the lowest-energy ground state of the electrons determines the structure and low-energy motions of the nuclei. The vast array of forms of matter – from the hardest material known, diamond carbon, to the soft lubricant, graphite carbon, to the many complex crystals and molecules formed by the elements of the periodic table – are largely manifestations of the ground state of the electrons. Motion of the nuclei, e.g., in lattice vibrations, in most materials is on a time scale much longer than typical electronic scales, so that the electrons may be considered to be in their instantaneous ground state as the nuclei move. This is the well-known adiabatic or Born–Oppenheimer approximation [90, 91] (see Appendix C).

Since the ground state of the electrons is an important part of electronic structure, a large part of current theoretical effort is devoted to finding accurate, robust methods to treat the ground state. To build up the essential features required in a theory, we need to understand the typical energies involved in materials. To be able to make accurate theoretical predictions, we need to have very accurate methods that can distinguish small energy differences between very different phases of matter. By far the most widespread approach for "first principles" quantitative calculations of solids is density functional theory, which is therefore a central topic of this book. In addition, the most accurate many-body method to find the properties of the ground state or thermal equilibrium is quantum Monte Carlo, which is treated in detail in the companion book [1] and only a few selected results are given as benchmarks in this book.

On the other hand, for given structures formed by nuclei, electronic excitations are the essence of the "electronic properties" of matter – including electrical conductivity, optical properties, thermal excitation of electrons, phenomena caused by extrinsic electrons in semiconductors, etc. These properties are governed by the spectra of excitation energies and the nature of the excited states. There are two primary types of excitation: addition or subtraction of single electrons, and excitations keeping the number of electrons constant. Excitations are not supposed to be given by Kohn–Sham DFT, but time-dependent DFT in principle is exact (and in practice often very good as brought out in Chapter 21) for excitations that preserve particle number like optical spectra, and there is a generalized Kohn–Sham approach (see Chapter 9) for addition and removal spectra. Green's function methods to treat excitations are covered in detail in [1], and they provide valuable insights and results referred to in this book.

Electronic excitations also couple to nuclear motion, which leads to effects such as electron–phonon interaction. This leads to broadening of electronic states and to potentially large effects in metals, since normal metals *always* have excitation energies at arbitrarily low energies, which therefore mix with low-energy nuclear excitations. The coupling can lead to phase transitions and qualitative new states of matter, such as the superconducting state. Here, we will consider the theory that allows us to understand and calculate electron–phonon interactions (for example in Chapter 20), but we will not deal explicitly with superconductivity itself.

2.2 Electronic Ground State: Bonding and Characteristic Structures

Stable structures of solids are most naturally classified in terms of the electronic ground state, which determines the bonding. Extensive discussion of characteristic structures can be found in many references, such as the texts listed at the end of this chapter. It is useful here to point out the characteristic types of bonding:

- Closed-shell systems, typified by rare gases and molecular solids, have electronic states qualitatively similar to those in the atom (or molecule), with only small broadening of the bands. Characteristic structures are close-packed solids for the rare gases and complex structures for solids formed from nonspherical molecules. In order to treat these systems it is essential to deal with weak van der Waals[1] bonding; developed functionals are described in Section 9.8.
- Ionic crystals are compounds formed from elements with a large difference in electronegativity. They can be characterized by charge transfer to form closed-shell ions, leading to structures with large anions in a close-packed arrangement, with small cations in positions to maximize Coulomb attraction, such as fcc NaCl and bcc CsCl.
- Metallic systems are conductors because there is no energy gap for electronic excitation. The bands can accept different numbers of electrons, leading to the ability of metals to form alloys among atoms with different valency and to the tendency for metals to adopt close-packed structures, such as fcc, hcp, and bcc. The homogeneous electron gas (Chapter 5) is an informative starting point for understanding condensed matter, especially the sp-bonded metals, but transition metals have partially filled d or f states with both atomic-like and band-like character that sometimes are ferromagnets and often are a challenge to theory.
- Covalent bonding involves a complete change of the electronic states, from those of isolated atoms or ions to well-defined bonding states in solids with an energy gap, illustrated by the crossover in Fig. 1.1. Directional covalent bonds lead to open structures, very different from the close packing typical of other types of bonding. A major success of quantitative theory is the description of semiconductors and the transition to more close-packed metallic systems under pressure.
- Hydrogen bonding is often identified as another type of bonding [98]. Hydrogen is unique because it has no core states; the proton is attracted to the electrons, with none of the repulsive terms that occur for other elements due to their cores. Hydrogen bonding

---

[1] Johannes van der Waals received the Nobel prize in 1910 for the equation of state for gases and liquids, first proposed in his thesis in Leiden in 1873 and developed by him in many contexts, for example, capillary action due to the weak attractive interactions. See [92] and his Nobel lecture [93] (available online at www.nobelprize.org). The explanation of the attractive force as due to polarizability was suggested by Debye in 1920, and a quantum theory was proposed by Wang in 1927 (see references in [94] and [95]); a systematic theory was developed in 1930 by Eisenschitz and London and [96, 97] for the dispersion or "London" force. We will use "dispersion" to denote the attraction via electromagnetic fields and "van der Waals" to denote weak interactions that may include other effects. See also the companion book [1].

is crucial in myriads of problems in biology and other fields that cannot be covered here, but the effects are illustrated by the interaction between different molecules in water (Section 2.10) and formation of sulfur hydride superconductors at high pressure (Section 2.6).

The bonding in a real material is, in general, a combination of types of bonding. For example, molecular crystals and so-called van der Waals heterostructures (see Section 2.12) involve strong covalent and ionic intramolecular or intralayer bonds and van der Waals or other weak bonding between the molecule or layers. Heteropolar covalent-bonded systems, such as BN, SiC, GaAs, etc., all have some ionic bonding, which goes hand in hand with a reduction of covalent bonding.

Electron Density in the Ground State

The electron density $n(\mathbf{r})$ plays a fundamental role in the theory and understanding of the system of electrons and nuclei. The density can be measured by scattering of X-rays and high-energy electrons. Except for the lightest atoms, the total density is dominated by the core. Therefore, determination of the density reveals several features of a material: (1) the core density, which is essentially atomic-like; (2) the Debye–Waller factor, which describes smearing of the average density due to thermal and zero-point motion (dominated by the cores); and (3) the change in density due to bonding and charge transfer.

A thorough study [99] of Si has compared experimental and theoretical results calculated using the LAPW method (Chapter 17) and different density functional approximations. The total density can be compared to the theoretical value using an approximate Debye–Waller factor, which yields information about the core density. The primary conclusion is that, compared to the local density approximation (LDA), the generalized-gradient approximation (GGA) (see Chapter 8) improves the description of the core density where there are large gradients; however, there is little change in the valence region where gradients are small. In fact, nonlocal Hartree–Fock exchange is more accurate for determining the core density. This is a general trend that is relevant for accurate theoretical description of materials and is discussed further in Chapter 10.

The covalent bonds are revealed by a maximum in the difference between the crystal density and that of a sum of superimposed neutral spherical atoms. This is shown in Fig. 2.1, which presents a comparison of experimental (left) with theoretical results for the LDA (middle) and GGA (right). From the figure it is apparent that the basic features are reproduced by both functionals; in addition, other calculations using LAPW [100] and pseudopotential [101, 102] methods are in good agreement. The conclusion is that the density can be measured and calculated accurately, with agreement in such detail that differences are at the level of the effects of anharmonic thermal vibrations (see [99] and references cited there).

Theory also allows one to break the density into contributions due to each band and to a decomposition in terms of localized Wannier functions (Chapter 23) that provide much

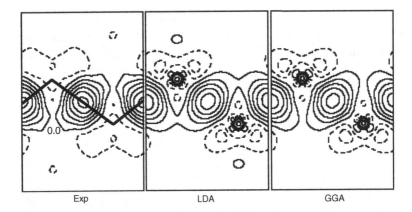

Figure 2.1. Electron density in Si shown as the *difference* of the total density from the sum of spherical atomic densities. The left-hand figure shows a contour plot of experimental (exp) measurements using electron scattering. The theoretical results were found using the linear augmented plane wave (LAPW) method (Chapter 17) and different density functionals (see text and Chapters 8 and 9). The difference density is in very good agreement with experimental measurements, with the differences in the figures due primarily to the thermal motion of the atoms in the experiment. LDA, local density approximation; GGA, generalized-gradient approximation. Provided by J. M. Zuo; similar to figure in [99].

more information than the density alone. For example, electric polarization of a crystal can be expressed in terms of Wannier functions even though it is not possible to determine the polarization from only the knowledge of the charge density in the bulk crystal. As described in Chapter 24, this has resolved a long-standing confusion in the literature and was an instrumental step in the progress leading up to the role of topology in electronic structure.

2.3 Volume or Pressure As the Most Fundamental Variable

The equation of state as a function of pressure and temperature is perhaps the most fundamental property of condensed matter. The stable structure at a given P and T determines all the other properties of the material. The basic quantities involved are the free energy $F(\Omega, T) = E(\Omega, T) - T S(\Omega, T)$, where the volume Ω and temperature T are the independent variables, or the Gibbs free energy $G(P, T) = H(P, T) - T S(P, T)$, where the pressure P and T are the independent variables and $H = E + P\Omega$ is the enthalpy. Most work considers only $T = 0$ because it is usually the dominant effect for solids and because it is much more difficult to calculate entropy, which requires a weighted integration over vibrational modes. In some cases thermal effects are added as a second step.[2] The total energy E at $T = 0$ as a function of volume Ω is the most convenient quantity for theoretical

---

[2] See Table 2.1 and Section 2.5 for examples. Of course, entropy is essential for liquids and is very important for problems like the phase diagrams for minerals under deep earth conditions in Section 2.10.

analysis because it is more straightforward to carry out electronic structure calculations at fixed volume. In essence, volume is a convenient "knob" that can be tuned to control the system theoretically. At $T = 0$ the fundamental quantities are energy E, pressure P, bulk modulus B,

$$E = E(\Omega) \equiv E_{\text{total}}(\Omega),$$

$$P = -\frac{dE}{d\Omega}, \tag{2.1}$$

$$B = -\Omega\frac{dP}{d\Omega} = \Omega\frac{d^2E}{d\Omega^2},$$

and higher derivatives of the energy. All quantities are for a fixed number of particles, e.g., in a crystal, E is the energy per cell of volume $\Omega = \Omega_{\text{cell}}$.

Comparison of theory and experiment is one of the touchstones of *ab initio* electronic structure theory, providing direct tests of the approximations made to treat electron–electron interactions. The first test is to determine the theoretical prediction for the equilibrium volume Ω^0, where E is minimum or $P = 0$, and the bulk modulus B is given by Eq. (2.1). Since Ω^0 and B can be measured with great accuracy (and extrapolated to $T = 0$), this is a rigorous test for the theory. One procedure is to calculate the energy E for several values of the volume Ω and fit to an analytic form, e.g., the Murnaghan equation of state [105]. The minimum gives the predicted volume Ω^0 and total energy, and the second derivative is the bulk modulus B. Alternatively, P can be calculated directly from the virial theorem or its generalization (Section 3.3), and B from response functions (Chapter 20).

During the 1960s and 1970s computational power and algorithms, using mainly atomic orbital bases (Chapter 15) or the augmented plane wave (APW) method (Chapter 16), made possible the first reliable self-consistent calculations of total energy as a function of volume for high-symmetry solids. Examples of calculations include ones for KCl [106, 107], alkali metals [108, 109], and Cu [110].[3] A turning point was the work of Janak, Moruzzi, and Williams [103, 111], who established the efficacy of the Kohn–Sham density functional theory as a practical approach to computation of the properties of solids. They used the Koringa–Kohn–Rostocker (KKR) method (Section 16.3) to calculate the equilibrium volume and bulk modulus for the entire series of transition metals using the local approximation, with results shown in Fig. 2.2. Except for a few cases where magnetic effects are essential, the calculated values are remarkably accurate – within a few percent of experiment. The overall shape of the curves has a very simple interpretation: the bonding is maximum at half-filling, leading to the maximum density, binding energy, and bulk modulus. Such comparisons of the predicted equilibrium properties with experimental values are now one of the routine tests of modern calculations.

---

[3] A theme of the work was comparison Slater average exchange with the Kohn–Sham formula (a factor of $2/3$ smaller). For example, Snow [110] made a careful comparison for Cu and found the lattice constant and other properties agreed with the experiment best for a factor 0.7225 instead of $2/3$, the value in density functional theory.

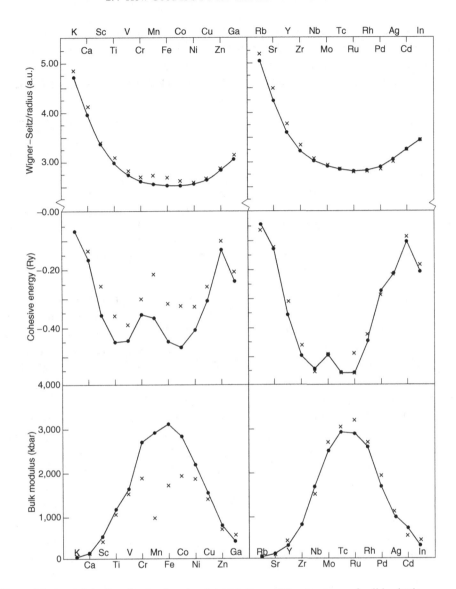

Figure 2.2. One of the first density functional calculations of the structures of solids: lattice constants (proportional to Wigner–Seitz radii), cohesive energy, and bulk moduli for the 3d and 4d series of transition metals, compared to experimental values denoted by points x. From Moruzzi, Janak, and Williams [103] (see also [104]).

2.4 How Good Is DFT for Calculation of Structures?

The title is meant to be thought provoking. Density functional theory is exact by construction (Chapter 6), but all functionals are approximations. The appropriate questions are: How good are useful, feasible approximations? For what classes of materials has a particular functional been tested?

Table 2.1. Calculated values of the lattice constant a and bulk modulus B compared to experiment for representative functionals and materials, selected from the results for 64 materials in [112]. MARE denotes the mean absolute relative error in % and MAX is the maximum relative error for the materials considered in [112]. Values for the magnetic moment of Fe are from [116].

Method	C a	C B	Si a	Si B	LiF a	LiF B	bcc Fe a	bcc Fe B	bcc Fe m	MARE a	MARE B	MAX a	MAX B
LDA	3.53	467	5.41	96	3.91	85	2.75	252	2.00	1.4	10.6	10.9	46.0
PBE	3.57	434	5.47	89	4.06	67	2.83	195	2.18	1.2	11.6	2.8	33.0
HSE06	3.55	469	5.43	98	4.00	78	2.90	151		0.8	8.6	3.2	35.0
SCAN	3.55	460	5.44	99	3.97	87	2.82	185	2.60	0.6	5.9	2.7	26.1
EXP	3.57	443	5.43	99	4.01	70	2.86	173	2.13				

One of the great successes of density functional theory is the prediction of structural properties of crystals. From the very first calculations with the original local density approximation, for example, as illustrated in Fig. 2.2, the results were a qualitative improvement over anything that had been available before. Since that time there have been great improvements in functionals that are the topic of Chapters 8 and 9. In addition, computational power and algorithms for calculation have improved so much that different methods for computation agree and the results are a true test of the functionals.

In Table 2.1 are listed the lattice constants and bulk moduli of a few selected crystals representing semiconductors, wide-gap covalent and ionic materials, and ferromagnetic transition metals, compared to experiment. These are taken from [112], which used the FHI-AIMS all-electron code (see [113] and Appendix R) to treat 64 different crystals with average performance (MARE) and maximum error (MAX). The values are in general agreement with many other references, which in total treat many hundreds of solids and thousands of molecules.[4] The functionals considered in the table are described in detail in Chapters 8 and 9: LDA, the local density approximation; PBE, the most widely used generalized gradient approximation; HSE06, a representative range-separated hybrid; and SCAN a functional of the density and the kinetic energy density called a meta functional.

One conclusion is that the more advanced functionals are all significant improvements over the LDA. The hybrid and meta functionals perform best in many ways. However, they require more complex codes and hybrid functionals are significantly more costly in computer time.[5] Meta functionals do not require as much computation as hybrids but they

---

[4] For example, over 1,000 materials are treated in [114], which concludes that SCAN leads to significantly more accurate predicted crystal volumes and mildly improved bandgaps as compared to PBE, but calculations are about five times slower than for PBE. Hybrid functionals are compared for many systems in [115].

[5] There is another very important consideration. A main purpose for the development of hybrid functionals is to improve the prediction of bandgaps, which is a major deficiency that plagues the LDA and GGA functionals. The improvement is stressed in many places in this book, such as the gaps shown in Fig. 2.24 and there are special considerations for solids illustrated by the optical spectra in Figs. 2.25, 2.26, and 21.5.

have not been as widely tested. Consequently, GGA functionals such as PBE are very widely used as a very successful improvement over the LDA for structural properties.

A general rule of thumb is that the lattice constants are predicted to within approximately 1% and the bulk moduli (also other elastic constants and vibrational frequencies) are predicted to within around 5%–10%. This is a remarkable achievement that has made a qualitative change in the way theory is used along with experiment and modern-day research. The agreement with experiment is sufficiently accurate that zero-point motion should be taken into account to compare to experiment; also it should be noted that there is uncertainty in experimental values extrapolated to $T = 0$.

However, it must be stressed that in every individual case great care must be taken to assure that the calculations are done properly and the conclusions are justified. The maximum errors listed in the last column of Table 2.1 are up to $\approx 6\%$ for the lattice constant a and approximately 35% for the bulk modules B for the more advanced functionals and more for the LDA. The largest errors tend to be for alkali metals and transition metals.

Each functional has deficiencies and there are no systematic proofs of when a functional works well and when it does not; there is only experience to justify the use in various classes of materials. Much of this book is devoted to analyzing classes materials and properties where we can expect a given functional to work well. Effort is made to point out cases in which there are problems and deficiencies, and the companion book [1] is devoted to many-body methods that are essential a many cases and provide benchmarks for functionals.

2.5 Phase Transitions under Pressure

Advances in experimental methods have made it possible to study matter over large ranges of pressures, sufficient to change the properties of ordinary materials completely; see, e.g., [117] and earlier references given there. In general, as the distance between the atoms is decreased, there is a tendency for all materials to transform to metallic structures, which are close packed at the highest pressures. Thus many interesting examples involve materials that have large-volume open structures at ordinary pressure and that transform to more close-packed structures under pressure.

Pressure different from zero (positive or negative!) is no problem for calculations, since the volume "knob" is easily turned to smaller or larger values. However, it is a great problem and a challenge of theory. It is an exacting requirement for a theory to predict the stability of phases with very different bonding, and it is only with the advent of density functional theory that it has been possible to make reliable predictions, and accurate prediction of stable structures and transition pressures provides a stringent test of different functionals.

Experiments are limited to structures that can actually be formed, but theory has no such restrictions. Understanding is gained by studying structures to quantify the reasons they are unfavorable and to find new structures that might be metastable. This is a double-edged sword: it is very difficult for a theorist to truly "predict" the most stable structures because of the difficulty in considering all possible structures. Nevertheless, great progress is being made with search methods discussed in Section 2.6.

The stability of phases of matter is determined by the free energy as a function of the temperature and pressure or temperature and volume. Fortunately for large classes of materials the most important effects are described by the energy at zero temperature, and the effects of temperature can be considered as a correction. At temperature $T = 0$, the condition for the stable structure at constant pressure P is that enthalpy $H = E + P\Omega$ be minimum. Transition pressures can be determined by calculating $E(\Omega)$ for different structures and using the Gibbs construction of tangent lines between the $E(\Omega)$ curves, the slope of which is the pressure for the transition between the phases.

Semiconductors are important examples that illustrate qualitative changes under pressure. The structures undergo various structural transitions and transformation from covalent open structures to metallic or ionic close-packed phases at high pressures [120, 121]. Figure 2.3 shows the energy versus volume for Si calculated [101] using the *ab initio* plane wave pseudopotential method and the local density approximation (LDA). This approach was pioneered by Yin and Cohen [101], a work that was instrumental in establishing the viability of theoretical predictions for stable structures of solids. The stable structure at $P = 0$ is cubic diamond as expected, and Si is predicted to transform to the β-Sn phase at the pressure indicated by the slope of the tangent line, ≈ 8 GPa. The right-hand figure includes phases labeled bc8 and st12, dense distorted metastable tetrahedral phases that are

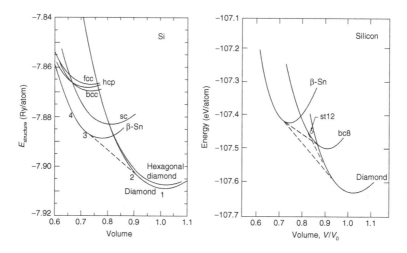

Figure 2.3. Energy versus volume for various structures of Si found using *ab initio* plane wave pseudopotential calculations. The left-hand figure, from the work of Yin and Cohen [101], is the first such fully self-consistent calculation, the success of which greatly stimulated the field of research. The tangent construction is indicated by the dashed line, the slope of which is the transition pressure. The right-hand figure is an independent calculation [118], which includes the dense tetrahedral phases bc8 and st12. The calculations find the phases to be metastable in Si but stable in C (see Fig. 19.2); similar results were found by Yin [119]. The calculations were done long before the widely used public codes were constructed. More recent calculations (e.g., in Fig. 2.4) are very similar for these phases and show that improved functionals tend to lead to higher transition pressures closer to experiment.

predicted to be almost stable; indeed they are well-known forms of Si produced upon release of pressure from the high-pressure metallic phases [122]. Many calculations have confirmed the general results and have considered many other structures [121, 123], including the simple hexagonal structure that was discovered experimentally [124, 125] and is predicted to be stable over a wide pressure range.

Similar calculations for carbon [118, 119, 126] correctly find the sign of the small energy difference between graphite and diamond at zero pressure and predict that diamond will undergo phase transitions similar to those in Si and Ge but at much higher pressures, ~3,000 GPa. Interestingly, the dense tetrahedral phases are predicted to be stable above ≈1,200 GPa, as is indicated in the phase diagram in Fig. 19.2.

The transitions also provide tests of functionals. The left side of Fig. 2.4 shows the equations of state for Si in the diamond and β-Sn structures using various functionals. Because the LDA favors dense structures it tends to predict transition pressures that are too low, and the improved functionals increase the transition pressure. The experimental transition pressure is ≈11 GPa, and it is clear that the pressure is underestimated by the LDA and improved significantly by the other functionals. However, in many cases it is difficult to compare directly with experiment because of hysteresis in the forward and backward transitions (often producing metastable bc8 structure as the pressure is decreased).

The stark contrast between the predictions of different functionals is illustrated for SiO_2 in the right side of Fig. 2.4. At low pressure there are many possible structures with

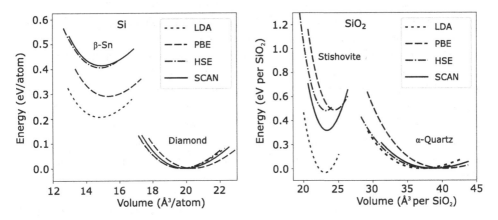

Figure 2.4. Energy versus volume for Si and SiO_2 for various functionals that are defined in Chapters 8 and 9. Transition pressures are determined common by tangent lines (not shown for simplicity), but it is clear that transition pressures are increased for all the other functionals compared to LDA. Comparison with experiment is discussed in the text for both Si and SiO_2. Left: for Si, calculated transition pressures are 7.1, 9.8, 13.3, 14.5 GPa, respectively, for LDA, PBE, HSE06, and SCAN functionals. Right: for SiO_2, results are shown for α-quartz and high-density stishovite phases. The LDA favors high density and even predicts the wrong structure at zero pressure, whereas the other functionals predict the correct low-pressure α-quartz structure. Modified from figures provided by Jefferson Bates; original references for Si [127, 128] and SiO_2 [127, 129].

networks of tetrahedrally coordinated Si typified by the well-known α-quartz structure. At high pressure there is a qualitative change in the bonding to a high-density stishovite structure with octahedrally coordinated Si atoms. The figure shows that the LDA predicts the stishovite structure to be most stable, i.e., a qualitatively incorrect prediction of the nature of SiO_2. The other functionals are qualitatively in agreement with experiment but the actual transition in SiO_2 is very sluggish and occurs to the intermediate coesite structure so that the experiment does not provide definitive tests of the functionals.

There have been many such calculations and much work combining theory and experiment for the whole range of semiconductors. The tendency of LDA to underestimate transition pressures has been brought out in reviews such as [120, 121]. Extensive tests of functionals for group IV, III–V, and II–VI compounds can be found in [128] for LDA, GGA, and the SCAN metafunctional, and in [115] for various hybrid functionals.

2.6 Structure Prediction: Nitrogen Solids and Hydrogen Sulfide Superconductors at High Pressure

Structure Searches

Electronic structure has reached the point where it is an indispensable part of one of the great goals of the physical sciences: to create new, useful materials and systems. Certainly a major factor is the capability to predict stable compositions and structures. The power of computers has made it possible to calculate the energies for enormous numbers of structures and to use a variety of approaches to search for materials with desired properties. Often this involves conditions that are not readily accessible experimentally. For example, predictions of stability under conditions in the mantle or core of the earth is of great importance for geophysics and can add to understanding of materials that exist under ordinary conditions. Essentially all work that attempts to calculate the energies in an *ab initio* manner with no empirical parameters uses density functional theory and the accuracy of the results depends on the ability of DFT to treat many different structures, compounds, stoichiometries, etc.under different conditions. Other methods such as quantum Monte Carlo can be used to study a few most favorable cases, but DFT is the method that makes large-scale searches possible.

A number of methods have been developed to search for stable or metastable structures.[6] Finding the most stable (lowest in energy or free energy) structure of a system with many atoms is a very difficult problem. In general, the potential energy surface as a function of positions of the atoms has many minima and a rugged landscape making it difficult to find the global minimum. A sobering message is the no-free-lunch theorem, which states that any search algorithm suffers the fate that if it performs better for one class of problems, it will be worse in some other classes [132]. Nevertheless, the set of reasonable possibilities

---

[6] Much of the discussion of search methods follows the review [130], which focuses on the random search methods and has an appendix with a summary of other methods. See also [131], which is about the role of structure prediction in materials discovery and design.

for a collection of atoms is much less than the set of all possible structures and there are methods proven to be useful.

One approach is simulated annealing that builds upon the early work of Kirkpatrick et al. [133]. For small systems thermal annealing is very useful for generating stable configurations, for example, the clusters in Fig. 2.20. For large systems like solids, however, barriers make it very difficult to generate transitions between structures.

In random search methods one starts with randomly generated structures, each of which is relaxed to find the minimum energy for that starting point. By limiting the random starting structures to ones that are "reasonable," e.g., without having atoms too close, Pickard and Needs [130] showed that their *ab initio* random structure searching (AIRS) method can successfully identify structures such as nitrogen solids. Their approach is to run the algorithm until the same lowest-energy structure has been found several times from different starting structures.

Particle swarm optimization (PSO) was developed by Kennedy and Eberhart in 1995 based on swarm behavior observed in fish and birds in nature.[7] The movement of a swarming particle (here it denotes a configuration of atoms) has a stochastic component and a deterministic component; it is attracted toward the position of the current global best of all the particles and its own best location in history, while at the same time it has a tendency to move randomly. There are many variations and one that has been used for structures of materials is CALYPSO (Crystal structure AnaLYsis by Particle Swarm Optimization) which was used in studies of H_2S structures under pressure.

Evolutionary algorithms are optimization techniques inspired by biological evolution, involving concepts such as reproduction, mutation and recombination, fitness, and selection [136]. They are widely used in many fields, and an example used in the materials community is the USPEX code (Universal Structure Predictor: Evolutionary Xtallography) [137] (original paper [138]) employed in finding structures of hydrogen sulfide materials.

Nitrogen

Molecular crystals are ideal to illustrate qualitative changes under pressure. A great example of the power of computational methods combined with intuition is nitrogen. The triple-bonded N_2 molecule is the most strongly bound of all diatomic molecules. At ordinary P and T nitrogen occurs only as N_2 gas or weakly bonded molecular solids and liquid. A great challenge for many years has been to create nonmolecular solid nitrogen. In the early 1990s Mailhiot, Yang, and McMahan [139] carried out calculations using a norm-conserving pseudopotential and the local density approximation, with results shown in Fig. 2.5 for the calculated energy versus volume for different structures. At low pressure there are various molecular phases, but at high pressures phases with higher coordination are more stable. That work found a new structure called "cubic gauche" (cg), perhaps the

---

[7] See, for example, the book [134]. The original reference by J. Kennedy and R. C. Eberhart [135] is a six-page paper with over 54,000 citations.

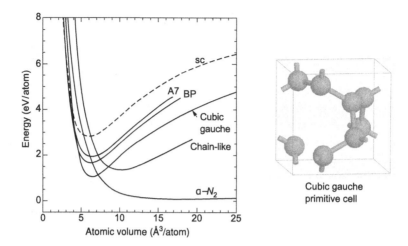

Figure 2.5. Total energy of nitrogen versus volume in molecular structures and the "cubic gauche" nonmolecular structure predicted at high pressure [139]. This is perhaps the first time DFT calculations have discovered a new structure never before known for any material; it was produced under pressure years later [140, 141].

first example of using DFT to predict a new structure that had never before been observed in any material. The cg structure is bcc with eight atoms per primitive cell as shown in Fig. 2.5; even though it has high symmetry, a glance at the configuration in Fig. 2.5 shows why it might be hard to guess.

The cubic gauche form of nitrogen was made experimentally years afterward by heating to a temperature above 2,000 K and pressure above 110 GPa in a laser-heated diamond cell [140, 141]. The structure was identified by X-ray scattering and it was found to have a bulk modulus greater than 300 GPa, characteristic of strong covalent solids, consistent with the theoretical prediction. Preparation of nonmolecular nitrogen at ambient pressure has been reported [142]; if indeed strongly bound solid nitrogen is metastable at low pressure the stored energy will be exceptional, more than five times that of the most powerfully energetic materials.

A later study using the random search method [143] found the cg structure to be lowest energy for a large pressure range and also identified a number of other stable structures. In some ways the behavior of nitrogen is analogous to carbon, which transforms under pressure from graphite with three neighbors double bonded to diamond with four neighbors and single bonds. The transition in nitrogen is from molecules where each atom has one neighbor with a triple bond to a three-dimensional crystal with threefold coordination and single bonds.

Hydrogen Sulfide Superconductors

There has been a long quest for metallic hydrogen, which has proven to be very difficult (see, e.g., [144]). Part of the impetus is the expectation that it would be a high-temperature

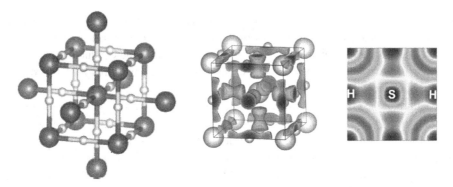

Figure 2.6. Left: Body-centered cubic structure of H_3S with sulfur atoms (large dark spheres) bonded to six hydrogens (small light spheres). Center and right: the electron localization function (ELF) in three-dimensions and in a (1 0 0) plane, which shows the bonding through the hydrogens. Provided by D. Duan, parts of figures in [146].

superconductor due to the high frequency of phonons and expected strong electron–phonon interaction. This has led to much work to find materials containing hydrogen that could have similar properties, in large part inspired by a paper by Niel Ashcroft [145]. A recounting of the story of studies of hydrogen-rich materials under pressure can be found in the review [117], and here we focus upon the discovery of superconductivity in hydrogen sulfides, where theory has played a key role.

Two theory papers published in 2014 set the stage for the discoveries. The paper by Li et al. [147] considered H_2S and optimized structures were identified using the particle swarm method CALYPSO structure search code. The result was that metallization occurs above 130 GPa with superconducting transition temperature predicted to be $\approx$80 K. The other paper by Duan et al. [146] considered the system $(H_2S)_2H_2$. This is one of the examples of closed-shell atoms or molecules that can absorb additional H_2 at high pressure to form H_2-containing stoichiometric compounds, and the structure of $(H_2S)_2H_2$ had already been established at relatively low pressure where it is still a molecular solid. The studies in [146] optimized the structures using an evolutionary algorithm method USPEX code and found a sequence of structures. At low pressure the structures of the molecular crystal were found in agreement with experiment. Above 111 GPa the H_2 molecules disappeared and H_3S molecules formed; there was a large volume collapse and the system became metallic. At 180 GPa the structure of H_3S molecules transformed smoothly into the high-symmetry bcc lattice shown at the left in Fig. 2.6 with hydrogen atoms symmetric between S atoms, and a predicted superconducting transition temperature between 191 and 204 K at 200 GPa.

The next year (2015) experimental results were reported by Drozdov et al. [148], who compressed H_2S and found that above $\approx$100 GPa, it becomes a superconductor with T_c that increases with pressure until it abruptly jumps to $\approx$200 K at pressures above 160 GPa. This work was inspired by the prediction of Li et al. [147], who studied H_2S; however, the results were actually consistent with the predictions of Duan et al. [146]. In addition, they found evidence for elemental sulfur and suggested the decomposition of H_2S to become more hydrogen rich. This was followed by synchrotron X-ray diffraction measurements reported in 2016 [149], which found that the structure of the superconducting phase is in good agreement with the predicted body-centered cubic (bcc) structure and confirmed the presence of elemental sulfur.

The nature of the bonding in the bcc phase of H_3S is revealed by the calculated electron localization function (ELF) shown in Fig. 2.6. As discussed in Section H.4, the ELF is derived from the kinetic energy density and it is defined in such a way that a large value indicates a region of space where there is a low probability of two electrons having the same spin–e.g., in a covalent bond with paired electrons of opposite spin. As indicated in Fig. 2.6, the value of ELF in the region of S–H bonds is large, which suggests a strong polar covalent bond; it is low between the nearest H atoms, indicating the absence of a covalent bonding, (i.e., no tendency for formation of H_2 molecules).

The calculation of the superconducting transition temperature in [146] and [147] was done with methods that have become standard. Although the theory of superconductivity is outside the scope of this volume, the basic ingredients in the theory of phonon-mediated superconductivity are the bread and butter of electronic structure (see Section 20.8): the electronic bands, the Fermi surface, the single-particle densities of states, phonon dispersion, and electron–phonon interactions. Calculations have also been done by others [150, 151] with general agreement that the high T_c is due to the high phonon frequencies but varying interpretations of the importance the electron–phonon interactions and the sharp peak in the electronic density of states at the Fermi energy.

In [146] the structures were derived in simulations that started from molecules of H_2S and H_2, with a fixed number of atoms. However, the ranges of stability of various hydrides and the transitions can be pinpointed most directly by calculations of the enthalpy $H = E + PV$ as a function of the pressure for each of the relevant stoichiometry and structures identified in the search. In Fig. 2.7 are the calculated enthalpies plotted as a function of the ratio $H_2/(S + H_2)$ in the range 0 (pure S) to 1 (pure H_2). The lines shown represent limits of stability for intermediate compounds; if the enthalpy is above the line, it is unstable to phase separation. For example, in the lower figures at 100 and 200 GPa, an H_2S compound has higher enthalpy than the same number of atoms separated into H_3S and elemental sulfur phases, and H_nS will separate into H_3S and H_2 for $n > 3$.

It is interesting to contrast the behavior of H_2S and H_2O. At low pressure H_2O molecules form structures with each O atom strongly bonded to two H atoms and weakly bonded to two other H atoms by hydrogen bonds, following ice rules. Under pressure, ice forms new structures including ice-X, in which the O atoms form a bcc lattice with four neighboring H atoms in tetrahedral positions midway between O atoms. It stays insulating and does not decompose up to pressures in the TPa range. See, for example, [153], which describes simulations of various phases and gives references to the literature.

Superhydrides. There are also other hydrogen-rich materials that are superconductors and theory plays a key role in identifying the most likely candidates. Perhaps the highest transition temperatures to date are in materials known as superhydrides. Two theoretical

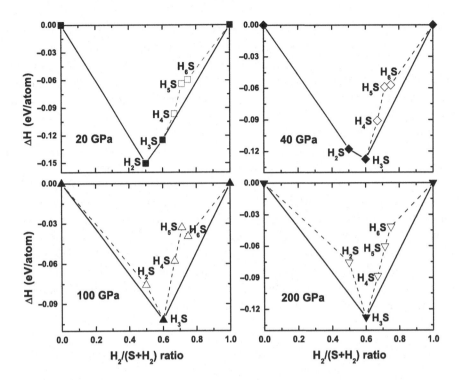

Figure 2.7. Phase diagram of the H–S system as a function of stoichiometry defined as $H_2/(S + H_2)$ so that 0 denotes pure S and 1 denotes pure H_2. Provided by D. Duan; parts of figures in [152].

papers have predicted that LaH_{10} and YH_{10} have a clathrate structure with 32 hydrogen atoms surrounding each La or Y atom [154, 155]. The hydrogen is so dense that the H–H distances are like that expected for atomic hydrogen at comparable pressures. Each paper reports DFT calculations and structure searches using the swarm-intelligence-based CALYPSO method (with random search methods in some cases [155]), with T_c ≈ 280 K for ≈ 200 GPA for LaH_{10} [154], and even higher T_c in YH_{10} [154, 155]. Subsequent experiments [156] found evidence for superconductivity above 260 K in LaH_{10} at ≈ 200 GPa.

2.7 Magnetism and Electron–Electron Interactions

Magnetic systems are ones in which the ground state has a spontaneous broken symmetry with spin and/or orbital moments of the electrons. In ferromagnets there is a net moment and in antiferromagnets there are spatially varying moments that average to zero by symmetry. The existence of a magnetic ground state is intrinsically a many-body effect caused by electron–electron interactions. Before the advent of quantum mechanics, it was recognized that the existence of magnetic materials was one of the key problems in physics, since it can be shown that within classical physics it is impossible for the energy of the system to be affected by an external magnetic field [72–74]. The solution was recognized in the earliest days of quantum mechanics, since a single electron has half-integral spin and interacting electrons can have net spin and orbital moments in the ground state. In open-shell atoms, this is summarized in Hund's rules that the ground state has maximum total spin and maximum orbital momentum consistent with the possible occupations of electrons in the states in the shell.

In condensed matter, the effects are often due to localized atomic-like moments and the correlation of moments on different sites leads to phase transitions and phenomena often called "strongly correlated." For that reason magnetic phenomena are treated more thoroughly in [1], which is devoted to the theory of interacting electrons. However, the average magnetization in an ordered state can be described by electrons in some average potential that is spin dependent. The effects of exchange and correlation among the electrons is replaced by an effective Zeeman field H_{Zeeman} represented by an added term in the Hamiltonian $m(\mathbf{r})V_m(\mathbf{r})$, where m is the spin magnetization $m = n^\uparrow - n^\downarrow$ and $V_m = \mu H_{\text{Zeeman}}$.[8] In analogy to the considerations of energy versus volume in Section 2.3, it is most convenient to find energy for fixed field V_m and the problem is to find the minimum of the energy and the susceptibility. The basic equations are

$$E - E(V_m) \equiv E_{\text{total}}(V_m),$$

$$m(\mathbf{r}) = -\frac{dE}{dV_m(\mathbf{r})}, \tag{2.2}$$

$$\chi(\mathbf{r},\mathbf{r}') = -\frac{dm(\mathbf{r})}{dV_m(\mathbf{r}')} = \frac{d^2 E}{dV_m(\mathbf{r})dV_m(\mathbf{r}')}.$$

If the electrons did not interact, the curvature of the energy χ would be positive with a minimum at zero magnetization, corresponding to bands filled with paired spins. However, exchange tends to favor aligned spins, so that $V_m(\mathbf{r})$ itself depends upon $m(\mathbf{r}')$ and can lead to a maximum at $m(\mathbf{r}') = 0$ and a minimum at nonzero magnetization, ferromagnetic if the average value $\bar{m}$ is nonzero antiferromagnetic otherwise. In general, $V_m(\mathbf{r})$ and $m(\mathbf{r}')$ must be found self-consistently.

The mean-field treatment of magnetism is a prototype for many problems in the theory of electronic structure. Magnetic susceptibility is an example of response functions described in Appendix D, where self-consistency leads to the mean-field theory expression (D.14); for magnetism the expressions for the magnetic susceptibility have the form first derived by Stoner [158, 159],

$$\chi = \frac{N(0)}{1 - I\,N(0)}, \tag{2.3}$$

where $N(0)$ is the density of states at the Fermi energy and the effective field has been expanded at linear order in the magnetization $V_m = V_m^{\text{ext}} + I\,m$ for the effective

---

[8] Density functional theory, discussed in the following chapters, shows that there exists a unique mean-field potential; however, no way is known of finding it exactly and there are only approximate forms at present.

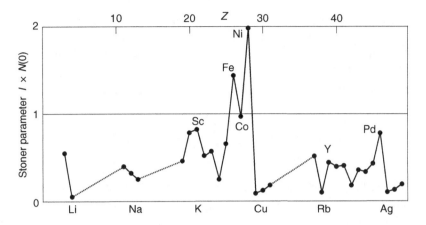

Figure 2.8. Stoner parameter for the elemental metals derived from densities of states and spin-dependent mean-field interactions from density functional theory. This shows the basic tendency for magnetism in the 3d elements Mn, Fe, Co, and Ni, due to the large density of states and interactions, both of which are consequences of the localized atomic-like nature of the 3d states discussed in the text. Data from Kübler and Eyert [157], originally from Moruzzi et al. [104].

interaction.[9] The denominator in Eq. (2.3) indicates a renormalization of the independent-particle susceptibility $\chi^0 = N(0)$, and an instability to magnetism is heralded by the divergence when the Stoner parameter $I N(0)$ equals unity. Figure 2.8 shows a compilation by Moruzzi et al. [104] of $I N(0)$ from separate calculations of the two factors I and $N(0)$ using density functional theory. Clearly, the theory is quite successful since the Stoner parameter exceeds unity only for the actual ferromagnetic metals Fe, Co, and Ni, and it is near unity for well-known enhanced paramagnetic cases like Pd. An example of the calculated average magnetization of Fe is given in Table 2.1.

Modern calculations can treat the spin susceptibility and excitations within density functional theory using either "frozen" spin configurations or response functions, with forms that are the same as for phonons described in the following section, Section 2.9, and Chapter 20. An elegant formulation of the former approach based on a Berry phase [160, 161] is described in Section 20.2. Examples of both Berry phase and response function approaches applied to Fe are given in Figs. 20.2 and 20.4.

2.8 Elasticity: Stress–Strain Relations

The venerable subject of stress and strain in materials has also been brought into the fold of electronic structure. This means that the origins of the stress–strain relations are traced back

---

[9] The Stoner parameter I $N(0)$ can be understood simply as the product of the second derivative of the exchange–correlation energy with respect to the magnetization (the average effect per electron) times the density of independent-particle electronic states at the Fermi energy (the number of electrons able to participate). This idea contains the essential physics of all mean-field response functions as exemplified in Appendix D and Chapters 20 and 21.

to the fundamental definition of stress in quantum mechanics, and practical equations have been derived that are now routine tools in electronic structure theory [102, 162]. Because the development of the theory has occurred in recent years, the subject is discussed in more detail in Appendix G.

The basic definition of the stress tensor $\sigma_{\alpha\beta}$ is the generalization of Eq. (2.1) to anisotropic strain,

$$\sigma_{\alpha\beta} = -\frac{1}{\Omega}\frac{\partial E_{\text{total}}}{\partial u_{\alpha\beta}}, \tag{2.4}$$

where $u_{\alpha\beta}$ is the symmetric strain tensor defined in Eq. (G.2). Likewise, the theoretical expressions are the generalization of the virial theorem for pressure to anisotropic stress [102, 162].

Figure 2.9 shows the first reported calculation [163] of stress in a solid from the electronic structure, which illustrates the basic ideas. This figure shows stress as a function of uniaxial strain in Si. The linear slopes yield two independent elastic constants, and nonlinear variations can be used to determine nonlinear constants. Linear elastic constants have been calculated for many materials, with generally very good agreement with experiment, typically ~5%–10%. Nonlinear constants are much harder to measure, so that the theoretical values are often predictions.

As an example of predictions of theory, Nielsen [164] has calculated the properties of diamond for general uniaxial and hydrostatic stresses, including second-, third-, and fourth-order elastic constants, internal strains, and other properties. The second-order constants agree with experiment to within ≈6% and the higher-order terms are predictions. At

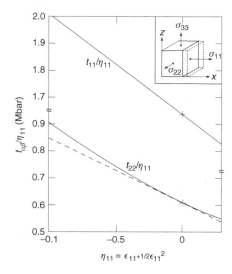

Figure 2.9. Stress in Si calculated as a function of strain in the (1 0 0) direction [163]. Two elastic constants can be found from the linear slopes; the nonlinear variation determines nonlinear constants. For nonlinear strains it is most convenient to use Lagrange stress and strain, $t_{\alpha\beta}$ and $\eta_{\alpha\beta}$ (see [163]), which reduce to the usual expression in the linear regime. From [163].

extremely large uniaxial stresses (4 Mb) the electronic bandgap collapses, and a phonon instability of the metallic diamond structure is found for compressions along the [110] and [111] crystal axes. This may be relevant for ultimate stability of diamond anvils in high-pressure experiments.

2.9 Phonons and Displacive Phase Transitions

A wealth of information about materials is provided by the vibrational spectra that are measured experimentally by infrared absorptions, light scattering, inelastic neutron scattering, and other techniques. The same holds for the response of the solid to electric fields, etc. Such properties are ultimately a part of electronic structure, since the electrons determine the changes in the energy of the material if the atoms are displaced or if external fields are applied. So long as the frequencies are low, the electrons can be considered to remain in their ground state, which evolves as a function of the displacements of the nuclei. The total energy can be viewed as a function of the positions of the nuclei $E(\{\mathbf{R}_I\})$ independent of the nuclear velocities. This is the adiabatic or Born–Oppenheimer regime (see Chapter 3 and Appendix C), which is an excellent approximation for lattice vibrations in almost all materials. In exact analogy to Eq. (2.1), the fundamental quantities are energy $E(\{\mathbf{R}_I\})$, forces on the nuclei $\mathbf{F}_I$, force constants C_{IJ},

$$E = E(\{\mathbf{R}_I\}) \equiv E_{\text{total}}(\{\mathbf{R}_I\}),$$

$$\mathbf{F}_I = -\frac{dE}{d\mathbf{R}_I}, \tag{2.5}$$

$$C_{IJ} = -\frac{d\mathbf{F}_I}{d\mathbf{R}_J} = \frac{d^2E}{d\mathbf{R}_I d\mathbf{R}_J},$$

and higher derivatives of the energy.

Quantitatively reliable theoretical calculations have added new dimensions to our understanding of solids, providing information that is not directly available from experiments. For example, except in a few cases, only frequencies and symmetries of the vibration modes are actually measured; however, knowledge of eigenvectors is also required to reconstruct the interatomic force constants C_{IJ}. In the past this has led to a plethora of models for the force constants that all fit the same data (see, e.g., [165, 166], and references therein). For example, large differences in eigenvectors were predicted for certain phonons in GaAs, and the issues were resolved only when reliable theoretical calculations became possible [167]. Modern theoretical calculations provide complete information directly on the force constants, which can serve as a database for simpler models and understanding of the nature of the forces. Furthermore, the same theory provides much more information: static dielectric constants, piezoelectric constants, effective charges, stress–strain relations, electron–phonon interactions, and much more.

As illustrated in the examples given here and in Chapter 20, theoretical calculations for phonon frequencies have been done for many materials, and agreement with experimental frequencies within $\approx 5\%$ is typical. Since there are no adjustable parameters in the theory, the agreement is a genuine measure of the success of current theoretical methods for such ground-state properties. This is an example where theory and experiment can work together, with experiment providing the crucial data and new discoveries, and the theory providing solid information on eigenvectors, electron–phonon interactions, and many other properties.

There are two characteristic approaches in quantitative calculations:

- Direct calculation of the total energy as a function of the positions of the atoms. This is often called the "frozen phonon" method.
- Calculations of the derivatives of the energy explicitly at any order. This is called the "response function" or "Green's function" method.

Frozen Phonons

The term "frozen phonons" denotes the direct approach in which the total energy and forces are calculated with the nuclei "frozen" at positions $\{\mathbf{R}_I\}$. This has the great advantage that the calculations use *exactly* the same computational machinery as for other problems: for example, the same program (with only slightly different input) can be used to calculate phonon dispersion curves (Chapter 20), surface and interface structures (Chapter 22), and many other properties. Among the first calculations were phonons in semiconductors, calculated in 1976 using empirical tight-binding methods [171] and again in 1979 using density functional theory and perhaps the first such calculation to find the small changes in energy [172]. Today these are standard applications of total energy methods.

Frozen Polarization and Ferroelectricity

The left side of Fig. 2.10 shows the calculated energy versus displacement of Ti atoms in $BaTiO_3$, where the origin corresponds to atoms in centrosymmetric positions of the perovskite structure shown in Fig. 4.8. If the energy had increased with displacement the structure would be stable and the curvature would determine the frequency of an optic mode. The negative curvature at the centrosymmetric position indicates an instability, and calculations for displacements in different directions are needed to determine the lowest energy minimum. From this information one can extract the various orders of the anharmonic terms; for example, the terms needed to construct a microscopic model (see, e.g., [173, 174]), which can be used to construct free energy models, and to study thermal phase transitions.

Incredible as it may seem, the problem of calculation of the electric polarization of a ferroelectric was only solved in the 1990s, despite the fact that expressions for the energy and forces have been known since the 1920s. As described in Chapter 24, the advance in recent years [175, 176] relates a change in polarization to a Berry phase [177] involving the change in *phases* of the electron wavefunctions. The theory provides practical methods for calculation of polarization in ferroelectrics and pyroelectrics, as well as effective charges and piezoelectric effects to all orders in lattice displacements and strains.

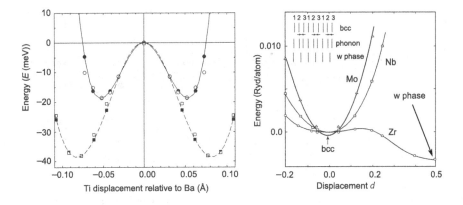

Figure 2.10. Frozen phonon calculations of energy versus displacement. Left: displacements of Ti atoms in $BaTiO_3$ is in two directions with dark symbols for calculations using local orbitals [168] (see Chapter 15) and open symbols from using full-potential LAPW methods [169] (Chapter 17). The centrosymmetric position is unstable, and the most stable minimum is for the rhombohedral direction resulting in a ferroelectric phase in agreement with experiment. Provided by R. Weht and J. Junquera (similar to the figure in [168]). Right: energy versus displacement for longitudinal displacements with wavevector $\mathbf{k} = \left(\frac{2}{3}, \frac{2}{3}, \frac{2}{3}\right)$ in the bcc structure [170]. For Mo and Nb the minimum energy is for bcc structure. The curvature agrees well with measured phonon frequencies and corresponds to a sharp dip in the phonon dispersion curves, which is a precursor to the phase transition that actually occurs in Zr. The minimum energy structure for Zr at low temperature is the "ω phase," which forms by displacements shown in the inset, with each third plane undisplaced and the other two planes forming a more dense bilayer. From [170].

Calculations with Supercells

The right side of Fig. 2.10 shows the results of calculations for three transition metals using a cell that corresponds to tripling the size of the cell [170]. For Mo and Nb, the curvature determines the frequency of the phonon at wavevector $\mathbf{k} = \left(\frac{2}{3}, \frac{2}{3}, \frac{2}{3}\right)$ in the bcc structure, but for Zr there is an instability to forming the ω phase where layers of atoms are displaced in the pattern shown in the figure. This is an example of instabilities that can occur in systems with strong electron–phonon interactions and narrow bands as found in many transition metals and discussed in Section 20.2.

Frozen phonon methods can also be used to calculate dispersion curves by taking a large enough supercell. The example in Fig. 20.1 shows this is feasible with cells that are manageable; however, the method based on response functions and Green's functions is more elegant and effective, and it is the method most widely used.

Linear (and Nonlinear) Response

Response function approaches denote methods in which the force constants are calculated based on expansions in powers of the displacements from equilibrium positions. This has the great advantage that it builds upon the theory of response functions (Appendix D), which

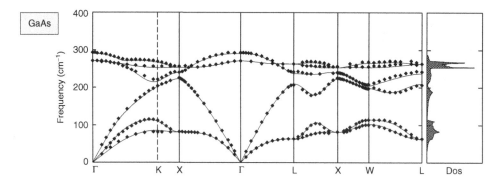

Figure 2.11. Phonon dispersion curves calculated for the semiconductor GaAs [182]. The points are from experiment and the curves from density functional theory using the response function method (Chapter 20). Similar agreement is found for the entire family of semiconductors. Calculations for many types of materials, e.g., in Figs. 20.3 and 20.4, have shown the wide applicability of this approach.

can be measured in experiments and was formulated [178–180] in the 1960s. Methods using Green's functions (see the review [181] and Chapter 20) have cast the expressions in forms much more useful for computation so that it is now possible to calculate phonon dispersion curves on an almost routine basis.

Figure 2.11 shows a comparison between experimental and theoretical phonon dispersion curves for GaAs [182]; such near-perfect agreement with experiment is found for many semiconductors using plane wave pseudopotential methods. In Chapter 20 examples of results for metals are shown, where the results are also impressive but the agreement with experiment is not as good for the transition metals. Similar results are found for many materials, agreeing to within ≈5% with experimental frequencies. The response function approach is also especially efficient for calculations of dielectric functions, effective charges, electron–phonon matrix elements, and other properties, as discussed in Chapter 20.

2.10 Thermal Properties: Solids, Liquids, and Phase Diagrams

One of the most important advances in electronic structure theory of recent decades is "quantum molecular dynamics" (QMD) pioneered by Car and Parrinello in 1985 [85] and often called "Car–Parrinello" simulations. As described in Chapter 19, QMD denotes classical molecular dynamics simulations for the nuclei, with the forces on the nuclei determined by solution of electronic equations as the nuclei move. By treating the *entire problem of electronic structure and motion of nuclei together*, this has opened the way for electronic structure to study an entire range of problems far beyond previous capabilities, including liquids and solids as a function of temperature beyond the harmonic approximation, thermal phase transitions such as melting, chemical reactions including molecules in solution, and many other problems. This work has stimulated many advances that are now standard features of electronic structure methods so that calculations on molecules and complex crystals routinely optimize the structure in addition to calculating the electronic structure.

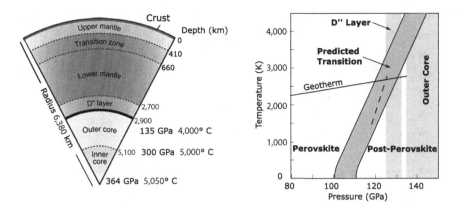

Figure 2.12. Left: structure of the interior of the earth, with pressure and temperature deep in the earth (adapted from figure at www.spring8.or.jp/). The different regions are determined from analysis of seismic waves, but experiments and theoretical calculations at high pressure and temperature are essential to determine the composition. The temperature follows the adiabatic geotherm shown at the right until it increases dramatically in the D'' layer at the mantle/outer core boundary. Right: predicted transition to a postperovskite structure corresponding to the D'' layer, with results that are significant for understanding the state of the lower mantle. The structure at high pressure is described in the text and the dark-shaded band is the estimated range of pressures based on the quasi-harmonic approximation with LDA and PBE GGA as lower and upper bounds. The slope from the Clausius–Clapeyron equation (see also Section 19.6) is consistent with the dashed line that was proposed earlier based on seismic data. Adapted from figure from [183].

Geophysics

One of the great challenges of geophysics is to understand the nature of the interior of the earth. As shown in the left side of Fig. 2.12, the pressure and temperature are estimated to be ≈ 135 GPa, $\approx 4{,}000$ K at the mantle/core boundary and ≈ 330 GPa, $\approx 5{,}000$ K at the boundary of the outer and inner core, conditions that are very difficult to reproduce in the laboratory. This is a great opportunity for simulations to make major contributions in areas described in sources such as [184] and the issue of *Reviews in Mineralogy and Geochemistry* on "Theoretical and Computational Methods in Mineral Physics: Geophysical Applications" [185] (https://pubs.geoscienceworld.org/rimg/issue/71/1).

The crust and the mantle are composed of silicate minerals, and $MgSiO_3$ in the high-pressure perovskite phase is believed to be the most abundant mineral in the earth's lower mantle. The possibility of a phase transition associated with the D'' layer has been controversial, and a combination of experiment and theory led to the discovery of a transition to a new postperovskite phase, which may have important implications for the lower mantle [183]. The simulations were a milestone for applications to minerals at high pressure including thermal effects by self-consistent calculations of lattice vibrations and the quasiharmonic approximation, with results for the transition shown at the right in Fig. 2.12. The line marked geotherm is the temperature/pressure relation in the earth and the calculations show that the transition occurs in the region of the D'' layer. The upper and

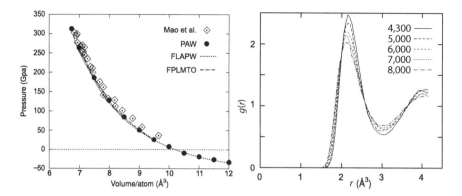

Figure 2.13. Left: pressure versus volume for hcp Fe at low temperature, which illustrates the agreement of calculations performed using ultrasoft pseudopotentials, the PAW method, and full-potential LMTO and LAPW calculations [186] and experiment [187]. Right: calculated radial density distribution $g(r)$ of liquid Fe as a function of temperature [186]. The density and the lowest temperatures $T \approx 5,000$ K are the values expected at the inner/outer core boundary in the earth. The~integral under the first peak is ≈ 12 atoms indicating a close-packed liquid at all pressures. Figures from [186].

lower bounds for the theoretical prediction are based on two calculations using the LDA and PBE GGA functionals, and the experience in many materials that calculated transition pressures tend to be low using the LDA and high using PBE. The new high-pressure phase has the $CaIrO_3$ structure with SiO_3 layers intercalated by eightfold coordinated Mg ions, and the slope from the Clausius–Clapeyron equation agrees with that inferred from seismic data. The conclusion is that the solid–solid transition to an anisotropic structure may be the explanation for the inhomogeneity and topography that are key for understanding the D'' layer in the deep lower mantle.

The core is made of Fe with other elements, and the simulations provide an example of the predictions that can be made. Before considering the calculations for the liquid, it is instructive to examine the equation of state for an Fe crystal at zero temperature. The left side of Fig. 2.13 shows the equation of state of hcp Fe up to pressures relevant for the core. The curves show that calculations made using ultrasoft pseudopotentials, the PAW method, and full-potential LMTO and LAPW calculations [186] are in agreement. They also agree with the experimental measurements made in diamond cells at low temperature [187], which gives confidence in the simulations of the liquid.

The radial density distribution for liquid Fe with fixed density $\rho = 10,700$ kg/m^3 (the value at the core/mantle boundary) is shown in the right side of Fig. 2.13 for several temperatures in which the calculated pressure ranges from 172 to 312 GPa [186]. This was done using the PAW method (Section 13.3) with forces calculated at each step of the molecular dynamics calculation, as in Section 19.2. The weight of the peak in $g(r)$ corresponds to slightly over 12 neighbors, i.e., to a close-packed liquid. As expected, the peak broadens with increasing temperature, with no abrupt transitions. The melting curve has been determined in several works, as discussed in [188].

An interesting case is carbon. Diamond is metastable at ordinary pressure but is formed under pressure deep in the earth. Simulations described in Section 19.6 find that crystalline diamond is the stable phase of carbon even under the conditions at the center of the earth. The stable structure under ordinary conditions is graphite. The many structures with threefold bonding that have been made are among the most fascinating of all materials and are central tropics in many places in this book.

Water and Aqueous Solutions

Certainly, water is the liquid most important for life. As a liquid, or in ice crystalline forms, it exemplifies the myriad of complex features due to hydrogen bonding [190–192]. Isolated molecules of H_2O and small clusters can be treated accurately by the methods of quantum chemistry, and many properties of the condensed states are described by potentials that have been fitted with great effort. QMD can play a special role in determining cases that are not understood from current experimental information. Examples include the actual atomic-scale nature of diffusion processes, which involves rearrangements of many molecules, the behavior of water under extreme conditions of pressure and temperature, and high-pressure phases of ice.

There are a vast number of QMD simulations of water and aqueous solutions. An example that illustrates the great advantage of methods that treat both the nuclei motion and the electronic states is the transfer of protons and accompanying transfer of electronic orbitals as shown in Fig. 2.14, which also illustrates the use of maximally localized Wannier functions (Chapter 23) to describe the electronic states. An example of a study of representative aqueous solutions containing solvated Na^+, K^+, and Cl^- ions is in [193], which has references to many earlier works.

A challenge is to describe intermolecular bonding in water accurately using density functional theory. A review by Gillan et al. in 2016 [197], "Perspective: How Good Is

Figure 2.14. "Snapshots" of the configurations of H and O atoms in a QMD simulation of water under high-pressure, high-temperature conditions [189]. A sequence of the motion of the atoms is shown from left to right, centered on one proton that transfers, and the Wannier function for the electronic state that transfers to form H^+ and $(H_3O)^-$. (The Wannier functions are defined by the "maximal localization" condition of Chapter 23.) Provided by E. Schwegler; essentially the same as fig. 2 of [189].

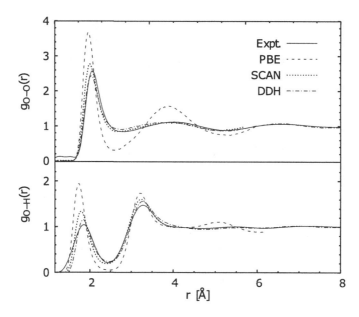

Figure 2.15. Radial density distribution functions for O–O and O–H in water at 300 K comparing results using different functionals described in the text: DDH [194], SCAN [195], PBE [193], and experiment [196]. Note that PBE leads to a $g(r)$ that is too structured, whereas the others are in much better agreement with experiment. Other functionals can also describe $g(r)$ with comparable accuracy [197]. Adapted from figure provided by Alex Gaiduk.

DFT for Water?," considers various configuration of molecules, ice, and simulations of the liquid; it gives many references and paints a picture that results are very sensitive to the approximations and improvements are required to unravel many issues. The local approximation gives bonding that is much too strong and some correlation functionals give bonding much too weak. As an example of recent work, the calculated radial density distributions for O–O and O–H distances are compared with experimental results in Fig. 2.15. Even though the PBE GGA functional is a significant improvement over LDA (not shown), it still leads to overstructure, i.e., correlations that are too strong with peaks sharper than experiment. The other functionals shown are much better: the SCAN meta functional (Section 9.4) and the DDH hybrid functional with screened long-range part (Section 9.3). Many other functionals have been tested (see the review [197]), for example, the effect of adding van der Waals functionals can improve the results [198].

There are many phases of ice, but for our purposes it is most relevant to point out that above $\approx$60 GPa [199] the molecular form of H_2O transforms to a strongly bonded structure where the hydrogen symmetric between oxygen atoms, which is a wide gap insulator as described by theoretical calculations [200]. It is interesting to note the difference from the hydrogen sulfides (see Section 2.6), where H_2S decomposes to H_3S and elemental sulfur under pressure. At high pressures H_3S also forms a structure with hydrogen in symmetric positions, but in this case it is metallic and a superconductor.

Reactions at Surfaces and Catalysis

A greater challenge yet is to describe chemical reactions catalyzed in solutions or on surfaces also in solution. As an example of the important role of theoretical calculations in understanding materials science and chemistry, QMD simulations [201, 202] have apparently explained a long-standing controversy in the Ziegler–Nata reaction that is a key step in the formation of polymers from the common alpha olefins, ethylene, and propylene. This is the basis for the huge chemical industry of polyethylene manufacture, used for "plastic" cups, grocery bags, the covers for CDs, etc. Since propylene is not as symmetric as ethylene, special care must be paid to produce a stereoregular molecular chain, where each monomer is bound to the next with a constant orientation. These high-quality polymers are intended for special use, e.g., for biomedical and space applications. The Ziegler–Nata process allows these cheap, harmless polymers to be made from common commercial gas, without strong acids, high temperatures, or other expensive procedures.

The process involves molecular reactions to form polymers at Ti catalytic sites on $MgCl_2$ supports; an example of a good choice for the support is made by cleaving $MgCl_2$ to form a (110) plane, as shown in Fig. 2.16. On this surface $TiCl_4$ sticks efficiently, giving a high density of active sites. The QMD simulations find that the relevant energetics, as well as the reactivity in the alkyl chain formation process, strongly depend on the local geometry. The dynamical approach follows the reaction pathway in an unbiased way during deposition of $TiCl_4$ and complex formation, which are energetically downhill. Constrained dynamics can then be used to determine free-energy profiles and estimate activation barriers in the alkene insertion processes. Steps in the insertion of a second ethylene molecule shown in the sequence in Fig. 2.16 offers insight into the chain growth process and the stereochemical character of the polymer, providing a complete picture of the reaction mechanism.

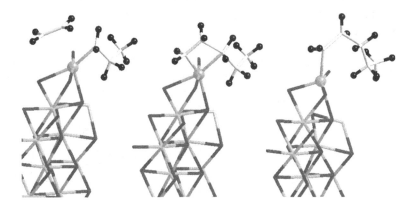

Figure 2.16. Simulation of the Ziegler–Nata reaction, which is an essential step in the production of polyethylene. Predicted steps in the main phases of the second insertion leading to the chain propagation: the π-complex (left), the transition state (middle), and insertion (right) of the ethylene molecule lengthening the polymer. Figure provided by M. Boero; essentially the same as fig. 11 of [201].

2.11 Surfaces and Interfaces

There is no infinite crystal in nature; every solid has a surface. More and more experiments can probe the details of the surface on an atomic scale, e.g., the scanning tunneling microscope and X-ray and electron diffraction from surfaces. Surface science is a vast subject that cannot be covered here, but a few examples are selected to illustrate the role of electronic structure theory and calculations, primarily in Chapter 22, which is devoted to surfaces, interfaces, and lower-dimensional systems, and in Chapters 25–28 on topological insulators where there are surface states with properties that were completely unknown only a few years ago. Reactions at surfaces of catalysts are particularly interesting and challenging for theory; an example is the Ziegler–Nata reaction described in Section 2.10.

Structure and Stoichiometry

It is sometimes stated that a surface is just one form of interface, an interface between a bulk solid and vacuum. But this is misleading. The so-called vacuum is really a very active, variable environment. Atoms at the surface can move and rebond and the stoichiometry can be changed by the overpressure of gasses. It is a major challenge to determine the structure, which may differ greatly from the bulk, for example, in semiconductors where the disruption of the strong covalent bonding leads to a variety of reconstructions of the surface. An example of reconstruction is the buckled dimer for Si or Ge (1 0 0) surfaces illustrated in Fig. 2.17. If the atoms remained in the positions as if they were in the bulk, there would be two dangling bonds, but the energy can be reduced by dimerizing to make a bond at the surface. This still leaves one dangling bond per surface atom, i.e., two half-filled states per dimer. The energy may be further reduced by a buckling as shown in the figure; one atom moves up and the other down to form a "lone pair" state primarily on one atom that is filled and an empty state primarily on the other atom. This final result in two surface bands in the bulk bandgap that are filled and empty respectively, as shown in the calculations in Fig. 22.4.

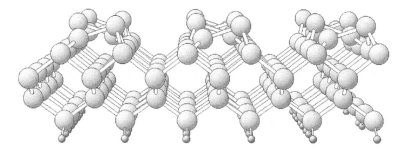

Figure 2.17. Illustration of the (1 0 0) surface of Si or Ge with buckled dimer reconstruction, where the surface atoms pair to make dimers and each dimer is buckled, i.e., one atom moves up and the other down. As explained in the text, the two atoms in the dimer are not equivalent and the electronic states of the dimer split with one filled and one empty, which finally result in two surface bands as shown in Fig. 22.4. Adapted from figure in [203].

Ionic semiconductors, such as III–V and II–VI crystals, present additional issues: not only must the total energy be compared for various possible reconstructions with the same numbers of atoms, but also one must compare anion and cation terminations with different numbers of atoms of each type, i.e., different stoichiometries. As an illuminating first step, the stoichiometry of the atoms for different types of structures can be predicted from simple electron counting rules [204, 205], i.e., charge compensation at the surface by the filling of all anion dangling bonds and the emptying of cation dangling bonds. The full analysis, however, requires that the surface energy be determined with reference to the chemical potential μ_I for each type of atom I, which can be controlled by the temperature and partial pressure in the gas (or other phase) in contact with the surface [206, 207], thereby allowing experimental control of the surface stoichiometry. Assuming the bulk, surface, and gas are in equilibrium,[10] the quantity to be minimized is not the free energy $E - T\,S$ but the grand potential [208],

$$\Omega = E - T\,S - \sum_I \mu_I N_I. \tag{2.6}$$

How can this be properly included in the theory? Fortunately, there are simplifications in a binary AB compound [208]:

- Pressure has negligible effect on the solid and the energy E_{AB} is close to its value at $T = 0$ (corrections for finite T can be made if needed).
- Assuming that the surface is in equilibrium with the bulk, there is only one free chemical potential, which can be taken to be μ_A since $\mu_A + \mu_B = \mu_{AB} \approx E_{AB}$.
- In equilibrium, the ranges of μ_A and μ_B are limited since each can never exceed the energy of the condensed pure element, $\mu_A \leq E_A$, and $\mu_B \leq E_B$.
- Therefore the allowed ranges of μ_A and μ_B can be determined by calculations for the elements and compound in the solid state, without any need to deal with gas phases.

This is sufficient for the theory to predict reconstructions of the surface *assuming equilibrium* as a function of the experimental conditions.

An instructive example is the study of various ZnSe (1 0 0) surface reconstructions [209] illustrated in Fig. 2.18. A $c(2 \times 2)$ reconstruction with half-monolayer coverage of twofold coordinated Zn atoms is stable in the Zn-rich limit. Under moderately Se-rich conditions, the surface adopts a (2×1) Se-dimer phase. In the extreme Se-rich limit, the theoretical calculations predicted a new structure with one and a half monolayer coverage of Se, proposed as an explanation for the high growth rates and migration observed in atomic layer epitaxy.

The structures in Fig. 2.18 also illustrate the importance of electrostatic effects in determining the pattern of a surface reconstruction that involves charge transfer within certain "building blocks." The preference for $c(2 \times 2)$ or (2×1) ordering of the building blocks is determined so that the electrostatic interaction (i.e., minimization of the surface Madelung energy – see Appendix F) is optimized, which was pointed out for GaAs (0 0 1) surfaces [210].

---

[10] Nonequilibrium involves kinetics that may allow other stoichiometries to be formed.

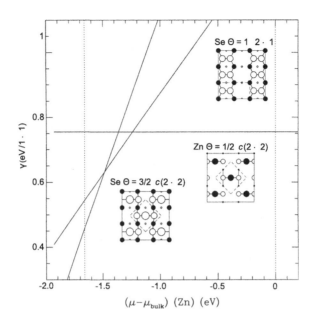

Figure 2.18. Energies of selected structures of the (1 0 0) surface of ZnSe as a function of the chemical potential (see text) calculated by the plane wave pseudopotential method. The Se-rich structure is an example of a theoretical prediction later found in experiment. Provided by A. Garcia and J. E. Northrup; essentially the same as in [209].

Surface States

Electronic states at surfaces are crucial for many properties of materials and they have a renewed relevance in electronic structure owing to the discovery of topological insulators. In Chapter 22 are two examples with comparison of theory and experiment. Gold is one of the classic examples of a metal with a Shockley surface state. Figure 22.3 illustrates the surface state in a way that brings out the importance of spin–orbit interaction and the Rashba effect. A powerful technique that allows photoemission measurements of states above the Fermi level, together with DFT calculations, show that the bands have nontrivial topological character even though gold is a metal. The theory also shows that if the spin–orbit interaction were larger, there would be a full gap with surface bands due to the topology. The results have simple interpretations in models in Chapter 27, which show similar evolution of the bands as the spin–orbit interaction is increased.

Surface states of semiconductors are strongly modified from the bulk because of the disruption of the covalent bonds, and there are many examples of reconstructions. One of the simpler reconstructions is the buckled dimer structure in Fig. 2.17 and the resulting bands for Ge are shown in Fig. 22.4. There are two dangling bonds for each dimer, which are split into two surface bands since buckling makes the two sites inequivalent, with the lower energy state filled to create a lone pair and a gap to the empty surface band.

Interfaces

Owing to their importance for technology, semiconductor interfaces have been prepared and characterized with great control. Of particular importance are the "band offsets" at an interface, which confine the carriers in semiconductor quantum devices [211]. Calculation of the offset involves two aspects of electronic structure. The energies of the states in each of the two materials is defined relative to a reference for that material; this is an intrinsic property independent of the interface. However, the reference energies are shifted by long-range Coulomb terms and relative energies in the two materials depends on the interface dipole (Section F.5), which requires a calculation of the interface. Sorting out these effects are topics of Sections 22.6 and 22.7. Studies of semiconductors, described briefly in Section 22.6, are a story of theory working with experiment, with a revision of previously held rules due to theoretical calculations [211–213]. More recent calculations are providing greater accuracy and confidence in the results.

One of the great advances in man-made structures is the growth of oxide interfaces with atomic-scale precision that rivals the well-known semiconductors. This has opened an entire field of two-dimensional systems with metallic bands, and many phenomena including magnetism and superconductivity. The spark for the field was the 2004 paper by Ohtomo and Hwang [214], which reported a high-mobility electron gas at the $LaAlO_3/SrTiO_3$ (LAO/STO) interface. There is much work with a variety of materials and Section 22.7 brings out some of the issues related to the origin of the carriers (the "polarization catastrophe" and ways to avoid it in Fig. 22.6) and the nature of the interface bands that have common features in many of the examples shown in Fig. 22.7.

2.12 Low-Dimensional Materials and van der Waals Heterostructures

There is great interest in two-dimensional materials formed as sheets a single atom thick, like graphene or BN, or a single cell of compounds like MoS_2, $MoSe_2$, WSe_2, and many others. They can exhibit properties completely different from usual three-dimensional materials. Graphene is the model system in many places in this book; it is the real two-dimensional material that is the strongest material known in the plane; and it has exceptional thermal and electronic transport, and many other properties. Other materials are metals, insulators, piezoelectrics, superconductors, magnets, etc. An example of the materials that may be of technological importance is single-layer MoS_2. Whereas bulk MoS_2 has an indirect gap and only weak optical absorption, a single layer has a direct gap and absorbs light strongly. This was predicted theoretically [216] and later realized experimentally in two works about the same time [217, 218]. In Figure 22.8 are shown the bands resulting from the calculation, which also provides a simple explanation for the effects.

One of the most promising developments in new materials this century is the ability to create an almost unlimited variety of new materials by stacking layers like Lego blocks, called "van der Waals heterostructures" since the layers are weakly bonded [219, 220]. Nevertheless, they are enough in contact to stabilize the solid and influence one another so that they form entirely new materials. Figure 2.19 shows the exquisite control with

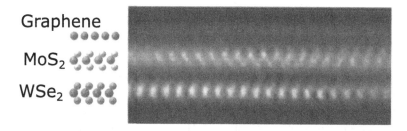

Figure 2.19. Electron microscope image of layers of graphene, MoS_2, and WSe_2 grown by chemical vapor deposition (CVD). This combination of layers was chosen to have large mismatch of mass densities and phonons, with the result that the thermal conductivity perpendicular to the layers is very low, less than air [215]. Provided by H. Zhang, S. Vaziri, and E. Pop.

which such structures can be made. In this case, graphene and single layers of MoS_2 and WSe_3 all have triangular Bravais lattices, as shown in Figs. 4.5 and 4.6, but they have very different properties. The structure in Fig. 2.19 was shown [215] to have extremely low thermal conductivity in the direction perpendicular to the layers, which is due to the large mismatch of mass density and phonon energies between the layers. The thermal resistance is greater than a layer of SiO_2 100 times as thick and may be useful for thermal isolation and routing of heat flow in nanoscale systems.

2.13 Nanomaterials: Between Molecules and Condensed Matter

In some ways clusters with nanometer size are just large molecules, but they share the property of condensed matter that clusters can have varying size. Yet they are small enough that their properties depend on finite-size quantization effects and by the fact that a large fraction of the atoms are in surface regions. Because the structure is extremely hard to determine directly from experiment, theory has a great role to play. This is exemplified by metallic clusters, the size of which can be varied from a few atoms to macroscopic dimensions. The observation of "magic numbers" for small clusters can be understood in terms of the filling of shells in a sphere [221, 222], but the details are more complex. The atomic-scale structures and optical spectra of such clusters are described in more detail in Chapter 21.

Semiconductor nanostructures have been of particular interest because confinement effects lead to large increases in the bandgaps and efficient light emission has been observed, even in Si, for which coupling to light is extremely weak in the bulk crystal. In the case of a semiconductor, broken bonds lead to reconstruction of the surface, and in the smallest clusters there is little resemblance to the bulk structures, as illustrated in Fig. 2.20. For example, in Si_{13} there is competition between a symmetric structure with 12 outer atoms surrounding a central atom and the low-symmetry structure found by Car–Parrinello methods and simulated annealing [223]. The symmetric structure was argued [224] to be stabilized by correlations not accounted for in the local approximation in density functional theory; however, quantum Monte Carlo calculations [225, 226] found the low-symmetry structure to be the most stable, in agreement with Car–Parrinello simulations.

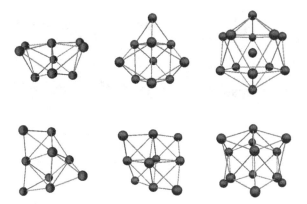

Figure 2.20. Atomic positions in competing Si_n clusters consisting of $n = 9$, 10, and 13 atoms. In each case, two very different structures are shown. The correct structure is not known from direct experiments, and theory plays an important role in sorting out likely candidates. Provided by J. Grossman; from among the cases studied in [225].

On the other hand, if the surface is terminated by atoms such as hydrogen or oxygen, which remove the dangling bonds, then the cluster is much more like a small, terminated piece of the bulk. The energy of the light emitted is increased by quantum confinement of the electron states in the cluster, and the emission strength is greatly increased by breaking of bulk selection rules due to the cluster size, shape, and detailed structure. This is an ideal case for combined theoretical and experimental work to interpret experiments and improve the desired properties. Calculations using time-dependent density functional theory (Section 21.2) are used as an illustration of the methods in Chapter 21, e.g., the variation of the gaps versus size, shown in Fig. 21.3.

Many of the materials that initiated the great interest in nanomaterials are the wealth of structures that can be made from carbon: fullerenes, C_{60}, C_{70}, ..., [227], nanotubes in the early 1990s [228], graphene [229], and graphene ribbons in the 2000s. They are extraordinary not only because of their exceptional properties but also because of their versatility and elegant simplicity. C_{60} is the most symmetric molecule in the sense that its point group (icosahedral) with 120 symmetry operations is the largest point group of the known molecules. A "buckyball" has the shape of a football (soccer ball in the USA), made of 60 carbon atoms arranged in 20 hexagons and 12 pentagons.[11] Interest in fullerenes increased dramatically with the discovery [230] of how to produce C_{60} in large enough quantities to make solids. In rapid succession it was found that intercalation of alkali–metal atoms in solid C_{60} leads to metallic behavior [231] and that some alkali-doped compounds (fullerides) are superconductors with transition temperatures surpassed only by the cuprates at the time.

---

[11] The name for the structures derives from R. Buckminster Fuller, professor at Southern Illinois University in Carbondale and a visionary engineer who conceived the geodesic dome.

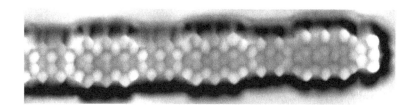

Figure 2.21. Graphene nonoribbon grown by a technique that uses self-assembled molecule precursors as a pattern [235]. This example is repeated segments of seven and nine atom widths that have different topological indices and a state at the end only for one case, as discussed in Section 26.7. Provided by Daniel Rizzo.

Nanotubes are made from graphene-like sheets (or multiple sheets) rolled into a tube [232–234] as illustrated in Fig. 14.6, which can form a great variety of semiconductors and metals. These are ideal systems in which the bands are beautifully described by theoretical rolling of the Brillouin zone of graphene. However, the curvature adds a coupling on the σ and π bonds not present in flat graphene sheets [232, 233] and large changes in the bands can occur in tubes with very small radii, as described in Sections 13.6 and 14.8.

Graphene is a single layer of atoms, the strongest material known in the plane, a model system that is an example of a simple band structure with a Dirac point (Section 14.7), one of the first topological insulators (Section 27.8), and it can form ribbons of varying widths. Figure 2.21 shows an STM image of a patterned graphene nanoribbon synthesized on an Ag substrate using a "bottom-up" procedure that provides atomically precise control through self-assembly of small-molecule precursors [235]. The repeated patten of seven- and nine-atom width segments illustrates the fine detail that can be engineered in nanoscale systems, and it provides a realization of a one-dimensional system that can be considered to be a crystalline topological insulator (see [236] and Section 26.7).

2.14 Electronic Excitations: Bands and Bandgaps

Electronic excitations can be grouped into two types: excited states with the same number N of electrons as the ground state, and single-particle excitations in which one electron is subtracted $N \rightarrow N - 1$ or added $N \rightarrow N + 1$. The former excitations determine the specific heat, linear response, optical properties, etc., whereas the latter are probed experimentally by tunneling and by photoemission or inverse photoemission.

The most important quantity for adding and removing electrons is the *fundamental gap*, which is the minimum difference between the energy for adding and subtracting an electron. The lowest gap is *not* an approximate concept restricted to independent-particle approximation. It is defined in a general many-body system as the difference in energy between adding an electron and removing one: if the ground state has N electrons, the fundamental gap is

$$E_{\text{gap}}^{\text{min}} = \min\{[E(N + 1) - E(N)] - [E(N) - E(N - 1)]\}. \tag{2.7}$$

Metals are systems in which the gap vanishes *and* the lowest-energy electron states are delocalized. On the other hand if the fundamental gap is nonzero or if the states are localized (due to disorder) the system is an insulator.

Angle- and Energy-Resolved Photoemission (ARPES)

The primary tool for direct observation of the spectrum of energies for removing an electron as a function of the crystal momentum **k** is angle-resolved photoemission [238, 239], shown schematically in Fig. 2.22. Because the electrons are restricted to a surface region, photoemission is a surface probe and care must be taken to separate surface and bulk information. The momentum of the excitation in the crystal parallel to the surface is determined by momentum conservation as illustrated in Fig. 2.22. The dispersion perpendicular to the surface can be mapped out from the dependence on the photon energy and information about the dispersion of the excited electron inside the crystal [238, 239]. In an independent-particle picture there are sharp peaks in the energies of the emitted electrons that are the eigenvalues or bands for the electrons as illustrated in Fig. 2.22. Weak interactions lead to small broadenings and shifts of the peaks, whereas strong interactions can lead to qualitative changes.

Angle-resolved photoemission was demonstrated as a quantitative experimental method in the late 1970s, as illustrated by the results for GaAs in Fig. 12.2. With dramatic improvements in resolution using synchrotron radiation, ARPES has become a powerful tool to measure the detailed dispersion and many-body effects for the one-electron removal spectrum in crystals. Inverse photoemission makes it possible to map out the electron addition spectrum. Schematically, it is inverse of ARPES shown in Fig. 2.22 but it is more difficult in practice. In many cases there are sharp peaks that describe well-defined bands

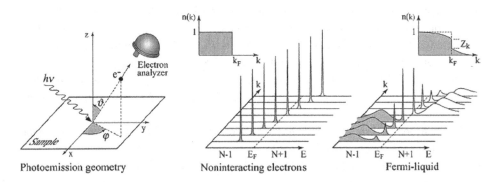

Figure 2.22. Schematic illustration of angle-resolved photoemission spectroscopy (ARPES). The middle figure shows an independent-particle spectrum of delta functions with occupation 1 or 0. The right figure exemplifies a spectrum for interacting electrons with "quasiparticle" peaks (i.e., peaks due to dressed one-particle excitations) that correspond to bands if the peaks are very sharp. However, there are also many-body effects that cause the panels to have fractional weight and "satellites" or "sidebands" due to additional excitations that are induced in the system. These are topics of [1] and are not addressed here. (Figure modified from [237].)

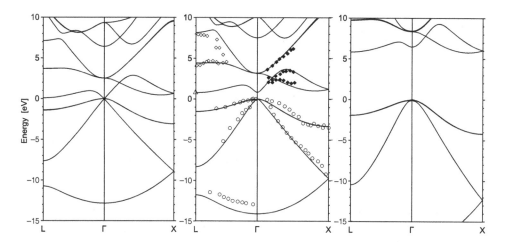

Figure 2.23. Bands for Ge from experiment (dots in the central panel) and theory using three different approximations. Left: the local density approximation where the infamous bandgap problem is so severe that it predicts Ge to be a metal. Right: the Hartree–Fock approximation with the well-known problem that the gaps are much too large. Center: the results using the HSE06 functional that is a combination of a density functional and Hartree–Fock-like exchange, as described in Section 9.3. Hybrid functionals often lead to bandgaps in much better agreement with experiment than the LDA or GGA functionals as shown in Fig. 2.24. Provided by J. Voss.

like those plotted as the points in the central Fig. 2.23 for Ge. In other cases the spectra are broader and show additional features due to interactions, as illustrated in the right side of Fig. 2.22. This is a primary topic in [1] and there are excellent reviews such as [240].

Theory and Experiment for Bands and Bandgaps

Despite the impressive agreement with experiment of many density functional theory calculations for ground state properties, the same calculations for insulators often lead to mediocre (or disastrous) predictions for excitations. Figure 2.23 illustrates the comparison of theory and experiment for Ge for both addition and removal spectra. This example is chosen because it illustrates a spectacular failure of the widely-used approach to interpret the eigenvalues of a Kohn–Sham calculation as the energies for adding and removing electrons.

The left-hand panel shows the results for the bands using the local density approximation which results in Ge predicted to be a metal, an extreme example of the "bandgap problem" which is found in many materials, as shown in Fig. 2.24. (The problem also occurs using a GGA functional such as PBE.) The central panel shows the results using a hybrid functional HSE06. The filled valence bands are almost unchanged but the conduction bands are lifted relative to the valence bands to open the gap to close to the experimental value. The improvement occurs in all the examples in Fig. 2.24 which are representative of many more materials.

The right-band panel in Fig. 2.23 shows the results of Hartree–Fock calculation for Ge. The gap is much too large, which is due to the large effect of exchange in a theory

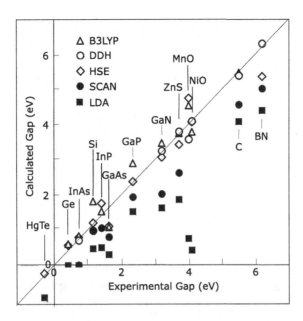

Figure 2.24. Bandgaps compared to experiment for various functionals described in Chapters 8 and 9. The LDA and meta-GGA functionals are denoted by closed symbols, and hybrid functionals by open symbols. In general, hybrid functionals are much better for gaps, but they require considerably more computational effort. Experiment, HSE06, and BLYP from [242]; LDA and SCAN from [243]; DDH from [244].

that explicitly excludes correlation. Even though it is a poor approximation, nevertheless, it is a proper theory in which it is correct to identify the eigenvalues with removal and addition energies.[12] A simplistic way to understand the results of the HSE06 calculation is that it is a mixture of a density functional and Hartree–Fock. However, there is much deeper reasoning. As explained in Chapter 9, the generalized Kohn–Sham (GKS) approach provides the framework for theories to describe more than just the ground state. Within that approach, the eigenvalues of the equations can properly be viewed as approximations to the energies for adding and removing electrons.

The fundamental gap is the key issue, and widely used approximate functionals in density functional theory lead to gaps (Eq. (2.7)) that are significantly below experimental values for essentially all materials. This is illustrated by results of calculations using the local density approximation (LDA) for a range of materials, shown by the dark square symbols in Fig. 2.24. The major effect of the improved functionals is to shift the empty bands almost rigidly up relative to the filled bands, as indicated in Fig. 2.23 for Ge and the left

---

[12] It is useful to note that many-body "GW" methods are essentially random phase approximation (RPA) calculations (see Section 5.5) for the quasiparticle self-energy, originally developed for the electron gas [241]. It can be considered to be a dynamically screened exchange that reduces the effect of the Hartree–Fock exchange. See [1] for extensive discussion.

side of Fig. 2.25 for GaAs. Improvements also occur for wide-gap materials not shown; for example, the gap in LiF is 14.2 (exp), 8.9 (LDA), 10.0 (SCAN), 11.5 (HSE[13]), and 16.1 (DDH).

Improvement in the theory of excitations in insulators is an active area of electronic structure research. In the generalized Kohn–Sham theory in Chapter 9, there are whole families of auxiliary systems that in principle lead to exact ground-state density and energy. The Kohn–Sham auxiliary system consists of independent particles, which means the exchange–correlation functional has to incorporate all aspects of exchange and correlation. If the auxiliary system has particles that interact in some way, then it could take some of the burden and make it easier to construct a functional that has to incorporate only the remaining aspects of exchange and correlation. The goal is a formulation that works at least as well as the usual Kohn–Sham approach for the ground state and also improves other properties.

2.15 Electronic Excitations and Optical Spectra

Dielectric functions and conductivity are among the most important response functions in condensed matter physics because they determine the optical properties of materials, electrical conductivity, and a host of technological applications. Optical spectra are perhaps the most widespread tool for studying the electronic excitations themselves. In addition, the response of the electrons to applied electric fields is directly related to the Coulomb interactions among the electrons. The phenomenological formulation of Maxwell's equations in the presence of polarizable or conducting media can be cast in terms of the complex frequency-dependent dielectric function $\epsilon(\omega)$ or conductivity $\sigma(\omega)$. The relations are summarized in Appendix E and the formulation in terms of electronic excitations is the subject of Chapter 21.

An excitation that does not change the number of electrons, e.g., an excitation created by absorbing a photon, can be viewed as the simultaneous addition of an electron and a hole, which can interact with one another. This is a traditional view that is very useful because it relates the excitation to the addition/removal spectra that can be measured separately and identifies the interaction as a quantity to be studied, e.g., in many-body pertubation theory as described in [1] and many other sources. However, it can also be viewed as an excitation of the density, which evokes the idea of a time-dependent (or frequency-dependent) density functional theory. This idea was used in early work like [245] and the theory of time-dependent density functional theory (TDDFT) was put on a firm footing by derivations [246] analogous to the theorem for static density functional theory. As described in Chapter 21 TDDFT describes excitation spectra in principle exactly, and in practice feasible approximations have proven to be extremely successful.

TDDFT is now widely used, especially in the chemistry community, for optical spectra of confined systems such as molecules and clusters [247–250]. Examples of metallic clusters

---

[13] Provided by J. Paier.

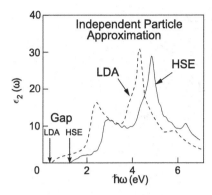 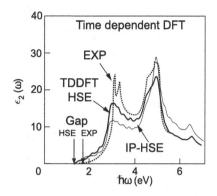

Figure 2.25. Optical spectra of GaAs comparing independent-particle and time-dependent DFT (TDDFT) for LDA and HSE functionals. The LDA has the well-known problem that the gap is much too small; the independent-particle spectrum (left) for the hybrid HSE06 functional shifted almost rigidly and the gap is closer to experiment. The gap is not changed in the TDFT spectra (right) but the weight is shifted to lower energy to be closer to experiment. LDA and experiment from [251]; HSE06 from [252].

and hydrogen-terminated Si clusters are described in Section 21.7, calculated assuming the usual adiabatic LDA functional. However, the adiabatic functional misses important physics, and the search for improved time-dependent functionals is a topic of much current research.

TDDFT is used less in condensed matter, where there are challenging issues for TDDFT. An example of optical absorption spectra is shown in Fig. 2.25 for the semiconductor GaAs. At the left is the independent-particle spectra, which is just the convolution of the filled and empty bands. The spectra for the LDA and the HSE06 hybrid functional illustrate a feature observed in many systems: the effect of the improved functional is mainly just a rigid shift of the empty states relative to the filled ones. The shift is sometimes called a scissor operation that has the effect of cutting the graph into two pieces and shifting the higher-energy part to increase the gap. At the right in Fig. 2.25 is shown the HSE06 independent-particle spectrum (the same as at the left) compared with TDDFT using the HSE functional. The first observation is that the gap does not change. As explained in Chapter 21 this is a general result, and TDDFT cannot be used to fix errors in the gap! The other observation is that the spectral weight is shifted by TDDFT so that it is closer to experiment, and we see that TDDFT has the potential to be very useful.

There is, however, a much more striking feature brought out by calculations for wide-gap insulators, as shown for LiF in Fig. 2.26. This is a case where the optical spectrum is dominated by excitons. The independent-particle spectrum is zero for any energy less than the gap of 14.2 eV, but the great peak is at lower energy and is completely different from the independent-particle spectrum. Furthermore, as explained in Chapter 21, the exciton peak cannot be described by any short-range functional like LDA, GGA, HSE, etc. Even though a functional like HSE06 improves the gap, it fails completely for excitons. The problem is illustrated for LDA in Fig. 2.26 and for HSE06 in the left panel of Fig. 21.5.

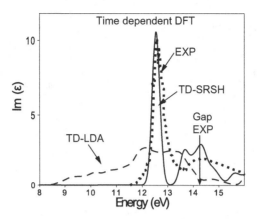

Figure 2.26. Optical spectra of LiF which is dominated by the exciton peak at 12.6 eV, which is more than 1 eV below the gap. The exciton is well described by the hybrid functional called SRSH, which has a long-range component as described in the text. That functional has been adjusted so the gap agrees with experiment, and the figure shows that there are two other exciton features below the gap. The TD-LDA has the well-known problem that the gap is much too low, but the most important point for this purpose is that it completely fails to describe the exciton peak. Any short-range functional (including hybrid HSE06 and related functionals) has the same failure to describe the dominant feature in the spectrum. See Section 21.8, especially Fig. 21.5, for a more complete discussion. Modified from a figure provided by S. Refaely-Abramson with information the same as in [253].

The difficulty in describing excitons in TDDFT is that the functional must have nonlocal character, which is very difficult to formulate in terms of the density only. Fortunately, hybrid functionals that have some fraction of long-range Hartree–Fock long exchange provide a practical solution. As shown in Fig. 2.26 such a functional can lead to exciton states below the gap and even provide quantitative descriptions of the spectrum, as discussed further in Chapter 21. In general one should not expect such quantitative agreement with experiment. The functional used for the calculations in Fig. 2.26 is a "tuned range-separated hybrid" that has been adjusted to fit the gap. Any similar function would lead to excitons very much like experiment, but the energy at which the peak occurs might not agree as well as the example in the figure.

It is ironic to resort to functionals that include long-range Hartree–Fock exchange, which is a disaster in metals (see Section 5.2), completely at variance with experiment as pointed out by Bardeen [254] in 1936. Yet the functional must have such terms in order to describe the spectra of insulators. How can this be overcome? One approach is to scale the part of the functional involving Hartree–Fock exchange by $1/\epsilon$ so that it vanishes for metals and is large in wide-gap insulators where the dielectric constant is small. This is the guiding principle for a class of functionals described in Section 9.3 and is incorporated in the functional used in the calculations in Fig. 2.26. Another example of this type of functional is used for the molecular dynamics simulations of water in Fig. 2.15.

2.16 Topological Insulators

The discovery of topological insulators in 2005–2006 by Kane and Mele [86, 87] and Bernevig and Zhang [88] was one of the greatest developments in the conceptual structure of condensed matter theory since the Bloch theorem. There is a long history of topological classification of electronic states, perhaps most famously in the Hall effect (QHE) (see Appendix Q) where Thouless and coworkers in 1982 showed that the precise integer multiples of the resistance is a consequence of topology and is given by a topological invariant. However, the QHE occurs only in the presence of a strong magnetic field with no need for quantitative calculations. The discovery of topological insulators, where spin–orbit interaction leads to related effects in the absence of a magnetic field, has brought topology squarely into electronic structure.

In this book, the theme is that the topology can be understood with only knowledge of band structure and spin–orbit interaction at the level of an undergraduate solid state physics textbook. The presentation in Chapters 25–28 builds upon the concepts of Berry phases. Appendix P shows that Berry phases can be understood in general terms using only the superposition principle of quantum mechanics, and it is only a few steps to Chern numbers that are the topological invariants. The topology essential for our purposes is for the electronic system as a whole as a function of momentum **k** in the Brillouin zone, which is brought out in Chapter 25. Another theme is that there are precedents in the work of Shockley in 1939 [42], Thouless in 1984 [256], and the theory of polarization by King-Smith and Vanderbilt in 1993 [175]. This background can be found in the theory of Shockley transition and surface states in Chapter 22, electron pumps in Chapter 26 and polarization in Chapter 24. The topological character of the states in graphene nanoribbons in Fig. 2.21 is described in Section 26.7.

There is qualitatively new physics in topological insulators, the topic of Chapters 27 and 28. By including spin and spin–orbit interaction entirely new effects were found and shown to be described by a topological invariant different from any of the previous works.

In this chapter the purpose is to illustrate experimental measurements that definitively demonstrate the effects. The cover of this book is an illustration of the surface state on Bi_2Se_3 observed by photoemission [255], and the left side of Fig. 2.27 shows the same data as a cut showing the dispersion of the surface state. The right side of the figure shows the theoretical calculation [257] on the same scale, which is described in more detail in Section 28.3. This is but one of the examples of topological character, which also occurs in gold (Section 22.4), HgTe/CdTe quantum wells (Section 27.7), graphene (Section 27.8), and many models in Chapters 26–28.

2.17 The Continuing Challenge: Electron Correlation

The competition between correlation due to interactions and delocalization due to kinetic energy leads to the most challenging problems in the theory of electrons in condensed

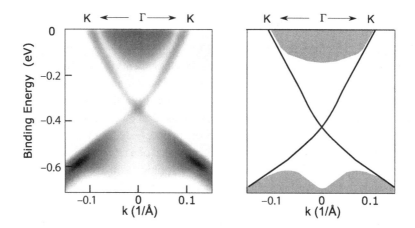

Figure 2.27. Left: ARPES measurement of the surface state of the topological insulator Bi_2Se_3, showing the Dirac point along with the continuum of bulk bands, which is the cross-section of the data shown on the cover of this book. Right: theoretical calculations taken from Fig. 28.5, plotted on the same energy and momentum scales as the experimental graph. There is no adjustment of the scales except that the theory has been positioned to approximately line up with the experimental Fermi energy for this sample, which is doped so that the Fermi energy is in the conduction band. The left figure was provided by Yulin Chen and is similar to figure 1b of [255].

matter. Correlations are responsible for metal–insulator transitions, the Kondo effect, heavy fermion systems, high-temperature superconductors, and many other phenomena (see, e.g., [240]). Low dimensionality leads to larger effects of correlation and new phenomena such as the quantum Hall effect. In one dimension, the Fermi liquid is replaced by a Luttinger–Tomanaga liquid in which the excitations are "holons" and "spinons." The broadening of quasiparticle peaks and sidebands in Fig. 2.22 requires explicit many-body methods.

These are the aspects of interactions that are the topics of [1]. However, it should be emphasized again that interactions must be taken into account in *any* theory that aspires to be quantitative. It is only because DFT incorporates effects of interactions with sufficient accuracy that it is now by far the dominant method in the theory of electrons in materials. It provides the wavefunctions used as the starting point for most quantitative many-body calculations. It is the method used to develop databases for materials by design and has many other uses.

As the theory of electronic structure becomes more powerful and more predictive, it is even more relevant to keep in mind the "big picture" of the possible consequences of many-body electron–electron interactions. Not only is a proper accounting of the effects of interactions essential for quantitative description of real materials, but also imaginative exploration of correlation can lead to exciting new phenomena qualitatively different from predictions of usual mean-field theories. As emphasized in the introduction, Section 1.4, the concepts are captured in the notion [81] of "more is different" with recent developments summarized in *More Is Different: Fifty Years of Condensed Matter Physics* [82].

SELECT FURTHER READING

The companion book [1] addresses many of the same issues, focusing on aspects that require many-body methods beyond density functional theory and the Kohn–Sham equations.

See the classic references at the end of Chapter 1.

More recent books that address the field of condensed matter as a whole:

Ashcroft, N. and Mermin, N., *Solid State Physics* (W. B. Saunders Company, New York, 1976).

Chaikin, P. N. and Lubensky, T. C., *Principles of Condensed Matter Physics* (Cambridge University Press, Cambridge, 1995).

Cohen, M. L. and Louie, S. G., *Fundamentals of Condensed Matter Physics* (Cambridge University Press, Cambridge, 2016).

Girvin S. M. and Yang K., *Modern Condensed Matter Physics* (Cambridge University Press, Cambridge, 2019).

Ibach, H. and Luth, H., *Solid State Physics: An Inroduction to Theory and Experiment* (Springer-Verlag, Berlin, 1991).

Kittel, C., *Introduction to Solid State Physics* (John Wiley & Sons, New York). Many editions.

Marder, M. *Condensed Matter Physics* (John Wiley and Sons, New York, 2000).

Books that focus upon electronic structure:

Giustino, F., *Materials Modelling Using Density Functional Theory: Properties and Predictions* (Oxford University Press, Oxford, 2014).

Harrison, W. A., *Elementary Electronic Structure* (World Publishing, Singapore, 1999).

Kaxiras, E., *Atomic and Electronic Structure of Solids* (Cambridge University Press, Cambridge, 2003).

Kaxiras, E. and Joannopoulos, J. D., *Quantum Theory of Materials*, 2nd rev. ed. (Cambridge University Press, Cambridge, 2019).

Kohanoff, J., *Electronic Calculations for Solids and Molecules: Theory and Computational Methods* (Cambridge University Press, Cambridge, 2003).

3

Theoretical Background

Summary

Our understanding of the electronic structure of matter is based on quantum mechanics and statistical mechanics. This chapter reviews fundamental definitions and expressions valid for many-body systems of interacting electrons and useful, simplified formulas that apply for noninteracting particles. This material is the foundation for subsequent chapters, which deal with the theory and practical methods for electronic structure.

3.1 Basic Equations for Interacting Electrons and Nuclei

The subject of this book is the ongoing progress toward describing properties of matter from theoretical methods firmly rooted in the fundamental equations. Thus our starting point is the hamiltonian for the system of electrons and nuclei,[1]

$$\hat{H} = -\frac{\hbar^2}{2m_e} \sum_i \nabla_i^2 - \sum_{i,I} \frac{Z_I e^2}{|\mathbf{r}_i - \mathbf{R}_I|} + \frac{1}{2} \sum_{i \neq j} \frac{e^2}{|\mathbf{r}_i - \mathbf{r}_j|}$$

$$- \sum_I \frac{\hbar^2}{2M_I} \nabla_I^2 + \frac{1}{2} \sum_{I \neq J} \frac{Z_I Z_J e^2}{|\mathbf{R}_I - \mathbf{R}_J|}, \tag{3.1}$$

where electrons are denoted by lowercase subscripts and nuclei – with charge Z_I and mass M_I – are denoted by uppercase subscripts. The challenge in the theory of electronic

---

[1] There are also other ingredients in the fundamental theory that are incorporated later, including spin–orbit interaction H_{SO} in Eq. (O.8) derived from the Dirac equation in Appendix O; electromagnetic fields that can be taken into account by adding a vector potential with the replacement $\mathbf{p} \rightarrow \boldsymbol{\pi} = \mathbf{p} - (e/c)\mathbf{A}$ (see Appendix E); and a Zeeman term in the hamiltonian $\mu_B \boldsymbol{\sigma} \cdot \mathbf{B}$, where $\mu_B = \hbar e/2m$ is the Bohr magneton. These are independent-electron terms that are often small and are ignored in much of this book; however, they are responsible for some of the most interesting phenomena in condensed matter including topological insulators.

structure is to develop methods that can treat the effects of the electron–electron interactions with sufficient accuracy that one can predict the diverse array of phenomena exhibited by matter starting from Eq. (3.1). It is most informative and productive to start with the fundamental many-body theory. Many expressions, such as the force theorem, are more easily derived in the full theory with no approximations. It is then straightforward to specialize to independent-particle approaches and the actual formulas needed for most of the following chapters.

There is only one term in Eq. (3.1) that can be regarded as small, the inverse mass of the nuclei $1/M_I$. A perturbation series can be defined in terms of this parameter, which is expected to have general validity for the full interacting system of electrons and nuclei. If we first set the mass of the nuclei to infinity, then the kinetic energy of the nuclei can be ignored. This is the Born–Oppenheimer or adiabatic approximation [90] defined in Appendix C, which is an excellent approximation for many purposes, e.g., the calculation of nuclear vibration modes in most solids [91, 180]. In other cases, it forms the starting point for perturbation theory in electron–phonon interactions, which is the basis for understanding electrical transport in metals, polaron formation in insulators, certain metal–insulator transitions, and the BCS theory of superconductivity. Thus we shall focus on the hamiltonian for the electrons, in which the positions of the nuclei are parameters.

Ignoring the nuclear kinetic energy, the fundamental hamiltonian for the theory of electronic structure can be written as

$$\hat{H} = \hat{T} + \hat{V}_{\text{ext}} + \hat{V}_{\text{int}} + E_{II}. \tag{3.2}$$

If we adopt Hartree atomic units $\hbar = m_e = e = 4\pi/\epsilon_0 = 1$, then the terms may be written in the simplest form. The kinetic energy operator for the electrons $\hat{T}$ is

$$\hat{T} = \sum_i -\frac{1}{2}\nabla_i^2; \tag{3.3}$$

$\hat{V}_{\text{ext}}$ is the potential acting on the electrons due to the nuclei,

$$\hat{V}_{\text{ext}} = \sum_{i,I} V_I(|\mathbf{r}_i - \mathbf{R}_I|); \tag{3.4}$$

$\hat{V}_{\text{int}}$ is the electron–electron interaction,

$$\hat{V}_{\text{int}} = \frac{1}{2}\sum_{i \neq j} \frac{1}{|\mathbf{r}_i - \mathbf{r}_j|}; \tag{3.5}$$

and the final term E_{II} is the classical interaction of nuclei with one another and any other terms that contribute to the total energy of the system but are not germane to the problem of describing the electrons. Here the effect of the nuclei upon the electrons is included in a fixed potential "external" to the electrons. This general form is still valid if the bare nuclear Coulomb interaction is replaced by a pseudopotential that takes into account effects of core electrons (except that the potentials are "nonlocal"; see Chapter 11). Also, other "external potentials," such as electric fields and Zeeman terms, can readily be included. Thus, for electrons, the hamiltonian, Eq. (3.2), is central to the theory of electronic structure.

Schrödinger Equation for the Many-Body Electron System

The fundamental equation governing a nonrelativistic quantum system is the time-dependent Schrödinger equation,

$$i\hbar \frac{d\Psi(\{\mathbf{r}_i\};t)}{dt} = \hat{H}\Psi(\{\mathbf{r}_i\};t), \tag{3.6}$$

where the many-body wavefunction for the electrons is $\Psi(\{\mathbf{r}_i\};t) \equiv \Psi(\mathbf{r}_1, \mathbf{r}_2, \ldots, \mathbf{r}_N; t)$, the spin is assumed to be included in the coordinate $\mathbf{r}_i$, and, of course, the wavefunction must be antisymmetric in the coordinates of the electrons $\mathbf{r}_1, \mathbf{r}_2, \ldots, \mathbf{r}_N$. The eigenstates of Eq. (3.6) can be written as $\Psi(\{\mathbf{r}_i\};t) = \Psi(\{\mathbf{r}_i\})e^{-i(E/\hbar)t}$. Even though it is not feasible to actually solve the equations for many interacting electrons, it is very useful and instructive to define properties such as the energy, density, forces, and excitations in the full many-body framework before making approximations.

For an eigenstate, the time-independent expression for any observable is an expectation value of an operator $\hat{O}$, which involves an integral over all coordinates,

$$\langle \hat{O} \rangle = \frac{\langle \Psi | \hat{O} | \Psi \rangle}{\langle \Psi | \Psi \rangle}. \tag{3.7}$$

The density of particles $n(\mathbf{r})$, which plays a central role in electronic structure theory, is given by the expectation value of the density operator $\hat{n}(\mathbf{r}) = \sum_{i=1,N} \delta(\mathbf{r} - \mathbf{r}_i)$,

$$n(\mathbf{r}) = \frac{\langle \Psi | \hat{n}(\mathbf{r}) | \Psi \rangle}{\langle \Psi | \Psi \rangle} = N \frac{\int d^3 r_2 \cdots d^3 r_N \sum_{\sigma_1} |\Psi(\mathbf{r}, \mathbf{r}_2, \mathbf{r}_3, \ldots, \mathbf{r}_N)|^2}{\int d^3 r_1 d^3 r_2 \cdots d^3 r_N |\Psi(\mathbf{r}_1, \mathbf{r}_2, \mathbf{r}_3, \ldots, \mathbf{r}_N)|^2}, \tag{3.8}$$

which has this form because of the symmetry of the wavefunction in all the electron coordinates. The total energy is the expectation value of the hamiltonian,

$$E = \frac{\langle \Psi | \hat{H} | \Psi \rangle}{\langle \Psi | \Psi \rangle} \equiv \langle \hat{H} \rangle = \langle \hat{T} \rangle + \langle \hat{V}_{\text{int}} \rangle + \int d^3 r \, V_{\text{ext}}(\mathbf{r}) n(\mathbf{r}) + E_{II}, \tag{3.9}$$

where the expectation value of the external potential has been explicitly written as a simple integral over the density function. The final term E_{II} is the electrostatic nucleus–nucleus (or ion–ion) interaction, which is essential in the total energy calculation but is only a classical additive term in the theory of electronic structure.

The eigenstates of the many-body hamiltonian are stationary points (saddle points or the minimum) of the energy expression (3.9) if Ψ is regarded as a variable trial function. These may be found by varying the ratio in Eq. (3.9) or by varying the numerator subject to the constraint of orthonormality ($\langle \Psi | \Psi \rangle = 1$), which can be done using the method of Lagrange multipliers,

$$\delta[\langle \Psi | \hat{H} | \Psi \rangle - E(\langle \Psi | \Psi \rangle - 1)] = 0. \tag{3.10}$$

This is equivalent to the well-known Rayleigh–Ritz principle[2]

$$\Omega_{RR} = \langle \Psi | \hat{H} - E | \Psi \rangle, \tag{3.11}$$

which is stationary at any eigensolution $|\Psi_m\rangle$.[3] Variation of the bra $\langle \Psi |$ leads to

$$\langle \delta \Psi | \hat{H} - E | \Psi \rangle = 0. \tag{3.12}$$

Since this must hold true for all possible $\langle \delta \Psi |$, this can be satisfied only if the ket $|\Psi\rangle$ satisfies the time-independent Schrödinger equation

$$\hat{H} |\Psi\rangle = E |\Psi\rangle. \tag{3.13}$$

In Exercise 3.1 it is shown that the same equations result from explicit variation of Ψ in Eq. (3.9) without Lagrange multipliers.

The ground-state wavefunction Ψ_0 is the state with the lowest energy, which can be determined, in principle, by minimizing the total energy with respect to all the parameters in $\Psi(\{\mathbf{r}_i\})$, with the constraint that Ψ must obey the particle symmetry and any conservation laws. Excited states are saddle points of the energy with respect to variations in Ψ.

Ground and Excited Electronic States

The distinction between ground and excited states pointed out in Chapter 2 is equally obvious when approached from the point of view of solving the many-body equations for the electrons. Except in special cases, the ground state must be treated by nonperturbative methods because the different terms in the energy equation are large and tend to cancel. The properties of the ground state include the total energy, electron density, and correlation functions. From the correlation functions one can derive properties that at first sight would not be considered to be ground-state properties, such as whether the material is a metal or an insulator. In any case, one needs to establish which state is the ground state, often comparing states that are very different in character but similar in energy.

On the other hand, excitations in condensed matter are usually small perturbations on the entire system. These perturbations can be classified into variations of the ground electronic state (e.g., small displacements of the ions in phonon modes) or true electronic excitations (e.g., optical electronic excitations). In both cases, perturbation theory is the appropriate tool. Using perturbation techniques, one can calculate excitation spectra and the real and imaginary parts of response functions. Nevertheless, even in this case one needs to know the ground state, since the excitations are perturbations on the ground state.

These approaches apply both in independent-particle and in many-body problems. The ground state is special in both cases and it is interesting that both density functional theory

---

[2] The method in its present form was developed by Ritz [258] in 1908, but the approach was used previously by Lord Rayleigh (John William Strutt) in several problems described in his 1877 book [259] and earlier work. The methods are described in many more texts, such as [260–262].

[3] This is an example of functional derivatives described in Appendix A for the case where the energy functional Eq. (3.9) is linear in both the bra $\langle \Psi |$ and the ket $|\Psi\rangle$ functions. Thus one may vary either or both at the same time with the same result.

and quantum Monte Carlo are primarily ground-state methods. The role of perturbation theory is rather different in independent-particle and many-body problems – in the latter, it plays a key role in the basic formulation of the problem in diagrammatic perturbation series and in suggesting the key ideas for summation of appropriate diagrams.

3.2 Coulomb Interaction in Condensed Matter

It is helpful to clarify briefly several points that are essential to a proper definition of energies in extended systems with long-range Coulomb interaction. For a more complete analysis see Appendix F. The key points are as follows:

- Any extended system must be neutral if the energy is to be finite.
- Terms in the energy must be organized in neutral groups for actual evaluation.

In Eq. (3.9) the most convenient approach is to identify and group together terms representing the classical Coulomb energies,

$$E^{CC} = E_{\text{Hartree}} + \int d^3r\, V_{\text{ext}}(\mathbf{r})n(\mathbf{r}) + E_{II}, \tag{3.14}$$

where E_{Hartree} is the self-interaction energy of the density $n(\mathbf{r})$ treated as a classical charge density

$$E_{\text{Hartree}} = \frac{1}{2} \int d^3r d^3r' \frac{n(\mathbf{r})n(\mathbf{r}')}{|\mathbf{r} - \mathbf{r}'|}. \tag{3.15}$$

Since E_{II} is the interaction among the positive nuclei and $\int d^3r\, V_{\text{ext}}(\mathbf{r})n(\mathbf{r})$ is the interaction of the electrons with the nuclei, Eq. (3.14) is a neutral grouping of terms so long as the system is neutral. The evaluation of classical Coulomb energies is an intrinsic part of quantitative electronic structure calculations; methods for dealing with long-range Coulomb interaction are described in Appendix F.

It then follows that the total energy expression, (3.9), can be written as

$$E = \langle \hat{T} \rangle + (\langle \hat{V}_{\text{int}} \rangle - E_{\text{Hartree}}) + E^{CC}, \tag{3.16}$$

where each of the three terms is well defined. The middle term in brackets, $\langle \hat{V}_{\text{int}} \rangle - E_{\text{Hartree}}$, is the difference between the Coulomb energies of interacting, correlated electrons with density $n(\mathbf{r})$ and that of a continuous classical charge distribution having the same density, which is defined to be the potential part of the exchange–correlation energy E_{xc} in density functional theory (see Sections 6.4 and 8.2, especially the discussion related to Eq. (8.3)).[4] Thus all long-range interactions cancel in the difference so that effects of exchange and correlation are short ranged. This is a point to which we will return in Chapter 7 and Appendix H.

---

[4] This definition differs from that given in many texts (see Section 3.7) as the difference from Hartree–Fock, since the Hartree–Fock density differs from the true density.

3.3 Force and Stress Theorems

Force (Hellmann–Feynman) Theorem

One of the beautiful theorems of physics is the "force theorem" for the force conjugate to any parameter in the hamiltonian. This is a very general idea perhaps formulated first in 1927 by Ehrenfest [263], who recognized that it is crucial for the correspondence principle of quantum and classical mechanics. He established the relevant relation by showing that the expression for force given below equals the expectation value of the operator corresponding to acceleration $\langle d^2 \hat{x}/dt^2 \rangle$. The ideas are implicit in the 1928 work of Born and Fock [264], and the explicit formulas used today were given by Güttiger [265] in 1932. The formulas were included in the treatises of Pauli [266] and Hellmann [267], the latter reformulating them as a variational principle in a form convenient for application to molecules. In 1939, when he was an undergraduate, Feynman [268] derived the force theorem and explicitly pointed out that the force on a nucleus is given strictly in terms of the charge density, independent of the electron kinetic energy, exchange, and correlation. Thus as an "electrostatic theorem," it should apparently be attributed to Feynman. The nomenclature "Hellmann–Feynman theorem" has been widely used, apparently originating with Slater [48]; however, we will use the term "force theorem."

The force conjugate to any parameter describing a system, such as the position of a nucleus $\mathbf{R}_I$, can always be written

$$\mathbf{F}_I = -\frac{\partial E}{\partial \mathbf{R}_I}. \tag{3.17}$$

From the general expression for the total energy Eq. (3.9), the derivative can be written using first-order perturbation theory (the normalization does not change and we assume $\langle \Psi | \Psi \rangle = 1$ for convenience),

$$-\frac{\partial E}{\partial \mathbf{R}_I} = -\left\langle \Psi \left| \frac{\partial \hat{H}}{\partial \mathbf{R}_I} \right| \Psi \right\rangle - \left\langle \frac{\partial \Psi}{\partial \mathbf{R}_I} | \hat{H} | \Psi \right\rangle - \left\langle \Psi | \hat{H} | \frac{\partial \Psi}{\partial \mathbf{R}_I} \right\rangle - \frac{\partial E_{II}}{\partial \mathbf{R}_I}. \tag{3.18}$$

Using the fact that at the exact ground-state solution the energy is extremal with respect to all possible variations of the wavefunction, it follows that the middle two terms in Eq. (3.18) vanish and the only nonzero terms come from the *explicit* dependence of the nuclear position. Furthermore, using the form of the energy in Eq. (3.9), it follows that the force depends only on the density n of the electrons and the other nuclei,

$$\mathbf{F}_I = -\frac{\partial E}{\partial \mathbf{R}_I} = -\int d^3 r\, n(\mathbf{r}) \frac{\partial V_{\text{ext}}(\mathbf{r})}{\partial \mathbf{R}_I} - \frac{\partial E_{II}}{\partial \mathbf{R}_I}. \tag{3.19}$$

Here $n(\mathbf{r})$ is the *unperturbed* density and the other nuclei are held fixed, as shown schematically in the left-hand side of Fig. I.1. Since each nucleus interacts with the electrons and other nuclei via Coulomb interactions, the right-hand side of Eq. (3.19) can be shown (Exercise 3.3) to equal the nuclear charge times the total electric field, which is the electrostatic theorem of Feynman. Thus even though the kinetic energy and internal interactions change as the nuclei move, all such terms cancel in the force theorem.

In the case of nonlocal potentials (such as pseudopotentials), the force cannot be expressed solely in terms of the electron density. However, the original expression is still valid and useful expressions can be directly derived from

$$-\frac{\partial E}{\partial \mathbf{R}_I} = -\left\langle \Psi \left| \frac{\partial \hat{H}}{\partial \mathbf{R}_I} \right| \Psi \right\rangle - \frac{\partial E_{II}}{\partial \mathbf{R}_I}. \tag{3.20}$$

Because the force theorem depends on the requirement that the electronic states are at their variational minimum, it follows that there must be a continuum of "force theorems" that corresponds to the addition of any linear variation in Ψ or n to the above expression. Although such terms vanish in principle, they can have an enormous impact upon the accuracy and physical interpretation of resulting formulas. The most relevant example in electronic structure is the case of core electrons: It is more physical and more accurate computationally to move the electron density in the core region along with the nucleus rather than holding the density strictly fixed. Methods to accomplish this are described in Appendix I and illustrated in Figure I.1.

Finally, there are drawbacks to the fact that expressions for the force theorem depend on the electronic wavefunction being an exact eigenstate. If the basis is not complete, or the state is approximated, then there may be additional terms. For example, if the basis is not complete and it changes as the positions of the nuclei move, then Pulay corrections [269] must be explicitly included so that the expression for the force given by the force theorem is identical to the explicit derivative of the energy (Exercise 3.4). Explicit expressions are given for use in independent-particle Kohn–Sham calculations in Section 7.5.

Stress (Generalized Virial) Theorem

A physically different type of variation is a scaling of space, which leads to the "stress theorem" [162, 163] for total stress. This is a generalization of the well-known virial theorem for pressure P, which was derived in the early days of quantum mechanics (see references in Appendix G). An elegant derivation was given by Fock [270] in terms of "Streckung des Grundgebietes" ("stretching of the ground state").

The stress is a generalized force for which the ideas of the force theorem can be applied. The key point is that for a system in equilibrium, the stress tensor $\sigma_{\alpha\beta}$ is minus the derivative of the energy with respect to strain $\epsilon_{\alpha\beta}$ per unit volume (see also Eq. (2.4))

$$\sigma_{\alpha\beta} = -\frac{1}{\Omega} \frac{\partial E}{\partial \epsilon_{\alpha\beta}}, \tag{3.21}$$

where α and β are the cartesian indices, and where strain is defined to be a scaling of space, $\mathbf{r}_\alpha \rightarrow (\delta_{\alpha\beta} + \epsilon_{\alpha\beta})\mathbf{r}_\beta$, where $\mathbf{r}$ is any vector in space including particle positions and translation vectors. The effect is to transform the wavefunction by scaling every particle coordinate [162],

$$\Psi_\epsilon(\{\mathbf{r}_i\}) = \det(\delta_{\alpha\beta} + \epsilon_{\alpha\beta})^{-1/2} \Psi(\{(\delta_{\alpha\beta} + \epsilon_{\alpha\beta})^{-1}\mathbf{r}_{i\beta}\}), \tag{3.22}$$

where the prefactor preserves the normalization. Since the wavefunction also depends on the nuclear positions (either explicitly, treating the nuclei as quantum particles, or implicitly, as parameters in the Born–Oppenheimer approximation discussed after Eq. (3.1)), so also must the nuclear positions be scaled. Of course, the wavefunction and the nuclear positions actually change in other ways if the system is compressed or expanded; however, this has no effect on the energy to first order because the wavefunction and the nuclear positions are at variational minima.

Substituting $\Psi_\epsilon(\{\mathbf{r}_i\})$ into expression (3.9) for the energy, changing variables in the integrations, and using Eq. (3.21) leads directly to the expression [162]

$$
\sigma_{\alpha\beta} = -\left\langle \Psi \left| \sum_k \frac{\hbar^2}{2m_k} \nabla_{k\alpha} \nabla_{k\beta} - \frac{1}{2} \sum_{k \neq k'} \frac{(\mathbf{x}_{kk'})_\alpha (\mathbf{x}_{kk'})_\beta}{x_{kk'}} \left(\frac{d}{dx_{kk'}} \hat{V} \right) \right| \Psi \right\rangle,
\tag{3.23}
$$

where the sum over k and k' denotes a double sum over all particles, nuclei, and electrons, where the interaction is a function of the distance $x_{kk'} = |\mathbf{x}_{kk'}|$. The virial theorem for pressure $P = -\sum_\alpha \sigma_{\alpha\alpha}$ is the trace of Eq. (3.23), which follows from isotropic scaling of space, $\epsilon_{\alpha\beta} = \epsilon\delta_{\alpha\beta}$. If all interactions are Coulombic and the potential energy includes all terms due to nuclei and electrons, the virial theorem leads to

$$
3P\Omega = 2E_{\text{kinetic}} + E_{\text{potential}},
\tag{3.24}
$$

where Ω is the volume of the system. The expression (3.24) is a general result valid in any system in equilibrium, classical or quantum, at any temperature, so long as all particles interact with Coulomb forces. Explicit expressions [102, 163] used in practical calculations in Fourier space are discussed in Appendix G.

3.4 Generalized Force Theorem and Coupling Constant Integration

In the previous section, the force on a nucleus I was shown to be given by the matrix element of the derivative of the hamiltonian with respect to position $\mathbf{R}_I$ because $\mathbf{R}_I$ can be considered to be a parameter. The same argument applies to any parameter, which we can denote by λ. Furthermore, finite energy differences between two states with values λ_1 and λ_2 can be calculated as an integral over a continuous variation of the hamiltonian from λ_1 to λ_2. This is also called an "adiabatic connection" following Harris [271] since it is a variation of the hamiltonian connecting the states of the system that is assumed to be in the ground state for each value of λ, i.e., it is an adiabatic variation.[5] The general expressions can be written

$$
\frac{\partial E}{\partial \lambda} = \left\langle \Psi_\lambda \left| \frac{\partial \hat{H}}{\partial \lambda} \right| \Psi_\lambda \right\rangle
\tag{3.25}
$$

[5] A clear derivation is given in [272] and an extensive description is given in [273].

and

$$\Delta E = \int_{\lambda_1}^{\lambda_2} d\lambda \frac{\partial E}{\partial \lambda} = \int_{\lambda_1}^{\lambda_2} d\lambda \left\langle \Psi_\lambda \left| \frac{\partial \hat{H}}{\partial \lambda} \right| \Psi_\lambda \right\rangle. \tag{3.26}$$

For example, if a parameter such as the charge squared of the electron e^2 in the interaction energy in the hamiltonian is scaled by $e^2 \to e^2\lambda$, then λ can be varied from 0 to 1 to vary the hamiltonian from the noninteracting limit to the fully interacting problem. Since the hamiltonian involves the charge only in the interaction term, and Eq. (3.5) is linear in e^2 (the nuclear term is treated separately as the "external potential"), it follows that the change in energy can be written

$$\Delta E = \int_0^1 d\lambda \langle \Psi_\lambda | V_{\mathrm{int}} | \Psi_\lambda \rangle, \tag{3.27}$$

where V_{int} is the full interaction term Eq. (3.5) and Ψ_λ is the wavefunction for intermediate values of the interaction[6] given by $e^2 \to e^2\lambda$. The disadvantage of this approach is that it requires the wavefunction at intermediate (unphysical) values of e; nevertheless, it can be very useful, e.g., in the construction of density functionals in Section 9.7.

3.5 Statistical Mechanics and the Density Matrix

From quantum statistical mechanics one can derive expressions for the energy U, entropy S, and free energy $F = U - TS$ at a temperature T. The general expression for F is

$$F = \mathrm{Tr}\hat{\rho} \left(\hat{H} + \frac{1}{\beta} \ln \hat{\rho} \right), \tag{3.28}$$

where $\hat{\rho}$ is the density matrix and $\beta = 1/k_B T$. Here Tr means trace over all the states of the system which have a fixed number of particles N. The final term is the entropy term, which is the log of the number of possible states of the system. A general property of the density matrix is that it is positive definite, since its diagonal terms are the density. The correct equilibrium density matrix is the positive-definite matrix that minimizes the free energy,

$$\hat{\rho} = \frac{1}{Z} e^{-\beta \hat{H}}, \tag{3.29}$$

with the partition function given by

$$Z = \mathrm{Tr} e^{-\beta \hat{H}} = e^{-\beta F}. \tag{3.30}$$

In a basis of eigenstates Ψ_i of $\hat{H}$, $\hat{\rho}$ has only diagonal matrix elements,

$$\rho_{ii} \equiv \langle \Psi_i | \hat{\rho} | \Psi_i \rangle = \frac{1}{Z} e^{-\beta E_i}; \quad Z = \sum_j e^{-\beta E_j}, \tag{3.31}$$

---

[6] The change in energy can be computed for any ground or excited state. States of different symmetry can be followed uniquely even if they cross. In many cases it is more efficient to solve a matrix equation (the size of the number of states of the same symmetry that are strongly mixed).

where ρ_{ii} is the probability of state i. Since the Ψ_i form a complete set, the operator $\hat{\rho}$ in Eq. (3.29) can be written

$$\hat{\rho} = \sum_i |\Psi_i\rangle \rho_{ii} \langle\Psi_i| \tag{3.32}$$

in Dirac bra and ket notation.

In the grand canonical ensemble, in which the number of particles is allowed to vary, the expressions are modified to include the chemical potential μ and the number operator $\hat{N}$. The grand potential Ω and the grand partition function Z are given by

$$Z = e^{-\beta\Omega} = \mathrm{Tr}\, e^{-\beta(\hat{H} - \mu\hat{N})}, \tag{3.33}$$

where now the trace is over all states with any particle number, and the grand density matrix operator is the generalization of Eq. (3.29),

$$\hat{\rho} = \frac{1}{Z} e^{-\beta(\hat{H} - \mu\hat{N})}. \tag{3.34}$$

All the equilibrium properties of the system are determined by the density matrix, just as they are determined by the ground-state wavefunction at $T = 0$. In particular, any expectation value is given by

$$\langle\hat{O}\rangle = \mathrm{Tr}\,\hat{\rho}\,\hat{O}, \tag{3.35}$$

which reduces to a ground-state expectation value of the form of Eq. (3.7) at $T = 0$. For the case of noninteracting particles, the general formulas reduce to the well-known expressions for fermions and bosons given in the next section.

3.6 Independent-Electron Approximations

There are two basic independent-particle approaches that may be classified as "non-interacting" and "Hartree–Fock." They are similar in that each assumes the electrons are uncorrelated except that they must obey the exclusion principle. However, they are different in that Hartree–Fock includes the electron–electron Coulomb interaction in the energy while neglecting the correlation that is introduced in the true wavefunction due to those interactions. In general, "noninteracting" theories have some effective potential that incorporates some effect of the real interaction, but there is no interaction term explicitly included in the effective hamiltonian. This approach is often referred to as "Hartree" or "Hartree-like," after D. R. Hartree [50], who included an average Coulomb interaction in a rather heuristic way.[7] More to the point of modern calculations, *all* calculations

---

[7] Historically, the first quantitative calculations on many-electron systems were carried out on atoms by D. R. Hartree [50], who solved, numerically, the equation for each electron moving in a central potential due to other electrons and the nucleus. Hartree defined a different potential for each electron because he subtracted a self-term for each electron that depended on its orbital. However, following the later development of the Hartree–Fock method [53], it is now customary to define the effective "Hartree potential" with an unphysical self-interaction term so that the potential is orbital independent. This unphysical term has no effect since it is canceled by the exchange term in Hartree–Fock calculations.

following the Kohn–Sham method (see Chapters 7–9) involve an auxiliary system with a noninteracting hamiltonian and an effective potential chosen to incorporate exchange and correlation effects approximately.

Noninteracting (Hartree-Like) Electron Approximation

With our broad definition, all noninteracting electron calculations involve the solution of a Schrödinger-like equation like Eqs. (1.1) and (1.2)

$$\hat{H}_{\text{eff}}\psi_i^\sigma(\mathbf{r}) = \left[-\frac{\hbar^2}{2m_e}\nabla^2 + V_{\text{eff}}^\sigma(\mathbf{r}) \right] \psi_i^\sigma(\mathbf{r}) = \varepsilon_i^\sigma \psi_i^\sigma(\mathbf{r}), \qquad (3.36)$$

where $V_{\text{eff}}^\sigma(\mathbf{r})$ is an effective potential that acts on each electron of spin σ at point $\mathbf{r}$.[8] The ground state for many noninteracting electrons is found by occupying the lowest eigenstates of Eq. (3.36) obeying the exclusion principle. If the hamiltonian is not spin dependent, then up and down spin states are degenerate and one can simply consider spin as a factor of two in the counting. Excited states involve the occupation of higher-energy eigenstates. There is no need to construct an antisymmetric wavefunction explicitly. Since the eigenstates of the independent-particle Schrödinger equation are automatically orthogonal, an antisymmetric wavefunction like Eq. (3.43) can be formed from a determinant of these eigenstates. It is then straightforward to show that *if the particles are noninteracting*, the relations reduce to the expressions given below for the energy, density, etc.(Exercise 3.6).

The solution of equations having the form of Eq. (3.36) is at the heart of the methods described in this volume. The basic justification of the use of such independent-particle equations for electrons in materials is density functional theory, which is the subject of Chapters 6–9. Most of the rest of the book is devoted to methods for solving the equations and applications to the properties of matter, such as predictions of structures, phase transitions, magnetism, elastic constants, phonons, piezoelectric and ferroelectric moments, topology of the band structure, and many other quantities.

At finite temperature it is straightforward to apply the general formulas of statistical mechanics given in the previous section to show that the equilibrium distribution of electrons is given by the Fermi–Dirac (or Bose–Einstein) expression (1.3) for occupation numbers of states as a function of energy (Exercise 3.7). The expectation value Eq. (3.35) is a sum over many-body states Ψ_j, each of which is specified by the set of occupation numbers $\{n_i^\sigma\}$ for each of the independent-particle states with energy ε_i^σ. Given that each n_i^σ can be either 0 or 1, with $\sum_i n_i^\sigma = N^\sigma$, it is straightforward (see Exercise 3.8) to show that Eq. (3.35) simplifies to

$$\langle \hat{O} \rangle = \sum_{i,\sigma} f_i^\sigma \langle \psi_i^\sigma | \hat{O} | \psi_i^\sigma \rangle, \qquad (3.37)$$

---

[8] Spin is introduced at this point because it is necessary to introduce a spin-dependent effective potential in order for the independent-particle equations to describe spin-polarized states. In order to include the spin–orbit interaction, Eq. (3.36) must be generalized to a single equation with two components as in Eq. (O.7) with a term like H_{SO} in Eq. (O.9), and sum over spin replaced by a trace.

where $\langle \psi_i^\sigma | \hat{O} | \psi_i^\sigma \rangle$ is the expectation value of the operator $\hat{O}$ for the one-particle state ψ_i^σ, and f_i^σ is the probability of finding an electron in state i, σ given in general by Eq. (1.3). The relevant case is the Fermi–Dirac distribution

$$f_i^\sigma = \frac{1}{e^{\beta(\varepsilon_i^\sigma - \mu)} + 1}, \tag{3.38}$$

where μ is the Fermi energy (or chemical potential) of the electrons. For example, the energy is the weighted sum of noninteracting particle energies ε_i^σ

$$E(T) = \langle \hat{H} \rangle = \sum_i^\sigma f_i^\sigma \varepsilon_i^\sigma. \tag{3.39}$$

Just as in the general many-body case, one can define a single-body density matrix operator

$$\hat{\rho} = \sum_i |\psi_i^\sigma\rangle f_i^\sigma \langle \psi_i^\sigma|, \tag{3.40}$$

in terms of which an expectation value Eq. (3.37) is $\langle \hat{O} \rangle = \mathrm{Tr} \hat{\rho} \hat{O}$ in analogy to Eq. (3.35). For example, in an explicit spin and position representation, $\hat{\rho}$ is given by

$$\rho(\mathbf{r}, \sigma; \mathbf{r}', \sigma') = \delta_{\sigma, \sigma'} \sum_i \psi_i^{\sigma *}(\mathbf{r}) f_i \psi_i^\sigma(\mathbf{r}'), \tag{3.41}$$

where the density is the diagonal part

$$n^\sigma(\mathbf{r}) = \rho(\mathbf{r}, \sigma; \mathbf{r}, \sigma) = \sum_i f_i^\sigma |\psi_i^\sigma(\mathbf{r})|^2. \tag{3.42}$$

Hartree–Fock Approximation

A standard method of many-particle theory is the Hartree–Fock method, which was first applied to atoms in 1930 by Fock [53]. In this approach one writes a properly antisymmetrized determinant wavefunction for a fixed number N of electrons and finds the single determinant that minimizes the total energy for the full interacting hamiltonian Eq. (3.2). If there is no spin–orbit interaction, the determinant wavefunction Φ can be written as a Slater determinant[9]

$$\Phi = \frac{1}{(N!)^{1/2}} \begin{vmatrix} \phi_1(\mathbf{r}_1, \sigma_1) & \phi_1(\mathbf{r}_2, \sigma_2) & \phi_1(\mathbf{r}_3, \sigma_3) & \cdots \\ \phi_2(\mathbf{r}_1, \sigma_1) & \phi_2(\mathbf{r}_2, \sigma_2) & \phi_2(\mathbf{r}_3, \sigma_3) & \cdots \\ \phi_3(\mathbf{r}_1, \sigma_1) & \phi_3(\mathbf{r}_2, \sigma_2) & \phi_3(\mathbf{r}_3, \sigma_3) & \cdots \\ \cdot & \cdot & \cdot & \cdots \\ \cdot & \cdot & \cdot & \cdots \end{vmatrix}, \tag{3.43}$$

---

[9] The determinant formulation had been realized by Dirac [24] before Slater's work, but the determinant of spin-orbitals is due to Slater [25], who believed this was his most popular work [48] because it replaced difficult group theoretical arguments by this simple form.

where the $\phi_i(\mathbf{r}_j, \sigma_j)$ are single-particle "spin-orbitals." If there is no spin–orbit interaction $\phi_i(\mathbf{r}_j, \sigma_j)$ can be written as a product of a function of the position $\psi_i^\sigma(\mathbf{r}_j)$ and a function of the spin variable $\alpha_i(\sigma_j)$. (Note that $\psi_i^\sigma(\mathbf{r}_j)$ is independent of spin σ in closed-shell cases. In open-shell systems, this assumption corresponds to the "spin-restricted Hartree–Fock approximation.") The spin-orbitals must be linearly independent and if in addition they are orthonormal the equations simplify greatly; it is straightforward to show (Exercise 3.10) that Φ is normalized to 1. Furthermore, if the hamiltonian is independent of spin or is diagonal in the basis $\sigma = |\uparrow\rangle; |\downarrow\rangle$, the expectation value of the hamiltonian Eq. (3.2), using Hartree atomic units, with the wavefunction Eq. (3.43) is given by (Exercise 3.11)

$$
\langle \Phi | \hat{H} | \Phi \rangle = \sum_{i,\sigma} \int d\mathbf{r}\, \psi_i^{\sigma *}(\mathbf{r}) \left[-\frac{1}{2}\nabla^2 + V_{\text{ext}}(\mathbf{r}) \right] \psi_i^\sigma(\mathbf{r}) + E_{II}
$$

$$
+ \frac{1}{2} \sum_{i,j,\sigma_i,\sigma_j} \int d\mathbf{r}\,d\mathbf{r}'\, \psi_i^{\sigma_i *}(\mathbf{r}) \psi_j^{\sigma_j *}(\mathbf{r}') \frac{1}{|\mathbf{r}-\mathbf{r}'|} \psi_i^{\sigma_i}(\mathbf{r}) \psi_j^{\sigma_j}(\mathbf{r}')
$$

$$
- \frac{1}{2} \sum_{i,j,\sigma} \int d\mathbf{r}\,d\mathbf{r}'\, \psi_i^{\sigma *}(\mathbf{r}) \psi_j^{\sigma *}(\mathbf{r}') \frac{1}{|\mathbf{r}-\mathbf{r}'|} \psi_j^\sigma(\mathbf{r}) \psi_i^\sigma(\mathbf{r}'). \tag{3.44}
$$

The first term groups together the single-body expectation values that involve a sum over orbitals, whereas the third and fourth terms are the direct and exchange interactions among electrons, which are double sums. We have followed the usual practice of including the $i = j$ "self-interaction," which is spurious but which cancels in the sum of direct and exchange terms. When this term is included, the sum over all orbitals gives the density and the direct term is simply the Hartree energy defined in Eq. (3.15). The "exchange" term, which acts only between same-spin electrons since the spin parts of the orbitals are orthogonal for opposite spins, is discussed below in Section 3.7 and in the chapters on density functional theory.

The Hartree–Fock approach is to minimize the total energy with respect to all degrees of freedom in the wavefunction with the restriction that it has the form Eq. (3.43). Since orthonormality was used to simplify the equations, it must be maintained in the minimization, which can be done by Lagrange multipliers as in Eqs. (3.10) to (3.13). If the spin functions are quantized along an axis, variation of $\psi_i^{\sigma *}(\mathbf{r})$ for each spin σ leads to the Hartree–Fock equations

$$
\left[-\frac{1}{2}\nabla^2 + V_{\text{ext}}(\mathbf{r}) + \sum_{j,\sigma_j} \int d\mathbf{r}'\, \psi_j^{\sigma_j *}(\mathbf{r}') \psi_j^{\sigma_j}(\mathbf{r}') \frac{1}{|\mathbf{r}-\mathbf{r}'|} \right] \psi_i^\sigma(\mathbf{r})
$$

$$
- \sum_j \int d\mathbf{r}'\, \psi_j^{\sigma *}(\mathbf{r}') \psi_i^\sigma(\mathbf{r}') \frac{1}{|\mathbf{r}-\mathbf{r}'|} \psi_j^\sigma(\mathbf{r}) = \varepsilon_i^\sigma \psi_i^\sigma(\mathbf{r}), \tag{3.45}
$$

where the exchange term is summed over all orbitals of the same spin including the self-term $i = j$, which cancels the unphysical self-term included in the direct term. If the exchange term is modified by multiplying and dividing by $\psi_i^\sigma(\mathbf{r})$, Eq. (3.45) can be written in a form analogous to Eq. (3.36) except that the effective hamiltonian is an operator that depends on the state

$$\hat{H}^i_{\text{eff}} \psi^\sigma_i (\mathbf{r}) = \left[-\frac{\hbar^2}{2m_e} \nabla^2 + \hat{V}^{i,\sigma}_{\text{eff}} (\mathbf{r}) \right] \psi^\sigma_i (\mathbf{r}) = \varepsilon^\sigma_i \psi^\sigma_i (\mathbf{r}), \tag{3.46}$$

with

$$\hat{V}^{i,\sigma}_{\text{eff}} (\mathbf{r}) = V_{\text{ext}}(\mathbf{r}) + V_{\text{Hartree}} (\mathbf{r}) + \hat{V}^{i,\sigma}_x (\mathbf{r}), \tag{3.47}$$

and the exchange term operator $\hat{V}_x$ is given by a sum over orbitals of the same spin σ

$$\hat{V}^{i,\sigma}_x (\mathbf{r}) = - \sum_j \int d\mathbf{r}' \psi^{\sigma*}_j (\mathbf{r}') \psi^\sigma_i (\mathbf{r}') \frac{1}{|\mathbf{r} - \mathbf{r}'|} \frac{\psi^\sigma_j (\mathbf{r})}{\psi^\sigma_i (\mathbf{r})}. \tag{3.48}$$

Note that this is a differential-integral equation for each orbital ψ^σ_i in terms of the exchange operator $\hat{V}^{i,\sigma}_x (\mathbf{r})$ that is an integral involving ψ^σ_i and all the other ψ^σ_j with the same spin. The term in square brackets is the Coulomb potential due to the "exchange charge density" $\sum_j \psi^{\sigma*}_j (\mathbf{r}') \psi^\sigma_i (\mathbf{r}')$ for the state i, σ. Furthermore, $\hat{V}^{i,\sigma}_x (\mathbf{r})$ diverges at points where $\psi^\sigma_i (\mathbf{r}) = 0$; this requires care in solving the equations but is not a fundamental problem since the product $\hat{V}^{i,\sigma}_x (\mathbf{r}) \psi^\sigma_i (\mathbf{r})$ has no singularity.

We will not discuss the solution of the Hartree–Fock equations in any detail since this is given in many texts [274, 275]. The only aspect that we consider here is the nature of the exchange term and issues for calculations. Unlike the case of independent Hartree-like equations, the Hartree–Fock equations can be solved directly only in special cases such as spherically symmetric atoms and the homogeneous electron gas. In general, one must introduce a basis, in which case the energy (3.44) can be written in terms of the expansion coefficients of the orbitals and the integrals involving the basis functions. Variation then leads to the Roothan and Pople–Nesbet equations widely used in quantum chemistry [274, 275]. In general, these are much more difficult to solve than the independent Hartree-like equations and the difficulty grows with size and accuracy since one must calculate N^4_{basis} integrals.

However, there are ways to reduce the amount of computation. The nonlocal exchange potential $\hat{V}^{i,\sigma}_x (\mathbf{r})$ acting on state i, σ involves the exchange charge density $\sum_j \psi^{\sigma*}_j (\mathbf{r}') \psi^\sigma_i (\mathbf{r}')$ involving the other orbitals j, σ. For localized basis functions, it is nonzero only where states i and j overlap, which decreases rapidly in general. For gaussians, the Coulomb integrals are analytic, and for numerical basis functions fast multiple methods [276] can be used. For large systems the calculation scales linearly with size, but there may be a large prefactor. For plane waves special considerations are needed as described in Section 13.4.

Koopmans' Theorem

What is the meaning of the eigenvalues of the Hartree–Fock equation (3.45)? Of course, Hartree–Fock is only an approximation to the energies for addition and removal of electrons, since all effects of correlation are omitted. Nevertheless, it is very valuable to have a rigorous understanding of the eigenvalues, which is provided by Koopmans' theorem[10] [277]:

> The eigenvalue of a filled (empty) orbital is equal to the change in the total energy Eq. (3.44) if an electron is subtracted from (added to) the system, i.e., decreasing (increasing) the size of the determinant by omitting (adding) a row and column involving a particular orbital $\phi_j(\mathbf{r}_i, \sigma_i)$, *keeping all the other orbitals the same.*

Koopmans' theorem can be derived by taking matrix elements of Eq. (3.45) with the normalized orbital $\psi_i^{\sigma*}(\mathbf{r})$ (see Exercise 3.18). For occupied states, the eigenvalues are lowered by the exchange term, which cancels the spurious repulsive self-interaction in the Hartree term. To find the energies for addition of electrons, one must compute empty orbitals of the Hartree–Fock equation (3.45). For these states there also is no spurious self-interaction since both the direct and the exchange potential terms in Eq. (3.45) involve only the occupied states. In general, the gaps between addition and removal energies for electrons are greatly overestimated in the Hartree–Fock approximation because of the neglect of relaxation of the orbitals and other effects of correlation.

ΔSCF Methods

In finite systems, such as atoms, it is possible to improve upon the use of the eigenvalues as approximate excitation energies. Significant improvement in the addition and removal energies result from the "delta Hartree–Fock approximation," in which one calculates total energy differences directly from Eq. (3.44), allowing the orbitals to relax and taking into account the exchange of an added electron with all the others. The energy difference approach for finite systems can be used in any self-consistent field method, hence the name "ΔSCF." Illustrations are given in Section 10.6.

3.7 Exchange and Correlation

The key problem of electronic structure is that the electrons form an interacting many-body system, with a wavefunction, in general, given by $\Psi(\{\mathbf{r}_i\}) \equiv \Psi(\mathbf{r}_1, \mathbf{r}_2, \ldots, \mathbf{r}_N)$, as discussed in Section 3.1. Since the interactions always involve pairs of electrons, two-body correlation functions are sufficient to determine many properties, such as the energy given by Eq. (3.9). Writing out the form for a general expectation value Eq. (3.7) explicitly, the joint probability $n(\mathbf{r}, \sigma; \mathbf{r}', \sigma')$ of finding electrons of spin σ at point $\mathbf{r}$ and of spin σ' at point $\mathbf{r}'$ is given by

---

[10] Tjalling Koopmans is famous in chemistry and physics for his theorem [277], but his major work was in economics, for which he was awarded a Nobel prize in 1975.

$$n(\mathbf{r},\sigma;\mathbf{r}',\sigma') = \left\langle \sum_{i \neq j} \delta(\mathbf{r} - \mathbf{r}_i)\delta(\sigma - \sigma_i)\delta(\mathbf{r}' - \mathbf{r}_j)\delta(\sigma' - \sigma_j) \right\rangle \tag{3.49}$$

$$= N(N-1) \sum_{\sigma_3,\sigma_4,\dots} \int d\mathbf{r}_3 \cdots d\mathbf{r}_N |\Psi(\mathbf{r},\sigma;\mathbf{r}',\sigma';\mathbf{r}_3,\sigma_3;\dots,\mathbf{r}_N,\sigma_N)|^2, \tag{3.50}$$

assuming Ψ is normalized to unity. For uncorrelated particles, the joint probability is just the product of probabilities so that the measure of correlation is $\Delta n(\mathbf{r},\sigma;\mathbf{r}',\sigma') = n(\mathbf{r},\sigma;\mathbf{r}',\sigma') - n(\mathbf{r},\sigma)n(\mathbf{r}',\sigma')$, so that

$$n(\mathbf{r},\sigma;\mathbf{r}',\sigma') = n(\mathbf{r},\sigma)n(\mathbf{r}',\sigma') + \Delta n(\mathbf{r},\sigma;\mathbf{r}',\sigma'). \tag{3.51}$$

It is also useful to define the normalized pair distribution,

$$g(\mathbf{r},\sigma;\mathbf{r}',\sigma') = \frac{n(\mathbf{r},\sigma;\mathbf{r}',\sigma')}{n(\mathbf{r},\sigma)n(\mathbf{r}',\sigma')} = 1 + \frac{\Delta n(\mathbf{r},\sigma;\mathbf{r}',\sigma')}{n(\mathbf{r},\sigma)n(\mathbf{r}',\sigma')}, \tag{3.52}$$

which is unity for uncorrelated particles so that correlation is reflected in $g(\mathbf{r},\sigma;\mathbf{r}',\sigma') - 1$. Note that all long-range correlation is included in the average terms so that the remaining terms $\Delta n(\mathbf{r},\sigma;\mathbf{r}',\sigma')$ and $g(\mathbf{r},\sigma;\mathbf{r}',\sigma') - 1$ are short range and vanish at large $|\mathbf{r} - \mathbf{r}'|$.

Exchange in the Hartree–Fock Approximation

The Hartree–Fock approximation (HFA) consists of neglecting all correlations *except* those required by the Pauli exclusion principle; however, the exchange term in Eq. (3.44) represents two effects: Pauli exclusion and the self-term that must be subtracted to cancel the spurious self-term included in the direct Coulomb Hartree energy. The effect is always to lower the energy, which may be interpreted as the interaction of each electron with a positive "exchange hole" surrounding it. The exchange hole $\Delta n_x(\mathbf{r},\sigma;\mathbf{r}',\sigma')$ is given by $\Delta n(\mathbf{r},\sigma;\mathbf{r}',\sigma')$ in the HFA, where Ψ in Eq. (3.50) is approximated by the single determinant wavefunction Φ of Eq. (3.43). If the single-particle spin-orbitals $\phi_i^\sigma = \psi_i^\sigma(\mathbf{r}_j) \times \alpha_i(\sigma_j)$ are orthonormal, it is straightforward (Exercise 3.13) to show that the pair distribution function can be written

$$n_{\text{HFA}}(\mathbf{r},\sigma;\mathbf{r}',\sigma') = \frac{1}{2!} \sum_{ij} \left| \begin{matrix} \phi_i(\mathbf{r},\sigma) & \phi_i(\mathbf{r}',\sigma') \\ \phi_j(\mathbf{r},\sigma) & \phi_j(\mathbf{r}',\sigma') \end{matrix} \right|^2, \tag{3.53}$$

and the exchange hole takes the simple form

$$\Delta n_{\text{HFA}}(\mathbf{r},\sigma;\mathbf{r}',\sigma') = \Delta n_x(\mathbf{r},\sigma;\mathbf{r}',\sigma') = -\delta_{\sigma\sigma'} \left| \sum_i \psi_i^{\sigma*}(\mathbf{r})\psi_i^\sigma(\mathbf{r}') \right|^2. \tag{3.54}$$

It is immediately clear from Eq. (3.51) and Eq. (3.54) that the exchange hole of an electron involves only electrons of the same spin and that the probability vanishes, as it must, for finding two electrons of the same spin at the same point $\mathbf{r} = \mathbf{r}'$. Note that from Eq. (3.54) and

Eq. (3.41), it follows that in the HFA $\Delta n_x(\mathbf{r},\sigma;\mathbf{r}',\sigma') = -\delta_{\sigma\sigma'}|\rho_\sigma(\mathbf{r},\mathbf{r}')|^2$, where $\rho_\sigma(\mathbf{r},\mathbf{r}')$ is the density matrix, which is diagonal in spin.

This is an example of the general property [278] that indistinguishability of particles leads to correlations, which in otherwise independent-particle systems can be expressed in terms of the first-order density matrix:

$$\Delta n_{\text{ip}}(\mathbf{x};\mathbf{x}') = \pm|\rho_\sigma(\mathbf{x},\mathbf{x}')|^2, \tag{3.55}$$

or

$$g_{\text{ip}}(\mathbf{x};\mathbf{x}') = 1 \pm \frac{|\rho_\sigma(\mathbf{x},\mathbf{x}')|^2}{n(\mathbf{x})n(\mathbf{x}')}, \tag{3.56}$$

where the plus (minus) sign applies for bosons (fermions) and $\mathbf{x}$ incorporates all coordinates including position $\mathbf{r}$ and spin (if applicable). Thus $\Delta n_{\text{ip}}(\mathbf{x};\mathbf{x}')$ is always positive for independent bosons and always negative for independent fermions.

There are stringent conditions on the exchange hole: (1) it can never be positive, $\Delta n_x(\mathbf{r},\sigma;\mathbf{r}',\sigma') \le 0$ (which means that $g_x(\mathbf{r},\sigma;\mathbf{r}',\sigma') \le 1$), and (2) the integral of the exchange hole density $\Delta n_x(\mathbf{r},\sigma;\mathbf{r}',\sigma')$ over all $\mathbf{r}'$ is exactly one missing electron per electron at any point $\mathbf{r}$. This is a consequence of the fact that if one electron is at $\mathbf{r}$, then that same electron cannot also be at $\mathbf{r}'$. It also follows directly from Eq. (3.54), as shown in Exercise 3.12. The exchange energy, the last term in Eq. (3.44), can be interpreted as the lowering of the energy due to each electron interacting with its positive exchange hole,

$$E_x = \left[\langle \hat{V}_{\text{int}}\rangle - E_{\text{Hartree}}(n)\right]_{\text{HFA}} = \frac{1}{2}\sum_{\sigma\sigma'}\int d^3r \int d^3r' \frac{\Delta n_x(\mathbf{r},\sigma;\mathbf{r}',\sigma')}{|\mathbf{r}-\mathbf{r}'|}. \tag{3.57}$$

In this form it is clear that the exchange energy cancels the unphysical self-interaction term in the Hartree energy.

The simplest example of an exchange hole is a one-electron problem, such as the hydrogen atom. There is, of course, no real "exchange" nor any issue of the Pauli exclusion principle, and it is easy to see that the "exchange hole" is exactly the electron density. Its integral is unity, as required by the sum rule, and the exchange energy cancels the spurious Hartree term. Because of this cancellation, the Hartree–Fock equation (3.45) correctly reduces to the usual Schrödinger equation for one electron in an external potential.

The next more complex case is a two-electron singlet such as the ground state of He. In this case (see Exercise 3.16) the two spins have identical spatial orbitals and the exchange term is minus one-half the Hartree term in the Hartree–Fock equation (3.44), so that the Hartree–Fock equation (3.45) simplifies to a Hartree-like equation of the form of Eq. (3.36) with V_{eff} a sum of the external (nuclear) potential plus one-half the Hartree potential.[11]

---

[11] This is exactly what D. R. Hartree did in his pioneering work [50]; however, his approach of subtracting a self-term for each electron is not the same as the more proper Hartree–Fock theory for more than two electrons.

For systems with many electrons the exchange hole must be calculated numerically, except for special cases. The most relevant for us is the homogeneous gas considered in the following section.

Beyond Hartree–Fock: Correlation

The energy of a state of many electrons in the Hartree–Fock approximation Eq. (3.44) is the best possible wavefunction made from a single determinant (or a sum of a few determinants in multireference Hartree–Fock [274] needed for degenerate cases). Improvement of the wavefunction to include correlation introduces extra degrees of freedom in the wavefunction and therefore always lowers the energy for any state, ground or excited, by a theorem often attributed to MacDonald [279]. The lowering of the energy is termed the "correlation energy" E_c.

This is not the only possible definition of E_c, which could also be defined as the difference from some other reference state. The definition in terms of the difference from Hartree–Fock is a well-defined choice in the sense that it leads to the smallest possible magnitude of E_c, since E_{HFA} is the lowest possible energy neglecting correlation. Another well-defined choice arises naturally in density functional theory, where E_c is also defined as the difference between the exact energy and the energy of an uncorrelated state as Eq. (3.44), but with the difference that the orbitals are required to give the *exact* density (see Section 3.2 and Chapter 7). In many practical cases this distinction appears not to be of great importance; nevertheless, it is essential to define the energies properly, especially as electronic structure methods become more and more powerful in their ability to calculate effects of correlation.

The effects of correlation can be cast in terms of the remaining part of the pair correlation function beyond exchange $n_c(\mathbf{r},\sigma;\mathbf{r}',\sigma')$ defined in terms of Eqs. (3.50) and (3.51) by

$$\Delta n(\mathbf{r},\sigma;\mathbf{r}',\sigma') \equiv n_{\mathrm{xc}}(\mathbf{r},\sigma;\mathbf{r}',\sigma') = n_x(\mathbf{r},\sigma;\mathbf{r}',\sigma') + n_c(\mathbf{r},\sigma;\mathbf{r}',\sigma'). \qquad (3.58)$$

Since the entire exchange–correlation hole obeys the sum rule that it integrates to one, the correlation hole $n_c(\mathbf{r},\sigma;\mathbf{r}',\sigma')$ must integrate to zero, i.e., it merely redistributes the density of the hole. In general, correlation is most important for electrons of opposite spin, since electrons of the same spin are automatically kept apart by the exclusion principle. For the ground state the correlation energy is always negative and any approximation should be negative. Excited states involve *energy differences* from the ground state, e.g., an exciton energy. Depending on the effects of correlation in the two states, the *difference* can be positive or negative.

The correlation energy is more complicated to calculate than the exchange energy because correlation affects both kinetic and potential energies. Both effects can be taken into account by a "coupling constant integration" using the methods of Section 3.4. The theory of interacting systems is beyond the scope of this book and the reader is referred to the companion volume [1] for an in-depth presentation. Nevertheless, it is essential to take correlation into account in realistic calculations, and the goal here is to show how to understand and use aspects that have an important role in present-day electronic structure theory and practical calculations. Coupling constant integration is discussed for the model system, the homogeneous gas in Chapter 5, and is a key aspect of ongoing developments of functionals for density functional theory (see especially Sections 8.2, 9.3, and 9.7).

SELECT FURTHER READING

The companion book [1] also presents the basic theory with an emphasis on the role of correlation beyond Hartree–Fock. In that book, Hartree–Fock plays a larger role because it is the starting point for the development of much of the theory.

See the list of texts at the end of Chapter 2.

Reference for Hartree–Fock:

Szabo, A. and Ostlund, N. S., *Modern Quantum Chemistry: Introduction to Advanced Electronic Structure Theory* (Dover, Mineola, New York, 1996).

Exercises

3.1 Show that the many-body Schrödinger equation (3.13) also results from explicit variation of the energy in Eq. (3.9) without use of Lagrange multipliers.

3.2 Show that the independent-particle Schrödinger equation (3.36) is a special case of the many-body solution. Show this first for one particle, then for many noninteracting particles.

3.3 As part of his undergraduate thesis, Feynman showed that the force theorem applied to a nucleus leads to the force being exactly the electric field at the given nucleus due to the charge density of the rest of the system (electrons and other nuclei) times the charge of the given nucleus. Derive this result from Eq. (3.19).

3.4 Derive the additional terms that must be included so that the expression for the force given by the force theorem is identical to the explicit derivative of the energy if the basis depends explicitly on the positions for the nuclei. Show that the contribution of these terms vanishes if the basis is complete.

3.5 Derive the stress theorem Eq. (3.23). Show that this equation reduces to the well-known virial theorem Eq. (3.24) in the case of isotropic pressure and Coulomb interactions.

3.6 Show that the relations for noninteracting particles given in the equations following Eq. (3.36) remain valid if a fully antisymmetric determinant wavefunction like Eq. (3.43) is created from the orbitals. Note that this holds *only if the particles are noninteracting*.

3.7 Derive the Fermi–Dirac distribution Eq. (3.38) for noninteracting particles from the general definition of the density matrix Eq. (3.32) using the fact that the sum over many-body states in Eq. (3.32) can be reduced to a sum over all possible occupation numbers $\{n_i^\sigma\}$ for each of the independent-particle states, subject to the conditions that each n_i^σ can be either 0 or 1, and $\sum_i n_i^\sigma = N^\sigma$.

3.8 Following Exercise 3.7, show that Eq. (3.35) simplifies to Eq. (3.37) for any operator in the independent-particle approximation.

3.9 Why is the independent particle density matrix Eq. (3.41) diagonal in spin? Is this always the case?

3.10 Show that the Hartree–Fock wavefunction Eq. (3.43) is normalized if the independent-particle orbitals are orthonormal.

3.11 Show that the Hartree–Fock wavefunction Eq. (3.43) leads to the exchange term in Eq. (3.44) and that the variational equation leads to the Hartree–Fock equation (3.45) if the independent-particle orbitals are orthonormal. Explain why the forms are more complicated if the independent-particle orbitals are not orthonormal.

3.12 Show explicitly from the definition Eq. (3.54) that the exchange hole around each electron always integrates to one missing electron. Show that, as stated in the text, this is directly related to the fact that "exchange" includes a self-term that cancels the unphysical self-interaction in the Hartree energy.

3.13 Derive the formulas for the pair distribution Eq. (3.53) and the exchange hole Eq. (3.54) for noninteracting fermions by inserting the Hartree–Fock wavefunction Eq. (3.43) into the general definition Eq. (3.50).

3.14 By expanding the 2×2 determinant in Eq. (3.53), (a) show that

$$\sum_{\sigma'} \int dr' \Delta n_x(\mathbf{r}, \sigma; \mathbf{r}', \sigma') = (N - 1)n(\mathbf{r}, \sigma), \tag{3.59}$$

where $n(\mathbf{r}, \sigma)$ is the density, and (b) derive the formula Eq. (3.54) for the exchange hole.

3.15 The relation Eq. (3.55) is a general property of noninteracting identical particles [278]. As shown in Eq. (3.54), $\Delta n_x(\mathbf{r}, \sigma; \mathbf{r}', \sigma')$ is always negative for fermions. Show that for bosons with a symmetric wavefunction, the corresponding exchange term is always positive.

3.16 Derive the results stated after Eq. (3.57) that (a) for a one-electron problem like hydrogen, the exchange term exactly cancels the Hartree term as it should and that (b) for the ground state of two electrons in a spin singlet state, e.g., in helium, the Hartree–Fock approximation leads to a V_{eff} sum of the external (nuclear) potential plus one-half the Hartree potential.

3.17 Following the exercise above, consider two electrons in a spin triplet state. Show that the situation is not so simple as for the singlet case, i.e., that in the Hartree–Fock approximation there must be two different functions V_{eff} for two different orbitals.

3.18 Derive Koopmans' theorem by explicitly taking matrix elements of the hamiltonian with an orbital to show that the eigenvalue is the same as the energy difference if that orbital is removed.

3.19 For adding electrons one must compute empty orbitals of the Hartree–Fock equation (3.45). Show that such an empty orbital does not experience a self-contribution to the exchange energy, whereas for a filled state there is a self-term in the exchange. Show that the same result for the addition energy is found if one explicitly includes the state in a calculation with one added electron but keeps the original orbitals unchanged.

3.20 In a finite system Hartree–Fock eigenfunctions have the (surprising) property that the form of the long-range decay of *all* bound states is the same, independent of binding energy. For example, core states have the same exponential decay as valence states, although the prefactor is smaller. Show that this follows from Eq. (3.45).

3.21 Show that all contributions involving i and j both occupied vanish in the expectation value Eq. (D.4).

3.22 Show that the correlation hole always integrates to zero, i.e., it rearranges the charge correlation. This does not require complex calculations beyond Hartree–Fock theory; all that is needed is to show that conservation laws must lead to this result.

3.23 As an example of the force theorem consider a one-dimensional harmonic oscillator with hamiltonian given by $-\frac{1}{2}(d^2/dx^2) + \frac{1}{2}Ax^2$, where A is the spring constant and the mass is set to unity. Using the exact solution for the energy and wavefunction, calculate the generalized force dE/dA by direct differentiation and by the force theorem.

4

Periodic Solids and Electron Bands

Summary

Classification of crystals and their excitations by symmetry is a general approach applicable to electronic states, vibrational states, and other properties. The first part of this chapter deals with translational symmetry, which has the same universal form in all crystals and which leads to the Bloch theorem that rigorously classifies excitations by their crystal momentum. (The discussion here follows Ashcroft and Mermin, [280], chapters 4–8.) The other relevant symmetries are time reversal and point symmetries. The latter depend on the specific crystal structure and are treated only briefly; detailed classification can be found in texts on group theory. Time-reversal symmetry is relatively simple to formulate but has profound consequences.

4.1 Structures of Crystals: Lattice + Basis

A crystal is an ordered state of matter in which the positions of the nuclei (and consequently all properties) are repeated periodically in space. It is completely specified by the types and positions of the nuclei in one repeat unit (primitive unit cell) and the rules that describe the repetition (translations).

- The positions and types of atoms in the primitive cell are called the basis. The set of translations, which generates the entire periodic crystal by repeating the basis, is a lattice of points in space called the Bravais lattice. Specification of the crystal can be summarized as

  Crystal structure = Bravais lattice + basis.

- The crystalline order is described by its symmetry operations. The set of translations forms a group because the sum of any two translations is another translation.[1] In addition

---

[1] A group is defined by the condition that the application of any two operations leads to a result that is another operation in the group. We will illustrate this with the translation group. The reader is referred to other sources for the general theory and the specific set of groups possible in crystals, e.g., books on group theory (see [281–283], and the comprehensive work by Slater [284].

there may be other point operations that leave the crystal the same, such as rotations, reflections, and inversions. This can be summarized as

$$\text{Space group} = \text{translation group} + \text{point group}.^2$$

The Lattice of Translations

First we consider translations, since they are intrinsic to all crystals. The set of all translations forms a lattice in space, in which any translation can be written as integral multiples of primitive vectors,

$$\mathbf{T}(\mathbf{n}) \equiv \mathbf{T}(n_1, n_2, \ldots) = n_1 \mathbf{a}_1 + n_2 \mathbf{a}_2 + \ldots, \tag{4.1}$$

where $\mathbf{a}_i$, $i = 1, \ldots, d$ are the primitive translation vectors and d denotes the dimension of the space. For convenience we write formulas valid in any dimension whenever possible and we define $\mathbf{n} = (n_1, n_2, \ldots, n_d)$.

In one dimension, the translations are simply multiples of the periodicity length a, $T(n) = na$, where n can be any integer. The primitive cell can be any cell of length a; however, the most symmetric cell is the one chosen symmetric about the origin $(-a/2, a/2)$ so that each cell centered on lattice point n is the locus of all points closer to that lattice point than to any other point. This is an example of the construction of the Wigner–Seitz cell.

The left-hand side of Fig. 4.1 shows a portion of a general lattice in two dimensions. Space is filled by the set of all translations of any of the infinite number of possible choices of the primitive cell. One choice of primitive cell is the parallelogram constructed from the two primitive translation vectors $\mathbf{a}_i$. This cell is often useful for formal proofs and for simplicity of construction. However, this cell is not unique since there are an infinite number of possible choices for $\mathbf{a}_i$. A more informative choice is the Wigner–Seitz cell, which is symmetric about the origin and is the most compact cell possible. It is constructed by drawing the perpendicular bisectors of all possible lattice vectors $\mathbf{T}$ and identifying the Wigner–Seitz cell as the region around the origin bounded by those lines.

In two dimensions there are special choices of lattices that have additional symmetry when the angles between the primitive vectors are $90°$ or 60. In units of the length a, the translation vectors are given by

$$
\begin{array}{llll}
& \text{square} & \text{rectangular} & \text{triangular} \\
\mathbf{a}_1 = & (1, 0) & (1, 0) & (1, 0), \\
\mathbf{a}_2 = & (0, 1) & (0, \frac{b}{a}), & \left(\frac{1}{2}, \frac{\sqrt{3}}{2}\right).
\end{array}
\tag{4.2}
$$

Examples of crystals having, respectively, square and triangular Bravais lattices are shown later in Fig. 4.5.

---

[2]   In some crystals the space group can be factorized into a product of translation and point groups; in others (such as the diamond structure) there are nonsymmorphic operations that can only be expressed as a combination of translation and a point operation.

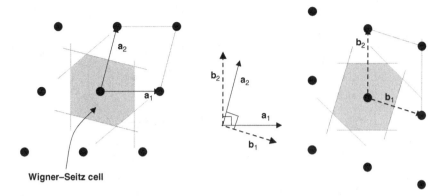

Figure 4.1. Real and reciprocal lattices for a general case in two dimensions. In the middle are shown possible choices for primitive vectors for the Bravais lattice in real space, a_1 and a_2, and the corresponding reciprocal lattice vectors, b_1 and b_2. In each case two types of primitive cells are shown, which when translated fill the two-dimensional space. The parallelogram cells are simple to construct but are not unique. The Wigner–Seitz cell in real space is uniquely defined as the most compact cell that is symmetric about the origin; the first Brillouin zone is the Wigner–Seitz cell of the reciprocal lattice.

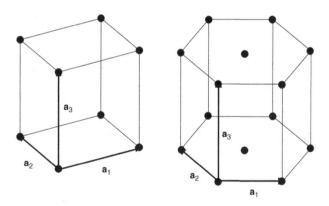

Figure 4.2. Simple cubic (left) and simple hexagonal (right) Bravais lattices. In the simple cubic case, the cell shown is the Wigner–Seitz cell and the Brillouin zone has the same shape. In the hexagonal case, the volume shown contains three atoms; the Wigner–Seitz cell is also a hexagonal prism rotated by 90° and 1/3 the volume. The reciprocal lattice is also hexagonal and rotated from the real lattice by 90°, and the Brillouin zone is shown in Fig. 4.10.

Figures 4.2–4.4 show examples of three-dimensional lattices that occur in many crystals. The primitive vectors can be chosen to be (in units of a)

$$
\begin{array}{ccccc}
 & \text{simple cubic} & \text{simple hex.} & \text{fcc} & \text{bcc} \\
\mathbf{a}_1 = & (1,0,0) & (1,0,0) & \left(0,\frac{1}{2},\frac{1}{2}\right) & \left(-\frac{1}{2},\frac{1}{2},\frac{1}{2}\right), \\
\mathbf{a}_2 = & (0,1,0) & \left(\frac{1}{2},\frac{\sqrt{3}}{2},0\right) & \left(\frac{1}{2},0,\frac{1}{2}\right) & \left(\frac{1}{2},-\frac{1}{2},\frac{1}{2}\right), \\
\mathbf{a}_3 = & (0,0,1) & \left(0,0,\frac{c}{a}\right) & \left(\frac{1}{2},\frac{1}{2},0\right) & \left(\frac{1}{2},\frac{1}{2},-\frac{1}{2}\right).
\end{array}
$$

(4.3)

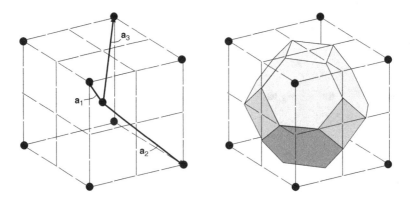

Figure 4.3. Body-centered cubic (bcc) lattice, showing one choice for the three lattice vectors. The conventional cubic cell shown indicates the set of all eight nearest neighbors at a distance $\frac{\sqrt{3}}{2}a$ around the central atom. (There are six second neighbors at distance a.) On the right-hand side of the figure is shown the Wigner–Seitz cell formed by the perpendicular bisectors of the lattice vectors (this is also the Brillouin zone for the fcc lattice).

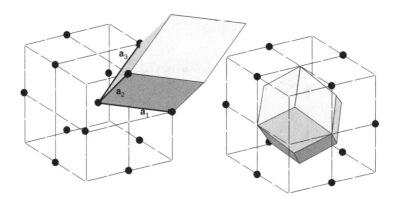

Figure 4.4. Face-centered cubic (fcc) lattice, drawn to emphasize the close packing of 12 neighbors around the central site. (The location of sites at face centers is evident if the cube is drawn with a lattice site at each corner and on each face of the cube.) Left: one choice for primitive lattice vectors and the parallelepiped primitive cell, which has lower symmetry than the lattice. Right: the symmetric Wigner–Seitz cell (which is also the Brillouin zone for the bcc lattice).

The body-centered cubic (bcc) and face-centered cubic (fcc) lattices are shown, respectively, in Figs. 4.3 and 4.4, each represented in the large conventional cubic cell (indicated by the dashed lines) with a lattice site at the center. All nearest neighbors of the central site are shown: 8 for bcc and 12 for fcc lattices. One choice of primitive vectors is shown in each case, but clearly other equivalent vectors could be chosen, and all vectors to the equivalent neighbors are also lattice translations. In the fcc case, the left-hand side of Fig. 4.4 shows one possible primitive cell, the parallelepiped formed by the primitive vectors. This is the simplest cell to construct; however, this cell clearly does not have cubic symmetry and other choices of primitive vectors lead to different cells. The Wigner–Seitz cells for each Bravais lattice, shown respectively in Figs. 4.3 and 4.4, are bounded by planes that are perpendicular bisectors of the translation vectors from the central lattice point. The Wigner–Seitz cell is particularly useful because it is the unique cell defined as the set of all points in space closer to the central lattice point than to any other lattice point; it is independent of the choice of primitive translations and it has the full symmetry of the Bravais lattice.

It is useful for deriving formal relations and for practical computer programs to express the set of primitive vectors as a square matrix $a_{ij} = (\mathbf{a}_i)_j$, where j denotes the cartesian component and i the primitive vector, i.e., the matrix has the same form as the arrays of vectors shown in Eqs. (4.2) and (4.3).

The volume of any primitive cell must be the same, since translations of any such cell fill all space. The most convenient choice of cell in which to express the volume is the parallelepiped defined by the primitive vectors. If we define Ω_{cell} as the volume in any dimension d (i.e., it has units (length)^d), simple geometric arguments show that $\Omega_{cell} = |a_1|$ ($d = 1$); $|\mathbf{a}_1 \times \mathbf{a}_2|$, ($d = 2$); and $|\mathbf{a}_1 \cdot (\mathbf{a}_2 \times \mathbf{a}_3)|$, ($d = 3$). In any dimension this can be written as the determinant of the $\mathbf{a}$ matrix (see Exercise 4.4),

$$\Omega_{cell} = \det(\mathbf{a}) = |\mathbf{a}|. \tag{4.4}$$

The Basis of Atoms in a Cell

The basis describes the positions of atoms in each unit cell relative to the chosen origin. If there are S atoms per primitive cell, then the basis is specified by the atomic position vectors τ_s, $s = 1, S$.

Two Dimensions

Two-dimensional cases are both instructive and relevant for important problems in real materials. For example, graphene, BN, MoS_2 and many other transition metal pnictides and chalcogenides can be made in actual single layers and they can be "stacked" to form the so-called van der Waals heterostructures described in Section 2.12. In addition there are three-dimensional crystals with layer structures that have strong bonding between the layers but nevertheless the electronic states near the Fermi energy are well described as two-dimensional planar systems with only weak hopping between the planes. For example, the three-dimensional structure of the high-temperature superconductor $YBa_2Cu_3O_7$ is shown in Fig. 17.3; however, the most interest is in the band confined to the CuO_2 square planes. These will serve as illustrative examples of simple bands in Chapter 14 and as notable examples of full density functional theory calculations.

The square lattice for CuO_2 planes is shown in Fig. 4.5. The lattice vectors are given above and the atomic position vectors are conveniently chosen with the Cu atom at the origin $\tau_1 = (0,0)$ and the other positions chosen to be $\tau_2 = (\frac{1}{2},0)a$ and $\tau_3 = (0,\frac{1}{2})a$. It is

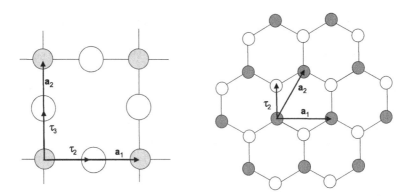

Figure 4.5. Left: square lattice for CuO_2 planes common to the cuprate high-temperature superconductors; there are three atoms per primitive cell. Right: honeycomb lattice for a single plane of graphite or hexagonal BN; the lattice is triangular and there are two atoms per primitive cell.

useful to place the Cu atom at the origin since it is the position of highest symmetry in the cell, with inversion, mirror planes, and fourfold rotation symmetry about this site.[3]

Graphene or a plane of hexagonal BN form a honeycomb lattice with a triangular Bravais lattice and two atoms per primitive cell, as shown on the right-hand side of Fig. 4.5. If the two atoms are the same chemical species, the structure is that of a plane of graphite. The primitive lattice vectors are $\mathbf{a}_1 = (1,0)a$ and $\mathbf{a}_2 = \left(\frac{1}{2}, \frac{\sqrt{3}}{2}\right)a$, where the nearest neighbor distance is $a/\sqrt{3}$. If one atom is at the origin, $\tau_1 = (0,0)$, one of the possible choices of τ_2 is $\tau_2 = (0, 1/\sqrt{3})a$ as shown in Fig. 4.5. It is also useful to define the atomic positions in terms of the primitive lattice vectors by $\tau_s = \sum_{i=1}^{d} \tau_{si}^L \mathbf{a}_i$, where the superscript L denotes the representation in lattice vectors. In this case, one finds $\tau_2 = \frac{2}{3}(\mathbf{a}_1 + \mathbf{a}_2)$ or $\tau_1^L = [0,0]$ and $\tau_2^L = \left[\frac{2}{3}, \frac{2}{3}\right]$.[4]

Single layers of many transition metal chalcogenides such as MoS_2, $MoSe_3$, and WSe_3 are of interest for possible technological applications and they can be grown in heterostructures as described in Section 2.12. For example, MoS_2 in the 2H structure is a semiconductor with a direct bandgap (see Fig. 22.8), which causes it to have strong optical absorption. The 2H structure is depicted in Fig. 4.6; the top view at the left looks the same as graphene and BN so that the Bravais lattice is also the triangular lattice, the same as a layer of close-packed spheres in Fig. 4.9. There is one Mo atom per primitive cell and we can choose the origin at the Mo site, i.e., $\tau_1 = 0$; however, there are two S atoms per cell above and below the Mo plane at positions τ_2 and τ_3, each with three Mo neighbors as indicated in the right side of Fig. 4.6. The 1T is similar except the two S atoms are not directly above one another but are at positions labeled B and C in Fig. 4.9.

---

[3] In any case where the origin can be chosen as a center of inversion, the Fourier transforms of all properties, such as the density and potential, are real. Also, all excitations can be classified into even and odd relative to this origin, and the fourfold rotational symmetry allows the roles of the five Cu d states to be separated.

[4] Simple reasoning shows that all covalently bonded crystals are expected to have more than one atom per primitive cell (see the examples of diamond and ZnS crystals and Exercise 4.8).

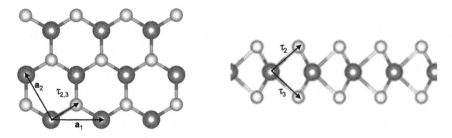

Figure 4.6. Crystal structure of a single MoS$_2$ layer in the 2H structure. The top view at the left shows that the Mo atoms (dark gray) form a triangular 2D Bravais lattice like graphene and BN in Fig. 4.5. The S atoms (light gray) each have 3 Mo neighbors like graphene and BN; however, they are above and below the Mo plane as shown in the side view at the right for the 2H structure. The 1T structure is the same but with the two S atoms in the alternate positions as described in the text.

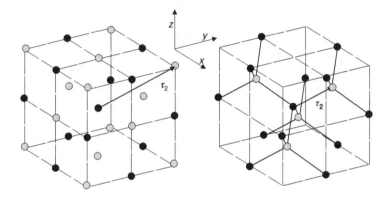

Figure 4.7. Two examples of crystals with a basis of two atoms per cell and fcc Bravais lattice. Left: rock salt (or NaCl) structure. Right: zinc-blende (cubic ZnS) structure. The positions of the atoms are given in the text. In the former case, a simple cubic crystal results if the two atoms are the same. In the latter case, if the two atoms are identical the resulting crystal has diamond structure.

Three Dimensions

NaCl and ZnS are two examples of crystals with the fcc Bravais lattice and a basis of two atoms per cell, as shown in Fig. 4.7. The primitive translation vectors are given in the previous section in terms of the cube edge a and illustrated in Fig. 4.4. For the case of NaCl, it is convenient to choose one atom at the origin $\tau_1 = (0,0,0)$, since there is inversion symmetry and cubic rotational symmetry around each atomic site, and the second basis vector is chosen to be $\tau_2 = \left(\frac{1}{2}, \frac{1}{2}, \frac{1}{2}\right)a$. In terms of the primitive lattice vectors, one can see from Fig. 4.7 that $\tau_2 = \sum_{i=1}^{d} \tau_{2i}^L \mathbf{a}_i$, where $\tau_2^L = \left[\frac{1}{2}, \frac{1}{2}, \frac{1}{2}\right]$. It is also easy to see that if the two atoms at positions τ_1 and τ_2 were the same, then the crystal would actually have a simple cubic Bravais lattice, with cube edge $a_{\text{sc}} = \frac{1}{2}a_{\text{fcc}}$.

A second example is the zinc-blende structure, which is the structure of many III–V and II–VI crystals such as GaAs and ZnS. This crystal is also fcc with two atoms per unit cell.

Although there is no center of inversion in a zinc-blende structure crystal, each atom is at a center of tetrahedral symmetry; we can place the origin at one atom, $\tau_1 = (0,0,0)a$, and $\tau_2 = \left(\frac{1}{4},\frac{1}{4},\frac{1}{4}\right)a$, as shown in Fig. 4.7, or any of the equivalent choices. Thus this structure is the same as the NaCl structure except for the basis, which in primitive lattice vectors is simply $\tau_2^L = \left[\frac{1}{4},\frac{1}{4},\frac{1}{4}\right]$. If the two atoms in the cell are identical, this is the diamond structure in which C, Si, Ge, and gray Sn occur. A bond center is the appropriate choice of origin for the diamond structure since this is a center of inversion symmetry. This can be accomplished by shifting the origin so that $\tau_1 = -\left(\frac{1}{8},\frac{1}{8},\frac{1}{8}\right)a$, and $\tau_2 = \left(\frac{1}{8},\frac{1}{8},\frac{1}{8}\right)a$; similarly, $\tau_1^L = -\left[\frac{1}{8},\frac{1}{8},\frac{1}{8}\right]$ and $\tau_2^L = \left[\frac{1}{8},\frac{1}{8},\frac{1}{8}\right]$.

The perovskite structure illustrated in Fig. 4.8 has chemical composition ABO_3 and occurs for a large number of compounds with interesting properties including ferroelectrics (e.g., $BaTiO_3$), Mott-insulator antiferromagnets (e.g., $CaMnO_3$), and alloys exhibiting metal–insulator transitions (e.g., $La_xCa_{1-x}MnO_3$). The crystal may be thought of as the CsCl structure with O on the edges. The environment of the A and B atoms is very different, with the A atoms having 12 O neighbors at a distance $a/\sqrt{2}$ and the B atoms having 6 O neighbors at a distance $a/2$. Thus these atoms play a very different role in the properties. Typically the A atom is a nontransition metal for which Coulomb ionic bonding favors the maximum number of O neighbors, whereas the B atom is a transition metal where the d states favor bonding with the O states. Note the contrast of the planes of B and O atoms with the CuO_2 planes in Fig. 4.5: although the planes are similar, each B atom in the cubic perovskites is in three intersecting orthogonal planes, whereas in the layered structures such as La_2CuO_4, the CuO_2 planes are clearly identified, with each Cu belonging to only one plane.

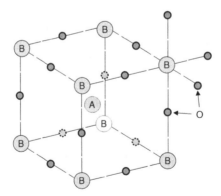

Figure 4.8. The perovskite crystal structure with the chemical composition ABO_3. This structure occurs for a large number of compounds with interesting properties, including ferroelectrics (e.g., $BaTiO_3$), antiferromagnets (e.g., $CaMnO_3$), and alloys (e.g., $Pb_xZr_{1-x}TiO_3$ and $La_xCa_{1-x}MnO_3$). The crystal may be thought of as cubes with A atoms at the center, B at the corners, and O on the edges. The environment of the A and B atoms is very different, the A atom having 12 O neighbors at a distance $a/\sqrt{2}$ and the B atoms having 6 O neighbors at a distance $a/2$. (The neighbors around one B atom are shown.) The Bravais lattice is simple cubic.

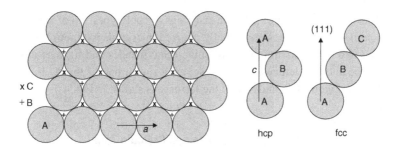

Figure 4.9. Stacking of close-packed planes to create close-packed three-dimensional lattices. Left: The only possible close packing in two dimensions is the hexagonal layer of spheres labeled A, with lattice constant a. Right: Three-dimensional stacking can have the next layers in either B or C positions. Of the infinite set of possible layer sequences, only the fcc stacking (... ABCABC ...) forms a primitive lattice; hexagonal close-packed (hcp) (... ABABAB ...) has two sites per primitive cell; all others have larger primitive cells.

Close-Packed Structures

In two dimensions there is only one way to make a "close-packed structure," defined as a structure in which hard spheres (or disks) can be placed with the maximum filling of space. That is the triangular lattice in Fig. 4.5 with one atom per lattice point. In the plane, each atom has six neighbors in a hexagonal arrangement, as shown in Fig. 4.9. All three-dimensional close-packed structures consist of such close-packed planes of atoms stacked in various sequences. As shown in Fig. 4.9, the adjacent plane can be stacked in one of two ways: if the given plane is labeled A, then two possible positions for the next plane can be labeled B and C.

The face-centered cubic structure (shown in Fig. 4.4) is the cubic close-packed structure, which can be viewed as the sequence of close-packed planes in the sequence ... ABCABC .... It has one atom per primitive cell, as may be seen by the fact that each atom has the same relation to all its neighbors, i.e., an A atom flanked by C and B planes is equivalent to a B atom flanked by A and C planes, etc. Specifically, if the lattice is a Bravais lattice then the vector from an atom in the A plane to one of its closest neighbors in the adjacent C plane must be a lattice vector. Similarly, twice that vector is also a lattice vector, as may be verified easily. The cubic symmetry can be verified by the fact that the close-packed planes may be chosen perpendicular to any of the [111] crystal axes.

The hexagonal closed-packed structure consists of close-packed planes stacked in a sequence ... ABABAB .... This is a hexagonal Bravais lattice with a basis of two atoms that are not equivalent by a translation. (This can be seen because – unlike the fcc case – twice the vector from an A atom to a neighboring B atom is *not* a vector connecting atoms. Thus the primitive cell is hexagonal as shown Fig. 4.2 with a equal to the distance between atoms in the plane and c the distance between two A planes. The ideal c/a ratio is that for packing of hard spheres, $c/a = \sqrt{8/3}$ [Exercise 4.11]). The two atoms in the primitive cell are equivalent by a combination of translation by $c/2$ and rotation by $\pi/6$, but this does not affect the analysis of the translation symmetry.

There are an infinite number of possible stackings or "polytypes" all of which are "close packed." In particular, polytypes are actually realized in crystals with tetrahedral bonding, like ZnS. The two simplest structures are cubic (zinc-blende) and hexagonal (wurtzite), based on the fcc and hcp lattices. In this case, each site in one of the A, B, or C planes corresponds to two atoms (Zn and S) and the fcc case is shown in Fig. 4.7.

4.2 Reciprocal Lattice and Brillouin Zone

Consider any function $f(\mathbf{r})$ defined for the crystal, such as the density of the electrons, which is the same in each unit cell,

$$f(\mathbf{r} + \mathbf{T}(n_1, n_2, \ldots)) = f(\mathbf{r}), \tag{4.5}$$

where $\mathbf{T}$ is any translation defined above. Such a periodic function can be represented by Fourier transforms in terms of Fourier components at wavevectors $\mathbf{q}$ defined in reciprocal space. The formulas can be written most simply in terms of a discrete set of Fourier components if we restrict the Fourier components to those that are periodic in a large volume of crystal Ω_{crystal} composed of $N_{\text{cell}} = N_1 \times N_2 \times \cdots$ cells. Then each component must satisfy the Born–von Karman periodic boundary conditions in each of the dimensions

$$\exp(i\mathbf{q} \cdot N_1 \mathbf{a}_1) = \exp(i\mathbf{q} \cdot N_2 \mathbf{a}_2) \ldots = 1, \tag{4.6}$$

so that $\mathbf{q}$ is restricted to the set of vectors satisfying $\mathbf{q} \cdot \mathbf{a}_i = 2\pi \frac{\text{integer}}{N_i}$ for each of the primitive vectors $\mathbf{a}_i$. In the limit of large volumes Ω_{crystal} the final results must be independent of the particular choice of boundary conditions.[5]

The Fourier transform is defined to be

$$f(\mathbf{q}) = \frac{1}{\Omega_{\text{crystal}}} \int_{\Omega_{\text{crystal}}} d\mathbf{r} f(\mathbf{r}) \exp(i\mathbf{q} \cdot \mathbf{r}), \tag{4.7}$$

which, for periodic functions, can be written

$$f(\mathbf{q}) = \frac{1}{\Omega_{\text{crystal}}} \sum_{n_1, n_2, \ldots} \int_{\Omega_{\text{cell}}} d\mathbf{r} \, f(\mathbf{r}) e^{i\mathbf{q} \cdot (\mathbf{r} + \mathbf{T}(n_1, n_2, \ldots))}$$

$$= \frac{1}{N_{\text{cell}}} \sum_{n_1, n_2, \ldots} e^{i\mathbf{q} \cdot \mathbf{T}(n_1, n_2, \ldots)} \frac{1}{\Omega_{\text{cell}}} \times \int_{\Omega_{\text{cell}}} d\mathbf{r} \, f(\mathbf{r}) e^{i\mathbf{q} \cdot \mathbf{r}}. \tag{4.8}$$

The sum over all lattice points in the middle line vanishes for all $\mathbf{q}$ except those for which $\mathbf{q} \cdot \mathbf{T}(n_1, n_2, \ldots) = 2\pi \times integer$ for all translations $\mathbf{T}$. Since $\mathbf{T}(n_1, n_2, \ldots)$ is a sum of integer multiples of the primitive translations $\mathbf{a}_i$, it follows that $\mathbf{q} \cdot \mathbf{a}_i = 2\pi \times integer$.

---

[5] Of course invariance to the choice of boundary conditions in the large-system limit must be proven. For short-range forces and periodic operators the proof is straightforward, but the generalization to Coulomb forces requires care in defining the boundary conditions on the potentials. The calculation of electric polarization is especially problematic and a satisfactory theory has been developed only within the past few years, as is described in Chapter 24.

The set of Fourier components $\mathbf{q}$ that satisfy this condition is the "reciprocal lattice." If we define the vectors $\mathbf{b}_i, i = 1, d$ that are reciprocal to the primitive translations $\mathbf{a}_i$, i.e.,

$$\mathbf{b}_i \cdot \mathbf{a}_j = 2\pi \delta_{ij} \tag{4.9}$$

– the only nonzero Fourier components of $f(\mathbf{r})$ are for $\mathbf{q} = \mathbf{G}$, where the $\mathbf{G}$ vectors are a lattice of points in reciprocal space defined by

$$\mathbf{G}(m_1, m_2, \ldots) = m_1 \mathbf{b}_1 + m_2 \mathbf{b}_2 + \ldots, \tag{4.10}$$

where the m_i, $i = 1, d$ are integers. For each $\mathbf{G}$, the Fourier transform of the periodic function can be written

$$f(\mathbf{G}) = \frac{1}{\Omega_{\text{cell}}} \int_{\Omega_{\text{cell}}} d\mathbf{r}\, f(\mathbf{r}) \exp(i\mathbf{G} \cdot \mathbf{r}). \tag{4.11}$$

The mutually reciprocal relation of the Bravais lattice in real space and the reciprocal lattice becomes apparent using matrix notation that is valid in any dimension. If we define square matrix $b_{ij} = (\mathbf{b}_i)_j$, exactly as was done for the a_{ij} matrix, then primitive vectors are related by

$$\mathbf{b}^T \mathbf{a} = 2\pi \mathbf{1} \rightarrow \mathbf{b} = 2\pi (\mathbf{a}^T)^{-1} \text{or} \mathbf{a} = 2\pi (\mathbf{b}^T)^{-1}. \tag{4.12}$$

It is also straightforward to derive explicit expressions for the relation of the $\mathbf{a}_i$ and $\mathbf{b}_i$ vectors; for example, in three dimensions, one can show by geometric arguments that

$$\mathbf{b}_1 = 2\pi \frac{\mathbf{a}_2 \times \mathbf{a}_3}{|\mathbf{a}_1 \cdot (\mathbf{a}_2 \times \mathbf{a}_3)|} \tag{4.13}$$

and cyclical permutations. The geometric construction of the reciprocal lattice in two dimensions is shown in Fig. 4.1.

It is easy to show that the reciprocal of a square (simple cubic) lattice is also a square (simple cubic) lattice, with dimension $\frac{2\pi}{a}$. The reciprocal of the triangular (hexagonal) lattice is also triangular (hexagonal) but rotated with respect to the crystal lattice. The bcc and fcc lattices are reciprocal to each other (Exercise 4.9). The primitive vectors of the reciprocal lattice for each of the three-dimensional lattices in Eq. (4.3) in units of $\frac{2\pi}{a}$ are given by

	simple cubic	simple hex.	fcc	bcc
$\mathbf{b}_1 =$	$(1, 0, 0)$	$\left(1, -\frac{1}{\sqrt{3}}, 0\right)$	$(1, 1, -1)$	$(0, 1, 1),$
$\mathbf{b}_2 =$	$(0, 1, 0)$	$\left(0, \frac{2}{\sqrt{3}}, 0\right)$	$(1, -1, 1)$	$(1, 0, 1),$
$\mathbf{b}_3 =$	$(0, 0, 1)$	$\left(0, 0, \frac{a}{c}\right)$	$(-1, 1, 1)$	$(1, 1, 0).$

$$\tag{4.14}$$

The volume of any primitive cell of the reciprocal lattice can be found from the same reasoning as used for the Bravais in real space. This is the volume of the first Brillouin zone Ω_{BZ} (see Section 4.2), which can be written for any dimension d in analogy to Eq. (4.4) as

$$\Omega_{BZ} = \det(\mathbf{b}) = |\mathbf{b}| = \frac{(2\pi)^d}{\Omega_{cell}}. \tag{4.15}$$

This shows the mutual reciprocal relation of Ω_{BZ} and Ω_{cell}. The formulas can also be expressed in the geometric forms $\Omega_{BZ} = |b_1|$ $(d = 1)$; $|\mathbf{b}_1 \times \mathbf{b}_2|$, $(d = 2)$; and $|\mathbf{b}_1 \cdot (\mathbf{b}_2 \times \mathbf{b}_3)|$, $(d = 3)$.

The Brillouin Zone

In this book, the term "Brillouin zone" or "BZ" is used in two ways. In some cases, especially for Fourier transforms and the proofs in the chapters on topology, it is most convenient to use a parallelepiped defined by the reciprocal lattice vectors. In general, however, it is most useful to use the customary convention that it means the Wigner–Seitz cell of the reciprocal lattice, which is defined by the planes that are the perpendicular bisectors of the vectors from the origin to the reciprocal lattice points. It is on these planes that the Bragg condition is satisfied for elastic scattering [280, 285]. For incident particles with wavevectors inside the BZ there can be no Bragg scattering. Construction of the BZ is illustrated in Figs. 4.1–4.4, and widely used notations for points in the BZ of several crystals are given in Fig. 4.10.

Useful Relations

Expressions for crystals often involve the lengths of vectors in real and reciprocal space, $|\tau + \mathbf{T}|$ and $|\mathbf{k} + \mathbf{G}|$ and the scalar products $(\mathbf{k} + \mathbf{G}) \cdot (\tau + \mathbf{T})$. If the vectors are expressed in a cartesian coordinate system, the expressions simply involve sums over each cartesian component. However, it is often more convenient to represent $\mathbf{T}$ and $\mathbf{G}$ by the integer multiples of the basis vectors, and positions τ and wave vectors $\mathbf{k}$ as fractional multiples of the basis vectors. It is useful to define lengths and scalar products in this representation, i.e., to define the "metric."

The matrix formulation makes it easy to derive the desired expressions. Any position vector τ with elements $\tau_1, \tau_2, \ldots,$ in cartesian coordinates can be written in terms of the primitive vectors by $\tau = \sum_{i=1}^{d} \tau_i^L \mathbf{a}_i$, where the superscript L denotes the representation in lattice vectors and τ^L has elements $\tau_1^L, \tau_2^L, \ldots,$ that are fractions of primitive translation vectors. In matrix form this becomes (here superscript T denotes transpose)

$$\tau = \tau^L \mathbf{a}; \quad \tau^L = \tau \mathbf{a}^{-1} = \frac{1}{2\pi} \tau \mathbf{b}^T, \tag{4.16}$$

where $\mathbf{b}$ is the matrix of primitive vectors of the reciprocal lattice. Similarly, a vector $\mathbf{k}$ in reciprocal space can be expressed as $\mathbf{k} = \sum_{i=1}^{d} k_i^L \mathbf{b}_i$ with the relations

$$\mathbf{k} = \mathbf{k}^L \mathbf{b}; \quad \mathbf{k}^L = \mathbf{k} \mathbf{b}^{-1} = \frac{1}{2\pi} \mathbf{k} \mathbf{a}^T. \tag{4.17}$$

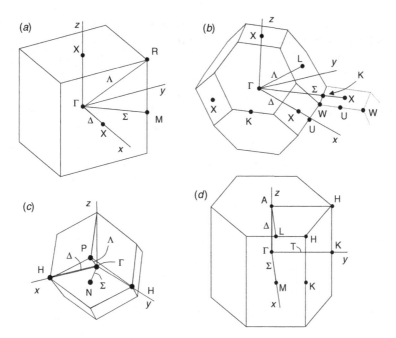

Figure 4.10. Brillouin zones for several common lattices: (*a*) simple cubic (sc), (*b*) face-centered cubic (fcc), (*c*) body-centered cubic (bcc), and (*d*) hexagonal (hex). High-symmetry points and lines are labeled according to Bouckaret, Smoluchowski, and Wigner; see also Slater [284]. The zone center ($\mathbf{k} = 0$) is designated Γ and interior lines by Greek letters; points on the zone boundary are designated by Roman letters. In the case of the fcc lattice, a portion of a neighboring cell is represented by dotted lines. This shows the orientation of neighboring cells that provides useful information, for example, that the line Σ from Γ to K continues to a point outside the first BZ that is equivalent to X.

The scalar product $(\mathbf{k} + \mathbf{G}) \cdot (\tau + \mathbf{T})$ is easily written in the lattice coordinates, using relation Eq. (4.9). If $\mathbf{T}(n_1, n_2, \ldots) = n_1 \mathbf{a}_1 + n_2 \mathbf{a}_2 + \cdots$ and $\mathbf{G}(m_1, m_2, \ldots) = m_1 \mathbf{b}_1 + m_2 \mathbf{b}_2 + \ldots$, then one finds the simple expression

$$(\mathbf{k} + \mathbf{G}) \cdot (\tau + \mathbf{T}) = 2\pi \sum_{i=1}^{d} (k_i^L + m_i)(\tau_i^L + n_i) \equiv 2\pi (\mathbf{k}^L + \mathbf{m}) \cdot (\tau^L + \mathbf{n}). \quad (4.18)$$

The relation in terms of the cartesian vectors is readily derived using Eqs. (4.16) and (4.17). On the other hand, the lengths are most easily written in the cartesian system. Using Eqs. (4.16) and (4.17) and the same vector notation as in Eq. (4.18), it is straightforward to show that lengths are given by

$$|\tau + \mathbf{T}|^2 = (\tau^L + \mathbf{n})\mathbf{a}\mathbf{a}^T (\tau^L + \mathbf{n})^T; \quad |\mathbf{k} + \mathbf{G}|^2 = (\mathbf{k}^L + \mathbf{m})\mathbf{b}\mathbf{b}^T (\mathbf{k}^L + \mathbf{m})^T \quad (4.19)$$

– i.e., $\mathbf{a}\mathbf{a}^T$ and $\mathbf{b}\mathbf{b}^T$ are the metric tensors for the vectors in real and reciprocal spaces expressed in their natural forms as multiples of the primitive translation vectors.

Finally, one often needs to find all the lattice vectors within some cutoff radius, e.g., in order to find the lowest Fourier components in reciprocal space or the nearest neighbors in real space. Consider the parallelepiped defined by all lattice points in real space $\mathbf{T}(n_1,n_2,n_3)$; $-N_1 \leq n_1 \leq N_1$; $-N_2 \leq n_2 \leq N_2$; $-N_3 \leq n_3 \leq N_3$. Since the vectors $\mathbf{a}_2$ and $\mathbf{a}_3$ form a plane, the distance in space perpendicular to this plane is the projection of $\mathbf{T}$ onto the unit vector perpendicular to the plane. This unit vector is $\hat{\mathbf{b}}_1 = \mathbf{b}_1/|\mathbf{b}_1|$ and, using Eq. (4.19), it is then simple to show that the maximum distance in this direction is $R_{max} = 2\pi \frac{N_1}{|\mathbf{b}_1|}$. Similar equations hold for the other directions. The result is a simple expression (Exercise 4.15) for the boundaries of the parallelepiped that bounds a sphere of radius R_{max},

$$N_1 = \frac{|\mathbf{b}_1|}{2\pi} R_{max}; N_2 = \frac{|\mathbf{b}_2|}{2\pi} R_{max}; \dots \tag{4.20}$$

In reciprocal space the corresponding condition for the parallelepiped that bounds a sphere of radius G_{max} is

$$M_1 = \frac{|\mathbf{a}_1|}{2\pi} G_{max}; M_2 = \frac{|\mathbf{a}_2|}{2\pi} G_{max}; \dots, \tag{4.21}$$

where the vectors range from $-M_i \mathbf{b}_i$ to $+M_i \mathbf{b}_i$ in each direction.

4.3 Excitations and the Bloch Theorem

The previous sections were devoted to properties of periodic functions in a crystal, such as the nuclear positions and electron density, that obey the relation Eq. (4.5), i.e., $f(\mathbf{r} + \mathbf{T}(n_1, n_2, \dots)) = f(\mathbf{r})$ for any translation of the Bravais lattice $\mathbf{T}(\mathbf{n}) \equiv \mathbf{T}(n_1, n_2, \dots) = n_1\mathbf{a}_1 + n_2\mathbf{a}_2 + \dots$, as defined in Eq. (4.1). Such periodic functions have nonzero Fourier components only for reciprocal space at the reciprocal lattice vectors defined by Eq. (4.10).

Excitations of the crystal do not, in general, have the periodicity of the crystal.[6] The subject of this section is the classification of excitations according to their behavior under the translation operations of the crystal. This leads to a Bloch theorem proved, in a general way, and applicable to all types of excitations: electrons, phonons, and other excitations of the crystal.[7] We will give explicit demonstrations for independent-particle excitations; however, since the general relations apply to any system, the theorems can be generalized to correlated many-body systems.

Consider the eigenstates of any operator $\hat{O}$ defined for the periodic crystal. Any such operator must be invariant to any lattice translation $\mathbf{T}(\mathbf{n})$. For example, $\hat{O}$ could be the hamiltonian $\hat{H}$ for the Schrödinger equation for independent particles,

---

[6] We take the Born–von Karman boundary conditions that the excitations are required to be periodic in the large volume $\Omega_{crystal}$ composed of $N_{cell} = N_1 \times N_2 \times \cdots$ cells, as was described previously in Eq. (4.6). See the footnote there regarding the proofs that the results are independent of the choice of boundary conditions in the thermodynamic limit of large size.

[7] The derivation here follows the "first proof" of the Bloch theorem as described by Ashcroft and Mermin [280]. It is rather formal and simpler alternative proofs are given in Chapters 12 and 14.

$$\hat{H}\psi(\mathbf{r}) = \left[-\frac{\hbar^2}{2m_e}\nabla^2 + V(\mathbf{r}) \right]\psi_i(\mathbf{r}) = \varepsilon_i\psi_i(\mathbf{r}). \tag{4.22}$$

The operator $\hat{H}$ is invariant to all lattice translations since $V_{\mathrm{eff}}(\mathbf{r})$ has the periodicity of the crystal[8] and the derivative operator is invariant to any translation.

Similarly, we can define translation operators $\hat{T}_{\mathbf{n}}$ that act on any function by displacing the arguments, e.g.,

$$\hat{T}_{\mathbf{n}}V(\mathbf{r}) = V[\mathbf{r} + \mathbf{T}(\mathbf{n})] = V(\mathbf{r} + n_1\mathbf{a}_1 + n_2\mathbf{a}_2 + \ldots). \tag{4.23}$$

Since the hamiltonian is invariant to any of the translations $\mathbf{T}(\mathbf{n})$, it follows that the hamiltonian operator commutes with each of the translations operators $\hat{T}_{\mathbf{n}}$,

$$\hat{H}\hat{T}_{\mathbf{n}} = \hat{T}_{\mathbf{n}}\hat{H}. \tag{4.24}$$

From Eq. (4.24) it follows that the eigenstates of $\hat{H}$ can be chosen to be eigenstates of *all* $\hat{T}_{\mathbf{n}}$ simultaneously. Unlike the hamiltonian, the eigenstates of the translation operators can be readily determined, independent of any details of the crystal; thus they can be used to "block diagonalize" the hamiltonian, rigorously classifying the states by their eigenvalues of the translation operators and thus leading to the "Bloch theorem" derived explicitly below.

The key point is that the translation operators form a simple group in which the product of any two translations is a third translation, so that the operators obey the relation,

$$\hat{T}_{\mathbf{n}_1}\hat{T}_{\mathbf{n}_2} = \hat{T}_{\mathbf{n}_1 + \mathbf{n}_2}. \tag{4.25}$$

Thus the eigenvalues $t_{\mathbf{n}}$ and eigenstates $\psi(\mathbf{r})$ of the operators $\hat{T}_{\mathbf{n}}$

$$\hat{T}_{\mathbf{n}}\psi(\mathbf{r}) = t_{\mathbf{n}}\psi(\mathbf{r}), \tag{4.26}$$

must obey the relations

$$\hat{T}_{\mathbf{n}_1}\hat{T}_{\mathbf{n}_2}\psi(\mathbf{r}) = t_{(\mathbf{n}_1 + \mathbf{n}_2)}\psi(\mathbf{r}) = t_{\mathbf{n}_1}t_{\mathbf{n}_2}\psi(\mathbf{r}). \tag{4.27}$$

By breaking each translation into the product of primitive translations, any $t_{\mathbf{n}}$ can be written in terms of a primitive set $t(\mathbf{a}_i)$

$$t_{\mathbf{n}} = [t(\mathbf{a}_1)]^{n_1}[t(\mathbf{a}_2)]^{n_2}\ldots. \tag{4.28}$$

Since the modulus of each $t(\mathbf{a}_i)$ must be unity (otherwise any function obeying Eq. (4.28) is not bounded), it follows that each $t(\mathbf{a}_i)$ can always be written

$$t(\mathbf{a}_i) = e^{i2\pi y_i}. \tag{4.29}$$

Since the eigenfunctions must satisfy periodic boundary conditions Eq. (4.6), $(t(\mathbf{a}_i))^{N_i} = 1$, so that $y_i = 1/N_i$. Finally, using the definition of the primitive reciprocal lattice vectors in Eq. (4.9), Eq. (4.28) can be written

$$t_{\mathbf{n}} = e^{i\mathbf{k}\cdot\mathbf{T}_{\mathbf{n}}}, \tag{4.30}$$

---

[8] The logic also holds if the potential is a nonlocal operator (as in pseudopotentials, with which we will deal later).

where

$$\mathbf{k} = \frac{n_1}{N_1}\mathbf{b}_1 + \frac{n_2}{N_2}\mathbf{b}_2 + \cdots \tag{4.31}$$

is a vector in reciprocal space. The range of $\mathbf{k}$ can be restricted to one primitive cell of the reciprocal lattice since the relation Eq. (4.30) is the same in every cell that differs by the addition of a reciprocal lattice vector $\mathbf{G}$ for which $\mathbf{G} \cdot \mathbf{T} = 2\pi \times integer$. Note that there are exactly the same number of values of $\mathbf{k}$ as the number of cells.

This leads us directly to the desired results:

1. **The Bloch theorem.**[9] From Eqs. (4.27), (4.30), and (4.31), one finds

$$\hat{T}_\mathbf{n}\psi(\mathbf{r}) = \psi(\mathbf{r} + \mathbf{T_n}) = e^{i\mathbf{k}\cdot\mathbf{T_n}}\psi(\mathbf{r}), \tag{4.32}$$

   which is the celebrated "Bloch theorem" that eigenstates of the translation operators vary from one cell to another in the crystal with the phase factor given in Eq. (4.32). The eigenstates of any periodic operator, such as the hamiltonian, can be chosen with definite values of $\mathbf{k}$, which can be used to classify any excitation of a periodic crystal. From Eq. (4.32) it follows that eigenfunctions with a definite $\mathbf{k}$ can also be written

$$\psi_\mathbf{k}(\mathbf{r}) = e^{i\mathbf{k}\cdot\mathbf{r}}u_\mathbf{k}(\mathbf{r}), \tag{4.33}$$

   where $u_\mathbf{k}(\mathbf{r})$ is periodic ($u_\mathbf{k}(\mathbf{r} + \mathbf{T_n}) = u_\mathbf{k}(\mathbf{r})$). Examples of Bloch states are shown in Fig. 4.11 and the Bloch theorem for independent-particle electron states in many different representations are given in Chapters 12–17.

2. **Bands of eigenvalues.** In the limit of a large (macroscopic) crystal, the spacing of the $\mathbf{k}$ points goes to zero and $\mathbf{k}$ can be considered a continuous variable. The eigenstates of the hamiltonian may be found separately for each $\mathbf{k}$ in one primitive cell of the reciprocal lattice. For each $\mathbf{k}$ there is a discrete set of eigenstates that can be labeled by an index i. This leads to bands of eigenvalues $\varepsilon_{i,\mathbf{k}}$ and energy gaps where there can be no eigenstates for any $\mathbf{k}$.

3. **Conservation of crystal momentum.** It follows from the analysis above that in a perfect crystal the wavevector $\mathbf{k}$ is conserved *modulo any reciprocal lattice vector* $\mathbf{G}$. Thus it is analogous to ordinary momentum in free space, but it has the additional feature that it is only conserved *within one primitive cell, usually chosen to be the Brillouin zone.* Thus two excitations at vectors $\mathbf{k}_1$ and $\mathbf{k}_2$ may have total momentum $\mathbf{k}_1 + \mathbf{k}_2$ outside the Brillouin zone at origin and their true crystal momentum should be reduced to the Brillouin zone around the origin by adding a reciprocal lattice vector. The physical process of scattering of two excitations by some perturbation is called "Umklapp scattering" [280].

4. **The role of the Brillouin zone (BZ).** All possible eigenstates are specified by $\mathbf{k}$ within any primitive cell of the periodic lattice in reciprocal space. However, the BZ is the cell of choice in which to represent excitations; its boundaries are the bisecting planes where Bragg scattering occurs and inside the Brillouin zone there are no such boundaries. Thus

---

[9] The properties of waves in periodic media were derived earlier by Flouquet in one dimension (see note in [280]) and is often referred to in the physics literature as the "Bloch–Flouquet theorem."

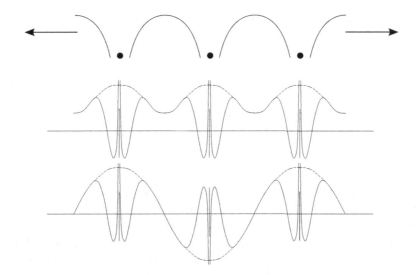

Figure 4.11. Schematic illustration of a periodic potential in a crystal extending to the left and right and two examples of Bloch states at $k = 0$ and at the zone boundary. The envelope is the smooth function that multiplies a periodic array of atomic-like 3s functions, chosen to be the same as in Fig. 11.2.

bands $\varepsilon_{i,\mathbf{k}}$ are analytic functions of $\mathbf{k}$ inside the BZ and nonanalytic dependence on $\mathbf{k}$ can occur only at the boundaries.

Examples of Brillouin zones for important cases are shown in Fig. 4.10 with labels for high-symmetry points and lines using the notation of Bouckaret, Smoluchowski, and Wigner (see also Slater [284]). The labels define the directions and points used in many figures given in the present work for electron bands and phonon dispersion curves.

5. **Integrals in k space.** For many properties, such as the counting of electrons in bands, total energies, etc., it is essential to sum over the states labeled by $\mathbf{k}$. The crucial point is that if one chooses the eigenfunctions that obey periodic boundary conditions in a large crystal of volume Ω_{crystal} composed of $N_{\text{cell}} = N_1 \times N_2 \times \cdots$ cells, as was done in the analysis of Eq. (4.6), then there is *exactly one value of* $\mathbf{k}$ *for each cell*. Thus in a sum over states to find an intrinsic property of a crystal expressed as "per unit cell" one simply has a sum over values of $\mathbf{k}$ divided by the number of values N_k. For a general function $f_i(\mathbf{k})$, where i denotes any of the discrete set of states at each $\mathbf{k}$, the average value per cell becomes

$$\bar{f}_i = \frac{1}{N_k} \sum_{\mathbf{k}} f_i(\mathbf{k}). \tag{4.34}$$

If one converts the sum to an integral by taking the limit of a continuous variable in Fourier space with a volume per $\mathbf{k}$ point of Ω_{BZ}/N_k,

$$\bar{f}_i = \frac{1}{\Omega_{\text{BZ}}} \int_{\text{BZ}} d\mathbf{k} \, f_i(\mathbf{k}) = \frac{\Omega_{\text{cell}}}{(2\pi)^d} \int_{\text{BZ}} d\mathbf{k} \, f_i(\mathbf{k}), \tag{4.35}$$

where Ω_{cell} is the volume of a primitive cell in real space.

6. **Equation for the periodic part of Bloch functions.** The Bloch functions $\psi_{i,\mathbf{k}}(\mathbf{r}) = e^{i\mathbf{k}\cdot\mathbf{r}}u_{i,\mathbf{k}}(\mathbf{r})$ are eigenfunctions of the hamiltonian operator $\hat{H}$. By inserting the expression for $\psi_{i,\mathbf{k}}(\mathbf{r})$ in terms of $u_{i,\mathbf{k}}(\mathbf{r})$, the equation becomes

$$e^{-i\mathbf{k}\cdot\mathbf{r}}\hat{H}e^{i\mathbf{k}\cdot\mathbf{r}}u_{i,\mathbf{k}}(\mathbf{r}) = \varepsilon_{i,\mathbf{k}}u_{i,\mathbf{k}}(\mathbf{r}). \tag{4.36}$$

For the hamiltonian in Eq. (4.22), the equation can be written

$$\hat{H}(\mathbf{k})u_{i,\mathbf{k}}(\mathbf{r}) = \left[-\frac{\hbar^2}{2m_e}(\nabla + i\mathbf{k})^2 + V(\mathbf{r})\right]u_{i,\mathbf{k}}(\mathbf{r}) = \varepsilon_{i,\mathbf{k}}u_{i,\mathbf{k}}(\mathbf{r}). \tag{4.37}$$

7. **Magnetic fields, spin, and spin–orbit interaction.** The generalization to spin-orbitals is treated briefly in the next section and described in various places in this book.
8. **Topology of the Bloch functions.** All of the points above were understood in the 1920s, but it was only starting in the 1980s that the topology of the Bloch functions was recognized by Thouless, Haldane, and others. Topology is a global property of the eigenfunctions as a function of $\mathbf{k}$ in the Brillouin zone that is discussed in Chapters 25–28). The new discoveries are in terms of the *phases* of the periodic part of Bloch functions, which lead to physically meaningful Berry phases. At this point it is important to emphasize that the Bloch theorem is still valid and the equations at any point $\mathbf{k}$ are exactly the same. Even though fabulous consequences are still being discovered, Eq. (4.37) is still valid and the methods for solution developed over the years still apply.

4.4 Time-Reversal and Inversion Symmetries

There is an additional symmetry that is present in all systems with no magnetic field. Since the hamiltonian is invariant to time reversal in the original time-dependent Schrödinger equation, it follows that the hamiltonian can always be chosen to be real. In a time-independent equation, such as Eq. (12.1), this means that if ψ is an eigenfunction, then ψ^* must also be an eigenfunction with the same real eigenvalue ε. According to the Bloch theorem, the solutions $\psi_{i,-\mathbf{k}}(\mathbf{r})$ can be classified by their wavevector $\mathbf{k}$ and a discrete band index i. If $\psi_{i,-\mathbf{k}}(\mathbf{r})$ satisfies the Bloch condition Eq. (4.32), then it follows that $\psi_{i,\mathbf{k}}^*(\mathbf{r})$ satisfies the same equation except with a phase factor corresponding to $-\mathbf{k}$. Thus there is never a need to calculate states at both $\mathbf{k}$ and $-\mathbf{k}$ in any crystal, $\psi_{i,-\mathbf{k}}(\mathbf{r})$ can always be chosen to be $\psi_{i,\mathbf{k}}^*(\mathbf{r})$, and the eigenvalues are equal $\varepsilon_{i,-\mathbf{k}} = \varepsilon_{i,\mathbf{k}}$. If in addition the crystal has inversion symmetry, then Eq. (4.37) is invariant under inversion since $V(-\mathbf{r}) = V(\mathbf{r})$ and $(\nabla + i\mathbf{k})^2$ is the same if we replace $\mathbf{k}$ and $\mathbf{r}$ by $-\mathbf{k}$ and $-\mathbf{r}$. Thus the periodic part of the Bloch function can be chosen to satisfy $u_{i,\mathbf{k}}(\mathbf{r}) = u_{i,-\mathbf{k}}(-\mathbf{r}) = u_{i,\mathbf{k}}^*(-\mathbf{r})$.

Spin–Orbit Interaction

So far we have ignored spin, considering only solutions for a single electron in a nonrelativistic hamiltonian. However, relativistic effects introduce a coupling of spin and spatial motion, i.e., the spin–orbit interaction derived in Appendix O. There it is shown that the equations can be written in terms of 2×2 matrices in terms of $\psi_{\uparrow,i,\mathbf{k}}(\mathbf{r})$ and $\psi_{\downarrow,i,\mathbf{k}}(\mathbf{r})$ where spin–orbit interaction leads to a diagonal term opposite for $\uparrow$ and $\downarrow$ and an off-diagonal spin-flip term. Time reversal leads to reversal of both spin and momentum so that a state $\psi_{i,\mathbf{k}}(\sigma,\mathbf{r})$ is transformed to $\psi_{i,-\mathbf{k}}(-\sigma,\mathbf{r})$, where σ is the spin variable. If there is time-reversal symmetry, Kramers theorem guarantees that states with reversed momentum and spin functions are degenerate and $\psi_{i,\mathbf{k}}(\sigma,\mathbf{r}) = \psi^*_{i,-\mathbf{k}}(-\sigma,\mathbf{r})$.

This leads to a conclusion that is a key to understanding topological insulators in Chapters 27 and 28. At certain $\mathbf{k}$ points in a crystal $\mathbf{k}$ and $-\mathbf{k}$ are the same ($\mathbf{k} = 0$) or related by a reciprocal lattice vector, and the Kramers theorem guarantees that the two spin states are degenerate. These are called TRIM (time-reversal invariant momentum) and their role is explained in Section 27.4. See especially Figs. 27.4 and 28.1, which show the TRIM points in two and three dimensions. Another example is the linear dispersion in the Rashba effect at a surface shown in Figs. O.1 and O.2. The states are degenerate at $\mathbf{k} = 0$ with linear dispersion that would not occur if there were no spin–orbit interaction. The effect can be viewed as if each spin state is in a magnetic field, opposite for the two spins so that overall there is time-reversal symmetry.

Symmetries in Magnetic Systems

The effects of a magnetic field can be included by modifying the hamiltonian in two ways: $\mathbf{p} \rightarrow (\mathbf{p} - \frac{e}{c}\mathbf{A})$, where $\mathbf{A}$ is the vector potential, and $\hat{H} \rightarrow \hat{H} + \hat{H}_{\text{Zeeman}}$, with $\hat{H}_{\text{Zeeman}} = g\mu\mathbf{H} \cdot \vec{\sigma}$. (See Appendix O for the elegant derivation by Dirac.) The latter term is easy to add to an independent-particle calculation in which there is no spin–orbit interaction; there are simply two calculations for different spins. The first term is not hard to include in localized systems like atoms where there are circulating currents. However, in extended condensed matter, there are quantitative effects that are macroscopic manifestations of quantum mechanics, including Landau diamagnetism, edge currents in quantum Hall systems (Appendix Q), and Chern insulators (Section 27.2). Topological insulators with time-reversal symmetry in Chapters 25–28, have related effects often called a quantum spin Hall effect.

In ferromagnetic systems there is a spontaneous breaking of time reversal symmetry. The ideas also apply to finite systems with a net spin, e.g., if there is an odd number of electrons. As far as symmetry is concerned, there is no difference from a material in an external magnetic field. However, the effects originate in the Coulomb interactions, which can be included in an independent-particle theory as an effective field (often a very large field). Such Zeeman-like spin-dependent terms are regularly used in independent-particle calculations to study magnetic solids such as spin-density functional theory. In Hartree–Fock calculations on finite systems, exchange induces such terms automatically; however, effects of correlation are omitted.

Antiferromagnetic solids are ones in which there is long-range order involving both space and time-reversal symmetries, e.g., a Neel state is invariant to a combination of translation and time reversal. States with such a symmetry can be described in an independent-particle approach by an effective potential with this symmetry breaking form. The broken symmetry leads to a larger unit cell in real space and a translation (or "folding") of the excitations into a smaller Brillouin zone compared with the nonmagnetic system. Antiferromagnetic solids are one of the outstanding classes of condensed matter in which many-body effects may play a crucial role. Mott insulators tend to be antiferromagnets, and metals with antiferromagnetic correlations often have large enhanced response functions. This is a major topic of the companion book [1], which addresses the many-body problem and issues that have confounded theorists since the early years of quantum mechanics.

4.5 Point Symmetries

This section is a brief summary needed for group theory applications. Discussion of group theory and symmetries in different crystal classes are covered in a number of texts and monographs. For example, Ashcroft and Mermin [280] give an overview of symmetries with pictorial representation, Slater [284] gives detailed analyses for many crystals with group tables and symmetry labels, and there are many useful books on group theory [281–283]. The codes for electronic structure for crystals in Appendix R all have some facilty to automatically generate and/or apply the group operations.

The total space group of a crystal is composed of the translation group and the point group. Point symmetries are rotations, inversions, reflections, and their combinations that leave the system invariant. In addition, there can be nonsymmorphic operations that are combinations with translations or "glides" of fractions of a crystal translation vector. The set of all such operations, $\{R_n, n = 1, \ldots, N_{\text{group}}\}$, forms a group. The operation on any function $g(\mathbf{r})$ of the full symmetry system (such as the density $n(\mathbf{r})$ or the total energy E_{total}) is

$$R_n g(\mathbf{r}) = g(R_n \mathbf{r} + \mathbf{t}_n), \tag{4.38}$$

where $R_n \mathbf{r}$ denotes the rotation, inversions, or reflections of the position $\mathbf{r}$ and $\mathbf{t}_n$ is the nonsymmorphic translation associated with operation n.

The two most important consequences of the symmetry operations for excitations can be demonstrated by applying the symmetry operations to the Schrödinger equation (4.22), with i replaced by the quantum numbers for a crystal, $i \rightarrow i, \mathbf{k}$. Since the hamiltonian is invariant under any symmetry operation R_n, the operation of R_n leads to a new equation with $\mathbf{r} \rightarrow R_i \mathbf{r} + \mathbf{t}_i$ and $\mathbf{k} \rightarrow R_i \mathbf{k}$ (the fractional translation has no effect on reciprocal space). It follows that the new function,

$$\psi_i^{R_i \mathbf{k}}(R_i \mathbf{r} + \mathbf{t}_i) = \psi_i^{\mathbf{k}}(\mathbf{r}); \text{ or } \psi_i^{R_i^{-1}\mathbf{k}}(\mathbf{r}) = \psi_i^{\mathbf{k}}(R_i \mathbf{r} + \mathbf{t}_i), \tag{4.39}$$

must also be an eigenfunction of the hamiltonian with the same eigenvalue $\varepsilon_i^{\mathbf{k}}$. This leads to two consequences:

- At "high-symmetry" $\mathbf{k}$ points, $R_i^{-1}\mathbf{k} \equiv \mathbf{k}$, so that Eq. (4.39) leads to relations among the eigenvectors at that $\mathbf{k}$ point, i.e., they can be classified according to the group representations. For example, at $\mathbf{k} = 0$, in cubic crystals all states have degeneracy 1, 2, or 3.

- One can define the "irreducible Brillouin zone" (IBZ), which is the smallest fraction of the BZ that is sufficient to determine all the information on the excitations of the crystal. The excitations at all other $\mathbf{k}$ points outside the IBZ are related by the symmetry operations. If a group operation $R_i^{-1}\mathbf{k}$ leads to a distinguishable $\mathbf{k}$ point, then Eq. (4.39) shows that the states at $R_i^{-1}\mathbf{k}$ can be generated from those at $\mathbf{k}$ by the relations given in Eq. (4.39), apart from a phase factor that has no consequence, and the fact that the eigenvalues must be equal,

$$\varepsilon_i^{R_i^{-1}\mathbf{k}} = \varepsilon_i^{\mathbf{k}}.$$

If there is time-reversal symmetry, the BZ can be reduced by at least a factor of 2 using relation of states at $\mathbf{k}$ and $-\mathbf{k}$; in a square lattice, the IBZ is $1/8$ the BZ, as illustrated in Fig. 4.12; in the highest-symmetry crystals (cubic), the IBZ is only $1/48$ the BZ.

Spin–Orbit Interaction

If there is spin–orbit interaction, the energy bands are labeled by irreducible representations of the double group for each crystal structure. The applications to crystals are worked out in many sources such as [281, 282, 284]. In this book we will only deal with cases such as the splitting of threefold degenerate p ($L = 1$) states at $\mathbf{k} = 0$ in a cubic crystal. The states split into twofold degenerate $J = 1/2$ and fourfold $J = 3/2$ states, which are illustrated for GaAs in Figs. 14.9 and 17.7. Spin–orbit interaction plays a crucial role in topological insulators in Chapters 27 and 28 and surface states in Chapter 22. We will not deal with the analysis in terms of the crystal symmetry, which can be found in the references.

4.6 Integration over the Brillouin Zone and Special Points

Evaluation of many quantities, such as energy and density, require integration over the BZ. There are two separate aspects of this problem:

- Accurate integration with a discrete set of points in the BZ. This is specific to the given problem and depends on having sufficient points in regions where the integrand varies rapidly. In this respect, the key division is between metals and insulators. Insulators have filled bands that can be integrated using only a few well-chosen points such as the "special points" discussed below. On the other hand, metals require careful integration near the Fermi surface for the bands that cross the Fermi energy where the Fermi factor varies rapidly.
- Symmetry can be used to reduce the calculations since all independent information can be found from states with $\mathbf{k}$ in the IBZ. This is useful in all cases with high symmetry, whether metals or insulators.

Special Points

The "special" property of insulators is that the integrals have the form of Eq. (4.34) where *the sum is over filled bands in the full BZ*. Since the integrand $f_i(\mathbf{k})$ is some function of the

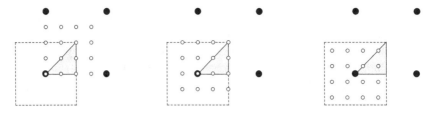

Figure 4.12. Grids for integration for a 2d square lattice, each with four times the density of the reciprocal lattice in each dimension. The left and center figures are equivalent with one point at the origin, and the six inequivalent points in the irreducible BZ shown in gray. Right: a shifted special point grid of the same density but with only three inequivalent points. Additional possibilities have been given by Moreno and Soler [286], who also pointed out that different shifts and symmetrization can lead to finer grids.

eigenfunctions $\psi_{i,\mathbf{k}}$ and eigenvalues $\varepsilon_{i,\mathbf{k}}$, it is a smoothly varying,[10] periodic function of $\mathbf{k}$. Thus $f_i(\mathbf{k})$ can be expanded in Fourier components,

$$f_i(\mathbf{k}) = \sum_{\mathbf{T}} f_i(\mathbf{T}) e^{i\mathbf{k}\cdot\mathbf{T}}, \qquad (4.40)$$

where $\mathbf{T}$ are the translation vectors of the crystal. The most important point is that the contribution of the rapidly varying terms at large $\mathbf{T}$ decreases exponentially so that the sum in Eq. (4.40) can be truncated to a finite sum. The proof [287] is related to transformations of the expressions to traces over Wannier functions (see Chapter 23) and the observation that the range of $f_i(\mathbf{T})$ is determined by the range of the Wannier functions.

Special points are chosen for efficient integration of smooth periodic functions.[11] The single most special point is the Baldereschi point [289], where the integration reduces to a single point. The choice is based upon (1) the fact that there is always some one "mean-value point" where the integrand equals the integral, and (2) use of crystal symmetry to find such a point approximately. The coordinates of the mean-value point for cubic lattices were found to be [289] simple cubic, $k = (\pi/a)(1/2, 1/2, 1/2)$; body-centered cubic, $k = (2\pi/a)(1/6, 1/6, 1/2)$; and face-centered cubic, $k = (2\pi/a)(0.6223, 0.2953, 1/2)$. Chadi and Cohen [290] have generalized this idea and have given equations for "best" larger sets of points.

The general method proposed by Monkhorst and Pack [287] is now the most widely used method because it leads to a uniform set of points determined by a simple formula valid for any crystal (given here explicitly for three dimensions):

$$\mathbf{k}_{n_1,n_2,n_3} \equiv \sum_{i}^{3} \frac{2n_i - N_i - 1}{2N_i} \mathbf{G}_i, \qquad (4.41)$$

---

[10] For an individual band the variation is not smooth at crossings with other bands; however, the relevant sums over all filled bands are smooth so long as all bands concerned are filled. This is always the case if the filled and empty bands are separated by a gap as in an insulator.

[11] In this sense, the method is analogous to Gauss–Chebyshev integration. (See [288], who found that Gauss–Chebyshev can be more efficient than the Monkhorst–Pack method for large sets of points.)

where $\mathbf{G}_i$ are the primitive vectors of the reciprocal lattice. The main features of the Monkhorst–Pack points are as follows:

- A sum over the uniform set of points in Eq. (4.41), with $n_i = 1, 2, \ldots, N_i$, *exactly integrates a periodic function that has Fourier components that extend only to $N_i \mathbf{T}_i$ in each direction*. (See Exercise 4.21. In fact, Eq. (4.41) makes a *maximum error* for higher Fourier components.)
- The set of points defined by Eq. (4.41) is a uniform grid in $\mathbf{k}$ that is a scaled version of the reciprocal lattice and offset from $\mathbf{k} = 0$. For many lattices, especially cubic, it is preferable to choose N_i to be even [287]. Then the set does *not involve the highest symmetry points*; it omits the $\mathbf{k} = 0$ point and points on the BZ boundary.
- The $N_i = 2$ set is the Baldereschi point for a simple cubic crystal (taking into account symmetry – see below). The sets for all cubic lattices are also the same as the offset Gilat–Raubenheimer mesh (see [291]).
- An informative tabulation of grids and their efficiency, together with an illuminating description is given by Moreno and Soler [286], who emphasized the generation of different sets of regular grids using a combination of offsets and symmetry.

The logic behind the Monkhorst–Pack choice of points can be understood in one dimension, where it is easy to see that the exact value of the integral,

$$I_1 = \int_0^{2\pi} dk \; \sin(k) = 0, \tag{4.42}$$

is given by the value of the integrand $f_1(k) = \sin(k)$ at the midpoint, $k = \pi$ where $\sin(k) = 0$. If one has a sum of two sin functions, $f_2(k) = A_1 \sin(k) + A_2 \sin(2k)$, then the exact value of the integral is given by a sum over two points:

$$I_2 = \int_0^{2\pi} dk \; f_2(k) = 0 = f_2(k = \pi/2) + f_2(k = 3\pi/2). \tag{4.43}$$

The advantage of the special point grids that do not contain the $\mathbf{k} = 0$ point is much greater in higher dimensions. As illustrated in Fig. 4.12 for a square lattice, an integration with a grid $4 \times 4 = 16$ times as dense as the reciprocal lattice can be done with only three inequivalent $\mathbf{k}$ points in the irreducible BZ (defined in the following subsection). This set is sufficient to integrate exactly any periodic function with Fourier components up to $\mathbf{T} = (4, 4) \times a$, where a is the square edge. The advantages are greater in higher dimensions.

Irreducible BZ

Integrals over the BZ can be replaced by integrals only over the IBZ. For example, the sums needed in the total energy (general expressions in Section 7.3 or specific ones for crystals, such as Eq. (13.1)) have the form of Eq. (4.34). Since the summand is a scalar, it must be invariant under each operation, $f_i(R_n\mathbf{k}) = f_i(\mathbf{k})$. It is convenient to define $w_\mathbf{k}$ to be the total number of *distinguishable* $\mathbf{k}$ points related by symmetry to the given $\mathbf{k}$ point in the IBZ (including the point in the IBZ) divided by the total number of points N_k. (Note that points on the BZ boundary related by $\mathbf{G}$ vectors are not distinguishable.) Then the sum Eq. (4.34) is equivalent to

$$\bar{f}_i = \sum_{\mathbf{k}}^{\text{IBZ}} w_{\mathbf{k}} f_i(\mathbf{k}). \tag{4.44}$$

Quantities such as the density can always be written as

$$n(\mathbf{r}) = \frac{1}{N_k} \sum_{\mathbf{k}} n_{\mathbf{k}}(\mathbf{r}) = \frac{1}{N_{\text{group}}} \sum_{R_n} \sum_{\mathbf{k}}^{\text{IBZ}} w_{\mathbf{k}} n_{\mathbf{k}}(R_n \mathbf{r} + \mathbf{t}_n). \tag{4.45}$$

Here points are weighted according to $w_{\mathbf{k}}$, just as in Eq. (4.44), and in addition the variable $\mathbf{r}$ is transformed in each term $n_{\mathbf{k}}(\mathbf{r})$. Corresponding expressions for Fourier components are given in Section 12.7.

Symmetry operations can be used to reduce the calculations greatly. Excellent examples are the Monkhorst–Pack meshes applied to cubic crystals, where there are 48 symmetry operations so that the IBZ is $1/48$ the total BZ. The set defined by $N_i = 2$ has $2^3 = 8$ points in the BZ, which reduces to 1 point in the IBZ. Similarly, $N_i = 4 \rightarrow 4^3 = 64$ points in the BZ reduces to 2 points; $N_i = 6 \rightarrow 6^3 = 216$ points in the BZ reduces to 10 points. As an example, for fcc the 2-point set is $(2\pi/a)(1/4, 1/4, 1/4)$ and $(2\pi/a)(1/4, 1/4, 3/4)$, which has been found to yield remarkably accurate results for energies of semiconductors, a fact that was very important in early calculations [171]. The 10-point set is sufficient for almost all modern calculations for such materials.

Interpolation Methods

Metals present an important general class of issues for efficient sampling of the desired states in the BZ. The Fermi surface plays a special role in all properties and the integration over states must take into account the sharp variation of the Fermi function from unity to zero as a function of $\mathbf{k}$. This plays a decisive role in all calculations of sums over occupied states for total quantities (e.g., the total electron density, energy, force, and stress in Chapter 7) and sums over both occupied and empty states for response functions and spectral functions (Chapter 20 and Appendix D).

In order to represent the Fermi surface, the tetrahedron method [292–295] is widely used. If the eigenvalues and vectors are known at a set of grid points, the variation between the grid points can always be approximated by an interpolation scheme using tetrahedra. This is particularly useful because tetrahedra can be used to fill all space for any grid. A simple case is illustrated on the left-hand side of Fig. 4.13, and the same construction can be used for other cases, e.g., an irregular grid that has more points near the Fermi surface and fewer points far from the Fermi surface where accuracy is not needed. The simplest procedure is a linear interpolation between the values known at the vertices, but higher-order schemes can also be used for special grids. Tetrahedron methods are very important in calculations for transition metals, rare earths, etc. where there are exquisite details of the Fermi surfaces that must be resolved.

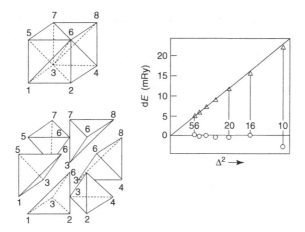

Figure 4.13. Example of generation of tetrahedra that fill the space between the grid points (left) and the results (right) of total energy calculations for Cu as a function of grid spacing Δ, comparing the linear method and the method of [292]. From [292].

One example is the method proposed by Blöchl [292] in which there is a grid of **k** points and tetrahedra that reduces to a special-points method for insulators. It also provides an interpolation formula that goes beyond the linear approximation of matrix elements within the tetrahedra, which can improve the results for metals. The use of a regular grid is helpful since the irreducible **k** points and tetrahedra can be selected by an automated procedure. An example of results for Cu metal is shown on the right-hand side of Fig. 4.13. Since the Fermi surface of Cu is rather simple, the improvement over simple linear interpolation may be surprising; it is due largely to the fact that the curvature of the occupied band crossing the Fermi energy is everywhere positive so that linear interpolation always leads to a systematic error.

4.7 Density of States

An important quantity for many purposes is the density of states (DOS) per unit energy E (and per unit volume Ω in extended matter),

$$\rho(E) = \frac{1}{N_k} \sum_{i,\mathbf{k}} \delta(\varepsilon_{i,\mathbf{k}} - E) = \frac{\Omega_{\text{cell}}}{(2\pi)^d} \int_{\text{BZ}} d\mathbf{k}\, \delta(\varepsilon_{i,\mathbf{k}} - E). \tag{4.46}$$

In the case of independent-particle states, where $\varepsilon_{i,\mathbf{k}}$ denotes the energy of an electron (or phonon), Eq. (4.46) is the number of independent-particle states per unit energy. Quantities like the specific heat involve excitations of electrons that do not change the number, i.e., an excitation from a filled to an empty state. Similarly, for independent-particle susceptibilities, such as general forms of χ^0 in Appendix D and the dielectric function given in Eq. (21.9), the imaginary part is given by matrix elements times a joint DOS, i.e., a double sum over bands i and j but a single sum over **k** due to momentum conservation, as a function of the energy difference $E = \varepsilon_j - \varepsilon_i$.

It is straightforward to show that the DOS has "critical points," or van Hove singularities [296], where $\rho(E)$ has analytic forms that depend only on the space dimension. In three dimensions, each band must have square root singularities at the maxima and minima and at saddle points in the bands. A simple example is illustrated later in Fig. 14.4 for a tight-binding model in one, two, and three dimensions, and a characteristic example for many bands in a crystal is shown in Fig. 16.12; an example for phonons is shown in Fig. 2.11.

SELECT FURTHER READING

Physically motivated discussion for many crystals with symmetry labels:

Ashcroft, N. and Mermin, N., *Solid State Physics* (W. B. Saunders Company, New York, 1976).

Cohen, M. L. and Louie, S. G., *Fundamentals of Condensed Matter Physics* (Cambridge University Press, Cambridge, 2016).

Girvin S. M. and Yang K., *Modern Condensed Matter Physics* (Cambridge University Press, Cambridge, 2019).

Slater, J. C., *Symmetry and Energy Bonds in Crystals* (Dover, New York, 1972), collected and reprinted version of *Quantum Theory of Molecules and Solids*, vol. 2 (1965).

Books on group theory with applications to crystals:

Heine, V., *Group Theory* (Pergamon Press, New York, 1960).

Lax, M. J., *Symmetry Principles in Solid State and Molecular Physics* (John Wiley & Sons, New York, 1974).

Tinkham, M., *Group Theory and Quantum Mechanics* (McGraw-HIll, New York, 1964).

Book with insightful problems with solutions:

Mihaly, L. and Martin, M. C., *Solid State Physics: Problems and Solutions*, 2nd ed. (Wiley-VCH, Berlin, 2009).

Exercises

4.1 Derive the expression for primitive reciprocal lattice in three dimensions given in Eq. (4.13).

4.2 For a two-dimensional lattice give an expression for primitive reciprocal lattice vectors that is equivalent to the one for three dimensions given in Eq. (4.13).

4.3 Show that for the two-dimensional triangular lattice the reciprocal lattice is also triangular and is rotated by $90°$.

4.4 Show that the volume of the primitive cell in any dimension is given by Eq. (4.4).

4.5 Find the Wigner–Seitz cell for the two-dimensional triangular lattice. Does it have the symmetry of a triangle or of a hexagon? Support your answer in terms of the symmetry of the triangular lattice.

4.6 Draw the Wigner–Seitz cell and the first Brillouin zone for the two-dimensional triangular lattice.

4.7 Consider a honeycomb plane of graphite in which each atom has three nearest neighbors. Give primitive translation vectors, basis vectors for the atoms in the unit cell, and reciprocal lattice primitive vectors. Show that the BZ is hexagonal.

4.8 Covalent crystals tend to form structures in which the bonds are *not* at 180°. Show that this means that the structures will have more than one atom per primitive cell.

4.9 Show that the fcc and bcc lattices are reciprocal to one another. Do this in two ways: by drawing the vectors and taking cross products and by explicit inversion of the lattice vector matrices.

4.10 Consider a body-centered cubic crystal, like Na, composed of an element with one atom at each lattice site. What is the Bravais lattice in terms of the conventional cube edge a? How many nearest neighbors does each atom have? How many second neighbors? Now suppose that the crystal is changed to a diatomic crystal like CsCl with all the nearest neighbors of a Cs atom being Cl, and vice versa. Now what is the Bravais lattice in terms of the conventional cube edge a? What is the basis?

4.11 Derive the value of the ideal c/a ratio for packing of hard spheres in the hcp structure.

4.12 Derive the formulas given in Eq. (4.12), paying careful attention to the definitions of the matrices and the places where the transpose is required.

4.13 Derive the formulas given in Eq. (4.18).

4.14 Derive the formulas given in Eq. (4.19).

4.15 Derive the relations given in Eqs. (4.20) and (4.21) for the parallelepiped that bounds a sphere in real and in reciprocal space. Explain the reason why the dimensions of the parallelepiped in reciprocal space involve the primitive vectors for the real lattice and vice versa.

4.16 Determine the coordinates of the points on the boundary of the Brillouin zone for fcc (X, W, K, U) and bcc (H, N, P) lattices.

4.17 Derive the formulas given in Eqs. (4.20) and (4.21). Hint: use the relations of real and reciprocal space given in the sentences before these equations.

4.18 Show that the expressions for integrals over the Brillouin zone Eq. (4.35), applied to the case of free electrons, lead to the same relations between density of one spin state n^σ and the Fermi momentum k_F^σ that was found in the section on homogeneous gas in Eq. (5.5). (From this one relation follow the other relations given after Eq. (5.5).)

4.19 In one dimension, dispersion can have singularities only at the end points where $E(k) - E_0 = A(k - k_0)^2$, with A positive or negative. Show that the singularities in the DOS form have the form $\rho(E) \propto |E - E_0|^{-1/2}$, as illustrated in the left panel of Fig. 14.4.

4.20 Show that singularities like those in Fig. 14.4 occur in three dimensions, using (4.46) and the fact that $E \propto Ak_x^2 + Bk_y^2 + Ck_z^2$ with A, B, C all positive (negative) at minima (maxima) or with different signs at saddle points.

4.21 The "special points" defined by Monkhorst and Pack are chosen to integrate periodic functions efficiently with rapidly decreasing magnitude of the Fourier components. This is a set of exercises to illustrate this property:
(a) Show that in one dimension the average of $f(k)$ at the k points $\frac{1}{4}\frac{\pi}{a}$ and $\frac{3}{4}\frac{\pi}{a}$ is exact if f is a sum of Fourier components $k + n\frac{2\pi}{a}$, with $n = 0, 1, 2, 3$, but that the error is maximum for $n = 4$.
(b) Derive the general form of Eq. (4.41).
(c) Why are uniform sets of points more efficient if they do it not include the Γ point?

(d) Derive the 2- and 10-point sets given for an fcc lattice, where symmetry has been used to reduce the points to the irreducible BZ.

4.22 The bands of any one-dimensional crystal are solutions of the Schrödinger equation (4.22) with a periodic potential $V(x + a) = V(x)$. The complete solution can be reduced to an informative analytic expression in terms of the scattering properties of a single unit cell and the Bloch theorem. This exercise follows the illuminating discussion by Ashcroft and Mermin [280], problem 8.1, and it lays a foundation for exercises that illustrate the pseudopotential concept (Exercises 11.2, 11.6, and 11.14) and the relation to plane wave, APW, KKR, and MTO methods, respectively, in Exercise 12.6, 16.1, 16.7, and 16.13.

An elegant approach is to consider a different problem first: an infinite line with $\tilde{V}(x) = 0$ except for a single cell in which the potential is the same as in a cell of the crystal, $\tilde{V}(x) = V(x)$ for $-a/2 < x < a/2$. At any positive energy $\varepsilon \equiv (\hbar^2/2m_e)K^2$, there are two solutions: $\psi_l(x)$ and $\psi_r(x)$ corresponding to waves incident from the left and from the right. Outside the cell, $\psi_l(x)$ is a given by $\psi_l(x) = e^{iKx} + re^{-iKx}$, $x < -\frac{a}{2}$, and $\psi_l(x) = te^{iKx}$, $x > a/2$, where t and r are transmission and reflection amplitudes. There is a corresponding expression for $\psi_r(x)$. Inside the cell, the functions can be found by integration of the equation, but we can proceed without specifying the explicit solution.

(a) The transmission coefficient can be written as $t = |t|e^{i\delta}$, with δ a phase shift that is related to the phase shifts defined in Appendix J as clarified in Exercise 11.2. It is well known from scattering theory that $|t|^2 + |r|^2 = 1$ and $r = \pm i|r|e^{i\delta}$, which are left as an exercise to derive.

(b) A solution $\psi(x)$ in the crystal at energy ε (if it exists) can be expressed as a linear combination of $\psi_l(x)$ and $\psi_r(x)$ evaluated at the same energy. Within the central cell all functions satisfy the same equation and $\psi(x)$ can always be written as a linear combination,

$$\psi(x) = A\psi_l(x) + B\psi_r(x), \quad -\frac{a}{2} < x < \frac{a}{2}, \tag{4.47}$$

with A and B chosen so that $\psi(x)$ satisfies the Bloch theorem for *some crystal momentum* k. Since $\psi(x)$ and $d\psi(x)/dx$ must be continuous, it follows that $\psi(\frac{a}{2}) = e^{ika}\psi(-\frac{a}{2})$ and $\psi'(\frac{a}{2}) = e^{ika}\psi'(-\frac{a}{2})$. Using this information and the forms of $\psi_l(x)$ and $\psi_r(x)$, find the 2×2 secular equation and show that the solution is given by

$$2t\cos(ka) = e^{-iKa} + (t^2 - r^2)e^{iKa}. \tag{4.48}$$

Verify that this is the correct solution for free electrons, $V(x) = 0$.

(c) Show that in terms of the phase shift, the solution, (4.48), can be written

$$|t|\cos(ka) = \cos(Ka + \delta), \quad \varepsilon \equiv \frac{\hbar^2}{2m_e}K^2. \tag{4.49}$$

(d) Analyze Eq. (4.49) to illustrate properties of bands and indicate which are special features of one dimension. (i) Since $|t|$ and δ are functions of energy ε, it is most convenient to fix ε and use Eq. (4.49) to find the wavevector k; this exemplifies the "root-tracing" method used in augmented methods (Chapter 16). (ii) There are necessarily bandgaps where there are no solutions, except for the free electron case. (iii) There is exactly one band of allowed states $\varepsilon(k)$ between each gap. (iv) The density of states, Eq. (4.46), has the form shown in the left panel in Fig. 14.4.

(e) Finally, discuss the problems with extending this approach to higher dimensions.

5

Uniform Electron Gas and sp-Bonded Metals

Summary

The simplest model system representing electrons in condensed matter is the homogeneous electron gas, in which the nuclei are replaced by a uniform positively charged background. This system is completely specified by the density n (or r_s, which is the average distance between electrons) and the spin density $n_\uparrow - n_\downarrow$ or the polarization $\zeta = (n_\uparrow - n_\downarrow)/n$. The homogeneous gas illustrates the problems associated with interacting electrons in condensed matter and is a prelude to the electronic structure of matter, which is governed by the combined effects of nuclei and electron interaction.

5.1 The Electron Gas

The homogeneous electron gas is the simplest system for illustrating key properties of interacting electrons and characteristic magnitudes of electronic energies in condensed matter. Since all independent-particle terms can be calculated analytically, this is an ideal model system for understanding the effects of correlation. In particular, the homogeneous gas best illustrates the issues of Fermi liquid theory [297, 298], which is the basis for our understanding of the "normal" (nonsuperconducting) state of real metals in terms of effective independent-particle approaches.

A homogeneous system is completely specified by its density $n = N_e/\Omega$, which can be characterized by the parameter r_s, defined as the radius of a sphere containing one electron on average,

$$\frac{4\pi}{3}r_s^3 = \Omega/N_e = \frac{1}{n}; \text{ or } r_s = \left(\frac{3}{4\pi n}\right)^{1/3}. \tag{5.1}$$

Thus r_s is a measure of the average distance between electrons. Table 5.1 gives values of r_s for valence electrons in a number of elements. The values shown are typical of characteristic electron densities in solids. For simple crystals, r_s is readily derived from the structure and

Table 5.1. Typical r_s values in elemental solids in units of the Bohr radius a_0. The valence is indicated by Z. The alkalis have bcc structure; Al, Cu, and Pb are fcc; the other group IV elements have diamond structure; and other elements have various structures. The values for metals are taken from [285] and [300]; precise values depend on temperature.

$Z = 1$	$Z = 2$	$Z = 1$	$Z = 2$	$Z = 3$	$Z = 4$
Li 3.23	Be 1.88			B	C 1.31
Na 3.93	Mg 2.65			Al 2.07	Si 2.00
K 4.86	Ca 3.27	Cu 2.67	Zn 2.31	Ga 2.19	Ge 2.08
Rb 5.20	Sr 3.56	Ag 3.02	Cd 2.59	In 2.41	Sn 2.39
Cs 5.63	Ba 3.69	Au 3.01	Hg 2.15	Tl	Pb 2.30

lattice constant; expressions for fcc and bcc, and the VI, III–V, and II–VI semiconductors are given in Exercises 5.1 and 5.2.

Of course, density is not constant in a real solid and it is interesting to determine the variation in density. In ordinary diamond-structure Si, there is a significant volume with low density (the open parts of the diamond structure). However, in the compressed metallic phase of Si with Sn structure, the variation in r_s is only $\pm \approx 20\%$. The distribution of local values of the density parameter r_s for valence electrons in Si can be found in [299].

The hamiltonian for the homogeneous system is derived by replacing the nuclei in Eq. (3.1) with a uniform positively charged background, which leads to

$$\hat{H} = -\frac{\hbar^2}{2m_e}\sum_i \nabla_i^2 + \frac{1}{2}\frac{4\pi}{\epsilon_0}\left[\sum_{i\neq j}\frac{e^2}{|\mathbf{r}_i - \mathbf{r}_j|} - \int d^3r d^3r' \frac{(ne)^2}{|\mathbf{r} - \mathbf{r}'|}\right]$$

$$\rightarrow -\frac{1}{2}\sum_i \nabla_i^2 + \frac{1}{2}\left[\sum_{i\neq j}\frac{1}{|\mathbf{r}_i - \mathbf{r}_j|} - \int d^3r d^3r' \frac{n^2}{|\mathbf{r} - \mathbf{r}'|}\right], \qquad (5.2)$$

where the second expression is in Hartree atomic units $\hbar = m_e = e = 4\pi/\epsilon_0 = 1$, where lengths are given in units of the Bohr radius a_0. The last term is the average background term, which must be included to cancel the divergence due to Coulomb interaction among the electrons. The total energy is given by

$$E = \langle\hat{H}\rangle = \langle\hat{T}\rangle + \langle\hat{V}_{\text{int}}\rangle - \frac{1}{2}\int d^3r d^3r' \frac{n^2}{|\mathbf{r} - \mathbf{r}'|}, \qquad (5.3)$$

where the first term is the kinetic energy of interacting electrons and the last two terms are the *difference* between the potential energy of the actual interacting electrons and the self-interaction of a classical uniform negative charge density, i.e., the exchange–correlation energy.[1] Note that the *difference* is well defined since there is a cancellation of the divergent Coulomb interactions, as discussed following Eq. (3.16).

In order to understand the interacting gas as a function of density, it is useful to express the hamiltonian Eq. (5.2) in terms of scaled coordinates $\tilde{\mathbf{r}} = \mathbf{r}/r_s$ instead of atomic units ($\mathbf{r}$ in units of a_0) assumed in the second expression in (5.2). Then Eq. (5.2) becomes (see Exercise 5.3 for the last term, which is essential for the expression to be well defined)

$$\hat{H} = \left(\frac{a_0}{r_s}\right)^2 \sum_i \left[-\frac{1}{2}\tilde{\nabla}_i^2 + \frac{1}{2}\frac{r_s}{a_0}\left(\sum_{j\neq i} \frac{1}{|\tilde{\mathbf{r}}_i - \tilde{\mathbf{r}}_j|} - \frac{3}{4\pi}\int d^3\tilde{\mathbf{r}} \frac{1}{|\tilde{\mathbf{r}}|} \right) \right], \tag{5.4}$$

where energies are in atomic units. This expression shows explicitly that one can view the system in terms of a scaled unit of energy (the Hartree scaled by $(a_0/r_s)^2$) and a scaled effective interaction proportional to r_s/a_0. In other words, the properties as a function of density r_s/a_0 are completely equivalent to a system at fixed density but with scaled electron–electron interaction $e^2 \rightarrow (r_s/a_0)e^2$ at fixed density and a scaled unit of energy.

5.2 Noninteracting and Hartree–Fock Approximations

In the noninteracting approximation, the solutions of Eq. (3.36) are eigenstates of the kinetic energy operator, i.e., normalized plane waves $\psi_{\mathbf{k}} = (1/\Omega^{1/2})e^{i\mathbf{k}\cdot\mathbf{r}}$ with energy $\varepsilon_{\mathbf{k}} = \frac{\hbar^2}{2m_e}k^2$. The ground state for a given density of up and down spin electrons is the determinant function Eq. (3.43) formed from the single-electron states with wavevectors inside the Fermi surface, which is a sphere in reciprocal space of radius k_F^σ, the Fermi wavevector for each spin σ. The value of k_F^σ is readily derived, since each allowed k state in a crystal of volume Ω is associated with a volume in reciprocal space $(2\pi)^3/\Omega$ (see Exercise 5.4 and Chapter 4.) Each state can contain one electron of each spin so that

$$\frac{4\pi}{3}\left(k_F^\sigma\right)^3 = \frac{(2\pi)^3}{\Omega}N_e^\sigma; \quad \text{i.e. } \left(k_F^\sigma\right)^3 = 6\pi^2 n^\sigma \quad \text{or} \quad k_F^\sigma = (6\pi^2)^{1/3}(n^\sigma)^{1/3}. \tag{5.5}$$

If the system is unpolarized, i.e., $n^\uparrow = n^\downarrow = n/2$, then $k_F = k_F^\uparrow = k_F^\downarrow$, where

$$(k_F)^3 = 3\pi^2 n; \quad \text{or} \quad k_F = (3\pi^2)^{1/3}n^{1/3} = \left(\frac{9}{4}\pi\right)^{1/3}/r_s. \tag{5.6}$$

The expression for the Fermi wavevector has the remarkable property that it also applies to interacting electron systems: the Luttinger theorem [47, 301] guarantees that the Fermi surface exists at the same k_F^σ as in the noninteracting case so long as there is no phase transition.

In the independent-particle approximation, Fermi energy E_{F0}^σ for each spin is given by

$$E_{F0}^\sigma = \frac{\hbar^2}{2m_e}\left(k_F^\sigma\right)^2 = \frac{1}{2}\left(k_F^\sigma a_0\right)^2 \rightarrow \frac{1}{2}\left(k_F^\sigma\right)^2, \tag{5.7}$$

---

[1] This can be derived from expression (3.16) for the energy, since in this case the total charge density (electrons + background) is everywhere zero, so that the final term in Eq. (3.16) vanishes.

Table 5.2. Characteristic energies for each spin σ for the homogeneous electron gas in the Hartree–Fock approximation: the Fermi energy E_{F0}^{σ}; kinetic energy T_0^{σ} and Hartree–Fock exchange energy per electron E_x^{σ}, which is negative; and the increase in band width in the Hartree–Fock approximation ΔW_{HFA}

Quantity	Expression	Atomic units
E_{F0}^{σ}	$\frac{\hbar^2}{2m_e}(k_F^{\sigma})^2$	$\frac{1}{2}(k_F^{\sigma})^2$
T_0^{σ}	$\frac{3}{5}E_F^{\sigma}$	$\frac{3}{5}E_F^{\sigma}$
$-E_x^{\sigma}$	$\frac{3e^2}{4\pi}k_F^{\sigma}$	$\frac{3}{4\pi}k_F^{\sigma}$
$\Delta W_{\text{HFA}}^{\sigma}$	$\frac{e^2}{\pi}k_F^{\sigma}$	$\frac{1}{\pi}k_F^{\sigma}$

Table 5.3. Useful expressions for the unpolarized homogeneous electron gas in terms of r_s in units of the Bohr radius a_0. See caption of Table 5.2 for definitions of energies.

Quantity	Expression	Atomic units	Common units
k_F	$\left(\frac{9}{4}\pi\right)^{1/3}/r_s$	$1.919, 158/r_s$	$3.626, 470/r_s$ (Ang.$^{-1}$)
E_{F0}	$\frac{1}{2}\left(\frac{9}{4}\pi\right)^{2/3}/r_s^2$	$1.841, 584/r_s^2$	$50.112, 45/r_s^2$ (eV)
T_0	$\frac{3}{5}E_F$	$1.104, 961/r_s^2$	$30.067, 47/r_s^2$ (eV)
$-E_x$	$\frac{3}{4\pi}\left(\frac{9\pi}{4}\right)^{1/3}/r_s$	$0.458, 165, 29/r_s$	$12.467, 311/r_s$ (eV)
ΔW_{HFA}	$\left(\frac{9}{4\pi^2}\right)^{1/3}/r_s$	$0.145, 838, 54/r_s$	$3.968, 4684/r_s$ (eV)

where the last expression is in atomic units with $a_0 = 1$. Useful relations for the Fermi wavevector and various energies are given in Tables 5.2 and 5.3.

The total kinetic energy per electron of a given spin in the ground state is given by integrating over the filled states

$$T_0^{\sigma} = \frac{\hbar^2}{2m_e}\frac{4\pi \int_0^{k_F^{\sigma}} dk\, k^4}{4\pi \int_0^{k_F^{\sigma}} dk\, k^2} = \frac{3}{5}E_{F0}^{\sigma} \tag{5.8}$$

(see Exercise 5.7 for one and two dimensions). Since the energy is positive, the homogeneous gas is clearly unbound in this approximation. The true binding in a material is provided by the added attraction to point nuclei and the attractive exchange and correlation energies.

Density Matrix

The density matrix in the homogeneous gas illustrates both the general expressions and the nature of the spatial dependence in many-electron systems. The general expression for independent fermions Eq. (3.41) simplifies for the homogeneous gas (for each spin) to

$$\rho(\mathbf{r}, \mathbf{r}') = \rho(|\mathbf{r} - \mathbf{r}'|) = \frac{1}{(2\pi)^3} \int d\mathbf{k} f(\varepsilon(k)) e^{i\mathbf{k} \cdot (\mathbf{r} - \mathbf{r}')}, \tag{5.9}$$

where $\varepsilon(k) = k^2/2$, which is just a Fourier transform of the Fermi function $f(\varepsilon(k))$. To evaluate the function it is convenient to transform the expression using a partial integration [302], yielding

$$\rho(r) = \frac{\beta}{(2\pi)^2} \frac{1}{r} \frac{d}{dr} \frac{1}{r} \frac{d}{dr} \int_{-\infty}^{\infty} dk \, \cos(kr) f'\left(\beta\left(\frac{1}{2}k^2 - \mu\right)\right). \tag{5.10}$$

This is a particularly revealing form that makes it clear why long-range oscillations in $r = |\mathbf{r} - \mathbf{r}'|$ must result from sharp variation in the derivative of the Fermi function $f'(\varepsilon)$, long known in Fourier transforms and attributed to Gibbs [303]. Since $f'(\varepsilon)$ approaches a delta function at low temperature, the range of $\rho(r)$ must increase as the temperature is reduced. At $T = 0$, $\rho(r)$ decays as $1/r^2$ [260],

$$\rho(r) = \frac{k_F^3}{3\pi^2} \left[3 \frac{\sin(y) - y \cos(y)}{y^3} \right], \tag{5.11}$$

with $y = k_F r$. The function in square brackets is defined to be normalized (Exercise 5.6) and is plotted in Fig. 5.1, where the decaying oscillatory form is evident, often called Friedel oscillations for charge and Ruderman–Kittel–Kasuya–Yosida oscillations for magnetic interactions [260, 280, 285]. Numerical results [302, 304] and simple analytic approximations [278, 302] can be found for $T \neq 0$, which show an exponential decay constant $\propto k_B T / k_F$.

Hartree–Fock Approximation

In the Hartree–Fock approximation, the one-electron orbitals are eigenstates of the non-local operator in Eq. (3.45). The solution of the Hartree–Fock equations in this case can be done analytically: the first step is to show that the eigenstates are plane waves, just as for noninteracting electrons (see Exercise 5.8). Thus the kinetic energy and the density matrix are the same as for noninteracting electrons, as they must be since the Hartree–Fock wavefunction contains no correlation beyond that required by the exclusion principle. The next step is to derive the eigenvalue for each k, which is $k^2/2$ plus the matrix element of the exchange operator Eq. (3.48). The integrals can be done analytically (the steps are outlined in Exercise 5.9 following [280, 297]), leading to

$$\varepsilon_k = \frac{1}{2}k^2 + \frac{k_F}{\pi} f(x), \tag{5.12}$$

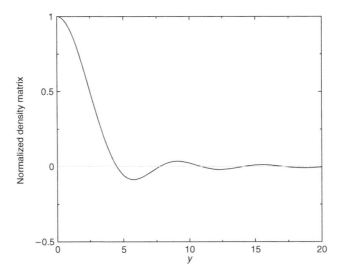

Figure 5.1. The dimensionless density matrix at $T = 0$ in the noninteracting homogeneous gas (the term in square brackets in Eq. (5.11) as a function of $y = k_F r$). The oscillations have spatial form governed by the Fermi wavevector k_F and describe charge around an impurity (Friedel oscillations) or magnetic interactions in a metal (Ruderman–Kittel–Kasuya–Yosida oscillations) [260, 280, 285]. See also Fig. 5.3, which shows the consequences for the pair correlation function.

where $x = k/k_F$ and

$$f(x) = -\left(1 + \frac{1 - x^2}{2x} \ln\left|\frac{1 + x}{1 - x}\right|\right). \tag{5.13}$$

(Note that the expression applies to each spin separately.)

The factor $f(x)$, shown in Fig. 5.2, is negative for all x; at the bottom of the band ($x = 0$), $f(0) = -2$, and at large x it approaches zero. Near the Fermi surface ($x = 1$), $f(x)$ varies rapidly and has a divergent slope; nevertheless the limiting value at $x = 1$ is well defined, $f(x \to 1) = -1$. Thus in the Hartree–Fock approximation, exchange increases the band width W by $\Delta W = k_F/\pi$. This holds separately for each spin, and in the unpolarized case, the factor can also be written $\Delta W = \left(\frac{9}{4\pi^2}\right)^{1/3}/r_s$ (see Table 5.3 and Exercise 5.10).

The Hartree–Fock eigenvalue relative to the Fermi energy, i.e., defined with $\varepsilon_k \equiv 0$ at $k = k_F$, can be written in scaled form,

$$\varepsilon_k = \frac{1}{2}k_F^2 \left\{(x^2 - 1) + \frac{2}{\pi k_F}[f(x) + 1]\right\}. \tag{5.14}$$

The expression in curly brackets is plotted on the right-hand side of Fig. 5.2 for several values of r_s. The broadening of the filled band due to interactions in the Hartree–Fock approximation is indicated by the value at $k = 0$, which is -1 for noninteracting electrons.

The singularity at the Fermi surface, first pointed out by Bardeen, [254], is a consequence of long-range Coulomb interaction and the existence of the Fermi surface where the separation of the occupied and empty states vanishes. The velocity at the Fermi surface

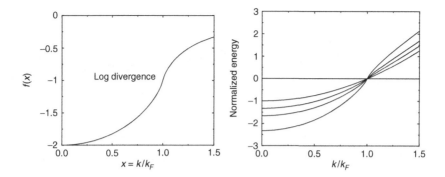

Figure 5.2. Left: the factor $f(x)$, Eq. (5.13), in the homogeneous gas that determines the dispersion $\varepsilon_{\mathrm{HFA}}(k)$ in the Hartree–Fock approximation. Right: $\varepsilon_{\mathrm{HFA}}(k)$ for three densities ($r_s = 1, 2, 4$) compared to the noninteracting case. The lowest density (largest r_s) is lowest at $k = 0$ and has the most visible singularity at the Fermi surface, $x = 1$. The normalized dimensionless eigenvalue is defined by the square brackets in Eq. (5.14), and $r_s = 0$ is the noninteracting limit $-1 + x^2$.

$d\varepsilon/dk$ diverges (Exercise 5.11), in blatant contradiction with experiment, where the well-defined velocities are determined by such measurements as specific heat and the de Haas–van Alfen effect [280, 285]. Thus this is an intrinsic failure of Hartree–Fock that carries over to any metal. The Hartree–Fock divergence, however, can be avoided either if there is a finite gap (i.e., in an insulator where Hartree–Fock is qualitatively correct and is widely applied in quantum chemistry) or if the Coulomb interaction is screened to be effectively short range. This is the *ansatz* – i.e., the assertion – of Fermi liquid theory: that the interactions are screened for low-energy excitations, leading to weakly interacting "quasiparticles," which is commonly justified by partial summation of diagrams in the random phase approximation (RPA) [297, 298].

The Exchange Energy and Exchange Hole

The exact total energy of the homogeneous gas is given by Eq. (5.3), which can be separated into the Hartree–Fock total energy, which is the sum of kinetic energy of independent electrons plus the exchange energy, and the remainder, termed the "correlation energy." As we have seen in Section 3.7, the exchange energy and exchange hole can be computed directly from the wavefunctions, which can be done analytically in this case. In addition, the exchange energy per electron is simply the average of the exchange contribution to the eigenvalue $\frac{k_F}{\pi} f(x)$ in Eq. (5.12) multiplied by 1/2 to take into account the fact that interactions should not be double counted. Using the fact that the average value of $f(x)$ is $-3/2$ (Exercise 5.12), it follows that the exchange energy per electron is

$$\epsilon_x^\sigma = E_x^\sigma / N^\sigma = -\frac{3}{4\pi} k_F^\sigma = -\frac{3}{4}\left(\frac{6}{\pi} n^\sigma\right)^{1/3}. \tag{5.15}$$

In the unpolarized case, one finds $\epsilon_x \equiv \epsilon_x^\uparrow = \epsilon_x^\downarrow = -\frac{3}{4\pi}\left(\frac{9\pi}{4}\right)^{1/3} / r_s$ and the explicit numerical relations in Table 5.3.

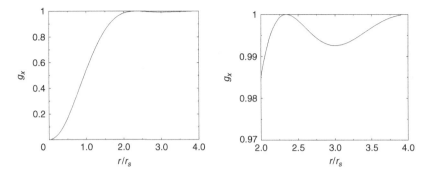

Figure 5.3. Exchange hole $g_x(r)$ in the homogeneous electron gas, Eq. (5.19) plotted as a function of r/r_s, where r_s is the average distance between electrons in an unpolarized system. The magnitude decreases rapidly with oscillation, as shown in the greatly expanded right-hand figure. Note the similarity to the calculated pair correlation function for parallel spins in Fig. 5.5.

For partially polarized cases, the exchange energy is just a sum of terms for the two spins, which can also be expressed in an alternative form in terms of the total density $n = n^\uparrow + n^\downarrow$ and fractional polarization,

$$\zeta = \frac{n^\uparrow - n^\downarrow}{n}. \tag{5.16}$$

It is straightforward to show that exchange in a polarized system has the form

$$\epsilon_x(n, \zeta) = \epsilon_x(n, 0) + [\epsilon_x(n, 1) - \epsilon_x(n, 0)] f_x(\zeta), \tag{5.17}$$

where

$$f_x(\zeta) = \frac{1}{2} \frac{(1 + \zeta)^{4/3} + (1 - \zeta)^{4/3} - 2}{2^{1/3} - 1}, \tag{5.18}$$

which is readily derived [305] from Eq. (5.15).

The exchange hole g_x defined in Eqs. (3.54) or (3.52) involves only electrons of the same spin and in a homogeneous system is a function only of the relative distance $|\mathbf{r}| = |\mathbf{r}_1 - \mathbf{r}_2|$, so that $g_x(\mathbf{r}_1, \sigma_1; \mathbf{r}_2, \sigma_2) = \delta_{\sigma_1, \sigma_2} g_x^{\sigma_1, \sigma_1}(|\mathbf{r}|)$. In the homogeneous gas, the form of the exchange hole can be calculated analytically in two ways (see Exercise 5.13): the definitions can be used directly [306] by inserting the plane wave eigenfunctions (normalized to a large volume Ω) and evaluating the resulting expression. Alternatively, $g_x(r)$ can be found from the general relation Eq. (3.56) of the pair correlation function and the density matrix in a noninteracting system[2] together with the density matrix $\rho(r)$ given by Eq. (5.11). For each spin, the hole can be given in terms of the dimensionless variable $y = k_F^\sigma r$ with the result

$$g_x^{\sigma, \sigma}(y) = 1 - \left[3 \frac{\sin(y) - y \cos(y)}{y^3} \right]^2, \tag{5.19}$$

which is shown graphically in Fig. 5.3. The exchange hole in the homogeneous gas illustrates the principle that for fermions the hole n^x must always be negative, i.e., $g_x^{\sigma, \sigma}$

---

[2] The arguments can be applied to any noninteracting particles [278]; for bosons the result is that $g_x(r) = 1 + |\rho(r)|^2/n^2$ is always greater than 1. See Section 3.7 and Exercise 3.15.

must be less than 1, and it approaches 1 as an inverse power law with the well-known Friedel oscillations due to the sharp Fermi surface.

5.3 Correlation Hole and Energy

"Screening" is the effect in a many-body system whereby the particles collectively correlate to reduce the net interaction among any two particles. For repulsive interactions, the hole (reduced probability of finding other particles) around each particle tends to produce a net weaker interaction strength, i.e., a lower total energy.

Thomas–Fermi Screening

The grandfather of models for screening is the Thomas–Fermi approximation for the electron gas, which is the quantum equivalent of Debye screening in a classical system. The screening is determined by analyzing the response of the gas to a static external charge density with Fourier component k. The response at wavevector k is determined by the change in energy of the electrons, which is a function of only density in the Thomas–Fermi approximation (Section 6.2). The result is that the long-range Coulomb interaction is screened to an exponentially decaying interaction, which in Fourier space can be written as

$$\frac{1}{k^2} \rightarrow \frac{1}{k^2 + k_{\mathrm{TF}}^2}, \tag{5.20}$$

where k_{TF} is the Thomas–Fermi screening wavevector (the inverse of the characteristic screening length). For an unpolarized system, k_{TF} is given by (see Exercise 5.14 and [280])

$$k_{\mathrm{TF}} = r_s^{1/2} \left(\frac{16}{3\pi^2} \right)^{1/3} k_F = \left(\frac{12}{\pi} \right)^{1/3} r_s^{-1/2}, \tag{5.21}$$

where r_s is in atomic units, i.e., in units of the Bohr radius a_0.

This is the simplest estimate for the characteristic length over which electrons are correlated, which is very useful in understanding the full results below in a homogeneous gas and estimates for real systems.

Correlation Energy

It is not possible to determine the correlation hole and energy analytically. The first quantitative form for the correlation energy of a homogeneous gas was proposed in the 1930s by Wigner [76, 307], as an interpolation between low- and high-density limits.[3] At low density the electrons form a "Wigner crystal" and the correlation energy is just the electrostatic energy of point charges on the body-centered cubic lattice. At the time, it was thought that the exchange energy per electron approached a constant in the high-density limit, and Wigner proposed the simple interpolation

---

[3] The first formula proposed by Wigner [76] was in error due to an incorrect expression for the low-density limit, as pointed out in [307].

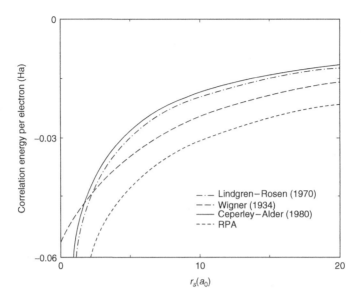

Figure 5.4. Correlation energy of an unpolarized homogeneous electron gas as a function of the density parameter r_s. The most accurate are the quantum Monte Carlo (QMC) calculations by Ceperley and Alder [311]; the curve labeled "Ceperley–Alder" is the interpolation formula fitted to the QMC results by Vosko, Wilk, and Nusair (VWN) [312]. The Perdew–Zunger (PZ) fit [313] is almost identical on this scale. In comparison are shown the Wigner interpolation formula, Eq. (5.22), the RPA (see text), and an improved many-body perturbation calculation taken from Mahan [306], where it is attributed to L. Lindgren and A. Rosen. Figure provided by H. Kim.

$$\epsilon_c = -\frac{0.44}{r_s + 7.8}\,(\text{in a.u.} = \text{Hartree}). \tag{5.22}$$

Correct treatment of correlation confounded many-body theory for decades until the work of Gellmann and Breuckner [308], who summed infinite series of diagrams to eliminate divergences that are present at each order and calculated the correlation energy exactly in the high-density limit, $r_s \to 0$. For an unpolarized gas ($n^\uparrow = n^\downarrow = n/2$), the result is [308, 309]

$$\epsilon_c(r_s) \to 0.311 \ \ln(r_s) - 0.048 + r_s(A \ \ln(r_s) + C) + \cdots, \tag{5.23}$$

where the ln terms are the signature of non-analyticity that causes so much difficulty. At low density the system can be considered a Wigner crystal with zero point motion leading to [298, 310]

$$\epsilon_c(r_s) \to \frac{a_1}{r_s} + \frac{a_2}{r_s^{3/2}} + \frac{a_3}{r_s^2} + \cdots, \tag{5.24}$$

There has been considerable work in the intervening years [306], including the well-known work of Hedin and Lundqvist [241] using the random phase approximation (RPA), which is the basis of much of our present understanding of excitations and other recent work such as self-consistent "GW" calculations [314]. The most accurate results for ground-state properties are found from quantum Monte Carlo (QMC) calculations that can treat interacting many-body systems [311, 315, 316], which are the benchmark for other methods. The QMC results for the correlation energy $\epsilon_c(r_s)$ per electron in an unpolarized gas are shown in Fig. 5.4 where they are compared with the Wigner interpolation formula, RPA, and improved many-body calculations of Lindgren and Rosen (results given in [306], p. 314). A more recent calculation [317] using the Bethe–Salpeter equation (BSE) finds energies slightly larger in magnitude than QMC similar to the Lindgren–Rosen curve in Fig. 5.4. A very important result is that for materials at typical solid densities ($r_s \approx 2 - 6$) the correlation energy is much smaller than the exchange energy; however, at very low densities (large r_s) correlation becomes more important and dominates in the regime of the Wigner crystal ($r_s > \approx 80$).

The use of the QMC results in subsequent electronic structure calculations relies upon parameterized analytic forms for $E_C(r_s)$ fitted to the QMC energies calculated at many values of r_s, mainly for unpolarized and fully polarized ($n^\uparrow = n$) cases, although some calculations have been done at intermediate polarization [315]. The key point is that the formulas fit the data well at typical densities and extrapolate to the high- and low-density limits, Eqs. (5.23) and (5.24). Widely used forms are due to Perdew and Zunger (PZ) [313] and Vosko, Wilkes, and Nussair (VWN) [312], which are given in Appendix B and are included in subroutines for functionals referred to there and available online.

The simplest form for the correlation energy as a function of spin polarization is the one made by PZ [313] that correlation varies the same as exchange

$$\epsilon_c(n, \zeta) = \epsilon_c(n, 0) + [\epsilon_c(n, 1) - \epsilon_c(n, 0)] f_x(\zeta), \tag{5.25}$$

where $f_x(\zeta)$ is given by Eq. (5.18). The slightly more complex form of VWN [312] has been found to be a slightly better fit to more recent QMC data [315].

It is also important for understanding the meaning of both exchange and correlation energies to see how they originate from the interaction of an electron with the exchange–correlation "hole" discussed in Section 3.7. The potential energy of interaction of each electron with its hole can be written

$$\epsilon_{xc}^{pot}(r_s) = E_{xc}^{pot}/N = \frac{1}{N}\left[\langle \hat{V}_{int}\rangle - E_{Hartree}(n)\right] = \frac{1}{2n}e^2 \int d^3r \frac{n_{xc}(|\mathbf{r}|)}{|\mathbf{r}|}, \tag{5.26}$$

where the factor $1/2$ is included to avoid double counting and we have explicitly indicated interaction strength e^2, which will be useful later. The exchange–correlation hole $n_{xc}(|\mathbf{r}|)$, of course, is spherically symmetric and is a function of density, i.e., of r_s. In the ground state, ϵ_{xc}^{pot} is negative since exchange lowers the energy if interactions are repulsive and correlation always lowers the energy. However, this is not the total exchange–correlation energy per electron ϵ_{xc} because the kinetic energy increases as the electrons correlate to lower their potential energy.

The full exchange–correlation energy including kinetic terms can be found in two ways: kinetic energy can be determined from the virial theorem [318] or from the "coupling constant integration formula" in Section 3.4. We will consider the latter as an example

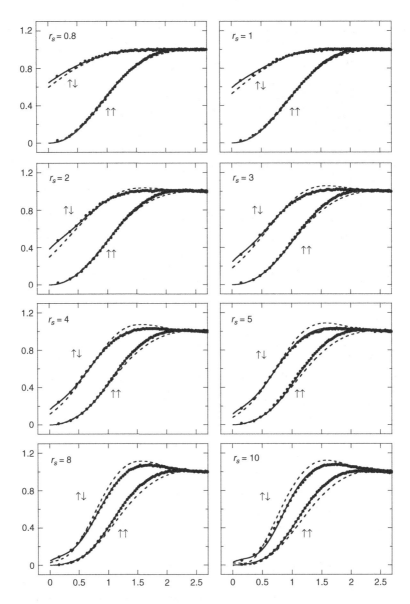

Figure 5.5. Spin-resolved normalized pair-correlation function $g_{xc}(r)$ for the unpolarized homogenous electron gas as a function of scaled separation r/r_s, for r_s varying from $r_s = 0.8$ to $r_s = 10$. Dots, QMC data of [319]; dashed line, Perdew–Wang model; solid line, coupling constant integrated form of [318]. From [318].

of the generalized force theorem, i.e., the coupling constant integration formula Eq. (3.27) in which the coupling constant e^2 is replaced by λe^2, which is varied from $\lambda = 0$ (the noninteracting problem) to the actual value $\lambda = 1$. Just as in Eq. (3.27), the derivative of the energy with respect to λ involves only the *explicit* linear variation of $\epsilon_{xc}^{pot}(r_s)$ in Eq. (5.26) with λ and there is no contribution from the implicit dependence of $n_{xc}^\lambda(|\mathbf{r}|)$ on λ since the energy is at a minimum with respect to such variations. This leads directly to the result that

$$\epsilon_{xc}(r_s) = \frac{1}{2n}e^2 \int d^3r \, \frac{n_{xc}^{av}(r)}{r}, \tag{5.27}$$

where $n_{xc}^{av}(r)$ is the coupling-constant-averaged hole

$$n_{xc}^{av}(r) = \int_0^1 d\lambda \, n_{xc}^\lambda(r). \tag{5.28}$$

The exchange–correlation hole has been calculated by quantum Monte Carlo methods at full coupling strength $\lambda = 1$, with results that are shown in Fig. 5.5 for various densities labeled r_s. By comparison with the exchange hole shown in Fig. 5.3, it is apparent that correlation is much more important for antiparallel spins than for parallel spins, which are kept apart by the Pauli principle. In general, correlation tends to reduce the long-range part of the exchange hole, i.e., it tends to cause screening.

Variation of the exchange–correlation hole with r_s can also be understood as variation with the strength of the interaction. As pointed out in Eq. (5.4), variation of e^2 from 0 to 1 at fixed density is equivalent to variation of r_s from 0 to the actual value. Working in scaled units r/r_s and $r_s \to \lambda r_s$, one finds

$$n_{xc}^{av}\left(\frac{r}{r_s}\right) = \int_0^1 d\lambda \, n_{xc}^\lambda\left(\frac{r}{\lambda r_s}\right). \tag{5.29}$$

Examples of the variation of the hole $n_{xc}(\frac{r}{\lambda r_s})$ are shown in Fig. 5.5 for various r_s for parallel and opposite spins in an unpolarized gas. Explicit evaluation of $\epsilon_{xc}(r_s)$ has been done using this approach in [318]. Note that this expression involves $\lambda r_s < r_s$ in the integrand, i.e., the hole for a system with density higher than the actual density. Exercises 5.15 and 5.16 deal with this relation, explicit shapes of the average holes for materials, and the possibility of making a relation that involves larger r_s (stronger coupling).

5.4 Binding in sp-Bonded Metals

The stage was set for understanding solids on a quantitative basis by Slater [320] and by Wigner and Seitz [54, 57] in the early 1930s. The simplest metals, the alkalis with one weakly bound electron per atom, are represented remarkably well by the energy of a homogeneous electron gas plus the attractive interaction with the positive cores. It was recognized that the ions were effectively weak scatterers even though the actual wavefunctions must have atomic-like radial structure near the ion. This is the precursor of the pseudopotential idea (Chapter 11) and also follows from the scattering analysis of Slater's APW method and the KKR approach (Chapter 16). Treating the electrons as a homogeneous gas, and adding the energies of the ions in the uniform background, leads to the expression for total energy per electron,

$$\frac{E_{total}}{N} = \frac{1.105}{r_s^2} - \frac{0.458}{r_s} + \epsilon_c - \frac{1}{2}\frac{\alpha}{r_s} + \epsilon_R, \tag{5.30}$$

where atomic units are assumed (r_s in units of a_0), and we have used the expressions in Table 5.3 for kinetic and exchange energies, and ϵ_c is the correlation energy per electron. The last two terms represent interaction of a uniform electron density with the ions: α is the Madelung constant for point charges in a background, and the final term is a repulsive correction due to the fact that the ion is not a point. Values of α are tabulated in Table F.1 for representative structures. The factor ϵ_R is due to core repulsion, which can be estimated using the effective model potentials in Fig. 11.3 that are designed to take this effect into account. This amounts to removing the attraction of the nucleus and the background in a core radius R_c around the ion

$$\epsilon_R = n2\pi \int_0^{R_c} dr\, r^2 \frac{e^2}{r} = \frac{3}{4\pi r_s^3} 2\pi e^2 R_c^2 = \frac{3}{2} \frac{a_0 R_c^2}{r_s^3} = \frac{3}{2} \frac{R_c^2}{r_s^3}, \tag{5.31}$$

where the last form is in atomic units.

Expression (5.30) contains much of the essential physics for the sp-bonded metals, as discussed in basic texts on solid state physics [280, 285, 300]. For example, the equilibrium value of r_s predicted by Eq. (5.30) is given by finding the extremum of Eq. (5.30). A good approximation is to neglect ϵ_c and to take $\alpha = 1.80$, the value for the Wigner–Seitz sphere that is very close to actual values in close-packed metals as shown in Table F.1 and Eq. (F.10). This leads to

$$\frac{r_s}{a_0} = 0.814 + \sqrt{0.899 + 3.31 \left(\frac{R_c}{a_0}\right)^2} \tag{5.32}$$

and improved expressions described in Exercise 5.17. Without the repulsive term, this leads to $r_s = 1.76$, which is much too small. However, a core radius $\approx 2a_0$ (e.g., a typical R_c in the model ion potentials shown in Fig. 11.3 and references given there) leads to a very reasonable $r_s \approx 4a_0$. The kinetic energy contribution to the bulk modulus is

$$B = \Omega \frac{d^2 E}{d\Omega^2} = \frac{3}{4\pi r_s} \frac{1}{9} \frac{d^2}{dr_s^2} \frac{1.105}{r_s^2} = \frac{0.176}{r_s^5} = \frac{51.7}{r_s^5} \text{Mbar}, \tag{5.33}$$

where a Mbar (=100 GPa) is a convenient unit. This sets a scale for understanding the bulk modulus in real materials, giving the right order of magnitude (often better) for materials ranging from sp-bonded metals to strongly bonded covalent solids.

5.5 Excitations and the Lindhard Dielectric Function

Excitations of a homogeneous gas can be classified into two types (see Section 2.14): electron addition or removal to create quasiparticles, and collective excitations in which the number of electrons does not change. The former are the bands for quasiparticles in Fermi liquid theory. How well do the noninteracting or Hartree–Fock bands shown in Fig. 5.2 agree with improved calculations and experiment? Figure 5.6 shows photoemission data for Na, which is near the homogeneous gas limit, compared to the noninteracting dispersion $k^2/2$. Interestingly, the bands are *narrower* than $k^2/2$, i.e., the opposite of what is predicted

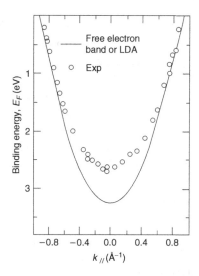

Figure 5.6. Experimental bands of Na determined from angle-resolved photoemission [322] compared to the simple $k^2/2$ dispersion for a non-interacting homogeneous electron gas at the density of Na, which is close to the actual calculated bands in Na. Such agreement is also found in other materials, providing the justification that density functional theory is a reasonable starting point for understanding electronic structure in solids such as the sp-bonded metals. From [322]; see also [323].

by Hartree–Fock theory. This is a field of active research in many-body perturbation theory to describe the excitations [321]. For our purposes, the important conclusion is that *the noninteracting case is a good starting point close to the measured dispersion*. This is germane to electronic structure of real solids, where it is found that Kohn–Sham eigenvalues are a reasonable starting point for describing excitations (see Section 7.7).

Excitations that do not change the particle number are charge density fluctuations (plasma oscillations) described by the dielectric function and spin fluctuations that are described by spin response functions. Expressions for the response functions are given in Chapters 20 and 21 and in Appendices D and E. The point of this section is to apply the expressions to a homogeneous system where the integrals can be done analytically. The discussion here follows Pines [297], sections 3–5, and provides examples that help to understand the more complex behavior of real inhomogeneous systems. In a homogeneous system, the dielectric function, Eqs. (E.8) and (E.11), is diagonal in the tensor indices and is an isotropic function of relative coordinates $\epsilon(|\mathbf{r} - \mathbf{r}'|, t - t')$, so that in Fourier space it is simply $\epsilon(q,\omega)$. Then, we have the simple interpretation that $\epsilon(q,\omega)$ is the response to an internal field,

$$\mathbf{D}(q,\omega) = \mathbf{E}(q,\omega) + 4\pi\mathbf{P}(q,\omega) = \epsilon(q,\omega)\mathbf{E}(q,\omega), \tag{5.34}$$

or, in terms of potentials,

$$\epsilon(q,\omega) = \frac{\delta V_{\text{ext}}(\mathbf{q},\omega)}{\delta V_{\text{test}}(\mathbf{q},\omega)} = 1 - v(q)\chi_n^*(\mathbf{q},\omega), \tag{5.35}$$

where $v(q) = \frac{4\pi e^2}{q^2}$ is the frequency-independent relation of the Coulomb potential at wavevector q to the electron density $n(q)$. No approximation has so far been made if χ^* is the full many-body response function (called the "proper" response function) to the internal electric field.

The well-known RPA [297] is the approximation where all interactions felt by the electrons average out because of their "random phases," except for the Hartree term, in which case each electron experiences an effective potential V_{eff} that is the same as that for a test charge V_{test}. Then $\chi_n^*(\mathbf{q}, \omega) = \chi_n^0(\mathbf{q}, \omega)$ and the RPA is an example of effective-field response functions treated in more detail in Section 21.4 and Appendix D. In a homogeneous gas, the expression for χ^0 given in Section 21.4 becomes an integral over states where $|\mathbf{k}| < k_F$ is occupied and $|\mathbf{k} + \mathbf{q}| > k_F$ is empty, which can be written

$$\chi_n^0(\mathbf{q}, \omega) = 4 \frac{1}{\frac{4\pi}{3} k_F^3} \int^{k=k_F} d\mathbf{k} \frac{1}{\varepsilon_k - \varepsilon_{|\mathbf{k}+\mathbf{q}|} - \omega + i\delta} \Theta(|\mathbf{k} + \mathbf{q}| - k_F). \tag{5.36}$$

The integral can be evaluated analytically for a homogeneous gas where $\varepsilon_k = \frac{1}{2} k^2$, leading to the Lindhard [324] dielectric function. The imaginary part can be derived as an integral over regions where the conditions are satisfied by $k < k_F$, $|\mathbf{k} + \mathbf{q}| > k_F$ and the real part of the energy denominator vanishes. The real part can be derived by a Kramers–Kronig transform Eq. (D.18), with the result (Exercise 5.18) [297],

$$\text{Im}\,\epsilon(q, \omega) = \frac{\pi}{2} \frac{k_{\text{TF}}^2}{q^2} \frac{\omega}{q v_F}, \omega < q v_F - \varepsilon_q,$$

$$= \frac{\pi}{4} \frac{k_{\text{TF}}^2}{q^2} \frac{k_F}{q} \left[1 - \frac{(\omega - \varepsilon_q)^2}{(q v_F)^2} \right], q v_F - \varepsilon_q < \omega < q v_F + \varepsilon_q,$$

$$= 0, \omega > q v_F - \varepsilon_q, \tag{5.37}$$

where v_F is the velocity at the Fermi surface and

$$\text{Re}\,\epsilon(q, \omega) = 1 + \frac{k_{\text{TF}}^2}{2q^2}$$

$$+ \frac{k_F k_{\text{TF}}^2}{4q^3} \times \left\{ \left[1 - \frac{(\omega - \varepsilon_q)^2}{(q v_F)^2} \right] \ln \left| \frac{\omega - q v_F - \varepsilon_q}{\omega + q v_F - \varepsilon_q} \right| + \omega \to -\omega \right\}. \tag{5.38}$$

The form of $\epsilon(q, \omega)$ for a homogeneous gas is shown in Fig. 5.7 for small q. The imaginary part of ϵ vanishes for $\omega > q v_F + \varepsilon_q$, so that there is no absorption above this frequency. The real part of the dielectric function vanishes at the plasmon frequency $\omega = \omega_p$, where $\omega_p^2 = 4\pi n_e e^2 / m_e$, with n_e the electron density. This corresponds to a pole in the inverse dielectric function $\epsilon^{-1}(q, \omega)$. The behavior of ϵ at the plasma frequency can be derived (Exercise 5.18) from Eq. (5.38) by expanding the logarithms, but the derivation is much more easily done using the general "f sum rule" given in Section E.3, together with the fact that the imaginary part of $\epsilon(q, \omega)$ vanishes at $\omega = \omega_p$.

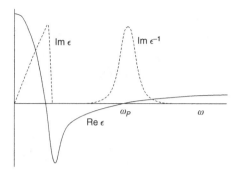

Figure 5.7. Lindhard dielectric function $\epsilon(k, \omega)$ for $k \ll k_{TF}$ given in Eq. (5.38) for a homogeneous electron gas in the random phase approximation (RPA). The imaginary part of ϵ is large only for low frequency. The frequency at which the real part of $\epsilon(k, \omega)$ vanishes corresponds to a peak in the imaginary part of $\epsilon^{-1}(k, \omega)$, which denotes the plasma oscillation at $\omega = \omega_P(k)$. For low frequencies, the real part approaches k_{TF}^2 / k^2, the same as the Thomas–Fermi form Eq. (5.20).

The Lindhard expression reveals many important properties that carry over qualitatively to solids. The low-frequency peak is still present in metals and is called the Drude absorption and there is generally additional broadening due to scattering [280, 285, 300]. In addition, the static screening Re $\epsilon(q, 0)$ has oscillations at twice the Fermi wavevector $q = 2k_F$, which lead to Friedel oscillations and the Kohn anomaly for phonons. Related affects carry over to response functions of solids (Appendix D and Chapter 20), except that $2k_F$ is replaced by anisotropic vectors that span the Fermi surface.

The primary difference in real materials is that there are also interband transitions that give nonzero imaginary ϵ above a threshold frequency. Examples of imaginary parts of $\epsilon(q \approx 0, \omega)$ for crystals are shown in Figs. 2.25 and 21.5. Interband absorption also causes a broadening of the plasmon peak in $\epsilon^{-1}(q, \omega)$, but, nevertheless, there still tends to be a dominant peak around the plasma frequency. Examples are given in Chapter 21, where the absorption of light by nanoscale clusters exhibits clearly the plasma-like peak.

SELECT FURTHER READING

The homogeneous electron gas is the testing ground for many-body theories and it has a large role in the companion book [1].

Texts and other sources with useful discussions:

Ashcroft, N. and Mermin, N. D., *Solid State Physics* (W. B. Saunders Company, New York, 1976).

Cohen, M. L. and Louie, S. G., *Fundamentals of Condensed Matter Physics* (Cambridge University Press, Cambridge, 2016).

Girvin S. M. and Yang K., *Modern Condensed Matter Physics* (Cambridge University Press, Cambridge, 2019).

Jones, W. and March, N. H., *Theoretical Solid State Physics*, vols. 1–2, (John Wiley & Sons, New York, 1976).

Mahan, G. D., *Many-Particle Physics*, 3rd edn. (Kluwer Academic/Plenum Publishers, New York, 2000).

Pines, D., *Elementary Excitations in Solids* (John Wiley & Sons, New York, 1964).
See the classic references by D. Pines and P. Nozières that have been reprinted [44, 298, 325].

Exercises

5.1 For fcc and bcc crystals with Z valence electrons per primitive cells, show that r_s is given, respectively, by

$$r_s = \frac{a}{2}\left(\frac{3}{2\pi Z}\right)^{1/3} \quad \text{and} \quad r_s = \frac{a}{2}\left(\frac{3}{\pi Z}\right)^{1/3}.$$

If r_s is in atomic units (a_0) and the cube edge a is in Å, then $r_s = 0.738 Z^{-1/3} a$ and $r_s = 0.930 Z^{-1/3} a$.

5.2 For semiconductors with eight valence electrons per primitive cell in diamond- or zinc-blende-structure crystals, show that $r_s = 0.369a$.

5.3 Argue that the expression for Coulomb interaction in large parentheses in Eq. (5.4) is finite due to cancellation of the two divergent terms. Show that the scaled hamiltonian given in Eq. (5.4) is indeed equivalent to the original hamiltonian Eq. (5.2).

5.4 Derive the relation Eq. (5.5) between the Fermi wavevector k_F^σ and the density n^σ for a given spin. Do this by considering a large cube of side L and requiring the wavefunctions to be periodic in length L in each direction (Born–von Karman boundary conditions).

5.5 Show that relation Eq. (5.6), between k_F and the density parameter r_s for an unpolarized gas, follows from the basic definition Eq. (5.5) (see also previous problem).

5.6 Show that Eq. (5.10) follows from Eq. (5.9) by carrying out the indicated differentiation and partial integration. Use this form to derive the $T = 0$ form, Eq. (5.11). Also show that the factor in brackets approaches unity for $y \to 0$.

5.7 Verify Eq. (5.8) for the kinetic energy of the ground state of a noninteracting electron gas. Note that in Eq. (5.8) the denominator counts the number of states and the numerator is the same integral but weighted by the kinetic energy of a state, so that this equation is independent of the number of spins. Derive the corresponding results for one and two dimensions.

5.8 Show that plane waves are eigenstates for the Hartree–Fock theory of a homogeneous electron gas – assuming the ground state is homogeneous, which may not be the case when interactions are included. Thus the kinetic energy is the same for Hartree–Fock theory as for noninteracting particles.

5.9 Derive Eq. (5.12) for eigenvalues in the Hartree–Fock approximation from the general definition in Eq. (3.48). Hint: the exchange integral for plane wave states has the form $-4\pi \sum_{\mathbf{k}'}^{k' < k_F} 1/|\mathbf{k} - \mathbf{k}'|^2$. This leads to the singular log form in three dimensions. For more details, see [280, 297].

5.10 Derive the broadening of the bands in the Hartree–Fock approximation from the unpolarized gas $\Delta W = (9/4\pi^2)^{1/3}/r_s$ using Eq. (5.12).

5.11 Derive analytically that the electron velocity $v = d\varepsilon/dk$ diverges at $k = k_F$ in the Hartree–Fock approximation. Argue that (1) this happens in *all* metals due to the Coulomb interaction and the Hartree–Fock approximation and (2) there is no divergence for short-range interactions.

5.12 Show that the average value of the factor $f(x)$ in Eq. (5.13) is $-3/4$, as stated before Eq. (5.15). Then, for the ground state of the homogeneous gas, verify the result for the exchange energy Eq. (5.15).

5.13 Show that Eq. (5.19) follows directly from evaluating the expressions in (3.54) or (3.52) by inserting the plane wave eigenfunctions (normalized to a large volume Ω) and evaluating the resulting expression. Alternatively, $g_x(r)$ can be found from the general relation Eq. (3.56) of the pair correlation function and the density matrix for noninteracting fermions [260, 278], $g_x(r) = 1 - |\rho(r)|^2/n^2$, where n is the density and the density matrix $\rho(r)$ is given by Eq. (5.11).

5.14 Consider a point charge in an otherwise uniform gas. Use the Thomas–Fermi (TF) approximation (Chapter 6) to derive the TF screening length Eq. (5.21). (Hint: assume the change in the density due to the impurity is $\delta n(r) \propto \exp(-k_{TF}r)/r$ and determine the decay constant k_{TF} from the TF equations expanded to linear order in $\delta n(r)$.)

5.15 Derive the expression for the exchange–correlation hole Eq. (5.29) in terms of the hole at larger densities (smaller r_s). Would there be an analogous form that involves an integral of λ from 1 to ∞, i.e., for larger r_s?

5.16 Using Fig. 5.5 sketch the shape of the average hole, Eq. (5.29), for antiparallel-spin electrons Al, Na, and Cs.

5.17 Derive the expression for the equilibrium r_s given in Eq. (5.32) from the expression for total energy and using $\alpha = 1.80$. In which direction will the predicted r_s change if correlation is included? Find the explicit expression using the Wigner interpolation formula for ϵ_c.

5.18 Derive the Lindhard expression for the dielectric function of a homogeneous gas Eq. (5.38). This is a tedious integral and the steps are given by Pines [297], p. 144.

6

Density Functional Theory: Foundations

E pluribus unum.

Summary

The fundamental tenet of density functional theory (DFT) is that *any* property of a system of many interacting particles can be viewed as a *functional* of the ground state density $n_0(\mathbf{r})$; that is, one scalar function of position $n_0(\mathbf{r})$, in principle, determines all the information in the many-body wavefunctions for the ground state and all excited states. The existence proofs for such functionals, given in the original works of Hohenberg and Kohn and of Mermin, are disarmingly simple. However, they provide no guidance whatsoever for constructing the functionals, and no exact functionals are known for any system of more than one electron. Density functional theory would remain a minor curiosity today if it were not for the construction of an auxiliary system by Kohn and Sham, which has provided a way to make useful, approximate functionals for real systems of many electrons. The subject of this chapter is density functional theory as a methodology for many-body systems; Chapter 7 describes the Kohn–Sham auxiliary independent-particle system and the formulation of the self-consistent Kohn–Sham equations; and Chapters 8 and 9 deal with widely used approximations for the exchange–correlation functional. Following chapters in this volume are devoted to algorithms for actual calculations, and applications to problems in atomic, molecular, and condensed matter physics.

6.1 Overview

The modern formulation of density functional theory originated in a famous paper by Hohenberg and Kohn in 1964 [83]. These authors showed that a special role can be assigned to the density of particles in the ground state of a quantum many-body system: the density can be considered to be a "basic variable" – i.e., that *all* properties of the system can be considered to be unique *functionals* of the ground-state density. Shortly thereafter in 1965,

Mermin [326] extended the Hohenberg–Kohn arguments to finite temperature canonical and grand canonical ensembles. Although the finite temperature extension has not been widely used, it illuminates both the generality of density functional theory and the difficulty of realizing the promise of exact density functional theory. Also in 1965 appeared the other classic work by Kohn and Sham [84], whose formulation of density functional theory has become the basis of much of present-day methods for treating electrons in atoms, molecules, and condensed matter.

In present-day literature, "density functional theory" and "DFT" usually mean the Kohn–Sham formulation, often blurring the distinction between Hohenberg–Kohn and Kohn–Sham. Here they are divided into different chapters to clearly distinguish between them. The goals of the chapters on density functional theory are to elucidate the fundamental ideas and current practices, to give the reader sufficient background to use density functional theory intelligently for real problems, and to expose potential pitfalls and possible avenues for future developments. The present chapter is concerned with the reformulation of the theory in interacting electrons in terms of functionals of the density. Sections 6.2 and 6.3 are devoted to background and the famous existence proofs by Hohenberg and Kohn; however, the less famous derivation by Levy and Lieb in Section 6.4 provides greater insight into the meaning of the functionals. Section 6.5 describes some of the extensions including the Mermin functional at nonzero temperature, current functionals, and issues related to electric polarization. Finally, some of the many intricacies and difficulties associated with density functional theory, as formulated by Hohenberg and Kohn, are the topics of Sections 6.6 and 6.7.

6.2 Thomas–Fermi–Dirac Approximation

The original density functional theory of quantum systems is the Thomas–Fermi method proposed in 1927 in separate papers by Thomas [327] and Fermi [328]. Although their approximation is not accurate enough for present-day electronic structure calculations, the approach illustrates the way density functional theory works. The approximation was to treat the kinetic energy of the system of electrons as an explicit functional of the density, idealized as noninteracting electrons in a homogeneous gas with density equal to the local density at any given point. Both Thomas and Fermi neglected exchange and correlation among the electrons; however, this was extended by Dirac [329] in 1930, who formulated the local approximation for exchange (see Sections 5.2 and 8.3) still in use today. This leads to the energy functional for electrons in an external potential $V_{ext}(\mathbf{r})$:

$$E_{TF}[n] = C_1 \int d^3r \, n(\mathbf{r})^{(5/3)} + \int d^3r \, V_{ext}(\mathbf{r})n(\mathbf{r})$$
$$+ C_2 \int d^3r \, n(\mathbf{r})^{4/3} + \frac{1}{2} \int d^3r d^3r' \frac{n(\mathbf{r})n(\mathbf{r}')}{|\mathbf{r} - \mathbf{r}'|}, \tag{6.1}$$

where the first term is the local approximation to the kinetic energy with $C_1 = \frac{3}{10}(3\pi^2)^{(2/3)} = 2.871$ in atomic units (see Section 5.2), the third term is the local exchange with $C_2 = -\frac{3}{4}(\frac{3}{\pi})^{1/3}$ (Eq. (5.15) for the case of equal up and down spins) and the last term is the classical electrostatic Hartree energy. (In Appendix H, improved approximations for inhomogeneous systems including gradients are given.)

The ground state density and energy can be found by minimizing the functional $E[n]$ in Eq. (6.1) for all possible $n(\mathbf{r})$ subject to the constraint on the total number of electrons

$$\int d^3r \, n(\mathbf{r}) = N. \tag{6.2}$$

Using the method of Lagrange multipliers (Exercise 6.1), the solution can be found by an unconstrained minimization of the functional

$$\Omega_{\text{TF}}[n] = E_{\text{TF}}[n] - \mu \left\{ \int d^3 r n(\mathbf{r}) - N \right\}, \tag{6.3}$$

where the Lagrange multiplier μ is the Fermi energy. For small variations of the density $\delta n(\mathbf{r})$, the condition for a stationary point is[1]

$$\int d^3r \left\{ \Omega_{\text{TF}}[n(\mathbf{r}) + \delta n(\mathbf{r})] - \Omega_{\text{TF}}[n(\mathbf{r})] \right\} \rightarrow$$

$$\int d^3r \left\{ \frac{5}{3} C_1 n(\mathbf{r})^{2/3} + V(\mathbf{r}) - \mu \right\} \delta n(\mathbf{r}) = 0, \tag{6.4}$$

where $V(\mathbf{r}) = V_{\text{ext}}(\mathbf{r}) + V_{\text{Hartree}}(\mathbf{r}) + V_x(\mathbf{r})$ is the total potential. Since Eq. (6.4) must be satisfied for any function $\delta n(\mathbf{r})$, it follows that the functional is stationary if and only if the density and potential satisfy the relation

$$\frac{1}{2}(3\pi^2)^{(2/3)} n(\mathbf{r})^{2/3} + V(\mathbf{r}) - \mu = 0. \tag{6.5}$$

Extensions to treat inhomogeneous systems more accurately require going beyond the local density approximation for the kinetic energy density. For example, the Weizsacker [330] correction, $\frac{1}{4}(\nabla n(\mathbf{r}))^2 / n(\mathbf{r})$, which is in essence a gradient correction in the same spirit as the gradient corrections in more modern density functionals in Section 8.5. In Section H.1 it is shown that this can be viewed as the kinetic energy density of a system of bosons with density $n(\mathbf{r})$, and a more accurate expression for fermions always increases the kinetic energy at each point $\mathbf{r}$.

The attraction of density functional theory is evident by the fact that one equation for the density is remarkably simpler than the full many-body Schrödinger equation that involves $3N$ degrees of freedom for N electrons. However, the Thomas–Fermi-type approach starts with approximations that are too crude, missing essential physics and chemistry, such as shell structures of atoms and binding of molecules [331]. Thus it falls short of the goal of a useful description of electrons in matter.

6.3 The Hohenberg–Kohn Theorems

The approach of Hohenberg and Kohn is to formulate density functional theory as an *exact theory of many-body systems*. The formulation applies to any system of interacting particles

---

[1] This is an example of functional equations described in Appendix A; see specifically Eq. (A.5).

$$V_{\text{ext}}(\mathbf{r}) \quad \overset{\text{HK}}{\Longleftarrow} \quad n_0(\mathbf{r})$$

$$\Downarrow \qquad\qquad \Uparrow$$

$$\Psi_i(\{\mathbf{r}\}) \quad \Rightarrow \quad \Psi_0(\{\mathbf{r}\})$$

Figure 6.1. Schematic representation of Hohenberg–Kohn theorem. The smaller arrows denote the usual solution of the Schrödinger equation where the potential $V_{\text{ext}}(\mathbf{r})$ determines all states of the system $\Psi_i(\{\mathbf{r}\})$, including the ground state $\Psi_0(\{\mathbf{r}\})$ and ground-state density $n_0(\mathbf{r})$. The long arrow labeled "HK" denotes the Hohenberg–Kohn theorem, which completes the circle.

in an external potential $V_{\text{ext}}(\mathbf{r})$, including any problem of electrons and fixed nuclei, where the hamiltonian can be written[2]

$$\hat{H} = -\frac{\hbar^2}{2m_e} \sum_i \nabla_i^2 + \sum_i V_{\text{ext}}(\mathbf{r}_i) + \frac{1}{2} \sum_{i \neq j} \frac{e^2}{|\mathbf{r}_i - \mathbf{r}_j|}. \tag{6.6}$$

Density functional theory is based on two theorems first proved by Hohenberg and Kohn [83]. Here we first present the theorems and the proofs along with discussion of the consequences; Section 6.4 contains the alternative formulation of Levy and Lieb, which is more general and gives a more intuitive definition of the functional. The relations established by Hohenberg and Kohn are illustrated in Fig. 6.1 and can be started as follows:

- **Theorem I.** For any system of interacting particles in an external potential $V_{\text{ext}}(\mathbf{r})$, the potential $V_{\text{ext}}(\mathbf{r})$ is determined uniquely, except for a constant, by the ground-state particle density $n_0(\mathbf{r})$.
  Corollary I. Since the hamiltonian is thus fully determined, except for a constant shift of the energy, it follows that the many-body wavefunctions for all states (ground and excited) are determined. *Therefore all properties of the system are completely determined given only the ground-state density $n_0(\mathbf{r})$.*
- **Theorem II.** A *universal functional* for the energy $E[n]$ in terms of the density $n(\mathbf{r})$ can be defined, valid for any external potential $V_{\text{ext}}(\mathbf{r})$. For any particular $V_{\text{ext}}(\mathbf{r})$, the exact ground-state energy of the system is the global minimum value of this functional, and the density $n(\mathbf{r})$ that minimizes the functional is the exact ground-state density $n_0(\mathbf{r})$.
  Corollary II. The functional $E[n]$ alone is sufficient to determine the exact ground-state energy and density. In general, excited states of the electrons must be determined by other means. Nevertheless, the work of Mermin (Section 6.5) shows that thermal equilibrium properties such as specific heat are determined directly by the free-energy functional of the density.

---

[2] The nuclei–nuclei interaction is added later; at this point it is not needed, except that care is needed to treat Coulomb interactions in extended systems (Section 3.2). Special considerations are required to include magnetic fields and there are subtle issues for electric fields in extended systems, see Section 6.5.

These assertions are so encompassing and the proofs are so simple that it is crucial for any practitioner in the field to understand the basis of the theorems and the limits of the logical consequences.

Proof of Theorem I: Density As a Basic Variable

The proofs of the Hohenberg–Kohn theorems are disarmingly simple. Consider first Theorem I, using the general expressions given in Eqs. (3.8) and (3.9) for density and energy in terms of the many-body wavefunction. Suppose that there were two different external potentials $V_{\text{ext}}^{(1)}(\mathbf{r})$ and $V_{\text{ext}}^{(2)}(\mathbf{r})$ that differ by more than a constant and that lead to the same ground-state density $n(\mathbf{r})$. The two external potentials lead to two different hamiltonians, $\hat{H}^{(1)}$ and $\hat{H}^{(2)}$, which have different ground-state wavefunctions, $\Psi^{(1)}$ and $\Psi^{(2)}$, that are hypothesized to have the same ground-state density $n_0(\mathbf{r})$. (It is straightforward to find different Ψs with the same density, as discussed below.) Since $\Psi^{(2)}$ is not the ground state of $\hat{H}^{(1)}$, it follows that

$$E^{(1)} = \langle \Psi^{(1)} | \hat{H}^{(1)} | \Psi^{(1)} \rangle < \langle \Psi^{(2)} | \hat{H}^{(1)} | \Psi^{(2)} \rangle. \tag{6.7}$$

The strict inequality follows if the ground state is nondegenerate, which we will assume here following the arguments of Hohenberg and Kohn.[3] The last term in Eq. (6.7) can be written

$$\langle \Psi^{(2)} | \hat{H}^{(1)} | \Psi^{(2)} \rangle = \langle \Psi^{(2)} | \hat{H}^{(2)} | \Psi^{(2)} \rangle + \langle \Psi^{(2)} | \hat{H}^{(1)} - \hat{H}^{(2)} | \Psi^{(2)} \rangle \tag{6.8}$$

$$= E^{(2)} + \int d^3r \left[V_{\text{ext}}^{(1)}(\mathbf{r}) - V_{\text{ext}}^{(2)}(\mathbf{r}) \right] n_0(\mathbf{r}), \tag{6.9}$$

so that

$$E^{(1)} < E^{(2)} + \int d^3r \left[V_{\text{ext}}^{(1)}(\mathbf{r}) - V_{\text{ext}}^{(2)}(\mathbf{r}) \right] n_0(\mathbf{r}). \tag{6.10}$$

On the other hand if we consider $E^{(2)}$ in exactly the same way, we find the same equation with superscripts (1) and (2) interchanged,

$$E^{(2)} < E^{(1)} + \int d^3r \left[V_{\text{ext}}^{(2)}(\mathbf{r}) - V_{\text{ext}}^{(1)}(\mathbf{r}) \right] n_0(\mathbf{r}). \tag{6.11}$$

Now if we add together Eqs. (6.10) and (6.11), we arrive at the contradictory inequality $E^{(1)} + E^{(2)} < E^{(1)} + E^{(2)}$. This establishes the desired result: there cannot be two different external potentials differing by more than a constant that give rise to the same nondegenerate ground-state charge density. The density uniquely determines the external potential to within a constant.

---

[3] This is not a necessary restriction. The proof can readily be extended to degenerate cases [332], which are also included in the alternative formulation by Levy [333–335] discussed in Section 6.4. Except in special cases the density of any one of the degenerate ground states uniquely determines the external potential. In the exercises is an example where two degenerate states have exactly the same density so that the expectation values of general operators cannot be unique functionals of the density. Even then, the expectation value of the energy is the same for all linear combinations of the degenerate states so that the Hohenberg–Kohn theorem is recovered.

The corollary follows since the hamiltonian is uniquely determined (except for a constant) by the ground-state density. Then, in principle, the wavefunction of any state is determined by solving the Schrödinger equation with this hamiltonian. Among all the solutions that are consistent with the given density, the unique ground-state wavefunction is the one that has the lowest energy.

Despite the appeal of this result, it is clear from the reasoning that no prescription has been given to solve the problem. Since all that was proved is that $n_0(\mathbf{r})$ uniquely determines $V_{\text{ext}}(\mathbf{r})$, we are still left with the problem of solving the many-body problem in the presence of $V_{\text{ext}}(\mathbf{r})$. For example, for electrons in materials, the external potential is the Coulomb potential due to the nuclei. The theorem only requires that the electron density uniquely determines the positions and types of nuclei, which can easily be proven from elementary quantum mechanics (see Exercise 6.6). At this level we have gained nothing: we are still faced with the original problem of many interacting electrons moving in the potential due to the nuclei.

Proof of Theorem II

The second theorem is just as easily proven once one has carefully defined the meaning of a functional of the density and restricted the space of densities. The original proof of Hohenberg–Kohn is restricted to densities $n(\mathbf{r})$ that are ground-state densities of the electron hamiltonian with *some* external potential V_{ext}. Such densities are called "*V*-representable." This defines a space of possible densities within which we can construct *functionals* of the density. (As discussed below in Section 6.4 it is possible to extend the range of validity of the functional.) Since all properties such as the kinetic energy, etc. are uniquely determined if $n(\mathbf{r})$ is specified, then each such property can be viewed as a functional of $n(\mathbf{r})$, including the total energy functional

$$E_{\text{HK}}[n] = T[n] + E_{\text{int}}[n] + \int d^3r \ V_{\text{ext}}(\mathbf{r})n(\mathbf{r}) + E_{II}$$

$$\equiv F_{\text{HK}}[n] + \int d^3r \ V_{\text{ext}}(\mathbf{r})n(\mathbf{r}) + E_{II}, \tag{6.12}$$

where E_{II} is the interaction energy of the nuclei (see Eq. (3.2) and related discussion). The functional $F_{\text{HK}}[n]$ defined in Eq. (6.12) includes all internal energies, kinetic and potential, of the interacting electron system,

$$F_{\text{HK}}[n] = T[n] + E_{\text{int}}[n], \tag{6.13}$$

which must be universal by construction because the kinetic energy and interaction energy of the particles are functionals only of the density.[4]

---

[4] Note that here "universal" means *the same for all electron systems*, independent of the external potential $V_{\text{ext}}(\mathbf{r})$. The Hohenberg–Kohn approach leads to different functionals for different particles depending on their masses and interactions. In this book the functionals described are for electrons, unless explicitly indicated otherwise. In fact there is another important application of the ideas of density functional theory in the theory of electronic

Now consider a system with the ground state density $n^{(1)}(\mathbf{r})$ corresponding to external potential $V_{\text{ext}}^{(1)}(\mathbf{r})$. Following the discussion above, the Hohenberg–Kohn functional is equal to the expectation value of the hamiltonian in the unique ground state, which has wavefunction $\Psi^{(1)}$

$$E^{(1)} = E_{\text{HK}}[n^{(1)}] = \langle \Psi^{(1)} | \hat{H}^{(1)} | \Psi^{(1)} \rangle. \tag{6.14}$$

Now consider a different density, say $n^{(2)}(\mathbf{r})$, which necessarily corresponds to a different wavefunction $\Psi^{(2)}$. It follows immediately that the energy $E^{(2)}$ of this state is greater than $E^{(1)}$, since

$$E^{(1)} = \langle \Psi^{(1)} | \hat{H}^{(1)} | \Psi^{(1)} \rangle < \langle \Psi^{(2)} | \hat{H}^{(1)} | \Psi^{(2)} \rangle = E^{(2)}. \tag{6.15}$$

Thus the energy given by Eq. (6.12) in terms of the Hohenberg–Kohn functional evaluated for the correct ground-state density $n_0(\mathbf{r})$ is indeed lower than the value of this expression for any other density $n(\mathbf{r})$.

It follows that if the functional $F_{\text{HK}}[n]$ was known, then by minimizing the total energy of the system, Eq. (6.12), with respect to variations in the density function $n(\mathbf{r})$, one would find the exact ground-state density and energy. This establishes Corollary II. Note that the functional only determines the ground-state properties; it does not provide any guidance concerning excited states.

6.4 Constrained Search Formulation of DFT

An alternative definition of a functional due to Levy [333–335] and Lieb [336–338] is very instructive because it

- extends the range of definition of the functional in a way that is formally more tractable and clarifies its physical meaning;
- provides an in-principle way to determine the exact functional; and
- leads to the same ground-state density and energy at the minimum as in the Hohenberg–Kohn analysis, and also applies for degenerate ground states.

The idea of Levy and Lieb (LL) is to define a *two-step* minimization procedure beginning with the usual general expression for the energy in terms of the many-body wavefunction Ψ given by Eq. (3.9). The ground state can be found, in principle, by minimizing the energy with respect to all the variables in Ψ. However, suppose one first considers the energy only for the class of many-body wavefunctions Ψ *that have the same density* $n(\mathbf{r})$. For any wavefunction, the total energy can be written

---

structure: the case of "noninteracting electrons" – i.e., fermions with the electron mass but with no interactions among themselves, which are the particles that explicitly enter the Kohn–Sham equations. It is advantageous to use the general ideas of density functionals in that case as well, and we will carefully indicate the distinction between the different use of the functionals for the Kohn–Sham equations.

$$E = \langle \Psi | \hat{T} | \Psi \rangle + \langle \Psi | \hat{V}_{\text{int}} | \Psi \rangle + \int d^3r \ V_{\text{ext}}(\mathbf{r}) n(\mathbf{r}). \tag{6.16}$$

Now if one minimizes the energy Eq. (6.16) over the class of wavefunctions with the same density $n(\mathbf{r})$, then one can define a unique lowest energy for that density

$$E_{\text{LL}}[n] = \min_{\Psi \to n(\mathbf{r})} [\langle \Psi | \hat{T} | \Psi \rangle + \langle \Psi | \hat{V}_{\text{int}} | \Psi \rangle] + \int d^3r \ V_{\text{ext}}(\mathbf{r}) n(\mathbf{r}) + E_{II}$$

$$\equiv F_{\text{LL}}[n] + \int d^3r \ V_{\text{ext}}(\mathbf{r}) n(\mathbf{r}) + E_{II}, \tag{6.17}$$

where the Levy–Lieb functional of the density is defined by

$$F_{\text{LL}}[n] = \min_{\Psi \to n(\mathbf{r})} \langle \Psi | \hat{T} + \hat{V}_{\text{int}} | \Psi \rangle. \tag{6.18}$$

In this form, $E_{\text{LL}}[n]$ is manifestly a functional of the density and the ground state is found by minimizing $E_{\text{LL}}[n]$.

The Levy–Lieb formulation is much more than just a restatement of the Hohenberg–Kohn functional, Eq. (6.12). First, Eq. (6.18) clarifies the meaning of the functional and provides a way to make an operational definition: *the minimum of the sum of kinetic plus interaction energies for all possible wavefunctions having the given density $n(\mathbf{r})$*. The LL functional also has important formal differences from the Hohenberg–Kohn functional; in particular, the LL functional in Eq. (6.18) is defined for *any* density $n(\mathbf{r})$ derivable from a wavefunction Ψ_N for N electrons. This is termed "N-representability" and the existence of such a wavefunction Ψ_N for any density satisfying simple conditions is known [339], as discussed in Section 6.6. In contrast, the Hohenberg–Kohn functional is defined only for densities that can be generated by some external potential; this is called "V-representability" and the conditions for such densities are not known in general. At the minimum of the total energy of the system in a given external potential, the Levy–Leib functional $F_{\text{LL}}[n]$ must equal the Hohenberg–Kohn functional defined in Eq. (6.13), since the minimum is a density that can be generated by an external potential. In addition, the LL form eliminates the restriction in the original proof of Hohenberg–Kohn to nondegenerate ground states; now one can do the search in the space of any one of a set of degenerate states.

Thus it has been established that a functional can be defined for any density (subject to certain conditions given below) and that by minimizing this functional one would find the exact density and energy of the true interacting many-body system. Just as for the original Hohenberg–Kohn proofs, however, we are faced with the cold fact that no method has been given to find the functional other than the original definition in terms of many-body wavefunctions. Nevertheless, as we shall see in the following chapter, the dependence of the functional on the kinetic and potential energies of the full, correlated many-body wavefunction points the way toward constructing approximate functionals that are of great utility in practical calculations and in understanding the effects of exchange and correlation among the electrons.

6.5 Extensions of Hohenberg–Kohn Theorems

Spin Density Functional Theory

The above analysis also shows how the Hohenberg–Kohn theorems can be generalized to several types of particles. The reason for the special role of the density and the external potential in the Hohenberg–Kohn theorems, rather than some other properties of the particles, is simply that these quantities enter the total energy Eq. (3.9) explicitly only through the simple bilinear integral term $\int d^3 r \, V_{\text{ext}}(\mathbf{r}) n(\mathbf{r})$. If there are other terms in the hamiltonian having this form, then each such pair of external potential and particle density will obey a Hohenberg–Kohn theorem.

The most relevant example for our purposes is a Zeeman term that is different for up and down spin fermions (i.e., a magnetic field that acts only on the spins, not on the orbital motion). This is in fact one of the important effects of an external magnetic field, so that this can be considered as a physically realistic approximation. Within this model, one can rigorously generalize all the above arguments to include two types of densities, the particle density $n(\mathbf{r}) = n(\mathbf{r}, \sigma = \uparrow) + n(\mathbf{r}, \sigma = \downarrow)$ and the spin density $s(\mathbf{r}) = n(\mathbf{r}, \sigma = \uparrow)$ $n(\mathbf{r}, \sigma = \downarrow)$. This leads to an energy functional

$$E = E_{\text{HK}}[n, s] \equiv E'_{\text{HK}}[n], \tag{6.19}$$

where in the last form it is assumed that $[n]$ denotes a functional of the density that depends on both position in space $\mathbf{r}$ and spin σ. "Spin density functional theory" is essential in the theory of atoms and molecules with net spins, as well as solids with magnetic order [305, 340, 341]. (Note that this does *not* include effects of a magnetic field on the orbital motion, which requires an extension to current functional theory [342–345].)

In the absence of external Zeeman fields, the lowest-energy solution may be spin polarized, i.e., $n(\mathbf{r}, \uparrow) \neq n(\mathbf{r}, \downarrow)$, which is analogous to the broken symmetry solution of unrestricted Hartree–Fock theory. (This must happen in a finite system with an odd number of electrons and also occurs in some atoms polarized to Hund's rules and in magnetic solids.) The spin functional is useful in these cases as well; however, the original Hohenberg–Kohn theorem remains valid and the ground state, in principle, is determined by the total ground-state density $n(\mathbf{r}) = n(\mathbf{r}, \uparrow) + n(\mathbf{r}, \downarrow)$ for any system where there is no spin-dependent external potential (see Exercise 6.8). The only modification of the statements of the theorems is to take into account the fact that the broken symmetry solution is necessarily degenerate.

Mermin Finite Temperature and Ensemble Density Functional Theory

The theorems of Hohenberg and Kohn for the ground state carry over to the equilibrium thermal distribution by constructing the density corresponding to the thermal ensemble. For each of the conclusions of Hohenberg and Kohn for the ground state, there exists a corresponding argument for a system in thermal equilibrium, as demonstrated by Mermin [326] shortly after the Hohenberg–Kohn paper. To show this, Mermin constructed a grand potential functional of the trial density matrices $\hat{\rho}$,

$$\Omega[\hat{\rho}] = \mathrm{Tr}\,\hat{\rho}\left[(\hat{H} - \mu\hat{N}) + \frac{1}{\beta}\ln\hat{\rho}\right], \tag{6.20}$$

whose minimum is the equilibrium grand potential

$$\Omega = \Omega[\hat{\rho}_0] = -\frac{1}{\beta}\ln\mathrm{Tr}\,e^{-\beta(\hat{H}-\mu\hat{N})}, \tag{6.21}$$

where $\hat{\rho}_0$ is the grand canonical density matrix

$$\hat{\rho}_0 = \frac{e^{-\beta(\hat{H}-\mu\hat{N})}}{\mathrm{Tr}\,e^{-\beta(\hat{H}-\mu\hat{N})}}. \tag{6.22}$$

The proof is completely analogous to the Hohenberg–Kohn proofs and uses only the minimum property of $\Omega[\hat{\rho}]$ and the fact that the energy depends on the external potential only through the term $\int V_{\text{ext}}(\mathbf{r})n(\mathbf{r})$. (The independent-particle version of the Mermin functional is given in Section 7.3.)

The Mermin theorem leads to even more powerful conclusions than the Hohenberg–Kohn theorems, namely that not only the energy but also the entropy, specific heat, etc. are functionals of the equilibrium density. However, the Mermin functional has not been widely applied. The simple fact is that it is much more difficult to construct useful, approximate functionals for the entropy (which involves sums over excited states) than for the ground-state energy. For example, in the Fermi liquid description of a metal the specific heat coefficient at low temperature is directly related to the effective mass at the Fermi surface. Thus the Mermin functional for the free energy must correctly describe the effective mass (with all its many-body renormalization) as well as the ground-state energy, whereas only the latter is required in the Hohenberg–Kohn functional.

The Hohenberg–Kohn theorems can also be generalized to other ensembles that are useful for aspects such as defining a functional of electron number as a continuous variable [346], whereas the original Hohenberg–Kohn theorems are formulated only for a ground state with a fixed integer number of electrons. The equilibrium thermal ensemble of Mermin at fixed chemical potential is an example where the number of electrons fluctuates around the average number given by the expectation value of the number operator $\hat{N}$. From ensemble theory, it also follows that there must be discontinuities in the derivative of the energy with respect to number at integer occupations or, in the case of solids, for filled bands. These are difficult properties to build into the functional and are absent in present-day approximate density functionals.

Current Density and Time-Dependent Density Functional Theory

The Hohenberg–Kohn theorems apply only to systems that are time-reversal invariant. If there is a magnetic field or time-dependent electric field, the hamiltonian involves terms of the form $V_{\text{ext}}(\mathbf{r})n(\mathbf{r})$ and $\mathbf{p}\cdot\mathbf{A}_{\text{ext}}$. Thus by exactly the same logic as the original Hohenberg–Kohn arguments, the properties must depend on both the density n and the current density $\mathbf{j} = -\frac{e}{m}\mathbf{p}$ [342, 343, 345]. However, the structure of the theory must be fundamentally different because there is no analogue of the variational principle for the ground-state energy or equilibrium-free energy.

The generalization of the Hohenberg–Kohn approach to time-dependent problems has been provided by Runge and Gross [246]. For a localized systems with simply connected geometry, the theory can be cast in terms of the time-dependent density since the current is determined by $\nabla \mathbf{j} = -dn/dt$. The result is that *given the initial wavefunction at one time t'*, the state at later times t is a functional of the time-dependent density $n(\mathbf{r}, t'')$ for all $t' \leq t'' \leq t$. This may be viewed as the formal construction of a density functional theory for excitations. Although the time-dependent functional must be quite intricate, there has been considerable progress within the Kohn–Sham approach, as described in Chapter 21.

In general, however, the theory must involve the current density. In particular, in a system with no boundaries, or is not simply connected, *the evolution is not a functional only of the density*. For example, in a uniform ring of charge, the density is unchanged if there is a net current, and the state is determined only if the current is specified [347]. Thus there is an essential link with current functionals and to properties such as the static electric polarization.

Electric Fields and Polarization

The issue of electric fields and polarization comes into play in extended systems. In infinite space, the potential due to an electric field $V(x) = Ex$ is unbounded; there is no lower bound to the energy and therefore there is no ground state. This is a famous problem [348, 349] in the theory of the dielectric properties of materials. However, if the ground state does not exist, the Hohenberg–Kohn theorems on the ground state do not apply [350].

Is there any way to include an electric field in density functional theory? This is a very subtle problem, and the answer is that in the presence of an electric field, one must apply some constraint, within which there is a stable ground state. In the case of molecules, this is routinely done by simply constraining the electrons to remain near the molecule. In a solid, however, the constraint is not so obvious. To the knowledge of the author, all proposals involve constraining the electrons to be in localized Wannier functions (Chapter 23) or equivalent conditions on Bloch functions. Since the energy contains a term $\mathbf{E} \cdot \mathbf{P}$, where $\mathbf{P}$ is the macroscopic polarization, the theory must become a "density polarization theory" (see [351, 352] and references cited there). An interesting point is that in a system with a net polarization at zero field $\mathbf{E} = 0$ (e.g., a ferroelectric) *the polarization is determined by the density alone* [351], i.e., the original Hohenberg–Kohn theorem applies. (But see Chapters 7 and 24 for the opposite conclusion in the Kohn–Sham approach.)

6.6 Intricacies of Exact Density Functional Theory

The challenge posed by the Hohenberg–Kohn theorems is how to make use of the reformulation of many-body theory in terms of functionals of the density. The theorems are in terms of unknown functionals of the density, and it is easy to show that these must be nonlocal functionals, depending simultaneously on $n(\mathbf{r})$ at different positions $\mathbf{r}$, which are difficult to cast in any simple form.

Allowed Densities for Electrons

There are a number of general questions related to the nature of the possible densities that are allowed for fermions, given only that they must integrate to the correct number of particles:

- Can one readily construct different wavefunctions Ψ that have the same density $n(\mathbf{r})$?
  Yes. An illuminating example is the homogeneous electron gas. All plane waves have the same uniform density, but only the choice of the lowest kinetic energy states gives the lowest-energy ground state for the noninteracting case. Interacting electrons also have the same uniform density even though the wavefunctions are correlated and thus quite different from a single determinant. The same logic can be applied to inhomogeneous cases, such as those discussed in Exercise 6.5.
- Is it possible to construct an antisymmetric wavefunction for fermions that can describe any possible density ("N-representability")?
  Yes, given a few restrictions on the density. As shown by Gilbert [339], it is possible to construct *any* density integrating to N total electrons of a given spin from a single Slater determinant of N one-electron orbitals, subject only to the condition that $n(\mathbf{r}) \geq 0$, and $\int |\nabla n(\mathbf{r})^{1/2}|^2$ is finite. In certain cases, explicit techniques exist for constructing such wavefunctions [273, 353], as described in Exercise 6.6.
- Is it possible to generate any such density as the ground state of some local external potential ("V-representability")?
  No. A number of "reasonable"-looking densities have been shown not to be the ground state for any potential V [334, 336]. Such densities are termed "non-V-representable." This applies to any linear combination of densities of a set of degenerate states; although the densities look "reasonable" they are not the ground state for the given number of electrons and any potential. An example is the spherically averaged density of an open-shell atom. If one weakens the question to ask if there are densities that cannot be generated by any smooth potential (one without delta functions) then one can find many counterexamples, e.g., any excited state density for single particles in finite systems. (The density of one electron in a 2s state in H is discussed in Exercise 6.7.)

Properties Obeyed by the Exact Density Functional Theory

The Hohenberg–Kohn arguments are very general for properties of interacting particle systems, yet special emphasis is on the ground state. Thus questions arise as to what properties of a material should be given correctly by the minimization of the Hohenberg–Kohn functional, if it were known exactly. *These examples make it clear how difficult it is to fulfill all the properties guaranteed by the Hohenberg–Kohn and Mermin theorems!*

Here we exclude magnetic fields and magnetic susceptibilities that require extension of density functional theory to include currents.

- Are excitation energies given correctly by the exact density functional theory?
  Yes. In principle, all properties are determined since the entire hamiltonian is determined.

- Are excitation energies given correctly by minimization of the exact Hohenberg–Kohn or Levy–Lieb functionals?

  No. The functional evaluated near the minimum provides no information about excitations. Some other method is needed to determine excitation energies.

- Is the exact specific heat versus temperature given correctly by the exact finite temperature Mermin functional?

  Yes. Even though the specific heat involves excitations from the ground state, nevertheless the thermal averages over these excitations must be a unique functional of the density and the temperature.

- Are static susceptibilities given correctly by the ground-state functional?

  Yes. Since the functional is universal, it is still valid in a perturbed system. All static susceptibilities are second derivatives of ground-state energies with respect to external fields. Thus they must be given correctly by the variation of the ground-state Hohenberg–Kohn functional as functions of external fields.[5]

- Is the exact Fermi surface of a metal given by the exact ground-state density functional theory?

  Yes. This is not a trivial question for two reasons. First, for the question to be meaningful, the many-body metal must have a well-defined Fermi surface; for the present purposes we assume this. Second, it is not a priori obvious that the Fermi surface is a ground-state property. One way to see that the Fermi surface is determined by ground-state properties is to consider susceptibilities to static perturbations. The exact density functional theory must lead to the correct Kohn anomalies and Friedel oscillations of the density far from an impurity, which depend in detail on the shape of the Fermi surface of the unperturbed metal.

- Must a Mott insulator (an insulator due to correlations among the electrons) be predicted correctly by the exact density functional theory?

  Yes. This follows from the above arguments on a metal in the special case where the Fermi surface vanishes.

6.7 Difficulties in Proceeding from the Density

The purpose of this section is to emphasize that density functional theory does *not* provide a way to understand the properties of a material merely by looking at the form of the density. Although the density is *in principle* sufficient, the relation is very subtle and no one has found a way to extract directly from the density any general set of properties, e.g., whether the material is a metal or an insulator. The key point is that the density is an allowed density of a quantum mechanical system; it is this fact that builds in the quantum effects.

---

[5] The dielectric susceptibility of an infinite system is a special case because the ground state is not strictly defined in the presence of a finite electric field. Nevertheless, susceptibilities can be calculated. Similarly a finite polarization is problematic. As discussed in Chapter 24, it is not given directly by the density; it is determined by expressions in terms of phases of the wavefunctions, which require some other method beyond the ground-state functional.

The difficulty can be illustrated by considering a case where the exact solution can be found – N noninteracting electrons in an external potential. This is the central problem in the Kohn–Sham approach to density functional theory, which is discussed in Chapter 7. In that case the *exact* Hohenberg–Kohn functional given by Eq. (6.12) is nothing other than the kinetic energy. In order to evaluate the kinetic energy exactly, the only way known is to revert to the usual expression in terms of a set of N wavefunctions. There is no known way to go directly from the density to the kinetic energy. The kinetic energy expressed in terms of wavefunctions has derivatives as a function of the number of electrons that are discontinuous at integer occupation numbers (see Exercise 6.11). From the virial theorem that relates kinetic and potential energies, it follows immediately that all parts of the exact functional (kinetic and potential) will vary in a *nonanalytic* manner as a function of the number of electrons. This is a property of a global integral of the density and is not simply determined from any aspect of the density only in some local region.

In the case of solids, the density is remarkably similar to sums of overlapping atom densities. For example, Fig. 2.1 shows the difference in density of electrons in silicon from superposed atoms, which is much smaller than the total density. In fact, the covalent bond is hard to distinguish in the total density. An ionic crystal is often considered as a sum of ions, but it is also well represented as the sum of neutral atoms [354]. This is possible because the negative anion is so large that its density extends around the positive cation, making the density similar to that of neutral atoms. Thus, even for well-known ionic crystals, it is not obvious how to extract pertinent information from the electron density. It is yet more difficult to distinguish metals from insulators (see Exercise 12.11 for an example).

This leads us to the Kohn–Sham approach, the success of which is based on the fact that it includes the kinetic energy of noninteracting electrons in terms of independent-particle wavefunctions, in addition to interaction terms explicitly modeled as functionals of the density. Because the kinetic energy is treated in terms of orbitals – *not* as an *explicit* functional of the density – it builds in the quantum properties that have no simple relation to the density. In the example of an ionic crystal, the key point is that the density is made up of fermions that obey the exclusion principle. It is this fact that leads to filling of four bands per cell and an insulating gap, which is the essence of this ionic crystal. So long as the true many-body solution is sufficiently close to the independent-particle formulation – e.g., the states must have the same symmetry – then the Kohn–Sham approach provides insightful guidance and powerful methods for electronic structure theory.

SELECT FURTHER READING

Original papers:

Hohenberg, P. and Kohn, W., "Inhomogeneous electron gas," *Phys. Rev.* 136:B864–871, 1964.

Kohn, W. and Sham, L. J., "Self-consistent equations including exchange and correlation effects," *Phys. Rev.* 140:A1133–1138, 1965.

Mermin, N. D., "Thermal properties of the inhomogeneous electron gas," *Phys. Rev.* 137:A1441–1443, 1965.

Reviews with historical perspective and Nobel prize lecture of Kohn:

Becke, A. D., "Perspective: Fifty years of density-functional theory in chemical physics," *J. Chem. Phys.* 140:301, 2014.

Burke, K., "Perspective on density functional theory," *J. Chem. Phys.* 136:150901 (2012).

Jones, R. O. and Gunnarsson, O., "The density functional formalism, its applications and prospect," *Rev. Mod. Phys.* 61:689–746, 1989.

Jones, R. O., "Density functional theory: Its origins, rise to prominence, and future," *Rev. Mod. Phys.* 87:897–923, 2015.

Kohn, W., "Nobel lecture: Electronic structure of matter wave functions and density functionals," *Rev. Mod. Phys.* 71:1253, 1999.

Book with extensive exposition:

Dreizler, R. M. and Gross, E. K. U., *Density Functional Theory: An Approach to the Quantum Many-Body Problem* (Springer, Berlin, 1990).

Engel, E. and Dreizler, R. M., *Density Functional Theory: An Advanced Course (Theoretical and Mathematical Physics* (Springer, New York, 2011).

Parr, R. G. and Yang, W., *Density-Functional Theory of Atoms and Molecules* (Oxford University Press, New York, 1989).

Sholl, D. and Stecklel, J. A., *Density Functional Theory: A Practical Introduction* (Wiley-Interscience, Hoboken, NJ, 2009).

Edited collections (see also many other volumes in the series):

Density Functional Methods in Physics, edited by R. M. Dreizler and J. da Providencia (Plenum, New York, 1985).

Density Functional Theory, edited by E. K. U. Gross and R. M. Dreizler (Plenum, New York, 1995).

A Primer in Density Functional Theory, edited by C. Fiolhais, F. Nogueira and M. Marques (Springer, New York, 2003).

More general books that discuss density functional theory:

Cohen, M. L. and Louie, S. G. *Fundamentals of Condensed Matter Physics* (Cambridge University Press, Cambridge, 2016).

Kaxiras, E., *Atomic and Electronic Structure of Solids* (Cambridge University Press, Cambridge, 2003).

Kaxiras, E. and Joannopoulos, J. D. *Quantum Theory of Materials*, 2nd rev. ed. (Cambridge University Press, Cambridge, 2019).

Kohanoff, J., *Electronic Calculations for Solids and Molecules: Theory and Computational Methods* (Cambridge University Press, Cambridge, 2003).

Exercises

6.1 Derive the Thomas–Fermi equation, (6.5), from the variational of the functional. Use the method of Lagrange multipliers as given in Eqs. (3.10) and used in (6.3).

6.2 See Exercise 5.14 for a problem involving the Thomas–Fermi (TF) approximation.

6.3 The simplest example of the Mermin theorem is the homogeneous gas. For a gas held at fixed volume, as the temperature is varied the density does not change. Describe the meaning of the Mermin functional in this case.

6.4 Theorem I of Hohenberg–Kohn shows that $n_0(\mathbf{r})$, *in principle*, uniquely determines all properties of the many-body system of electrons, including ground and excited states. We have argued that, for example, the electron density uniquely determines the positions and types of nuclei, which then defines the complete hamiltonian and therefore, *in principle*, determines all properties. Show explicitly that only the density and its derivatives near the nuclei are sufficient to establish the proof in this case.

6.5 In one dimension it is possible to construct orthonormal independent-particle orbitals that describe *any* density that satisfies simple positivity and continuity conditions. See Exercise 7.9.

6.6 Following the approach of Section 6.6, show that it is not possible to construct the density of the 2s state of hydrogen (one electron in the potential of a proton) as the ground-state density of any smooth potential, i.e., one without delta functions.

6.7 Consider the lowest-energy state of Li with three electrons, which may be $1s^2 2s$ or one of the degenerate states $(1s)^2 2p^0$, $1s^2 2p^-$, or $1s^2 2p^+$. The densities of the last two states are identical, so that the density does not determine the state. Show that, nevertheless, the energy is the same for any combination of these states so that the energy is still a functional of the density as needed for the Hohenberg–Kohn functional.

6.8 In this problem you are asked to show that in the absence of an external magnetic field, the total density is, in principle, enough to determine all the properties of the system even if it is spin polarized. To do this, consider the system in a Zeeman field $\mathbf{h} \cdot \sigma$ that distinguishes between σ parallel and antiparallel to $\mathbf{h}$. Show that if $\mathbf{h}$ is reversed, the new solution will have exactly the same density but with σ reversed. Using this fact show that you can reach the desired conclusion.

6.9 Suppose particles can be divided into two types (e.g., spins) of density n_1 and n_2 with internal energy $E_{int}[n_1, n_2]$. If the external potential acts on n_1 and n_2 equally, the total energy can be written $E_{total} = E[n_1, n_2] + \int V_{ext} n$, where $n = n_1 + n_2$. Show that E_{total} is a functional only of n. Do this in three ways: (a) using arguments similar to the original arguments of Hohenberg and Kohn; (b) the Levy–Lieb constrained search method; and (c) formal solution by variational equations in terms of n and $\sigma = n_1 - n_2$.

6.10 Consider a many-body hamiltonian $\hat{H} = \hat{H}_{int} + V_{ext}$, where $\hat{H}_{int}$ denotes all intrinsic internal kinetic and interaction terms and V_{ext} is the external potential. Show that the external potential $V_{ext}(\mathbf{r})$ is determined to within a constant, given $\hat{H}_{int}$ and *any* eigenfunction Ψ_i. Hint: solve for $V_{ext}(\mathbf{r})$ using the Schrödinger equation. (Note a specific example of a determinant wavefunction is considered as an exercise in Chapter 7.)

6.11 Show that in a finite system the kinetic energy must be a nonanalytic function of the density n with derivatives that are discontinuous at integer occupations. Hint: it is sufficient to show the result in an independent-particle example (see Exercise 7.5) with an argument that the result must also apply to many-body cases. Generalize this argument to all properties of the system and to solids with an insulating gap.

7

The Kohn–Sham Auxiliary System

If you don't like the answer, change the question.

Summary

Density functional theory is the most widely used method today for electronic structure calculations because of the approach proposed by Kohn and Sham in 1965: *to replace the original many-body problem by an auxiliary independent-particle problem*. The auxiliary system explicitly takes into account important parts of the problem including the nuclear potentials, the average repulsive Coulomb interactions, and the kinetic energy of independent fermions; the difficult problem of interacting, correlated electrons are incorporated in an exchange–correlation functional $E_{xc}[n]$. In principle, this provides a method to calculate exact ground-state properties of many-body systems using independent-particle methods; in practice, approximate forms for $E_{xc}[n]$ have proved to be remarkably successful. This chapter is devoted to defining the Kohn–Sham system and solving the self-consistent Kohn–Sham equations, which apply for any exchange–correlation functional. Progress in finding successful approximations to the $E_{xc}[n]$ functional is the topic of the following two chapters.

7.1 Replacing One Problem with Another

The Kohn–Sham approach is to replace the difficult interacting many-body system obeying the hamiltonian Eq. (3.1) with a different *auxiliary system* that can be solved more easily. Since there is no unique prescription for choosing the simpler auxiliary system, this is an *ansatz*[1] that rephrases the issues. Kohn and Sham proposed that the ground-state density of the original interacting system is equal to that of some chosen noninteracting system. This leads to independent-particle equations for the noninteracting system that can be considered exactly soluble (in practice by numerical means) with all the difficult

---

[1] *Ansatz*: attempt, approach. A mathematical assumption, especially about the form of an unknown function, which is made in order to facilitate solution of an equation or other problem (*Oxford English Dictionary*).

many-body terms incorporated into an *exchange–correlation functional of the density*. By solving the equations, one finds the ground-state density and energy of the original interacting system with the accuracy limited only by the approximations in the exchange–correlation functional.

Indeed, the Kohn–Sham approach has led to very useful approximations that are now the basis of most calculations that attempt to make "first-principles" or *ab initio* predictions for the properties of condensed matter and large molecular systems. The local density approximation (LDA) or various generalized-gradient approximations (GGAs) described below are remarkably accurate, most notably for "wide-band" systems, such as the group IV and II–V semiconductors, sp-bonded metals like Na and Al, insulators like diamond, NaCl, and molecules with covalent and/or ionic bonding. It also appears to be successful for many cases in which the electrons have stronger effects of correlations, such as transition metals. However, these approximations fail for many strongly correlated cases including the copper oxide planar materials, which are antiferromagnetic insulators for exactly half-filled bands, whereas the LDA or present GGA functionals find them to be metals [240]. This leads to the present situation in which there is great interest in utilizing and improving the density functional approach: to build upon the many successes of current approximations and to overcome the known deficiencies and failures in strongly correlated electron systems.

Here we will consider the Kohn–Sham *ansatz* for the ground state, which is by far the most widespread way in which the theory has been applied. However, in the big picture *this is only the first step*. The fundamental theorems of density functional theory (Chapter 6) show that *in principle* the ground-state density determines *everything*. A great challenge in present theoretical work is to develop methods for calculating excited state properties. We will return to these issues at the end of this chapter and in Chapter 9, but for the moment we will be concerned only with the theory of the ground state.

The Kohn–Sham construction of an auxiliary system rests upon two assumptions:

1. The exact ground-state density can be represented by the ground-state density of an auxiliary system of noninteracting particles. This is called *"noninteracting-V-representability"*; although there are no rigorous proofs for real systems of interest, we will proceed assuming its validity. This leads to the relation of the actual and auxiliary systems shown in Fig. 7.1.

2. The auxiliary hamiltonian is chosen to have the usual kinetic operator and an effective *local* potential $V_{\text{eff}}^{\sigma}(\mathbf{r})$ acting on an electron of spin σ at point $\mathbf{r}$. The local form is essential in order for there to be a one-to-one correspondence of the potential.[2] As in Chapter 6, we assume that the external potential $\hat{V}_{\text{ext}}$ is spin independent;[3] nevertheless,

---

[2] The generalized Kohn–Sham approach in Chapter 9 involves nonlocal orbital-dependent functionals. This is indeed outside the framework of the Kohn–Sham method as defined here, but it is interesting to note that the original paper of Kohn and Sham also proposed an alternative approach with a Hartree–Fock-like orbital-dependent operator for exchange, as in Eq. (3.45).

[3] Spin–orbit interaction is ignored as this point. It is a relativistic effect that can be included in the usual nonrelativistic equations as a term H_{SO} as described in Appendix O, and it is not included in the exchange–correlation functional.

$$
\begin{array}{ccccccc}
V_{\text{ext}}(\mathbf{r}) & \overset{\text{HK}}{\Longleftarrow} & n_0(\mathbf{r}) & \overset{\text{KS}}{\Longleftrightarrow} & n_0(\mathbf{r}) & \overset{\text{HK}_0}{\Longrightarrow} & V_{\text{KS}}(\mathbf{r}) \\
\Downarrow & & \Uparrow & & \Uparrow & & \Downarrow \\
\Psi_i(\{\mathbf{r}\}) & \Rightarrow & \Psi_0(\{\mathbf{r}\}) & & \psi_{i=1,N_e}(\mathbf{r}) & \Leftarrow & \psi_i(\mathbf{r})
\end{array}
$$

Figure 7.1. Schematic representation of Kohn–Sham auxiliary system (compare to Fig. 6.1). The notation HK_0 denotes the Hohenberg–Kohn theorem applied to the noninteracting problem. The arrow labeled KS provides the connection in both directions between the many-body and the auxiliary independent-particle systems, so that the arrows connect any point to any other point. Therefore, in principle, solution of the independent-particle Kohn–Sham problem determines *all properties* of the full many-body system.

except in cases that are spin symmetric, the auxiliary effective potential $V_{\text{eff}}^\sigma(\mathbf{r})$ must depend on spin in order to give the correct density for each spin.

The actual calculations are performed on the auxiliary independent-particle system defined by the auxiliary hamiltonian (using Hartree atomic units $\hbar = m_e = e = 4\pi/\epsilon_0 = 1$)

$$
\hat{H}_{\text{aux}}^\sigma = -\frac{1}{2}\nabla^2 + V^\sigma(\mathbf{r}). \tag{7.1}
$$

The expressions must apply for all $V^\sigma(\mathbf{r})$ in some range, in order to define functionals for a range of densities. For a system of $N = N^\uparrow + N^\downarrow$ independent electrons obeying this hamiltonian, the ground state has one electron in each of the N^σ orbitals $\psi_i^\sigma(\mathbf{r})$ with the lowest eigenvalues ϵ_i^σ of the hamiltonian Eq. (7.1). The density of the auxiliary system is given by sums of squares of the orbitals for each spin

$$
n(\mathbf{r}) = \sum_\sigma n(\mathbf{r}, \sigma) = \sum_\sigma \sum_{i=1}^{N^\sigma} |\psi_i^\sigma(\mathbf{r})|^2, \tag{7.2}
$$

the independent-particle kinetic energy T_s can be expressed in two ways,

$$
T_s = -\frac{1}{2}\sum_\sigma \sum_{i=1}^{N^\sigma} \langle \psi_i^\sigma |\nabla^2| \psi_i^\sigma \rangle = \frac{1}{2}\sum_\sigma \sum_{i=1}^{N^\sigma} \int d^3r |\nabla \psi_i^\sigma(\mathbf{r})|^2, \tag{7.3}
$$

and we define the classical Coulomb interaction energy of the electron density $n(\mathbf{r})$ interacting with itself (the Hartree energy defined in Eq. (3.15)) as

$$
E_{\text{Hartree}}[n] = \frac{1}{2}\int d^3r d^3r' \frac{n(\mathbf{r})n(\mathbf{r}')}{|\mathbf{r} - \mathbf{r}'|}. \tag{7.4}
$$

The Kohn–Sham approach to the full interacting many-body problem is to rewrite the Hohenberg–Kohn expression for the ground-state energy functional Eq. (6.12) in the form

$$
E_{\text{KS}} = T_s[n] + \int d\mathbf{r} V_{\text{ext}}(\mathbf{r})n(\mathbf{r}) + E_{\text{Hartree}}[n] + E_{II} + E_{\text{xc}}[n]. \tag{7.5}
$$

Here $V_{\text{ext}}(\mathbf{r})$ is the external potential due to the nuclei and any other external fields (assumed to be independent of spin) and E_{II} is the interaction between the nuclei (see Eq. (3.2)). Thus the sum of the terms involving V_{ext}, $E_{Hartree}$, and E_{II} forms a neutral grouping that is well defined (see Section 3.2). The independent-particle kinetic energy T_s is given explicitly as a functional of the orbitals; however, T_s for each spin σ must be a unique functional of the density $n(\mathbf{r}, \sigma)$ by application of the Hohenberg–Kohn arguments applied to the independent-particle hamiltonian Eq. (7.1); see Exercise 7.4.

All many-body effects of exchange and correlation are grouped into the exchange–correlation energy E_{xc}. Comparing the Hohenberg–Kohn, Eqs. (6.12) and (6.19), and Kohn–Sham, Eq. (7.5), expressions for the total energy (recall that the auxiliary density $n(\mathbf{r}, \sigma)$ of Eq. (7.2) is required to equal the true density for each spin σ) shows that E_{xc} can be written in terms of the Hohenberg–Kohn functional Eq. (6.13) as

$$E_{xc}[n] = F_{HK}[n] - (T_s[n] + E_{Hartree}[n]), \tag{7.6}$$

or in the more revealing form

$$E_{xc}[n] = \langle \hat{T} \rangle - T_s[n] + \langle \hat{V}_{int} \rangle - E_{Hartree}[n]. \tag{7.7}$$

Here $[n]$ denotes a functional of the density $n(\mathbf{r}, \sigma)$, which depends on both position in space $\mathbf{r}$ and spin σ. One can see that $E_{xc}[n]$ must be a functional since the right-hand sides of the equations are functionals. The latter equation shows explicitly that E_{xc} is just the difference of the kinetic and the internal interaction energies of the true interacting many-body system from those of the fictitious independent-particle system with electron–electron interactions replaced by the Hartree energy.

If the universal functional $E_{xc}[n]$ defined in Eq. (7.7) (or $\epsilon_{xc}([n], \mathbf{r})$ in Eq. (8.1)) were known, then the exact ground-state energy and density of the many-body electron problem could be found by solving the Kohn–Sham equations for independent particles. To the extent that an approximate form for $E_{xc}[n]$ describes the true exchange–correlation energy, the Kohn–Sham method provides a feasible approach to calculating the ground-state properties of the many-body electron system.

7.2 The Kohn–Sham Variational Equations

Solution of the Kohn–Sham auxiliary system for the ground state can be viewed as the problem of minimization with respect to either the density $n(\mathbf{r}, \sigma)$ or the effective potential $V_{eff}^{\sigma}(\mathbf{r})$ (see Section 9.5). Since T_s Eq. (7.3) is explicitly expressed as a functional of the orbitals but all other terms are considered to be functionals of the density, one can vary the wavefunctions and use the chain rule to derive the variational equation[4]

$$\frac{\delta E_{KS}}{\delta \psi_i^{\sigma*}(\mathbf{r})} = \frac{\delta T_s}{\delta \psi_i^{\sigma*}(\mathbf{r})} + \left[\frac{\delta E_{ext}}{\delta n(\mathbf{r}, \sigma)} + \frac{\delta E_{Hartree}}{\delta n(\mathbf{r}, \sigma)} + \frac{\delta E_{xc}}{\delta n(\mathbf{r}, \sigma)} \right] \frac{\delta n(\mathbf{r}, \sigma)}{\delta \psi_i^{\sigma*}(\mathbf{r})} = 0, \tag{7.8}$$

[4] Note that even if E_{xc} is explicitly represented as a functional of the wavefunctions (as in the optimized effective potential OEP method, Section 9.5), one does *not* use $\delta E_{xc}/(\delta \psi_i^{\sigma*}(\mathbf{r}))$, which would lead to nonlocal potential operators. See Chapter 9 for a generalized approach.

subject to the orthonormalization constraints

$$\langle \psi_i^\sigma | \psi_j^{\sigma'} \rangle = \delta_{i,j} \delta_{\sigma,\sigma'}. \tag{7.9}$$

This is equivalent to the Rayleigh–Ritz principle [258, 259] and the general derivation of the Schrödinger equations in (3.10)–(3.12), except for the explicit dependence of $E_{Hartree}$ and E_{xc} on n.

Using expressions (7.2) and (7.3) for $n^\sigma(\mathbf{r})$ and T_s, which give

$$\frac{\delta T_s}{\delta \psi_i^{\sigma*}(\mathbf{r})} = -\frac{1}{2} \nabla^2 \psi_i^\sigma(\mathbf{r}); \quad \frac{\delta n^\sigma(\mathbf{r})}{\delta \psi_i^{\sigma*}(\mathbf{r})} = \psi_i^\sigma(\mathbf{r}), \tag{7.10}$$

and the Lagrange multiplier method for handling the constraints Eqs. (3.10)–(3.13), this leads to the Kohn–Sham Schrödinger-like equations:

$$(H_{KS}^\sigma - \varepsilon_i^\sigma) \psi_i^\sigma(\mathbf{r}) = 0, \tag{7.11}$$

where the ε_i are the eigenvalues, and H_{KS} is the effective hamiltonian (in Hartree atomic units)

$$H_{KS}^\sigma(\mathbf{r}) = -\frac{1}{2} \nabla^2 + V_{KS}^\sigma(\mathbf{r}), \tag{7.12}$$

with

$$V_{KS}^\sigma(\mathbf{r}) = V_{ext}(\mathbf{r}) + \frac{\delta E_{Hartree}}{\delta n(\mathbf{r}, \sigma)} + \frac{\delta E_{xc}}{\delta n(\mathbf{r}, \sigma)}$$

$$= V_{ext}(\mathbf{r}) + V_{Hartree}(\mathbf{r}) + V_{xc}^\sigma(\mathbf{r}). \tag{7.13}$$

The meaning of the functional derivatives in the definitions of the Kohn–Sham potential, Eqs. (7.8) and (7.13), is described in Appendix A along with illustrative examples.

Equations (7.11)–(7.13) are the famous Kohn–Sham equations, with the resulting density $n(\mathbf{r}, \sigma)$ and total energy E_{KS} given by Eqs. (7.2) and (7.5). The equations have the form of independent-particle equations with a potential that must be found self-consistently with the resulting density. These equations are independent of any approximation to the functional $E_{xc}[n]$, and would lead to the exact ground-state density and energy for the interacting system if the exact functional $E_{xc}[n]$ were known. Furthermore, it follows from the Hohenberg–Kohn theorems (see Exercise 7.3) that the ground-state density uniquely determines the potential at the minimum (except for a trivial constant), so that there is a unique Kohn–Sham potential $V_{eff}^\sigma(\mathbf{r})|_{min} \equiv V_{KS}^\sigma(\mathbf{r})$ associated with any given interacting electron system.[5]

---

[5] It is straightforward to add terms in the Kohn–Sham equations in (7.12) and (7.13) to account for spin–orbit interaction interactions (see Appendix O) and cases where the spin is not quantized along the same axis at all points in space. The latter case is called "noncollinear spin" and the equations must be written in terms of the spin density matrix $\rho^{\alpha\beta}(\mathbf{r}) = \sum_i f_i \psi_i^{\alpha*}(\mathbf{r}) \psi_i^\beta(\mathbf{r})$, so that the Kohn–Sham hamiltonian Eq. (7.12) becomes a 2×2 matrix. These make the equations more complicated but are otherwise the same so long as the E_{xc} functional is not modified (see Section 8.3). Examples of calculations can be found in [355–358] and in Fig. 19.4. However, the equations are formally exact only if E_{xc} is a functional of the density matrix.

The extraordinary success of the Kohn–Sham approach is due to the separation of the problem of interacting electron into two parts: tractable equations that are considered in this chapter, and the difficult exchange–correlation term $E_{xc}[n]$ and $V_{xc}^{\sigma}(\mathbf{r})$ that is the topic of the following two chapters. Because of the difficulty of the original problem the latter are extraordinarily difficult to determine. There are no exact solutions, but it has been found that it is possible to find very useful approximations. The success of the methods depends totally upon the faithfulness of the functionals to describe the interacting many-body system, and they are given special attention in the following two chapters, Chapters 8 and 9.

7.3 Solution of the Self-Consistent Coupled Kohn–Sham Equations

The Kohn–Sham equations are summarized in the flow diagram in Fig. 7.2. They are a set of Schrödinger-like independent-particle equations that must be solved subject to the condition that the effective potential $V_{\text{eff}}^{\sigma}(\mathbf{r})$ and the density $n(\mathbf{r}, \sigma)$ are consistent. The explicit reference to spin will be dropped except where needed, and notation V_{eff} and n will be assumed to designate both space and spin dependence (of course, the potential for each spin depends on the densities for both spins). An actual calculation utilizes a numerical procedure that successively changes V_{eff} and n to approach the self-consistent solution. The computationally intensive step in Fig. 7.2 is "solve KS equation" for a given potential V_{eff}. This is the subject of the following chapters. Here this step is considered a "black box" that uniquely solves the equations for a given input V^{in} to determine an output density n^{out}, i.e., $V^{\text{in}} \rightarrow n^{\text{out}}$. Conversely, for a given form of the xc functional, any density n determines a potential V_{eff}, as shown in the second box. (This is the same as Eq. (7.13) and examples of specific expressions are given in Section 8.6.)

The problem is that except at the exact solution, the input and output potentials and densities do not agree. To arrive at the solution one operationally defines a new potential $n^{\text{out}} \rightarrow V^{\text{new}}$, which can then start a new cycle with V^{new} as the new input potential. Clearly, the procedure shown in Fig. 7.2 can be made into the iterative progression

$$V_i \rightarrow n_i \rightarrow V_{i+1} \rightarrow n_{i+1} \rightarrow \cdots, \tag{7.14}$$

where i labels the step in the iteration. The progression converges with a judicious choice of the new potential in terms of the potential or density found at the previous step (or steps).

Methods for reaching self-consistency are described in Section 7.4. However, it is best to first examine the nature of various possible total energy functionals. The expressions are needed for the final calculation of the energy and, in addition, the behavior of any of the functionals near the correct solution provides the basis for analysis of the convergence characteristics using that functional.

Total Energy Functionals

The Kohn–Sham equations are derived by minimizing the energy E_{KS} in Eq. (7.8); however, there are various choices for functionals, all of which have the same minimum energy

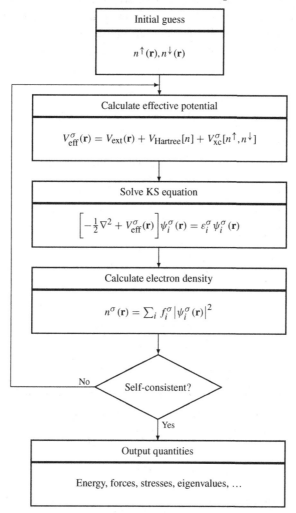

Figure 7.2. Schematic representation of the self-consistent loop for solution of the Kohn–Sham equations. In a spin-polarized system there are two such loops that must be iterated simultaneously, with the potential for each spin $V_{\text{eff}}^{\sigma}[n^{\sigma}, n^{\sigma}]$ a functional of the density of both spins. In the generalized Kohn–Sham approach the form of the loop is the same even though the potential is a nonlocal orbital-dependent operator as described in Chapter 9.

solution of the Kohn–Sham equations but behave differently away from the minimum. In particular, it is not essential to regard the density as the independent variable in the equations; different functionals can be found by a Legendre transformation to change the independent and dependent variables, as is familiar in thermodynamics. In terms of the Kohn–Sham equations, this means the behavior is a functional of the difference of input and output quantities $\Delta V = V^{\text{out}} - V^{\text{in}}$ and $\Delta n = n^{\text{out}} - n^{\text{in}}$, where n^{out} is the resulting density from solving the Schrödinger-like equation with the potential V^{in}. *It is essential to utilize correct variational expressions in order to have the desired variational properties.*

The original expression for the Kohn–Sham energy functional is given by Eq. (7.5), which is repeated here, with the grouping of all the potential terms to define $E_{pot}[n]$,

$$E_{KS} = T_s[n] + E_{pot}[n], \tag{7.15}$$

$$E_{pot}[n] = \int d\mathbf{r} V_{ext}(\mathbf{r}) n(\mathbf{r}) + E_{Hartree}[n] + E_{II} + E_{xc}[n]. \tag{7.16}$$

The first three terms on the right-hand side of the second equation together form a neutral grouping equal to the classical Coulomb interaction E^{CC} in Eq. (3.14). Since the eigenvalues of the Kohn–Sham equations are given by

$$\varepsilon_i^\sigma = \langle \psi_i^\sigma | H_{KS}^\sigma | \psi_i^\sigma \rangle, \tag{7.17}$$

the kinetic energy can be expressed as

$$T_s = E_s - \sum_\sigma \int d\mathbf{r} \, V^{\sigma,in}(\mathbf{r}) n^{out}(\mathbf{r}, \sigma), \tag{7.18}$$

where

$$E_s = \sum_\sigma \sum_{i=1}^{N^\sigma} \varepsilon_i^\sigma. \tag{7.19}$$

The advantages of this formulation are that the eigenvalues are available in actual calculations, and, furthermore, E_s in Eq. (7.19) is itself a functional. It is the ground-state energy of a noninteracting electron system, for which the Hohenberg–Kohn theorems, the force theorem, etc. all apply in a particularly simple way.

The Kohn–Sham Functional of the Potential $E_{KS}[V]$

Although the Kohn–Sham energy Eq. (7.15) is, *in principle*, a functional of the density, it is operationally a functional of the input potential $E_{KS}[V^{in}]$, as indicated in Fig. 7.2. (Here V denotes the potential for each spin, $V^\sigma(\mathbf{r})$.) At any stage of a Kohn–Sham calculation when the energy is not at the minimum, V^{in} determines all the quantities in the energy. This is clearly shown if we write E_{KS} from Eq. (7.15) as

$$E_{KS}[V^{in}] = E_s[V^{in}] - \sum_\sigma \int d\mathbf{r} \, V^{\sigma,in}(\mathbf{r}) n^{out}(\mathbf{r}, \sigma) + E_{pot}[n^{out}], \tag{7.20}$$

where the first two terms on the right-hand side are a convenient way of calculating the independent-particle kinetic energy as in Eq. (7.18), and E_{pot} is the sum of potential terms given in Eq. (7.16) evaluated for $n = n^{out}$. Since E_s is the sum of eigenvalues, Eq. (7.19), and $n^{out}(\mathbf{r}, \sigma)$ is the output density, each determined directly by the potential $V^{\sigma,in}(\mathbf{r})$, clearly the energy is a functional of V^{in}. Of course, E_{KS} formally can be regarded as a functional of n^{out}, since there is a one-to-one relation of the output density and the input potential (except for a trivial constant in V^{in}); however, the Kohn–Sham equations provide no way of choosing n^{out} except as an output determined by a potential.

The solution of the Kohn–Sham equations is for the potential V^{in} that minimizes the energy, Eq. (7.20). Then $V^{in} = V_{KS}$, the output density n^{out} is the ground-state density n^0, and the potential and density are consistent with the relation in Eq. (7.13). The functional $E_{KS}[V^{in}]$ is variational and all other potentials lead to energies that are higher by an amount that is quadratic in the error $V^{in} - V_{KS}$. Near the minimum energy solution, the error in the energy must also be quadratic in the error in the density $\delta n = n^{out} - n^0$, so that

$$E_{KS}[V^{in}] = E_{KS}[V_{KS}] + \frac{1}{2} \sum_{\sigma,\sigma'} \int d\mathbf{r} d\mathbf{r}' \left[\frac{\delta^2 E_{KS}}{\delta n(\mathbf{r},\sigma)\delta n(\mathbf{r}',\sigma')} \right]_{n^0} \delta n(\mathbf{r},\sigma)\delta n(\mathbf{r}',\sigma'), \quad (7.21)$$

where the second term is always positive.

Explicit Functionals of the Density

As shown by Harris [359], Weinert et al. [360], and Foulkes and Haydock [361], one can choose different expressions for the total energy functional that are given *explicitly* in terms of the density. The functional is cast in terms of the density n^{in} that, via Eq. (7.13), determines the input potential $V[n^{in}] \equiv V_{n^{in}}$, which in turn leads directly to the sum of eigenvalues, the first term on the right-hand side of Eq. (7.20). The energy is then defined by evaluating the functional $E_{pot}[n^{in}]$ in Eq. (7.16) in terms of the chosen *input* density $n^{in}(\mathbf{r},\sigma)$ (instead of the output density $n^{out}(\mathbf{r},\sigma)$ as in the Kohn–Sham functional),

$$E_{HWF}[n^{in}] \equiv E_s[V_{n^{in}}] - \sum_{\sigma} \int d\mathbf{r}\, V_{n^{in}}^{\sigma}(\mathbf{r})n^{in}(\mathbf{r},\sigma) + E_{pot}[n^{in}]. \quad (7.22)$$

The stationary properties of this functional can be understood straightforwardly following the arguments of Foulkes [361]. For a given input density n^{in} and potential $V_{n^{in}}$, the difference in the two expressions for the energy involves only the potential terms

$$E_{KS}[V^{in}] - E_{HWF}[n^{in}] = \sum_{\sigma} \int d\mathbf{r}\, V_{n^{in}}^{\sigma}(\mathbf{r}) \left[n^{out}(\mathbf{r},\sigma) - n^{in}(\mathbf{r},\sigma) \right]$$

$$+ \left[E_{pot}[n^{out}] - E_{pot}[n^{in}] \right]. \quad (7.23)$$

Near the correct solution where $\Delta n = n^{out} - n^{in}$ is small, one can expand the difference in Eq. (7.23) in powers of Δn. The linear terms cancel (which follows from the fact that $V_{n^{in}}^{\sigma}(\mathbf{r}) = [\delta E_{pot}/(\delta n(\mathbf{r},\sigma))]_{n^{in}}$; see Exercise 7.15) so that the lowest-order terms are

$$E_{KS}[V^{in}] - E_{HWF}[n^{in}] \approx \frac{1}{2} \sum_{\sigma,\sigma'} \int d\mathbf{r} d\mathbf{r}'\, K(\mathbf{r},\sigma;\mathbf{r}',\sigma')_{n^{in}} \Delta n(\mathbf{r},\sigma)\Delta n(\mathbf{r}',\sigma'), \quad (7.24)$$

where the kernel K is defined to be

$$K(\mathbf{r},\sigma;\mathbf{r}',\sigma') \equiv \frac{\delta^2 E_{\text{Hxc}}[n]}{\delta n(\mathbf{r},\sigma)\delta n(\mathbf{r}',\sigma')}$$

$$= \frac{1}{|\mathbf{r}-\mathbf{r}'|}\delta_{\sigma,\sigma'} + \frac{\delta^2 E_{\text{xc}}[n]}{\delta n(\mathbf{r},\sigma)\delta n(\mathbf{r}',\sigma')}, \qquad (7.25)$$

evaluated for $n = n^{\text{in}}$. (Note that K has been defined in terms of $E_{\text{Hxc}}[n] \equiv E_{\text{Hartree}}[n] + E_{\text{xc}}[n]$; the other terms in $E_{\text{pot}}[n]$ do not contribute since they are constant or linear in n.) Since the differences in the energies are quadratic in the errors in the density, it follows that at the exact solution where $\Delta n(\mathbf{r},\sigma) = 0$, the functional $E_{HWF}[n^{\text{in}}]$ equals the usual Kohn–Sham energy and it is stationary. However, it is *not variational*, which can be seen from Eq. (7.24). Since the kernel K tends to be positive (see below), then $E_{HWF}[n^{\text{in}}]$ is lower than $E_{KS}[V^{\text{in}}]$. Thus even though $E_{KS}[V^{\text{in}}]$ is always above the Kohn–Sham energy, $E_{HWF}[n^{\text{in}}]$ may be lower by an amount that is second order in the error $\Delta n(\mathbf{r},\sigma)$.

The primary advantage of the explicit functional of the density Eq. (7.22) is that, for densities near the correct solution, it can accurately approximate the true Kohn–Sham energy. In particular, it is often an excellent approximation to stop the calculation after one calculation of eigenvalues with *no self-consistency*: in this case one does not even need to calculate the output density. This approach is remarkably successful if $n(\mathbf{r})$ is approximated by a sum of atomic densities [172, 359, 361–363]. Perhaps the first example was a calculation of phonon frequencies [172]. Foulkes has used this as a conceptual basis for the success of empirical tight-binding models where the energy is given strictly by sums of eigenvalues plus additional terms that can be accounted for in this approach (see Section 14.11 on total energies in tight-binding methods). In addition, it is particularly simple to calculate the energy relative to neutral atoms in terms of the *difference in the density from a sum of neutral atoms*. This yields directly desirable physical quantities, as described in Section F.4.

In a full self-consistent calculation the functional Eq. (7.22) is useful at each step of the iteration in Fig. 7.2. It is now standard to calculate both energies, Eqs. (7.20) and (7.22), at each step in the iteration. The KS functional of the potential is variational, but the nonvariational functional of the density energy is usually closer to the true energy for reasons explained in Section 7.4. It is also very useful to calculate both energies and treat the difference as a measure of the lack of self-consistency during a calculation.

It is tempting to assume that the explicit density functional Eq. (7.22) is a *maximum* as a function of density. However, this is not the case in general because the second-derivative functional $K(\mathbf{r},\sigma;\mathbf{r}',\sigma')$ in Eq. (7.25) is *not* guaranteed to be positive definite [364–366]. From the definition of K in Eq. (7.25), the first term is positive definite since it is due to the repulsive Hartree term. One might expect that the second attractive term would never overcome the repulsion. However, approximations such as the LDA violate this condition since the extreme local $\delta(|\mathbf{r}-\mathbf{r}'|)$ behavior leads to large negative contributions for short wavelength density variations.

Generalized Functionals of V and n, $E[V; n]$

It is also possible to define functionals of the density and potential varied independently, as pointed out by a number of authors [361, 363, 367, 368]. We will denote n and V by n^{in} and V^{in} to emphasize that *both* are independent input functions. The expression is exactly the same as Eq. (7.22), except that V^{in} is regarded as an independent function so that the expression can be written

$$E[V^{in}, n^{in}] = E_s[V^{in}] - \sum_\sigma \int V^{\sigma, in}(\mathbf{r}) n^{in}(\mathbf{r}, \sigma) d\mathbf{r} + E_{pot}[n^{in}]. \tag{7.26}$$

The first term is solely a functional of V^{in}, the last term is a functional only of n^{in}, and the only coupling of V^{in} and n^{in} is through the simple bilinear second term. The properties of the functional can be seen clearly following the description by Methfessel [363]. Considering variations around any V^{in} and n^{in}, to linear order

$$\delta E[V^{in}, n^{in}] = \sum_\sigma \int \left[V_{n^{in}}^\sigma(\mathbf{r}) - V^{\sigma, in}(\mathbf{r}) \right] \delta n(\mathbf{r}, \sigma) d\mathbf{r}$$

$$+ \sum_\sigma \int \left[n_{V^{in}}^{out}(\mathbf{r}, \sigma) - n^{in}(\mathbf{r}, \sigma) \right] \delta V^\sigma(\mathbf{r}) d\mathbf{r}, \tag{7.27}$$

where $V_{n^{in}}^\sigma(\mathbf{r}) = \left[\frac{\delta E_{pot}}{\delta n(\mathbf{r}, \sigma)} \right]_{n^{in}}$ is the potential determined by the input density (as used in Eq. (7.22)), and $n_{V^{in}}^{out}(\mathbf{r}, \sigma)$ is the output density determined by the potential V^{in} (as used in Eq. (7.20)). Since the terms in brackets vanish at self-consistency, the functional is stationary and the value equals the Kohn–Sham energy $E_{KS}[V^{KS}]$.

It is also straightforward to show [363] that for any fixed density n^{in}, the stationary point of $E[V^{in}, n^{in}]$ as a function of V^{in} is in fact a *global maximum* as a function of V^{in}, at which point the value of $E_s[V^{max}] - \sum_\sigma \int V^{\sigma, max}(\mathbf{r}) n^{in}(\mathbf{r}, \sigma) d\mathbf{r}$ equals the Kohn–Sham kinetic energy functional $T_s[n^{in}]$. Although the maximum property may seem surprising, it follows from inequalities similar to the Hohenberg–Kohn arguments and it can be understood from Eq. (7.27), which shows that

$$\frac{\delta E}{\delta V^\sigma(\mathbf{r})} = n^{out}(\mathbf{r}, \sigma) - n^{in}(\mathbf{r}, \sigma) \Rightarrow \frac{\delta^2 E}{\delta V^\sigma(\mathbf{r}) \delta V^{\sigma'}(\mathbf{r}')} = \frac{\delta n^{out}(\mathbf{r}, \sigma)}{\delta V^{\sigma'}(\mathbf{r}')}. \tag{7.28}$$

The eigenvalues of this functional are always negative since the density decreases where the potential is increased [363]. The curvature of E as a functional of n^{in} is given by the kernel Eq. (7.25), which involves only the potential terms $E_{Hxc}[n]$ since the other terms are constant or linear. As explained following Eq. (7.25), E tends to be a minimum as a functional of n^{in}; however, this is not guaranteed and only with constraints on the density variations is the solution a minimum [363].

The importance of the stationarity is that one can approximate both V^{in} and n^{in}. For example, one can choose convenient forms for the potentials, such as spherical muffin-tin-type potentials often used in augmented methods. If one carries out the Kohn–Sham calculation exactly for this potential, of course this is just a restatement of the variational property of $E_{KS}[V]$. The generalized functional shows that the errors in the energy are still quadratic if the density is also approximated using convenient functional forms. This can be used to advantage in calculations as illustrated in [363].

Free Energy Functionals

Introducing temperature has many potential benefits:

- Direct calculation of thermal quantities: entropy S, free energy $F = E - TS$, etc.
- The density matrix becomes shorter range as the temperature increases, which can be used to advantage, e.g., in order-N methods (Chapter 18).
- Smearing the occupation makes calculations for metals less sensitive to numerical approximations.

Expressions for the energy are given by any of the previous functionals with the sum of single-particle energies $E_s \to E_s(T)$ generalized to finite T as in Eq. (3.39). The entropy is given by the single particle form of the Mermin finite temperature functional Eq. (6.20),

$$S = -\left[\sum_i f_i \ln f_i + \sum_i (1 - f_i) \ln(1 - f_i)\right],\tag{7.29}$$

where f_i denotes the occupation number $f(\varepsilon_i - \mu)$.

These formulas can be used as a clever way to calculate $E(T = 0)$. The simple idea is that $E(T)$ increases quadratically with T, whereas $F(T)$ decreases quadratically. A combination of the two, $E + F$ (see Exercise 7.17), can cancel the quadratic terms and give an expression equal to $E(T = 0)$ with only quartic corrections. For example, this has been used by Gillan [369] to calculate the vacancy energy in Al using a calculation actually done at a temperature of 10,000 K. The high temperature greatly simplifies the calculations by reducing the finite size effects in the calculation.

In iterative methods (Appendix M), one is seeking to find the solution for both the potential and the wavefunctions at the same time, i.e., the wavefunctions are not consistent with the potential, as is assumed in the above expressions. As shown in [370], one can generalize the Fermi function f_i to a matrix f_{ij}, which is constrained to have eigenvalues in the range $[0, 1]$. Then the density is given by

$$n(\mathbf{r}) = \sum_{ij} f_{ij} \psi_i^*(\mathbf{r}) \psi_j(\mathbf{r}),\tag{7.30}$$

and the grand energy functional Eq. (6.20) is generalized to

$$\tilde{\Omega}[V^{\text{in}}, n^{\text{in}}, T, \mu] = E[V^{\text{in}}, n^{\text{in}}]_0 + \mu(N_0 - \text{Tr}[f])$$
$$+ k_B T \text{Tr}\left[f \ln f + (1 - f) \ln(1 - f)\right].\tag{7.31}$$

This form is particularly useful in iterative methods where it can speed the convergence in metals by effectively allowing for unitary transformations of the wavefunctions that are problematic because they correspond to low-energy "slow modes" of the electronic system.

The most complete expression for a generalized functional is found by including temperature T via the Mermin functional (see Section 6.5) and the chemical potential μ to allow variation in particle number. Then, as shown by Nicholson et al. [368], one can define a grand functional,

$$\Omega[V^{in}, n^{in}, T, \mu] = E[V^{in}, n^{in}, T]_0 + \mu\left(N_0 - \sum_i f_i\right)$$

$$+ k_B T\left[\sum_i f_i \ln f_i + \sum_i (1 - f_i) \ln(1 - f_i)\right]. \tag{7.32}$$

This functional is stationary with respect to V^{in}, n^{in}, μ, T, and the form of the occupation function $f(\varepsilon)$.

7.4 Achieving Self-Consistency

A key problem is the choice of procedure for updating the potential V^{σ} or the density n^{σ} in each loop of the Kohn–Sham equations illustrated in Fig. 7.2. Obviously one can vary either V^{σ} or n^{σ}, but it is simpler to describe in terms of n^{σ}, which is unique, whereas V^{σ} is subject to shift by a constant. (The spin index σ is omitted below for simplicity.) This section is devoted to the basic ideas of linear mixing, dielectric screening, and numerical methods; there are many variations and combinations and there is no attempt to review all methods.

The simplest approach is *linear mixing*, estimating an improved density input n^{in}_{i+1} at step $i + 1$ as a fixed linear combination of n^{in}_i and n^{out}_i at step i,

$$n^{in}_{i+1} = \alpha n^{out}_i + (1 - \alpha)n^{in}_i = n^{in}_i + \alpha(n^{out}_i - n^{in}_i). \tag{7.33}$$

This is a good choice in the absence of other information and is essentially moving in an approximate "steepest descent" direction for minimizing the energy.

Why cannot one simply take the output density at one step as the input to the next? What are the limits on α? How can one do better? The answers lie in linear analysis of the behavior near the minimum [371, 372].[6] As in Eq. (7.21), let us define the deviation from the correct density to be $\delta n \equiv n - n_{KS}$ at any step in the iteration. Then near the solution, the error in the output density to linear order in the error in the input is given by

$$\delta n^{out}[n^{in}] = n^{out} - n_{KS} = (\tilde{\chi} + 1)(n^{in} - n_{KS}), \tag{7.34}$$

where

$$\tilde{\chi} + 1 = \frac{\delta n^{out}}{\delta n^{in}} = \frac{\delta n^{out}}{\delta V^{in}} \frac{\delta V^{in}}{\delta n^{in}}. \tag{7.35}$$

Here $\delta n^{out}/\delta V^{in}$ is a response function defined to be χ^0 in Eq. (D.6) and $\delta V^{in}/\delta n^{in}$ is K defined in Eq. (7.25). Thus the needed function $\tilde{\chi}$ can be calculated and is closely related

---

[6] The description here follows that of Pickett in [372].

to other uses of response functions. The best choice for the new density is one that would make the error zero, i.e., $n_{i+1}^{in} = n_{KS}$. Since n_i^{out} and n_i^{in} are known from step i, if $\tilde{\chi}$ is also known, then Eq. (7.34) can be solved for n_{KS},

$$n_{KS} = n_i^{in} - \tilde{\chi}^{-1}\left(n_i^{out} - n_i^{in}\right). \tag{7.36}$$

If Eq. (7.36) were exact, this would be the answer and the iterations could stop; since it is not exact this gives the best input for the next iteration.

Although Eq. (7.36) is a more complex integral equation, it bears a strong resemblance to the linear-mixing equation (7.33). If we resolve the response function $\tilde{\chi}$ into eigenfunctions $\tilde{\chi}(\mathbf{r},\mathbf{r}') = \sum_m \chi_m f_m(\mathbf{r}) f_m(\mathbf{r}')$, the eigenvalues χ_m give the optimal α for the change in density resolved into the density eigenvectors $f_m(\mathbf{r})$. Furthermore, the radius of convergence of the linear-mixing scheme is determined by the maximum eigenvalue $\tilde{\chi}_{max}^{-1} = 1/\tilde{\chi}_{min}$ of the matrix $\tilde{\chi}^{-1}$. If a constant α is used, it is straightforward to show [372] that the maximum error at iteration i varies as $(1 - \alpha\tilde{\chi}_{max}^{-1})^i$, so that the iterations converge only if $\alpha < 2/\tilde{\chi}_{max}^{-1} = 2\tilde{\chi}_{min}$ (see Exercises 7.21 and 13.3).

Physically, the response of the system is a measure of the polarizability. Linear mixing with large α works well for strongly bound, rigid systems, such as wide-gap insulators. However, convergence can be very difficult to achieve for "soft cases," for which metal surfaces are an especially difficult example. Convergence algorithms using the response kernel K have been proposed [373] for such cases. In these examples, it is most useful to analyze the response in Fourier space, which is done in Section 13.2 in the chapter on plane waves.

Numerical Mixing Schemes

The difficulty with the analysis in terms of the response kernel $\tilde{\chi}$ (or K) is that in real problems, it can be found only by calculations (similar to those for response functions; see Appendix D and Chapter 20) that are more costly than many iterations of a standard minimization algorithm. It can be much more efficient to adopt methods from the numerical literature that build up the information on the Jacobian J (the second-derivative matrix) of the system automatically rather than using physical arguments. In fact, the matrix $\tilde{\chi}$ is the Jacobian J, but in this section we will use the notation J to be consistent with commonly used notation (see Appendix L).

General numerical approaches for reaching a consistent solution include the Broyden method [374] described in Appendix L.[7] and the RMM-DIIS approach described in Section M.7. In the Broyden method the desired quantity, the inverse Jacobian J^{-1} itself, is built up as the iterations proceed. Starting with an approximate form, J^{-1} is improved at each iteration in a way so that the change in density for step $i + 1$ is made in a direction orthogonal to all previous directions. (This is the general idea in all numerical methods that generate a "Krylov subspace" – see Appendices L and M.) The magnitude of the step is chosen to be such that it would give the result of step i projected onto the subspace generated thus far. (Note the similarity of this last requirement with solution Eq. (7.36) using $\tilde{\chi}^{-1}$;

---

[7] This method was first used in solid-state calculations by Bendt and Zunger [375] and described in more detail by Srivastava [376].

the difference is that in the Broyden method only partial information is known about the Jacobian at any step i.) Thus the Broyden method combines the "best of both worlds" to make an automatic method that generates the needed parts of the Jacobian as the calculation proceeds, with essentially no added cost above that encountered in simple linear mixing.

At each iteration i the input density for the next step is given by an equation analogous to Eq. (7.36) except that $\tilde{\chi}$ is replaced by the approximate Jacobian J_i

$$n_{i+1}^{\text{in}} = n_i^{\text{in}} - J_i^{-1}\left(n_i^{\text{out}} - n_i^{\text{in}}\right), \tag{7.37}$$

and J_i^{-1} is improved at each step by Expression (L.24). This can be used directly if the Jacobian matrix is small, i.e., if there are only a few components of the density for which convergence is a problem. An example is given in Section 13.2 in the chapter on plane waves. Srivastava [376] has shown how to avoid storage of the Jacobian matrices by writing the predicted change $\delta n_{i+1}^{\text{in}}$ in terms of a sum over all the previous steps involving only the initial J_0^{-1}. A modified Broyden method was proposed by Vanderbilt and Louie [380] and adapted by Johnson [381] to also incorporate Srivistava's improvements [376]. The basic equation is given in Eq. (L.25) with discussion of the weights given in the original and subsequent papers.

An example of the power of the Broyden method using this approach is shown in Fig. 7.3 for the density at a (1 0 0) surface of W using an LAPW method (Chapter 17). The quantity shown is the "distance" d, which is the norm of the residual

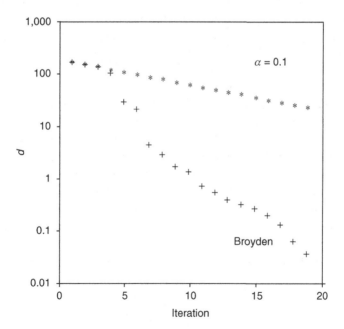

Figure 7.3. Convergence of the density for a W(1 0 0) surface (see Eq. (7.38) for the definition of d) versus iteration number for linear mixing and the Broyden method. From [377]. There are many more recent variations; see Section 13.2 and assessments in references such as [378] and [379].

$$d = \frac{1}{\Omega_{\text{cell}}} \int_{\Omega_{\text{cell}}} d^3 r (n^{\text{out}} - n^{\text{in}})^2, \tag{7.38}$$

plotted for linear mixing with $\alpha = 0.1$ and for Broyden with $J_0 = \alpha \mathbf{1}$. There are many variations and improved methods, too many to mention, and the reader is advised that this is an important consideration that can greatly affect the time required for a calculation. See, for example, references such as [378] and [379] and the discussion in Section 13.2.

7.5 Force and Stress

It is straightforward to see that the usual form (Section 3.3) of the force theorem holds in density functional calculations. The essential point is that the energy is at a variational minimum (or saddle point in generalized functionals) with respect to the density at the self-consistent solution. Thus changes in the density as a nucleus is moved do not contribute to the first-order derivatives. The result follows from the Hohenberg–Kohn expression for the total energy, Eq. (6.12), or any of the expressions in Section 7.3. Since the only terms that depend *explicitly* on the positions of the nuclei are the interaction E_{II} and the external potential, one immediately finds

$$\mathbf{F}_I = -\frac{\partial E}{\partial \mathbf{R}_I} = -\int d\mathbf{r}\, n(\mathbf{r}) \frac{\partial V_{\text{ext}}(\mathbf{r})}{\partial \mathbf{R}_I} - \frac{\partial E_{II}}{\partial \mathbf{R}_I}, \tag{7.39}$$

which is the "electrostatic theorem" for the forces due to Feynman [268] and given in Eq. (3.19). For nonlocal pseudopotentials, the force is only formally a function of the density; operationally it is defined in terms of the Kohn–Sham wavefunctions, with the general expression given in Eq. (3.20) and explicit plane wave expressions given in Eq. (13.3).

There are many possible alternative expressions for forces since any linear variation of the density can be added to Eq. (7.39) with no change in the result. The main point is very simple and is illustrated in Fig. I.1: the usual force theorem involves a nucleus moving relative to *all the electrons* as shown on the left-hand side of Fig. I.1. It is more appropriate in many actual calculations (especially ones involving core electrons) to move part of the density along with the nucleus as illustrated in the middle part of the figure. The resulting equations can be made very simple through clever choices, as described in Appendix I.

In actual calculations, there are two factors that can affect the use of the force theorem Eq. (7.39): (1) explicit dependence of the basis on the positions of the atoms, and (2) errors due to non-self-consistency. Both factors can be addressed by considering the nature of the terms omitted in going from Eqs. (3.18) to (3.19). The middle terms in Eq. (3.18) that involve variations of the wavefunctions can be written in the independent-particle case as

$$\mathbf{F}_I^{(2)} = -2\text{Re} \sum_i \int d\mathbf{r} \frac{\partial \psi_i^*}{\partial \mathbf{R}_I} \left[\frac{1}{2} \nabla^2 + V_{\text{KS}} - \varepsilon_i \right] \psi_i$$

$$- \int d\mathbf{r} \left[V_{\text{KS}} - V^{\text{in}} \right] \frac{\partial n}{\partial \mathbf{R}_I}, \tag{7.40}$$

where the term involving ε_i is due to the orthonormality constraint just as in the derivation of the Kohn–Sham equation. Here V_{KS} is defined to be the self-consistent Kohn–Sham potential for the given basis set, and V^{in} is the non-self-consistent input potential that leads to the wavefunctions ψ_i.

Since the ψ_i are the eigenstates of the hamiltonian with potential V^{in}, the first term in Eq. (7.40) is zero if the changes in the ψ_i maintain orthonormality when the atom is displaced. This happens in two cases: (1) if the basis is independent of the atom positions (as in plane waves), or (2) if the basis is complete. However, this term is nonzero if the basis is tied to the atoms (as in atom-centered orbitals) and the basis is incomplete. This contribution, often called the Pulay correction term [269], is straightforward – but often tedious – to include in a calculation. Only if it is included will the force be equal to the change in total energy per unit displacement. One of the great advantages of plane waves is that they are manifestly zero even if the basis is not complete.

The last term in Eq. (7.40) is the contribution due to the lack of self-consistency in the solution. This is a more serious concern for forces than for the energy since the energy is variational (errors are second order), whereas the force expression is not. Strategies can be devised for approximate inclusion of such terms at any stage in the self-consistency iterations, even though the final potential V_{KS} is not known. These methods are based on essentially the same logic as those for achieving self-consistency discussed in Section 7.4, where the goal is to find the optimum choice of potential at the next step.

Stress

Stress and strain are important concepts in characterizing the states of condensed matter; however, general expressions in terms of the ground-state wavefunction have been formulated only in the 1980s [102, 162]. There are a number of subtle issues and complications, so that a separate appendix, Appendix G, is devoted to the definition of stress and strain and to the resulting formulas that can be used in various applications.

The main results are that the stress tensor is the generalization of pressure to all the independent components of dilation and shear, and the "stress theorem" provides a way to calculate all components of the stress tensor from the ground-state wavefunction as a generalization of the virial theorem for pressure. In condensed matter, the state of the system is specified by the forces on each atom and by the macroscopic stress, which is an independent variable. The conditions for equilibrium are (1) the total force vanishes on each atom, and (2) the macroscopic stress equals the externally applied stress. This is well established in classical simulations [382] (e.g., the Parrinello–Rahman [383] and variable metric methods [384]) and is now an integral part of electronic structure calculations [385] in which one relaxes the structure by minimizing with respect to both the positions of the atoms in a unit cell and the size and shape of the cell.

7.6 Interpretation of the Exchange–Correlation Potential V_{xc}

The exchange–correlation potential $V_{xc}^{\sigma}(\mathbf{r})$ is the functional derivative of E_{xc}; it is *not* a potential that can be identified with interactions between particles, and it behaves in ways that seem paradoxical. The difference from an ordinary potential is brought out if it is expressed in terms of the exchange–correlation energy density $\epsilon_{xc}([n], \mathbf{r})$ defined in Eq. (8.1),

$$V_{xc}^{\sigma}(\mathbf{r}) = \epsilon_{xc}([n],\mathbf{r}) + n(\mathbf{r})\frac{\delta\epsilon_{xc}([n],\mathbf{r})}{\delta n(\mathbf{r},\sigma)}. \tag{7.41}$$

The second term (sometimes called the "response potential" [386]) is due to the *change* in the exchange–correlation hole with density. In an insulator, this derivative is discontinuous at a bandgap where the nature of the states changes discontinuously as a function of n. This leads to a "derivative discontinuity" whereby the Kohn–Sham potential for *all* the electrons in a crystal changes by a constant amount when a single electron is added [387, 388]. Thus even in the exact Kohn–Sham theory, the difference between the highest occupied and lowest unoccupied eigenvalues should *not* equal the actual bandgap. Similarly, there can be a shift in absolute energies of states of one molecule due to the presence of another molecule far away [389].

The behavior of the Kohn–Sham potential as a function of density seems paradoxical. How can adding one electron shift the potential for all the other electrons in a solid? The answer is in the definition of the functional and the behavior can be understood from examination of the kinetic energy. The great advance of the Kohn–Sham approach over the Thomas–Fermi approximation is the incorporation of orbitals to define the kinetic energy. In terms of orbitals, it is easy to see that the kinetic energy T_s for independent particles in Eq. (7.3) changes discontinuously in going from an occupied to an empty band, since the $\psi_i^{\sigma}(\mathbf{r})$ are different for different bands. In terms of the density this means the formal density functional $T_s[n]$ has discontinuous derivatives at densities that correspond to filled bands. This is a direct consequence of quantum mechanics and is not paradoxical; the real problem is that it is difficult to incorporate into an explicit density functional. It is likewise straightforward to see that the true exchange–correlation functional must change discontinuously. None of these properties is incorporated in any of the simple explicit functionals of the density, such as the local density or gradient approximations (Sections 8.3 and 8.5); however, they occur naturally (and are not paradoxical) in terms of orbital-dependent formulations in Chapter 9.

A different way to see the properties is to note that the Kohn–Sham potential V_{KS} is *defined* by the requirement that it yield the exact charge density. This is an exacting requirement that must be accomplished by the properties of V_{xc}, $V_{KS}^{\sigma}(\mathbf{r}) = V_{ext}(\mathbf{r}) + V_{Hartree}(\mathbf{r}) + V_{xc}^{\sigma}(\mathbf{r})$, where $V_{ext}(\mathbf{r})$ is known and $V_{Hartree}(\mathbf{r})$ is a simple explicit functional of the density. Thus one way to determine $V_{xc}^{\sigma}(\mathbf{r})$ is the requirement that $V_{KS}^{\sigma}(\mathbf{r})$ lead to the exact density. Conversely, the application of the Hohenberg–Kohn theorem to the Kohn–Sham noninteracting system implies that the exact density can be fit by only one $V_{xc}^{\sigma}(\mathbf{r})$, which is unique except for an additive constant.

7.7 Meaning of the Eigenvalues

It is often said that Kohn–Sham eigenvalues have no physical meaning. Indeed, the eigenvalues are *not* the energies to add or subtract electrons from the interacting many-body system. There is only one exception [390]: the highest eigenvalue in a finite system, which is minus the ionization energy, $-I$. The asymptotic long-range density of a bound system is governed by the occupied state with highest eigenvalue; since the density is assumed to be exact, so must the eigenvalue be exact. No other eigenvalue is guaranteed to be correct by the Kohn–Sham construction.

Nevertheless, the eigenvalues have a well-defined meaning within the theory and they can be used to construct physically meaningful quantities. One approach is the development of perturbation expressions for excitation energies starting from the Kohn–Sham eigenfunctions and eigenvalues. This can take the form of a functional [391] or it can be an operational definition, such as an explicit many-body calculation that uses the Kohn–Sham eigenfunctions and eigenvalues as input. The latter is actually done in quantum Monte Carlo and many-body perturbation approaches discussed in detail in [1]. For example, in fixed-node diffusion Monte Carlo, the resulting energies are determined by the nodes of the many-body trial function. If the trial function is taken to be a determinant made of Kohn–Sham orbitals, each result is operationally a functional of the Kohn–Sham potential.

Within the Kohn–Sham formalism itself, the eigenvalues have a definite mathematical meaning, often known as the Slater–Janak theorem [392]. The eigenvalue is the derivative of the total energy with respect to occupation of a state

$$\varepsilon_i = \frac{dE_{\text{total}}}{dn_i} = \int d\mathbf{r} \frac{dE_{\text{total}}}{dn(\mathbf{r})} \frac{dn(\mathbf{r})}{dn_i}. \tag{7.42}$$

For a noninteracting system this is trivial. However, for the Kohn–Sham problem it raises interesting points. The exchange–correlation energy is a functional of the density and the derivative of the potential terms in $dE_{\text{total}}/dn(\mathbf{r})$ in Eq. (7.42) is the effective potential $V_{\text{xc}}(\mathbf{r})$ in Eq. (7.41). As pointed out following that equation, $V_{\text{xc}}(\mathbf{r})$ contains a "response part" that is the derivative of $\epsilon_{\text{xc}}([n], \mathbf{r})$ with respect to $n(\mathbf{r})$. This can vary discontinuously between states giving rise to jumps in eigenvalues that are at first surprising. This is the well-known "bandgap discontinuity" [387, 388].

Thus it follows that for the critical problem of the gap in an insulator, the eigenvalues of the *ground state Kohn–Sham potential should not be the correct gap*, at least in principle. However, the magnitude of the discontinuity has not been established and there is active research especially using "optimized effective potentials" (Section 9.5) to clarify the issues regarding electron addition and removal energies.

7.8 Intricacies of Exact Kohn–Sham Theory

This section asks similar questions of Kohn–Sham theory as were asked in Section 6.6 of Hohenberg–Kohn density functional theory. In some cases the answers are the same and will be abbreviated here, but in other cases the difference in the answers is fundamental for understanding practical forms of density functional theory.

Allowed Densities for Electrons

Since the Hohenberg–Kohn theorems also apply to independent-particle problems, the reasoning of Section 6.6 shows the following:

- One can construct different wavefunctions ψ_i that have the same density $n(\mathbf{r})$.

- An antisymmetric wavefunction for fermions can describe any possible density ("*N*-representability") with some analyticity conditions.
- It is *not* possible to generate any reasonable density as the ground state of some local external potential ("*V*-representability"). One example is a linear combination of densities of a set of degenerate states. A second is the density corresponding to an excited state of a potential, which cannot be the ground state of another potential if it is required not to have singularities. (The example of a 2s state in H is discussed in Exercise 6.6).

The new question is this:

- For any ground-state density of an *interacting* electron system, is it possible to reproduce the density exactly as the ground-state density of a *noninteracting* electron system ("noninteracting *V*-representability")?

  The answer is not known. This is the Kohn–Sham *ansatz*, which is the basis for the entire industry, but it has never been proven in general. It is obviously true for the homogeneous gas; it can be demonstrated easily for any one- or two-electron problem (see Exercises 7.2 and 7.12); and it has been shown by Kohn and Sham [84] for small deviations from the homogeneous gas (Exercise 7.10); but to the knowledge of the author, there are no general proofs. Nevertheless, results of calculations appear very reasonable and detailed tests have shown that it is possible to fit the best numerical densities in many cases. We will follow the standard practice and proceed under the *assumption* that it is either valid or is good enough to be worth all this effort. The definition of "exact Kohn–Sham theory" followed here is that it is *exact – assuming that it exists.*

Properties Obeyed by "Exact Kohn–Sham Theory"

The Kohn–Sham approach places even heavier emphasis on the ground state than the Hohenberg–Kohn theorems. The only properties guaranteed to be correct by construction in the exact Kohn–Sham theory are the density and the energy. Thus questions arise as to what properties of a material should be given correctly by Kohn–Sham theory if the exchange–correlation functional were known exactly.

There are restrictions on the applicability of the Kohn–Sham approach. The questions and answers below assume that it is possible to represent any reasonable density even though it is not proven and that there is no magnetic field or spin–orbit interaction. Generalizations of the theory are considered in the following sections.

These are difficult questions. The answers given here are the opinions of the author with short explanations of the reasoning. *They are meant to encourage the reader to probe more deeply into the theory and answer the questions for him/herself.*

- Is the ground-state spin density correct in Kohn–Sham theory?

  Yes, in the spin-density theory where a spin-dependent effective potential is introduced specifically to give the correct density and spin density. Noncollinear spin functionals

(Section 8.3) allow the proper rotation invariance, which is broken in theories that fix only the z-component of the spin.

- Are static charge and spin susceptibilities given correctly by the ground-state functional? Yes. Static susceptibilities are second derivatives of ground-state energy with respect to external fields. Since the functional is valid for all potentials, the derivatives must be correct.[8]
- Is the macroscopic polarization in a crystal given correctly by the Kohn–Sham theory in terms of the density $n(\mathbf{r})$ in the bulk of the crystal?
  No. It is now understood that the static polarization can be determined by the *phases* of the wavefunctions (see Chapter 24), which may not be correct since the Kohn–Sham wavefunctions are auxiliary functions not necessarily meaningful.
- Is the exact Fermi surface of a metal given by eigenvalues in the exact Kohn–Sham theory?
  Not known to the author. In a metal, susceptibilities have Kohn anomalies at extremal spanning vectors of the Fermi surface, but it is not clear this uniquely determines the Fermi surface. According to [393] a local potential cannot describe at least some Fermi surfaces; however, it may be possible to describe in terms of functionals that are extremely nonlocal functionals of the density.[9]
- Must a Mott insulator – an insulator due to correlations among the electrons – be predicted correctly by the eigenvalues in the exact Kohn–Sham theory?
  The answer depends upon the meaning of "Mott insulator," which is used in various ways. If a Mott insulator means *any* system that becomes an insulator and violates the Luttinger theorem, there are arguments that it must have some type of quantum order, which may require a new density (or other field) to be introduced, perhaps analogous to the pair density for a superconductor described in the next section. If the question is restricted to systems like NiO, perhaps the initial material studied by Mott (see Chapter 1), these are just ordered states that should be given by the exact spin-density theory. This is discussed more fully in [1].
- Are excitation energies given correctly by the eigenvalues of the Kohn–Sham equations?
  No. The eigenvalues are not the true energies for adding or subtracting electrons, nor for neutral excitations (see Section 7.7). Even if the generalized Kohn–Sham approach (Chapter 9) provides new approaches, certainly entire continuous spectra cannot be described by eigenvalues.
- Is any excitation energy given correctly by an eigenvalue of the Kohn–Sham equations?
  Yes. The highest eigenvalue in a finite system must be correct [390] because that state dominates the long-range tail of the density, which is defined to be correct.
- Is the exact specific heat versus temperature given correctly by the exact finite temperature Mermin functional?

---

[8] The dielectric susceptibility is a special case and care must be taken to describe the electric polarization properly. There is a term outside the usual Kohn–Sham theory related to the following question and described in Chapter 24.

[9] This may be analogous to the problem of describing excitations in insulators in time-dependent density functional theory, which requires a nonlocal potential or a very nonlocal functional (Section 21.8).

Yes. Even though the specific heat involves excitations from the ground state, nevertheless the thermal averages over these excitations must be a unique functional of the density
and the temperature. However, it is more difficult to derive the exchange–correlation
functional as function of temperature.

- Is it possible to determine excitation energies by any means using the Kohn–Sham
  theory?
  Yes. This question is in the spirit of the Hohenberg–Kohn existence proofs. Since the
  Kohn–Sham density is exact by construction, it follows from the Hohenberg–Kohn
  theorems that *all properties are determined* since the entire hamiltonian is determined.
  Thus there should be some way to use the Kohn–Sham potential and eigenfunctions to
  determine all excitations exactly, but this requires a theory beyond the naive use of
  Kohn–Sham eigenvalues. One approach is to use the eigenstates as the basis for
  many-body calculations, which is literally done in configuration interaction, Monte
  Carlo, and many-body perturbation theory calculations such as the GW and BSE as
  described in [1] and other references. Other formulations bring certain excitations into
  the fold of the Kohn–Sham approach itself, most importantly, time-dependent Kohn–
  Sham theory.

7.9 Time-Dependent Density Functional Theory

The Kohn–Sham approach replaces the many-body problem with an independent-particle
problem, in which the effective potential depends on the density. As discussed in Section 7.7, the eigenvalues of the Kohn–Sham equations are independent-particle eigenvalues
that do not correspond to true electron removal or addition energies. Similarly, eigenvalue
differences do not correspond to excitation energies.

How can the Kohn–Sham approach properly describe excitations? The answer is to return
to the formulation in terms of the interacting density. In the full many-body problem,
excitations are most readily described in terms of the response functions, i.e., the response of
the system to external perturbations. The excitation energies in the response in Eq. (D.2) are
the exact many-body excitation energies. Following the analysis of frequency-dependent
dynamical response functions in Appendix D, the exact density response function has
poles as a function of frequency ω at the exact excitation energies. Therefore, the goal
is to construct a theory of the dynamical density response function within the Kohn–Sham
framework.

Such a theory exists: "time-dependent Kohn–Sham density functional theory" (TDDFT)
is a generalization of the original static Kohn–Sham method that was put on a firm
theoretical basis by Runge and Gross [246] and is described in more detail in Chapter 21.
The theory appears to be remarkably simple and straightforward, and it is now widely
used as a standard tool for calculation of spectra, especially in the chemistry community.
However, it has many subtle aspects; the apparent simplicity is because of the adiabatic
approximation that uses the standard ground-state functionals. There are fundamental
issues and active research to improve the functionals and go beyond this approximation,
as described, for example, in [247, 394, 395].

7.10 Other Generalizations of the Kohn–Sham Approach

The overarching guiding principle of the Kohn–Sham approach is the replacement of the full many-body problem with a simpler problem. In the usual Kohn–Sham theory of Eq. (7.1), the simpler problem is a system of noninteracting particles chosen to reproduce *only* the correct ground-state density and energy. In this framework, the eigenvalues and eigenfunctions do not correspond to actual excitations, except the highest eigenvalue of a localized system. However, this is *not* essential: the density is supposed to determine *everything*. A general approach for requiring that the auxiliary system reproduce the density and some other quantity has been outlined by Jansen [396]. Why not require that other properties of the Kohn–Sham system are equal to the exact values? For example, the ground-state energy and density and *also the bandgap*?

We have already used the extension that includes the spin density as well as number density, which serves as one model for other generalizations. An example is the density functional theory of superconductivity introduced in [397] and turned into a practical method in [398] and [399]. The essential ingredient is the addition of an anomalous pair density that is analogous to the spin density. The functional is constructed in terms of the electron–phonon interaction, the same as the well-known previous theories; nevertheless, there is an important consequence that the density functional naturally includes the effect of the so-called Coulomb interaction term $\mu*$, which is a parameter in previous methods. Another example is density polarization theory [350–352] pointed out in Chapter 24.

A much more sweeping modification is called "generalized Kohn–Sham theory" (GKS) which was formulated by Seidl et al. [400] and is the topic of much of Chapter 9. The original Kohn–Sham paper mentioned the possibility of an alternative approach in which the potential is not required to be local. The GKS approach can be cast in terms of functionals of the wavefunctions, with operators similar to the Hartree–Fock nonlocal exchange. This is a theoretical basis for hybrid functionals of the density and the wavefunctions, and other related functionals in Chapter 9, which provide much improved results for bandgaps and optical excitations.

SELECT FURTHER READING

For general references for DFT, see the list at the end of Chapter 6.

Applications of Kohn–Sham DFT in materials with an introduction to the theory include the following:

Giustino, F., *Materials Modelling Using Density Functional Theory: Properties and Predictions* (Oxford University Press, Oxford, 2014).

Kaxiras, E. and Joannopoulos, J. D. *Quantum Theory of Materials*, 2nd rev. ed. (Cambridge University Press, Cambridge, 2019).

Kohanoff, J., *Electronic Calculations for Solids and Molecules: Theory and Computational Methods* (Cambridge University Press, Cambridge, 2003).

For selected references on time-dependent DFT, see the list at end of Chapter 21.

Exercises

7.1 For any one-electron problem, one can readily determine whether or not any given density is a possible ground-state density. Using the known properties of solutions of the Schrödinger equation, give a sufficient set of conditions that any function must satisfy in order to guarantee that it is the ground-state density of some potential. See Exercise 7.7 for an example of an allowed density and Exercise 6.6 for a function that is not an allowed ground-state density.

7.2 For any density $n(\mathbf{r})$ that is allowed (see Exercise 7.1) and integrates to one electron, show that the Kohn–Sham potential $V_{\text{eff}}^{\sigma}(\mathbf{r})|_{\min} \equiv V_{\text{KS}}^{\sigma}(\mathbf{r})$ is unique except for an arbitrary constant, and give an explicit algorithm for constructing $V_{\text{KS}}^{\sigma}(\mathbf{r})$ from $n(\mathbf{r})$. See Exercise 7.7 for an explicit example of an allowed density.

7.3 Generalize the arguments of Exercise 7.2 to show that $V_{\text{KS}}^{\sigma}(\mathbf{r})$ is unique except for an arbitrary constant for a noninteracting Kohn–Sham system of any integer number of electrons.

7.4 For any noninteracting Kohn–Sham system, use the result of Exercise 7.3 to show that the kinetic energy T_s for each spin σ must be a unique functional of the density $n(\mathbf{r}, \sigma)$ for that spin. Generalize the argument to show that *all* properties of the system are uniquely determined by the density.

7.5 Based on the result of Exercise 7.4, show that in a finite system with discrete states the kinetic energy functional $T_s[n]$ must be a nonanalytic function of the density n with derivatives that are discontinuous at integer occupations. Hint: use the known solutions of the Schrödinger equation, ψ_i which are different for each i. Generalize this argument to all properties of the system and to filled bands in the case of a solids.

7.6 As an example of the fact that arbitrary densities *cannot* be constructed from the lowest eigenstates of a noninteracting hamiltonian, see Exercise 6.6. Use this example as the basis for constructing a general argument that it is not possible to construct any density from a determinant formed from the lowest N eigenvectors of a noninteracting particle problem.

7.7 As an example of the explicit construction of a potential determined by the density, find the one-dimensional potential $V(x)$ that gives the density $A \exp(-\alpha x^2)$, where normalization constant A is chosen so that the density corresponds to one electron. Express the answer in terms of α.

7.8 For a one-electron radial problem it is straightforward to find the unique Kohn–Sham potential that will lead to any radial density with no nodes. (The Schrödinger equation in radial coordinates is given in Section 10.1.)
(a) Find the potential $V_{\text{KS}}(r)$ that gives the hydrogen atom density.
(b) Find the potential for a gaussian density $A \exp(-\alpha r^2)$, where A is a normalization constant chosen so that the density integrates to one (see also Exercise 7.7.). Express the answer in terms of α.

7.9 This problem is an example of explicit construction of orthonormal independent-particle orbitals that describe *any* density of N particles and, furthermore, that there are many such choices for the same density. This example is for one dimension and is taken from p. 55 of [273]. For a density $n(x)$ and $s(x) \equiv n(x)/N$ given in the range $x_1 \leq x \leq x_2$, define the set of functions

$$\psi_k(x) = [s(x)]^{1/2} \exp[i2\pi k q(x)], \tag{7.43}$$

with $q(x) \equiv \int_{x_1}^{x} s(x')dx'$ and k = integers or half-integers. Show that the orbitals satisfy the desired conditions since each has the same density $s(x)$ and the orbitals are orthonormal. Show that it follows that an infinite number of such choices can be made.

7.10 Show that to lowest order, small deviations from the homogeneous density can be reproduced by noninteracting fermions. Hint: use the fact that to lowest order, any change in the density is linear in the potential.

7.11 Consider an independent-particle hamiltonian $\hat{H} = \hat{H}_{int} + V_{ext}$ for which the wavefunction for any state i is a single determinant Φ_i and the subscript "int" denotes all internal terms. Then the total energy can be written $E_{tot} = E_{int}[\Phi] + \int d^3\mathbf{r} V_{ext}(\mathbf{r}) n(\mathbf{r})$. Show that the external potential $V_{ext}(\mathbf{r})$ is determined to within a constant given $\hat{H}_{int}$ and any eigenfunction Φ_i, not only the ground state. (Hint: solve for $V_{ext}(\mathbf{r})$ using the Schrödinger equation.) Explain why it is more difficult numerically to find $V_{ext}(\mathbf{r})$ from the wavefunction for an excited state than for the ground state.

7.12 For a two-electron problem in a singlet state, it is straightforward to find the Kohn–Sham potential that will lead to any density with no nodes. The purpose of this exercise is to emphasize the relation to the one-electron case in Exercise 7.8 by constructing the potential $V_{KS}(r)$ for the following cases:
(a) A density that is twice that of the H atom
(b) A gaussian density $A \exp(-\alpha r^2)$, where A is chosen so that the density integrates to two electrons

7.13 Project: using an atomic program (such as the one discussed in conjunction with Chapter 10) one can find the density of a closed-shell atom and the Kohn–Sham potential.
(a) This exercise is to invert the problem: construct a minimization program to find the potential $V(r)$ that will produce that density and show that it is the same potential. This is essential for the potential to be unique.
(b) Now modify the density by multiplying by a gaussian and normalizing. For this density find the potential.

7.14 In actual calculations one can determine the energy from either of the two functionals Eqs. (7.21) or (7.22). Describe how it can be useful to compute both. Which is expected to be closest to the actual converged result before convergence is reached? Which is a true variational bound? Can the difference be used as a measure of convergence?

7.15 As posed before Eq. (7.24), derive the expressions for the linear terms and thus the form of Eq. (7.24).

7.16 Fill in the steps to show that Eq. (7.26) defines a functional that is indeed extremal at the correct solution for independent variations of potential and density.

7.17 On general thermodynamic grounds, show that $E(T)$ increases quadratically with T, whereas $F(T)$ decreases quadratically. Thus a linear combination of $E(T)$ and $F(T)$ can be chosen in which the quadratic terms cancel. Using the expressions for $E(T)$ and $F(T)$ that follow from the occupation numbers, find the value of α for which $\alpha E(T) + (1 - \alpha)F(T) = E(T = 0)$ with corrections $\propto T^4$.

7.18 Complete the arguments to show that Eq. (7.32) is extremal at the correct solution for independent variations of all the quantities: V^{in}, n^{in}, μ, T, and the form of the occupation function $f(\varepsilon)$.

7.19 Show that the form of the electronic entropy $\sum_i f_i \ln f_i + \sum_i (1 - f_i) \ln(1 - f_i)$ presented in Eq. (7.32) in fact follows from the general many body from in terms of the density matrix given by Mermin in Eq. (6.20).

7.20 Show that $\tilde{\chi}$ in Eq. (7.34) is given by

$$\tilde{\chi} + 1 = \frac{\delta n^{\text{out}}}{\delta n^{\text{in}}} = \frac{\delta n^{\text{out}}}{\delta V^{\text{in}}} \frac{\delta V^{\text{in}}}{\delta n^{\text{in}}}, \tag{7.44}$$

where $\delta n^{\text{out}}/\delta V^{\text{in}}$ is a response function defined to be χ^0 in Eq. (D.6) and the last term $\delta V^{\text{in}}/\delta n^{\text{in}}$ is K defined in Eq. (7.25). Thus the needed function $\tilde{\chi}$ can be calculated and is closely related to other uses of response functions.

7.21 Derive the constraint on the α parameter in the simple linear-mixing scheme in terms of the response function; i.e., that the iterations converge only if $\alpha < 2/\tilde{\chi}_{\text{max}}^{-1} = 2\tilde{\chi}_{\text{min}}$. See also Exercise 13.3.

7.22 Derive the two terms in the corrections to the force given in Eq. (7.40) for a self-consistent independent-particle method, starting from the general form, Eq. (3.18). The self-consistency adds the second term that is not present is the general case where the hamiltonian never changes. Hint: derive this term from the original definition of the force as a derivative of the total energy.

8

Functionals for Exchange and Correlation I

It is rain that grows flowers, not thunder.

Rumi

Summary

Density functional theory is the most widely used method today for electronic structure calculations because of the success of practical, approximate forms for the exchange–correlation energy as a functional of the density $E_{xc}[n]$. The first part of this chapter is devoted to the basic understanding of the way exchange and correlation affect the energy, which is determined by the exchange–correlation hole, i.e., the decrease in probability of finding electrons near one another due to the exclusion principle and repulsive Coulomb interaction. This is followed by two of the most widely used functionals, the local density approximation (LDA) and the semilocal generalized-gradient approximations (GGAs), and important features are illustrated by a few selected results on atoms and the H_2 molecule. Orbital-dependent functionals (hybrid, meta functionals of the kinetic energy, and other approaches) as well as other more advanced functionals and nonlocal van der Waals functionals are considered in Chapter 9.

8.1 Overview

The genius of the Kohn–Sham approach is the construction of an auxiliary system described by tractable independent-particle equations with all the difficulties of the full many-body interacting electron problem contained in the exchange–correlation energy $E_{xc}[n]$ that is a functional of the electron density. This is an exact formulation for the ground-state properties of the full interacting many-body system. However, the exact functional $E_{xc}[n]$ must be very complex and the Kohn–Sham construction would be only a footnote if that were the only result. The reason that density functional theory is by far the dominant method for practical calculations for materials is because it has proven to be possible to find approximations for exchange–correlation functional $E_{xc}[n]$ that are remarkably successful. This chapter and the following one are devoted to these approximations.

The central quantity that determines the exchange–correlation energy $E_{xc}[n]$ is the exchange–correlation hole described in general terms in Section 8.2. The "hole" denotes the decrease in the probability of finding another electron in the region around each electron, which is due to Pauli exclusion principle and correlation caused by the repulsive Coulomb interaction. For large classes of materials, a vast amount of experience has found that the resulting exchange–correlation energy can be reasonably approximated by a local functional (Section 8.3) and Section 8.4 is devoted to justification in terms of the exchange–correlation hole. The next step is the semilocal generalized gradient approximation in Section 8.5. The local and semilocal approximations are the first two steps up the ladder of functionals in Fig. 9.1.

On the other hand, there are important classes of materials and phenomena that cannot be described by local correlations, for example, van der Waals dispersion interactions, which are due to correlated fluctuations of dipoles on different atoms or molecules. Section 9.8 is an example of how such nonlocal behavior can be cast in terms of a functional of the density.

A natural progress in the development is functionals of the wavefunction in addition to the density. This requires a generalized form of the Kohn–Sham approach, which has proven to be very successful for describing single-particle excitations as well as the ground state, and in some cases they are derived in a systematic way that provides additional insights. This and other advanced developments are the topics of the following chapter.

8.2 E_{xc} and the Exchange–Correlation Hole

In the Kohn–Sham approach, the only quantities needed are the independent-particle kinetic energy, the Hartree energy and the exchange–correlation energy $E_{xc}[n]$ as a functional of the density. In order to carry out a DFT calculation all one needs is a computational method like those in Part IV and the formula for a functional. However, if one wants to understand the reasoning that has gone into the creation of a functional, or to create a new functional, then it is essential to understand the aspects of exchange and correlation that are needed. It is useful to express the energy E_{xc} in the form

$$E_{xc}[n] = \int d\mathbf{r}\, n(\mathbf{r})\epsilon_{xc}([n], \mathbf{r}), \tag{8.1}$$

where $\epsilon_{xc}([n], \mathbf{r})$ is an energy per electron at point $\mathbf{r}$ that depends on the density $n(\mathbf{r}, \sigma)$ in some neighborhood of point $\mathbf{r}$.[1] Since the Hartree energy includes the average Coulomb interactions, the potential energy of interaction per electron $\epsilon_{xc}([n], \mathbf{r})$ involves the joint probability function for each pair of electrons minus the Hartree term, as described in Section 3.7. The exchange term can be written as in Eq. (3.57) and there is an analogous term for correlation, so that the exchange–correlation interaction energy energy can be written

---

[1] For simplicity we consider systems with equal spin up and down. Spin-polarized systems are most straightforwardly treated by introducing two spin densities. To the knowledge of the author, there are no functionals that include the effect of spin–orbit interaction in exchange and correlation.

$$\epsilon_{xc}^{int}([n], \mathbf{r}) = \frac{1}{2} \int d^3 r' \frac{n_{xc}(\mathbf{r}, \mathbf{r}')}{|\mathbf{r} - \mathbf{r}'|}, \qquad (8.2)$$

where $n_{xc}(\mathbf{r}, \mathbf{r}')$ is the exchange–correlation hole around an electron at point $\mathbf{r}$, as described in Section 3.7 summed over parallel ($\sigma = \sigma'$) and antiparallel ($\sigma \neq \sigma'$) spins. There are several important aspects that can be seen at this point:

- Since the Coulomb interaction is isotropic only the spherical average of $n_{xc}(\mathbf{r}, \mathbf{r}')$ as a function of $\mathbf{r}'$ is relevant for the energy.
- $n_{xc}(\mathbf{r}, \mathbf{r}')$ is a sum of exchange and correlation terms. The exchange hole $n_x(\mathbf{r}, \mathbf{r}') \leq 0$ is never positive and it integrates to one missing electron. The correlation hole $n_c(\mathbf{r}, \mathbf{r}')$ integrates to zero since it only reflects the change in relative positions of electrons.
- This is *not* the entire exchange–correlation energy. As far as exchange is concerned this is the whole story, but correlation also changes the kinetic energy. The exchange–correlation functional must include the difference from the independent-particle kinetic energy, as brought out in Eq. (7.7).

An expression for $\epsilon_{xc}([n], \mathbf{r})$ including both interaction and kinetic terms can be found using the "coupling constant integration formula" described in the theoretical background, Section 3.4, which is also called an "adiabatic connection" [271]. In this case, the electronic charge is varied from zero (the noninteracting case) to the actual value (one in atomic units used here), with the added constraint that the density must be kept constant during this variation. Then all other terms remain constant and the change in energy is given by

$$E_{xc}[n] = \int_0^{e^2} d\lambda \langle \Psi_\lambda | \frac{dV_{int}}{d\lambda} | \Psi_\lambda \rangle - E_{Hartree} = \frac{1}{2} \int d^3 r n(\mathbf{r}) \int d^3 r' \frac{\bar{n}_{xc}(\mathbf{r}, \mathbf{r}')}{|\mathbf{r} - \mathbf{r}'|}, \qquad (8.3)$$

where $\bar{n}_{xc}(\mathbf{r}, \mathbf{r}')$ is the coupling-constant-averaged hole

$$\bar{n}_{xc}(\mathbf{r}, \mathbf{r}') = \int_0^1 d\lambda n_{xc}^\lambda(\mathbf{r}, \mathbf{r}'). \qquad (8.4)$$

Together with Eqs. (8.1), (8.3) shows that the exchange–correlation density $\epsilon_{xc}([n], \mathbf{r})$ can be written as

$$\epsilon_{xc}([n], \mathbf{r}) = \frac{1}{2} \int d^3 r' \frac{\bar{n}_{xc}(\mathbf{r}, \mathbf{r}')}{|\mathbf{r} - \mathbf{r}'|}. \qquad (8.5)$$

This is an important result that shows that the exact exchange–correlation energy can be understood in terms of the potential energy due to the exchange–correlation hole averaged over the interaction from $e^2 = 0$ to $e^2 = 1$. For $e^2 = 0$ the wavefunction is just the independent-particle Kohn–Sham wavefunction so that $n_{xc}^0(\mathbf{r}, \sigma, \mathbf{r}', \sigma') = n_x(\mathbf{r}, \sigma, \mathbf{r}', \sigma')$, where the exchange hole is known from Eq. (3.54). Since the density everywhere is required to remain constant as λ is varied, clearly $\epsilon_{xc}([n], \mathbf{r})$ is implicitly a functional of the density in all space. Thus $E_{xc}[n]$ can be considered as an interpolation between the exchange-only and the full correlated energies at the given density $n(\mathbf{r}, \sigma)$.

Analysis of the nature of the averaged hole $\bar{n}_{xc}(\mathbf{r},\mathbf{r}')$ is one of the primary approaches for developing improved approximations for $E_{xc}[n]$. In particular, the exchange–correlation hole obeys a sum rule that its integral must be unity, as shown in Section 3.7. The sum rule is satisfied for any case that is derived from an actual electron hamiltonian and it places constraints on any approximate forms that may be proposed [272]. This and other sum rules [401] are among the primary guidelines for systematic improvement of functionals.

8.3 Local (Spin) Density Approximation (LSDA)

Homogeneous Gas

The exchange–correlation hole in the homogeneous electron gas has been presented in Chapter 5. The results are relevant here because they present representative cases from weak to strong correlation and they are the basis for the local density approximation. In the independent-particle approximation there is no correlation; the hole is purely the exchange hole involving electrons of the same spin given by Eq. (5.19) and shown in Fig. 5.3. At full coupling strength the hole has been calculated by quantum Monte Carlo methods, with results shown in Fig. 5.5. The average hole is some mean between the two, which can also be found by an appropriate average of the holes from high density (where correlation is negligible) to the actual density. The key point is that Fig. 5.5 allows one to have a feeling for the radial shapes and the characteristic extent of the exchange–correlation hole.

Examples of exchange–correlation hole are also shown in Fig. 8.5 for two densities corresponding to the highest and lowest densities of the valence states in silicon.

Local (Spin) Density Approximation

Already in their seminal paper, Kohn and Sham pointed out that solids can often be considered as close to the limit of the homogeneous electron gas. In that limit, it is known that the effects of exchange and correlation are local in character, and they proposed making the local density approximation (LDA) (or more generally the local spin density approximation (LSDA)), in which the exchange–correlation energy is simply an integral over all space with the exchange–correlation energy density at each point assumed to be the same as in a homogeneous electron gas with that density,[2]

$$E_{xc}^{LSDA}[n^{\uparrow},n^{\downarrow}] = \int d^3 r\, n(\mathbf{r})\epsilon_{xc}^{hom}(n^{\uparrow}(\mathbf{r}),n^{\downarrow}(\mathbf{r}))$$

$$= \int d^3 r\, n(\mathbf{r})[\epsilon_x^{hom}(n^{\uparrow}(\mathbf{r}),n^{\downarrow}(\mathbf{r})) + \epsilon_c^{hom}(n^{\uparrow}(\mathbf{r}),n^{\downarrow}(\mathbf{r}))]. \qquad (8.6)$$

---

[2] The functional in Eq. (8.6) is defined for spin quantized along the same axis at every point in space, but this is not essential. In the local approximation, the same functional applies to a system with a noncollinear spin density where the axis rotates as a function of position [305] (see the footnote after Eq. (7.13)). In gradient approximations there are added terms in the functional.

The LSDA can be formulated in terms of either two spin densities $n^\uparrow(\mathbf{r})$ and $n^\downarrow(\mathbf{r})$, or the total density $n(\mathbf{r})$ and the fractional spin polarization $\zeta(\mathbf{r})$ defined in Eq. (5.16),

$$\zeta(\mathbf{r}) = \frac{n^\uparrow(\mathbf{r}) - n^\downarrow(\mathbf{r})}{n(\mathbf{r})}. \tag{8.7}$$

The LSDA is the most general local approximation and is given explicitly by Eqs. (5.17) and (5.18) for exchange and by approximate (or fitted) expressions given in Section 5.3 for correlation. For unpolarized systems the LDA is found simply by setting $n^\uparrow(\mathbf{r}) = n^\downarrow(\mathbf{r}) = n(\mathbf{r})/2$.

The local approximation is considered to be the first rung on the ladder in Fig. 9.1. The only information needed is the exchange–correlation energy of the homogeneous gas as a function of density. The exchange energy of the homogeneous gas is given by the simple analytic form in Eq. (5.15), and the correlation energy has been calculated to great accuracy with Monte Carlo methods [311]. Variations of exchange and correlation energies with density are discussed in Chapter 5 (where they are compared with insightful approximations), and explicit analytic forms fitted to the numerical results are given in Appendix B. As long as there are no further approximations in the calculations, the results of LDA and LSDA calculations can be considered as tests of the local approximation itself; the local approximation lives or dies depending on how the answers agree with experiment (or with in some cases many-body calculations that can be considered essentially exact).

8.4 How Can the Local Approximation Possibly Work As Well As It Does?

Much of this part of the book is devoted to improvements over the local approximation. But it is instructive to first consider the astounding successes of the local approximation! Since it is derived from the electron gas, how can it possibly be relevant for an atom that is nothing like an electron gas? How can it possibly produce such accurate results that lattice constants of crystals are predicted to within a few percent?

There are reasons that may help to explain the successes. One is that the hole obeys all the sum rules since it is the exact hole for some external potential, even if it is not the actual one [341]. Thus the hole satisfies constraints imposed by the sum rules, which are difficult to satisfy if one makes arbitrary approximations. Furthermore, the detailed shape of the hole need not be correct since only the spherical average of the xc hole enters the energy. However, such arguments must be backed up to draw anything more than just qualitative reasoning.

Atoms

Atoms are an extreme example of finite, inhomogeneous systems, often with wide bandgaps, that are very different from an infinite homogeneous gas. The one-electron problem, the hydrogen atom, is the worst in many ways. In this case there is no correlation and Hartree–Fock is exact for the ground-state energy with exact cancellation of the exchange and Hartree energies. This is a test of the LSDA where there is correlation like in

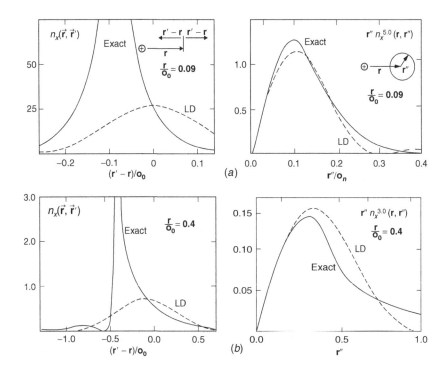

Figure 8.1. Exchange hole in an Ne atom. Left: $n(\mathbf{r}, \mathbf{r}')$ plotted for two values of $|\mathbf{r}|$ as a function of $|\mathbf{r}' - \mathbf{r}|$ along a line through the nucleus and compared to the local density approximation. The origin is centered on an electron a distance $|r|$. All quantities are in units of the Bohr radius, a_0. Right: the spherical average as a function of the relative distance, which shows the close resemblance to the local density approximation. From Gunnarsson et al. [402].

a homogeneous gas. The result shown in Table 9.1 is a small (fictitious) correlation energy that tends to correct the error in the exchange energy, leading to ≈7% error in the binding energy. A similar error is found for the other atoms in Table 9.1. The other functionals are much better, but given the crudeness of the LDA this level of accuracy is quite remarkable.

In order to get a physical picture of how the approximation can be so good, we need to look at the exchange–correlation hole that determines the energy. In an atom the hole associated with an electron depends on the electron position and is nonspherical if the electron is not at the atom center; however, for the energy, only the spherical average is needed. The consequences are illustrated in Fig. 8.1 for the exchange hole $n_x(\mathbf{r}, \mathbf{r}')$ in a neon atom taken from [402]. For the two representative cases shown at the left, the hole is extremely nonspherical, and yet the spherical average, shown on the right, is quite similar to the hole in the homogeneous electron gas with density equal to the local density at the point chosen. Thus even in this case the local approximation is remarkably good for the energy. Such agreement is evidence that the local approximation can be applied to many systems even if they are very inhomogeneous. In addition, it is a good sign for the success of improvements that involve gradients in Section 8.5 and other approaches in Chapter 9.

Two-Electron Problems: He and H_2

The neutral He atom and H_2 molecule are the simplest two-electron systems, which nevertheless exemplify issues related to many of the most important problems in condensed matter physics. The exchange and correlation energies for He are also given in Table 9.1. The good agreement for the LYP functional is not surprising since it was constructed using He as a starting point; however, the quality of the results is impressive for the other functionals.

The neutral H_2 molecule is a two-electron system for which there are essentially exact quantum Monte Carlo calculations [403]. As shown in Fig. 8.2 it is truly remarkable that the LDA is almost indistinguishable near the minimum and at shorter bond lengths. However, a close examination of the figure reveals a failure at large bond length. The curve for LSDA has a kink where the lowest-energy solution changes from the restricted spin singlet to the unrestricted broken symmetry solution with spin up on one site and spin down on the other. The reason is that at large distances there is a strong correlation between the electrons, with a greatly reduced probability of finding two electrons near the same atom at one time, compared to the probability of $1/4$, which would occur in the noninteracting case. The way this is accomplished in the DFT calculation is to break the symmetry, but this is unphysical

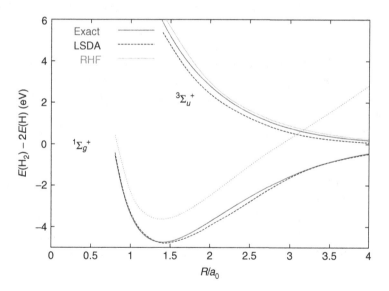

Figure 8.2. Energy versus separation R for an H_2 molecule, comparing LSDA (unrestricted) and Hartree–Fock (restricted RHF) with the exact energies from [403]. The two sets of curves are for the spin singlet (bonding) and triplet (antibonding) states. The most remarkable result is the accuracy of the LSDA near the minimum, whereas the Hartree–Fock curve is too high since it omits correlation. At large R the unrestricted LSDA has a broken symmetry solution that approaches the usual spin-polarized isolated-atom LSDA limit. The triplet Hartree–Fock energy approaches the exact isolated-atom limit, $E \equiv 0$, for large R, but the singlet approaches the wrong limit in the restricted approximation. Figure provided by O. E. Gunnarsson.

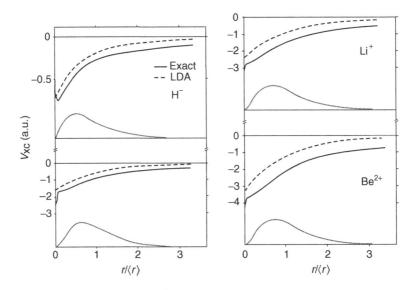

Figure 8.3. Exchange–correlation potential in two-electron ions, comparing the exact V_{XC} with the LDA. Each is derived from an essentially exact density. Note that the LDA potential is too high, leading to an eigenvalue that is also too high. This can be easily understood from the fact that the exact potential has an attractive $1/r$ form at long range that is missing in the LDA. From Almbladh and Pedroza [404].

since the correct solution is always a spin singlet. This is an example of an intrinsic failure of any method that assumes the spin of each electron is quantized along an axis up or down, not only the local approximation, and it can be fixed only by a method that properly treats a singlet wavefunction.

How do the eigenvalues compare with experimental removal energies? Since the highest eigenvalue is exact in an exact Kohn–Sham calculation, this is a test of approximate functionals. As illustrated in Fig. 8.3, there is a large effect in finite systems due to the long-range form of the potential [404]. The self-interaction that occurs in approximate functionals has the effect of adding a spurious repulsive term that raises the eigenvalues and makes states too weakly bound. Proper treatment of the nonlocal exchange eliminates this effect and makes the states more strongly bound. Similar consequences of nonlocal exchange are found in calculations on many-electron atoms as illustrated in Section 10.5.

Solids

There are very few quantitative calculations of the exchange and correlation holes in solids, but an informative example is shown in Figs. 8.4 and 8.5 for Si determined by a quantum Monte Carlo simulation with a chosen variational wavefunction [405]. The figures show separately the exchange and correlation holes, demonstrating the basic fact that the exchange dominates over correlation. It is so large because the main effect of the exchange term is to remove the spurious self-interaction in the Hartree interactions. There is a sum rule

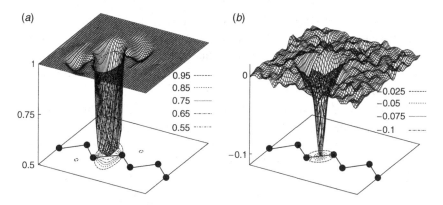

Figure 8.4. Exchange (a) and coupling-constant-averaged correlation hole (b) for an electron at the bond center in Si, calculated by a variational Monte Carlo method. Note the much smaller scale for the correlation hole. From Hood et al. [405].

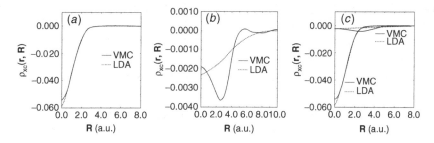

Figure 8.5. Spherical and coupling-constant-averaged exchange–correlation hole in Si, calculated as in Fig. 8.4, compared with the LDA approximation. Left: hole around an electron at the bond center. Middle: hole around an electron at the interstitial hole. Right: a comparison to scale. From Hood et al. [405].

that the exchange hole integrates to one, whereas the correlation hole integrates to zero since correlation only affects relative positions. Despite the fact that the hole varies greatly from high-density bond-center regions to low-density interstitial regions, the spherical average is given reasonably well by the local density approximation. However, the large difference in the interstitial region shown in Fig. 8.5 indicates possible sources of inaccuracies. Since the hole obeys a sum rule, the deepening at short range due to correlation must be offset by a decrease at large range, i.e., screening that effectively decreases the range of correlation.

8.5 Generalized-Gradient Approximations (GGAs)

The success of the LSDA has led to the development of various generalized-gradient approximations (GGAs), which are a step up to the second rung of the DFT ladder in Fig. 9.1. They lead to marked improvements over LSDA for many cases, and GGAs can now provide the accuracy required for density functional theory to be the standard for large classes of solids and molecules. In this section we briefly describe some of the physical ideas that are the foundation for construction of GGAs, and we focus especially on the PBE functional as a representative GGA functional that is widely used. For example, the PBE functional was chosen for benchmarking the codes most used in the condensed matter community [406].

The first step beyond the local approximation is a functional of the magnitude of the gradient of the density $|\nabla n^\sigma|$ as well as the value n at each point. Such a "gradient expansion approximation" (GEA) was suggested in the original paper of Kohn and Sham and carried out by Herman et al. [407] and others. The low-order expansion of the exchange and correlation energies is known [408]; however, the GEA does *not* lead to consistent improvement over the LSDA. It violates the sum rules and other relevant conditions [407] and, indeed, often leads to worse results. The basic problem is that gradients in real materials are so large that the expansion breaks down.

The term *generalized*-gradient expansion (GGA) denotes a variety of ways proposed for functions that modify the behavior at large gradients in such a way as to preserve desired properties. It is convenient [409] to define the functional as a generalized form of Eq. (8.6),

$$E_{xc}^{GGA}[n^\uparrow, n^\downarrow] = \int d^3 r n(\mathbf{r}) \epsilon_{xc}(n^\uparrow, n^\downarrow, |\nabla n^\uparrow|, |\nabla n^\downarrow|, \dots)$$

$$\equiv \int d^3 r n(\mathbf{r}) \epsilon_x^{hom}(n) F_{xc}(n^\uparrow, n^\downarrow, |\nabla n^\uparrow|, |\nabla n^\downarrow|, \dots), \qquad (8.8)$$

where F_{xc} is dimensionless and $\epsilon_x^{hom}(n)$ is the exchange energy of the unpolarized gas given in Table 5.3.

For exchange, it is straightforward to show (Exercise 8.1) that there is a "spin-scaling relation,"

$$E_x[n^\uparrow, n^\downarrow] = \frac{1}{2}\left[E_x[2n^\uparrow] + E_x[2n^\downarrow]\right], \qquad (8.9)$$

where $E_x[n]$ is the exchange energy for an unpolarized system of density $n(\mathbf{r})$. Thus for exchange we need to consider only the spin-unpolarized $F_x(n, |\nabla n|)$. It is natural to work in terms of dimensionless reduced-density gradients of mth order that can be defined by

$$s_m = \frac{|\nabla^m n|}{(2k_F)^m n} = \frac{|\nabla^m n|}{2^m (3\pi^2)^{m/3} (n)^{(1+m/3)}}. \qquad (8.10)$$

Since the Fermi momentum is given by $k_F = 3(2\pi/3)^{1/3} r_s^{-1}$, s_m is proportional to the mth-order variation in density normalized to the average distance between electrons r_s. The explicit expression for the first gradients can be written (Exercise 8.2)

$$s_1 \equiv s = \frac{|\nabla n|}{(2k_F)n} = \frac{|\nabla r_s|}{2(2\pi/3)^{1/3} r_s}. \qquad (8.11)$$

The lowest-order terms in the expansion of F_x have been calculated analytically [408, 409]

$$F_x = 1 + \frac{10}{81} s_1^2 + \frac{146}{2025} s_2^2 + \cdots. \qquad (8.12)$$

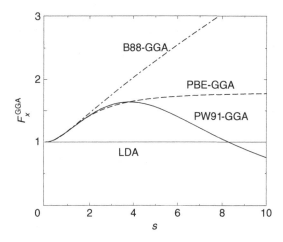

Figure 8.6. Exchange enhancement factor F_x as a function of the dimensionless density gradient s for various GGAs. (From H. Kim; similar to fig. 1 of [409] but for a larger range of s). Note that in the relevant range for most materials, $0 < s \lesssim 3$, the magnitude of the exchange is increased by a factor ≈ 1.3–1.6. (This provides an a posteriori reason why there was some success in using the constant factor $4/3$ in Slater's average local exchange.)

In Fig. 8.6 are shown the exchange enhancement factor $F_x(n,s)$ for three widely used functionals: Becke (B88) [410]; Perdew and Wang (PW91) [411]; and Perdew, Burke, and Enzerhof (PBE) [412]. As shown in the figure, one can divide the GGA into two regions: (i) small s ($0 < s < 3$) and (ii) large s ($s > 3$) regions. In region (i), which is relevant for most physical applications, different F_x have nearly identical shapes, which is the reason that different GGAs give similar improvement for many conventional systems with small-density gradient contributions. Most importantly, $F_x \geq 1$, so all the GGAs lead to an exchange energy lower than the LDA. Typically, there are more rapidly varying density regions in atoms than in condensed matter, which leads to greater lowering of the exchange energy in atoms than in molecules and solids. This results in the reduction of binding energy, correcting the LDA overbinding and improving agreement with experiment, which is one of the most important characteristics of GGAs [413].

Note that for s in the range up to ≈ 3, which is appropriate form many materials, the average enhancement is roughly $4/3$, making the average exchange similar to that proposed by Slater (see Section 9.5), although for very different reasons. Perhaps this accounts for the improvement that has often been found in calculations that use the factor $4/3$ or an adjustable factor called "$X\alpha$" which tends to be between 1 and $4/3$.

In region (ii), the different limiting behaviors of F_x result from choosing different physical conditions for $s \to \infty$. In B88, $F_x^{\text{B88}}(s) \sim s/\ln(s)$ was chosen to give the correct exchange energy density ($\epsilon_x \to -1/2r$) [410]. In PW91, choosing $F_x^{\text{PW91}}(s) \sim s^{-1/2}$ satisfies the Lieb–Oxford bound (see [412]) and the nonuniform scaling condition that must be satisfied if the functional is to have the proper limit for a thin layer or a line [411]. In PBE, the nonuniform scaling condition was dropped in favor of a simplified

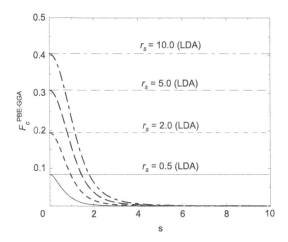

Figure 8.7. Correlation enhancement factor F_c as a function of the dimensionless density gradient s for the PBE functional. The actual form is given in Appendix B. Other functionals are qualitatively similar. (See caption of Fig. 8.6.)

parametcrization with $F_x^{PBE}(s) \sim$ const. [412]. The fact that different physical conditions lead to very different behaviors of F_x in region (ii) not only reflects the lack of knowledge of the large-density gradient regions but also an inherent difficulty of the density gradient expansion in this region: even if one form of GGA somehow gives the correct result for a certain physical property while others fail, it is not guaranteed that the form is superior for other properties in which different physical conditions prevail.

Correlation is more difficult to cast in terms of a functional, but its contribution to the total energy is typically much smaller than the exchange. The lowest-order gradient expansion at high density has been determined by Ma and Brueckner [414] (see [412]) to be

$$F_c = \frac{\epsilon_c^{LDA}(n)}{\epsilon_x^{LDA}(n)}(1 - 0.21951 s_1^2 + \cdots). \tag{8.13}$$

For large-density gradients the magnitude of correlation energy decreases and vanishes as $s_1 \to \infty$. This decrease can be qualitatively understood since large gradients are associated with strong confining potentials that increase level spacings and reduce the effect of interactions compared to the independent-electron terms. As an example of a GGA for correlation, Fig. 8.7 shows the correlation enhancement factor F_c^{PBE} for the PBE functional, which is almost identical to that for the PW91. The actual analytic form for the PBE correlation is given in Appendix B.

There are now many GGA functionals that are used in quantitative calculations, especially in chemistry (see, e.g., [413] and [415], which gives an overview and extensive assessment of 200 density functionals). A few examples can give a sense of how the functionals are constructed. In many cases correlation is treated using the Lee–Yang–Parr (LYP) [416] functional, which was derived from the orbital-dependent Colle–Salvetti functional [417]. That functional was in turn derived for the He atom and parameterized to fit atoms with more electrons. The BLYP functional is the combination of YLP for correlation and the Becke exchange functional [410], which was adjusted to approximate the Hartree–Fock exchange in closed-shell atoms. In a more theoretical approach, Krieger and coworkers [418] constructed a functional based on many-body calculations [419] of an artificial "jellium with a gap" problem that attempts to incorporate the effect of a gap into a functional. Many other functionals have parameters adjusted to fit to molecular data. Selected explicit forms can be found in [273, 413, 415, 420].

8.6 LDA and GGA Expressions for the Potential $V_{xc}^{\sigma}(r)$

The part of the Kohn–Sham potential due to exchange and correlation $V_{xc}^{\sigma}(\mathbf{r})$ is defined by the functional derivative in Eqs. (7.13) or (7.41). The potential can be expressed more directly for LDA and GGA functionals, Eqs. (8.6) and (8.8), since they are expressed in terms of functions (*not functionals*) of the local density of each spin $n(\mathbf{r}, \sigma)$ and its gradients at point $\mathbf{r}$. Explicit forms are given in Appendix B.

In the LDA, the form is very simple,

$$\delta E_{xc}[n] = \sum_{\sigma} \int d\mathbf{r} \left[\epsilon_{xc}^{hom} + n \frac{\partial \epsilon_{xc}^{hom}}{\partial n^{\sigma}} \right]_{\mathbf{r},\sigma} \delta n(\mathbf{r}, \sigma), \tag{8.14}$$

so that the potential,

$$V_{xc}^{\sigma}(\mathbf{r}) = \left[\epsilon_{xc}^{hom} + n \frac{\partial \epsilon_{xc}^{hom}}{\partial n^{\sigma}} \right]_{\mathbf{r},\sigma}, \tag{8.15}$$

involves only ordinary derivatives of $\epsilon_{xc}^{hom}(n^{\uparrow}, n^{\downarrow})$. Here the subscript $\mathbf{r}, \sigma$ means the quantities in square brackets are evaluated for $n^{\sigma} = n(\mathbf{r}, \sigma)$. The LDA exchange terms are particularly simple: since $\epsilon_x^{hom}(n^{\sigma})$ scales as $(n^{\sigma})^{1/3}$ it follows that

$$V_x^{\sigma}(\mathbf{r}) = \frac{4}{3} \epsilon_x^{hom}(n(\mathbf{r}, \sigma)). \tag{8.16}$$

The correlation potential depends on the form assumed, with selected examples given in Appendix B.

In the GGA one can identify the potential by finding the change $\delta E_{xc}[n]$ to linear order in δn and $\delta \nabla n = \nabla \delta n$,

$$\delta E_{xc}[n] = \sum_{\sigma} \int d\mathbf{r} \left[\epsilon_{xc} + n \frac{\partial \epsilon_{xc}}{\partial n^{\sigma}} + n \frac{\partial \epsilon_{xc}}{\partial \nabla n^{\sigma}} \nabla \right]_{\mathbf{r},\sigma} \delta n(\mathbf{r}, \sigma). \tag{8.17}$$

The term in square brackets might be considered to be the potential; however, it does not have the form of a local potential because of the last term, which is a differential operator.

There are three approaches to handling the last term. The first is to find a local $V_{xc}^{\sigma}(\mathbf{r})$ by partial integration (see Appendix A) of the last term in square brackets to give

$$V_{xc}^{\sigma}(\mathbf{r}) = \left[\epsilon_{xc} + n \frac{\partial \epsilon_{xc}}{\partial n^{\sigma}} - \nabla \left(n \frac{\partial \epsilon_{xc}}{\partial \nabla n^{\sigma}} \right) \right]_{\mathbf{r},\sigma}. \tag{8.18}$$

This form is commonly used; however, it has the disadvantage that it requires higher derivatives of the density that can lead to pathological potentials and numerical difficulties. This may happen near the nucleus where the density varies rapidly or in the outer regions of atoms where it is very small (see Exercise 8.3).

A second approach is to use the operator form Eq. (8.17) directly by modifying the Kohn–Sham equations [421]. Using the fact that the density can be written in terms of the wavefunctions ψ_i, the matrix elements of the operator can be written (for simplicity we omit the variables $\mathbf{r}$ and σ)

$$\langle \psi_j | \hat{V}_{xc} | \psi_i \rangle = \int \left[\tilde{V}_{xc} \psi_j^* \psi_i + \psi_j^* \mathbf{V}_{xc} \cdot \nabla \psi_i + (\mathbf{V}_{xc} \cdot \nabla \psi_j^*) \psi_i \right], \qquad (8.19)$$

where $\tilde{V}_{xc} = \epsilon_{xc} + n(\partial \epsilon_{xc}/\partial n)$ and $\mathbf{V}_{xc} = n(\partial \epsilon_{xc}/\partial \nabla n)$. This form is numerically more stable; however, it requires inclusion of the additional vector operator in the Kohn–Sham equation, which may significantly increase the computational cost; for example, in plane wave approaches four Fourier transforms are required instead of one.

Finally, a different approach proposed by White and Bird [422] is to treat E_{xc} strictly as a function of the density; the gradient terms are *defined* by an operational definition in terms of the density. Then Eq. (8.17) can be written using the chain rule as

$$\delta E_{xc}[n] = \sum_{\sigma} \int d\mathbf{r} \left[\epsilon_{xc} + n \frac{\partial \epsilon_{xc}}{\partial n^{\sigma}} \right]_{\mathbf{r},\sigma} \delta n(\mathbf{r}, \sigma)$$

$$+ \sum_{\sigma} \int \int d\mathbf{r} d\mathbf{r}' n(\mathbf{r}) \left[\frac{\partial \epsilon_{xc}}{\partial \nabla n^{\sigma}} \right]_{\mathbf{r},\sigma} \frac{\delta \nabla n(\mathbf{r}')}{\delta n(\mathbf{r})} \delta n(\mathbf{r}, \sigma), \qquad (8.20)$$

where $(\delta \nabla n(\mathbf{r}')/\delta n(\mathbf{r}))$ denotes a functional derivative (which is independent of spin). For example, on a grid, the density for either spin is given only at grid points $n(\mathbf{r}_m)$ and the gradient at $\nabla n(\mathbf{r}_m)$ is determined by the density by a formula of the form

$$\nabla n(\mathbf{r}_m) = \sum_{m'} \mathbf{C}_{m-m'} n(\mathbf{r}_{m'}), \qquad (8.21)$$

so that

$$\frac{\delta \nabla n(\mathbf{r}_m)}{\delta n(\mathbf{r}_{m'})} \rightarrow \frac{\partial \nabla n(\mathbf{r}_m)}{\partial n(\mathbf{r}_{m'})} = \mathbf{C}_{m-m'}. \qquad (8.22)$$

(Note that each $\mathbf{C}_{m''} = \{C_{m''}^x, C_{m''}^y, C_{m''}^z\}$ is a vector in the space coordinates.) In a finite difference method, the coefficients $\mathbf{C}_{m''}$ are nonzero for some finite range; in a Fourier transform method, the $\mathbf{C}_{m''}$ follow simply by noting that

$$\nabla n(\mathbf{r}_m) = \sum_{\mathbf{G}} i \mathbf{G} n(\mathbf{G}) e^{i\mathbf{G}\cdot\mathbf{r}_m} = \frac{1}{N} \sum_{\mathbf{G},m'} i \mathbf{G} e^{i\mathbf{G}\cdot(\mathbf{r}_m - \mathbf{r}_{m'})} n(\mathbf{r}_{m'}). \qquad (8.23)$$

Finally, varying $n(\mathbf{r}_m, \sigma)$ in the expression for E_{xc} and using the chain rule leads to

$$V_{xc}^{\sigma}(\mathbf{r}_m) = \left[\epsilon_{xc} + n \frac{\partial \epsilon_{xc}}{\partial n} \right]_{\mathbf{r}_m, \sigma} + \sum_{m'} \left[n \frac{\partial \epsilon_{xc}}{\partial |\nabla n|} \frac{\nabla n}{|\nabla n|} \right]_{\mathbf{r}_{m'}, \sigma} \mathbf{C}_{m'-m}. \qquad (8.24)$$

This form reduces the numerical problems associated with Eq. (8.18) without a vector operator as in Eq. (8.19). Note that $V_{xc}^{\sigma}(\mathbf{r}_m)$ is a nonlocal function of $n(\mathbf{r}_{m'}, \sigma)$, the form of which depends on the way the derivative is calculated. This is an advantage in actual calculations because it ensures consistency between E_{xc} and V_{xc}. The method can be extended to other bases by specifying the derivative in the appropriate basis.

8.7 Average and Weighted Density Formulations: ADA and WDA

An approach to the generalization of the local density approximation proposed by Gunnarsson et al. [402] is to construct a nonlocal functional that depends on the density in some region around each point $\mathbf{r}$. The original proposals were designed to provide a natural extension of the local functional in a way that satisfies the sum rules. This led to two approaches, the average density approximation (ADA) and the weighted density approximation (WDA) [402]. In the ADA, the exchange–correlation hole Eq. (8.4) and energy Eq. (8.5) are approximated by the corresponding quantity for a homogeneous gas of average density $\bar{n}^{\sigma}$ instead of the local density $n(\mathbf{r}, \sigma)$. This leads to

$$E_{xc}^{ADA}[n^{\uparrow}, n^{\downarrow}] = \int d^3 r\, n(\mathbf{r}) \epsilon_{xc}^{hom}(\bar{n}^{\uparrow}(\mathbf{r}), \bar{n}^{\downarrow}(\mathbf{r})), \qquad (8.25)$$

where

$$\bar{n}(\mathbf{r}) = \int d^3 r'\; w(\bar{n}(\mathbf{r}); |\mathbf{r} - \mathbf{r}'|)\, n(\mathbf{r}') \qquad (8.26)$$

is a nonlocal functional of the density for each spin separately. The weight function w can be chosen in several ways. Gunnarsson et al. [402] originally proposed a form based on the linear response of the homogeneous electron gas and given in tabular form. The WDA is related but differs in the way the weighting is defined.

Tests have shown that there are advantages of the ADA and WDA, but there have not been extensive studies. A clear superiority over the ordinary LDA and GGAs is that in the limit where a three-dimensional system approaches two dimensional (e.g., in a confined electron gas in semiconductor quantum wells) the nonlocal functionals are well behaved whereas most of the LDA and GGA functionals diverge [423] (see Exercise 8.4). On the other hand, the ADA and WDA functionals suffer from the serious difficulty that core electrons distort the weighting in an unphysical way, so that any reasonable weighting must involve some shell decomposition to separate the effects on core and valence electrons.

8.8 Functionals Fitted to Databases

There are many other functionals that cannot be described here. In particular, functionals fitted to databases are extensively used in chemistry. It is not feasible here to cover the large number of such functionals even if they are very useful. An example of widely used functionals is the M06 suite developed by Zhao and Truhlar [424] and the performance of many other functionals can be found in the review by Mardirossian and Head-Gordon [415]. As a general rule, one might expect that functionals fit to a database work better for similar materials than functionals derived from theoretical arguments and idealized systems like the homogeneous gas. However, they may not work as well when applied to other materials that are significantly different and may not have as much predictive power for wide ranges of problems.

A promising area for future developments is the use of "machine learning" techniques, which may provide ways to discover new functionals by finding patterns that are not apparent and not suggested by traditional chemical and physical intuition. An early paper in the field is [425]; more recent work can be found by searching for the citations to this paper.

SELECT FURTHER READING

See the list at the end of Chapter 6 for general references on density functional theory.

Overviews and reviews:

Casida, M. E., in *Recent Developments and Applications of Density Functional Theory*, edited by J. M. Seminario (Elsevier, Amsterdam, 1996), p. 391.

Cohen, A. J., Mori-Snchez, P., and Yang, W., "Challenges for density functional Theory" *Chem. Rev.* 112:289–320, 2012.

Koch, W. and Holthausen, M. C., *A Chemist's Guide to Density Functional Theory* (Wiley-VCH, Weinheim, 2001).

Mardirossian, N. and Head-Gordon, M., "Thirty years of density functional theory in computational chemistry: An overview and extensive assessment of 200 density functionals," *Mol. Phys.* 115:2315–2372, 2017.

In addition to an extensive assessment of many functionals, the authors describe their approach "to use a combinatorial design strategy that can be loosely characterised as 'survival of the most transferable.'"

Perdew, J. P. and Burke, K., "Comparison shopping for a gradient-corrected density functional," *Int. J. Quant. Chem.* 57:309–319, 1996.

Towler, M. D., Zupan, A., and Causa, M., "Density functional theory in periodic systems using local gaussian basis sets," *Comp. Phys. Commun.* 98:181–205, 1996. (Summarizes explicit formulas for functionals.)

Exercises

8.1 Derive the "spin-scaling relation" Eq. (8.9). From this it follows that in the homogeneous gas, one needs only the exchange in the unpolarized case.

8.2 (a) Show that the expression for the dimensionless gradients $s_1 = s$ in Eq. (8.10) can be written in terms of r_s as Eq. (8.11).
(b) Find the form of the second gradient s_2 in terms of r_s.

8.3 Use the known form of the density near a nucleus to analyze the final term in Eq. (8.18) near the nucleus. Show that the term involves higher-order derivatives of the density that are singular at the nucleus.

(a) Argue that such a potential is unphysical using the facts that the exact form of the exchange potential is known and correlation is negligible compared to the divergent nuclear potential.

(b) Show that, nevertheless, the result for the total energy is correct since it is just a transformation of the equations.

(c) Finally, discuss how the singularity can lead to numerical difficulties in actual calculations.

8.4 Show that if a three-dimensional system is compressed in one direction so that the electrons are confined to a region that approaches a two-dimensional plane, the density diverges and the LDA expression for the exchange energy approaches negative infinity. Show that this is unphysical and that the exchange energy should approach a finite value that depends on the area density. Argue that this is not necessarily the case for a GGA but that the unphysical behavior can be avoided only by stringent conditions on the form of the GGA.

8.5 Problem on a diatomic molecule that demonstrates the breaking of symmetry in mean-field solutions such as LSDA.

(a) Prove that the lowest state is a singlet for two electrons in any local potential.

(b) Show this explicitly for the two-site Hubbard model with two electrons.

(c) Carry out the unrestricted HF calculation for the two-site Hubbard model with two electrons. Show that for large U the lowest-energy state has broken symmetry.

(d) Computational exercise (using available codes for DFT calculations): Carry out the same set of calculations for the hydrogen molecule in the LSDA. Show that the lowest-energy state changes from the correct symmetric singlet to a broken symmetry state as the atoms are pulled apart.

(e) Explain why the unrestricted solution has broken symmetry in parts (c) and (d), and discuss the extent to which it represents correct aspects of the physics even though the symmetry is not correct.

(f) Explain how to form a state with proper symmetry using the solutions of (c) and (d) and a sum of determinants.

8.6 Compute the exact exchange potential as a function of radius r in an H atom using the exact wavefunction in the ground state. This can be done with the formulas in Chapter 10 and numerical integration. Compare with the LDA approximation for the exchange potential using the exact density and Expression (8.16) (note the system is full-spin polarized). Show the comparison explicitly by plotting the potentials as a function of radius. Justify the different functional forms of the potentials at large radius in the two cases.

8.7 The hydrogen atom is also a test case for correlation functionals; of course, correlation should be zero in a one-electron problem. Calculate the correlation potential using the approximate forms given in Appendix B (or the simpler Wigner interpolation form). Is the result close to zero? Does the correlation potential tend to cancel the errors in the local exchange approximation?

9

Functionals for Exchange and Correlation II

When it rains it pours.

Summary

The previous chapter is devoted to local density and generalized gradient approximations, which are successful in many ways and are still the bread and butter of many applications to materials. However, this is not the end of the story. It is the start of ongoing research to use physical insights and theory to construct improved functionals that can treat other properties, such as van der Waals dispersion interactions. Excitations can be treated within the framework of generalized Kohn–Sham methods in which the exchange–correlation functional depends on the wavefunctions and response functions, including hybrid functionals with nonlocal exchange, "meta" functionals that involve the kinetic energy density, and SIC and DFU+U methods designed to treat localized d and f orbitals with strong interactions. A theme throughout this book is to consider a set of representative functionals and how they perform for different types of materials and properties. At the end of the chapter is a comparison for atoms and a list of pointers to other places in the book that illustrate applications to solids.

9.1 Beyond the Local Density and Generalized Gradient Approximations

There is much activity and a great variety of approaches to develop new functionals that go beyond the local density and generalized gradient approximations. The efforts are fueled by the fact that there is an ever-increasing payoff for development of functionals that are more accurate for important classes of materials and make possible application to new systems. Because there is no one approach to development of functionals, this has led to a plethora of methods with increasingly lengthy acronyms as more features are incorporated, as described in reviews such as [415, 426, 427]. Many of the functionals have been created by fitting to large databases of molecules and these are widely used, especially in chemistry. It is not feasible to cover all the developments; here the focus is on a representative set of functionals that are based on theoretical considerations or using information from model systems.

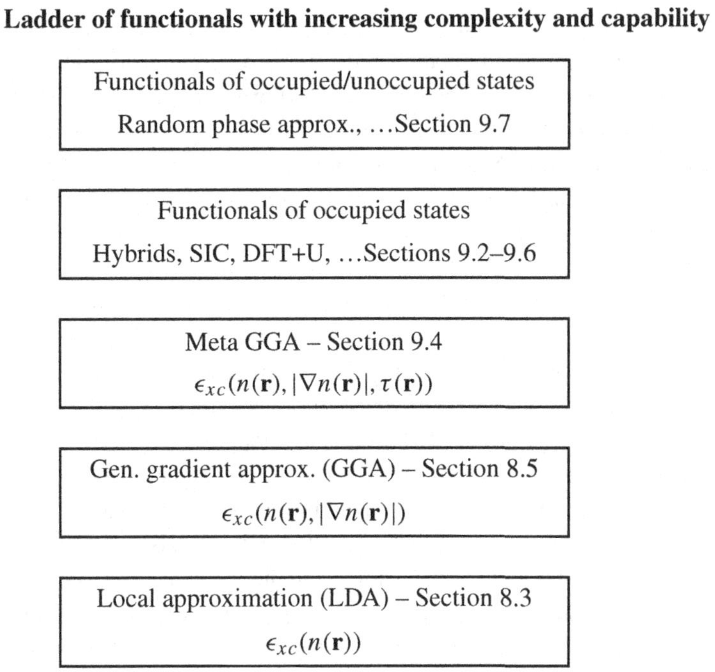

Ladder of functionals with increasing complexity and capability

Functionals of occupied/unoccupied states

Random phase approx., ...Section 9.7

Functionals of occupied states

Hybrids, SIC, DFT+U, ...Sections 9.2–9.6

Meta GGA – Section 9.4

$\epsilon_{xc}(n(\mathbf{r}), |\nabla n(\mathbf{r})|, \tau(\mathbf{r}))$

Gen. gradient approx. (GGA) – Section 8.5

$\epsilon_{xc}(n(\mathbf{r}), |\nabla n(\mathbf{r})|)$

Local approximation (LDA) – Section 8.3

$\epsilon_{xc}(n(\mathbf{r}))$

Figure 9.1. From bottom to top: classes of functionals with increasing complexity, with the goal of increased capability and accuracy that offset the increased computational requirements. They are arranged as rungs on a ladder that has been termed a "Jacob's ladder" after the biblical story of a ladder reaching to heaven, i.e., extending toward the exact formulation. The van der Waals functionals in Section 9.8 can be placed at various levels. They are derived using many of the features in the more advanced functionals, but they have been formulated as functionals only of the density.

The progression of functionals has been organized on the "density functional ladder" or "Jacob's ladder" [428] that categorizes the classes of functionals described in this chapter, as illustrated in Fig. 9.1. This helps to avoid being overwhelmed by the hundreds of choices by developing a logical, instructive path with typical functionals for each rung of the ladder. The higher rungs can be cast in terms of "generalized Kohn–Sham theory" that provides new avenues for approaches to the problems. Development of functionals is an ongoing field of research and it is not appropriate to try to describe any method in detail. The aim is to bring out the central features in a way that will still be useful as new ideas emerge in the future.

9.2 Generalized Kohn–Sham and Bandgaps

The great advance of the Kohn–Sham approach over Thomas–Fermi-type methods is that the kinetic energy Eq. (7.3) is explicitly expressed as a functional of the independent-particle orbitals ψ_i. It is implicitly a functional of the density, since the orbitals are determined by the potential $V_{KS}^\sigma(\mathbf{r})$, which in turn is a functional of the density $n^\sigma(\mathbf{r})$; however, the functional dependence on n must be highly nontrivial, nonanalytic, and nonlocal. In particular, derivatives of the Kohn–Sham kinetic energy dT_s/dn are discontinuous functions of n at densities corresponding to filled shells. This is the way in which shell structure occurs in the Kohn–Sham approach, whereas it is missing in Thomas–Fermi-type approximations. An intrinsic part of the Kohn–Sham construction in Section 7.1 is that $V_{KS}^\sigma(\mathbf{r})$ is a *local* potential that is a function only of the position and spin; otherwise there cannot be a one-to-one relation of the potential and the density $n^\sigma(\mathbf{r})$.

There is another approach called "generalized Kohn–Sham theory" (GKS) in which the potential can be a nonlocal operator. The action of an operator can be expressed in a basis so that it acts differently on different wavefunctions, i.e., it is "orbital dependent." An example is the Hartree–Fock equations in Eqs. (3.45)–(3.47), where the action of the exchange term operator $\hat{V}_x$ on a wavefunction at point $\mathbf{r}$ depends on an integral over that same wavefunction over all space. The idea of a nonlocal Hartree–Fock-like exchange term is already in the original Kohn–Sham paper, but it was only later that Seidl and coworkers [400] proposed that a family of possible auxiliary systems could be defined with the only requirement that the wavefunction be a single determinant of single-body orbitals $\psi_i(\mathbf{r})$. The Kohn–Sham system of noninteracting particles is one choice; another is the Hartree–Fock functional of the orbitals ψ_i that incudes interactions in an average sense; and there are many possible choices of systems that include interactions in some way. Thus there are whole ranges of auxiliary systems that in principle lead to the exact density and ground-state energy. To the knowledge of the author there are no formal proofs that the generalized theory is exact for other properties; however, it provides a framework for addressing other properties.

Why would one want to construct a theory that appears to be more difficult? The most direct answer is that it provides a greater range of possibilities. It is potentially more accurate since it can incorporate various physical properties in a form that goes beyond Kohn–Sham and yet is feasible to calculate. The deeper reason is that in the Kohn–Sham approach the E_{xc} functional has to take into account all the effects of exchange and correlation. If some of the effects can be taken into account by a different auxiliary system, then it may be helpful in constructing the functional. It turns out that for many problems, the added difficulty in solving the equations is worthwhile because the functionals are more accurate and take into account features that are very difficult to build into a functional of only the density.

In many ways GKS is a natural progression: the improvement of Kohn–Sham over Thomas-Fermi was to introduce an orbital-dependent kinetic energy, and the next step is to improve exchange–correlation functionals by expressing E_{xc} explicitly in terms of the independent-particle orbitals ψ_i. For example, the true E_{xc} functional must have discontinuities at filled shells [387, 388, 429], which is essential for a correct description of energy gaps. The "bandgap problem" results from the use of the eigenvalues, even though it is fundamentally unjustified. The generalized theory provides the framework to properly consider the eigenvalues as meaningful, even if they are approximate. There are solid theoretical arguments that a generalized theory can take into account physical effects which are missing in the original Kohn–Sham approach and which can include at least part of the discontinuity using practical functionals [400]. The many-author paper by Perdew et al.

[430] provides an instructive overview of gaps in a generalized theory and the foundations of GKS are discussed in [427, 431].

There are two primary types of functionals of the wavefunction placed on the third and fourth rungs of the ladder in Fig. 9.1. Hybrid functionals on the fourth rung are discussed first because they involve concepts that are well established – a combination of Hartree–Fock and DFT – and are very important in a large fraction of present-day calculations. Calculations using hybrids can match (or improve) the success of LDA and GGAs for ground state properties and also lead to better estimates for the bands, bandgaps, and excitations. Meta-GGA denotes functionals of the kinetic energy density as well as the density. They are on the third rung of the ladder because they involve only a single added function calculated from information already present in a Kohn–Sham DFT calculation. However, meta-GGAs have not been used as much and are not as, well tested as hybrids, and they are presented later in Section 9.4.

9.3 Hybrid Functionals and Range Separation

Hybrid functionals are a combination of explicit density-dependent and orbital-dependent Hartree–Fock (or Hartree–Fock-like) functionals.[1] These are often the method of choice in the chemistry community (see, e.g., [413, 415, 426]) because they lead to marked improvements over semilocal functionals and they were readily incorporated in existing codes since Hartree–Fock and DFT are each well-developed methods for molecules. For solids, Hartree–Fock calculations are more difficult, and the use of hybrids was very limited until efficient methods were developed. More important, the Hartree–Fock approximation is problematic for extended solids where it has disastrous consequences for metals, as discussed in Chapter 5. On the other hand, in insulators it leads to quantitative improvements in many cases and the long-range interaction is essential for excitons (see Chapter 21). One approach is range separation in which a Hartree–Fock-like exchange is included in different ways for the short- and long-range parts of the Coulomb interaction. Here the goal is to bring out the basic ideas and give representative examples; more details and in-depth analysis can be found in review articles such as [415, 426, 427] and sources listed at the ends of this chapter and Chapter 8.

The most striking improvements made possible by hybrid functionals are bandgaps and excitation energies. If the eigenvalues are treated as energies for adding or removing electrons, standard density-dependent functionals like LDA and GGA tend to greatly underestimate gaps whereas Hartree–Fock tends to give gaps that are much too large. An example is the bands of Ge shown in Fig. 2.23 where the LDA leads to a metal and Hartree–Fock to an insulator with a gap much too large, whereas the hybrid is in much better agreement with experiment. Figure 2.24 shows the improvement for a range of materials.

---

[1] The arguments in this section relate to the exchange. What about correlation? The reason for so much emphasis on exchange is that it is usually the larger effect and the approximations are often more important quantitatively. However, there must ultimately be a reckoning with the difficult problem of correlation!

There are several approaches to assign the fraction α of the Hartree–Fock and $1 - \alpha$ of the Kohn–Sham components of the hybrid, which broadly can be classified as fitting to experimental data versus theoretical considerations. One approach to justify the choice is to use the coupling constant integration formulation in Eq. (8.3), along with information at the end points and the dependence as a function of the coupling constant λ. In particular, at $\lambda = 0$ the energy of the uncorrelated wavefunction is the Hartree–Fock exchange energy, which is easily expressed in terms of the exchange hole that can be calculated from the orbitals (the fourth term in Eq. (3.44)). Becke [432] argued that the potential part of the LDA or GGA functional is most appropriate at full coupling $\lambda = 1$ and suggested that the integral Eq. (8.3) can be approximated by assuming a linear dependence on λ, which leads to the "half-and-half" form that is the average of the density and Hartree-Fock functionals. A closer look at the variation as a function of λ led Perdew, Ernzerhof, and Burke [433] to conclude it is nonlinear and they proposed the form

$$E_{xc}^{PBE0} = E_{xc}^{PBE} + \frac{1}{4}(E_x^{HF} - E_x^{PBE}), \qquad (9.1)$$

with $\alpha = 1/4$ mixing of Hartree–Fock exchange and with correlation energy given by the density functional unchanged. Their reasoning applies to any functional and the "1/4" hybrid with the PBE exchange–correlation functional, called PBE0 [412], is widely used.

There are many parameterized forms for hybrid functionals (see, e.g., [415]). An example is "B3LYP," which uses the Becke B88 exchange functional [410] and the LYP correlation [416],

$$E_{xc} = E_{xc}^{LDA} + a_0(E_x^{HF} - E_x^{LDA}) + a_x(E_x^{B88} - E_x^{LDA}) + a_c(E_c^{LYP} - E_c^{LDA}), \qquad (9.2)$$

with coefficients $a_0 = 0.2$, $a_x = 0.72$, $a_c = 0.81$. These were empirically adjusted to fit atomic and molecular data, but it turns out that the values are close to $a_x = 1 - a_0, a_c = 1.0$ in which case Eq. (9.2) reduces to the much simpler form in Eq. (9.1) with a_0 that is in a range around 1/4 depending on which GGA functional is used.

Variation of the Long-Range Exchange Hybrid Functionals

There are several motivations to go beyond the form of hybrids that use a fixed fraction of Hartree–Fock exchange. (See also related arguments for range separation.) As it stands, any functional with nonzero α cannot be used for a metal. The long-range part of the exchange must be eliminated completely since it gives unphysical results at the Fermi surface, as described in Section 5.2. In addition, it has been observed that larger values of α lead to improved gaps for materials with large gaps and smaller values of α for ones with small gaps. Furthermore, it seems physically reasonable that there should be larger effects of the exchange in wider gap materials. There have been a number of proposals to make α inversely proportional to the dielectric constant. Thus it vanishes for a metal and is small in materials like Si where $\epsilon_0 = 12$ but is large in materials like LiF where $\epsilon_0 = 1.9$, with results that are described in Chapter 21. In [434] can be found physical reasoning supported by many-body perturbation theory based on works over the decades on screening the long-range fields in solids. A PBE0-like form with $\alpha = 1/\epsilon$ was proposed as a nonempirical "dielectric-dependent hybrid" (DDH) functional in which ϵ is calculated self-consistently was proposed in [244], which provided extensive tests. The DDH functional is used in calculations for water as shown in Fig. 2.15 and bandgaps in Fig. 2.24.

Range-Separated Hybrid Functionals

Range separation denotes schemes in which the exchange energy is separated into contributions due to short- and long-range parts of the Coulomb interaction. Instead of scaling the Hartree–Fock exchange by the dielectric constant, a different approach is based on the idea that screening is dependent on the distance with little screening at short distance and reaching the macroscopic value at large distance. This idea is used in various functionals [415], and we consider the HSE [435–437] and related approaches that are most used in condensed matter applications. A convenient division into short- and long-range parts, adopted in the HSE functional and other works, is the same as used in the Ewald transformation in Section F.2,

$$\frac{1}{r} = \frac{1 - \mathrm{erf}(\eta r)}{r} + \frac{\mathrm{erf}(\eta r)}{r}, \tag{9.3}$$

where erf is the error function and η (called ω in [435]) is the range parameter. The various ways to choose a functional are generalizations of expressions like Eq. (9.1) to apply separately to the short- and long-range parts,

$$E_{\mathrm{xc}} = E_{\mathrm{xc}}^{\mathrm{DFT}} + a(E_x^{\mathrm{HF,SR}} - E_x^{\mathrm{DFT,SR}}) + b(E_x^{\mathrm{HF,LR}} - E_x^{\mathrm{DFT,LR}}) \tag{9.4}$$

where DFT stands for any of the local or semilocal density functionals such as PBE.

Short-Range-Only Functionals

For solids a widely used choice is the HSE functional [437] in which the long-range Hartree–Fock is completely eliminated ($b = 0$ in Eq. (9.4)),

$$E_{\mathrm{xc}} = E_{\mathrm{xc}}^{\mathrm{PBE}} + \frac{1}{4}(E_x^{\mathrm{HF,SR}} - E_x^{\mathrm{PBE,SR}}), \tag{9.5}$$

where a is chosen to be $1/4$ the same as Eq. (9.1). This form has the great advantage that it can be applied to metals, and it reduces to the PBE GGA for $\eta \to \infty$ and the PBE0 hybrid for $\eta \to 0$.

The range-separation length η is a parameter chosen to be $0.11/a_0$ in HSE06 in order to provide the best description of a range of materials and properties. HSE06 is one of the characteristic functionals selected in this book, with results shown for many examples: lattice constants and bulk moduli in Table 2.1; phase transition pressures in Fig. 2.4; bandgaps in Fig. 2.24 and various examples; atoms in Table 9.1; and optical spectra in Chapter 21. The choice $\eta = 0.11/a_0$ works well for semiconductors and many other cases, but it is found to be less accurate for wider bandgap insulators. Indeed on physics grounds one expects that effects are more localized in wide-gap materials, i.e., larger η in Eq. (9.3).

Range Separation with Screened Long-Range Exchange

There are fundamental reasons to include some degree of the long-range part of the exchange in insulators, in addition to the fact that hybrid functionals are very successful for quantitative calculations in molecules and insulating solids. Optical spectra calculated using time-dependent perturbation theory are very successful and widely used for molecules. Even though short-range forms like HSE seem to work well for materials like semiconductors, they can never describe excitons in a solid! It is only if there is some fraction of long-range exchange that the functionals can produce a bound state below the gap, as illustrated in Fig. 21.5 for LiF.

Nonempirical Hybrid Functionals

At this point we have range-separated functionals with two types of parameters that characterize the range and the magnitude of the long-range exchange respectively. There are good physical arguments for the variation of each one for different materials. Does this mean choosing a different functional for each material? If the functional is fitted for each material, this would be an enormous step backward from the goal of a universal functional for all systems. It would go against the history of density functional theory, where approximate functionals have been developed and their quality judged by the degree to which a single function could apply to many materials.

One approach to take advantage of the positive features of the range-separated functionals is to develop the theory in such a way that the parameters η and α are predicted by the theory itself, i.e., an internally consistent approach with no empirical input. An example is the dielectric-dependent hybrid (DDH) functional [244]) with long-range exchange reduced by the dielectric function ϵ, which is calculated self-consistently. A related approach involving the dielectric function with no free parameters is in [438]. Such functionals can be viewed as adding a very nonlocal functional of the density – a single parameter α that depends on a global property of the system. (Note, however, that there are issues for systems with different dielectric functions in different regions, e.g., an interface between different materials, a surface or a molecule in vacuum.) Another example is the van der Waals functional in [439] (see Section 9.8), which uses well-known formulas in terms of polarizabilities that are finally expressed in terms of the density determined in the calculation.

In this context the term "tuned" has been introduced as a general approach [440] to finding such functionals who gave a specific approach to finding η using the ionization potential and the fact that in an exact Kohn–Sham theory it is equal to the highest occupied eigenvalue. This has the benefit of dealing with the perplexing problems of charge transfer between weakly coupled systems that depend on the relative eigenvalues. However, there is a difficulty in a solid where the ionization potential is not a well-defined intrinsic property.

9.4 Functionals of the Kinetic Energy Density: Meta-GGAs

The third rung of the DFT ladder in Fig. 9.1 is functionals of the kinetic energy density $t(\mathbf{r}) = n(\mathbf{r})\tau(\mathbf{r})$ of noninteracting particles as well as the particle density,

$$E_{xc}^{GGA\,\tau} = \int d^3 r n(\mathbf{r}) \epsilon_{xc}(n^\uparrow, n^\downarrow, |\nabla n^\uparrow|, |\nabla n^\downarrow|, \tau^\uparrow, \tau^\downarrow), \tag{9.6}$$

where each term is evaluated at position $\mathbf{r}$. The kinetic energy density $t(\mathbf{r})$ is defined in Section H.1, and the problem is that it is not unique. Two possibilities are given in Eq. (H.8), which can be written

$$t^{\sigma(1)}(\mathbf{r}) = -\frac{1}{2} \sum_{i=1}^{N} \psi_i^{\sigma*}(\mathbf{r}) \nabla^2 \psi_i^\sigma(\mathbf{r}) \quad \text{or} \quad t^{\sigma(2)}(\mathbf{r}) = \frac{1}{2} \sum_{i=1}^{N} |\nabla \psi_i^\sigma(\mathbf{r})|^2, \tag{9.7}$$

where spin σ is indicated explicitly for clarity. It might seem that there is a fundamental problem in using $\tau^\sigma(\mathbf{r})$ to construct a functional since there is not a unique expression. However, in Section H.1 it is shown that it can be considered to be a sum of two terms $t^\sigma(\mathbf{r}) = t_n^\sigma(\mathbf{r}) + t_x^\sigma(\mathbf{r})$ (see Eq. (H.9)) where $t_x^\sigma(\mathbf{r}) = n^\sigma(\mathbf{r})\tau_x^\sigma(\mathbf{r})$ is a well-defined, unique positive-definite function, and all the nonuniqueness is in the term $t_n^\sigma(\mathbf{r})$ that depends solely on the density $n^\sigma(\mathbf{r})$ and its derivatives at point $\mathbf{r}$. Either form for $t(\mathbf{r})$ in Eq. (9.7) can be used to define the unique function $\tau_x^\sigma(\mathbf{r})$; here we give explicit expressions using the second form (see [415] for works that use the other choice),

$$\tau_x^\sigma(\mathbf{r}) = \frac{1}{2n^\sigma(\mathbf{r})} \sum_{i=1}^{N} |\nabla \psi_i^\sigma(\mathbf{r})|^2 - \frac{1}{8} \left[\frac{\nabla n^\sigma(\mathbf{r})}{n^\sigma(\mathbf{r})} \right]^2 \equiv \tau^\sigma(\mathbf{r}) - \tau_W^\sigma(\mathbf{r}), \tag{9.8}$$

where we have dropped the superscript (2) to define $\tau^\sigma(\mathbf{r})$ and $\tau_W^\sigma(\mathbf{r})$ is the Weizsacker expression [330] for the kinetic energy (see also Eq. (H.15)), which has been used in many contexts as a correction to the Thomas–Fermi approximation.

There are good reasons [441] to expect $\tau_x^\sigma(\mathbf{r})$ to be an appropriate variable with which to express the functional. As summarized in Section H.1, $\tau_x^\sigma(\mathbf{r})$ has a clear physical meaning, the curvature of the exchange hole, which can be shown to be the relative kinetic energy of pairs of electrons. In fact, the properties of the exchange hole including its curvature have been used in creating the GGA functionals [410–412, 442] in Section 8.5, based on the physical reasoning that the large gradient regime is cut off by the extent of the exchange hole. In addition, $\tau_x^\sigma(\mathbf{r})$ is the basis for the electron localization function (ELF) defined in Eq. (H.21) which is used to analyze the electronic structure.

We have not completely escaped from the problem of nonuniqueness since there are still two possibilities for the part of the kinetic energy denoted $\tau_n^\sigma(\mathbf{r})$, the Weizsacker form or one in terms of second derivatives as shown in Eq. (H.14). However, the key point is that either expression for $\tau_n^\sigma(\mathbf{r})$ involves only the density $n^\sigma(\mathbf{r})$ and its derivatives. This is the reason that functionals of the kinetic energy should also involve gradients of the the density (hence the appellation "meta-GGA") and the reason that the combined functional can be physically meaningful.

A meta-GGA functional is an example of an orbital-dependent functional that leads to a generalized Kohn–Sham method like other orbital-dependent functionals (see Section 9.2). However, it depends explicitly only on local variables and it leads to a differential operator in the Kohn–Sham equations in contrast to Hartree–Fock where the exchange is a nonlocal integral operator (see Eq. (3.45)). The Kohn–Sham equations can be derived by variation of the functional with respect the wavefunctions, which leads an operator with the form [243]

$$\hat{V}_{xc,\sigma}^{\tau}(\mathbf{r})\psi_i^{\sigma}(\mathbf{r}) = \frac{1}{2}\nabla \cdot \left[\frac{\delta E_{xc}}{\delta \tau^{\sigma}}(\mathbf{r})\nabla \psi_i^{\sigma}(\mathbf{r})\right]. \tag{9.9}$$

This term alters the form of the Kohn–Sham equations, but it can be included in ways similar to Eq. (8.19) since it is a derivative operator. One can generate a local optimized effective potential as in Section 9.5, but the resulting equations are more difficult than using the operator in Eq. (9.9).

The SCAN Meta-GGA Functional

Functionals involving the kinetic energy density have been derived in the 1980s, such as [442], and there are more recent developments that were stimulated by the approach in [443]. Here we describe the functional [444] denoted by the acronym SCAN, which stands for "strongly constrained and appropriately normed" where "constrained" means that it is required to obey certain conditions (including ones previously applied to GGAs) and "appropriate norms" are systems that are argued to be appropriate to use in constructing the functional (for example, the homogeneous gas is used to define the LDA). The SCAN functional is constructed in terms of the dimensionless variable $\alpha(\mathbf{r}) = \tau_x^{\sigma}(\mathbf{r})/\tau^{unif}(n(\mathbf{r}))$, where $\tau^{unif}(n(\mathbf{r}))$ is the kinetic energy density of the electron gas with density $n(\mathbf{r})$. The form of exchange energy as a function of the density, gradients and α was determined by the requirement that it satisfies 17 different constraints and it fits as well as possible information from several systems (see the supplementary material for [444]): (a) uniform and slowly varying densities, (b) jellium surface energy, (c) H atom, (d) the He atom and the limit of large atomic number for the rare-gas atoms, plus compressed Ar_2, and (e) the $Z \to \infty$ limit of the two-electron ion. This list may seem arbitrary, but the first two are the same as used for LDA and GGA functionals, the H atom is a one-electron systems already used in a previous kinetic energy functional [443], and the others add new information on inhomogeneous systems with strong gradients in different geometries. Since there is no information about bonded systems in the construction of the functional, the application to molecules and crystals is a test of the predictive power of the functional. This is in contrast to functionals that are fitted to data sets of molecules, which are expected to perform very well for similar systems but may not be as good when applied to different systems.

The SCAN functional has been tested in a number of applications including bandgaps of solids in Fig. 2.24, lattice constants in Table 2.1, phase transitions under pressure in Fig. 2.4, simulations of water in Fig. 2.15, and atoms in Table 9.1. These are a sample of much more extensive tests for atoms and molecules in the original paper [444], equilibrium structures [112], phase transitions under pressure [128], simulations of water [195] and applications in combination with a van der Waals functional [445].

9.5 Optimized Effective Potential

The generalized Kohn–Sham theory in Section 9.2 leads to a nonlocal potential. Is there a way to formulate the original Kohn–Sham theory not with a local potential but with an orbital-dependent functional $E_{xc}[\{\psi_i\}]$? In fact there is a long history of such methods that predates the work of Kohn and Sham, apparently first formulated in a short paper [446] by Sharp and Horton in 1953 as the problem of finding "that potential, the same for all electrons, such that when ... given a small variation, the energy of the system remains stationary." This approach has come to be known as the optimized effective potential (OEP) method [447–449]. The key point is that if one considers orbitals ψ_i that are determined by a potential through the usual independent-particle Schrödinger equation, then it is straightforward, in principle, to define the energy functional of the potential V,

$$E_{OEP}[V] = E[\{\psi_i[V]\}]. \tag{9.10}$$

The OEP method is fully within the Kohn–Sham approach since it is just the optimization of the potential V that appears in the very first Kohn–Sham equation (7.1). Furthermore, as emphasized in Section 7.3, the usual Kohn–Sham expressions are operationally functionals of the potential; the OEP is merely an orbital formulation of the general idea. The OEP method has been applied primarily to the Hartree–Fock exchange functional, which is straightforward to write in terms of the orbitals (the fourth term in Eq. (3.44)), which is called "exact exchange" or "EXX." However, the OEP approach is more general and applicable to orbital-dependent kinetic energy and correlation functionals as well.

The variational equation representing the minimization of energy Eq. (9.10) can be written using the density formalism as an intermediate step. Since the potential V acts equally on all orbitals, it follows that

$$V_{xc}^{\sigma,OEP}(\mathbf{r}) = \frac{\delta E_{xc}^{OEP}}{\delta n^{\sigma}(\mathbf{r})}, \tag{9.11}$$

which can be written (see Exercise 9.2) using the chain rule [251, 449] as

$$V_{xc}^{\sigma,OEP}(\mathbf{r}) = \sum_{\sigma'} \sum_{i=1}^{N^{\sigma'}} \int d\mathbf{r}' \frac{\delta E_{xc}^{OEP}}{\delta \psi_i^{\sigma'}(\mathbf{r}')} \frac{\delta \psi_i^{\sigma'}(\mathbf{r}')}{\delta n^{\sigma}(\mathbf{r})} + c.c. \tag{9.12}$$

$$= \sum_{\sigma'} \sum_{i=1}^{N^{\sigma'}} \int d\mathbf{r}' \int d\mathbf{r}'' \left[\frac{\delta E_{xc}^{OEP}}{\delta \psi_i^{\sigma'}(\mathbf{r}')} \frac{\delta \psi_i^{\sigma'}(\mathbf{r}')}{\delta V^{\sigma',KS}(\mathbf{r}'')} + c.c. \right] \frac{\delta V^{\sigma',KS}(\mathbf{r}'')}{\delta n^{\sigma}(\mathbf{r})},$$

where $V^{\sigma',KS}$ is the total potential in the independent-particle Kohn–Sham equations that determine the $\psi_i^{\sigma'}$. Each term has a clear meaning and can be evaluated from well-known expressions:

- The first term is an orbital-dependent nonlocal (NL) operator that can be written

$$\frac{\delta E_{xc}^{OEP}}{\delta \psi_i^{\sigma'}(\mathbf{r})} \equiv V_{i,xc}^{\sigma',NL}(\mathbf{r}')\psi_i^{\sigma'}(\mathbf{r}'). \tag{9.13}$$

For example, in the exchange-only approximation, $V_{i,xc}^{\sigma',NL}(\mathbf{r}')$ is the orbital-dependent Hartree–Fock exchange operator Eq. (3.48).

- The second term can be evaluated by perturbation theory,[2]

$$\frac{\delta \psi_i^{\sigma'}(\mathbf{r}')}{\delta V^{\sigma',KS}(\mathbf{r}'')} = G_{i,0}^{\sigma'}(\mathbf{r}',\mathbf{r}'')\psi_i^{\sigma'}(\mathbf{r}''), \tag{9.14}$$

where the Green's function for the noninteracting Kohn–Sham system is given by (see Eq. (D.6), which is written here with spin explicitly indicated)

$$G_{i,0}^{\sigma}(\mathbf{r},\mathbf{r}') = \sum_{\substack{j \neq i}}^{\infty} \frac{\psi_j^{\sigma}(\mathbf{r})\psi_j^{\sigma*}(\mathbf{r}')}{\varepsilon_{\sigma i} - \varepsilon_{\sigma j}}. \tag{9.15}$$

- The last term is the inverse of a response function χ_0 given by

$$\chi_0^{\sigma,KS}(\mathbf{r},\mathbf{r}') = \frac{\delta n^{\sigma}(\mathbf{r})}{\delta V^{\sigma',KS}(\mathbf{r}'')} = \sum_{i=1}^{N^{\sigma}} \psi_i^{\sigma*}(\mathbf{r})G_{i,0}^{\sigma}(\mathbf{r},\mathbf{r}')\psi_i^{\sigma}(\mathbf{r}'), \tag{9.16}$$

where we have used a chain rule and the fact that n is given by the sum of squares of orbitals, Eq. (7.2).

The integral form of the OEP equations (see Exercise 9.2) can be found by multiplying Eq. (9.13) by $\chi_0^{\sigma}(\mathbf{r},\mathbf{r}')$ and integrating:

$$\sum_{i=1}^{N^{\sigma}} \int d\mathbf{r}' \psi_i^{\sigma*}(\mathbf{r}') \left[V_{xc}^{\sigma,OEP}(\mathbf{r}') - V_{i,xc}^{\sigma,NL}(\mathbf{r}') \right] G_{i,0}^{\sigma}(\mathbf{r}',\mathbf{r})\psi_i^{\sigma}(\mathbf{r}) + c.c. = 0. \tag{9.17}$$

This form shows the physical interpretation that $V_{xc}^{\sigma,OEP}(\mathbf{r})$ is a particular weighted average of the nonlocal orbital-dependent potentials.

The integral form is the basis for useful approximations for which the potential can be given explicitly, e.g., as proposed by Krieger, Li, and Iafrate (KLI) [449–452]. Although KLI gave a more complete derivation, a heuristic derivation [446, 449, 450] is to replace the energy denominator in the Green's function by a constant $\Delta\varepsilon$. Then the value of $\Delta\varepsilon$ drops out of Eqs. (9.17) and (9.15) becomes

$$G_{i,0}^{\sigma}(\mathbf{r},\mathbf{r}') \rightarrow \sum_{\substack{j \neq i}}^{\infty} \frac{\psi_j^{\sigma}(\mathbf{r})\psi_j^{\sigma*}(\mathbf{r}')}{\Delta\varepsilon} = \frac{\delta(\mathbf{r}-\mathbf{r}') - \psi_i^{\sigma}(\mathbf{r})\psi_i^{\sigma*}(\mathbf{r}')}{\Delta\varepsilon}. \tag{9.18}$$

---

[2] Note that G_0 and the derivative in Eq. (9.14) are diagonal in spin since they involve the noninteracting Kohn–Sham system.

As discussed in Exercise 9.5, the KLI approximation leads to the simple form

$$V_{xc}^{\sigma,KLI}(\mathbf{r}) = V_{xc}^{\sigma,S}(\mathbf{r}) + \sum_{i=1}^{N^\sigma} \frac{n_i^\sigma(\mathbf{r})}{n^\sigma(\mathbf{r})} \left[\bar{V}_{i,xc}^{\sigma,KLI} - \bar{V}_{i,xc}^{\sigma,NL} \right], \tag{9.19}$$

where $V_{xc}^{\sigma,S}(\mathbf{r})$ is the density-weighted average proposed by Slater [453]

$$V_{xc}^{\sigma,S}(\mathbf{r}) = V_{xc}^{\sigma}(\mathbf{r}) + \sum_{i=1}^{N^\sigma} \frac{n_i^\sigma(\mathbf{r})}{n^\sigma(\mathbf{r})} \bar{V}_{i,xc}^{\sigma,NL}, \tag{9.20}$$

and the $\bar{V}$ are expectation values

$$\bar{V}_{i,xc}^{\sigma,KLI} = \langle \psi_i^\sigma | V_{xc}^{\sigma,KLI} | \psi_i^\sigma \rangle \quad \text{and} \quad \bar{V}_{i,xc}^{\sigma,NL} = \langle \psi_i^\sigma | V_{i,xc}^{\sigma,NL} | \psi_i^\sigma \rangle. \tag{9.21}$$

Finally, by taking matrix elements of Eq. (9.19), the equations become a set of linear equations for the matrix elements $\bar{V}_{i,xc}^{\sigma,KLI}$, which can be solved readily. The KLI approximation, including only exchange, has been shown to be quite accurate in many cases [449].

Slater Local Approximation for Exchange

An interesting detour is the difference between the Kohn–Sham formula for the exchange potential Eq. (8.16) and the local form Slater had proposed earlier [453] based on his approach of finding a local potential that is a weighted average of the nonlocal Hartree–Fock exchange operators Eq. (9.20). By averaging the *exchange potential* of the homogeneous gas, Slater found $V_x = 2\epsilon_x$, rather than the factor 4/3 in Eq. (8.16) found by Kohn and Sham from the derivative of the exchange energy. In the context of the nonlocal exchange energy functional, it is not immediately clear which is the better approximation to carry over to an inhomogeneous system. Only recently has this issue been resolved [448, 451] by careful treatment of the reference for the zero of energy in transferring the potential from the gas to an inhomogeneous system and the second factor in Eq. (9.19). The result is the Kohn–Sham form Eq. (8.16).

It has been observed that the Slater local approximation for exchange often gives eigenvalues in better agreement with experiment than does the Kohn–Sham form. This has led to the "Xα" approximation with an adjustable constant. In hindsight, this can be justified in part by the fact that gradient corrections lead to typical increases of similar magnitude, as shown in Fig. 8.6.

9.6 Localized-Orbital Approaches: SIC and DFT+U

Many of the most interesting problems in condensed-matter physics involve materials in which the electrons tend to be localized and strongly interacting, such as transition metal oxides and rare earth elements and compounds with partially occupied d and f states.[3] Often the usual functionals, such as the LDA and GGAs, find the system to be a metal whereas it is in fact a magnetic insulator. Even if the ground state is correct, the Kohn–Sham bands may be very far from the true spectra. Various methods have been developed to extend the functional approach to treat localized d and f orbitals differently from the more diffuse s and p band-like states. A fundamental problem is that there is not a unique way to identify local orbitals in an extended system. Nevertheless, there are often very reasonable choices and in many cases the results do not depend on the details. The challenge is to use the methods in ways that bring insight and understanding – and often quantitative results – for interesting problems!

Self-Interaction Correction (SIC)

SIC denotes methods that add corrections to attempt to correct for the unphysical self-interaction in the Hartree potential. This interaction of an electron with itself in the Hartree interaction is canceled in the exact treatments of exchange in Hartree–Fock and EXX discussed in Section 9.5. However, this is not the case for approximations to E_{xc}, and the errors can be significant since these terms involve large Coulomb interactions. There is a long history to such approaches, the first by Hartree himself [50] in his calculations on atoms. As noted in Section 3.6, Hartree defined a different potential for each occupied state by subtracting a self-term due to the charge density of that state, literally the first SIC. An example of density functional calculations for d → s promotion energies for electrons in the 3d transition metal series is shown in Fig. 10.2, which shows a striking improvement when SIC is included. For an extended state in a solid, the correction vanishes since the interaction scales inversely with the size of the region in which the state is localized. Thus, in extended systems the value depends on the choice of the localized orbital much like the DFT+U method.

An approach to extended systems has been developed in which a functional is defined with self-terms subtracted; minimization of the functional in an unrestricted manner allows the system of electrons to minimize the total energy by delocalizing the states (in a crystal, this is the usual Kohn–Sham solution with vanishing correction) or by localizing some or all of the states to produce a different solution [313, 454]. This approach has an intuitive appeal in that it leads to atomic-like states in systems like transition metal oxides and rare earth systems, where the electrons are strongly interacting. Studies using the SIC and related approaches have led to an improved description of the magnetic state and magnetic order in transition metal oxides [455], high T_c materials [456], and 4f occupation in rare earth compounds [457], which are discussed in more detail in [1].

---

[3] Extensive discussion of the interesting phenomena and theoretical methods can be found in [1], especially in chapters 19 and 20. Calculation of the interaction U from many-body methods and subtleties regarding the use of the "U" term are also discussed there.

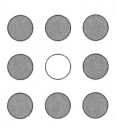
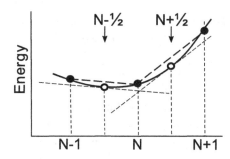

Figure 9.2. Schematic for calculation of interaction parameter U for electrons on a site (the central site show at the left) in a molecule or solid. The calculation is the same as for an atom in Section 10.6 except that the localized function must be chosen by some criterion and the interaction is screened by the surrounding system. The energy varies with the occupation N as shown at the right: black dots denote $E(N)$, the slope of heavy dashed lines are the addition energies ΔE which vary discontinuously, and the light dashed lines indicate the eigenvalues at half-occupation which approximate ΔE (Slater transition rule). The interaction U is the difference in eigenvalues as given in Eq. (10.22). The same approach can be used to determine exchange energies J and other terms.

Adding Hubbard-Like Interactions: DFT+U

The quaint acronym "LDA+U" is often used to denote methods that involve LDA- or GGA-type calculations coupled with an additional orbital-dependent interaction [458, 459], but we use the more general term "DFT+U." The additional interaction is usually considered only for highly localized atomic-like orbitals on the same site, i.e., of the same form as the "U" interaction in Hubbard models [460, 461]. The effect of the added term is to shift the localized orbitals relative to the other orbitals, which attempts to correct errors known to be large in the usual LDA or GGA calculations. For example, the promotion energies in transition metal atoms in Fig. 10.2 illustrate the fact that the relative energies shift depending on the approximation for exchange. The other effect occurs in partially filled d and f states, where occupation of one orbital raises the energy of the other orbitals, with the result that it favors magnetic states. Because the effects are critical for many problems involving 3d transition metal oxides and other materials, DFT+U calculations are an essential part of present-day methods.

One way to calculate the on-site interaction U parameter is the "constrained density functional" approach, which is an adaptation of the ΔSCF calculations for atoms in Section 10.6 and shown schematically in Fig. 9.2. For an isolated system like an atom it is straightforward to calculate the energy for different integer occupations of the states and determine energies to add, subtract, or promote electrons from differences of the total energy like in Eqs. (10.21) and (10.22). Since these involve ground-state energies the energy differences are in principle exact, unlike the Kohn–Sham eigenvalues. In addition, those equations also define the way that the Slater transition rule can be used to calculate the energy differences from the eigenvalue of the state at one-half integer occupation. As described in Section 10.6 the results can be quite accurate for atoms.

Many examples of "DFT+U" calculations are given in [1] and [459]. The prototypical examples are the transition metal oxides. Perhaps the best-known examples are the parent compounds of the CuO superconductors, which are found to be nonmagnetic metals in the usual LDA and GGA calculations (see Chapter 17), whereas "DFT+U" calculations find the correct antiferromagnetic insulator solution [459]. The usual spin density theory for materials like MnO and NiO finds the correct spin states and an energy gap, but the value of the gap is much too small as shown in Fig. 2.24. The gap is much better with hybrid functionals; however, it is often easier and more intuitive to correct the gap by a "U" term that increases the gap between the filled and empty 3d states and shifts the states relative to the oxygen states to agree much better with experiment.

Generalized Koopmans' Functionals

A different formulation of the problem has been made starting from the fact that for a localized state the electron addition and removal energies as a function of electron number are the straight lines that join with discontinuous slope (piecewise linear) as depicted in Fig. 9.2. The linearity means that the energy of an orbital is independent of the occupation of that orbital itself, which is the very definition of having no self-interaction, and the discontinuities occur at the points where the orbitals change.[4] This occurs at integer occupations and the energy of an orbital is shifted, as indicated in Fig. 9.2, by interaction with electrons in the other orbitals. This condition that the addition energies in a localized system are piecewise linear has been termed a generalization of Koopmans' theorem [462]; the original Koopmans' theorem (see Section 3.6) is that the addition energy for a electron in a particular state in Hartree–Fock is equal to its eigenvalue if all the states are rigid and do not relax. The generalization is the requirement that the energy of a state does not depend on its own occupation but may depend on the form and occupation of other orbitals.

This has been used to develop "Koopmans-compliant" functionals [463] that correct the smooth curvature in a continuous functional by adding a term to make it piecewise linear. Such a functional has features like SIC and DFT+U; however, it is more general in the sense that there may be different ways to choose the orbitals. For an extended system the eigenstates of the independent-particle hamiltonian are delocalized and the consequence of the occupation of one orbital vanishes as the size increases. The proposal in [463] is to define orbitals that are linear combinations of occupied states, which can vary from delocalized in which corrections vanish to localized with the piecewise linear behavior of the Koopmans-compliant functional, and to choose the degree of localization by minimizing the energy. The concepts have much in common with the earlier work on the unrestricted SIC method [313, 454], but it is more general and has been applied to a wide range of semiconductors and insulators [463].

---

[4] Note that here an orbital denotes the spatial state and spin, e.g., a d shell consists of 10 orbitals. The idea of the energy as a function of N can be realized by considering the localized system in contact with a reservoir with Fermi energy μ. As μ varies the energy of the orbital relative to μ varies linearly; its occupation changes abruptly when the energy crosses μ but its energy does not change.

9.7 Functionals Derived from Response Functions

The next rung on the DFT ladder is to include information about the excited states. In the previous sections we have seen how to treat the exchange energy in the exact exchange and hybrid functional methods. However, in those methods correlation is not calculated from theory but is included as an LDA or GGA functional, where it is derived from some other source, such as calculations on two-electron atoms, the homogeneous gas, etc. Great insight and ingenuity was needed to invent the functionals of the density that have proven so useful; however, there are no systematic derivations that allow one to understand what is included, what is not, and what are steps toward further improvement.

The approach in this section is very different. In many ways it is a systematic application of exact relations and approximations using perturbation theory. Thus it is more straightforward to understand, but the price to be paid is that the calculations are much more involved and computationally intensive. The methods build on the many-body theory of excitations, which is covered in much more detail in the companion book [1]. However, it is not essential to go into the details of the many-body methods because work over the decades has built up much experience and approximations that can be understood using only methods developed in the previous sections and knowledge of response functions in Appendix D and the relation to the dielectric function in Section E.4.

The theoretical formulation is called the "adiabatic coupling fluctuation dissipation (ACFD) theorem," which can in understood in two parts. The first is the coupling-constant integration, which is also termed an "adiabatic connection" between a noninteracting problem and the full interacting system, with the interaction scaled by a λ that varies from 0 to 1 (see Section 3.4) and the density constrained to be constant. This is an exact relation for the exchange–correlation energy as explained in Sections 5.3 and 8.2, and it is used in an approximate way to create hybrid functionals in Section 9.3. The other part is "fluctuation dissipation theorem," which is a well-known relationship given in Appendix D between the correlated fluctuations in a system to the response to a driving force (a response function). The result is that the correlation energy E_c can be calculated by an integral over λ and frequency of the density–density response function $\chi_\lambda(\omega)$ that depends on λ. It is in this sense that ground-state correlation energy is related to the frequency-dependent excitations of the system; the integral over all frequencies is needed to find the correlation at the same time that determines E_c.

The result is the ACFD theorem:[5]

$$E_c = -\frac{1}{2\pi} \int_0^1 d\lambda \int_0^\infty du \, \mathrm{Tr} \left[v_c \left(\chi_\lambda(iu) - \chi_0(iu) \right) \right],$$

$$= -\frac{1}{2\pi} \int_0^1 d\lambda \int dr \, dr' v_c(r - r') \int_0^\infty du \left[\chi_\lambda(r', r; iu) - \chi_0(r', r; iu) \right], \quad (9.22)$$

---

[5] Clear discussions can be found in the early references [272, 464] and the reviews [465, 466], which also give many references, and there is extensive discussion of the RPA in [1]. Here we consider only the case with equal spin up and down; correlation involves both spins and there are nontrivial changes for a polarized system.

where the first line reveals the elegant simplicity of the relation, and the second line is written out in real space. In Eq. (9.22) u is an imaginary frequency, v_c is the Coulomb interaction, χ_0 is the response function for the independent-particle Kohn–Sham auxiliary system given in Eq. (9.24) below, and χ_λ is the response of a system with scaled interaction λv_c, and with ground-state density required to be the same as λ changes. This is an exact relation, but of course χ_λ for an interacting system is not known. When χ_λ is approximated in the RPA, $\chi_\lambda = \chi_0/(1 - \lambda v_c \chi_0)$, the λ-integration can be done analytically and expressed as a logarithm or equivalently as a power series:

$$E_c^{RPA} = \frac{1}{2\pi} \int_0^\infty du \, \text{Tr}\left[\ln(1 - v_c\chi_0) - v_c\chi_0\right] = \frac{-1}{2\pi} \int_0^\infty du \sum_{n=2}^\infty \frac{1}{n} \text{Tr}\left[(v_c\chi_0)^n\right].$$
(9.23)

The calculations can be done since $\chi_0(\omega)$ is given in terms of independent-particle wavefunctions. The general form is in Section D.4 and it is instructive to note that the frequency-dependent expressions needed here are analogous to the static OEP equations. Expressed in real space $\chi_0(\mathbf{r}, \mathbf{r}', \omega)$ is given by the generalization of the static χ Eqs. (9.15) and (9.16) (see also Section D.4),

$$\chi_0(\mathbf{r}, \mathbf{r}', \omega) = 2 \sum_{i=1}^{occ} \sum_j^{empty} \frac{\psi_i(\mathbf{r})\psi_j(\mathbf{r})\psi_j(\mathbf{r}')\psi_i(\mathbf{r}')}{\varepsilon_i - \varepsilon_j + \omega + i\eta},$$
(9.24)

where the factor of 2 is for spin, and there are similar expressions in momentum space. The calculations are much more computationally intensive than those for standard DFT since they involve sums over excited states.[6] In general the wavefunctions and eigenvalues are determined by a nonlocal operator, similar to Hartree–Fock or a hybrid method. Alternatively, the energy in Eq. (9.23) can be calculated from the output of Kohn–Sham calculation or as a self-consistent procedure to minimize the total energy for wavefunctions that are generated by a local Kohn–Sham potential, where the latter amounts to finding an optimized effective potential (OEP) for the RPA formulation. References for methods and many results can be found in reviews [465, 466], and extensive benchmarks for RPA and related functionals for solids in [467].

One of the great advantages of the RPA is that it can treat dispersion interactions (see Section 9.8) since the interaction between two separated systems is due only to the long-range Coulomb interaction. The first term in the expansion in Eq. (9.23) is second order in the polarizability and it includes the leading term for weak interactions between pairs of atoms $\propto 1/R^6$, even though the coefficient C_{6AB} is not exact in the RPA. Examples of applications are given in [466] where there is also a demonstration that for weak interactions, the system can be treated efficiently as coupled effective harmonic oscillators. The RPA formalism also provides the basis for one of the methods to develop the van der Waals functional in Section 9.8. The second-order expansion used in [468] is closely related

---

[6] There are more efficient methods that do not require explicit sums over excited states as described in Sections 20.4 and 21.5, but they are still more costly than DFT with standard density functionals.

to the power series expansion in Eq. (9.23) and, in order to construct a density functional, those methods use a single-pole approximation at a plasma frequency that is determined by the density of electrons.

9.8 Nonlocal Functionals for van der Waals Dispersion Interactions

A major deficiency of all local and semilocal functionals is that they cannot describe the van der Waals interaction that is ubiquitous in weakly bonded atomic and molecular systems. Aside from the average Coulomb interaction that is included in the nuclear and Hartree terms, the longest-range interaction is purely a correlation effect due to quantum fluctuations of the electric dipole on one atom or molecule that induces dipoles on another, oriented such that the energy is lowered. The attraction between two atoms or molecules is often called the dispersion or London interaction, where the longest-range part of the interaction can be expressed as C_6/R^6 (see Exercise 9.6). The challenge in density functional theory is to develop a nonlocal functional that captures the effects of the long-range correlation on the energy, describes the intermediate-range interactions, and merges seamlessly into the regime where shorter-range effects are described by well-developed functionals. This is essential to describe important cases such as molecules on metal surfaces that require a method that can handle many types of bonding within the same framework.

The different approaches can be grouped into two types. One starts with long-range pair-wise interactions between atoms that are modified at short range in such a way that they can be added to standard density functionals with little or no change in the short-range behavior. The other starts with the homogeneous gas and develops functionals that describe nonlocal correlation in such a way that it describes the long-range dispersion interactions in systems that are so inhomogeneous that they separate into nonoverlapping regions. Of course, there are connections between these approaches since each strives to have the same behavior at large distance, and each is designed to take advantage of the successes of local and semilocal approximations for metallic, covalent, and ionic bonding.

The van der Waals functionals could be grouped with the functionals in Chapter 8 since they are functionals only of the density. They are presented here because the underlying phenomena and the theory are fundamentally related to polarizability, which is a response function that depends on the occupied and unoccupied states, the topic of Section 9.7.

Pair-Wise Additive Contributions to the Energy

The theory of the dispersion interactions can be formulated in terms of the Casimir–Polder [469] formula for the two non-overlapping atoms with frequency-dependent polarizability $\alpha(\omega)$,

$$C_{6AB} = \frac{3}{\pi} \int_0^\infty du \, \alpha_A(iu)\alpha_B(iu), \tag{9.25}$$

where iu is an imaginary frequency.[7] Expressions for higher-order terms (C_8/R^8, etc.) are given by algebraic expressions in terms of C_6 and the fourth moment $\langle r^4 \rangle$ of the dipole distribution of each atom [470]. In order to have a form that can be added to a standard DFT functional, the divergence at short distance must be eliminated, which can be accomplished by introducing a damping function $f(R)$ so that the interaction becomes a sum over pairs of atoms of terms with the form

$$E_{disp} = -\frac{1}{2}\sum_{AB}\frac{C_{6AB}}{(R_{AB})^6}f(R_{AB}), \tag{9.26}$$

where $f(R)$ varies smoothly from 0 at $R = 0$–1 at large R. Convenient forms are Pade [470] and Fermi [439] functions, which are determined by the characteristic range and the rapidity of the variation.

One approach is to carry out the integral in Eq. (9.25) directly for each pair of atoms. This has been developed by Grimme and coworkers [470] using $\alpha(\omega)$ calculated by time-dependent DFT, which has proven very useful for atoms, molecules, and clusters (see Sections 21.2 and 21.4). It is most desirable to find the constituent $\alpha(\omega)$ for atoms in an environment similar to that in the molecule or solid, and the choice in [470] is a hydride molecule in which the atom is attached to hydrogens so that it forms a closed-shell system, which has been done for all 94 elements from H to Pu. It turns out that the C_8/R^8 term is quite important at intermediate range and should be included explicitly or incorporated in the form of the damping function $f(R)$. In order to apply the results as an add-on to DFT functionals, the approach in [470] is to adjust the damping function $f(R)$ with two parameters that are different for each functional, and it been shown to be very useful as a general method to treat systems from weakly interacting molecules to strongly bonded solids.

There are other approaches that aim to develop simpler more universal forms using approximations to the frequency dependence of the polarizability. It is instructive to consider the method of Tkatchenko and Scheffler (TS) [439] where the atomic realizability $\alpha(\omega)$ is approximated by a single pole at an effective frequency ω_0, so that

$$\alpha(\omega) = \frac{\alpha^0\omega_0^2}{\omega_0^2 - \omega^2} \quad \text{and} \quad \alpha(iu) = \frac{\alpha^0\omega_0^2}{\omega_0^2 + u^2}. \tag{9.27}$$

The integral is readily done, and one consequence is the relation (Exercise 9.7)

$$C_{6AB} = \frac{2C_{6AA}C_{6BB}}{\frac{\alpha_B^0}{\alpha_A^0}C_{6AA} + \frac{\alpha_A^0}{\alpha_B^0}C_{6BB}}, \tag{9.28}$$

which is equivalent to formulas derived by London [97] and Slater-Kirkwood [95]. This vastly simplifies the problem and has been tested for large data sets for molecules. The

---

[7] Imaginary frequencies are often used in many-body theory where they provide compact expressions like Eq. (9.25). For our purposes we only need to know that it is a straightforward integral if there is an analytic form for $\alpha(\omega)$, e.g., in the model in Eq. (9.27).

only inputs are the homopolar C_{AA} and C_{BB}, which in turn are determined by only two parameters per atom, the static polarizability α^0 and the effective frequency ω_0.

The approach of TS is to turn this into a functional of the density by using the relation of polarizability and volume[8] and the Hirshfeld partitioning [472] of the density $n(\mathbf{r})$ to assign a volume for each atom in the molecule or solid. Values of α^0, ω_0 and the homopolar C_6 are taken from previous work for a set of closed-shell atoms in free space, and the values of C_6 for each pair of atoms in a molecule or solid is found from the difference of the volumes from free atoms. The damping function $f(R_{AB})$ is chosen to be a Fermi function with a form that is fitted to a set of data on selected systems, when it is used as an addition to the PBE functional. Although this route to a functional may seem to be based on many approximations, each one is tested separately on well-known systems and the resulting form has been applied to many problems.

Functionals Based on Theories of Nonlocal Correlation

A different approach has been developed by Langreth and Lundqvist and their coworkers in a series of papers (see [468] and references to earlier work given there). The goal is to formulate a general theory of the correlation energy as a sum of local and nonlocal terms,

$$E_c[n] = E_c^{LDA}[n] + E_c^{nl}[n], \tag{9.29}$$

where $E_c^{LDA}[n]$ is the local density expression in terms of the correlation energy for the homogeneous gas, and the nonlocal part can be expressed as

$$E_c^{nl} = \frac{1}{2} \int d^3 r d^3 r' n(\mathbf{r}) \phi(\mathbf{r},\mathbf{r}') n(\mathbf{r}'), \tag{9.30}$$

where $\phi(\mathbf{r},\mathbf{r}')$ is a functional of the density.[9] In order to describe dispersion interactions between two systems with densities that do not overlap, $\phi(\mathbf{r},\mathbf{r}')$ must include the long-range Coulomb interaction and incorporate information on the polarizability of each system. The challenge is to find a single-function $\phi(\mathbf{r},\mathbf{r}')$ that describes the correlation energy for systems that range from the homogeneous gas (where ϕ must vanish) to van der Waals bonding solely in terms of the density.

The methods to find useful approximations for E_c^{nl} and $\phi(\mathbf{r},\mathbf{r}')$ build on the long history of many-body methods to treat correlation, which have been developed to the point where there are simple forms that provide approximate but time-tested methods. The quintessential example is the random phase approximation (RPA), which is described in detail in the companion book [1], and the most relevant aspects can be found in Section 9.7. Calculation of the correlation energy can be done with the adiabatic coupling dissipation fluctuation

---

[8] Exercise 9.8 is to show that polarizability has units of volume and that the polarizability of a conducting sphere of radius R is equal to its volume. It follows from the general expression in (21.7) that an effective volume can be defined for atoms and molecules. Ways to determine the effective volume are described, for example, in [471].

[9] As it stands, Eq. (9.30) scales as N^2, where N is the number of atoms; however, methods [473] have been developed using FFTs to reduce the computation to $\propto N \ln N$.

(ACFD) method as described in Section 9.7, which leads to the correct long-range $1/R^6$ form of the dispersion interaction. The value of the coefficient C_6 is not exact, but there is much experience in using the RPA to calculate accurate practicabilities $\alpha(\omega)$. At the other extreme is the homogeneous electron gas described in Section 5.3. Figure 5.4 shows that the RPA is not very accurate, but it provides a good description of the *change* in the correlation energy as a function of density, which is the information needed to calculate corrections to the LDA.

The problem is that the expressions in Section 9.7 involve integrals over dynamical density response functions $\chi(\omega)$, whereas the goal is a functional of only the ground-state density. It has proven to be possible to find very useful forms by building on previous experience with RPA in solids. It is useful to recall the relation to the dielectric function in Eq. (E.15), which can be written in schematic form as $\epsilon^{-1}(\omega) = 1 - v_c\chi(\omega)$ where $\chi(\omega) = (\chi_0(\omega)/(1 - v_c\chi_0(\omega))$ in the RPA. The inverse function $\epsilon^{-1}(\omega)$ describes charge response and has peaks at the plasma frequency; for many purposes it is well approximated by a single pole at the plasma frequency ω_P, which is determined by the electron density in analogy to Eq. (9.27) and the relation of the polarizability to volume in atoms.

The problem is more difficult in a solid since ω_P varies with momentum and the dielectric function is a matrix in momenta $\mathbf{q}$ and $\mathbf{q}'$. The formulation proposed in [468], based on earlier works, is a single-pole approximation in terms of $S = 1 - \epsilon^{-1} = v_c\chi$ that can be expressed as

$$S_{\mathbf{q},\mathbf{q}'}(\omega) = \int d^3r e^{(\mathbf{q}-\mathbf{q}')\cdot\mathbf{r}} \frac{\omega_P(\mathbf{r})}{(\omega_{\mathbf{q}}(\mathbf{r}) + \omega)(\omega_{\mathbf{q}'}(\mathbf{r}) - \omega)}, \tag{9.31}$$

where $\omega_P(\mathbf{r}) = 4\pi n(\mathbf{r})e^2/m$ is the long-wavelength plasma frequency for a uniform system with density $n(\mathbf{r})$, i.e., a local approximation, and $\omega_{\mathbf{q}}(\mathbf{r})$ and $\omega_{\mathbf{q}'}(\mathbf{r})$ have parameterized forms that interpolate between long and short wavelength limits, and they are assumed to depend only on the density and its gradients at the point $\mathbf{r}$. The nonlocal part of the correlation energy $E_c^{nl}[n]$ can be found by expanding the RPA formula in Eq. (9.23) to second order in S [468],

$$E_c^{nl} = \frac{1}{4\pi}\int_0^\infty du \, \text{Tr}\left[S^2 - \left(\frac{\nabla S \cdot \nabla v_C}{4\pi}\right)^2\right], \tag{9.32}$$

which vanishes for a uniform system. For nonoverlapping densities, integration by parts leads to $\nabla^2 v_c \propto 1/R^3$ and thus a $1/R^6$ dependence of the last term in Eq. (9.32) [474]. Various functionals can be made by different choices for the $\omega_{\mathbf{q}}(\mathbf{r})$, for example, the simple analytic form called VV10 due to Vydrov and Voorhis [475], which can be expressed in the form of Eq. (9.30) with

$$\phi(\mathbf{r},\mathbf{r}') = \frac{3e^4}{2m^2 g g'(g + g')}, \tag{9.33}$$

where g and g' are analytic functions of the density and its gradient as points $\mathbf{r}$ and $\mathbf{r}'$ respectively.

It is not appropriate to attempt to compare functionals here because there are so many combinations of short-range and long-range dispersion functionals. An extensive review by

Mardirossian and Head-Gordon [415] compares the performance of 200 functionals applied to large data sets of molecules. Among all the functionals considered they concluded that the most promising overall is the VV10 form combined with a range-separated hydrid meta-GGA. However, it is important to consider also the computational effort required and to examine carefully how well a chosen functional performs when applied to specific types of systems and phenomena.

9.9 Modified Becke–Johnson Functional for V_{xc}

A remarkably simple effective local potential for exchange was derived by Becke and Johnson (BJ) [476]. Its value at a point $\mathbf{r}$ depends only on densities evaluated at that same point and yet it is a good approximation to the potential derived from the nonlocal Fock operator. The goal was to describe the shell structure that is characteristic of the exact exchange potential and also to reproduce the uniform gas limit and be correct for hydrogenic systems (the same requirements as used in other functionals such as SCAN in Section 9.4). This was accomplished by including a term that involves the kinetic energy density $t(\mathbf{r})$, which is chosen to be the positive symmetric form defined in Eq. (9.7); the way it incorporates the shell structure is discussed in Section H.4 and illustrated in Fig. H.2. The BJ form was subsequently modified by Tran and Blaha [477] to add a parameter c giving the form

$$V_x^{MBJ}(\mathbf{r}) = c V_x^{BR}(\mathbf{r}) + (3c - 2)\frac{1}{\pi}\sqrt{\frac{5}{12}}\sqrt{\frac{2t(\mathbf{r})}{n(\mathbf{r})}}, \qquad (9.34)$$

where $V_x^{BR}(\mathbf{r})$ is the local potential defined by Becke and Roussel [442], which is close to the Slater potential (see Eq. (9.20)), and $c = 1$ corresponds to the original BJ potential.

Since $V_x^{MBJ}(\mathbf{r})$ is a semilocal potential it is much less expensive to calculate than hybrid functionals. Note, however, the formulation provides a potential, not an energy functional. There is no way to minimize a total energy and the procedure is to use another functional, such as LDA or a GGA, to find a self-consistent solution for the density and eigenfunctions, which are then used to construct the MBJ potential $V_x^{MBJ}(\mathbf{r})$. in Eq. (9.34), and finally to compute the bands with $V_x^{MBJ}(\mathbf{r})$.

By choosing a value of c in the range 1.1–1.3, it was found that the resulting gaps are in remarkable agrement with experiment; however, the band widths are too narrow. Extensive comparisons with other methods and experiment can be found in [477] and [478]. For example, the gaps in Ge and NiO shown in Figs. 2.23 are 0 and $\approx$0.4eV in the LDA, but the MBJ potential starting from the LDA yields gaps 0.85 and 4.16 respectively [477] in good agreement with experiment and the hybrid HSE calculations shown in the figures.

9.10 Comparison of Functionals

It is instructive to examine the consequences of different approximations for the exchange–correlation functional in various ways. In this section are results for a few selected atoms, chosen because they have nondegenerate ground states, which avoids the complications of open-shell systems. Comparisons for solids are given in many places in this book, which are summarized in a list near the end of this section.

One-Electron Problems: Hydrogen

The results for hydrogen are a special case. For any one-electron problem, Hartree–Fock is exact. In the calculation there are Hartree and exchange terms, each of which is unphysical since there are no electron–electron interactions, but they cancel exactly, leading to the exact solution. Exact exchange (EXX) involves the Hartree–Fock orbital-dependent exchange functional as described in, Section 9.5. For one electron, it is the same as Hartree–Fock and so it is also exact. However, one-particle problems are severe tests for functionals of the density, such as the LDA and GGAs, and for the other functionals, except for BLYP and SCAN, which are constructed to be exact for hydrogen. The functionals are designed to deal with many electrons, such as the homogeneous gas, and their application to a one-particle problem introduces unphysical nonzero values for the correlation energy and a local or other approximation for exchange that does not cancel the Hartree term. The results for various functionals are given in Table 9.1 and the most obvious result is that there are large errors in the LSDA that are considerably improved by the GGA and hybrid functionals. Note that the separate errors in exchange and correlation in the LSDA tend to cancel, so that the final LSDA energy is ≈0.48 Ha, in surprisingly good agreement with the exact value, 0.5 Ha. Although there is no such cancellation for the GGAs, their final results for the total energy are much improved over LSDA.

Many-Electron Atoms

For more than one electron, the quantity called exchange is not a physically measurable energy. It is usually defined to be the Hartree–Fock value and correlation is defined to be the remaining energy beyond Hartree–Fock. There are quantitative differences (usually very small) for exact exchange EXX; here it is chosen since it may be considered more appropriate for comparison of functionals. The results shown in Table 9.1[10] show the calculated values for exchange and correlation energies separately, and one sees immediately two important conclusions:

- Exchange is the dominant effect. Even though the fractional error in correlation may be larger, it is usually the error in the exchange that determines the final accuracy of the functional.

---

[10] It should be noted that the values are not all taken from the same reference and they are computed using different definitions, but the differences are not large and do not change the primary conclusions. The results for LDA, PBE, and BLYP are from [479] and used the same EXX density and wavefunctions to calculate the energies for each functional. The SCAN calculations used Hartree–Fock wavefunctions, which should be very close to EXX but are not identical. The HSE calculations were self-consistent with HSE wavefunctions, which lead to differences that are very small but significant for the detailed comparisons. A correction has been included, the same as the correction needed to bring the PBE calculations into agreement with the values from [479] quoted in the table, which is justified by the fact that HSE and PBE are closely related.

Table 9.1. Magnitude of exchange and correlation energies in Ha for selected spherically symmetric atoms for functionals described in the text. "Exact" denotes exact exchange (EXX) and correlation energies taken from [479]. Other energies are from [479], calculated with the same EXX density, except HSE06 (provided by Johannes Voss) and SCAN (provided by Biswajit Santra). See footnote in the text concerning the differences. The last row is the mean absolute relative error (MARE).

Atom	Exact	LSDA	PBE	HSE06	BLYP	SCAN
			Exchange			
H	0.3125	0.2680	0.3059	0.3088	0.3098	0.3125
He	1.0258	0.8840	1.0136	1.0193	1.0255	1.0306
Be	2.6658	2.3124	2.6358	2.6464	2.6578	2.6602
N	6.6044	5.908	6.5521	6.5764	6.5961	6.4114
Ne	12.1050	11.0335	12.0667	12.098	12.1378	12.1636
MARE %	0	12.1	1.1	0.61	0.32	0.82
			Correlation			
H	0.0000	0.0222	0.0060	0.0055	0.0000	0.0000
He	0.0420	0.1125	0.0420	0.0409	0.0438	0.0379
Be	0.0950	0.2240	0.0856	0.0881	0.0945	0.0827
N	0.1858	0.4268	0.1799	0.1775	0.1919	0.1809
Ne	0.3929	0.7428	0.3513	0.3432	0.3835	0.3448
MARE %	0	130	59	6.8	2.7	9.4
			Total Exchange-Correlation			
H	0.3125	0.2902	0.3143	0.3104	0.3098	0.3125
He	1.0678	0.9965	1.0602	1.0553	1.0663	1.0685
Be	2.7608	2.5364	2.7345	2.7349	2.7439	2.7429
N	6.7902	6.3348	6.7539	6.7384	6.7766	6.5923
Ne	12.498	11.7763	12.441	12.4069	12.5043	12.5084
MARE %	0	6.9	0.85	0.65	0.31	0.74

- There are very large errors for LDA, more than 100% in the correlation energy! The other functionals are much better and hybrid functionals are the most accurate. For hybrids this is understandable since they involve the Hartree–Fock exchange in addition to the improvements due to the GGA, for example, HSE is built on the PBE functional plus a fraction of Hartree–Fock exchange. The SCAN functional also incorporates properties of GGA functionals and adds additional constraints.

It is often stated that errors in exchange and correlation tend to cancel. However, except for LDA where errors are very large, this is not a major factor and it is not always the case.

Pointers to Places in This book with Comparisons of Functionals for Solids

- Lattice constants, bulk moduli for selected elements and compounds in Table 2.1 extracted from a set of 64 solids in [112], and the magnetization of Fe.
- Phase transitions under pressure for Si and SiO_2 in Section 2.5.
- Phase transition in $MgSiO_3$ potentially important for the lower mantel of the Earth comparing LDA and PBE in Fig. 2.12.
- Radial density distribution function for water in Fig. 2.15.
- Band structure of Ge in Fig. 2.23, which compares LDA, HSE and Hartree–Fock.
- Plot of gaps for selected elements in Fig. 2.24 comparing LDA, SCAN, and the hybrid functionals BLYP, HSE, and DDH.
- Optical spectra comparing LDA and HSE in the independent-particle approximation are shown in the left side of Fig. 2.25. Time-dependent DFT in Figs. 2.26 and 21.5 shows that long range functionals are essential to describe excitons.
- Bands near the Γ point for CdTe and the inverted bands of HgTe, comparing PBE and HSE are shown in Fig. 27.10.

There are many places where much more extensive comparison of functionals can be found. One referred to earlier is the compendium by Mardirossian and Head-Gordon [415], which gives an overview and extensive assessment of 200 density functionals. An assessment of functionals applied to an important problem, water in many form – molecular interactions, liquid water, ices, etc. – can be found in [197].

SELECT FURTHER READING

The functionals in this chapter rely heavily on many-body theory, amd many of the aspects are discussed in depth in the companion book [1].

See references at the end of Chapters 6 and 7 for general references for density functional theory and at the end of Chapter 8 for references on functionals. Almost all the reading on functionals includes hybrid functionals, which are widely used.

Reviews that foucus on functionals in this chapter:

Anisimov, V. I., Aryasetiawan, F., and Lichtenstein, A. I., "First principles calculations of the electronic structure and spectra of strongly correlated systems: The LDA + U method," *J. Phys.: Condensed Matter* 9:767–808, 1997.

Becke, A. D., "Perspective: Fifty years of density-functional theory in chemical physics," *J. Chem. Phys.* 140:301, 2014.

Grabo, T., Kreibich, T., Kurth, S., and Gross, E. K. U., "Orbital functionals in density functional theory: The optimized effective potential method," in *Strong Coulomb Correlations in Electronic Structure: Beyond the Local Density Approximation*, edited by V. I. Anisimov (Gordon & Breach, Tokyo, 1998).

Kummel, S. and Kronik, L., "Orbital-dependent density functionals: Theory and applications," *Rev. Mod. Phys.* 80:3–60, 2008.

Exercises

9.1 It is interesting to note that the construction of a kinetic energy functional of the density is a "fermion problem." See Section H.1 and Exercise H.1 and H.2. For noninteracting bosons, construct explicitly a practical, exact density functional theory.

9.2 Derive the general OEP expression, (9.11), using the chain rule, and show that it leads to the compact integral expression, (9.17).

9.3 Write out explicit expressions for the inversion of the response function needed in Eq. (9.11) by expressing the response function in a basis. Consider appropriate bases for two cases: a radially symmetric atom (with the potential and density on a one-dimensional radial grid) and a periodic crystal (with all quantities represented in Fourier space).

9.4 An impediment in actual application of the OEP formula, Eq. (9.11) is the fact that the response function is singular. Show that this is the case since a constant shift in the potential causes no change in the density. Describe how such a response function can be inverted. Hint: one can define a nonsingular function by projecting out the singular part. This may be most transparent in the case of a periodic crystal where trouble arises from a constant potential that is undetermined.

9.5 Show that the approximation, Eq. (9.18), substituted into the integral equation, (9.17), leads to the KLI form, Eq. (9.19), and discuss the ways in which this is a much simpler expression than the integral equation, (9.17).

9.6 Derive the power law for the London dispersion interaction, using a model with two oscillators, each with a charged particle attached to a spring, that are coupled by the Coulomb interaction. The longest-range term results from using the lowest nonzero term in a perturbation expansion. Is this sufficient to establish the result for any problem approximated by a single pole?

9.7 Derive the relation in Eq. (9.28) using the expression for a coefficient in Eq. (9.25) and the single-pole approximation in Eq. (9.27).

9.8 Show that the polarizability in Eq. (21.7) has units of volume. Show also that the polarizability of a perfectly conducting sphere is equal to its volume. This is derived in [480] and other books, but it is a good exercise to do it yourself.

9.9 See Exercise 8.6 and 8.7 for exchange and correlation (fictitious) terms in the LDA for H. Compare with the results in Table 9.1. Discuss the extent to which the results are improved in the hybrid functionals.

PART III

IMPORTANT PRELIMINARIES ON ATOMS

10

Electronic Structure of Atoms

Summary

This chapter is concerned with the issues of solving the self-consistent Kohn–Sham and Hartree–Fock equations in the simplest geometry. We will *not* be concerned with the intricate details of the states of many-electron atoms but only those aspects relevant to our primary goal, the electronic structure of condensed matter and molecules. Studies of the atom illustrate the concepts and are directly relevant for the following sections since they are the basis for construction of *ab initio* pseudopotentials (Chapter 11) and the augmentation functions (Chapter 16) that are at the heart of the augmented plane wave (APW), the linear combination of muffin-tin orbitals (LMTO), and KKR methods. Furthermore, we shall see that calculations on atoms and atomic-like radial problems are extremely useful in qualitative understanding of many aspects of condensed matter, including band widths, equilibrium volume, and bulk modulus, as discussed in Section 10.7.

10.1 One-Electron Radial Schrödinger Equation

We start with the case of the hydrogenic one-electron atom. Although this is treated in many texts, it is useful to establish notation that will be used in many chapters. If there is no spin–orbit coupling, the wavefunction can be decoupled into a product of space and spin. (The relativistic Dirac equations and spin–orbit interaction are treated in Section 10.3 and Appendix O.) Since the potential acting on the electron is spherically symmetric $V_{\text{ext}}(\mathbf{r}) = V_{\text{ext}}(r) = -Z/r$, the spatial part of the orbital can be classified by angular momentum ($L = \{l, m_l\}$)

$$\psi_{lm}(\mathbf{r}) = \psi_l(r)Y_{lm}(\theta,\phi) = r^{-1}\phi_l(r)Y_{lm}(\theta,\phi), \tag{10.1}$$

where the normalized spherical harmonics are given by

$$Y_{lm}(\theta,\phi) = \sqrt{\frac{2l+1}{4\pi}\frac{(l-m)!}{(l+m)!}} P_l^m[\cos(\theta)]e^{im\phi}, \tag{10.2}$$

with $P_l^m(x)$ denoting the associated Legendre polynomials defined in Section K.2. The traditional labels for angular momenta are s, p, d, f, g, $\ldots$, for $l = 0, 1, 2, 3, \ldots$, and explicit formulas for the first few functions are given in Eq. (K.10).

Using the form of the Laplacian in spherical coordinates,

$$\nabla^2 = \frac{1}{r^2}\frac{\partial}{\partial r}\left(r^2\frac{\partial}{\partial r}\right) + \frac{1}{r^2\sin(\theta)}\frac{\partial}{\partial\theta}\left[\sin(\theta)\frac{\partial}{\partial\theta}\right] + \frac{1}{r^2\sin^2(\theta)}\left(\frac{\partial^2}{\partial\phi^2}\right), \tag{10.3}$$

the wave equation can be reduced to the radial equation (Exercise 10.1) for principal quantum number n

$$-\frac{1}{2r^2}\frac{d}{dr}\left[r^2\frac{d}{dr}\psi_{n,l}(r)\right] + \left[\frac{l(l+1)}{2r^2} + V_{\text{ext}}(r) - \varepsilon_{n,l}\right]\psi_{n,l}(r) = 0, \tag{10.4}$$

or

$$-\frac{1}{2}\frac{d^2}{dr^2}\phi_{n,l}(r) + \left[\frac{l(l+1)}{2r^2} + V_{\text{ext}}(r) - \varepsilon_{n,l}\right]\phi_{n,l}(r) = 0. \tag{10.5}$$

The equations can be solved for bound states with the boundary conditions $\phi_{n,l}(r)$, $\psi_{n,l}(r) \to 0$ for $r \to \infty$, and $\phi_{n,l}(r) \propto r^{l+1}$ and $\psi_{n,l}(r) \propto r^l$ for $r \to 0$, and subject to the normalization condition

$$\int_0^\infty dr\, \phi_{n,l}(r)^2 = 1. \tag{10.6}$$

For a one-electron atom, the well-known analytic solutions have eigenvalues independent of l given by

$$\varepsilon_{n,l} = -\frac{1}{2}\frac{Z^2}{n^2} \tag{10.7}$$

in Hartree atomic units.

Logarithmic Grid

It is convenient to have regular grids in numerical algorithms; however, for atoms a higher density of radial points is needed near the origin and only a low density in the outer region. In the program of Herman and Skillman [481] this is accomplished by doubling the grid density several times as one proceeds toward the nucleus. This is simple in concept but leads to a complicated algorithm. Another choice is to use a logarithmic grid with $\rho \equiv \ln(r)$, which is suggested by the hydrogenic orbital, which has amplitude $\propto \exp(-Zr)$, where

Z is the atomic number. If we define $\tilde{\phi}_l(\rho) = r^{1/2}\psi(r)$, then the radial equation (10.5) becomes (see Fischer [482] and Exercise 10.2)

$$\left\{ \frac{-\hbar^2}{2m_e}\frac{d^2}{d\rho^2} + \frac{l}{2}\left(l+\frac{1}{2}\right)^2 + r^2[V_{\text{ext}}(r) - \varepsilon_{n,l}] \right\} \tilde{\phi}_l(\rho) = 0, \tag{10.8}$$

This has the disadvantage of transforming the interval $0 \le r \le \infty$ to $-\infty \le \rho \le \infty$. In practice, one can treat an inner region $0 \le r \le r_1$ with a series expansion [482]

$$\phi_l(r) \propto r^{l+1}\left[1 - \frac{Zr}{l+1} + \alpha r^2 + O(r^3) \right], \tag{10.9}$$

where α is given in Section 6.2 of [482]. The boundary r_1 is chosen so that $\rho_1 = \ln(Zr)$ is constant for all atoms. Then the outer region $\rho_1 \le \rho \le \infty$ can be treated on a regular grid in the variable ρ.

The atomic equations can be solved on a regular grid in r or ρ following the flow chart for a Kohn–Sham calculation given in Fig. 7.2. The radial equations can be solved using the Numerov method described in Appendix L, Section L.1. Excellent atomic programs exist, often built upon the one written originally by Herman and Skillman [481]. The ideas are described in great detail by Slater [483, 484] and by Fischer [482] in the Hartree–Fock approximation, and a simplified description is given by Koonin and Meredith [485]. Some of the codes for atomic calculations are listed in Appendix R.

10.2 Independent-Particle Equations: Spherical Potentials

The Kohn–Sham equations for a general problem have been given in Eqs. (7.11)–(7.13), which are independent-particle equations with a potential that must be determined self-consistently. The same form applies to the Hartree–Fock equations (3.45) or (3.46), except that the exchange potential Eq. (3.48) is state dependent. In each case, the solution requires solving independent-particle equations having the same form as the one-electron equation (10.4) or (10.5), except that V_{ext} is replaced by some effective potential V_{eff} that must be determined self-consistently.

For closed-shell systems, such as rare-gas atoms, all the filled states are spin paired and the charge density $n(r)$,

$$n(r) = \sum_{n,l}^{\text{occupied}} (2l+1)|\psi_{n,l}(r)|^2 = \sum_{n,l}^{\text{occupied}} (2l+1)|\phi_{n,l}(r)|^2/r^2, \tag{10.10}$$

has spherical symmetry. The potential

$$V_{\text{eff}}(r) = V_{\text{ext}}(r) + V_{\text{Hartree}}(r) + V_{\text{xc}}(r) \tag{10.11}$$

is obviously spherically symmetric in the Kohn–Sham approach. In the Hartree–Fock case, the last term is the orbital-dependent exchange $\hat{V}_x^{n,l}(\mathbf{r})$, but it is not difficult to show (Exercise 10.5) that matrix elements of $\hat{V}_x$ are independent of m and σ and lead to an effective radial potential for each n, l.

Thus the independent-particle states can be rigorously classified by the angular momentum quantum numbers $L = \{l, m_l\}$ and there is no net spin. This leads to the simplest-case radial equations with eigenvectors ϕ_{l,m_l} independent of spin and eigenvalues independent of m_l. The resulting radial equation for $\phi_{n,l}(r)$, analogous to Eq. (10.5), is

$$-\frac{1}{2}\frac{d^2}{dr^2}\phi_{n,l}(r) + \left[\frac{l(l+1)}{2r^2} + V_{\text{eff}}(r) - \varepsilon_{n,l}\right]\phi_{n,l}(r) = 0, \qquad (10.12)$$

which can be solved for bound states with the same boundary conditions and normalization as for the one-electron atom.

Achieving Self-Consistency

The general form for solution of self-consistent equations has been given in Section 7.4. In a closed-shell case the effective potential is spherically symmetric (Exercise 10.5). In most cases, strongly bound states pose no great problems and the linear-mixing algorithm, Eq. (7.33), is usually sufficient. (A value of $\alpha < 0.3$ will converge for most cases but may have to be reduced for heavy atoms.) However, weakly bound "floppy" states may need the more sophisticated methods described in Section 7.4. and systems with near degeneracies (e.g., energies of 3d and 4s states in transition metal atoms) may present special problems, since the order of states may change during the iterations, so that filling the states according to the minimum energy principle leads to abrupt changes in the potential. Since this principle really applies only at the minimum, often there is a simple solution; however, in some cases, there is no stable solution.

An essential part of the problem is the calculation of the Hartree or Coulomb potential. There are two approaches: solution of the Poisson equation [485] or analytic formulas that can be written down for the special case of wavefunctions that are radial functions times spherical harmonics [484]. The former approach has the advantage that the Poisson equation,

$$\frac{d^2}{dr^2}V_{\text{Hart}}(r) = -4\pi n(r), \qquad (10.13)$$

has the form of the second-order equation (L.1) and can be solved by numerical methods similar to those used for the Schrödinger equation [485]. The latter approach involves expressions that are applicable to open-shell atoms (Section 10.4) and special cases of the Fock integrals needed in for Hartree–Fock calculations [484].

The exchange–correlation potential $V_{\text{eff}}(r)$ depends on the type of independent-particle approximation. An explicit expressions for the state-dependent exchange potential $V_x^{n,l}(r)$ in the Hartree–Fock approximation is given in the following section. (The exchange potential is purely radial for a closed-shell atom and is shown in Exercise 10.5.) The OEP formulation of the energy is exactly the same as for Hartree–Fock but the potential is required to be $V_x(r)$ independent of the state. In approximations such as the LDA and GGAs, explicit expressions are given in Appendix B. Examples of results for selected spherically symmetric atoms with various functionals are given in Table 9.1 and are discussed further in Section 10.5.

10.3 Spin–Orbit Interaction

Relativistic effects are essential for heavy atoms. Fortunately, they originate deep inside the core, and they act on each electron independently, so that it is sufficient to solve the relativistic equations in a spherical atomic geometry. The results carry over to molecules and solids essentially unchanged. Appendix O is devoted to the Dirac equation and the conclusion that, for cases where the effects are not too large, they can be included in the ordinary nonrelativistic Schrödinger equation by adding a spin–orbit interaction term in the hamiltonian (see Eq. (O.8))

$$H_{SO} = \frac{e\hbar}{4m^2c^2}(\mathbf{p} \times \nabla V) \cdot \boldsymbol{\sigma} \tag{10.14}$$

and other scalar relativistic terms that shift the energies of the states. This may appear to be an innocuous addition to the hamiltonian and it is typically much smaller than other terms; however, it has major qualitative consequences that are different from anything that can be produced by a potential. The spin–orbit interaction opens gaps that cannot occur otherwise, and it leads to such fascinating phenomena as topological insulators in Chapters 25–28.

In an actual calculation on solids or molecules, relativistic effects can be included directly within the augmentation methods (Chapter 16), where the calculation in the spherical regions is analogous to that for an atom or in local orbitals centered on the nuclei (Chapters 14 and 15). Relativistic effects can be built into pseudopotentials as described in Section 11.5 by generating them using relativistic atomic calculations; the pseudopotentials can then be used in a nonrelativistic Schrödinger equation to determine the valence states including relativistic effects [486, 487].

10.4 Open-Shell Atoms: Nonspherical Potentials

The term "open shell" denotes cases where the spins are not paired and/or the angular momentum states $m = -l, \ldots, l$, are not completely filled for a given l. Then the proper classification is in terms of "multiplets," with given total angular momentum $J = \{j, m_j\}$ that are linear combinations of the space $(L = \{l, m_l\})$ and spin $(S = \{s, m_s\})$ variables. In general, one must deal with multiple-determinant wavefunctions. Even though the external potential (the nuclear potential for an atom) has spherical symmetry, the effective independent-particle potential $V_{\text{eff}}(\mathbf{r})$ does not, since it depends on the occupations of the orbitals. The only simplification is that one can choose the axes of quantization so that $V_{\text{eff}}(\mathbf{r}) = V_{\text{eff}}(r, \theta)$ has cylindrical symmetry. Fortunately, a method due to Slater [453] shows how to reduce all the needed calculations to purely radial calculations, by using symmetry to choose appropriate multiplets. There are no general rules, but an extensive compilation of cases is given in appendix 21 of [484].[1] In the atom one needs to consider only the cases where V_{eff} is purely radial $V_{eff}(r)$ or cylindrical $V_{eff}(r,\theta)$.

For open-shell problems there are a set of approximations with various degrees of accuracy:

- Restricted: Treat the problem as spherical (derive $V_{eff}(r)$ by a spherical average over any nonspherical terms) and independent of spin (average over spin states so that orbitals are the same for each spin state). This is the correct form for closed-shell, spin $= 0$ systems and, with care, can be viewed as an approximation for open shells.
- Spin unrestricted: Treat as spherical but allow the potential and the orbitals to depend on the spin. This is the correct form for half-filled shells with maximum spin so that they are closed shell for each spin separately. This case has been treated in the section on closed shells.
- Unrestricted: Treat the full problem in which only the total m_l and m_s are good quantum numbers. In this case, Slater's method [484] can be used to simplify the problem.

Equations for Open-Shell Cases

For the fully unrestricted case, additional complications arise from the electron–electron interaction terms. If we chose an axis then the density $n(r,\theta)$ and the potential $V_{xc}(r,\theta)$ have cylindrical symmetry. In the Kohn–Sham approach, the wavefunctions are expanded in spherical harmonics and angular integrals with $V_{xc}(r,\theta)$ must be done numerically because the nonlinear relation of $V_{xc}(\mathbf{r})$ to $n(\mathbf{r})$ means that an expansion of $V_{xc}(r,\theta)$ in spherical harmonics has no maximum cutoff in L. Also the Coulomb potential has multi-pole moments and the solution of the Poisson equation is not as simple as in the spherical case.

For the open-shell case, the Hartree–Fock equations are actually simpler because the exchange term can be expanded in a finite sum of spherical harmonics. In order to calculate the matrix elements of the electron–electron interaction, one can use the well-known expansion [480] in terms of the spherical harmonics (Appendix K), which allows the factorization into terms involving $\mathbf{r}_1$ and $\mathbf{r}_2$

$$\frac{1}{|\mathbf{r}_1 - \mathbf{r}_2|} = 4\pi \sum_{l=0}^{\infty} \sum_{m=-l}^{l} \frac{1}{2l+1} \frac{r_<^l}{r_>^{l+1}} Y_{-lm}^*(\theta_2,\phi_2) Y_{-lm}(\theta_1,\phi_1), \tag{10.15}$$

where $r_<$ and $r_>$ are the lesser and the greater of r_1 and r_2. Using this expression, matrix elements involving the orbitals i,j,r,t can be written

$$\left\langle i\,j\left|\frac{1}{r_{12}}\right|r\,t\right\rangle = \delta(\sigma_i,\sigma_r)\,\delta(\sigma_j,\sigma_t)\,\delta(m_i+m_j,m_r+m_t)$$

$$\times \sum_{k=0}^{k_{max}} c^k(l_i,m_i;l_r,m_r)\,c^k(l_t,m_t;l_j,m_j)\,R^k(i,j;r,t), \tag{10.16}$$

---

[1] In general one needs nondiagonal Lagrange multipliers $\varepsilon_{n,l;n',l'}$ [484], but these terms appear to be small [484].

where

$$R^k(i,j;r,t) = \int_0^\infty \int_0^\infty \phi_{n_i,l_i}^\dagger(r_1)\phi_{n_j,l_j}^\dagger(r_2)\frac{r_<^k}{r_>^{k+1}}\phi_{n_r,l_r}(r_1)\phi_{n_t,l_t}(r_2)dr_1 dr_2. \tag{10.17}$$

The δ functions in Eq. (10.16) reflect the fact that the interaction is spin independent and conserves the z component of the angular momentum. The angular integrals can be done analytically resulting in the Gaunt coefficients $c^k(l,m;l',m')$, which are given explicitly in [484] and Appendix K. Fortunately, the values of R^k are only needed for a few values of k; the maximum value is set by the vector-addition limits $|l - l'| \le k_{max} \le |l + l'|$, and, furthermore, $c^k = 0$ except for $k + l + l' = $ even.

The radial Hartree–Fock equations are derived by functional derivatives of the energy with respect to the radial functions $\psi_{n,l,m,\sigma}(r)$. If we define a function[2]

$$Y^k(n_i,l_i;n_j,l_j;r) = \frac{1}{r^k}\int_0^r dr' \, \phi_{n_i,l_i}^\dagger(r')\phi_{n_j,l_j}(r')r'^k + r^{k+1}\int_r^\infty dr' \, \phi_{n_i,l_i}^\dagger(r')\phi_{n_j,l_j}(r')\frac{1}{r'^{k+1}}$$

$$\tag{10.18}$$

and use the relation between the Gaunt and Clebsch–Gordan coefficients Eq. (K.17), then the Hartree potential is given by

$$V_{\text{Hartree}}(r) = \sum_{\sigma=\uparrow,\downarrow}\sum_{j=1,N_\sigma}\sum_{k=0}^{\min(2l_i,2l_j)}(-1)^{m_i+m_j}\frac{(2l_i+1)(2l_j+1)}{(2k+1)^2}\frac{Y^k(n_j,l_j;n_j,l_j;r)}{r}$$

$$\times C_{l_i0,l_i0}^{k0} C_{l_j0,l_j0}^{k0} C_{l_im_i,l_i-m_i}^{k0} C_{l_jm_j,l_j-m_j}^{k0}, \tag{10.19}$$

and the exchange potential acting on state n_i, l_i, σ_i can be written

$${}^\bullet V_x(r) = -\sum_{\sigma=\uparrow,\downarrow}\delta(\sigma,\sigma_i)\sum_{j=1,N_\sigma}\sum_{k=|l_i-l_j|}^{l_i+l_j}\frac{(2l_i+1)(2l_j+1)}{(2k+1)^2}$$

$$\times \left[C_{l_i0,l_j0}^{k0} C_{l_im_i,l_j-m_j}^{km_i-m_j}\right]^2 \frac{Y^k(n_j,l_j;n_j,l_j;r)}{r}\frac{\phi_{n_j,l_j}(r)}{\phi_{n_i,l_i}(r)}, \tag{10.20}$$

where $C_{j_1m_1,j_2m_2}^{j_3m_3}$ are the Clebsch–Gordan coefficients defined in Appendix K.

10.5 Example of Atomic States: Transition Elements

Examples for selected spherically symmetric atoms are given in Table 9.1 using various functionals for exchange and correlation, respectively. Three atoms shown there (He, Be, and Ne) are closed shell and the other two (H and N) are half-filled shells in which the spatial wavefunction is spherically symmetric. The latter are called "spin unrestricted" and require spin functionals with separate potentials for $V_{\text{eff}}^\sigma(r)$ for spin up and down. The results show that the local approximation works remarkably well, considering that it is derived

---

[2] This follows the definition of Slater [483], p. 180.

from the homogeneous gas, and that GGAs in general improve the overall agreement with experiment.

Hydrogen is the special case where the one-electron solution for the ground-state energy is exact. This is satisfied in any theory that has no self-interactions, including Hartree–Fock and exact exchange (EXX). The results in the tables indicate the error in the other functionals in this limit. The accuracy is quite remarkable, especially for functionals derived from the homogeneous gas, which supports the use of the functionals for the entire range from homogeneous solids to isolated atoms. Nevertheless, there are important errors. In particular, there are large effects on the eigenvalues due to the long-range asymptotic form of the potential. The form is correct in Hartree–Fock and EXX calculations that take into account the nonlocal exchange but is incorrect in local and GGA approximations. The effects are large in one- and two-electron cases, shown explicitly in Section 9.10, but are smaller in heavier atoms with many electrons.

As an example of a many-electron atom, the wavefunctions for spin-polarized Mn are shown in Fig. 10.1, calculated without relativistic corrections and spin–orbit coupling. The states of this transition metal element illustrate the difference between the loosely bound outer 4s states and the more localized 3d states. The atom is in the $3d^{5\uparrow}4s^2$ state and the right-hand side of the figure shows the up and down wavefunctions for the outer states. Since the d shell is filled for one spin (Hund's first rule for maximum spin), the atom is in a spherically symmetric spatial state. Note that the 3d states are actually in the same spatial region as the strongly bound "semicore" 3s and 3p states.

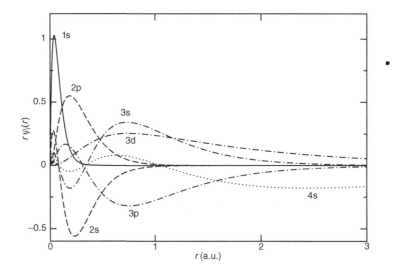

Figure 10.1. Radial wavefunction $\phi_l(r) = r\psi_l(r)$ for Mn in the $3d^{5\uparrow}4s^2$ state showing all the orbitals. Note that the 4s states are much more delocalized than the 3d states even though they have similar energies. In contrast, the maximum in the 3d is close to that of the 3s and 3p, even though these are much lower in energy and are called "semicore" states. Since the atom is spin polarized, the orbital shapes depend on the spin. There is a clear effect on the 3d and 4s, which is not shown for simplicity.

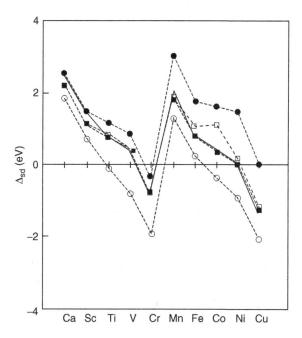

Figure 10.2. Promotion energies for d → s electrons in the 3d transition metal series. Shown are the energies $\Delta_{sd} \equiv E_{total}[3d^{n-1}4s^1] - E_{total}[3d^{n-2}4s^2]$ from experiment (solid line) and from calculated total energy differences with different functionals. This shows that LSDA (open circles), typically tends to underestimate the promotion energy, i.e., the d states are underbound, whereas full nonlocal exchange (solid circles) tends to have the opposite effect. Including correction for self-interaction (open squares) and screening, the exchange (solid squares) improves the results considerably. Such effects carry over to solids as well. From [488].

Atomic calculations can be used to gain insight into practical aspects of density functional theory and how it can be expected to work in solids. Transition metals provide an excellent example because the d states retain much of their atomic character in the solid. For example, the relative energies of the 3d and 4s states can be expected to carry over to the solid. Figure 10.2 shows the promotion energies for transferring a d electron to an s state. The energies plotted are the experimental promotion energies and the calculated values with different functionals. The calculated energies are *total energy differences* $\Delta_{sd} \equiv E_{total}[3d^{n-1}4s^1] - E_{total}[3d^{n-2}4s^2]$, not differences of eigenvalues, since Δ_{sd} is a better measure of the true energy for promotion than the eigenvalue difference. The primary point to notice is that there are opposite tendencies for the local approximation (LSDA) and for exact nonlocal exchange (EXX), which is very close to Hartree–Fock. The former leads to underbinding of the d relative to the s state, whereas the latter leads to overbinding. The example of screened exchange is one version of hybrid functionals (Section 9.3) that tend to give results intermediate between Hartree–Fock and LDA.

Three conclusions can be drawn from the atomic calculations that are very important for applications in molecules and solids:

- Typical density functional theory calculations using LDA or GGA functionals can be expected to give errors in the relative positions of bands, especially bands of different character, such as localized 3d versus delocalized 4s states. The errors can be of the order of electron volts.
- Hybrid functionals (Section 9.3) improve the accuracy of many results, such as bandgaps as shown in Fig. 2.24. They are widely applied in chemistry where implementation is straightforward as a mixture of Hartree–Fock and density functional calculations. In a solid Hartree–Fock is more difficult, but methods have been developed explicitly for the purpose of using hybrid functionals.
- The results for transition metals illustrate the large difference between valence orbitals that are delocalized (s and p) and the d states that are much more localized. The effects of exchange and correlation are much more important in highly localized orbitals, leading to effects of strong correlation in transition metal systems. The essence of the self-interaction correction (SIC) and the "DFT+U" approach (Section 9.6) is to include orbital-dependent interactions that describe this large difference. In atoms the states of each symmetry are well defined; however, in a solid there are choices in how to define localized orbitals. Nevertheless, these approaches can provide significantly improved descriptions of strongly correlated systems like transition metal oxides [459].

10.6 Delta-SCF: Electron Addition, Removal, and Interaction Energies

In localized systems, electron excitation, addition, and removal energies can all be calculated as *energy differences* $\Delta E_{12} = E_2 - E_1$ for a transition between states 1 and 2, instead of eigenvalues calculated for state 1 or 2. This is known as "ΔSCF" and in self-consistent field methods, it produces more accurate results since the energy difference includes effects of relaxation of all the orbitals. Following the Slater transition state argument ([489], p. 51), the energy difference can be approximated by the eigenvalue calculated at the occupation *halfway between the two states*. For example, an electron removal energy is the eigenvalue when $1/2$ an electron is missing in the given state; a transition energy is the eigenvalue difference calculated when $1/2$ an electron is transferred between the two states,

$$\Delta E(N \rightarrow N-1) = E(N-1) - E(N) \approx \epsilon_i\left(N - \frac{1}{2}\right), \qquad (10.21)$$

where i denotes a particular state and $N - \frac{1}{2}$ means the density is $n(\mathbf{r}) - \frac{1}{2}|\psi_i(\mathbf{r})|^2$. See Exercise 10.8 for a statement of the ideas involved in the arguments and their proof. A schematic figure for the variation of the energy with occupation N is shown in the right side of Fig. 9.2.

The ΔSCF or transition state methods can be used in atomic calculations and compared with experiment. For example, in [490] it was shown that results from both ΔSCF and transition state calculations using LDA are in good agreement with experiment for the first and second ionization energies of 3d electrons in Cu. In addition, it is straightforward to calculate interaction energies as energy differences. An effective interaction energy that includes relaxation of the orbitals is given by the difference of first and second ionization energies, which in terms of the transition state rule can be written

$$U \equiv [E(N-1) - E(N)] - [E(N-2) - E(N-1)] \approx \epsilon_i\left(N - \frac{1}{2}\right) - \epsilon_i\left(N - \frac{3}{2}\right).$$

(10.22)

The LDA calculation [490] for the 3d states of a Cu atom in free space yields $U_{av} = E(d^{10},s^1) + E(d^8,s^1) - 2E(d^9,s^1) = 15.88$ eV compared to the experimental value 16.13 eV for the average of the multiplet energies for the three occupations. See Exercise 10.11 for discussion of the interpretation and suggested exercises.

In Section 9.6 the same approach is applied to a localized state in a molecule or solid, but in such cases it is not obvious how to identify the localized orbital whose occupation can be varied, and one must take into account the screening by the rest of the system. Here it is useful to point out the way that an atomic calculation can be used to estimate the screened U in a solid. The addition of a charge in a d or f state in a solid is screened by the other electrons; within a short distance the charge is reduced to $\approx \pm 1/\epsilon$ where the dielectric constant $\epsilon \gg 1$. This can be mimicked in an atom if the charge in the d or f state is compensated by an electron in an s state [491]. For example, in a Cu atom the energy differences $E(d^{10},s^0) + E(d^8,s^2) - 2E(d^9,s^1)$ can be interpreted as a screened interaction, with the values 3.96 eV found in an LDA calculation [490] compared to 4.23 eV in experiment. Similar values are found for all 3d transition metals. For the lanthanides the result is $\approx$ 6–7 eV [491, 492]. These are remarkably close to the values for screened effective interactions found in calculations for 3d and 4f states in solids, as discussed in Section 9.6 and in greater detail in [1].

10.7 Atomic Sphere Approximation in Solids

In a solid the wavefunctions tend to be atomic-like near each atom. This results from the full calculations are described in Part IV, but it is instructive to see that qualitative (sometimes quantitative) information about electronic bands, pressure, and energy of simple solids can be derived from calculations with spherical symmetry, analogous to the usual atomic calculations with only one difference: a change of boundary conditions to mimic the extreme limits of each band in a solid. Such calculations are also instructive because they are very closely related to the radial atomic-like calculations used in augmented plane wave (APW), linear combination of muffin-tin orbitals (LMTO), and KKR methods of Chapters 16 and 17.

The basic ideas are in the original work of Wigner and Seitz [54], extended by Andersen [495] to describe the width of bands formed by states with a given angular momentum. The environment of an atom in a close-packed solid is mimicked by boundary conditions on an atomic sphere, i.e., approximating the Wigner–Seitz cell as a sphere. As indicated in Fig. 10.3, for each angular momentum l the free-atom boundary conditions are replaced by the condition that the wavefunction be zero at the boundary (the highest-energy

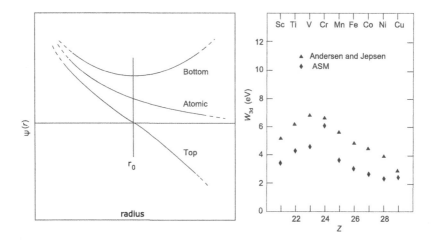

Figure 10.3. Left: Schematic figure of the radial wavefunction in the atomic sphere approximation (ASA) to a solid, where r_0 is the Wigner–Seitz radius, roughly $1/2$ the distance to a neighboring atom. The lines denote the different boundary conditions that correspond to the "bottom" lowest-energy bonding-type state, the "top" highest-energy antibonding-type state, and the usual free-atom state. The difference in energy from bottom to top is an estimate of the band width in a solid. Right: estimates of the d-band width for the 3d transition metals from Eq. (10.25) (called "ASM") compared to full calculations using the LMTO method (Chapter 17) by Andersen and Jepsen [493]. From Straub and Harrison [494].

antibonding-type state that corresponds to the top of the band) or have zero derivative at the boundary (the lowest-energy bonding-type state that corresponds to the bottom of the band). The difference is the band width W_l for angular momentum l.

This simple picture contains the important ingredients for understanding and semiquantitative prediction of band widths in condensed matter. The width is directly related to the magnitude, the wavefunction, and its slope at the boundary. Thus the width varies from narrow-band atomic-like to wide-band delocalized states exactly as indicated in the first figure of this book, Fig. 1.1, as there is increased overlap of the atomic states. Furthermore, it is reasonably accurate as illustrated on the right-hand side of Fig. 10.3 from the work of Straub and Harrison [455] using Eq. (10.25). It gives an estimate that is particularly good in close-packed systems but also is a good starting point for liquids and even dense plasmas [496–499] One can also calculate properties like the pressure, as discussed in Section I.3. Thus a single standard atomic calculation and Eq. (10.25) provide a great start to understanding electronic structure of condensed matter!

Explicit expressions for the band widths can be derived from the radial equation, (10.4), for the wavefunction $\psi_{n,l}(r)$. The band formed for each state n, l is considered separately, and we can drop the subscripts to simplify the equations. Consider any two solutions $\psi^1(r)$ and $\psi^2(r)$ with eigenvalues ε^1 and ε^2 obtained with two different specific boundary conditions. Let the equation for $\psi^1(r)$ be multiplied by $r^2\psi^2(r)$ and integrated from $r = 0$ to the boundary $r = r_0$, and similarly for the equation for $\psi^2(r)$. Integrating by parts and subtracting the equations leads to [494] (Exercise 10.12),

$$-\frac{1}{2}r_0^2\left(\psi^2\frac{d\psi^1}{dr}-\psi^1\frac{d\psi^2}{dr}\right)_{r-r_0}=(\varepsilon^1-\varepsilon^2)\int_0^{r_0}dr\,r^2\psi^1\psi^2. \qquad (10.23)$$

If we let $\psi^2(r)$ be the solution for the top of the band ($\psi^2(r_0)=0$) and $\psi^1(r)$ for the bottom ($d\psi^1(r)/dr=0$ at $r=r_0$), then this equation can be written

$$W\equiv\varepsilon^2-\varepsilon^1=-\frac{1}{2}\frac{r_0^2\left(\psi^1\frac{d\psi^2}{dr}\right)_{r=r_0}}{\int_0^{r_0}dr\,r^2\psi^1\psi^2}. \qquad (10.24)$$

This gives the width W for each n,l in terms of the two solutions with the boundary conditions described above.

Finally, a simple, insightful expression for the band width as a function of the Wigner–Seitz radius r_0 can be derived [494] from a single atomic calculation with the usual boundary conditions. As suggested by the interpretation of the bottom and top of the band as bonding and antibonding, as illustrated in Fig. 10.3, the value of the bonding function at r_0 is approximately twice the value of the atomic function ψ^a, and similarly for the slope. Further approximating the product to be $\psi^1\psi^2\approx[\psi^a]^2$ leads to the very simple expression (see Exercise 10.12)

$$W\approx-2\frac{r_0^2\left(\psi^a\frac{d\psi^a}{dr}\right)_{r=r_0}}{\int_0^{r_0}dr\,r^2(\psi^a)^2}. \qquad (10.25)$$

The right-hand side of Fig. 10.3 shows the band widths for d-bands in transition metals calculated from this simple formula compared to full calculations using the LMTO method (Chapter 17).

SELECT FURTHER READING

Theory and calculations for atoms:

Fischer, C. F., *The Hartree–Fock Method for Atoms: A Numerical Approach* (John Wiley & Sons, New York, 1977).

Slater, J. C., *Quantum Theory of Atomic Structure*, vols. 1–2 (McGraw-Hill, New York, 1960).

Tinkham, M., *Group Theory and Quantum Mechanics* (McGraw-Hill, New York, 1964).

Early numerical work:

Herman, F. and Skillman, S., *Atomic Structure Calculations* (Prentice-Hall, Engelwood Cliffs, NJ, 1963).

Relativistic theory:

Koelling, D. D. and Harmon, B. N., "A technique for relativistic spin-polarized calculations," *J. Phys. C* 10:3107–3114, 1977.

Kübler, J., *Theory of Itinerant Electron Magnetism* (Oxford University Press, Oxford, 2001).

Kübler, J. and Eyert, V., "Electronic structure calculations," in *Electronic and Magnetic Properties of Metals and Ceramics*, edited by K. H. J. Buschow (VCH-Verlag, Weinheim, Germany, 1992), p. 1.

MacDonald, A. H., Pickett, W. E., and Koelling, D., "A linearised relativistic augmented-plane-wave method utilising approximate pure spin basis functions," *J. Phys. C: Solid State Phys.* 13:2675–2683, 1980.

Instructive methods for calculations:
Koonin, S. E. and Meredith, D. C., *Computational Physics* (Addison Wesley, Menlo Park, CA, 1990).

Exercises

10.1 Show explicitly that the wave equation can indeed be written in the form of Eq. (10.4).

10.2 Derive the form of the radial equation, (10.8) in terms of the transformed variable $\rho \equiv \ln(r)$. Give two reasons why a uniform grid in the variable ρ is an advantageous choice for an atom.

10.3 Show that the Hartree–Fock equations are exact for the states of H. Show that the change in energy computed by *energy differences* gives exact excitations but that the eigenvalues do not.

10.4 Show that the OEP equations are exact for the states of H, just like Hartree–Fock. But unlike Hartree–Fock the eigenvalues give exact excitation energies.

10.5 Show that the general Hartree–Fock equations simplify in the closed-shell case so that the exchange potential is spherically symmetric.

10.6 Show that for the ground state of He the general Hartree–Fock equations simplify to the very simple problem of one electron moving in the average potential of the other, with both electrons required to be in the same spatial orbital.

10.7 There are results that emerge in relativistic quantum mechanics that may be surprising. For example, show that there is a 2p state that has nonzero expectation value at the origin, whereas it is zero in the nonrelativistic theory.

10.8 The Slater transition state argument ([489], p. 51) is based on two facts. First, an eigenvalue is the derivative of the energy with respect to the occupation of the given state (the "Janak theorem"), and, second, that the eigenvalue varies with occupation and can be represented in a power series.
(a) Using these facts derive the "halfway" rule.
(b) Argue that one can derive the "halfway" rule based purely on the fact that one wants a result that is symmetric between the two states.
(c) Derive the explicit expression (10.21) for electron removal.

10.9 Solve the Schrödinger equation in Section 10.1 for a particle in a spherical box of radius R. If the boundary conditions are that $\psi = 0$ at $r = R$, show that the solutions are $\psi(r) = \sin kr/r$ and derive the eigenvalues and normalization factors for the states with the three lowest energies. Show that all energies scale as $\propto 1/R^2$.

10.10 Derive the pressure $-dE/d\Omega$ from the expression for the energy in the problem above. Show that this is equivalent to the expression for the pressure in a spherical geometry given in Eq. (I.8).

10.11 The expression (10.22) provides a way to calculate interactions.
(a) Show that these are "effective" in the sense that orbital relations are included and are *exact* if the energies $E(N)$, $E(N-1)$, and $E(N-2)$ are exact.

(b) Derive Expression (10.22) using the same arguments as in Exercise 10.8.

(c) Use an atomic code to calculate the first and second ionization energies of 3d electrons in Cu. The difference is the effective d–d interaction. A better measure of the net effect in a solid is to calculate the difference $E(3d^9) - E(3d^{84}s^1)$. Compare your results with those of [490]. In this case, the effective d–d interaction is decreased because the added s electron "screens" the change in charge of the d state. As argued in [492] this is close to the screening that occurs in a solid; hence, the screened interaction is the appropriate effective interaction in the solid.

10.12 Following the arguments given in conjunction with Eqs. (10.23)–(10.25), derive the approximate expressions (10.24) and (10.25) for the band width. The full argument requires justifying the argument that this corresponds to the maximum band width in a solid and deriving the explicit expression using the linearized formulas for energy as a function of boundary condition.

10.13 The wavefunction for atomic hydrogen can be used to estimate hydrogen band widths at various states, using the approximate form of Eq. (10.25). Apply this approach to the H_2 molecule to calculate bonding/antibonding splitting and compare these with the results shown in Fig. 8.2. Use this expression to derive a general argument for the functional form of the splitting as a function of proton separation R at large R. Evaluate explicitly at the equilibrium R and compare with Fig. 8.2. Calculate the band width expected for hydrogen at high density ($r_s = 1.0$) where it is expected to be stable as a close-packed crystal with 12 neighbors. (The result can be compared with the calculations in Exercises 12.13 and 13.4.)

10.14 Use an atomic code (possibly modified to have different boundary conditions) to calculate the band widths for elemental solids using the approach described in Section 10.7. As an example consider 3d and 4s bands in fcc Cu. Compare these with the bands shown in Fig. 16.4.

11

Pseudopotentials

Summary

The fundamental idea of a "pseudopotential" is the replacement of one problem with another. The primary application in electronic structure is to replace the strong Coulomb potential of the nucleus and the effects of the tightly bound core electrons by an effective ionic potential acting on the valence electrons. A pseudopotential can be generated in an atomic calculation and then used to compute properties of valence electrons in molecules or solids, since the core states remain almost unchanged. Furthermore, the fact that pseudopotentials are not unique allows the freedom to choose forms that simplify the calculations and the interpretation of the resulting electronic structure. The advent of *ab initio* "norm-conserving" and "ultrasoft" pseudopotentials has led to accurate calculations that are the basis for much of the current research and development of new methods in electronic structure, as described in the following chapters.

Many of the ideas originated in the orthogonalized plane wave (OPW) approach that casts the eigenvalue problem in terms of a smooth part of the valence functions plus core (or core-like) functions. The OPW method has been brought into the modern framework of total energy functionals by the projector augmented wave (PAW) approach that uses pseudopotential operators but keeps the full core wavefunctions.

11.1 Scattering Amplitudes and Pseudopotentials

It is useful to first consider scattering properties of a localized spherical potential at energy ε, which can be formulated concisely in terms of the phase shift $\eta_l(\varepsilon)$, which determines the effect on the wavefunction outside the localized region. This is a central concept for many phenomena in physics, such as scattering cross-sections in nuclear and particle physics, resistance in metals due to scattering from impurities, and electron states in crystals described by phase shifts in the augmented plane wave and multiple scattering KKR methods (Chapter 16). One of the early examples of this idea is illustrated in Fig. 11.1, taken from papers by Fermi and coworkers. Exacty the same figure appears in two papers,

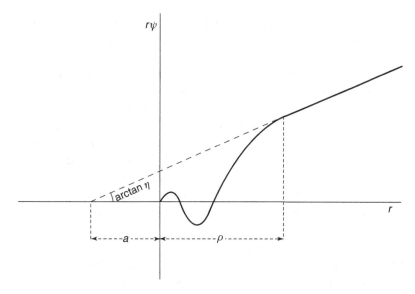

Figure 11.1. Radial wavefunction $\phi = r\psi$ for low-energy scattering as illustrated in a figure from the 1934 and 1935 papers of Fermi and coworkers for low-energy electron scattering from atoms [65] and neutron scattering from nuclei [500]. The nodes in the wavefunction near the origin show that the potential is attractive and strong enough to have bound states. The cross-section for scattering from the localized potential is determined by the phase shift (or equivalently the extrapolated scattering length as discussed in Exercise 11.1) and is the same for a weaker pseudopotential with the same phase shift modulo 2π.

one on low-energy electron scattering from atoms [65] and the other on low-energy neutron scattering from nuclei [500]. The incident plane wave is resolved into spherical harmonics as in Fig. J.1, and the figure shows the radial wavefunction for one angular momentum l in a scattering state with a small positive energy. The closely spaced nodes in the wavefunction near the origin indicate that the kinetic energy is large, i.e., that there is a strong attractive potential. In fact, there must be lower-energy bound states (with fewer nodes) to which the scattering state must be orthogonal. The same effect on the wavefunction outside the core can be produced by a weaker potential that has no bound states.

It is instructive to consider the changes in the wavefunction $\phi = r\psi$ outside the scattering region as a function of the scattering potential. If there were no potential, i.e., phase shift $\eta_l(\varepsilon) = 0$, then Eq. (J.4) leads to $\phi \propto r j_l(\kappa r)$, which extrapolates to zero at $r = 0$. In the presence of a potential the wavefunction outside the central region is also a free wave but phase shifted as in Eq. (J.4). A weak potential leads to a small phase shift $\eta < 2\pi$. If the potential is made more attractive, the phase shift increases with a new bound state formed for each integer multiple 2π. From the explicit solution Eq. (J.4), it is clear that the wavefunction outside the central region is exactly the same for any potential that gives the same phase shift $\eta_l(\varepsilon)$ modulo any multiple of 2π. In particular, the scattering in Fig. 11.1 can be reproduced at the given energy ε by a weak potential that has no bound states and a scattering state with no nodes. For example, one can readily find a square well with the same scattering properties at this energy (see Exercise 11.2). The aim of pseudopotential theory is to find useful pseudopotentials that faithfully represent the scattering over a desired energy range.

This is the essential idea for pseudopotentials, to describe the low-energy states – the valence states that are responsible for bonding – without explicitly treating the tightly bound core states. In a solid or molecule the potential is not zero outside the core region, but it is slowly varying and the same approach can be used to replace the strong potential with a weaker one. Perhaps the first use of pseudopotentials in solids was by Hellmann [66, 67] in 1935, who developed an effective potential for scattering of the valence electrons from the ion cores in metals and formulated a theory for binding of metals that is remarkably similar to present-day pseudopotential methods. The potentials, however, were not very weak [501], so that the calculations were not very accurate using perturbation methods available at the time.

Interest in pseudopotentials in solids was revived in the 1950s by Antoncik [502, 503] and Phillips and Kleinman [504], who showed that the orthogonalized plane wave (OPW) method of Herring [64, 505] (see Section 11.2) can be recast in the form of equations for the valence states only that involve only a weaker effective potential. Their realization that the band structures of sp-bonded metals and semiconductors could be described accurately by a few empirical coefficients led to the basic understanding of a vast array of properties of sp-bonded metals and semiconductors. Excellent descriptions of the development of pseudopotentials before 1970 can be found in the review of Heine and Cohen [506, 507] and in the book *Pseudopotentials in the Theory of Metals* by Harrison [508].

Today modern pseudopotential calculations are based on *ab initio* approaches in which the pseudopotential is derived from calculations on an atom without any fitting to properties of a molecule or solid. "Norm-conserving" potentials (Sections 11.4–11.8) are in a sense a return to the model potential concepts of Fermi and Hellmann, but with important additions. The requirement of norm conservation is the key step in making *accurate, transferable* pseudopotentials, which is essential so that a pseudopotential constructed in one environment (usually the atom) can faithfully describe the valence properties in different environments including atoms, ions, molecules, and condensed matter.[1] The basic principles are given in some detail in Section 11.4 because they are closely related to scattering phase shifts (Appendix J), the augmentation approaches of Chapter 16, and the properties of the wavefunctions needed for linearization and given explicitly in Section 17.2. Section 11.5 is devoted to the generation of "semilocal" potentials $V_l(r)$ that are l-dependent, i.e., act differently upon different angular momenta l. In Section 11.8 we describe the transformation to a separable, fully nonlocal operator form that is often advantageous.

---

[1] Of course, there is some error due to the assumption that the cores do not change. Many tests have shown that this is an excellent approximation in atoms with small, deep cores. Errors occur in cases with shallow cores and requiring high accuracy.

This approach has been extended by Blöchl [509] and Vanderbilt [510], who showed that one can make use of auxiliary localized functions to define "ultrasoft pseudopotentials" (Section 11.11). By expressing the pseudofunction as the sum of a smooth part and a more rapidly varying function localized around each ion core (formally related to the original OPW construction [64] and the Phillips–Kleinman–Antoncik transformation), the accuracy of norm-conserving pseudopotentials can be improved, while at the same time making the calculations less costly (although at the expense of added complexity in the programs).

The projector augmented wave (PAW) formulation (Section 11.12) is a reformulation of the OPW method into a form that is particularly appropriate for density functional theory calculations of total energies and forces. The valence wavefunctions are expressed as a sum of smooth functions plus core functions, which leads to a generalized eigenvalue equation just as in the OPW approach. Unlike pseudopotentials, however, the PAW method retains the entire set of all-electron core functions along with smooth parts of the valence functions. Matrix elements involving core functions are treated using muffin-tin spheres as in the augmented methods (Chapter 16). As opposed to augmented methods, however, the PAW approach maintains the advantage of pseudopotentials that forces can be calculated easily.

The concept of a pseudopotential is not limited to reproducing all-electron calculations within independent-particle approximations, such as the Kohn–Sham density functional approach. In fact, the original problem of "replacing the effects of core electrons with an effective potential" presents a larger challenge: can this be accomplished in a true many-body theory taking into account the fact that all electrons are indistinguishable? Although the details are beyond the scope of the present chapter, Section 11.13 provides the basic issues and ideas for construction of pseudopotentials that describe the effects of the cores *beyond the independent electron approximation*.

11.2 Orthogonalized Plane Waves (OPWs) and Pseudopotentials

Orthogonalized plane waves (OPWs), introduced by Herring [64, 505] in 1940, were the basis for the first quantitative calculations of bands in materials other than sp-bonded metals (see, e.g., [59, 511, 512] and the review by Herman [68]). The calculations of Herman and Callaway [59] for Ge, done in the 1950s, are shown in Fig. 1.3; similarly, OPW calculations provided the first theoretical understanding that Si is an indirect bandgap material with the minimum of the conduction band near the X ($\mathbf{k} = (1,0,0)$) zone boundary point [513, 514]. Combined with experimental observations [515], this work clarified the nature of the bands in these important materials. The OPW method is described in this chapter because it is the direct antecedent of modern pseudopotential and projector augmented wave (PAW) methods.

The original OPW formulation [64] is a very general approach for construction of basis functions for valence states with the form

$$\chi_{\mathbf{q}}^{\text{OPW}}(\mathbf{r}) = \frac{1}{\Omega} \left\{ e^{i\mathbf{q}\cdot\mathbf{r}} - \sum_j \langle u_j | \mathbf{q} \rangle u_j(\mathbf{r}) \right\}, \tag{11.1}$$

where

$$\langle u_j | \mathbf{q} \rangle \equiv \int d\mathbf{r} u_j(\mathbf{r}) e^{i\mathbf{q} \cdot \mathbf{r}}, \tag{11.2}$$

from which it follows that $\chi_{\mathbf{q}}^{\text{OPW}}$ is orthogonal to each function u_j. The functions $u_j(\mathbf{r})$ are left unspecified but are required to be localized around each nucleus.

If the localized functions u_j are well chosen, Eq. (11.1) divides the function into a smooth part plus the localized part, as illustrated on the left-hand side of Fig. 11.2. In a crystal a smooth function can be represented conveniently by plane waves, hence the emphasis on plane waves in the original work. In the words of Herring [64]:

> This suggests that it would be practical to try to approximate [the eigenfunction in a crystal] by a linear combination of a few plane waves, plus a linear combination of a few functions centered about each nucleus and obeying wave equations of the form[2]

$$\frac{1}{2} \nabla^2 u_j + \left(E_j - V_j \right) u_j = 0. \tag{11.3}$$

The potential $V_j = V_j(r)$ and the functions u_j are to be chosen to be optimal for the problem. With this broad definition present in the original formulation [64], the OPW approach is the prescience of all modern pseudopotential and PAW methods. As is clear in the sections below, those methods involve new ideas and clever choices for the functions and operations on the functions. This has led to important advances in electronic structure that have made many of the modern developments in the field possible. For present purposes it is useful to consider the orthogonalized form for the valence states in an atom, where the state is labeled by angular momentum lm, and, of course, the added functions must also have the same lm. Using the definitions Eqs. (11.1) and (11.2), it follows immediately that the general OPW-type relation takes the form

$$\psi_{lm}^v(\mathbf{r}) = \tilde{\psi}_{lm}^v(\mathbf{r}) + \sum_j B_{lmj} u_{lmj}(\mathbf{r}), \tag{11.4}$$

where ψ_{lm}^v is the valence function, $\tilde{\psi}_{lm}^v$ is the smooth part, and all quantities can be expressed in terms of the original OPWs by Fourier transforms:

$$\psi_{lm}^v(\mathbf{r}) = \int d\mathbf{q} c_{lm}(\mathbf{q}) \chi_{\mathbf{q}}^{\text{OPW}}(\mathbf{r}), \tag{11.5}$$

$$\tilde{\psi}_{lm}^v(\mathbf{r}) = \int d\mathbf{q} c_{lm}(\mathbf{q}) e^{i\mathbf{q} \cdot \mathbf{r}}, \tag{11.6}$$

$$B_{lmj} = \int d\mathbf{q} c_{lm}(\mathbf{q}) \langle u_j | \mathbf{q} \rangle. \tag{11.7}$$

A schematic example of a 3s valence state and the corresponding smooth function is illustrated in Fig. 11.2.

---

[2] This is the original equation except that the factor of $1/2$ was not included since Herring's equation was written in Rydberg atomic units.

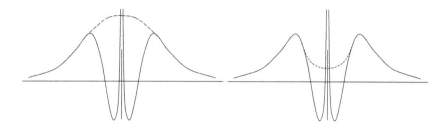

Figure 11.2. Schematic example of a valence function that has the character of a 3s orbital near the nucleus (which is properly orthogonal to the 1s and 2s core states) and two examples of smooth functions (dashed lines) that equal the full wavefunction outside the core region. Left: the dashed curve illustrates the smooth part of the valence function $\tilde{\psi}$ defined by OPW-like equations Eqs. (11.4) and (11.6). Right: a smooth pseudofunction ψ_l^{PS} that satisfies the norm-conservation condition Eq. (11.21). In general, ψ_l^{PS} is not as smooth as $\tilde{\psi}$.

It is also illuminating to express the OPW relation Eq. (11.4) as a transformation

$$|\psi_{lm}^v\rangle = \mathcal{T}|\tilde{\psi}_{lm}^v\rangle. \tag{11.8}$$

This is nothing more than a rewritten form of Eq. (11.4), but it expresses in compact form the idea that a solution for the smooth function $\tilde{\psi}_{lm}^v$ is sufficient; one can always recover the full function ψ_{lm}^v using a linear transformation denoted $\mathcal{T}$ in Eq. (11.8). This is exactly the form used in the PAW approach in Section 11.12.

The simplest approach is to choose the localized states to be core orbitals $u_{lmi} = \psi_{lmi}^c$, i.e., to choose the potential in Eq. (11.3) to be the actual potential (assumed to be spherical near the nucleus), so that ψ_{lmi}^c are the lowest eigenstates of the hamiltonian

$$H\psi_{lmi}^c = \varepsilon_{li}^c \psi_{lmi}^c. \tag{11.9}$$

Since the valence state ψ_{lm}^v must be orthogonal to the core states ψ_{lmi}^c, the radial part of $\psi_l^v(r)$ must have as many nodes as there are core orbitals with that angular momentum. One can show (Exercise 11.3) that the choice of $u_{li} = \psi_{li}^c$ leads to a smooth function $\tilde{\psi}_l^v(r)$ that has no radial nodes, i.e., it is indeed smoother than $\psi_l^v(r)$. Furthermore, often the core states can be assumed to be the same in the molecule or solid as in the atom. This is the basis for the actual calculations [68] in the OPW method.

There are several relevant points to note. As is illustrated on the left-hand side of Fig. 11.2, an OPW is like a smooth wave with additional structure and reduced amplitude near the nucleus. The set of OPWs is not orthonormal and each wave has a norm less than unity (Exercise 11.4)

$$\langle \chi_{\mathbf{q}}^{OPW} | \chi_{\mathbf{q}}^{OPW} \rangle = 1 - \sum_j |\langle u_j | \mathbf{q} \rangle|^2. \tag{11.10}$$

This means that the equations for the OPWs have the form of a generalized eigenvalue problem with an overlap matrix.

The Pseudopotential Transformation

The pseudopotential transformation of Phillips and Kleinman [504] and Antoncik [502, 503] (PKA) results if one inserts Expression (11.4) for $\psi_i^v(\mathbf{r})$ into the original equation for the valence eigenfunctions

$$\hat{H}\psi_i^v(\mathbf{r}) = \left[-\frac{1}{2}\nabla^2 + V(\mathbf{r}) \right] \psi_i^v(\mathbf{r}) = \varepsilon_i^v \psi_i^v(\mathbf{r}), \tag{11.11}$$

where V is the total effective potential, which leads to an equation for the smooth functions, $\tilde{\psi}_i^v(\mathbf{r})$,

$$\hat{H}^{\text{PKA}}\tilde{\psi}_i^v(\mathbf{r}) \equiv \left[-\frac{1}{2}\nabla^2 + \hat{V}^{\text{PKA}} \right] \tilde{\psi}_i^v(\mathbf{r}) = \varepsilon_i^v \tilde{\psi}_i^v(\mathbf{r}). \tag{11.12}$$

Here

$$\hat{V}^{\text{PKA}} = V + \hat{V}^R, \tag{11.13}$$

where $\hat{V}^R$ is a nonlocal operator that acts on $\tilde{\psi}_i^v(\mathbf{r})$ with the effect

$$\hat{V}^R\tilde{\psi}_i^v(\mathbf{r}) = \sum_j \left(\varepsilon_i^v - \varepsilon_j^c \right) \langle \psi_j^c | \tilde{\psi}_i^v \rangle \psi_j^c(\mathbf{r}). \tag{11.14}$$

Thus far this is nothing more than a formal transformation of the OPW expression, (11.11). The formal properties of the transformed equations suggest both advantages and disadvantages. It is clear that $\hat{V}^R$ is repulsive since Eq. (11.14) is written in terms of the energies $\varepsilon_i^v - \varepsilon_j^c$, which are always positive. Furthermore, a stronger attractive nuclear potential leads to deeper core states so that Eq. (11.14) also becomes more repulsive. This tendency was pointed out by Phillips and Kleinman and Antoncik and derived in a very general form as the "cancellation theorem" by Cohen and Heine [516]. Thus $\hat{V}^{\text{PKA}}$ is much weaker than the original $V(\mathbf{r})$, but it is a more complicated nonlocal operator. In addition, the smooth pseudofunctions $\tilde{\psi}_i^v(\mathbf{r})$ are *not orthonormal* because the complete function ψ_i^v also contains the sum over core orbitals in Eq. (11.4). Thus the solution of the pseudopotential equation (11.12) is a generalized eigenvalue problem.[3] Furthermore, since the core states are still present in the definition, Eq. (11.14), this transformation does not lead to a "smooth" pseudopotential.

The full advantages of the pseudopotential transformation are realized by using *both* the formal properties of pseudopotential $\hat{V}^{\text{PKA}}$ and the fact that the same scattering properties can be reproduced by different potentials. Thus the pseudopotential can be chosen to be both smoother and weaker than the original potential V by taking advantage of the nonuniqueness of pseudopotentials, as discussed in more detail in following sections.

Even though the potential operator is a more complex object than a simple local potential, the fact that it is weaker and smoother (i.e., it can be expanded in a small number of Fourier

---

[3] "Norm-conserving" potentials described in Section 11.4 remove this complication; however, nonorthogonality is resurrected in "ultrasoft" pseudopotentials, which are formally similar to the operator construction described here (see Section 11.11).

components) has great advantages, conceptually and computationally. In particular, it immediately resolves the apparent contradiction (see Chapter 12) that the valence bands $\varepsilon_{n\mathbf{k}}^v$ in many materials are nearly-free-electron-like, even though the wavefunctions $\psi_{n\mathbf{k}}^v$ must be very non-free-electron-like since they must be orthogonal to the cores. The resolution is that the bands are determined by the secular equation for the smooth, nearly-free-electron-like $\tilde{\psi}_{n\mathbf{k}}^v$ that involves the weak pseudopotential $\hat{V}^{\mathrm{PKA}}$ or $\hat{V}^{\mathrm{model}}$.

11.3 Model Ion Potentials

Based on the foundation of pseudopotentials in scattering theory, and the transformation of the OPW equations and the cancellation theorem, the theory of pseudopotentials has become a fertile field for generating new methods and insight for the electronic structure of molecules and solids. There are two approaches: (1) to define ionic pseudopotentials, which leads to the problem of interacting valence-only electrons, and (2) to define a *total pseudopotential* that includes effects of the other valence electrons. The former is the more general approach since the ionic pseudopotentials are more transferable with a single ion potential applicable for the given atom in different environments. The latter approach is very useful for describing the bands accurately if they are treated as adjustable empirical potentials; historically, empirical pseudopotentials have played an important role in the understanding of electronic structures [506, 507], and they reappear in Section 12.6 as a useful approach for understanding bands in a plane wave basis.

Here we concentrate on ionic pseudopotentials and the form of model potentials that give the same scattering properties as the pseudopotential operators of Eqs. (11.13) and Eq. (11.14) or more general forms. Since a model potential replaces the potential of a nucleus and the core electrons, it is spherically symmetric and each angular momentum l, m can be treated separately, which leads to nonlocal l-dependent model pseudopotentials $V_l(r)$. The qualitative features of l-dependent pseudopotentials can be illustrated by the forms shown in Fig. 11.3. Outside the core region, the potential is Z_{ion}/r, i.e., the combined Coulomb potential of the nucleus and core electrons. Inside the core region the potential is

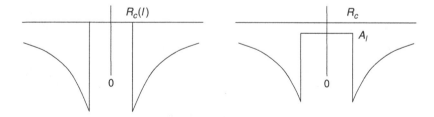

Figure 11.3. Left: "empty core" model potential of Ashcroft [517] in which the potential is zero inside radius $R_C(l)$, which is different for each l. Right: square well model potential with value A_l inside a cutoff radius R_C, proposed by Abarenkov and Heine [518] and fit to atomic data by Animalu and Heine [519, 520] (see also by Harrison [508]). The fact that the potentials are weak, zero, or even positive inside a cutoff radius R_C is an illustration of the "cancellation theorem" [516].

expected to be repulsive [516] to a degree that depends on the angular momentum l, as is clear from the analysis of the repulsive potential in Eq. (11.14).

The dependence on l means that, in general, a pseudopotential is a nonlocal operator that can be written in "semilocal" (SL) form

$$\hat{V}_{SL} = \sum_{lm} |Y_{lm}\rangle V_l(r) \langle Y_{lm}|, \tag{11.15}$$

where $Y_{lm}(\theta,\phi) = P_l(\cos(\theta))e^{im\phi}$. This is termed semilocal (SL) because it is nonlocal in the angular variables but local in the radial variable: when operating on a function $f(r,\theta',\phi')$, $\hat{V}_{SL}$ has the effect

$$\left[\hat{V}_{SL}f\right]_{r,\theta,\phi} = \sum_{lm} Y_{lm}(\theta,\phi)V_l(r) \int d(\cos\theta')d\phi' Y_{lm}(\theta',\phi')f(r,\theta',\phi'). \tag{11.16}$$

All the information is in the radial functions $V_l(r)$ or their Fourier transforms, which are defined in Section 12.4. An electronic structure involves calculation of the matrix elements of $\hat{V}_{SL}$ between states ψ_i and ψ_j

$$\langle\psi_i|\hat{V}_{SL}|\psi_j\rangle = \int dr\ \psi_i(r,\theta,\phi)\left[\hat{V}_{SL}\psi_j\right]_{r,\theta,\phi}. \tag{11.17}$$

(Compare this with Eq. (11.43) for a fully nonlocal separable form of the pseudopotential.)

There are two approaches to the definition of potentials:

- Empirical potentials fitted to atomic or solid state data. Simple forms are the "empty core" [517] and square well [518–520] models illustrated in Fig. 11.3. In the latter case, the parameters were fit to atomic data for each l and tabulated for many elements by Animalu and Heine [519, 520] (tables given also by Harrison [508]).[4]
- *Ab initio* potentials constructed to fit the valence properties calculated for the atom. The advent of "norm-conserving" pseudopotentials provided a straightforward way to make such potentials that are successfully transferrable to calculations on molecules and solids.

11.4 Norm-Conserving Pseudopotentials (NCPPs)

Pseudopotentials generated by calculations on atoms (or atomic-like states) are termed *ab initio* because they are *not fitted to experiment*. The concept of "norm conservation" has a special place in the development of *ab initio* pseudopotentials; at one stroke it simplifies

---

[4] Construction of such model potentials from atomic information presents a conceptual problem: the potential represents the effects of $\hat{V}^{PKA}$, which depends on the valence eigenvalue ε_i^v in the atom, which is defined relative to a reference energy equal to zero at infinity; however, the goal is to apply the pseudopotentials to infinite solids, where this is not a well-defined reference energy, and to molecules, where the levels are shifted relative to the atom. How can one relate the eigenvalues of the two types of systems? This was a difficult issue in the original pseudopotentials that was resolved by the conditions for construction of "norm-conserving" pseudopotentials described in Section 11.4.

the application of the pseudopotentials and it makes them more accurate and transferable. The latter advantage is described below, but the former can be appreciated immediately. In contrast to the PKA approach (Section 11.2) (where the equations were formulated in terms of the smooth part of the valence function $\tilde{\psi}_i^v(\mathbf{r})$ to which another function must be added, as in Eq. (11.4)), norm-conserving pseudofunctions $\psi^{PS}(\mathbf{r})$ are normalized and are solutions of a model potential chosen to reproduce the valence properties of an all-electron calculation. A schematic example is shown on the right-hand side of Fig. 11.2, which illustrates the difference from the unnormalized smooth part of the OPW. In the application of the pseudopotential to complex systems, such as molecules, clusters, solids, etc., *the valence pseudofunctions satisfy the usual orthonormality conditions* as in Eq. (7.9),

$$\langle \psi_i^{\sigma,PS} | \psi_j^{\sigma',PS} \rangle = \delta_{i,j} \delta_{\sigma,\sigma'}, \tag{11.18}$$

so that for the Kohn–Sham equations have the same form as in Eq. (7.11),

$$(H_{KS}^{\sigma,PS} - \varepsilon_i^{\sigma}) \psi_i^{\sigma,PS}(\mathbf{r}) = 0, \tag{11.19}$$

with $H_{KS}^{\sigma,PS}$ given by Eq. (7.12) and Eq. (7.13), and the external potential given by the pseudopotential specified in the section following.

Norm-Conservation Condition

Quantum chemists and physicists have devised pseudopotentials called, respectively, "shape-consistent" [521, 522] and "norm-conserving" [523].[5] The starting point for defining norm-conserving potentials is the list of requirements for a "good" *ab initio* pseudopotential given by Hamann, Schluter, and Chiang (HSC) [523]:

1. All-electron and pseudovalence eigenvalues agree for the chosen atomic reference configuration.
2. All-electron and pseudovalence wavefunctions agree beyond a chosen core radius R_c.
3. The logarithmic derivatives of the all-electron and pseudo-wavefunctions agree at R_c.
4. The integrated charge inside R_c for each wavefunction agrees (norm conservation).
5. The *first energy derivative* of the logarithmic derivatives of the all-electron and pseudowavefunctions agrees at R_c, and therefore for all $r \geq R_c$.

From points 1 and 2 it follows that the NCPP equals the atomic potential outside the "core region" of radius R_c; this is because the potential is uniquely determined (except for a constant that is fixed if the potential is zero at infinity) by the wavefunction and the energy ε, which need not be an eigenenergy. Point 3 follows since the wavefunction $\psi_l(r)$ and its radial derivative $\psi_l'(r)$ are continuous at R_c for any smooth potential. The dimensionless logarithmic derivative D is defined by

$$D_l(\varepsilon, r) \equiv r \psi_l'(\varepsilon, r) / \psi_l(\varepsilon, r) = r \frac{d}{dr} \ln \psi_l(\varepsilon, r), \tag{11.20}$$

also given in Eq. (J.5).

---

[5] Perhaps the earliest work was that of Topp and Hopfield [524].

Inside R_c the pseudopotential and radial pseudo-orbital ψ_l^{PS} differ from their all-electron counterparts; however, point 4 requires that the integrated charge,

$$Q_l = \int_0^{R_c} dr\, r^2 |\psi_l(r)|^2 = \int_0^{R_c} dr\, \phi_l(r)^2, \tag{11.21}$$

be the same for ψ_l^{PS} (or ϕ_l^{PS}) as for the all-electron radial orbital ψ_l (or ϕ_l) for a valence state. The conservation of Q_l insures that (a) the total charge in the core region is correct, and (b) the normalized pseudo-orbital is equal[6] to the true orbital outside of R_c (in contrast to the smooth orbital of Eq. (11.6), which equals the true orbital outside R_c only if it is not normalized). Applied to a molecule or solid, these conditions ensure that the normalized pseudo-orbital is correct in the region outside R_c between the atoms where bonding occurs, and that the potential outside R_c is correct as well since the potential outside a spherically symmetric charge distribution depends only on the total charge inside the sphere.

Point 5 is a crucial step toward the goal of constructing a "good" pseudopotential: one that can be generated in a simple environment like a spherical atom and then used in a more complex environment. In a molecule or solid, the wavefunctions and eigenvalues change and a pseudopotential that satisfies point 5 will reproduce the changes in the eigenvalues to linear order in the change in the self-consistent potential. At first sight, however, it is not obvious how to satisfy the condition that the *first energy derivative* of the logarithmic derivatives $dD_l(\varepsilon, r)/d\varepsilon$ agree for the pseudo- and the all-electron wavefunctions evaluated at the cutoff radius R_c and energy ε_l chosen for the construction of the pseudopotential of angular momentum l.

The advance due to HSC [523] and others [521, 522] was to show that point 5 is implied by point 4. This "norm-conservation condition" can be derived straightforwardly, e.g., following the derivation of Shirley et al. [526], which uses relations due to Luders [527] (see Exercises 11.8 and 11.9 for intermediate steps). The radial equation for a spherical atom or ion, Eq. (10.12), which can be written

$$-\frac{1}{2}\phi_l''(r) + \left[\frac{l(l+1)}{2r^2} + V_{\text{eff}}(r) - \varepsilon\right]\phi_l(r) = 0, \tag{11.22}$$

where a prime means derivative with respect to r, can be transformed by defining the variable $x_l(\varepsilon, r)$

$$x_l(\varepsilon, r) \equiv \frac{d}{dr}\ln\phi_l(r) = \frac{1}{r}[D_l(\varepsilon, r) + 1]. \tag{11.23}$$

It is straightforward to show that Eq. (11.22) is equivalent to the nonlinear first-order differential equation,

$$x_l'(\varepsilon, r) + [x_l(\varepsilon, r)]^2 = \frac{l(l+1)}{r^2} + 2[V(r) - \varepsilon]. \tag{11.24}$$

---

[6] Equality can be strictly enforced only for local functionals, not for nonlocal cases as in Hartree–Fock and EXX potentials. For example, see [525].

Differentiating this equation with respect to energy gives

$$\frac{\partial}{\partial \varepsilon} x_l'(\varepsilon, r) + 2 x_l(\varepsilon, r) \frac{\partial}{\partial \varepsilon} x_l(\varepsilon, r) = -1. \tag{11.25}$$

Combining this with the relation valid for any function $f(r)$ and any l,

$$f'(r) + 2 x_l(\varepsilon, r) f(r) = \frac{1}{\phi_l(r)^2} \frac{\partial}{\partial r} [\phi_l(r)^2 f(r)], \tag{11.26}$$

multiplying by $\phi_l(r)^2$ and integrating, one finds at radius R

$$\frac{\partial}{\partial \varepsilon} x_l(\varepsilon, R) = -\frac{1}{\phi_l(R)^2} \int_0^R dr \phi_l(r)^2 = -\frac{1}{\phi_l(R)^2} Q_l(R), \tag{11.27}$$

or in terms of the dimensionless logarithmic derivative $D_l(\varepsilon, R)$

$$\frac{\partial}{\partial \varepsilon} D_l(\varepsilon, R) = -\frac{R}{\phi_l(R)^2} \int_0^R dr \phi_l(r)^2 = -\frac{R}{\phi_l(R)^2} Q_l(R). \tag{11.28}$$

This shows immediately that if ϕ_l^{PS} has the same magnitude as the all-electron function ϕ_l at R_c and obeys norm conservation (Q_l the same), then the first energy derivative of the logarithmic derivative $x_l(\varepsilon, R)$ and $D_l(\varepsilon, R)$ is the same as for the all-electron wavefunction. This also means that the norm-conserving pseudopotential has the same scattering phase shifts as the all-electron atom to linear order in energy around the chosen energy ε_l, which follows from Expression (J.6), which relates to $D_l(\varepsilon, R)$ and the phase shift $\eta_l(\varepsilon, R)$.[7]

11.5 Generation of *l*-Dependent Norm-Conserving Pseudopotentials

Generation of a pseudopotential starts with the usual all-electron atomic calculation as described in Chapter 10. Each state l, m is treated independently except that the total potential is calculated self-consistently for the given approximation for exchange and correlation and for the given configuration of the atom. The next step is to identify the valence states and generate the pseudopotentials $V_l(r)$ and pseudo-orbitals $\psi_l^{\mathrm{PS}}(r) = r \phi_l^{\mathrm{PS}}(r)$. The procedure varies with different approaches, but in each case one first finds a total "screened" pseudopotential acting on valence electrons in the atom. This is then "unscreened" by subtracting from the total potential the sum of Hartree and exchange–correlation potentials $V_{H\mathrm{xc}}^{\mathrm{PS}}(r) = V_{\mathrm{Hartree}}^{\mathrm{PS}}(r) + V_{\mathrm{xc}}^{\mathrm{PS}}(r)$

$$V_l(r) \equiv V_{l,\mathrm{total}}(r) - V_{H\mathrm{xc}}^{\mathrm{PS}}(r), \tag{11.29}$$

where $V_{H\mathrm{xc}}^{\mathrm{PS}}(r)$ is defined for the valence electrons in their pseudo-orbitals. Further aspects of "unscreening" are deferred to Section 11.6.

---

[7] This relation is very important and used in many contexts: in Appendix J it is seen to be the Friedel sum rule, which has important consequences for resistivity due to impurity scattering in metals. In Chapter 17 it is used to relate the band width to the value of the wavefunction at the boundary of a sphere.

It is useful to separate the ionic pseudopotential into a local (l-independent) part of the potential plus nonlocal terms

$$V_l(r) = V_{\text{local}}(r) + \delta V_l(r). \tag{11.30}$$

Since the eigenvalues and orbitals are required to be the same for the pseudo and the all-electron case for $r > R_c$, each potential $V_l(r)$ equals the local (l-independent) all-electron potential, and $V_l(r) \to -\frac{Z_{\text{ion}}}{r}$ for $r \to \infty$. Thus $\delta V_l(r) = 0$ for $r > R_c$ and all the long-range effects of the Coulomb potential are included in the local potential $V_{\text{local}}(r)$. Finally, the "semilocal" operator Eq. (11.15) can be written as

$$\hat{V}_{\text{SL}} = V_{\text{local}}(r) + \sum_{lm} |Y_{lm}\rangle \delta V_l(r) \langle Y_{lm}|. \tag{11.31}$$

Even if one requires norm conservation, there is still freedom of choice in the form of $V_l(r)$ in constructing pseudopotentials. There is no one "best pseudopotential" for any given element – there may be many "best" choices, each optimized for some particular use of the pseudopotential. In general, there are two overall competing factors:

- Accuracy and transferability generally lead to the choice of a small cutoff radius R_c and "hard" potentials, since one wants to describe the wavefunction as well as possible in the region near the atom.
- Smoothness of the resulting pseudofunctions generally leads to the choice of a large cutoff radius R_c and "soft" potentials, since one wants to describe the wavefunction with as few basis functions as possible (e.g., plane waves).

Here we will try to present the general ideas in a form that is the basis of widely used methods, with references to some of the many proposed forms that cannot be covered here.

An example of pseudopotentials [523] for Mo is shown in Fig. 11.4. A similar approach has been used by Bachelet, Hamann, and Schlüter (BHS) [528] to construct pseudopotentials for all elements from H to Po, in the form of an expansion in gaussians with tabulated coefficients. These potentials were derived starting from an assumed form of the potential and varying parameters until the wavefunction has the desired properties, an approach also used by Vanderbilt [529]. A simpler procedure is that of Christiansen et al. [521] and Kerker [530], which defines a pseudowavefunction $\phi_l^{\text{PS}}(r)$ with the desired properties for each l and numerically inverts the Schrödinger equation to find the potential $V_l(r)$ for which $\phi_l^{\text{PS}}(r)$ is a solution with energy ε. The wavefunction outside the radius R_c is the same as the true function and at R_c it is matched to a parameterized analytic function. Since the energy ε is fixed (often it is the eigenvalue from the all-electron calculation, but this is not essential), it is straightforward to invert the Schrödinger equation for a nodeless function $\phi_l^{\text{PS}}(r)$ for each l separately, yielding

$$V_{l,\text{total}}(r) = \varepsilon - \frac{\hbar^2}{2m_e} \left[\frac{l(l+1)}{2r^2} - \frac{\frac{d^2}{dr^2}\phi_l^{\text{PS}}(r)}{\phi_l^{\text{PS}}(r)} \right]. \tag{11.32}$$

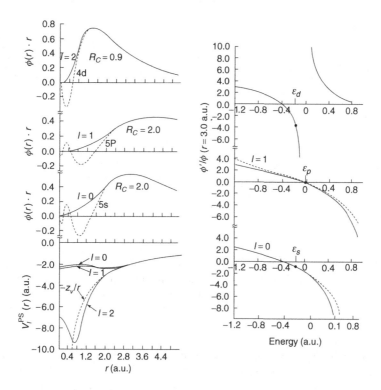

Figure 11.4. Example of norm-conserving pseudopotentials, pseudofunctions, and logarithmic derivative for the element Mo. Left bottom: $V_l(r)$ in Rydbergs for angular momentum $l = 0, 1, 2$ compared to Z_{ion}/r (dashed). Left top: all-electron valence radial functions $\phi_l(r) = r\psi_l(r)$ (dashed) and norm-conserving pseudofunctions. Right: logarithmic derivative of the pseudopotential compared to the full-atom calculation; the points indicate the energies, ε, where they are fitted. The derivative with respect to the energy is also correct due to the norm-conservation condition Eq. (11.27). From [523].

The analytic form chosen by Kerker is $\phi_l^{PS}(r) = e^{p(r)}$, $r < R_c$, where $p(r) =$ polynomial to the fourth power with coefficients fixed by requiring continuous first and second derivatives at R_c and norm conservation.

One of the important considerations for many uses is to make the wavefunction as smooth as possible, which allows it to be described by fewer basis functions, e.g., fewer Fourier components. For example, the BHS potentials [528] are a standard reference for comparison; however, they are generally harder and require more Fourier components in the description of the pseudofunction than other methods. Troullier and Martins [531] have extended the Kerker method to make it smoother by using a higher-order polynomial and matching more derivatives of the wavefunction. A comparison of different pseudopotentials for carbon is given in Fig. 11.5, showing the forms both in real and reciprocal space. The one-dimensional radial transforms $V_l(q)$ (or "form factors") for each l are defined in Section 12.4; these are the functions that enter directly in plane wave calculations and their extent in Fourier space determines the number of plane waves needed for convergence.

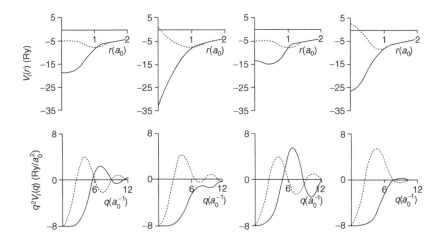

Figure 11.5. Comparison of pseudopotentials for carbon (dotted line for s and solid line for p) in real space and reciprocal space, illustrating the large variations in potentials that are all norm conserving and have the same phase shifts at the chosen energies. In order from left to right, generated using the procedures of Troullier and Martins [531]; Kerker [530]; Hamann, Schlüter, and Chiang [523]; Vanderbilt [529]. From Troullier and Martins [531].

A number of authors have proposed ways to make smoother potentials to reduce the size of calculations. One approach [532, 533] is to minimize the kinetic energy of the pseudofunctions explicitly for the chosen core radius. This can be quantified by examination of the Fourier transform and its behavior at large momentum q, as described in Section 11.7. Optimization of the potentials can be done in the atom, and the results carry over to molecules and solids, since the convergence is governed by the form inside the core radius.

Relativistic Effects

Effects of special relativity can be incorporated into pseudopotentials, since they originate deep in the interior of the atom near the nucleus, and the consequences for valence electrons can be easily carried into molecular or solid-state calculations. This includes shifts due to scalar relativistic effects and spin–orbit interactions. The first step is generation of a pseudopotential from a relativistic all-electron calculation on the atom for both $j = l + 1/2$ and $j = l - 1/2$. From the two potentials we can define [372, 528]

$$V_l = \frac{l}{2l+1} \left[(l+1)V_{l+1/2} + lV_{l-1/2} \right], \tag{11.33}$$

$$\delta V_l^{so} = \frac{2}{2l+1} \left[V_{l+1/2} - V_{l-1/2} \right]. \tag{11.34}$$

Scalar relativistic effects are included in the first term and the spin–orbit effects are included in a short-range nonlocal term [486, 487],

$$\delta \hat{V}_{SL}^{so} = \sum_{lm} |Y_{lm}\rangle \delta V_l^{so}(r) \mathbf{L} \cdot \mathbf{S} \langle Y_{lm}|. \tag{11.35}$$

11.6 Unscreening and Core Corrections

In the construction of *ab initio* pseudopotentials there is a straightforward one-to-one relation of the valence pseudofunction and the *total* pseudopotential. It is then a necessary step to "unscreen" to derive the bare-ion pseudopotential, which is transferable to different environments. However, the process of "unscreening" is not so straightforward. If the effective exchange–correlation potential were a linear function of density (as is the Hartree potential $V_{Hartree}$) there would be no problem, and Eq. (11.29) could be written as

$$V_{l,\text{total}} = V_l(r) + V_{Hartree}([n^{PS}], \mathbf{r}) + V_{xc}([n^{PS}], \mathbf{r}), \tag{11.36}$$

where the notation $[n^{PS}]$ means the quantity is evaluated as a functional of the pseudodensity n^{PS}. This is true for the Hartree potential, but the fact that V_{xc} is a nonlinear functional of n (and may also be nonlocal) leads to difficulties and ambiguities. (Informative discussions can be found in [534] and [525].)

Nonlinear Core Corrections

So long as the exchange–correlation functional involves only the density or its gradients at each point, then the unscreening of the potential in the atom can be accomplished by defining the effective exchange–correlation potential in Eq. (11.29) as

$$\tilde{V}_{xc}(\mathbf{r}) = V_{xc}([n^{PS}], \mathbf{r}) + \left[V_{xc}([n^{PS} + n^{core}], \mathbf{r}) - V_{xc}([n^{PS}], \mathbf{r}) \right]. \tag{11.37}$$

The term in square brackets is a core correction that significantly increases the transferability of the pseudopotential [534]. There are costs, however: the core charge density must be stored along with the pseudodensity and the implementation in a solid must use $\tilde{V}_{xc}(\mathbf{r})$ defined in Eq. (11.37), and the rapidly varying core density would be a disadvantage in plane wave methods. The second obstacle can be overcome [534] using the freedom of choice inherent in pseudopotentials by defining a smoother "partial core density" $n^{core}_{partial}(r)$ that can be used in Eq. (11.37). The original form proposed by Louie, Froyen, and Cohen [534] is[8]

$$n^{core}_{partial}(r) = \begin{cases} \dfrac{A \sin(Br)}{r}, & r < r_0, \\ n^{core}(r), & r > r_0, \end{cases} \tag{11.38}$$

where A and B are determined by the value and gradient of the core charge density at r_0, a radius chosen where n^{core} is typically one to two times $n^{valence}$. The effect is particularly large for cases in which the core is extended (e.g., the 3d transition metals where the 3p "core" states strongly overlap the 3d "valence" states) and for magnetic systems where there may be a large difference between up and down valence densities even though the fractional difference in total density is much smaller. In such cases, description of spin-polarized

---

[8] The form in Eq. (11.38) has a discontinuity in the second derivative r_0, which causes difficulties in conjunction with GGA functionals. This problem is readily remedied by using a more flexible functional form.

configurations can be accomplished with a spin-independent ionic pseudopotential, with no need for separate spin-up and spin-down ionic pseudopotentials.

Nonlocal E_{xc} Functionals

There is a complication in "unscreening" in cases where the E_{xc} functional is intrinsically nonlocal, as in Hartree–Fock and exact exchange (EXX). In general it is not possible to make a potential that keeps the wavefunctions outside a core radius the same as in the original all-electron problem because the nonlocal effects extend to all radii. The issues are discussed thoroughly in [525].

11.7 Transferability and Hardness

There are two meanings to the word "hardness." One meaning is a measure of the variation in real space, which is quantified by the extent of the potential in Fourier space. In general, "hard" potentials describe the properties of the localized rigid ion cores and are more transferable from one material to another; attempts to make the potential "soft" (i.e., smooth) have tended to lead to poorer transferability. There is considerable effort to make accurate, transferable potentials that do not extend far in Fourier space. What we care about is the extent of the pseudowavefunction (not the potential, even though they are related) where the Fourier transform of the wavefunction for each l is given by

$$\psi_l(q) = \int_0^\infty dr \; j_l(qr) \, r^2 \psi_l(r).$$ (11.39)

In a calculation with plane waves up to a cutoff q_c a quantitative measure of the error proposed by Rappe et al. [532] is the residual kinetic energy, i.e., the kinetic energy above the cutoff,

$$E_l^r(q_c) = \int_{q_c}^\infty dq \; q^4 \, |\psi_l(q)|^2,$$ (11.40)

and the idea is to minimize $E_l^r(q_c)$ with the constraint of continuity, slope, and second derivative at the pseudopotential cutoff radius r_c. Pseudopotentials generated in this way were termed "optimized" [532]. This approach was extended by Hamann [535], who developed a procedure to expand the wavefunction in a set of functions that are used to satisfy continuity conditions up to an arbitrary number of derivatives.[9]

The second meaning is a measure of the ability of the valence pseudoelectrons to describe the response of the system to a change in the enviroment properly [536–538]. We have already seen that norm conservation guarantees that the electron states of the atom have the correct first derivative with respect to change in energy. This meaning of "hardness" is a measure of the faithfulness of the response to a change in potential. Potentials can be tested

---

[9] The auxiliary functions are used only for the purpose of generating the pseudopotential.

versus spherical perturbations (change of charge, state, radial potential) using the usual spherical atom codes. Goedecker and Maschke [536] have given an insightful analysis in terms of the response of the charge density in the core region; this is relevant since the density is the central quantity in density functional theory and the integrated density is closely related to norm-conservation conditions. Also tests with nonspherical perturbations ascertain the performance with relevant perturbations, in particular, the polarizability in an electric field [538].

Tests in Spherical Boundary Conditions

We have seen in Section 10.7 that some aspects of solids are well modeled by imposing different spherical boundary conditions on an atom or ion. A net consequence is that the valence wavefunctions tend to be more concentrated near the nucleus than in the atom. How well do pseudopotentials derived for an isolated atom describe such situations? The answer can be found directly using computer programs for atoms and pseudoatoms; examples are given in the exercises. These are the types of tests that should be done *whenever generating a new pseudopotential.*

11.8 Separable Pseudopotential Operators and Projectors

It was recognized by Kleinman and Bylander (KB) [539] that it is possible to construct a *separable* pseudopotential operator, i.e., $\delta V(\mathbf{r}, \mathbf{r}')$ written as a sum of products of the form $\Sigma_i f_i(\mathbf{r}) g_i(\mathbf{r}')$. KB showed that the effect of the semilocal $\delta V_l(r)$ in Eq. (11.30) can be replaced, to a good approximation, by a separable operator $\delta \hat{V}_{\mathrm{NL}}$ so that the total pseudopotential has the form

$$\hat{V}_{\mathrm{NL}} = V_{\mathrm{local}}(r) + \sum_{lm} \frac{|\psi_{lm}^{\mathrm{PS}} \delta V_l\rangle \langle \delta V_l \psi_{lm}^{\mathrm{PS}}|}{\langle \psi_{lm}^{\mathrm{PS}} | \delta V_l | \psi_{lm}^{\mathrm{PS}}\rangle}, \tag{11.41}$$

where the second term written explicitly in coordinates, $\delta \hat{V}_{\mathrm{NL}}(\mathbf{r}, \mathbf{r}')$, has the desired separable form. Unlike the semilocal form Eq. (11.15), it is fully nonlocal in angles θ, ϕ and radius r. When operating on the reference atomic states ψ_{lm}^{PS}, $\delta \hat{V}_{\mathrm{NL}}(\mathbf{r}, \mathbf{r}')$ acts the same as $\delta V_l(r)$, and it can be an excellent approximation for the operation of the pseudopotential on the valence states in a molecule or solid.

The functions $\langle \delta V_l \psi_{lm}^{\mathrm{PS}}|$ are *projectors* that operate on the wavefunction

$$\langle \delta V_l \psi_{lm}^{\mathrm{PS}} | \psi \rangle = \int d\mathbf{r} \, \delta V_l(r) \psi_{lm}^{\mathrm{PS}}(\mathbf{r}) \psi(\mathbf{r}). \tag{11.42}$$

Each projector is localized in space, since it is nonzero only inside the pseudopotential cutoff radius where $\delta V_l(r)$ is nonzero. This is independent of the extent of the functions $\psi_{lm}^{\mathrm{PS}} = \psi_{lm}(r) P_l(\cos(\theta)) e^{im\phi}$, which have the extent of atomic valence orbitals or can even be nonbound states.

The advantage of the separable form is that matrix elements require only products of projection operations Eq. (11.42)

$$\langle \psi_i | \delta \hat{V}_{NL} | \psi_j \rangle = \sum_{lm} \langle \psi_i | \psi_{lm}^{PS} \delta V_l \rangle \frac{1}{\langle \psi_{lm}^{PS} | \delta V_l | \psi_{lm}^{PS} \rangle} \langle \delta V_l \psi_{lm}^{PS} | \psi_j \rangle. \tag{11.43}$$

This can be contrasted with Eq. (11.17), which involves a radial integral for each pair of functions ψ_i and ψ_j. This leads to savings in computations that can be important for large calculations. However, it does lead to an additional step, which may lead to increased errors. Although the operation on the given atomic state is unchanged, the operations on other states at different energies may be modified, and care must be taken to ensure that there are no artificial "ghost states" introduced. (As discussed in Exercise 11.12, such ghost states at low energy are expected when $V_{local}(r)$ is attractive and the nonlocal $\delta V_l(r)$ are repulsive. This choice should be avoided [540].)

It is straightforward to generalize to the case of spin–orbit coupling using the states of the atom derived from the Dirac equation with total angular momentum $j = l \pm \frac{1}{2}$ [487, 539]. The nonlocal projections become

$$\hat{V}_{NL}^{j=l\pm\frac{1}{2}} = V_{local}(r) + \sum_{lm} \frac{|\psi_{l\pm\frac{1}{2},m}^{PS} V_{l\pm\frac{1}{2}} \rangle \langle \delta V_{l\pm\frac{1}{2}} \psi_{l\pm\frac{1}{2},m}^{PS}|}{\langle \psi_{l\pm\frac{1}{2},m}^{PS} | \delta V_{l\pm\frac{1}{2}} | \psi_{l\pm\frac{1}{2},m}^{PS} \rangle}. \tag{11.44}$$

The KB construction can be modified to generate the separable potential directly without going through the step of constructing the semilocal $V_l(r)$ [510]. Following the same procedure as for generating the norm-conserving pseudopotential, the first step is to define pseudofunctions $\psi_{lm}^{PS}(\mathbf{r})$ and a local pseudopotential $V_{local}(r)$ that are equal to the all-electron functions outside a cutoff radius $r > R_c$. For $r > R_c$, $\psi_{lm}^{PS}(\mathbf{r})$ and $V_{local}(r)$ are chosen in some smooth fashion as was done in Section 11.5. If we now define new functions

$$\chi_{lm}^{PS}(\mathbf{r}) \equiv \left\{ \varepsilon_l - \left[-\frac{1}{2}\nabla^2 + V_{local}(r) \right] \right\} \psi_{lm}^{PS}(\mathbf{r}), \tag{11.45}$$

it is straightforward to show that $\chi_{lm}^{PS}(\mathbf{r}) = 0$ outside R_c and that the operator

$$\delta \hat{V}_{NL} = \sum_{lm} \frac{|\chi_{lm}^{PS}\rangle \langle \chi_{lm}^{PS}|}{\langle \psi_{lm}^{PS} | \delta V_l | \psi_{lm}^{PS} \rangle} \tag{11.46}$$

has the same properties as the KB operator Eq. (11.41), i.e., ψ_{lm}^{PS} is a solution of $\hat{H}\psi_{lm}^{PS} = \varepsilon_l \psi_{lm}^{PS}$ with $\hat{H} = -\frac{1}{2}\nabla^2 + V_{local} + \delta \hat{V}_{NL}$.

11.9 Extended Norm Conservation: Beyond the Linear Regime

Two general approaches have been proposed to extend the range of energies over which the phase shifts of the original all-electron potential are described. Shirley and coworkers [526] have given general expressions that must be satisfied for the phase shifts to be correct to arbitrary order in a power series expansion in $(\varepsilon - \varepsilon_0)^N$ around the chosen energy ε_0.

A second approach is easier to implement and is the basis for further generalizations in the following Sections and in Section 17.9. The construction of the projectors can be done at any energy ε_s and the procedure can be generalized to satisfy the Schrödinger equation at more than one energy for a given l, m [509, 510]. (Below we omit superscript PS and subscript l, m for simplicity.) If pseudofunctions ψ_s are constructed from all-electron calculations at different energies ε_s, one can form the matrix $B_{s,s'} = \langle \psi_s | \chi_{s'} \rangle$, where the χ_s are defined by Eq. (11.45). In terms of the functions $\beta_s = \sum_{s'} B_{s,s'}^{-1} \chi_{s'}$, the generalized nonlocal potential operator can be written

$$\delta \hat{V}_{\mathrm{NL}} = \sum_{lm} \left[\sum_{s,s'} B_{s,s'} | \beta_s \rangle \langle \beta_{s'} | \right]_{lm} . \tag{11.47}$$

It is straightforward to show (Exercise 11.13) that each ψ_s is a solution of $\hat{H} \psi_s = \varepsilon_s \psi_s$. With this modification, the nonlocal separable pseudopotential can be generalized to agree with the all-electron calculation to arbitrary accuracy over a desired energy range.

The transformation Eq. (11.47) exacts a price; instead of the simple sum of products of projectors in Eq. (11.43), matrix elements of Eq. (11.47) involve a matrix product of operators. For the spherically symmetric pseudopotential, the matrix is $s \times s$ and is diagonal in l, m. (A similar idea is utilized in Section 17.9 to transform the equations for the general problem of electron states in a crystal.)

11.10 Optimized Norm-Conserving Potentials

Although the "ultrasoft" potentials in the next section were developed much earlier, it is logical to first consider "optimized norm-conserving Vanderbilt pseudopotentials" (ONCV) developed by Hamann [535]. The name explains the method: "Vanderbilt" denotes that the potentials involve multiple projectors as defined in Eq. (11.47) and norm conserving means that the pseudofunctions are normalized. Thus they can be used just like other norm-conserving potentials, except that the separable operator in Eq. (11.41) becomes a sum of operators. This is only a minor complication in codes that are designed to use separable potentials. Finally, "optimized" means that great care has been taken to make the potentials soft using the condition of minimizing the residual kinetic energy in Eq. (11.40). The set of projectors for each angular momentum l must be optimized, and it was in this context that a new approach was devised in terms of a set of auxiliary functions to satisfy multiple continuity constraints [535].

The ONCV potentials are an important extension to the family of norm-conserving potentials. The fact that only one projector was used in the Kleinman–Bylander approach leads to inaccuracies (sometimes quite severe) in many cases and the potential needs to be very hard (requiring many plane waves) to converge. The ONCV potentials address both these difficulties, with higher accuracy due to the multiple projectors for the same reason this occurs in the ultrasoft potentials. The renewed focus on softness has allowed the ONCV potentials to be competitive with ultrasoft potentials.

Tests of the potentials in [535] show they are in excellent agreement with all-electron calculations, including cases where the single projector methods make serious errors, such as K, Cu, and $SrTiO_3$, which have shallow semicore states. Further details of the potentials and extensive tests are in [541] entitled "The PseudoDojo: Training and Grading a 85 Element Optimized Norm-Conserving Pseudopotential Table."

There are great advantages of norm-conserving potentials because they are normalized and do not involve auxiliary functions like the ultrasoft potentials. For example, they can be used in quantum Monte Carlo calculations where the accuracy of the pseudopotential is the key limiting factor.

11.11 Ultrasoft Pseudopotentials

One goal of pseudopotentials is to create pseudofunctions that are as "smooth" as possible and yet are accurate. For example, in plane wave calculations the valence functions are expanded in Fourier components, and the cost of the calculation scales as a power of the number of Fourier components needed in the calculation (see Chapter 12). The approach in the previous sections is to maximize "smoothness" by minimizing the range in Fourier space needed to describe the valence properties to a given accuracy. "Norm-conserving" pseudopotentials achieve the goal of accuracy, usually at some sacrifice of "smoothness."

A different approach known as "ultrasoft pseudopotentials" reaches the goal of accurate calculations by a transformation that reexpresses the problem in terms of a smooth function and an auxiliary function around each ion core that represents the rapidly varying part of the density. Although the equations are formally related to the OPW equations and the Phillips–Kleinman–Antoncik construction given in Section 11.2, ultrasoft pseudopotentials are a practical approach for solving equations beyond the applicability of those formulations. We will focus on examples of states that present the greatest difficulties in the creation of accurate, smooth pseudofunctions: valence states at the beginning of an atomic shell, 1s, 2p, 3d, etc. For these states, the OPW transformation has no effect since there are no core states of the same angular momentum. Thus the wavefunctions are nodeless and extend into the core region. Accurate representation by norm-conserving pseudofunctions requires that they are at best only moderately smoother than the all-electron function (see Fig. 11.6).

The transformation proposed by Blöchl [509] and Vanderbilt [510] rewrites the nonlocal potential in Eq. (11.47) in a form involving a smooth function $\tilde{\phi} = r\tilde{\psi}$ that is *not norm conserving*. (We follow the notation of [510], omitting the labels PS, l, m, and σ for simplicity.) The difference in the norm equation (11.21), from that norm-conserving function $\phi = r\psi$ (either an all-electron function or a pseudofunction) is given by

$$\Delta Q_{s,s'} = \int_0^{R_c} dr\, \Delta Q_{s,s'}(r), \tag{11.48}$$

where

$$\Delta Q_{s,s'}(r) = \phi_s^*(r)\phi_{s'}(r) - \tilde{\phi}_s^*(r)\tilde{\phi}_{s'}(r). \tag{11.49}$$

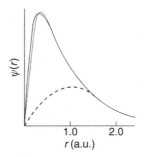

Figure 11.6. 2p radial wavefunction $\psi(r)$ for oxygen treated in the LDA, comparing the all-electron function (solid line), a pseudofunction generated using the Hamann–Schluter–Chiang approach ([523] (dotted line), and the smooth part of the pseudofunction $\tilde{\psi}$ in the "ultrasoft" method (dashed line). From [510].

A new nonlocal potential that operates on the $\tilde{\psi}_{s'}$ can now be defined to be

$$\delta \hat{V}_{NL}^{US} = \sum_{s,s'} D_{s,s'} |\beta_s\rangle\langle\beta_{s'}|, \tag{11.50}$$

where

$$D_{s,s'} = B_{s,s'} + \varepsilon_{s'} \Delta Q_{s,s'}. \tag{11.51}$$

For each reference to atomic states s, it is straightforward to show that the smooth functions $\tilde{\psi}_s$ are the solutions of the *generalized eigenvalue problem*

$$\left[\hat{H} - \varepsilon_s \hat{S}\right] \tilde{\psi}_s = 0, \tag{11.52}$$

with $\hat{H} = -\frac{1}{2}\nabla^2 + V_{local} + \delta \hat{V}_{NL}^{US}$ and $\hat{S}$ an overlap operator,

$$\hat{S} = \hat{\mathbf{1}} + \sum_{s,s'} \Delta Q_{s,s'} |\beta_s\rangle\langle\beta_{s'}|, \tag{11.53}$$

which is different from unity only inside the core radius. The eigenvalues ε_s agree with the all-electron calculation at as many energies s as desired. The full density can be constructed from the functions $\Delta Q_{s,s'}(r)$, which can be replaced by a smooth version of the all-electron density.

The advantage of relaxing the norm-conservation condition $\Delta Q_{s,s'} = 0$ is that each smooth pseudofunction $\tilde{\psi}_s$ can be formed independently, with only the constraint of matching the value of the functions $\tilde{\psi}_s(R_c) = \psi_s(R_c)$ at the radius R_c. Thus it becomes possible to choose an R_c much larger than for a norm-conserving pseudopotential while maintaining the desired accuracy by adding the auxiliary functions $\Delta Q_{s,s'}(r)$ and the overlap operator $\hat{S}$. An example of the unnormalized smooth function for the 2p state of oxygen is shown in Fig. 11.6, compared to a much more rapidly varying norm-conserving function.

In a calculation that uses an "ultrasoft pseudopotential" the solutions for the smooth functions $\tilde{\psi}_i(\mathbf{r})$ are orthonormalized according to

$$\langle \tilde{\psi}_i | \hat{S} | \tilde{\psi}_{i'} \rangle = \delta_{i,i'}, \tag{11.54}$$

and the valence density is defined to be

$$n_v(\mathbf{r}) = \sum_i^{\text{occ}} \tilde{\psi}_i^*(\mathbf{r}) \tilde{\psi}_{i'}(\mathbf{r}) + \sum_{s,s'} \rho_{s,s'} \Delta Q_{s,s'}(\mathbf{r}), \tag{11.55}$$

where

$$\rho_{s,s'} = \sum_i^{\text{occ}} \langle \tilde{\psi}_i | \beta_{s'} \rangle \langle \beta_s | \tilde{\psi}_i \rangle. \tag{11.56}$$

The solution is found by minimizing the total energy

$$E_{\text{total}} = \sum_i^{\text{occ}} \langle \tilde{\psi}_n | -\frac{1}{2} \nabla^2 + V_{\text{local}}^{\text{ion}} + \sum_{s,s'} D_{s,s'}^{\text{ion}} |\beta_s\rangle \langle \beta_{s'} \, || \tilde{\psi}_n \rangle$$

$$+ \, E_{\text{Hartree}}[n_v] + E_{II} + E_{\text{xc}}[n_v], \tag{11.57}$$

which is the analog of Eqs. (7.5) and (7.16), except that now the normalization condition is given by Eq. (11.54).[10] If we define the "unscreened" bare-ion pseudopotential by $V_{local}^{ion} \equiv V_{local} - V_{Hxc}$, where $V_{Hxc} = V_H + V_{xc}$, and similarly $D_{s,s'}^{ion} \equiv D_{s,s'} - D_{s,s'}^{Hxc}$ with

$$D_{s,s'}^{Hxc} = \int d\mathbf{r} V_{Hxc}(\mathbf{r}) \Delta Q_{s,s'}(r), \tag{11.58}$$

this leads to the generalized eigenvalue problem

$$\left[-\frac{1}{2} \nabla^2 + V_{\text{local}} + \delta \hat{V}_{\text{NL}}^{\text{US}} - \varepsilon_i \hat{S} \right] \tilde{\psi}_i = 0, \tag{11.59}$$

where $\delta \hat{V}_{\text{NL}}^{\text{US}}$ is given by the sum over ions of Eq. (11.50). Fortunately, such a generalized eigenvalue problem is not a major complication with iterative methods (see Appendix M).

11.12 Projector Augmented Waves (PAWs): Keeping the Full Wavefunction

The projector augmented wave (PAW) method is a general approach to solution of the electronic structure problem that reformulates the OPW method, adapting it to modern techniques for calculation of total energy, forces, and stress. The original derivations of the method were in the 1990s [542–544] and more recent work has made an efficient method in which the core states are updated consistently [545]. Like the "ultrasoft" pseudopotential method, it introduces projectors and auxiliary localized functions. The PAW approach also defines a functional for the total energy that involves auxiliary functions and it uses

---

[10] Note that one can add a "non-linear core correction" in E_{xc} just as in other pseudopotential methods.

advances in algorithms for efficient solution of the generalized eigenvalue problem like
Eq. (11.59). However, the difference is that the PAW approach keeps the full all-electron
wavefunction in a form similar to the general OPW expression given earlier in Eq. (11.1);
since the full wavefunction varies rapidly near the nucleus, all integrals are evaluated as
a combination of integrals of smooth functions extending throughout space plus localized
contributions evaluated by radial integration over muffin-tin spheres, as in the augmented
plane wave (APW) approach of Chapter 16.

Here we only sketch the basic ideas of the definition of the PAW method for an
atom, following [542]. Further developments for calculations for molecules and solids are
deferred to Section 13.3. Just as in the OPW formulation, one can define a smooth part
of a valence wavefunction $\tilde{\psi}_i^v(\mathbf{r})$ (a plane wave as in Eq. (11.1) or an atomic orbital as
in Eq. (11.4)) and a linear transformation $\psi^v = \mathcal{T}\tilde{\psi}^v$ that relates the set of all-electron
valence functions $\psi_j^v(\mathbf{r})$ to the smooth functions $\tilde{\psi}_i^v(\mathbf{r})$. The transformation is assumed to
be unity except with a sphere centered on the nucleus, $\mathcal{T} = \mathbf{1} + \mathcal{T}_0$. For simplicity, we omit
the superscript v, assuming that the ψs are valence states, and the labels i, j. Adopting the
Dirac notation, the expansion of each smooth function $|\tilde{\psi}\rangle$ in partial waves m within each
sphere can be written (see Eqs. (J.1) and Eq. (16.4))

$$|\tilde{\psi}\rangle = \sum_m c_m |\tilde{\psi}_m\rangle, \tag{11.60}$$

with the corresponding all-electron function,

$$|\psi\rangle = \mathcal{T}|\tilde{\psi}\rangle = \sum_m c_m |\psi_m\rangle. \tag{11.61}$$

Hence the full wavefunction in all space can be written

$$|\psi\rangle = |\tilde{\psi}\rangle + \sum_m c_m \left\{ |\psi_m\rangle - |\tilde{\psi}_m\rangle \right\}, \tag{11.62}$$

which has the same form as Eqs. (11.4) and (11.8).

If the transformation $\mathcal{T}$ is required to be linear, then the coefficients must be given by a
projection in each sphere

$$c_m = \langle \tilde{p}_m | \tilde{\psi}\rangle, \tag{11.63}$$

for some set of projection operators $\tilde{p}$. If the projection operators satisfy the biorthogonality
condition,

$$\langle \tilde{p}_m | \tilde{\psi}_{m'}\rangle = \delta_{mm'}, \tag{11.64}$$

then the one-center expansion $\sum_m |\tilde{\psi}_m\rangle\langle\tilde{p}_m|\tilde{\psi}\rangle$ of the smooth function $\tilde{\psi}$ equals $\tilde{\psi}$ itself.

The resemblance of the projection operators to the separable form of pseudopotential
operators (Section 11.8) is apparent. Just as for pseudopotentials, there are many possible

choices for the projectors with examples given in [542] of smooth functions for $\tilde{p}(\mathbf{r})$ closely related to pseudopotential projection operators. The difference from pseudopotentials, however, is that the transformation $\mathcal{T}$ still involves the full all-electron wavefunction

$$\mathcal{T} = 1 + \sum_m \left\{ |\psi_m\rangle - |\tilde{\psi}_m\rangle \right\} \langle \tilde{p}_m|. \tag{11.65}$$

Furthermore, the expressions apply equally well to core and valence states so that one can derive all-electron results by applying the expressions to all the electron states.

The general form of the PAW equations can be cast in terms of the transformation Eq. (11.65). For any operator $\hat{A}$ in the original all-electron problem, one can introduce a transformed operator $\tilde{A}$ that operates on the smooth part of the wavefunctions

$$\tilde{A} = \mathcal{T}^\dagger \hat{A} \mathcal{T} = \hat{A} + \sum_{mm'} |\tilde{p}_m\rangle \left\{ \langle \psi_m | \hat{A} | \psi_{m'} \rangle - \langle \tilde{\psi}_m | \hat{A} | \tilde{\psi}_{m'} \rangle \right\} \langle \tilde{p}_{m'}|, \tag{11.66}$$

which is very similar to a pseudopotential operator as in Eq. (11.41). Furthermore, one can add to the right-hand side of Eq. (11.66) any operator of the form

$$\hat{B} - \sum_{mm'} |\tilde{p}_m\rangle \langle \tilde{\psi}_m | \hat{B} | \tilde{\psi}_{m'} \rangle \langle \tilde{p}_{m'}|, \tag{11.67}$$

with no change in the expectation values. For example, one can remove the nuclear Coulomb singularity in the equations for the smooth function, leaving a term that can be dealt with in the radial equations about each nucleus.

The expressions for physical quantities in the PAW approach follow from Eqs. (11.65) and (11.66). For example, the density is given by[11]

$$n(\mathbf{r}) = \tilde{n}(\mathbf{r}) + n^1(\mathbf{r}) - \tilde{n}^1(\mathbf{r}), \tag{11.68}$$

which can be written in terms of eigenstates labeled i with occupations f_i as

$$\tilde{n}(\mathbf{r}) = \sum_i f_i |\tilde{\psi}_i(\mathbf{r})|^2, \tag{11.69}$$

$$n^1(\mathbf{r}) = \sum_i f_i \sum_{mm'} \langle \tilde{\psi}_i | \tilde{p}_m \rangle \psi_m^*(\mathbf{r}) \psi_{m'}(\mathbf{r}) \langle \tilde{p}_{m'} | \tilde{\psi}_i \rangle, \tag{11.70}$$

and

$$\tilde{n}^1(\mathbf{r}) = \sum_i f_i \sum_{mm'} \langle \tilde{\psi}_i | \tilde{p}_m \rangle \tilde{\psi}_m^*(\mathbf{r}) \tilde{\psi}_{m'}(\mathbf{r}) \langle \tilde{p}_{m'} | \tilde{\psi}_i \rangle. \tag{11.71}$$

The last two terms are localized around each atom, and the integrals can be done in spherical coordinates with no problems from the strong variations near the nucleus, as in augmented methods. Section 13.3 is devoted to the PAW method and expressions for other quantities in molecules and condensed matter.

---

[11] The equations are modified if the core functions are not strictly localized in the augmentation spheres [546].

11.13 Additional Topics

Operators with Nonlocal Potentials

The nonlocal character of pseudopotentials leads to complications that the user should be aware of. One is that the usual relation of momentum and position matrix elements does not hold [547, 548]. The analysis at Eq. (20.33) shows that for nonlocal potentials the correct relation is

$$[H, \mathbf{r}] = i \frac{\hbar}{m_e} \mathbf{p} + [\delta V_{nl}, \mathbf{r}], \tag{11.72}$$

where δV_{nl} denotes the nonlocal part of the potential. The commutator can be worked out using the angular projection operators in δV_{nl} [547, 548].

Reconstructing the Full Wavefunction

In a pseudopotential calculation, only the pseudowavefunction is determined directly. However, the full wavefunction is required to describe many important physical properties, e.g., the Knight shift and the chemical shift measured in nuclear resonance experiments [549, 550]. These provide extremely sensitive probes of the environment of a nucleus and the valence states, but the information depends critically on the perturbations of the core states. Other experiments, such as core-level photoemission and absorption, involve core states directly.

The OPW and PAW methods provide the core wavefunctions. Is it possible to reconstruct the core wavefunctions from a usual pseudopotential calculation? The answer is yes, within some approximations. The procedure is closely related to the PAW transformation Eq. (11.65). For each scheme of generating *ab initio* pseudopotentials, one can formulate an explicit way to reconstruct the full wavefunctions given the smooth pseudofunction calculated in the molecule or solid. Such reconstruction has been used, e.g., by Mauri and coworkers, to calculate nuclear chemical shifts [550, 551].

Pseudohamiltonians

A pseudohamiltonian is a more general object than a pseudopotential; in addition to changing the potential, *the mass is varied to achieve the desired properties of the valence states*. Since the pseudohamiltonian is chosen to represent a spherical core, the *pseudo–kinetic energy operator* is allowed only to have a mass that can be different for radial and tangential motion and whose magnitude can vary with radius [552]. Actual pseudohamiltonians derived thus far have assumed that the potential is local [552–554]. If such a form can be found, it will be of great use in Monte Carlo calculations, where the nonlocal operators are problematic [552, 553]; however, it has so far not proven possible to derive pseudohamiltonians of general applicability.

Beyond the Single-Particle Approximation

It is also possible to define pseudopotentials that describe the effects of the cores *beyond the independent-electron approximation* [521, 555–557]. At first sight, it seems impossible to define a hamiltonian for valence electrons only, omitting the cores, when all electrons are identical. However, a proper theory can be constructed relying on the fact that all low-energy excitations can be mapped one to one onto a valence-only problem. In essence, the outer valence electrons can be viewed as *quasiparticles* that are renormalized by the presence of the core electrons. Further treatment is beyond the scope of the present work, but extensive discussion and actual pseudopotentials can be found in [555, 556].

SELECT FURTHER READING

Overview and perspective on earlier work on norm-conserving and ultrasoft potentials can be found in:

Hamann, D. R., "Optimized norm-conserving Vanderbilt pseudopotentials," *Phys. Rev. B* 88: 085117, 2013.

For history and early work see, for example:

Heine, V., in *Solid State Physics*, edited by H. Ehrenreich, F. Seitz, and D. Turnbull (Academic, New York, 1970), p. 1.

Exercises

11.1 Consider s-wave ($l = 0$) scattering in the example illustrated in Fig. 11.1. Using Formula Eq. (J.4) for the radial wavefunction ψ, with the definition $\phi = r\psi$, and the graphical construction indicated in Fig. 11.1, show that the scattering length approaches a well-defined limit as $\kappa \to 0$, and find the relation to the phase shift $\eta_0(\varepsilon)$.

11.2 The pseudopotential concept can be illustrated by a square well in one dimension with width s and depth $-V_0$. (See also Exercises 11.6 and 11.14; the general solution for bands in one dimension in Exercise 4.22; and relations to the plane wave, APW, KKR, and MTO methods, respectively, in Exercises 12.6, 16.1, 16.7, and 16.13.)

A plane wave with energy $\varepsilon > 0$ traveling to the right has a reflection coefficient r and transmission coefficient t (see Exercise 4.22).

(a) By matching the wavefunction at the boundary, derive r and t as a function of V_0, s, and ε. Note that the phase shift δ is the shift of phase of the transmitted wave compared to the wave in the absence of the well.

(b) Show that the same transmission coefficient t can be found with different V_0' and/or s' at a chosen energy ε_0.

(c) Combined with the analysis in Exercise 4.22, show that a band in a one-dimensional crystal is reproduced approximately by the modified potential. The bands agree exactly at energy $\varepsilon_k = \varepsilon_0$ and have errors linear in $\varepsilon_k - \varepsilon_0$ plus higher-order terms at other energies at most $\propto \varepsilon_k - \varepsilon_0$.

11.3 Following Eq. (11.9) it is stated that if $u_{li} = \psi^c_{li}$ in the OPW, then the smooth function $\tilde{\psi}^v_l(\mathbf{r})$ has no radial nodes. Show that this follows from the definition of the OPW.

11.4 Verify Expression (11.10) for the norm of an OPW. Show this means that different OPWs are not orthonormal and each has norm less than unity.

11.5 Derive the transformation from the OPW equation (11.11) to the pseudopotential equation (11.12) for the smooth part of the wavefunction.

11.6 Consider the one-dimensional square well defined in Exercise 11.2. There (and in Exercise 4.22) the scattering was considered in terms of left and right propagating waves ψ_l and ψ_r. However, pseudopotentials are defined for eigenstates of the symmetry. In one dimension the only spatial symmetry is inversion so that all states can be classified as even or odd. Here we construct a pseudopotential; the analysis is also closely related to the KKR solution in Exercise 16.7.

(a) Using linear combinations of ψ_l and ψ_r, construct even and odd functions, and show they have the form

$$\psi^+ = e^{-ik|x|} + (t+r)e^{ik|x|},$$
$$\psi^- = \left[e^{-ik|x|} + (t-r)e^{ik|x|}\right] \text{sign}(x). \tag{11.73}$$

(c) From the relation of t and r given in Exercise 4.22, show that the even and odd phase shifts are given by

$$e^{2i\eta^+} \equiv t + r = e^{i(\delta+\theta)},$$
$$e^{2i\eta^-} \equiv t - r = e^{i(\delta-\theta)}, \tag{11.74}$$

where $t = |t|e^{i\delta}$ and $\theta \equiv \cos^{-1}(|t|)$.

(d) Repeat the analysis of Exercise 11.2 and show that the band of a one-dimensional crystal at a given energy ε is reproduced by a pseudopotential if *both* phase shifts $\eta^+(\varepsilon)$ and $\eta^-(\varepsilon)$ are correct.

11.7 Find the analytic formulas for the Fourier transforms of a spherical square well potential $V(r) = v_0, r < R_0$, and a gaussian potential $V(r) = A_0 \exp -\alpha r^2$, using the expansion of a plane wave in spherical harmonics.

11.8 Show that the radial Schrödinger equation can be transformed to the nonlinear first-order differential equation (11.24).

11.9 Show that Eq. (11.26) indeed holds for any function f and that this relation leads to Eq. (11.27) with the choice $f(r) = (\partial/\partial\varepsilon)x_l(\varepsilon,r)$. To do this use the fact that $\phi = 0$ at the origin so that the final answer depends only on $f(R)$ and $\phi(R)$ at the outer radius.

11.10 Show that the third condition of norm conservation (agreement of logarithmic derivatives of the wavefunction) ensures that the potential is continuous at R_c.

11.11 Computational exercise (using available codes for pseudopotential calculations): Generate a "high-quality" (small R_c) pseudopotential for Si in the usual atomic ground state $3s^2 3p^2$. Check that the eigenvalues are the same as the all-electron calculation.

(a) Use the same pseudopotential to calculate the eigenvalues in various ionization states $+1$, $+2$, $+3$, $+4$. How do the eigenvalues agree with the all-electron results?

(b) Repeat for a poorer-quality (larger R_c) pseudopotential. Is the agreement worse? Why or why not?

(c) Carry out another set of calculations for a "compressed atom" – i.e., confined to a radius $\approx \frac{1}{2}$ the nearest neighbor distance. (This may require changes in the code.) Calculate the changes in eigenvalues using the all-electron code and using the same pseudopotential, i.e., one derived from the "compressed atom." How do they agree?

(d) Nonlinear core correlation corrections can also be tested. In many generation codes, the corrections can simply be turned on or off. One can also calculate explicitly the exchange–correlation energy using the pseudo and the entire density. The largest effects are for spin-polarized transition metals, e.g., Mn $3d^{5\uparrow}$ compared to $3d^{4\uparrow} 3d^{1\downarrow}$.

11.12 Show that unphysical "ghost states" can occur at low energies as eigenvalues of the hamiltonian with the nonlocal potential operator Eq. (11.41) if $V_{\text{local}}(r)$ is chosen to be large and negative (attractive) so that the nonlocal $\delta V_l(r)$ must be large and positive. Hint: consider the limit of a very large negative $V_{\text{local}}(r)$ acting on a state that is orthogonal to $\phi_l(r)$.

11.13 Show that each ψ_s is a solution of $\hat{H} \psi_s = \varepsilon_s \psi_s$ if the "ultrasoft" potential is constructed using Eq. (11.47).

11.14 The square well in one dimension considered in Exercises 11.2 and 11.6 illustrates ideas of the OPW and pseudopotential methods and also shows close relations to other methods (see Exercise 11.2). In this example we consider a bound state with $\varepsilon < 0$, but similar ideas apply for $\varepsilon > 0$ (Exercise 11.2).

(a) A deep well has states analogous to core states with $\varepsilon_c \ll 0$. Consider a well with width $s = 2a_0$ and depth $-V_0 = -12Ha$. Solve for the two lowest "core" states using the approximation that they are bound states of an infinite well. Solve for the third "valence" state by matching the wavefunction.

(b) Construct a generalized OPW-like valence state using the definition $\psi^v(x) = \tilde{\psi}^v(x) + \sum_j B_j u_j(x)$, analogous to (11.4). Rather than using the expressions in Fourier space, it is easiest to use the definition $B_j = \langle u_j | \tilde{\psi}^v \rangle$. The overlap B_j is zero for one of the "core" states; give the reason and generalize the argument to apply to core states of an atom in three dimensions. Show that the "smooth state" $\tilde{\psi}^v$ is indeed smoother than the original ψ^v.

(c) Construct the PKA pseudopotential analogous to Eq. (11.13) and show that its operation on $\tilde{\psi}^v$ is effectively that of a weaker potential.

(d) Construct a model potential with the same width s but weaker potential V_0' that has the same logarithmic derivative at the "valence" energy ε. Is this potential norm-conserving?

(e) Construct a norm-conserving potential, which can be done by first finding a nodeless norm-conserving wavefunction and inverting it as in Eq. (11.32). If the form of the wavefunction is analytic, e.g., a polynomial inside the well, all steps can be done analytically.

(f) Write a computer code to integrate the one-dimensional Schrödinger equation and evaluate the logarithmic derivative as a function of energy near ε and compare the results for the original problem with the pseudopotentials from parts (d) and (e).

(g) Transform the potential to a separable form as in Section 11.8. There is only one projector since only one state is considered. Show that for a symmetric well in one dimension the general form involves only two projectors for even and odd functions.

(h) Generate an "ultrasoft" potential and the resulting generalized eigenvalue problem analogous to Eq. (11.59). Discuss the relation to the OPW method and PKA form of the potential.

(i) Generate a PAW function and show the relation to the OPW and APW methods (part (b) above and Exercise 16.1).

PART IV

DETERMINATION OF ELECTRONIC STRUCTURE: THE BASIC METHODS

Overview of Chapters 12–18

The best revenge is massive success.

Frank Sinatra

There are three basic approaches to the calculation of independent-particle electronic states in materials. There are no fundamental disagreements. All agree when applied carefully and taken to convergence, as demonstrated by the heroic comparison of methods in [406]. Nevertheless, each of the approaches has particular values:

- Each method provides instructive, complementary ways to understand electronic structure.
- Each method has its advantages: each is most appropriate for a range of problems where it can provide insightful results with effective use of computational power.
- Each method has its pitfalls: the user beware. It is all too easy to make glaring errors or over-interpret results if the user does not understand the basics of the methods.
- There are well-developed codes available for each of the methods. In Appendix R is a list of a few widely-used codes that are used for examples in this book.

The purpose of the short introduction is to put the methods into perspective, identifying their main features, differences and similarities.

1. **Plane wave and grid methods** provide general approaches for solution of differential equations, including the Schrödinger and Poisson equations. At first sight, plane waves and grids are very different, but in fact each is an effective way of representing smooth functions. Furthermore, grids are involved in modern efficient plane wave calculations that use fast Fourier transforms.

   Chapter 12 is devoted to the basic concepts and methods of electronic structure. Plane waves are presented first because of their simplicity and because Fourier transforms provide a simple derivation of the Bloch theorem. Since plane waves are eigenfunctions of the Schrödinger equation with constant potential, they are the natural basis for description of bands in the nearly-free-electron approximation which provides important insight into band structures of many materials including sp-bonded metals, semiconductors, etc. Pseudopotentials are intertwined with plane wave methods because they allow calculations to be done with a feasible number of plane waves. The basic ideas can be understood in terms of empirical pseudopotentials which lead up to full DFT calculations using accurate first-principles pseudopotentials in the following chapter. Real-space grids and multiresolution methods are well-developed methods used in many fields that can be vary efficient and useful for many problems, especially finite systems and other cases where very different resolution is needed in different regions.

Chapter 13 is devoted to self-consistent *ab initio* methods that utilize plane waves (and/or grids) to solve the Kohn–Sham equations. Because of the simplicity of plane waves, they are often the basis of choice for development of new methods, such as Car-Parrinello quantum molecular dynamics simulations (Chapter 19), efficient iterative methods (Appendix M), and other innovations. Three approaches have made high accuracy calculations feasible and effective for many problems. "Norm-conserving" potentials and the ONCR extensions with multiple projectors provide accurate solutions with pseudofunctions that are orthonormal solutions of ordinary differential equations. With ultrasoft pseudopotentials, the problem is cast in terms of localized spherical functions and smooth wavefunctions that obey a generalized eigenvalue equation with an OPW-type hamiltonian. The PAW formulation completes the transformation by expressing the wavefunctions as a sum of smooth functions plus core functions, just as in the OPW approach. Unlike pseudopotentials, the PAW method keeps the entire set of all-electron core functions and the smooth parts of the valence functions. Matrix elements involving core functions are treated using muffin-tin spheres as in augmented methods (Chapter 16). Nevertheless, the ultrasoft and PAW methods maintain the advantage of pseudopotentials that forces can be calculated easily.

Real space methods are not as widely used in electronic structure at the time of this writing, but they have many advantages. Their use in both pseudopotential and all-electron methods is described in Chapter 12.

2. **Localized atomic(-like) orbitals (LCAO)** provide a basis that captures the essence of the atomic-like features of solids and molecules. They provide a satisfying, localized description of electronic structure widely used in chemistry, in "order-N" methods (Chapter 18), and in constructing useful models.

Chapter 14 defines the orbitals and presents basic theory. In particular, local orbitals provide an illuminating derivation (indeed the original derivation of Bloch in 1928) of the Bloch theorem. The semiempirical tight-binding method, associated with Slater and Koster, is particularly simple and instructive since one needs only the matrix elements of the overlap and hamiltonian. Tables of tight-binding matrix elements can be used to determine electronic states, total energies, and forces with very fast, simple calculations.

Chapter 15 is devoted to methods for full calculations done with localized basis functions such as gaussians, Slater-type orbitals, and numerical radial atomic-like orbitals. Calculations can vary from quick (and often dirty) to highly refined with many basis orbitals per atom. Even in the latter case, the calculations can be much smaller than with plane waves or grids. However, compared to general methods involving plane waves and grids, it is harder to reach convergence and greater care is needed in constructing basis functions of sufficient quality.

3. **Atomic sphere methods** are the most general methods for precise solution of the Kohn–
   Sham equations. The basic idea is to divide the electronic structure problem, providing
   efficient representation of atomic-like features that are rapidly varying near each nucleus
   and smoothly varying functions between the atoms.

   Chapter 16 is devoted to the original methods in which smooth functions are "augmented" near each nucleus by solving the Schrödinger equation in the sphere at each
   energy and matching to the outer wavefunction. The resulting APW and KKR methods
   are very powerful, but suffer from the fact that they require solution of non-linear
   equations. The Green's function KKR method is particularly elegant, providing local
   information as well as global information such as the Fermi surface. The non-linearity
   does not present any problem in a Green's function approach; however, it is difficult
   to extend the KKR approach beyond the muffin-tin potential approximation. Muffin-
   tin orbitals (MTOs) are localized, augmented functions that can lead to physically
   meaningful descriptions of electronic states in terms of a *minimal basis*, including the
   concept of "canonical bands," described in terms of structure constants and a very few
   "potential parameters."

   Chapter 17 deals with the advance that has made augmented methods much more
   tractable and useful: the "L" methods that make use of linearization of the equations
   around reference energies. This allows any of the augmented methods to be written in the
   familiar form of a secular equation linear in energy involving a hamiltonian and overlap
   matrix. The simplification has led to further advances, e.g., the development of full-
   potential methods, so that LAPW provides the most precise solutions of the Kohn–Sham
   equations available today. The LMTO approach describes electronic states in terms of
   a reduced linear hamiltonian with basis functions that are localized and designed to
   provide understanding of the electronic states. LMTO involves only a small basis and can
   be cast in the form of an *ab initio* orthogonal tight-binding hamiltonian with all matrix
   elements derived from the fundamental Kohn–Sham hamiltonian. It is also possible to
   go beyond linearization and a methodology is provided by the "NMTO" generalization
   to order N.

   Linear scaling O(N) methods in Chapter 18 are in a different category from the
   previous methods. The central idea is that any of those methods can be adapted to solve
   the Kohn–Sham equations in a time linearly proportional to the size of the system instead
   of cubic scaling of the standard eigenvalue methods (which can be reduced to square scaling
   in some methods). However, this applies only for large systems and requires special efforts
   to ensure accuracy. The most appropriate methods involve localized functions; however,
   even plane waves can be adapted by using functions confined to localized regions.

12

Plane Waves and Grids: Basics

Summary

Plane waves and grids provide general methodologies for solution of differential equations including the Schrödinger and Poisson equations; in many ways they are very different and in other ways they are two sides of the same coin. Plane waves are especially appropriate for periodic crystals, where they provide intuitive understanding as well as simple algorithms for practical calculations. Methods based on grids in real space are most appropriate for finite systems and are prevalent in many fields of science and engineering. We introduce them together because modern electronic structure algorithms use both plane waves and grids with fast Fourier transforms.

This chapter is organized first to give the general equations in a plane wave basis and a transparent derivation of the Bloch theorem, complementary to the one given in Chapter 4. The remaining sections are devoted to relevant concepts and useful steps, such as nearly-free-electron approximation and empirical pseudopotentials, that reveal the characteristic properties of electronic bands in materials. This lays the ground work for the full solution of the density functional equations using *ab initio* nonlocal pseudopotentials given in Chapter 13. The last sections are devoted to real-space finite difference grid methods and multiresolution methods including finite elements and wavelets.

12.1 The Independent-Particle Schrödinger Equation in a Plane Wave Basis

The eigenstates of any independent-particle Schrödinger-like equation in which each electron moves in an effective potential $V_{eff}(\mathbf{r})$,[1] such as the Kohn–Sham equations, satisfy the eigenvalue equation

$$\hat{H}_{eff}(\mathbf{r})\psi_i(\mathbf{r}) = \left[-\frac{\hbar^2}{2m_e}\nabla^2 + V_{eff}(\mathbf{r}) \right] \psi_i(\mathbf{r}) = \varepsilon_i \psi_i(\mathbf{r}). \qquad (12.1)$$

---

[1] The derivations in this section also hold if the potential is a nonlocal operator acting only on valence electrons (as for a nonlocal pseudopotential) or is energy dependent (as in the APW method). See Exercise 12.8.

In a solid (or any state of condensed matter) it is convenient to require the states to be normalized and obey periodic boundary conditions in a large volume Ω that is allowed to go to infinity. (Any other choice of boundary conditions will give the same result in the large Ω limit [91].) Using the fact that any periodic function can be expanded in the complete set of Fourier components, an eigenfunction can be written

$$\psi_i(\mathbf{r}) = \sum_{\mathbf{q}} c_{i,\mathbf{q}} \times \frac{1}{\sqrt{\Omega}} \exp(i\mathbf{q} \cdot \mathbf{r}) \equiv \sum_{\mathbf{q}} c_{i,\mathbf{q}} \times |\mathbf{q}\rangle, \tag{12.2}$$

where $c_{i,\mathbf{q}}$ are the expansion coefficients of the wavefunction in the basis of orthonormal plane waves $|\mathbf{q}\rangle$ satisfying

$$\langle \mathbf{q}'|\mathbf{q}\rangle \equiv \frac{1}{\Omega} \int_{\Omega} d\mathbf{r} \, \exp(-i\mathbf{q}' \cdot \mathbf{r})\exp(i\mathbf{q} \cdot \mathbf{r}) = \delta_{\mathbf{q},\mathbf{q}'}. \tag{12.3}$$

Inserting Eq. (12.2) into Eq. (12.1), multiplying from the left by $\langle \mathbf{q}'|$ and integrating as in Eq. (12.3) leads to the Schrödinger equation in Fourier space

$$\sum_{\mathbf{q}} \langle \mathbf{q}'|\hat{H}_{eff}|\mathbf{q}\rangle c_{i,\mathbf{q}} = \varepsilon_i \sum_{\mathbf{q}} \langle \mathbf{q}'|\mathbf{q}\rangle c_{i,\mathbf{q}} = \varepsilon_i c_{i,\mathbf{q}'}. \tag{12.4}$$

The matrix element of the kinetic energy operator is simply

$$\langle \mathbf{q}'| - \frac{\hbar^2}{2m_e}\nabla^2|\mathbf{q}\rangle = \frac{\hbar^2}{2m_e}|q|^2\delta_{\mathbf{q},\mathbf{q}'} \rightarrow \frac{1}{2}|q|^2\delta_{\mathbf{q},\mathbf{q}'}, \tag{12.5}$$

where the last expression is in Hartree atomic units. For a crystal, the potential $V_{eff}(\mathbf{r})$ is periodic and can be expressed as a sum of Fourier components (see Eqs. (4.7) to (4.11))

$$V_{eff}(\mathbf{r}) = \sum_{m} V_{eff}(\mathbf{G}_m) \exp(i\mathbf{G}_m \cdot \mathbf{r}), \tag{12.6}$$

where $\mathbf{G}_m$ are the reciprocal lattice vectors, and

$$V_{eff}(\mathbf{G}) = \frac{1}{\Omega_{\text{cell}}} \int_{\Omega_{\text{cell}}} V_{eff}(\mathbf{r}) \exp(-i\mathbf{G} \cdot \mathbf{r})d\mathbf{r}, \tag{12.7}$$

with Ω_{cell} the volume of the primitive cell. Thus the matrix elements of the potential

$$\langle \mathbf{q}'|V_{eff}|\mathbf{q}\rangle = \sum_{m} V_{eff}(\mathbf{G}_m)\delta_{\mathbf{q}'-\mathbf{q},\mathbf{G}_m} \tag{12.8}$$

are nonzero only if $\mathbf{q}$ and $\mathbf{q}'$ differ by some reciprocal lattice vector $\mathbf{G}_m$.

Finally, if we *define* $\mathbf{q} = \mathbf{k} + \mathbf{G}_m$ and $\mathbf{q}' = \mathbf{k} + \mathbf{G}_{m'}$ (which differ by a reciprocal lattice vector $\mathbf{G}_{m''} = \mathbf{G}_m - \mathbf{G}_{m'}$), then the Schrödinger equation for any given $\mathbf{k}$ can be written as the matrix equation

$$\sum_{m'} H_{m,m'}(\mathbf{k})c_{i,m'}(\mathbf{k}) = \varepsilon_i(\mathbf{k})c_{i,m}(\mathbf{k}), \tag{12.9}$$

where[2]

$$H_{m,m'}(\mathbf{k}) = \langle \mathbf{k} + \mathbf{G}_m | \hat{H}_{eff} | \mathbf{k} + \mathbf{G}_{m'} \rangle = \frac{\hbar^2}{2m_e} |\mathbf{k} + \mathbf{G}_m|^2 \delta_{m,m'} + V_{eff}(\mathbf{G}_m - \mathbf{G}_{m'}).$$

(12.10)

Here we have labeled the eigenvalues and eigenfunctions $i = 1, 2, \ldots$, for the discrete set of solutions of the matrix equations for a given $\mathbf{k}$. Equations (12.9) and (12.10) are the basic Schrödinger equations in a periodic crystal, leading to the formal properties of bands derived in the next section as well as to the practical calculations that are the subject of the remainder of this chapter.

12.2 Bloch Theorem and Electron Bands

The fundamental properties of bands and the Bloch theorem have been derived from the translation symmetry in Section 4.3; this section provides an alternative, simpler derivation[3] in terms of the Fourier analysis of the previous section.

1. **The Bloch theorem.** Each eigenfunction of the Schrödinger equation, (12.9), for a given $\mathbf{k}$ is given by Eq. (12.2), with the sum over $\mathbf{q}$ restricted to $\mathbf{q} = \mathbf{k} + \mathbf{G}_m$, which can be written

$$\psi_{i,\mathbf{k}}(\mathbf{r}) = \sum_m c_{i,m}(\mathbf{k}) \times \frac{1}{\sqrt{\Omega}} \exp(i(\mathbf{k} + \mathbf{G}_m) \cdot \mathbf{r}) = \exp(i\mathbf{k} \cdot \mathbf{r}) \frac{1}{\sqrt{N_{\text{cell}}}} u_{i,\mathbf{k}}(\mathbf{r}),$$

(12.11)

where $\Omega = N_{\text{cell}} \Omega_{\text{cell}}$ and

$$u_{i,\mathbf{k}}(\mathbf{r}) = \frac{1}{\sqrt{\Omega_{\text{cell}}}} \sum_m c_{i,m}(\mathbf{k}) \exp(i\mathbf{G}_m \cdot \mathbf{r}),$$

(12.12)

which has the periodicity of the crystal. This is the Bloch theorem also stated in Eq. (4.33): any eigenvector is a product of $\exp(i\mathbf{k} \cdot \mathbf{r})$ and a periodic function. Since we require $\psi_{i,\mathbf{k}}(\mathbf{r})$ to be orthonormal over the volume Ω, then $u_{i,\mathbf{k}}(\mathbf{r})$ are orthonormal in one primitive cell, i.e.,

$$\frac{1}{\Omega_{\text{cell}}} \int_{\text{cell}} d\mathbf{r} u^*_{i,\mathbf{k}}(\mathbf{r}) u_{i',\mathbf{k}}(\mathbf{r}) = \sum_m c^*_{i,m}(\mathbf{k}) c_{i',m}(\mathbf{k}) = \delta_{i,i'},$$

(12.13)

where the final equation means the $c_{i,m}(\mathbf{k})$ are orthonormal vectors in the discrete index m of the reciprocal lattice vectors.

2. **Bands of eigenvalues.** Since the Schrödinger equation, (12.9), is defined for each $\mathbf{k}$ separately, each state can be labeled by the wavevector $\mathbf{k}$ and the eigenvalues and eigenvectors for each $\mathbf{k}$ are independent unless they differ by a reciprocal lattice vector.

---

[2] The effective potential $V_{eff}(\mathbf{G}_m - \mathbf{G}_{m'})$ must be generalized for nonlocal potentials to depend on all the variables $V_{eff}(\mathbf{K}_m, \mathbf{K}_{m'})$, where $\mathbf{K}_m = \mathbf{k} + \mathbf{G}_m$ (see Section 12.4).

[3] This derivation follows the "second proof" of the Bloch theorem given by Ashcroft and Mermin [280].

In the limit of large volume Ω, the $\mathbf{k}$ points become a dense continuum and the eigenvalues $\varepsilon_i(\mathbf{k})$ become continuous *bands*. At each $\mathbf{k}$ there are a discrete set of eigenstates labeled $i = 1, 2, \ldots$, that may be found by diagonalizing the hamiltonian, Eq. (12.10), in the basis of discrete Fourier components $\mathbf{k} + \mathbf{G}_m$, $m = 1, 2, \ldots$.

3. **Conservation of crystal momentum.** Since any state can be labeled by a well-defined $\mathbf{k}$, it follows that $\mathbf{k}$ is *conserved* in a way analogous to ordinary momentum in free space; however, in this case $\mathbf{k}$ is *conserved modulo addition of any reciprocal lattice vector* $\mathbf{G}$. In fact, it follows from inspection of the Schrödinger equation, (12.9), with the hamiltonian, Eq. (12.10), that the solutions are periodic in $\mathbf{k}$, so that all unique solutions are given by $\mathbf{k}$ in one primitive cell of the reciprocal lattice.

4. **The role of the Brillouin zone.** Since all possible eigenstates are specified by the wavevector $\mathbf{k}$ within any one primitive cell of the periodic lattice in reciprocal space, the question arises: is there a "best choice" for the cell? The answer is "yes." The first Brillouin zone (BZ) is the uniquely defined cell that is the most compact possible cell, and it is the cell of choice in which to represent excitations. It is unique among all primitive cells because its boundaries are the bisecting planes of the $\mathbf{G}$ vectors where Bragg scattering occurs (see Section 4.2). Inside the Brillouin zone there are no such boundaries: the bands must be continuous and analytic inside the zone. The boundaries are of special interest since every boundary point is a $\mathbf{k}$ vector for which Bragg scattering can occur; this leads to special features, such as zero group velocities due to Bragg scattering at the BZ boundary. The construction of the BZ is illustrated in Figs. 4.1, 4.2, 4.3, and 4.4, and widely used notations for points in the BZ of several crystals are given in Fig. 4.10.

5. **Integrals in k space.** For many properties such as the counting of electrons in bands, total energies, etc., it is essential to integrate over $\mathbf{k}$ throughout the BZ. As pointed out in Section 4.3, an intrinsic property of a crystal expressed "per unit cell" is an average over $\mathbf{k}$, i.e., a sum over the function evaluated at points $\mathbf{k}$ divided by the number of values N_k, which in the limit is an integral. For a function $f_i(\mathbf{k})$, where i denotes the discrete band index, the average value is

$$\bar{f}_i = \frac{1}{N_k} \sum_{\mathbf{k}} f_i(\mathbf{k}) \rightarrow \frac{\Omega_{\text{cell}}}{(2\pi)^d} \int_{\text{BZ}} d\mathbf{k} f_i(\mathbf{k}), \qquad (12.14)$$

where Ω_{cell} is the volume of a primitive cell in real space and $(2\pi)^d / \Omega_{\text{cell}}$ is the volume of the BZ. Specific algorithms for integration over the BZ are described in Section 4.6.

12.3 Nearly-Free-Electron Approximation

The nearly-free-electron approximation (NFEA) is the starting point for understanding bands in crystals. Not only is it a way to illustrate the properties of bands in periodic crystals, but the NFEA quantitatively describes bands for many materials. In the homogeneous gas, described in Chapter 5, the bands are simply the parabola $\varepsilon(\mathbf{q}) = (\hbar^2/2m_e)|\mathbf{q}|^2$. The first step in the NFEA is to plot the free-electron bands in the BZ of the given crystal. The bands

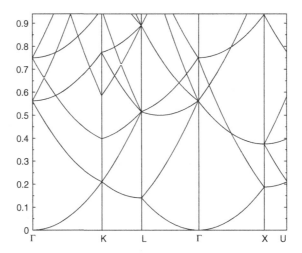

Figure 12.1. Free-electron bands plotted in the BZ of an fcc crystal. The BZ is shown in Fig. 4.10, which defines the labels. Compare this with the actual bands of Al in Fig. 16.6 that were calculated using the KKR method. Al is an ideal case where the bands are well explained by a weak pseudopotential [506–508, 560].

are still the simple parabola $\varepsilon(\mathbf{q}) = (\hbar^2/2m_e)|\mathbf{q}|^2$, but they are plotted as a function of $\mathbf{k}$ where $\mathbf{q} = \mathbf{k} + \mathbf{G}_m$, with $\mathbf{k}$ restricted to the BZ. Thus for each Bravais lattice, the free-electron bands have a characteristic form for lines in the Brillouin zone, with the energy axis scaled by $\Omega^{-2/3}$, where Ω is the volume of the primitive cell. By this simple trick we can plot the bands that result from the Schrödinger equation, (12.9), for a vanishing potential.

An example of a three-dimensional fcc crystal is shown in Fig. 12.1. The bands are degenerate at high symmetry points like the zone center, since several **G** vectors have the same modulus. The introduction of a weak potential on each atom provides a simple way of understanding NFEA bands, which are modified near the zone boundaries. An excellent example is Al, for which bands are shown in Fig. 16.6, compared to the free-electron parabolic dispersion. The bands are very close to free-electron-like, yet the Fermi surface is highly modified by the lattice effects because it involves bands very near zone boundary points where degeneracies are lifted and there are first-order effects on the bands. The bands have been calculated using many methods: the KKR method is effective since one is expanding around the analytic free-electron Green's function (outside the core) [558]; the OPW [559] and pseudopotential methods [504] make use of the fact that for weak effective scattering only a few plane waves are needed. Computer programs available online (see Appendix R) can be used to generate the bands and understand them in terms of the NFEA using only a few plane waves. See Exercises 12.4, 12.7, 12.11, and 12.12.

The fcc NFEA bands provide an excellent illustration of the physics of band structures. For sp-bonded metals like Na and Al, the NFEA bands are very close to the actual bands (calculated and experimental). The success of the NFEA directly demonstrates the fact that the bands can in some sense be considered "nearly free" even though the states must actually be very atomic-like with structure near the nucleus so that they are properly orthogonal to the core states. The great beauty of the pseudopotential, APW, and KKR methods is that they provide a very simple explanation in terms of the weak scattering properties of the atom even though the potential is strong.

12.4 Form Factors and Structure Factors

An important concept in the Fourier analysis of crystals is the division into "structure factors" and "form factors." For generality, let the crystal be composed of different species of atoms each labeled $\kappa = 1, n_{\text{species}}$, and for each κ there are n^{κ} identical atoms at positions $\tau_{\kappa, j}, j = 1, n^{\kappa}$ in the unit cell. Any property of the crystal, e.g., the potential, can be written

$$V(\mathbf{r}) = \sum_{\kappa=1}^{n_{\text{species}}} \sum_{j=1}^{n^{\kappa}} \sum_{\mathbf{T}} V^{\kappa}(\mathbf{r} - \tau_{\kappa, j} - \mathbf{T}), \tag{12.15}$$

where $\mathbf{T}$ denotes the set of translation vectors. It is straightforward (Exercise 12.2) to show that the Fourier transform of Eq. (12.15) can be written as

$$V(\mathbf{G}) \equiv \frac{1}{\Omega_{\text{cell}}} \int_{\Omega_{\text{cell}}} V(\mathbf{r}) \exp(i\mathbf{G} \cdot \mathbf{r}) d\mathbf{r} = \sum_{\kappa=1}^{n_{\text{species}}} \frac{\Omega^{\kappa}}{\Omega_{\text{cell}}} S^{\kappa}(\mathbf{G}) V^{\kappa}(\mathbf{G}), \tag{12.16}$$

where the **structure factor** for each species κ is

$$S^{\kappa}(\mathbf{G}) = \sum_{j=1}^{n^{\kappa}} \exp(i\mathbf{G} \cdot \tau_{\kappa, j}) \tag{12.17}$$

and the **form factor** is[4]

$$V^{\kappa}(\mathbf{G}) = \frac{1}{\Omega^{\kappa}} \int_{\text{allspace}} V^{\kappa}(\mathbf{r}) \exp(i\mathbf{G} \cdot \mathbf{r}) d\mathbf{r}. \tag{12.18}$$

The factors in Eqs. (12.16)–(12.17) have been chosen so that $V^{\kappa}(|\mathbf{G}|)$ is defined in terms of a "typical volume" Ω^{κ} for each species κ, so that $V^{\kappa}(|\mathbf{G}|)$ is independent of the crystal. In addition, the structure factor is defined so that $S^{\kappa}(\mathbf{G} = 0) = n^{\kappa}$. These are arbitrary – but convenient – choices; other authors may use different conventions.

Equation (12.16) is particularly useful in cases where the potential is a sum of spherical potentials in real space,

$$V^{\kappa}(\mathbf{r} - \tau_{\kappa, j} - \mathbf{T}) = V^{\kappa}(|\mathbf{r} - \tau_{\kappa, j} - \mathbf{T}|). \tag{12.19}$$

This always applies for nuclear potentials and bare ionic pseudopotentials. Often it is also a reasonable approximation for the total crystal potential as the sum of spherical

---

[4] Note the difference from Eqs. (4.11), between (12.18) and where for the latter the integral is over the cell instead of all space; Exercise 12.3 shows the equivalence of the expressions.

potentials around each nucleus.[5] Using the well-known expansion of plane waves in spherical harmonics, Eq. (J.1), Eq. (12.18) can be written as [102, 372, 561]

$$V^\kappa(\mathbf{G}) = V^\kappa(|\mathbf{G}|) = \frac{4\pi}{\Omega^\kappa} \int_0^\infty dr\, r^2\, j_0(|\mathbf{G}|r) V^\kappa(r). \qquad (12.20)$$

For a nuclear potential, $V^\kappa(\mathbf{G})$ is simply

$$V^\kappa_{\text{nucleus}}(|\mathbf{G}|) = \frac{4\pi}{\Omega^\kappa} \frac{-Z^\kappa_{\text{nucleus}} e^2}{|\mathbf{G}|^2}, \quad \mathbf{G} \neq 0,$$

$$= 0, \quad \mathbf{G} = 0, \qquad (12.21)$$

where the divergent $\mathbf{G} = 0$ term is treated separately, as discussed in Section 3.2 and Appendix F. For a bare pseudopotential, the potential form factor Eq. (12.20) is the transform of the pseudopotential $V_l(\mathbf{r})$, given in Chapter 11. Again the $\mathbf{G} = 0$ term must be treated carefully. One procedure is to calculate the potential and total energy of point ions of charge Z^κ in a compensating background that represents the $\mathbf{G} = 0$ Fourier component of the electron density. In that case, there is an additional contribution that arises from the fact that the ion is not a point charge [562],

$$\alpha^\kappa = \int 4\pi r^2 dr \left[V^\kappa_{\text{local}}(r) - \left(-\frac{Z^\kappa}{r} \right) \right]. \qquad (12.22)$$

Each ion contributes a constant term in the total energy (see Eq. (13.1) below) equal to $(N_e/\Omega)\alpha^\kappa$, where N_e/Ω is the average electron density.

The generalization of Eq. (12.16) to nonlocal potentials $V^\kappa_{\text{NL}}(\mathbf{r}, \mathbf{r}')$ is straightforward. For each $\mathbf{k}$ and basis vectors $\mathbf{G}_m$ and $\mathbf{G}_{m'}$, it is convenient to define $\mathbf{K}_m = \mathbf{k} + \mathbf{G}_m$ and $\mathbf{K}_{m'} = \mathbf{k} + \mathbf{G}_{m'}$. The structure factor $S(\mathbf{G})$ still depends only on $\mathbf{G} = \mathbf{K}_m - \mathbf{K}_{m'} = \mathbf{G}_m - \mathbf{G}_{m'}$, but the matrix elements of the semilocal form factor are more complicated since the matrix elements depends on two arguments. Using the fact that the spherical symmetry of the nonlocal operator guarantees that it can be written as a function of the magnitudes $|\mathbf{K}_m|$, $|\mathbf{K}_{m'}|$ and the angle θ between $\mathbf{K}_m$ and $\mathbf{K}_{m'}$, the matrix elements of the semilocal form factor Eq. (11.15), are (Exercise 12.9)

$$\delta V^\kappa_{NL}(\mathbf{K}_m, \mathbf{K}_{m'}) = \frac{4\pi}{\Omega^\kappa} \sum_l (2l+1) P_l(\cos(\theta)) \int_0^\infty dr\, r^2\, j_l(|\mathbf{K}_m|r) j_l(|\mathbf{K}_{m'}|r) \delta V^\kappa_l(r).$$

$$(12.23)$$

This formula has the disadvantage that it must be evaluated for each $|\mathbf{K}_m|$, $|\mathbf{K}_{m'}|$, and θ, i.e., for a three-dimensional object. In order to treat this in a computationally efficient manner, one can discretize this function on a grid and interpolate during an actual calculation.

The separable Kleinman–Bylander form, Eq. (11.41), is simpler because it is a sum of products of Fourier transforms. Each Fourier transform is a one-dimensional function of

---

[5] Many studies have verified that the total potential $V(\mathbf{r})$ is close to the sum of neutral atom potentials. This is especially true for examples like transition metals where the environment of each atom is nearly spherical. See Chapter 16.

$|\mathbf{K}_m|$ (and the same function of $|\mathbf{K}_{m'}|$), which is much more convenient. The form in k space is analogous to that in real space [372, 539]. (Here we denote the azimuthal quantum number as m_l to avoid confusion with the index m for basis functions $\mathbf{G}_m$.)

$$\delta V_{NL}^{\kappa}(\mathbf{K}_m, \mathbf{K}_{m'}) = \sum_{lm_l} \frac{Y_{lm_l}^*(\hat{\mathbf{K}}_m) T_l^*(|\mathbf{K}_m|) \times T_l(|\mathbf{K}_{m'}|) Y_{lm_l}(\hat{\mathbf{K}}_{m'})}{\langle \psi_{lm}^{PS} | \delta V_l | \psi_{lm}^{PS} \rangle}, \qquad (12.24)$$

where $T_l(q)$ is the Fourier transform of the radial function $\psi_l^{PS}(r)\delta V_l(r)$. The simplicity of this form has led to its widespread use in calculations. Furthermore it is straightforward to extend to "ultrasoft" potentials that involve additional projectors (see Section 11.11).

12.5 Approximate Atomic-Like Potentials

A first step in including the effects of the nuclei is to assume that the potential is a sum of atomic-like potentials. This gives all the qualitative features of the bands and often gives semiquantitative results. One procedure is simply to use the potential directly from an atomic calculation; another is to assume the potential has some simple analytic form. For example, if we approximate the electrons as nearly-free-electron-like then the total potential due to the nuclei and electrons to first order in perturbation theory is given by

$$V_{total}(\mathbf{G}) \approx V_{screened}(\mathbf{G}) \equiv V_{bare}(\mathbf{G})/\epsilon(\mathbf{G}), \qquad (12.25)$$

where V_{bare} is a bare nuclear or ionic potential and $\epsilon(\mathbf{G})$ is the screening function. In the NFE limit, the screening is evaluated for the homogeneous gas, so it is isotropic $\epsilon(|\mathbf{G}|)$ and a reasonable approximation is the Thomas–Fermi screening, where ϵ can be written

$$\epsilon(|\mathbf{G}|) = \frac{|\mathbf{G}|^2}{|\mathbf{G}|^2 + k_0^2}, \qquad (12.26)$$

using Eqs. (5.20) and (5.21), where k_0 is dependent only on the electron density (i.e., r_s). Furthermore, since the screening is linear in this approximation, the total potential is a sum of spherical screened nuclear or ionic potential which are neutral and atomic-like.

This approach was instrumental in the early work on *ab initio* pseudopotentials, e.g., the Heine–Abarenkov potentials [506, 508, 518] that are derived from atomic data and have been very successfully used in solids with an approximate screening function such as Eq. (12.26). A simple, instructive example is hydrogen at high pressure, i.e., high density or small $r_s \approx 1$. This corresponds to about 10 GPa, pressures that can be found in the interiors of the giant planets. At such densities, hydrogen is predicted to form a monatomic crystal with nearly-free-electron bands. Since the "bare" potential is just $\propto 1/|\mathbf{G}|^2$, it is easy to work out the screened potential in the Thomas–Fermi approximation. Exercise 12.13 calculates the appropriate form factors, estimates band structure in perturbation theory, carries out calculations using available programs (or by writing one's own), and compares with fully self-consistent calculations.

This approximation is sufficient to illustrate two points. First, the total potential near each nucleus is very well approximated by a spherical atomic-like form. This is widely used in augmented methods such as APW, KKR, and LMTO that treat the region around the nucleus using spherical coordinates (Chapters 16 and 17). Second, the approximation demonstrates the problems with the straightforward application of plane waves. Except for the lowest Z elements, materials with core electrons require huge numbers of plane waves (see Exercise 12.10). This is why pseudopotentials (Chapter 11) are so intimately related to the success of plane wave methods.

12.6 Empirical Pseudopotential Method (EPM)

Even though the general ideas of pseudopotentials have been known for many years [65–67], and model potentials close to those used in modern work were already applied to solids as early as the 1930s [66, 67], the modern use of pseudopotentials started with the work of Phillips and Kleinman [504], and Antonchik [502, 503]. Those authors realized that the band structure of sp-bonded metals and semiconductors could be quantitatively described by a few numbers: the values of the spherical atomic-like potentials at a few lowest reciprocal lattice vectors. By fitting to experimental data, a few parameters could be used to describe a tremendous amount of data related to the band structure, effective masses and bandgaps, optical properties, etc. The "empirical pseudopotential" method has been described in detail by Heine and Cohen [506, 507], who showed the connections to the underlying theory. Applications to metals are covered thoroughly by Harrison [508], and a very complete exposition of the method and results for semiconductors has been given by Cohen and Chelikowsky [561].

The EPM method has played an important role in understanding electronic structure, especially for the sp-bonded metals and semiconductors. As an example, Fig. 12.2 shows the bands of GaAs measured [563] by photoemission spectroscopy are compared with EPM bands calculated [564] many years before. The agreement with the photoemission data is nearly perfect for this nonlocal pseudopotential that was adjusted to fit the bandgaps, effective masses, and optical spectra [561]. Comparison of Fig. 12.2 with Fig. 2.23 shows the agreement with inverse photoemission and more recent many-body calculations, and the fact that the adjusted EPM provides a better description of the bands than do LDA calculations. The pseudocharge density has been calculated for many materials [561]: as illustrated in Fig. 12.3, the results show the basic features of the chemical bonding and the nature of individual states.

The method is more than just a fitting procedure if one makes the approximation that the total potential is a sum of spherical potentials that *have analytic form* and are *transferable between different structures*. Although this is an approximation, it has been tested in many cases and, at least, provides semiquantitative results. With the assumption of transferability, the EPM method can readily be applied to calculations for many structures and for properties like electron–phonon interactions (see, e.g., [565]), where the distorted lattice is simply viewed as a different structure.

The simplicity of the EPM makes possible calculations not feasible using *ab initio* pseudopotentials. It is a great advantage to have an analytic representation since it can

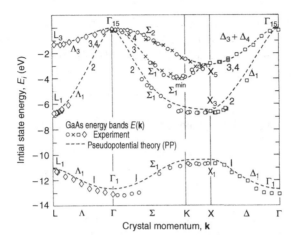

Figure 12.2. Experimental energy bands of GaAs measured by photoemission in [563] compared to empirical pseudopotential calculations [564]. The pseudopotential was fitted earlier to independent optical data, so this is a test of the transferability of information within an independent-particle theory. Spin–orbit interaction is small on this scale and is not included. From [563].

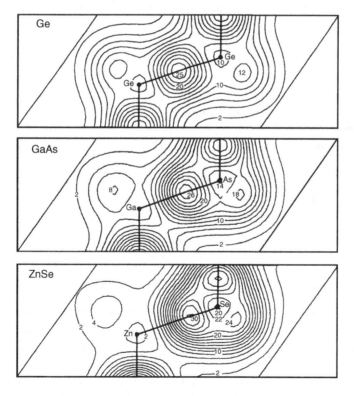

Figure 12.3. Theoretical calculations of the valence charge density of semiconductors showing the formation of the covalent bond and the progression to more ionic character in the series Ge, GaAs, and ZnSe. The results have the same basic features as full self-consistent calculations such as that for Si in Fig. 2.1. From [105].

be used for any structure. For example, [566, 567] report calculations of the electronic structure of pyramidal quantum dots containing up to 250,000 atoms, using spin–orbit-coupled, nonlocal, empirical pseudopotentials and with results that differ from those found using the effective-mass approximation.

12.7 Calculation of Electron Density: Introduction of Grids

One of the most important operations is the calculation of the density of electrons n. The general form for a crystal treated in independent-particle theory, e.g., Eqs. (3.42) or (7.2), can be written as

$$n(\mathbf{r}) = \frac{1}{N_k} \sum_{\mathbf{k},i} f(\varepsilon_{i,\mathbf{k}})n_{i,\mathbf{k}}(\mathbf{r}), \text{ with } n_{i\mathbf{k}}(\mathbf{r}) = |\psi_{i,\mathbf{k}}(\mathbf{r})|^2, \tag{12.27}$$

which is an average over $\mathbf{k}$ points (see Eq. (12.14)), with i denoting the bands at each $\mathbf{k}$ point (including the spin index σ) and $f(\varepsilon_{i,\mathbf{k}})$ denoting the Fermi function. For a plane wave basis, Expression (12.11) for the Bloch functions leads to

$$n_{i,\mathbf{k}}(\mathbf{r}) = \frac{1}{\Omega} \sum_{m,m'} c_{i,m}^*(\mathbf{k})c_{i,m'}(\mathbf{k}) \exp(\mathrm{i}(\mathbf{G}_{m'} - \mathbf{G}_m) \cdot \mathbf{r}) \tag{12.28}$$

and

$$n_{i,\mathbf{k}}(\mathbf{G}) = \frac{1}{\Omega} \sum_{m,m''} c_{i,m}^*(\mathbf{k})c_{i,m''}(\mathbf{k}), \tag{12.29}$$

where m'' denotes the $\mathbf{G}$ vector for which $\mathbf{G}_{m''} \equiv \mathbf{G}_m + \mathbf{G}$.

The symmetry operations R_n of the crystal can be used as in Sections 4.5 and 4.6 to find the density in terms only of the $\mathbf{k}$ points in the IBZ,

$$n(\mathbf{r}) = \frac{1}{N_k} \sum_{i,\mathbf{k}} n_{i,\mathbf{k}}(\mathbf{r}) = \frac{1}{N_{\mathrm{group}}} \sum_{R_n} \sum_{\mathbf{k}}^{\mathrm{IBZ}} w_{\mathbf{k}} \sum_i f(\varepsilon_{i,\mathbf{k}})n_{i,\mathbf{k}}(R_n\mathbf{r} + \mathbf{t}_n), \tag{12.30}$$

and

$$n(\mathbf{G}) = \frac{1}{N_{\mathrm{group}}} \sum_{R_n} \exp(\mathrm{i}R_n\mathbf{G} \cdot \mathbf{t}_n) \sum_{\mathbf{k}}^{\mathrm{IBZ}} w_{\mathbf{k}} \sum_i f(\varepsilon_{i,\mathbf{k}})n_{i,\mathbf{k}}(R_n\mathbf{G}). \tag{12.31}$$

The phase factor due to the translation $\exp(\mathrm{i}R_n\mathbf{G} \cdot \mathbf{t}_n)$ follows from Eq. (12.28).

Despite the simplicity of Eq. (12.29), it is not the most efficient way to calculate the density $n(\mathbf{r})$ or $n(\mathbf{G})$. The problem is that finding all the Fourier components using Eq. (12.29) involves a double sum, i.e., a convolution in Fourier space that requires N_G^2 operations, where N_G is the number of $\mathbf{G}$ vectors needed to describe the density. For large systems this becomes very expensive. On the other hand, if the Bloch states are known on a grid of N_R points in real space, the density can be found simply as a square, in N_R operations. The trick is to use a fast Fourier transform (FFT) that allows one to transform from one space to the other in $N \log N$ operations, where $N = N_R = N_G$. The flowchart,

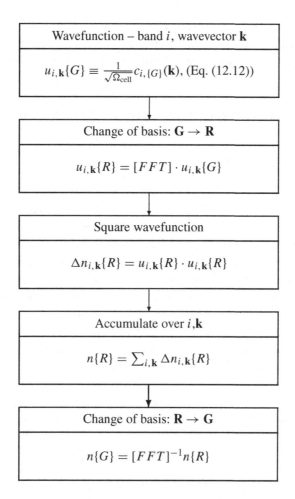

Figure 12.4. Calculation of the density using Fourier transforms and grids. The notation $\{G\}$ and $\{R\}$ denotes the sets of N **G** vectors and N grid points **R**. Since the fast Fourier transform (FFT) scales as $N \ln N$, the algorithm is faster than the double sum needed to calculate $n\{G\}$ that scales as N^2. In addition, the result is given in both real and reciprocal space, needed for calculation of the *exchange–correlation* and Hartree terms. The algorithm is essentially the same as used in iterative methods (Appendix M).

Fig. 12.4, illustrates the algorithm, and the general features for all such operations are described in Section M.11. A great advantage is that $n(\mathbf{r})$ is needed to find $\epsilon_{xc}(\mathbf{r})$ and $V_{xc}(\mathbf{r})$. The inverse transform can be used to find $n(\mathbf{G})$, which can be used for solving the Poisson equation in Fourier space.

It is relevant to note that the density n requires Fourier components that extend twice as far in each direction as those needed for the wavefunction ψ because $n \propto |\psi|^2$. Also the FFT requires a regular grid in the form of a parallelepiped, whereas the wavefunction cutoff is generally a sphere with $(1/2)|\mathbf{k} + \mathbf{G}|^2 < E_{\text{cutoff}}$. Thus the number of points in the FFT grid for density $N = N_R = N_G$ is roughly an order of magnitude larger than the number N_G^{wf} of **G** vectors in the basis for the wavefunctions. *Nevertheless, the FFT approach is much more efficient for large systems* since the number of operations scales as $N \log N$.

12.8 Real-Space Methods I: Finite Difference and Discontinuous Galerikin Methods

Since the Kohn–Sham equations are a set of coupled second-order differential equations, it is natural to ask: Why not use real-space methods widely employed in many areas of numerical analysis, such as finite element, finite difference, multigrid, wavelets, etc.? There are reasons they have not been developed as much as other methods for electronic structure applications. There are problems involving characteristic features of atoms, namely that the potentials are highly peaked around the nuclei, where they are nearly spherically symmetric with core states that are almost inert and yet must be treated accurately. Great accuracy is required to calculate the relevant energies and it is very useful to have methods that take advantage of the structure of the equations. In crystals the periodicity suggests plane waves and pseudopotentials make possible calculations with feasible number of plane waves; iterative methods that use fast Fourier transforms (FFTs) are so efficient that it is hard for other methods to compete. Forces can be calculated straightforwardly since the plane waves are independent of the atom positions. For molecules, methods that use gaussians have been extensively developed and have the great feature that there are simple analytic expressions for Coulomb matrix elements.

Nevertheless, there are advantages to the real-space methods. By using pseudopotentials and related methods the atomic-like potentials are sufficiently smooth so that grids do not have to be extremely fine. In fact, the iterative plane methods are so efficient only because they use an FFT to transform parts of the problem to a regular grid in real space. This is shown explicitly in the flowcharts Fig. 12.4 for operations involving the density and Fig. M.2 for multiplication of the hamiltonian times the wavefunction. The plane wave method can be turned inside out to consider it as a real-space method on a uniform grid that uses plane waves only as a convenient way to handle the kinetic energy and to store the information in the basis of plane waves, which may be much smaller than the number of grid points (see the discussion in the previous section).

Viewing the problem as a real-space grid method opens the door to new possibilities of ways to utilize the many methods developed in real space. Here we can only skim the surface of this large field and bring out some of the salient features and methods for electronic structure, with references to some of the original works. More complete descriptions can be found in the review by Beck [568] and the papers given in this and the following sections, which give references to the more recent literature.

Finite Difference

In a finite difference (FD) method, the kinetic energy laplacian operator is approximated from values of the function at a set of grid points. For example, the FD method of

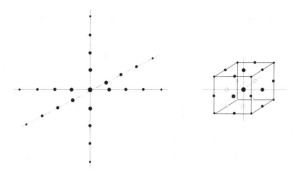

Figure 12.5. Two examples of stencils for finite difference calculation of the laplacian. The size of the points represents the weights schematically. Left: orthogonal "cross" with 25 points [570]. Right: more compact cube of 27 points that has been used with the Mehrstellen (Numerov) method [576].

Chelikowsky et al. (see original papers [569, 570], more recent reviews [571, 572], and works such as [573]) uses higher-order expansions for the kinetic energy laplacian operator, separable in the x, y, z orthogonal components. For a uniform orthogonal three-dimensional (3D) grid with points (x_i, y_j, z_k), the mth order approximation is

$$\left[\frac{\partial^2 \psi}{\partial x^2}\right]_{x_i, y_j, z_k} = \sum_{-m}^{m} C_m \psi(x_i + mh, y_j, z_k) + O(h^{2m+2}), \qquad (12.32)$$

where h is the grid spacing and m is a positive integer. As illustrated on the left-hand side of Fig. 12.5, the laplacian at the central point is computed in terms of values of the function on the "cross" of points along the axes; the size of the dots for the 25 points represents the decreasing magnitude of C_m. This approximation is accurate to $O(h^{2m+2})$ assuming ψ can be approximated by a polynomial in h.[6] Algorithms are available to compute the coefficients C_m for any grid to arbitrary order in h [574]. Expansion coefficients for $m = 1$ to 6 for a uniform grid are given in table 1 of [570]. There have been many applications to finite clusters, such as optical properties in Chapter 21, where Figs. 21.2 and 21.3 show spectra for hydrogen-terminated silicon systems from SiH_4 up to large clusters calculated using FD methods [575]. Examples of codes are OCTOPUS, PARSEC, and SPARC listed in Appendix R.

A different discretization uses the "Mehrstellen" operator, which is an extension of the Numerov method (Section L.1) to higher dimensions (see [577], p., 164 as cited in [576]). As illustrated on the right-hand side of Fig. 12.5, the 27 points are more compact in space than the 25-point cross. This is an advantage, especially for finite systems where the more extended "cross" leads to larger boundary effects and linear-scaling order N algorithms (Chapter 18), where localization is essential.

---

[6] The method can readily be extended to nonorthogonal systems and nonuniform grids, but at the price of having to compute many different sets of C.

Since the number of grid points is much larger than the number of eigenstates needed, methods are needed to extract the low-energy states. One such work is termed "spectrum slicing," which uses Chebyshev polynomials to filter out selected energy ranges that can be solved independently [578].

Discontinuous Galerkin (DG) Construction of Basic Functions

Discontinuous basis functions may seem strange as a way to describe wavefunctions; however, they are not strange at all. A wavefunction expressed as values on a grid of points can be considered as a set of delta functions or as the value of the wavefunction in each volume surrounding each point, which are discontinuous, nonoverlapping functions, each representing a region. If there is a local potential, the matrix elements are purely diagonal since the potential only multiplies the function in each region. The only matrix elements between regions is due to the kinetic energy; in the finite difference method it is expressed in terms of the difference between regions, i.e., the discontinuities at the boundaries. The advantage is that the hamiltonian is very simple and very sparse; the disadvantage is that many grid points are required to accurately represent wavefunctions in realistic problems.

The discontinuous Galerkin (DG) method is well developed in numerical analysis and it has been applied to the Kohn–Sham problem [579] and developed further in the context of grid methods [580]. In the application to the Kohn–Sham problem with wavefunctions expressed on a grid, the idea is to define regions that contain several grid points. The regions are defined to be nonoverlapping and to cover all space so functions in different regions are automatically orthogonal; by allowing the functions to be discontinuous they have complete freedom to describe the wavefunction within each region and there are hamiltonian matrix elements only to other regions within a small neighborhood. The advantage that only a few functions for each region (not all the possible functions) are needed to find the low-energy solutions, so that the hamiltonian is expressed in a much smaller basis. The disadvantage is that the hamiltonian is not as simple and the DG wavefunctions in each region must be calculated.

The approach in [580] is illustrated in Fig. 12.6, which shows a region enclosed by the dark square that contains 16 grid points. All space can be filled with nonoverlapping regions by repetition of the box. For each of these regions the Kohn–Sham problem is solved in a larger box illustrated by the dashed lines, for example, with the boundary condition that the wavefunctions are zero at the boundary. As illustrated at the right, the basis is defined by simply truncating the functions. In general, the result is a set of functions that are discontinuous in value and slope. It is shown in [580] that one can find accurate solutions that span the space of the occupied states with a number of functions much smaller than the number of grid points. The hamiltonian matrix is thus much smaller and still sparse for large systems since there are matrix elements only between nearby basis functions, each of which is strictly localized. The price to pay is that the functions must be calculated in each region, but there is a great advantage that each local calculation is independent and can be done in parallel, and there can be a great speedup of the overall calculation using parallel computers. In addition, this part of the calculation is linear scaling in the size of the

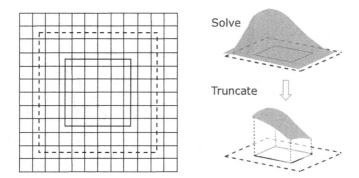

Figure 12.6. Schematic illustration of the construction of the basis for the discontinuous Galerkin (DG) method. At the left is a grid that is divided into regions of 16 sites with one of the regions indicated by the dark box. In the approach in [580] the functions are calculated in the larger region, indicated by the dashed lines, and a set of the low energy functions are truncated to form the basis as shown at the right. The functions can be continuous as shown or discrete values on the grid points depending on the way the Kohn–Sham equation is solved in the local regions.

system for large systems, and the construction can be a practical approach to $O(N)$ methods in Chapter 18.

12.9 Real-Space Methods II: Multiresolution Methods

Multiresolution denotes the ability to describe all regions with desired accuracy: regions where there are strong variations and high resolution is needed, and others where less resolution is required. One approach is the multigrid method, which is generally credited to Brandt [581] in 1977 and developed by him and others. It can be used with any algorithm for solution of the differential equations and it works by cycling up and down between levels of resolution to use the speed of coarse functions while adding corrections due to fine functions. Regions that require fine detail are detected autometically. There are codes for electronic structure calculations that use the multigrid approach [576, 582–584], for example, the RMGDFT method (Appendix R) based on the Mehrstellen form for the laplacian (see Fig. 12.5). It has been applied to periodic and nonperiodic problems including a simulation of water comparable to that described in Section 2.10 and transport through molecular nanojunctions [576, 582, 584].

Finite Elements

Finite element (FE) methods are widely used in many fields from civil engineering to quantum mechanics. Finite elements form a localized basis in which variational calculations can be done, unlike the finite difference (FD) method, which approximates the laplacian. The FE method employs a basis of strictly local piecewise polynomials, where each overlaps only its immediate neighbors. Because the basis consists of polynomials, the method is completely general and systematically improvable, like plane wave methods.

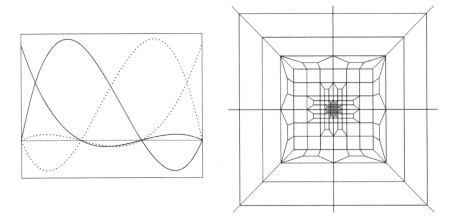

Figure 12.7. Left: examples of cubic polynomial finite element basis functions, which are strictly zero outside the range shown. Right: multiresistant using finite elements to zoom in on the nucleus in a calculation for metallic aluminum. Provided by V. Gavini

Examples in one dimension are the cubic polynomial functions shown in Fig. 12.7. Matrix elements of operators are integrals $\langle m|\hat{O}|m'\rangle$ just as for any other basis.

Introduction to the use in electronic structure problems can be found in the reviews by Beck [568] and Pask and Sterne [585], where one can also find references to classic texts on FE analysis. There are ongoing efforts to make available FE methods for electronic structure, for example, the DFT-FE codes (Appendix R) described in [586]. The methods have been applied to realistic problems such as all-electron calculations for aluminum. The ability to add fine resolution is illustrated in the right side of Fig. 12.7, which shows the locations of finite elements that vary in scale from coarse in the regions between the nuclei where less resolution is needed to very fine resolution around an Al nucleus. An application that shows the power of the methods is the calculation of electron–nuclear spin interactions in molecules and solids using the DFT-FE code [587].

Wavelets

There are many forms for localized wavelike functions that are termed wavelets. However, one type is very special: Daubechies[7] wavelets have the remarkable properties that they are localized in both real and Fourier space, and they are orthonormal at the same level and between levels at all levels of resolution. A function is represented in terms of scaling functions translated on a regular grid. Starting with a set of scaling functions at the coarsest level, one can generate finer and finer levels by adding wavelets. They do not have an analytic form but they can be generated by recursion so that matrix elements

---

[7] They were developed by Ingrid Daubechies, a professor of mathematics at Duke who was the first woman to be awarded the Fields Medal, often considered the Nobel prize in mathematics. Her 1992 work [589] on wavelets has over 30,000 citations as of 2019.

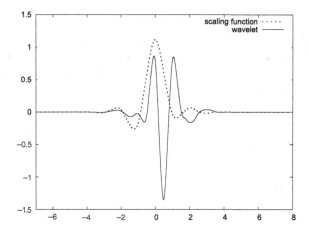

Figure 12.8. Daubechies wavelet and scaling function at 16 order. There is no analytic formula, but it can represent polynomials up to 16th order and matrix elements can be calculated recursively. Provided by S. Mohr; similar to figure in [588]

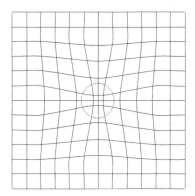

Figure 12.9. Schematic figure of adaptive coordinates defined by a smooth transformation that maps a regular grid onto a nonuniform set of points. The transformation can be thought of as working in curved space. The example shows a transformation to provide greater resolution near an atom.

can be computed efficiently. Low-order wavelets were used in early electronic structure calculations [590, 591], and there is a full DFT code called BigDFT (Appendix R), which use smoother 16-order Daubechies wavelets shown in Fig. 12.8. The codes are developed for periodic systems, molecules, and clusters and have been applied to complex systems like molecular dynamics of water clusters [588, 592].

Adaptive Curvilinear Coordinates

A different approach suggested by Gygi [593] and Hamann [594] is to warp the grid using a smooth transformation as illustrated in Fig. 12.9. The transformation can be defined in terms of a smooth set of basis functions that map the regular points $\mathbf{r}_i$ specifying the grid to the points $\mathbf{r}'_i(\mathbf{r}_i)$. For example, the transformation can be specified in terms of plane waves. The method can be adaptive in the sense that one can make an algorithm to determine where more resolution is needed and adjust the adaption. In any case, the resulting equations expressed on the regular grid $\mathbf{r}_i$ have the form of operators in curved space. An alternative approach uses a local set of fixed transformations [595, 596] around each atom that overlap to form the complete transformation; these are easy to visualize and are equivalent to a form of global transformation [597].

SELECT FURTHER READING

Basic aspects of band-structure methods:

See list at the end of Chapter 2.

Ziman, J. M., *Principles of the Theory of Solids* (Cambridge University Press, Cambridge, 1989).

The empirical pseudopotential method:

Cohen, M. L. and Heine, V., in *Solid State Physics*, edited by H. Ehrenreich, F. Seitz, and D. Turnbull (Academic Press, New York, 1970), p. 37.

Cohen, M. L. and Chelikowsky, J. R., *Electronic Structure and Optical Properties of Semiconductors*, 2nd ed. (Springer-Verlag, Berlin, 1988).

Heine, V., in *Solid State Physics*, edited by H. Ehrenreich, F. Seitz, and D. Turnbull (Academic Press, New York, 1970), p. 1.

Grid methods:

Beck, T. L., "Real-space mesh techniques in density-functional theory," *Rev. Mod. Phys.* 72, 1041–1080 (2000).

Pask, J. E. and Sterne, P. A., "Finite element methods in *ab initio* electronic structure calculations," *Model. Simul. Mater. Sc.*, 13, R71–R96, 2005.

Exercises

12.1　See many excellent problems (and solutions) on the nearly-free-electron approximation in the book by Mihaly and Martin [598].

12.2　Show that the Fourier transform of Eq. (12.15) leads to the expression in terms of form and structure factors given in Eq. (12.16).

12.3　Show the equivalence of Expressions (4.11) and (12.18), which express the final Fourier component in two ways, one an integral over the cell and the other as a structure factor times an integral for one unit only but over all space.

12.4　Plot the bands for a nearly-free-electron system in one dimension if the lattice constant is a.
(a) First plot the bands using analytic expressions for the energy in the free-electron limit.
(b) Then qualitatively sketch the changes if there is a small lattice potential.
(c) Use an empirical pseudopotential program, such as the ones listed in Appendix R, or write your own to calculate the bands for a pure sine wave potential $V(x) = V_0 \sin(2\pi x/a)$. This is the Mathieu potential for which there are solutions; check your results with known results.

12.5 Consider a one-dimensional crystal with potential $V(x) = V_0 \cos(2\pi x/a)$ as in Exercise 12.4. In this exercise make the simplifying approximation that a state with wavevector k is the solution of the 2×2 hamiltonian

$$\begin{vmatrix} \frac{k^2}{2} - \varepsilon(k) & V_0 \\ V_0 & \frac{(k-G)^2}{2} - \varepsilon(k) \end{vmatrix} = 0, \qquad (12.33)$$

where $G = 2\pi/a$. Give the analytic expressions for the bands $\varepsilon(k)$ and the periodic part of the Bloch functions $u_k(x)$. If there are two electrons per cell, give the expression for the density $n(x)$ as an integral over k. Evaluate the density using a grid of "special" k points (Section 4.6). Note that more points are required for an accurate answer if V_0 is small. Plot the lowest two bands and the electron density for the case where $V_0 = \frac{1}{4}(\pi/a)^2$ in atomic units. (See Exercise 23.13 Wannier functions and Exercise 24.10 for polarization using a variation of this model.)

12.6 Consider a one-dimensional crystal with a square well potential that in the cell at the origin has the form $V(x) = V_0$ for $-s/2 < x < s/2$ and $V = 0$ otherwise. The potential is repeated periodically in one dimension with $V(x + Na) = V(x)$, with cell length $a > s$. (See also Exercises 11.2, 11.6, 11.14; the general solution for bands in one dimension in Exercise 4.22; and relations to the APW, KKR, and MTO methods, respectively, in Exercises 16.1, 16.7, and 16.13.)
(a) First find the Fourier transform of the potential $V(G)$.
(b) Computational exercise: Construct a computer (it is a matter of setting up the hamiltonian and diagonalizing) or use an available empirical pseudopotential code (see Appendix R) to solve for the bands. As an explicit example, choose $a = 4$, $s = 2$, and $V_0 = 0.2$ in atomic units and choose a sufficient number of plane waves so that the solution is accurate.
(c) Compare the results with the solutions in Exercise 16.1 in which the bands are found by matching the wavefunctions at the boundary, i.e., a simple example of the APW method. Of course, the result must be the same as derived by other methods: compare and contrast the plane approach with the general solution for any potential in one dimension given in Exercise 4.22.

12.7 Find the bands for Al using a simple empirical pseudopotential. One source is the paper by Segall [558] that shows bands similar to those in Fig. 16.6 calculated with $V(111) = 0.0115Ha$ and $V(111) = 0.0215Ha$ and mass $m^* = 1.03m$. (The last can be included as a scaling factor.) Use the NFEA to calculate energies at the X point analytically. Use an empirical pseudopotential program (see notes on codes in Exercise 12.12.) to generate full bands.

12.8 Show that the derivations in Section 12.1 also hold for nonlocal potentials as given in Eq. (12.24).

12.9 Derive the semilocal and separable forms of the pseudopotential in Eqs. (12.23) and (12.24). Hint: use the definitions of the potential operators in real space in Chapter 11 and the expansion of a plane wave in spherical harmonics, Eq. (J.1).

12.10 Pseudopotentials are used because calculations with the full nuclear Coulomb potential are very expensive for heavy atoms of nuclear charge Z. Derive the power law with which the number of plane waves needed scales with Z. Do this by using perturbation theory for very high Fourier components, where the matrix element is given by $V(G)$ and the energy denominator is approximately given by the kinetic energy. Argue that screening is not effective for high Fourier components.

12.11 Project: Use an empirical pseudopotential program (see notes on codes in Exercise 12.12) to find the bands and charge densities of Si in the diamond structure at the lattice constant 10.26 a_0. The bands should be insulating and the bonds should be visible in the charge density.
(a) Verify that the minimum along the Δ direction (see Fig. 4.10) is qualitatively the same as in experiment, which is given in many texts, e.g., [285].
(b) Now compress the system until it is metallic (this can only be done in theory; in reality it transforms). Can you tell when the system becomes a metal just from the density? In principle, if you had the exact functional, what aspect of the density would be the signature of the insulator–metal transition?
(c) Do a similar calculation replacing the Si atoms with Al, still in the diamond structure with lattice constant 10.26 a_0. (Of course this is a theoretical structure.) There are three Al electrons/atom, i.e., six electrons per cell – and it turns out to be a metal. Show that it must be metallic *without doing the calculation*. Does the density plot look a lot like Si? Can you find any feature that shows it is a metal?

12.12 Project: Use an empirical pseudopotential program to find the bands for GaAs. (See the codes listed in Appendix R. Figure 14.9 was calculated using online tools at nanohob.org, which also has empirical pseudopotential codes.)
(a) Verify that it has a direct gap at Γ.
(b) Displace the atoms in the unit cell a small amount along the (111) direction. Check the splitting of the top of the valence band at γ. Is the splitting what you expect?
(c) Repeat with the displacement in the (100) direction.

12.13 This exercise is to work out the form factor for the screened H potential in the Thomas–Fermi approximation and to calculate the bands for fcc H at very high density, $r_s = 1.0$.
(a) Estimate the deviation of the bands from the free-electron parabola by calculating the gaps at the X and L points of the BZ in lowest non-zero-order perturbation theory.
(b) Carry out calculations using an empirical pseudopotential program (see Exercise 12.12) and compare with the results from perturbation theory.
(c) Compare with the simple expression for the band width in Exercise 10.13 and with fully self-consistent band structure results as described in Exercise 13.4.

13

Plane Waves and Real-Space Methods: Full Calculations

Summary

The subject of this chapter is the role of plane waves and grids in modern electronic structure calculations, which builds on the basic formulation of Chapter 12. Plane waves have played an important role from the early OPW calculations to widely used methods involving *ab initio* pseudopotentials and related methods in Chapter 11. Plane waves continue to be the basis of choice for many new developments, such as quantum molecular dynamics simulations (Chapter 19), owing to the simplicity of operations. Efficient iterative methods (Appendix M) have made it feasible to apply plane waves to large systems, and methods such as "ultrasoft" pseudopotentials and projector augmented waves (PAWs) reduce the number of plane waves required to treat difficult cases such as materials containing transition metals. Real-space grids are an intrinsic part of efficient plane wave calculations and there is a growing development of real-space methods, including finite difference, multigrids, finite elements, and wavelets described in Chapter 12.

Plane waves are by far the method most used in present-day calculations for electronic structure, which is made possible by the combination of two things: (1) efficient methods using fast Fourier transforms (FFTs) and iterative methods in the previous chapter and Appendix M and (2) accurate *ab initio* pseudopotentials and related methods described in Chapter 11. Because of the simplicity of operations with plane waves or real-space grids, it is straightforward to combine the parts to create methods for full self-consistent Kohn–Sham calculations. This is a short chapter where needed aspects are worked out; at the end are pointers to places in other chapters where the methods are applied.

The previous chapter has brought out some of the real-space approaches that have important advantages. However, they are not as well developed or tested; references to ongoing work is given there and we can only describe a few applications.

13.1 *Ab initio* Pseudopotential Method

Expressions for Total Energy, Force, and Stress in Fourier Space

The starting point for derivation of the full Kohn–Sham theory is the total energy and the Kohn–Sham equations for which general expressions have been given in Chapter 7; the subject of this section is the derivation of explicit expressions in reciprocal space. For example, the variational expression for energy (Eqs. (7.5) or (7.20)) in terms of the output wavefunctions and density can be written [102, 372, 562, 599]

$$
E_{\text{total}}[V_{\text{eff}}] = \frac{1}{N_k} \sum_{\mathbf{k},i} w_{k,i} \left\{ \sum_{m,m'} c_{i,m}^*(\mathbf{k}) \left[\frac{\hbar^2}{2m_e} |\mathbf{K}_m|^2 \delta_{m,m'} + V_{\text{ext}}(\mathbf{K}_m, \mathbf{K}_{m'}) \right] c_{i,m'}(\mathbf{k}) \right\}
$$
$$
+ \sum_{\mathbf{G}} \epsilon_{\text{xc}}(\mathbf{G}) n(\mathbf{G}) + \frac{1}{2} 4\pi e^2 \sum_{\mathbf{G} \neq 0} \frac{n(\mathbf{G})^2}{G^2} + \gamma_{\text{Ewald}} + \left(\sum_{\kappa} \alpha_\kappa \right) \frac{N_e}{\Omega}. \quad (13.1)
$$

Since E_{total} is the total energy per cell, the average over $\mathbf{k}$ and sum over bands is the same as for the density in Eq. (12.27). Similarly, the sums can be reduced to the IBZ just as in Eq. (4.44). The potential terms involve $\mathbf{K}_m \equiv \mathbf{k} + \mathbf{G}_m$; the xc term is the total exchange–correlation energy; and the final three terms are considered below. Alternatively, one can use Eq. (7.20) for the energy, in which the eigenvalues replace the term in square brackets in Eq. (13.1). As discussed in Chapter 7, the form in Eq. (13.1) is manifestly a functional of V_{eff}, which determines each term (except the final two terms that depend only on the structure and number of electrons).

Correct treatment of the Coulomb terms is accomplished by *consistently* separating out the $\mathbf{G} = 0$ components in the potential and the total energy. The Hartree term in Eq. (13.1) is the Coulomb interaction of the electrons with themselves *excluding the divergent term due to the average electron density*. Similarly, *the $\mathbf{G} = 0$ Fourier component of the local potential is defined to be zero* in Eq. (13.1). Both these terms are included in the Ewald term γ_{Ewald}, which is the energy of point ions in a compensating background (see Appendix F, Eq. (F.6)), i.e., this term includes the ion–ion terms as well as the interactions of the average electron density with the ions and with itself. *Only by combining the terms together is the expression well defined.* The final term in Eq. (13.1), is a contribution due to the non-Coulombic part of the local pseudopotential (see Eq. (12.22)), where $\frac{N_e}{\Omega}$ is the average electron density.

Following the analysis of Section 7.3, one can define a functional[1]

$$
\tilde{E}_{\text{total}} = \frac{1}{N_k} \sum_{\mathbf{k},i} w_{k,i} \varepsilon_i + \sum_{\mathbf{G}} [\epsilon_{\text{xc}}(\mathbf{G}) - V_{\text{xc}}(\mathbf{G})] n(\mathbf{G})
$$
$$
+ \left[\gamma_{\text{Ewald}} - \frac{1}{2} 4\pi e^2 \sum_{\mathbf{G} \neq 0} \frac{n(\mathbf{G})^2}{G^2} \right] + \left(\sum_{\kappa} \alpha_\kappa \right) \frac{N_e}{\Omega}, \quad (13.2)
$$

---

[1] The electron Coulomb term on the second line cancels the double counting in the eigenvalues. The terms are arranged in two neutral groupings: the difference of the ion and the electron terms in the square bracket and the sum of eigenvalues that are the solution of the Kohn–Sham equation with a neutral potential.

where all terms involve the *input density* $n \equiv n^{in}$. This expression is not variational but instead is a saddle point as a function of n^{in} around the consistent solution $n^{out} = n^{in}$. It is very useful because it often converges faster to the final consistent energy so that it is useful at every step of a self-consistent calculation. Furthermore, it is the basis for useful approximations, e.g., stopping after one step and never evaluating any output quantity other than the eigenvalues [172, 359, 361–363].

The force on any atom $\tau_{\kappa, j}$ can be found straightforwardly from the "force theorem" or "Hellmann–Feynman theorem" given in Section 3.3. For this purpose, Expression (13.1) is the most useful and the explicit expression for (3.20) in Fourier components can be written

$$F_j^\kappa = -\frac{\partial E}{\partial \tau_{\kappa, j}} = -\frac{\partial \gamma_{Ewald}}{\partial \tau_{\kappa, j}} - i \sum_m \mathbf{G}_m e^{i(\mathbf{G}_m \cdot \tau_{\kappa, j})} V_{local}^\kappa (\mathbf{G}_m) n(\mathbf{G}_m)$$

$$\frac{-i}{N_k} \sum_{\mathbf{k}, i} w_{k,i} \sum_{m,m'} c_{i,m}^*(\mathbf{k}) \left[\mathbf{K}_{m,m'} e^{i(\mathbf{K}_{m,m'} \cdot \tau_{\kappa, j})} \delta V_{NL}^\kappa (\mathbf{K}_m, \mathbf{K}_{m'}) \right] c_{i,m'}(\mathbf{k}),$$

$$(13.3)$$

where the Ewald contribution is given in Eq. (F.11). Here the external pseudopotential has been separated into the local part, which contains the long-range terms, and the short-range nonlocal operator $\delta V_{ext}^\kappa (\mathbf{K}_m, \mathbf{K}_{m'})$, with $\mathbf{K}_{m,m'} \equiv \mathbf{K}_m - \mathbf{K}_{m'}$. The expression for stress in Fourier components is given in Section G.3.

Solution of the Kohn–Sham Equations

The Kohn–Sham equation is given by Eqs. (12.9) and (12.10) with the local and nonlocal parts of the pseudopotential specified by the formulas of Section 12.4. Consistent with the definitions above, the local part of the potential in the Kohn–Sham equation can be written straightforwardly as the Fourier transform of the external local potential Eq. (12.16), Hartree, and xc potentials in Eq. (7.13),

$$V_{KS, local}^\sigma (\mathbf{G}) = V_{local}(\mathbf{G}) + V_{Hartree}(\mathbf{G}) + V_{xc}^\sigma (\mathbf{G}), \qquad (13.4)$$

where all $\mathbf{G} = 0$ Fourier components are omitted. The $\mathbf{G} = 0$ term represents the average potential, which is only a shift in the zero of energy that has no consequence for the bands, since the zero of energy is arbitrary in an infinite crystal [213, 254, 600]. The full potential is Eq. (13.4) plus the nonlocal potential Eqs. (12.23) or (12.24).

The equations are solved by the self-consistent cycle shown in Fig. 7.2, where the solution of the equations for a fixed potential is the same as for a non-self-consistent EPM calculation. The new steps that must be added are as follows:

- Calculation of the output density $n^{out}(\mathbf{G})$
- Generation of a new input density $n^{in}(\mathbf{G})$, which leads to the new effective potential
- After self-consistency is reached, calculation of the total energy (Eqs. (13.1), (13.2), or related variational formulas using the expressions of Section 7.3), forces, stress, etc.

13.2 Approach to Self-Consistency and Dielectric Screening

The plane waves framework affords a simple case in which to discuss the approach to self-consistency, bringing out issues addressed in Section 7.4. The simplest approach – that works very well in many cases – is linear mixing:

$$V_{i+1}^{\sigma,\text{in}}(\mathbf{G}) = \alpha V_i^{\sigma,\text{out}}(\mathbf{G}) + (1 - \alpha)V_i^{\sigma,\text{in}}(\mathbf{G}). \tag{13.5}$$

Choice of α by trial and error is often sufficient since the same value will apply to many similar systems.

In order to go further and analyze the convergence, one can treat the region near convergence, where the error in the output density or potential is proportional to the error in the input potential δV^{in}. Using the definition of the dielectric function, the error in the output potential is given by[2]

$$\delta V^{\text{out}}(\mathbf{G}) = \sum_{\mathbf{G}'} \epsilon(\mathbf{G}, \mathbf{G}')\delta V^{\text{in}}(\mathbf{G}'). \tag{13.6}$$

(Note that this *does not apply to the* $\mathbf{G} = 0$ *component*, which is fixed at zero.) It follows that the error in the output density $\delta n^{\text{out}}(\mathbf{G}) = \delta V^{\text{out}}(\mathbf{G})(G^2/4\pi e^2)$ is also governed by the dielectric function, and the kernel χ in Eq. (7.34) is related by $\chi(\mathbf{G}, \mathbf{G}') = \epsilon(\mathbf{G}, \mathbf{G}')G'^2/G^2$. In general the dielectric function approaches unity for large $\mathbf{G}$ or $\mathbf{G}'$; however, it may be much larger than unity for small wavevectors. For example, for Si, $\epsilon \approx 12$ for small wavevectors, so that *the error in the output potential (or density) is 12 times larger than the error in the input!* For a metal, the problem may be worse since ϵ diverges.

How can the iterations reach the solution? There are two answers. First, for crystals with small unit cells, this is *not a problem* because all the $\mathbf{G} \neq 0$ components of the potential are for large values of $|\mathbf{G}|$, and the $\mathbf{G} \equiv 0$ is taken care of in combination with the Ewald term (see Appendix F). It is only if one has small nonzero components that problems arise. This happens for large unit cells and is called the "charge sloshing problem." It is worst for cases like metal surfaces, where the charge can "slosh" to and from the surface with essentially no cost in energy. In such cases the change is in the right direction but one must take only small steps in that direction. If a linear mixing formula is used, then the mixing of the output must be less than $1/\epsilon(G_{\text{min}})$ for convergence.

The relation to the dielectric function also suggests an improved way to reach convergence. It follows from Eq. (13.6) that the exact potential can be reached after one step (see also Eqs. (7.34) and (7.36)) by solving the equation

$$\delta V^{\text{out}}(\mathbf{G}) \equiv V^{\text{out}}(\mathbf{G}) - V_{\text{KS}}(\mathbf{G}) = \sum_{\mathbf{G}'} \epsilon(\mathbf{G}, \mathbf{G}') \left[V^{\text{in}}(\mathbf{G}') - V_{\text{KS}}(\mathbf{G}') \right] \tag{13.7}$$

for the converged Kohn–Sham potential $V_{\text{KS}}(\mathbf{G})$. The input and output potentials are known from the calculation; however, the problem is that it is very difficult to find $\epsilon(\mathbf{G}, \mathbf{G}')$ – a more difficult problem than solving the equations. Nevertheless, approximate forms for

---

[2] Note the similarity to the Thomas–Fermi expression, (12.25). The reason that we have ϵ here instead of ϵ^{-1} is that here we are considering the response to an înternal field.

$\epsilon(\mathbf{G}, \mathbf{G}')$ such as the diagonal Thomas–Fermi form, Eq. (12.26), can be used to improve the convergence [373]. One can also take advantage of the fact that a linear mixing leads to exponential convergence (or divergence) near the solution; by fitting three points to an exponential, the solution for an infinite number of steps can be predicted [601].

From a numerical point of view the dielectric matrix (or the second derivatives defined in Chapter 7) are nothing more than the Jacobian. Since it is in general not known, approximations (such as approximate dielectric functions) are really "preconditioners" as discussed in the chapter on iterative methods and in Appendix M. This leads to practical approaches that involve combinations of the methods, For example, for high Fourier components the dielectric function is near unity and nearly diagonal, so one can use Eq. (13.5) with $\alpha \approx 0.5$ to 1; for low Fourier components one can use general numerical approaches to build up the Jacobian iteratively as the calculations proceed, e.g., the Broyden-type methods described in Section 7.4 or the RMM-DIIS method (see Section M.7). For some cases there are large problems caused by charge sloshing, i.e., large response of the system to errors. This can be analyzed in terms of the dielectric function and model functions are very useful in reaching convergence.[3]

13.3 Projector Augmented Waves (PAWs)

The projector augmented wave (PAW) method [542–545] described in Section 11.12 is analogous to pseudopotentials in that it introduces projectors acting on smooth valence functions $\tilde{\psi}^v$ that are the primary objects in the calculation. It also introduces auxiliary localized functions like the "ultrasoft" pseudopotential method. However, the localized functions actually keep all the information on the core states like the OPW and APW (see Chapter 16) methods. Thus many aspects of the calculations are identical to pseudopotential calculations; e.g., all the operations on smooth functions with FFTs, generation of the smooth density, etc. are the same. However, since the localized functions are rapidly varying, augmentation regions around each nucleus (like the muffin-tin spheres in Chapter 16) are introduced and integrals within each sphere are done in spherical coordinates.

The expressions given in Section 11.12 apply here also. The linear transformation to the all-electron valence functions $\psi^v = \mathcal{T}\tilde{\psi}^v$ is assumed to be a sum of nonoverlapping atom-centered contributions $\mathcal{T} = 1 + \sum_{\mathbf{R}} \mathcal{T}_{\mathbf{R}}$, each localized to a sphere denoted Ω_{vecr}. If the smooth wavefunction is expanded in spherical harmonics inside each sphere, omitting the labels v and i as in Eq. (11.60),

$$|\tilde{\psi}\rangle = \sum_m c_m |\tilde{\psi}_m\rangle, \qquad (13.8)$$

---

[3] An example is the combination of dielectric screening and RMM-DIIS residual minimization, which is described in the online information on optimization in the VASP code.

with the corresponding all-electron function,

$$|\psi\rangle = \mathcal{T}|\tilde{\psi}\rangle = \sum_m c_m |\psi_m\rangle. \tag{13.9}$$

Hence the full wavefunction in all space can be written

$$|\psi\rangle = |\tilde{\psi}\rangle + \sum_{\mathbf{R}m} c_{\mathbf{R}m} \left\{ |\psi_{\mathbf{R}m}\rangle - |\tilde{\psi}_{\mathbf{R}m}\rangle \right\}. \tag{13.10}$$

The biorthogonal projectors $\langle \tilde{p}_{\mathbf{R}m}|$ in each sphere are the same as defined in Eq. (11.62) since the spheres are nonoverlapping. In much of the work using PAW the core functions are "frozen" and assumed to be the same as in reference atomic-like systems. In [545] are efficient ways to update the core functions self-consistently in the calculations.

Thus the expressions carry over with the generalization to many spheres, for example, the density given by Eqs. (11.68)–(11.71). Here it is particularly relevant to give the form for the total energy, from which follow the basic Kohn–Sham equations and expressions for forces, etc. [542, 544]. Like the density, the energy can be written as a sum of three terms:

$$E_{\text{total}} = \tilde{E}_{\text{total}} + E_{\text{total}}^1 + \tilde{E}_{\text{total}}^1, \tag{13.11}$$

where $\tilde{E}$ denotes the energy due to the smooth functions evaluated in Fourier space or a grid that extends throughout space, $\tilde{E}^1$ denotes the same terms evaluated only in the spheres on radial grids, and E^1 the energy in the spheres with the full functions. The classical Coulomb terms are straightforward in the sense that they are given directly by the density; however, they can be rearranged in different ways to improve convergence of the Coulomb sums. In the PAW approach, an additional density is added to both auxiliary densities in $\tilde{n}(\mathbf{r})$ and $\tilde{n}^1(\mathbf{r})$ so that the multipole moments of the terms $n^1(\mathbf{r}) - \tilde{n}^1(\mathbf{r})$ in Eq. (11.68) vanish. Thus the electrostatic potential due to these terms vanishes outside the augmentation spheres around each atom, just as is accomplished in full-potential LAPW methods [602]. A discussion of different techniques for the additional density terms [542, 544] is given in [544]. The expression for E_{xc} also divides into three terms with each involving the total density evaluated in the different regions [544]. It is not hard to derive the expressions for the Kohn–Sham equations by functional derivatives of the total energy and a detailed account can be found in [542].

It is advantageous that expressions for the total energy are closely related in the ultrasoft and the PAW formulations, differing only in the choice of auxiliary functions and technical aspects. Thus the expressions for forces and stress are also essentially the same. In particular, the large intra-atomic terms do not enter the derivatives and forces can be derived by derivatives of the structure constants [603]. Stress can also be derived as referred to in [544].

13.4 Hybrid Functionals and Hartree–Fock in Plane Wave Methods

Hybrid functionals that have a fraction of Hartree–Fock exchange are increasingly important in electronic structure calculations because they are often much more accurate than local or semilocal functionals, especially for bandgaps. However, there are difficulties to treat the nonlocal exchange in Hartree–Fock using plane waves. In methods with localized orbitals the calculations are straightforward and the exchange term decreases rapidly with distance as described in Section 3.6. In a plane wave basis, the problem is the $1/q^2$ singularity of the Coulomb interaction, which must be addressed in order to use hybrid functionals. There are two main approaches: methods to treat the divergence using analytical functions, and transformation to localized orbitals. For a short-range hybrid like HSE (see Section 9.3) there is no problem in principle; nevertheless, efficient ways to handle the exchange term are useful.

The Hartree–Fock exchange energy and potential are given in Eqs. (3.44)–(3.48), which are expressed in real space. In a plane wave basis matrix elements of the exchange term can be written as

$$\hat{V}_x^\sigma(\mathbf{k}, \mathbf{G}_m; \mathbf{k}, \mathbf{G}_{m'}) = -\frac{4\pi e^2}{\Omega} \sum_{j, \mathbf{k}''} \sum_{\mathbf{G}_{m''}} \frac{c_{j,\mathbf{k}''}^*(\mathbf{G}_m + \mathbf{G}_{m''}) c_{j,\mathbf{k}''}(\mathbf{G}_{m'} + \mathbf{G}_{m''})}{|\mathbf{k} - \mathbf{k}'' - \mathbf{G}_{m''}|^2}, \quad (13.12)$$

where $c_{j,\mathbf{k}''}(\mathbf{G})$ is the coefficient of the Bloch function j with momentum $\mathbf{k}''$ (see Section 12.1). Comparison with Eq. (12.7) for a local potential shows immediately that Eq. (13.12) is much more complicated. Since the calculation is done on a discrete set of $\mathbf{k}$-points, the sum in Eq. (13.12) does not properly account for the integral over the divergence in the $\mathbf{G}_{m''} = 0$ term. Gygi and Baldereschi [604] proposed a method to handle the divergence for $|\mathbf{k} - \mathbf{k}''| \to 0$ by adding a function $c_{j,\mathbf{k}}^*(\mathbf{G}_m) c_{j,\mathbf{k}}(\mathbf{G}_{m'}) F(\mathbf{k} - \mathbf{k}'')$ to the $\mathbf{G}_{m''} = 0$ term in Eq. (13.12) that cancels the divergence, so that the sum is over terms that vary smoothly. The final result is given by subtracting the integral over the same terms calculated analytically. This approach, combined with techniques using FFTs [605], was developed in [606] to make an efficient method that scales almost linearly with the number of plane waves and quadratically with the number of Bloch functions $j, \mathbf{k}$. For a large calculation with periodic boundary conditions (only $\mathbf{k} = 0$) this means a quadratic scaling in the size.

A different approach is to transform the Bloch functions into localized functions, i.e., Wannier or related localized functions, as described in Chapter 23. The problem is thus transformed to be the same as for localized basis functions described in Section 3.6. A linear-scaling method has been created in [607], which uses maximally localized functions that are generated using the minimization method in Section 23.3. The calculations require treating all pairs of overlapping Wannier functions, but the efficiency of the calculations can be improved significantly by methods such as the fast multiple approach used in [607]. An example of an application is a simulation of water in [198].

13.5 Supercells: Surfaces, Interfaces, Molecular Dynamics

Because plane waves are inherently associated with periodic systems, an entire approach has been developed: to create "supercells" that are large enough to treat problems that are intrinsically nonperiodic. The problem is made artificially periodic and all the usual plane

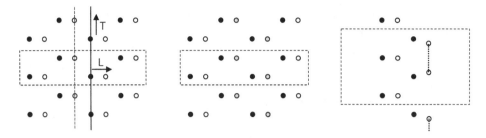

Figure 13.1. Examples of the use of "supercells" for the case of zinc blende crystals with the long axis in the [1 0 0] direction and the vertical axis in the [1 1 0] direction. The unit cell is shown by the dashed boundary. Left: perfect crystal with one plane of atoms displaced; as described in Section 20.2, calculation of forces on other planes, e.g., the one shown by the dashed line, allows calculation of longitudinal (L) or transverse (T) phonon dispersion curves in the [1 0 0] direction. Middle: interface between two crystals, e.g., GaAs–AlAs, which allows calculation of band offsets, interface states, etc. Right: two surfaces of a slab created by removing atoms, leaving a vacuum spacer. Complicated reconstructions often lead to increased periodicity in the surface and larger unit cells as indicated.

wave methods apply. Despite the obvious disadvantage that many plane waves are required, the combination of efficient iterative methods (Appendix M) and powerful computers makes this an extremely effective approach. From a more fundamental point of view, the variation in calculated properties with the size of the "supercell" is an example of finite-size scaling, which is a powerful method to extrapolate to the large-system limit. This is the opposite of the approach in Chapter 18 to treat localized properties; each has advantages and disadvantages. Of course, other methods can be used as well.[4]

Schematic examples of supercells in one dimension are shown in Fig. 13.1: a crystal with a displaced plane, a superlattice formed by different atoms, and a slab with vacuum on either side. In fact, the periodic cell made by repeating the cell shown is potentially a real physical problem of interest in its own right. In addition, the limit of infinite horizontal cell size can be used in various ways to represent the physical limit of an isolated plane of displaced atoms, an isolated interface, and a pair of surfaces. Although each case may be better treated by some other method, the fact that a plane wave code can treat all cases with only changing a few lines of input information can be used to describe the limits accurately is an enormous advantage.

It is essential to treat the long-range Coulomb interaction carefully. In methods that use periodic boundary conditions, it is not hard to calculate terms that behave as $1/|\mathbf{k} + \mathbf{G}|^2$, but it is often very tricky to interpret them properly! See Section F.5 and Appendix F.

"Frozen" Phonons and Elastic Distortions

An example of the use of supercells in calculating phonon properties is given in Section 20.2. Any particular cell can be used to calculate all the vibrational properties at

---

[4] Real-space approaches can treat such systems without assuming artificial periodicity; however, it is only for the surface that it is straightforward to impose correct boundary conditions.

the discrete wavevectors corresponding to the reciprocal lattice vectors of the supercell. This is very useful for many problems, and a great advantage is that nonlinear, anharmonic effects can be treated with the same method (see, e.g., Fig. 2.10) as well as phase transitions such as in $BaTiO_3$ and Zr in Fig. 2.10. Furthermore, it is possible [608–610] to derive full dispersion curves from the direct force calculations on each of the atoms in a supercell, like those shown on the left-hand side of Fig. 13.1. The requirement is that the cell must be extended to twice the range of the forces. An example of phonon dispersion curves in GaAs in the [100] direction is shown in Fig. 20.1. The inverse dielectric constant ϵ^{-1} and the effective charges Z_I^* can also be calculated from the change in the potentials due to an induced dipole layer in the supercell [608] or by finite wavevector analysis [611].

Interfaces

Interfaces between bulk materials are a major factor in determining the properties of materials: grain boundaries in metals that determine ductility; heterostructures that are very important in creating man-made semiconductor devices; oxide interfaces that form atomically thin electronic states; and so forth. In general, the different bonding of atoms at the interface leads to displacements of atoms and often superstructures; because the positions of the atoms may be crucial and it is very difficult to determine the structure experimentally, it makes a qualitative difference to have methods to reliably determine the structures. Superlattices provide a way to do the calculations in an unbiased way for many different structures. The "polar catastrophe" is a particular problem that is exemplified for oxide interfaces in Fig. 22.6. Representative examples for semiconductors and oxide interfaces are in Chapter 22.

Surfaces

Even though surfaces are interfaces of a solid with vacuum, surface science is a field in itself because surfaces are experimentally accessible and have untold number of uses. At a surface atoms have much more degrees of freedom and examples of surface structures determined by plane wave total energy calculations are given in Section 2.11 for ZnSe, which illustrate issues that occur in polar semiconductors. Catalysis is a great challenge for electronic structure methods to play a major role in sorting out the mechanisms and developing new catalysts.

Molecular Dynamics

Molecular dynamics (MD) in condensed matter is a venerable area of simulations, where boundary conditions have been studied in depth. Methods in which forces are calculated by calculating the electronic state at each step (quantum MD in Chapter 19) are far more computationally intensive than with atomic force models. Cells are smaller, boundary conditions are much more important, and periodic boundary conditions are almost always the best. Plane waves have been the method of choice for the great majority of quantum MD calculations, which have made a qualitative change in the types of problems in the realm of electronic structure, with examples in Chapters 2 and 19.

13.6 Clusters and Molecules

For finite systems such as clusters and molecules real-space methods are an obvious choice using open boundary conditions. In the case of plane waves, it is essential to construct a supercell in each dimension in which the system is localized. Since the supercell must be large enough so that any spurious interactions with the images are negligible, which means that many plane waves must often be used, and Coulomb interactions require special care; nevertheless, it may still be an effective way of solving the problem.

Calculations that employ grids are featured in Chapter 21 on excitations, where finite difference methods [569, 570] have been used effectively for total energy minimization and time-dependent DFT studies of finite systems from atoms to clusters of hundreds of atoms [575]. An application of multigrid methods is a boron nitride–carbon nanotube junction [612].

Two examples of plane wave calculations in one-dimensional systems involve a patterned graphene nanoribbon in Section 26.7 and a small-diameter carbon nanotube in Section 22.9. Figure 22.9 shows the bands for the nanotube, where the generality of plane waves led to discovery of an effect that is very different from a simple picture of tubes as rolled graphene. In the graphene nanoribbon, tight-binding models are often sufficient, but the plane wave calculations provide quantitative results with no parameters.

13.7 Applications of Plane Wave and Grid Methods

Because plane waves are so pervasive, many applications are given in other chapters, listed here along with a few salient comments:

- prediction of structures and calculation of equations of state (the pioneering work of Yin and Cohen [101, 599] for the phase transition of Si under pressure shown in Fig. 2.3; prediction [139] of a new structure of nitrogen under pressure in Fig. 2.5; prediction of the structure and transition temperature of sulfur hydride superconductors (Figs. 2.6 and 2.7); phase transition in $MgSiO_3$ and the equation of state of Fe at pressures and temperatures found deep in the earth in Figs. 2.12 and 2.13; thermal simulations of water in Fig. 2.15, the catalysis problem shown schematically in Fig. 2.16, and the phase diagram of carbon at high temperature and pressure in Fig. 19.2)
- phonons in Fig. 2.11 and Chapter 20.
- effective charges and spontaneous polarization in Section 24.5.
- tests of various functionals in the phase transitions of Si and SiO_2 under pressure in Fig. 2.4
- band structures and excitations of many materials including Ge in Fig. 2.23, GaAs and LiF in Figs. 2.25 and 2.26, MoS_2 in Fig. 22.8, HgTe and CdTe in Fig. 27.10, and Bi_2Se_3 in Fig. 28.4

- surface states of gold in Fig. 22.3 and Bi_2Se_3 in Fig. 28.5, and end states of a patterned graphene nanoribbon in Section 26.7
- excitations in molecules and clusters, using plane waves for small metal clusters in Fig. 21.1 and C_{60} in Fig. 21.4, and the finite difference grid method in Figs. 21.2 and 21.3

SELECT FURTHER READING

General references are given in Chapter 12. For developments with plane waves:

Kohanoff, J., *Electronic Calculations for Solids and Molecules: Theory and Computational Methods* (Cambridge University Press, Cambridge, 2003).

Payne, M. C., Teter, M. P., Allan, D. C., Arias, T. A., and Joannopoulos, J. D., "Iterative minimization techniques for *ab initio* total-energy calculations: molecular dynamics and conjugate gradients," *Rev. Mod. Phys.* 64:1045–1097, 1992.

Pickett, W. E., "Pseudopotential methods in condensed matter applications," *Comput. Phys. Commun.* 9:115, 1989.

Singh, D. J., *Planewaves, Pseudopotentials, and the APW Method* (Kluwer Academic Publishers, Boston, 1994), and references therein.

Srivastava, G. P. and Weaire, D., "The theory of the cohesive energy of solids," *Adv. Phys.* 36:463–517, 1987.

PAW method:

Blöchl, P. E., "Projector augmented-wave method," *Phys. Rev. B* 50:17953–17979, 1994.

Holzwarth, N. A. W., Tackett, A. R., and Matthews, G. E., "A projector augmented wave (PAW) code for electronic structure calculations, part I: ATOMPAW for generating atom-centered functions," *Comp. Phys. Commun.* 135:329–347, 2001.

Holzwarth, N. A. W., Tackett, A. R., and Matthews, G. E., "A projector augmented wave (PAW) code for electronic structure calculations, part II: PWPAW for periodic solids in a plane wave basis," *Comp. Phys. Commun.* 135:348–376, 2001.

Kresse, G. and Joubert, D., "From ultrasoft pseudopotentials to the projector augmented-wave method," *Phys. Rev. B* 59:1758–1775, 1999.

Grid methods:

Beck, T. L., "Real-space mesh techniques in density-functional theory," *Rev. Mod. Phys.* 72:1041–1080, 2000.

Also see other references cited in Chapter 12.

Exercises

13.1 There are excellent open-source codes available online for DFT calculations using plane waves (see Appendix R). Two well-known ones are ABINIT and quantumESPRESSO, which have large user groups. These codes can be used in the calculations for metallic H in Eq. (13.4) and other problems in the book. The codes often have tutorials similar to the exercises and also many examples for materials.

13.2 Show that the Eq. (7.20) leads to the expression (13.2) written in Fourier components. In particular, show that the groupings of terms lead to two well-defined neutral groupings: the difference of the ion and the electron terms in the square bracket and the sum of eigenvalues that are the solution of a the Kohn–Sham equation with a neutral potential.

13.3 Derive the result that the α parameter in the linear-mixing scheme Eq. (13.5) must be less than $1/\epsilon(G_{min})$ for convergence. Show that this is a specific form of the general equations in Section 7.4 and is closely related to Exercise 7.21. In this case it is assumed that ϵ_{max} occurs for $G = G_{min}$. Discuss the validity of this assumption. Justify it in the difficult extreme case of a metal surface as discussed in Section 13.2.

13.4 This exercise is to calculate the band structure of metallic H at high density ($r_s = 1$ is a good choice) in the fcc structure and to compare with (1) the free-electron bands expected for that density and (2) bands calculated with the Thomas–Fermi approximation for the potential (Exercise 12.13). Use the Coulomb potential for the proton and investigate the number of plane waves required for convergence. (There is no need to use a pseudopotential at high density, since a feasible number of planes is sufficient.) Comparison with the results of Exercise 12.13 can be done either by comparing the gaps at the X and L points of the BZ in lowest non-zero-order perturbation theory, or by carrying out the full band calculation with the Thomas–Fermi potential.

14

Localized Orbitals: Tight-Binding

Summary

Localized functions afford a satisfying description of electronic structure and bonding in an intuitive localized picture. They are widely used in chemistry and have been revived in recent years in physics for efficiency in large simulations, especially "order-N" methods (Chapter 18). The tight-binding method is particularly simple and instructive since the basis is not explicitly specified and one needs only the matrix elements of the overlap and the hamiltonian. This chapter starts with a definition of the problem of electronic structure in terms of localized orbitals and considers various illustrative examples in the tight-binding approach. Two-band models are illustrated for graphene and are the basis for much of the analysis of Shockley surface states, topological insulators, and other phenomena in Chapters 22–28. Many of the concepts and forms carry over to full calculations with localized functions that are the subject of the following chapter, Chapter 15.

The hallmark of the approaches considered in this chapter and the next is that the wavefunction is expanded in a linear combination of fixed, energy-independent orbitals, each associated with a specific atom in the molecule or crystal. For example, the linear combination of atomic orbitals (LCAO) formulation denotes a basis of atomic or modified atomic-like functions. Such a basis provides a natural, physically motivated description of electronic states in materials; in fact, possibly the first theory of electrons in a crystal was the tight-binding[1] method developed by Bloch [35] in 1928. The history of this approach is summarized nicely by Slater and Koster [613], who point out that the seminal work of Bloch considered only the simplest s-symmetry function and the first to consider a basis of different atomic orbitals were Jones, Mott, and Skinner [614] in 1934.

---

[1] Here "tight-binding" means "highly localized atomic states," whereas it has taken different meaning as a semiempirical method with two-center matrix elements following the Slater-Koster approach (Section 14.4) in more recent years.

We will highlight three ways in which the local orbital, or tight-binding, formulation plays an important role in electronic structure:

- Of all the methods, perhaps tight-binding provides the simplest understanding of the fundamental features of electronic bands. In particular, this provided the original derivation of the Bloch theorem [35], which will also suffice for us to derive the theorem yet again in Section 14.1.
- Viewed as simply models for electrons on a lattice expressed in terms of a localized basis, this is the formulation of many of the most used models in physics, such as the Hubbard model for interacting electron problems (not explored here but studied in detail in the companion book [1]) and the two-band models that are the basis for understanding topological insulators in Chapters 25–28.
- Viewed as useful models to describe electronic bands and total energies, empirical tight-binding methods can provide both accurate results and valuable insights for real materials. In this approach, one assumes a form for the hamiltonian and overlap matrix elements without actually specifying anything about the orbitals except their symmetry. The values of the matrix elements may be derived approximately or may be fitted to experiment or other theory. This is generically called "tight-binding" and is widely used as a fitting procedure or as a quick (and sometimes dirty) electronic structure method. As such it is the method of choice for development of many ideas and new methods, e.g., "order-N" techniques in Chapter 18. This is the subject of later sections in this chapter.
- Finally, local orbital methods are not restricted to such simplifications. They can be used as a basis to carry out a full self-consistent solution of independent-particle equations. Analytic forms, especially gaussians, are extensively used in chemistry, where standard basis sets have been developed. Alternatively, one can use atomic-like orbitals with all integrals calculated numerically. Localized orbital methods are powerful, general tools and are the subject of Chapter 15.

14.1 Localized Atom-Centered Orbitals

A local orbital basis is a set of orbitals $\chi_\alpha(\mathbf{r} - \mathbf{R}_I)$, each associated with an atom at position $\mathbf{R}_I$. In order to simplify notation, we will let m denote both α and site I, so that $m = 1, \ldots, N_{\text{basis}}$ labels all the states in the basis, which can also be written $\chi_m(\mathbf{r} - \mathbf{R}_m)$.[2] In a crystal, the atoms in a unit cell are at positions $\tau_{\kappa,j}$, where $\tau_{\kappa,j}$ is the position of $j = 1, \ldots, n^\kappa$ atoms of type κ. The composite index $\{\kappa, j, \alpha\} \to m$ allows the entire basis to be specified by $\chi_m(\mathbf{r} - (\tau_m + \mathbf{T}))$, where $\mathbf{T}$ is a translation vector. The matrix elements of the hamiltonian of a state m in the cell at the origin and state m' in the cell labeled by translation vector $\mathbf{T}$ is

$$H_{m,m'}(\mathbf{T}) = \int d\mathbf{r} \chi_m^*(\mathbf{r} - \tau_m)\hat{H}\chi_{m'}[\mathbf{r} - (\tau_{m'} + \mathbf{T})], \qquad (14.1)$$

---

[2] Here the subscript m is a generic label for a basis function as in other chapters; when used in combination with l, m denotes the azimuthal quantum number, e.g., in Section 14.2.

which applies to any orbitals m and m' in cells separated by the translation $\mathbf{T}$, since the crystal is translationally invariant. Similarly, the overlap matrix is given by

$$S_{m,m'}(\mathbf{T}) = \int d\mathbf{r}\chi_m^*(\mathbf{r} - \tau_m)\chi_{m'}[\mathbf{r} - (\tau_{m'} + \mathbf{T})]. \tag{14.2}$$

The Bloch theorem for the eigenstates can be derived by defining a basis state with wavevector $\mathbf{k}$,

$$\chi_{m\mathbf{k}}(\mathbf{r}) = A_{m\mathbf{k}} \sum_{\mathbf{T}} e^{i\mathbf{k}\cdot\mathbf{T}} \chi_m[\mathbf{r} - (\tau_m + \mathbf{T})], \tag{14.3}$$

where $A_{m\mathbf{k}}$ is a normalization factor (Exercise 14.4). The analysis proceeds much like the derivation of the Bloch theorem in a plane wave basis in Sections 12.1 and 12.2, except that here the wavevector $\mathbf{k}$ is restricted to the Brillouin zone. This is sufficient since the phase factor $e^{i\mathbf{k}\cdot\mathbf{T}}$ in Eq. (14.3) is unchanged if a reciprocal lattice vector is added. Using the translation invariance of the hamiltonian, it is straightforward (Exercise 14.3) to show that matrix elements of the hamiltonian with basis functions $\chi_{m\mathbf{k}}$ and $\chi_{m'\mathbf{k}'}$ are nonzero only for $\mathbf{k} = \mathbf{k}'$, with

$$H_{m,m'}(\mathbf{k}) = \int d\mathbf{r}\chi_{m\mathbf{k}}^*(\mathbf{r})\hat{H}\chi_{m'\mathbf{k}}(\mathbf{r}) = \sum_{\mathbf{T}} e^{i\mathbf{k}\cdot\mathbf{T}} H_{m,m'}(\mathbf{T}), \tag{14.4}$$

and

$$S_{m,m'}(\mathbf{k}) = \int d\mathbf{r}\chi_{m\mathbf{k}}^*(\mathbf{r})\chi_{m'\mathbf{k}}(\mathbf{r}) = \sum_{\mathbf{T}} e^{i\mathbf{k}\cdot\mathbf{T}} S_{m,m'}(\mathbf{T}). \tag{14.5}$$

Since the hamiltonian conserves $\mathbf{k}$, an eigenfunction of the Schrödinger equation in a basis always can be written in the form

$$\psi_{i\mathbf{k}}(\mathbf{r}) = \sum_m c_m(\mathbf{k})\,\chi_{m\mathbf{k}}(\mathbf{r}), \tag{14.6}$$

and the secular equation for wavevector $\mathbf{k}$ is

$$\sum_{m'} \left[H_{m,m'}(\mathbf{k}) - \varepsilon_i(\mathbf{k}) S_{m,m'}(\mathbf{k}) \right] c_{i,m'}(\mathbf{k}) = 0. \tag{14.7}$$

This has the same form as Eq. (12.9) except that in Eq. (14.7) the orbitals are not assumed to be orthonormal. The only fundamental sense in which local orbitals are different from any other basis is that the locality of $\chi_m(\mathbf{r} - (\tau_m + \mathbf{T}))$ is expected to cause $H_{m,m'}(\mathbf{T})$ and $S_{m,m'}(\mathbf{T})$ to decrease and become negligible for large distances $|\tau_m - (\tau_{m'} + \mathbf{T})|$.

14.2 Matrix Elements with Atomic-Like Orbitals

Much can be gained from consideration of the symmetries of the basis orbitals and the crystal or molecule. This is the basis for tight-binding approaches and continues to be essential in full calculations (Chapter 15). An appropriate choice for basis functions is a set of atomic-like functions centered on the atom sites, i.e., functions with the same symmetry

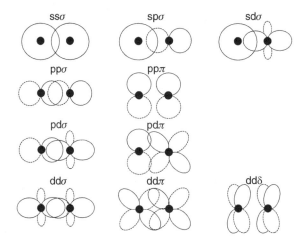

Figure 14.1. Schematic figures of local orbitals indicating all possible overlap and two-center hamiltonian matrix elements for s, p, and d orbitals, which are classified by the angular momentum about the axis with the notation σ ($m = 0$), π ($m = 1$), and δ ($m = 2$). The orbitals shown are the real combinations of the angular momentum eigenstates. Positive and negative lobes are denoted by solid and dashed lines, respectively. Note that the sign of the p orbitals must be fixed by convention; here and in Table 14.1 the positive p_x lobe is along the positive x-axis, etc.

as for an atom in free space but may have radial dependence modified from the atom. On each site κ, j the basis functions can be written as radial functions multiplied by spherical harmonics,

$$\chi_\alpha(\mathbf{r}) \rightarrow \chi_{nlm}(\mathbf{r}) = \chi_{nl}(r)Y_{lm}(\hat{\mathbf{r}}), \tag{14.8}$$

where n indicates different functions with the same angular momentum. Real basis functions can also be defined using the real angular functions $S_{lm}^+ = \frac{1}{\sqrt{2}}(Y_{lm} + Y_{lm}^*)$ and $S_{lm}^- = \frac{1}{\sqrt{2}i}(Y_{lm} - Y_{lm}^*)$ defined in Eq. (K.11). These are useful for visualization and in actual calculations, but the Y_{lm} are most convenient for symmetry analysis. Examples of real s, p, and d orbitals are given in Fig. 14.1.

The matrix elements, Eqs. (14.1) and (14.2), can be divided into one-, two-, and three-center terms. The simplest is the overlap matrix S in Eq. (14.2), which involves only one center if the two orbitals are on the same site ($\mathbf{T} = 0$ and $\tau_m = \tau_{m'}$) and two centers otherwise. The hamiltonian matrix elements in Eq. (14.1) consist of kinetic and potential terms with

$$\hat{H} = -\frac{1}{2}\nabla^2 + \sum_{\mathbf{T}\kappa j} V^\kappa[|\mathbf{r} - (\tau_{\kappa j} + \mathbf{T})|], \tag{14.9}$$

where the first term is the usual kinetic energy and the second is the potential decomposed into a sum of spherical terms centered on each site κ, j in the unit cell.[3] The kinetic

---

[3] This decomposition can always be done formally. Often $V^\kappa(|\mathbf{r}|)$ can be approximated as spherical atomic-like potentials associated with atom of type κ.

part of the hamiltonian matrix element always involves one or two centers. However, the potential terms may depend on the positions of other atoms; they can be divided into the following:

- One center, where both orbitals and the potential are centered on the same site. These terms have the same symmetry as an atom in free space. Spin–orbit interaction is a one-center term that is considered in Section 14.3.
- Two center, where the orbitals are centered on different sites and the potential is on one of the two. These terms have the same symmetry as other two-center terms.
- Three center, where the orbitals and the potential are all centered on different sites. These terms can also be classified into various symmetries based on the fact that three sites define a triangle.
- A special class of two-center terms with both orbitals on the same site and the potential centered on a different site. These terms add to the one-center terms above but depend on the crystal symmetry.

Two-Center Matrix Elements

Two-center matrix elements play a special role in calculations with local orbitals and are considered in more detail here. The analysis applies to all overlap terms and to any hamiltonian matrix elements that involve only orbitals and potentials on two sites. For these integrals the problem is the same as for a diatomic molecule in free space with cylindrical symmetry. The orbitals can be classified in terms of the azimuthal angular momentum about the line between the centers, i.e., the value of m with the axis chosen along the line, and the only nonzero matrix elements are between orbitals with the same $m = m'$. If $K_{lm,l'm'}$ denotes an overlap or two-center hamiltonian matrix element for states lm and $l'm'$, then in the standard form with orbitals quantized about the axis between the pair of atoms, the matrix elements are diagonal in mm' and can be written $K_{lm,l'm'} = K_{ll'm}\delta_{m,m'}$. The quantities $K_{ll'm}$ are independent matrix elements that are irreducible, i.e., they cannot be further reduced by symmetry. By convention the states are labeled with l or l' denoted by s, p, d, ..., and $m = 0, \pm 1, \pm 2, \ldots$, denoted by $\sigma, \pi, \delta, \ldots$, leading to the notation $K_{ss\sigma}, K_{sp\sigma}, K_{pp\pi}, \ldots$.

Figure 14.1 illustrates the orbitals for the nonzero σ, π, and δ matrix elements for s, p, and d orbitals. The orbitals shown are the real basis functions $S_{lm}^{\pm}$ defined in Eq. (K.11) as combinations of the $\pm m$ angular momentum eigenstates. These are oriented along the axes defined by the line between the neighbors and two perpendicular axes. All states except the s state have positive and negative lobes, denoted by solid and dashed lines, respectively. States with odd l are odd under inversion, and their sign must be fixed by convention. Typically one chooses the positive lobe along the positive axis of the displacement vector from the site denoted by the first index to the site denoted by the second index. For example, in Fig. 14.1, the $K_{sp\sigma}$ matrix element in the top center has the negative lobe of the p function oriented toward the s function. Interchange of the indices leads to $K_{ps\sigma} = -K_{sp\sigma}$ and, more generally, to $K_{ll'm} = (-1)^{l+l'} K_{l'lm}$ (Exercise 14.5).

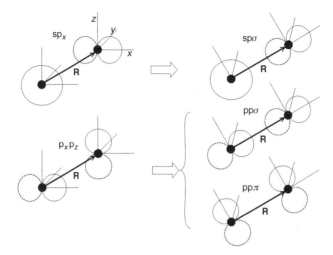

Figure 14.2. Schematic figures for two-center matrix elements of s and $p_i = \{p_x, p_y, p_z\}$ orbitals for atoms separated by displacement vector **R**. Matrix elements are related to σ and π integrals by the transformation to a combination of orbitals that are aligned along **R** and perpendicular to **R**. The top figure illustrates the transformation to write a real matrix element K_{s,p_x} in terms of $K_{sp\sigma}$: the s orbital is unchanged and the p_x orbital is written as a sum of the σ orbital, which is shown, and the π orbitals, which are not shown because there is no spπ matrix element. The lower figure illustrates the transformation needed to write K_{p_x,p_z} in terms of $K_{pp\sigma}$ and $K_{pp\pi}$. The coefficients of the transformation for all s and p matrix elements are given in Table 14.1. Matrix elements for arbitrary angular momenta can be found using the rotation matrix method described in Appendix N.

In general the displacement vector **R** is not oriented along the cartesian axes and the functions must be transformed to utilize the standard irreducible form of the matrix elements. Examples of two-center matrix elements of s and $p_i = \{p_x, p_y, p_z\}$ orbitals are shown in Fig. 14.2. Each of the orbitals on the left-hand side can be expressed as a linear combination of orbitals that have the standard form oriented along the rotated axes, as shown on the right. An s orbital is invariant and a p orbital is transformed to a linear combination of p orbitals. The only nonzero matrix elements are the σ and π matrix elements, as shown. The top row of the figure illustrates the transformation of the p_x orbital needed to write the matrix element K_{s,p_x} in terms of $K_{sp\sigma}$ and the bottom row illustrates the relation of K_{p_x,p_z} to $K_{pp\sigma}$ and $K_{pp\pi}$. The relations for s and p matrix elements are given in Table 14.1, and all other s and p matrix elements are related by symmetry. Expressions for d orbitals are given in many places including [354, 613], and [615]; arbitrary angular momenta can be treated using the procedures in Appendix N.

Three-Center Matrix Elements

The hamiltonian matrix elements, in general, depend on the presence of other atoms, resulting in three-center or multicenter matrix elements. Such terms are discussed in Chapter 15 since they arise naturally in the integrals required in a local orbital basis. In this

Table 14.1. Two-center matrix elements for real orbitals s and p_x, p_y, p_z, where the vector between sites is $\hat{\mathbf{R}} \equiv \{x, y, z\}$ (see Fig. 14.2)

Element	Expression
$K_{s,s}$	$K_{ss\sigma}$
K_{s,p_x}	$x^2 K_{sp\sigma}$
K_{p_x,p_x}	$x^2 K_{pp\sigma} + (1 - x^2) K_{pp\pi}$
K_{p_x,p_z}	$xz(K_{pp\sigma} - K_{pp\pi})$

chapter we consider only the "empirical tight-binding" or "semiempirical tight-binding" approaches that involve only the matrix elements $H_{m,m'}(\mathbf{T})$ and $S_{m,m'}(\mathbf{T})$ expressed in a parameterized form, without an explicit representation for the basis orbitals.

The only rigorous result that one can apply immediately to the nature of matrix elements Eqs. (14.1) and (14.2) is that they must obey crystal symmetry. This is often very helpful in reducing the number of parameters to a small number for a high-symmetry crystal as is done in the tables of Papaconstantopoulos [616] for many crystalline metals. In this form, the tight-binding method is very useful for interpolation of results of more expensive methods [616].

14.3 Spin–Orbit Interaction

The spin–orbit interaction is a relativistic effect derived in Appendix O, where it is shown that the effects can be described using the nonrelativistic Schrödinger equation with an added term

$$\hat{H}_{SO} = \frac{\hbar^2}{4M^2 c^2} \frac{1}{r} \frac{dV}{dr} \mathbf{L} \cdot \sigma. \tag{14.10}$$

Because $\hat{H}_{SO}$ is a velocity-dependent operator, it has effects that cannot be reproduced by any potential. Since the effect is largest near the nucleus where the potential is spherically symmetric, it can be take into account by a term in the hamiltonian centered on each atom, which can be written as a matrix in the basis of ↑ and ↓ spin states,

$$H_{SO} = \begin{bmatrix} H_{SO}(\uparrow, \uparrow) & H_{SO}(\uparrow, \downarrow) \\ H_{SO}(\downarrow, \uparrow) & H_{SO}(\downarrow, \downarrow) \end{bmatrix}, \tag{14.11}$$

with $H_{SO}(\downarrow\downarrow) = -H_{SO}(\uparrow\uparrow)$ and $H_{SO}(\uparrow\downarrow) = H_{SO}^\dagger(\downarrow\uparrow)$. Thus the size of the hamiltonian is doubled, and in general the eigenstates are linear combinations of the ↑ and ↓ spin states.

In electronic structure methods that deal with wavefunctions in the continuum, such as plane waves and local orbitals, the problem is defined by taking matrix elements of Eq. (14.10). In methods with a basis of orbitals with definite angular momentum L centered on the atoms, the spin–orbit interaction can be included as an on-site term for each atom that is determined by one parameter for each $L > 0$. (Strictly, this is an approximation and there are other terms if the orbitals on an atom overlap the nuclei on neighboring atoms.)

This has the same symmetry as in an atom, i.e., each state with angular momentum L is split into states with $J = L \pm 1/2$, but the quantitative values may be different since the wavefunctions in the solid are not the same as in the atom. In calculations for a crystal it is often more convenient to work with a basis specified in cartesian coordinates, for example, p states p_x, p_y and p_z. In this basis the hamiltonian in Eq. (14.11) has the form

$$
H_{SO}(\uparrow\uparrow) = \begin{bmatrix} 0 & -i\zeta & 0 \\ i\zeta & 0 & 0 \\ 0 & 0 & 0 \end{bmatrix}, \text{ and } H_{SO}(\downarrow\uparrow) = \begin{bmatrix} 0 & 0 & \zeta \\ 0 & 0 & -i\zeta \\ -\zeta & i\zeta & 0 \end{bmatrix}, \tag{14.12}
$$

where $\zeta = \frac{\hbar e}{4m^2c^2}\langle p_x|(V'(r)/r)(i\mathbf{p} \times \mathbf{r})_z|p_y\rangle$, which is a real number. It is especially simple to include in an sp tight-binding model where ζ is a parameter fit to experiment or to a calculation for the solid. The only thing one needs to do is to double the matrix size, compared to a case without spin–orbit interaction, and add H_{SO} in the p components of the hamiltonian at each site.

14.4 Slater–Koster Two-Center Approximation

Slater and Koster (SK) [613] developed the widely used [354, 615] approach that bears their name. They proposed that the hamiltonian matrix elements be approximated with the two-center form and fitted to theoretical calculations (or empirical data) as a simplified way of describing and extending calculations of electronic bands. Within this approach, all matrix elements have the same symmetry as for two atoms in free space given in Fig. 14.2 and Table 14.1. This is a great simplification that leads to an extremely useful approach to understanding electrons in materials. Of all the methods for treating electrons, the SK approach provides the simplest, most illuminating picture of electronic states. In addition, more accurate treatments involving localized orbitals (Chapter 15) are often best understood in terms of matrix elements having SK form plus additional terms that modify the conclusions in quantitative ways.

Slater and Koster gave extensive tables for matrix elements with the symmetry of atom-centered s, p, and d states, and analytic formulas for bands in several crystal structures. Explicit expressions for the s and p matrix elements are given in Table 14.1 and illustrated in Fig. 14.2. The power of using such models to capture qualitative effects such as topology of band structures are illustrated in Chapters 25–28. Examples of quantitative descriptions for band structures in Section 14.10 illustrate useful information that can be derived simple, pedagogical models that also can be used to description of large complicated systems, including the bands, total energies, and forces for relaxation of structures and molecular dynamics. These different applications have very different requirements that often lead to different choices of SK parameters.

For quantitative description of bands, the parameters are usually designed to fit selected eigenvalues for a particular crystal structure and lattice constant. For example, the extensive tables derived by Papaconstantopoulos [616] are very useful for interpolation of results of more expensive methods. It has been pointed out by Stiles [617] that for a fixed ionic configuration, effects of multicenter integrals can be included in two-center terms that can be generated by an automatic procedure. This makes it possible to describe any band structure accurately with a sufficient number of matrix elements in SK form. However, the two-center matrix elements are not transferable to different structures.

On the other hand, any calculation of total energies, forces, etc. requires that the parameters be known *as a function of the positions of the atoms*. Thus the choices are usually compromises that attempt to fit a large range of data. Such models are fit to structural data and, in general, are only qualitatively correct for the bands. Since the total energy depends only on the occupied states, the conduction bands may be poorly described in these models. Of particular note, Harrison [354, 615] has introduced a table that provides parameters for any element or compound. The forms are chosen for simplicity, generality, and ability to describe many properties in a way that is instructive and useful, albeit approximate. The basis is assumed to be orthonormal, i.e., $S_{mm'} = \delta_{mm'}$. The diagonal hamiltonian matrix elements are given in a table for each atom. Any hamiltonian matrix element for orbitals on neighboring atoms separated by a distance R is given by a factor times $1/R^2$ for s and p orbitals and $1/R^{l+l'}$ for $l > 1$. The form for s and p orbitals comes from scaling arguments on the homogeneous gas [615] and the form for higher angular momenta is taken from muffin-tin orbital theory (Section 16.7).

Many other SK parameterizations have been proposed, each tailored to particular elements and compounds. Examples are given in Sections 14.5–14.11, chosen to illustrate various aspects of electronic structure calculations in the present and other chapters. Care must be used in applying the different parameterizations to the appropriate problems.

14.5 Tight-Binding Bands: Example of a Single s Band

This section and following ones are concerned with electronic bands calculated using tight-binding with the SK two-center form for the hamiltonian that can be worked out analytically, with further examples in exercises. The later sections of this chapter give examples of applications to bands in materials and more complex problems, such as the electronic structure of nanotubes.

s-Bands in Line, Square, and Cubic Bravais Lattices

The simplest possible example of bands is for s-symmetry states on each site I in a Bravais lattice so that there is only one band. As a further simplification, we consider the case of orthogonal basis states and nonzero hamiltonian matrix elements $\langle I|\hat{H}|I'\rangle \equiv t$ only if I and I' are nearest neighbors. The on-site term can be chosen to be zero, $\langle 0|\hat{H}|0\rangle = 0$. There are three cases (line, square, and cubic lattices) that can be treated together. For the cubic lattice with spacing a the general expressions (14.4) and (14.7) reduce to

$$\varepsilon(\mathbf{k}) = H(\mathbf{k}) = 2t\left[\cos(k_x a) + \cos(k_y a) + \cos(k_z a)\right]. \qquad (14.13)$$

The bands for the square lattice in the x, y-plane are given by this expression, omitting the k_z term; for a line in the x-direction, only the k_x term applies.

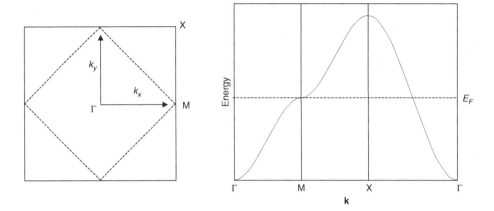

Figure 14.3. Tight-binding bands in the square lattice with only an s state on each site and nearest neighbor interactions. The left figure shows the BZ and the dashed line shows the Fermi surface in the case of a half-filled band. The right figure shows the bands with **k** along the lines between the high-symmetry points.

This simple example leads to useful insights. In particular, the bands are symmetric about $\varepsilon(\mathbf{k}) = 0$ in the sense that every state at $+\varepsilon$ has a corresponding state at $-\varepsilon$. This can be seen by plotting the bands in two ways: first in the usual Brillouin zone centered on $\mathbf{k} = 0$, and second in a cell of the reciprocal lattice centered on $\mathbf{k} = (\pi/a, \pi/a, \pi/a)$. Since $\cos(k_x a - \pi) = -\cos(k_x a)$, etc., it follows that the bands have exactly the same shape except that the sign of the energy is changed,

$$\varepsilon(\mathbf{k}) = -\varepsilon(\mathbf{k} - (\pi/a, \pi/a, \pi/a)). \tag{14.14}$$

The same arguments apply to the line and square: the line has a simple cosine band and the bands for a square lattice are illustrated in Fig. 14.3. The densities of states (DOS) for one, two, and three dimensions are shown in Fig. 14.4. The shapes can be found analytically in this case, which is the subject of Exercise 14.6.

There are several remarkable consequences in the case of the square. The energy $\varepsilon(\mathbf{k}) = 0$ at a zone face $\mathbf{k} = (\pi/a, 0)$, which is easily verified using Eq. (14.13) and omitting the k_z term. This is a saddle point since the slope vanishes and the bands curve upward and downward in different directions as shown in Fig. 14.3. This leads to a density of states with a logarithmic divergence at $\varepsilon = 0$ (Exercise 14.7). Furthermore, for a half-filled band (one electron per cell), the Fermi surface is at energy $\varepsilon(\mathbf{k}) = 0$. This leads to the result shown in Fig. 14.4 that the Fermi surface is a square (Exercise 14.7) rotated by $\pi/4$ with half the volume of the Brillouin zone, and the density of states diverges at $\varepsilon = E_F$ as shown in Fig. 14.3. If there are second-neighbor interactions, the symmetry of the bands in $\pm\varepsilon$ is broken and the Fermi surface is no longer square.

Nonorthogonal Orbitals

The solution of the tight-binding equations in terms of non-orthogonal orbitals can be found simply in terms of the overlap matrix S using Eq. (14.7). The matrix elements of S can be

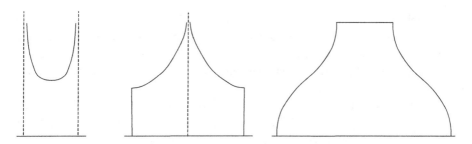

Figure 14.4. Schematic densities of states (DOS) for an s band in a one-dimensional (1D) line, a two-dimensional (2D) square, and a three-dimensional (3-D) simple cubic lattice with nearest neighbor interactions t. The bands are symmetric about the center and the width of each segment is $4t$, i.e., the width of the 1-D DOS is $4t$, that of the 2-D DOS is $8t$ divided into two parts, and the 3D band has width $12t$ divided into three parts of equal width. The special property of the square lattice in this leads to a logarithmic singularity at the band center. See also Exercise 14.6 for further information.

parameterized in the same way as the hamiltonian, with the added benefit that the two-center form is rigorous and each orbital is normalized so that $S_{mm} = 1$. The effect can be illustrated by line, square, or simple cubic lattices, with nearest neighbor overlap defined to be s. Then the solution for the bands, Eq. (14.13), is generalized to

$$\varepsilon(\mathbf{k}) = \frac{H(\mathbf{k})}{S(\mathbf{k})} = \frac{2t\left[\cos(k_x) + \cos(k_y a) + \cos(k_z a)\right]}{1 + 2s\left[\cos(k_x a) + \cos(k_y a) + \cos(k_z a)\right]}. \tag{14.15}$$

The effect of nonzero s is discussed in Exercises 14.16 and 14.17. In this case, the symmetry about $\varepsilon = 0$ is broken, so that the conclusions on bands and the Fermi surface no longer apply. In fact s has an effect like longer-range hamiltonian matrix elements, indeed showing strictly infinite range but rapid exponential decay.

Nonorthogonal orbitals play an essential role in realistic tight-binding models. As discussed more completely in Section 14.12, it is never rigorously consistent to cut off the hamiltonian matrix elements while assuming orthogonal orbitals. This is a manifestation of the well-known properties of Wannier functions (Chapter 23) and the fact that Wannier functions are very environment dependent. On the other hand, nonorthogonal functions can be much more useful because they are more transferable between different environments. This is illustrated by examples in Sections 14.12 and 23.4.

14.6 Two-Band Models

Two-band models play a special role in electronic structure of crystals. The hamiltonian can be expressed in the form

$$H(k) = \begin{bmatrix} E_1(\mathbf{k}) & t(\mathbf{k}) \\ t^*(\mathbf{k}) & E_2(\mathbf{k}) \end{bmatrix}, \tag{14.16}$$

where the diagonal and off-diagonal terms have different interpretation in different models. In one dimension the lowest two bands can always be represented by Bloch states that are linear combinations of an even and an odd state per cell. This was used by by Shockley to draw very general conclusions about the bands including a transition in the bulk band structure and emergence of a surface state. This is described in Chapter 22, where it is the origin of a class of surface states often observed on metals, and in Chapter 26, where it is a forerunner of topological insulators.

Because the models are so useful in the chapters on topology, description of the models is deferred to Chapter 26. However, it is appropriate to emphasize here that the models are not only for surfaces and topological insulators! They are instructive models for physical problems. Two characteristic models are shown in Figs. 26.2 and Fig. 26.3. The first case has two sites, each with one state, which can represent an ionic crystal with different energies on the two sites, or a molecular crystal with strong and weak hopping. The second has one site with even and odd states, which is the natural model to describe covalent bonding. See Exercise 26.2 for examples. These and other models all share the same mathematical structure and can be mapped into one another as brought out in Chapter 26. In this chapter, the two-band model is illustrated by graphene in two dimensions.

14.7 Graphene

Graphene is one the most fascinating materials in nature and in theoretical models. It can be made easily by the "Scotch tape method" to peel off one layer from graphite; it is the thinnest stable material known (one atomic layer) and the strongest material known (in the two-dimensional layer); it has exceptionally large thermal conductivity and ballistic electron transport; and it is the basis for a plethora of structures like nanotubes and ribbons with a fabulous array of properties. It is also one of the best examples where a very simple model provides quantitative results, as well as qualitative new effects including nontrivial topology!

Graphene has the planar honeycomb structure shown in Fig. 4.5. The Brillouin zone is a hexagon that is the same as for the three-dimensional hexagonal zone in Fig. 4.10 with $k_z = 0$; it is also shown in Fig. 27.12, where the corners of the zone are labeled K and K', which are related by inversion. For a single, flat graphene sheet, the π states are odd in reflection in the plane and symmetry forbids coupling to the σ bands that are well below and well above the Fermi energy (in the absence of spin–orbit interaction). Since graphene has two atoms per cell, the π states are an example of the two-site model with the same energy of each site that can be set to zero for convenience, and hopping matrix element $t_{pp\pi} \equiv t$ between nearest neighbors. The bands $\varepsilon(\mathbf{k})$ are given by the determinant equation (see Exercise 14.20 and [233])

$$\left| \hat{H}(\mathbf{k}) - \varepsilon(\mathbf{k}) \right| = \begin{vmatrix} -\varepsilon(\mathbf{k}) & t(\mathbf{k}) \\ t^*(\mathbf{k}) & -\varepsilon(\mathbf{k}) \end{vmatrix} = 0, \tag{14.17}$$

with

$$t(\mathbf{k}) = t \left[e^{ik_y a/\sqrt{3}} + 2e^{-ik_y a/2\sqrt{3}} \cos\left(k_x \frac{a}{2}\right) \right], \tag{14.18}$$

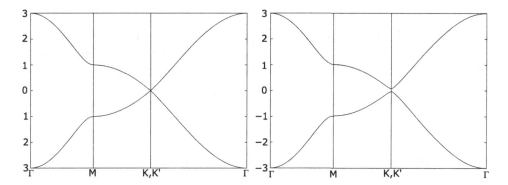

Figure 14.5. The π bands for graphene without and with spin–orbit interaction. At the left is shown only the π bands without spin–orbit interaction showing the Dirac linear dispersion with no gap at the K points. At the right is shown the bands that are modified by adding the s and other p states with spin–orbit interaction, which shifts the bands and opens the gap. (The magnitude of the spin–orbit interaction is chosen to be much larger than the actual value for graphene in order to make the effect visible.)

where a is the lattice constant, t is the nearest neighbor hopping matrix element, and x and y are the horizontal and vertical directions in Fig. 4.5. This is readily solved to yield the band energies

$$\varepsilon(\mathbf{k}) = \pm|t(\mathbf{k})| = \pm t \left[1 + 4\cos\left(k_y \frac{\sqrt{3}a}{2}\right)\cos\left(k_x \frac{a}{2}\right) + 4\cos^2\left(k_x \frac{a}{2}\right) \right]^{1/2}. \quad (14.19)$$

The resulting bands are shown in the left side of Fig. 14.5 in units where the hopping matrix element is set to $t = 1$. This shows the remarkable feature that the bands touch with zero gap at the K and K' points. These are called Dirac points, since the dispersion is linear like a relativistic particle with no mass. But it should be kept in mind that this is only a useful analogy: it is not a relativistic effect, the velocities are much less than the speed of light and the dispersion is not strictly linear for momenta a finite distance from the K and K' points.

Spin–Orbit Interactions

It might seem surprising that there is an effect of spin–orbit interaction on the π bands. If there were only a single π state on each atom, like in the model described by Eq. (14.19), there would be no angular momentum and no effect. However, in actual graphene the other p states are present and are part of the bonding and antibonding states that are well below and above the π bands, outside the energy range shown in the figure. There is an effect because the spin–orbit interaction mixes the three p states, but it is extremely small for two reasons: carbon is a light atom with small spin–orbit interaction and it is a second-order effect due to mixing with the other p states. The right side of Fig. 14.5 shows the bands including a spin–orbit interaction that is chosen to be much larger than the actual value to make the effect more visible. Note that a point K' is related to K by time-reversal symmetry such that it has opposite momentum and opposite spin state. The eigenvalues are the same as at K but since the spins are opposite this is referred to as a negative gap in the sense that at K the valence band is for one spin (say $\uparrow$) and the bottom of the conduction band is $\downarrow$, whereas at K' the spins are reversed. This is an essential aspect of graphene as a topological insulator in Section 27.8.

14.8 Nanotubes

Nanotubes are ideal for illustrating the use of tight-binding to reveal the most vital information about the electronic structure in a simple, illuminating way. Carbon nanotubes were discovered in 1991 by Iijima [228] and recognized to be nanoscale versions of "microtubes," long tubular graphitic structures that had been grown using iron particles for catalysis [234]. In a perfect single-wall tube, each carbon atom is equivalent and is at a junction of three hexagons. The various ways a sheet of graphene (i.e., a single honeycomb-structure plane of carbon) can be rolled into a tube leads to an enormous variety of semiconductors and metals [232–234] that originate in the way the Dirac points in graphene are modified by the structure of the nanotube.

The structures of nanotubes are defined in terms of a graphene layer as shown in Fig. 14.6. The vector indicated connects atoms in the layer that are equivalent in the tube, i.e., the tube is defined by rolling the plane of graphene to bring those points together. The tube axis is perpendicular to the vector. The convention is to label the vector with multiples of graphene translation vectors, $\mathbf{a}_1$ and $\mathbf{a}_2$, defined as in Fig. 4.5. The example shown is for a $(6, 1)$ tube, which denotes $(6 \times \mathbf{a}_1, 1 \times \mathbf{a}_2)$ and which defines the chiral tube shown on the right. Special examples of "zigzag" $(n, 0)$ and "armchair" (n, n) tubes are shown in Fig. 14.7. These are not chiral but have very different properties due to the underlying atomic structure. See Exercise 14.21.

The first approximation is to assume that the bands are unchanged from graphene and the only effect is that certain states are allowed by the boundary condition. The condition

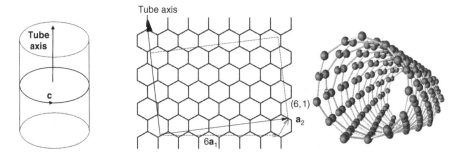

Figure 14.6. Rolling of a graphene sheet to form a nanotube. On the left is shown a tube with circumference indicated by the circle **c**. The middle and right figures show a graphene translation vector $\mathbf{c} = n\mathbf{a}_1 + m\mathbf{a}_2$ with $n = 6$ and $m = 1$ and the chiral $(6, 1)$ tube formed by rolling to join the site related by **c**. The basis vectors $\mathbf{a}_1$ and $\mathbf{a}_2$ shown are the same as in Fig. 4.5 and in [233, 618]; in this notation the example shown is a $(6, 1)$ nanotube. Provided by J. Kim

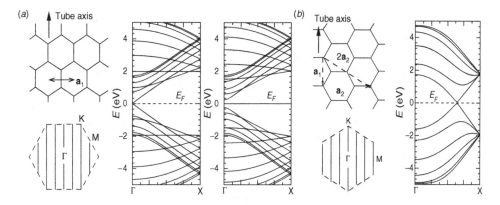

Figure 14.7. Structures, Brillouin zones, and bands of the "zigzag" (*a*) and "armchair" (*b*) nanotubes. The bands are calculated using the orthogonal tight-binding model of Xu et al. [619]. The zigzag tubes shown are denoted $(13,0)$ and $(12,0)$; the latter is insulating and the former has a small gap that is due to curvature. However, for smaller tubes, the curvature can induce large effects, including band overlap leading to metallic tubes as shown in Fig. 22.9 and discussed in the text. The armchair (n,n) tubes are always metallic since the lines of allowed states with k along the tube always include Γ and one of the K points. (*b*) illustrates the bands for a $(3,3)$ tube. In each case the bands of the tube are plotted in the one-dimensional BZ denoted $\Gamma \to X$. Provided by J. Kim

on allowed functions is that they must be single valued in the circumferential direction but have Bloch boundary conditions along the tube axis. This leads to allowed **k** vectors, which are shown as lines in Fig. 14.7. The resulting bands have been analyzed in general [232–234] with the simple conclusion that there is a gap between filled and empty states unless the allowed k lines pass through the point K. If they do include K, e.g., the armchair tube, then the interesting situation of one-dimensional metallic bands arises.

The next approximation is to include expected effects due to curvature [232, 233]. The simplest depends only on symmetry: curvature makes bonds along the axis inequivalent to those at an angle to the axis. Therefore, the k point where the bands touch moves away from the point K opening a small gap, as shown in Fig. 14.7 for the $(13, 0)$ zigzag tube. The gap is expected to increase for small tubes with larger curvature. On the other hand, there is no effect upon the band crossing at point K, which is along the tube axis for the armchair tube, so that it remains metallic in all cases.

However, small changes from graphene are not the whole story. As shown in Fig. 22.9, calculations [620] on small tubes have shown that the bands can be qualitatively changed from the graphene-like states considered thus far. The reason is that in small tubes there is large mixing of π states with a σ antibonding state that pushes a band below the Fermi level. This leads to the prediction [620] that small-diameter nanotubes can be metallic due to band overlap, even in cases where analogy to graphene would expect an insulator. This effect is also found in local orbital calculations [621] and in the improved tight-binding model of [622], which has been fitted to LDA calculations of many properties (eigenvalues, total energies, phonons, etc.) of carbon in various coordinations and geometries. This is an example of a tight-binding method that allows fast calculations that span the range from small to large tubes, as well as other carbon structures.

There is an interesting relation to graphene topological insulators. As the circumference of a tube is increased to be much larger than the length it becomes a ribbon: the zigzag and armchair tubes map directly onto the zigzag and armchair ribbons in Section 27.8. For example, the zero-gap tubes in Fig. 14.7 are the equivalent of the points where the bulk spectrum has zero gaps in Fig. 27.14, and as the tube circumference increases the other low-energy states in Fig. 14.7 approach zero and become the edge state in Fig. 27.14. The spin–orbit interaction would lead to a gap and circulating end states!

Boron Nitride Nanotubes

Nanotubes of boron nitride have been proposed theoretically [623] and later made experimentally [624]. Structures for the tubes are allowed if they maintain the B–N equal stoichiometry, and the tubes always have a large gap due to the difference between the B and N atoms. Thus the electronic properties are very different from carbon nanotubes and BN tubes hold the potential to act like one-dimensional semiconductors in the III–V family. Like other III–V materials they exhibit piezoelectric and pyroelectric effects, but in this case the one dimensionality leads to extreme anisotropy and novel electric polarization and piezoelectric effects [612, 625].

14.9 Square Lattice and CuO$_2$ Planes

The problem of an s band in a square lattice has a particularly noteworthy application in the case of the cuprate high-temperature superconductors.[4] Figure 4.5 shows the square lattice structure of CuO$_2$ planes that is the common feature of these materials, e.g., each of the planes in the bilayer in YBa$_2$Cu$_3$O$_7$ shown in Fig. 17.3. Extensive calculations, exemplified by the bands presented in Fig. 17.4, have shown that the primary electronic states at the Fermi energy are a single band formed from Cu d and O p states. The band has the same symmetry as d$_{x^2-y^2}$ states centered on each Cu (where x and y are in directions toward the neighboring Cu atoms). This can be understood in terms of the Cu and O states shown in Fig. 14.8. The three states per cell form a bonding, a nonbonding, and an antibonding combination, with the antibonding band crossing the Fermi level. In fact, the single antibonding band has the same symmetry as a Cu d$_{x^2-y^2}$ band with an effective hamiltonian matrix element (Exercise 14.15) so that the problem is equivalent to a model with one d$_{x^2-y^2}$ state per Cu, as shown on the right-hand side of Fig. 14.8. This highly schematic figure is supported by detailed calculations of the one-band orbital shown in Fig. 17.8. The orbital has d$_{x^2-y^2}$ symmetry about the Cu site and it

---

[4] This is a well-known case [240] where the simple LDA and GGA functionals predict a metal at half-filling, whereas the real solution is an antiferromagnetic insulator. Nevertheless, the metallic state created by doping appears to be formed from the band as described here.

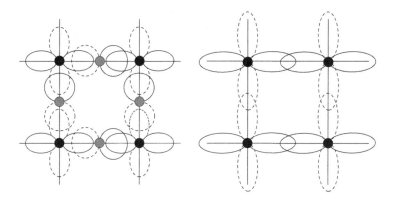

Figure 14.8. Tight-binding models representing electronic states in the square lattice structure of a CuO plane. On the left is shown the three-band model representing one Cu $d_{x^2-y^2}$ and two O p states per cell. As discussed in the text, the most relevant states are the antibonding combination of Cu and O states that are equivalent to the model shown on the right with modified orbitals that have $d_{x^2-y^2}$ symmetry around each Cu site. The effective orbitals are more extended, as shown on the right; actual calculated orbitals shown in Fig. 17.8 are more realistic and show extended shape with $d_{x^2-y^2}$ symmetry. Finally, the band is isomorphic to a single s band because, by symmetry, all hamiltonian matrix elements have the same symmetry; e.g., the nearest neighbor elements all have the same sign, as is evident from the right-hand figure.

is extended in the directions along the Cu–O bonds, with large amplitude on the O sites. If the orbitals are required to be orthonormal, like Wannier functions, then each orbital must also extend to the neighboring Cu sites.

Finally, the problem is isomorphic to a single s band; this occurs because nearest neighbor $d_{x^2-y^2}$ states always have lobes of the same sign ($++$ or $--$) so that the matrix elements are equal for all four neighbors, exactly as for s symmetry states. Thus the simplest model for the bands is a single s band, with dispersion shown in Fig. 14.4, and a square Fermi surface at half-filling. In fact, there are second-neighbor interactions that modify the bands and the calculated Fermi surface [490, 626]. The single band resulting from the orbital in Fig. 17.8 is shown in Fig. 17.9. It accurately describes the actual band, and its dispersion is significantly different from the nearest neighbor model due to longer-range matrix elements in a realistic model.

14.10 Semiconductors and Transition Metals

In this section two examples illustrate different applications of the tight-binding approach. The left side of Fig. 14.9 shows the bands of GaAs calculated using nanoHUB (nanohub.org), which provides databases and codes that can be run online to simulate the electronic structure of semiconductor nanostructure devices. This example is for bulk GaAs using a large s, p, d, s∗ basis with parameters carefully fit to experiment. The calculations show the direct gap at Γ and the effect of spin–orbit interaction that splits the valence band at Γ into Γ_7 and Γ_8 components, the same as for CdTe and HgTe in Chapter 27 on topological insulators.

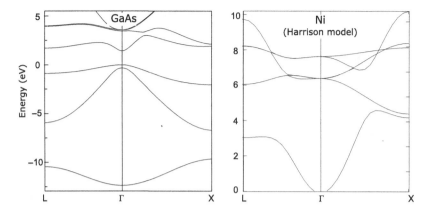

Figure 14.9. Left: band structure of GaAs calculated using the online tools at nanoHUB (nanohub.org) and a s, p, d, s∗ basis with parameters fit to experiment and results good enough for device simulations. Compare with the LDA bands in Fig. 17.7, where the gap is too small. Right: bands for the transition metal Ni calculated with Harrison's universal parameters [354, 615]. This is an example of tight-binding used to find approximate results that reveal characteristic features of the narrow d bands and wide s bands, as shown by comparing with LDA calculations in Figs. 16.4 and 16.5.

Simpler models with minimal s, p basis are not nearly so accurate. For example, calculations using the parameters in Harrison's "universal" table [354, 615] yield valence bands in reasonable agreement with experiment but the conduction bands may be qualitatively incorrect. The conduction bands can be greatly improved, e.g., for Si the correct indirect gap can be fit by including a second s symmetry state (called s∗) at high energy [627]. This is a way to mimic the effect found in full calculations that it is the admixture of Si 3d states that affects the shape of the conduction bands. The detailed fitting illustrated for GaAs in Fig. 14.9 uses both d and s∗ states. Description of bands using extended tight-binding models for many elements is given in the book [628].

The bands of transition metals have features that are dominated by localized d states. Tight-binding is a very natural approach for these states, as is also emphasized in the tight-binding LMTO method (Sections 16.7 and 17.6). The right side of Fig. 14.9 shows the bands of Ni calculated using Harrison's parameters [354, 615]. The bands have the right features, compared to full calculations (see Figs. 16.4 and 16.5) with narrow d bands and a wide s band, but the bands are somewhat too narrow and the model is inadequate to describe the other bands at higher energy. Nevertheless, such an approach is very valuable because it can describe the main features of entire classes on materials. Improved tight-binding descriptions of transition metal systems can be found in [629] and in the extensive handbook of band structures [628].

14.11 Total Energy, Force, and Stress in Tight-Binding

The total energy in any self-consistent method, such as the Kohn–Sham approach, can be written as in Eqs. (7.20) and (7.22) expressed as a sum of eigenvalues (Eq. (7.19)) plus the interaction of the nuclei and a correction needed to avoid double counting the interactions

$$E_{total} = \sum_i \varepsilon_i f(\varepsilon_i) + F[n]. \tag{14.20}$$

Here $f(\varepsilon_i)$ is the Fermi function and i labels eigenstates including the spin index; in a crystal, the sum is over all bands and $\mathbf{k}$ in the BZ. In terms of E_{pot} in Eq. (7.16), $F[n]$ is given by

$$F[n] \equiv E_{pot}[n] - \int d\mathbf{r}\, V_{KS}(\mathbf{r}) n(\mathbf{r})$$

$$= E_{II} - E_{Hartree}[n] + \int d\mathbf{r}\, [\epsilon_{xc}(\mathbf{r}) - V_{xc}(\mathbf{r})] n(\mathbf{r}), \tag{14.21}$$

where V_{KS} is the Kohn–Sham potential taken to be spin independent for simplicity.

In the tight-binding method, the parameterized hamiltonian matrix elements lead to the eigenvalues ε_i. How can the second term be included in such an approach? How can it be approximated as a function of the positions of the nuclei, even though the full theory defines $F[n]$ as a complicated functional of the density? An elegant analysis of the problem has been given by Foulkes and Haydock [361] based on the expression for the energy, Eq. (7.22). They used the variational properties of that functional and the choice of the density as a sum of neutral spherical atom densities, which is a good approximation [359, 360, 630]. It immediately follows that the difference of the Hartree and ion–ion terms in Eq. (14.21) is a sum of short-range, pair-wise interactions between atoms. The exchange–correlation term in Eq. (14.21) is also short range and it can be approximated as a pair potential. Thus we are led to the approximation that F can be expressed as a sum of terms $f(|\mathbf{R}_I - \mathbf{R}_J|)$ that depend only on distances to near neighbors, so that the total energy in the tight-binding approach can be expressed as

$$E_{total} = \sum_{i=1}^{N} \sum_{m,m'} c_{i,m}^* H_{m,m'} c_{i,m'} + \sum_{I<J} f(|\mathbf{R}_I - \mathbf{R}_J|), \tag{14.22}$$

where the eigenvectors are given by

$$\psi_i = \sum_m c_{i,m} \chi_m. \tag{14.23}$$

Finally, defining the density matrix,

$$\rho_{m,m'} = \sum_{i=1}^{N} c_{i,m}^* c_{i,m'}, \tag{14.24}$$

the energy can be written

$$E_{total} = \sum_{m,m'} \rho_{m,m'} H_{m,m'} + \sum_{I<J} f(|\mathbf{R}_I - \mathbf{R}_J|) = \text{Tr}\{\hat{\rho}\hat{H}\} + \sum_{I<J} f(|\mathbf{R}_I - \mathbf{R}_J|). \tag{14.25}$$

The added functions F or f can be found by fitting to additional information related to the total energy, e.g., the elastic constants or phonon frequencies. There can be no unique form of F, however, because of the fundamental ambiguity in separating the two terms in Eq. (14.20). An arbitrary function of the nuclear positions can be added to the hamiltonian matrix elements, shifting all eigenvalues rigidly with no change in the physics. The function F must be chosen consistent with the choice in the matrix elements. One choice of F is a sum of pair potentials, as is done in many models such as that of Harrison [354, 615] and models mentioned below. A different approach to is define eigenvalues that include a shift due to repulsive affects [631],

$$\varepsilon'_i \equiv \varepsilon_i + F/N_e. \tag{14.26}$$

Then the total energy is simply

$$E_{\text{total}} = \sum_{i=1}^{N} \varepsilon'_i = \text{Tr}\{\hat{\rho}\hat{H}'\}, \tag{14.27}$$

and the challenge is to parameterize the tight-binding matrix elements to describe both the eigenvalues and the total energy. Considerable success has been demonstrated with this form (see Section 14.12). In any case, the results depend on the availability of calculated and/or experimental energies and the adequacy of the forms chosen to represent different structures.

Forces can be found by taking derivatives of the energy using the force theorem just as in other methods. In tight-binding, the matrix elements are considered as functions of the positions of the nuclei and the expression follows from the condition that the energy is variational w.r.t. the density matrix $\hat{\rho}$ (Exercise 14.22). Taking the derivative of Eq. (14.25) with respect to the position of atom I leads to

$$\mathbf{F}_I = -\text{Tr}\left\{\hat{\rho}\frac{\partial \hat{H}}{\partial \mathbf{R}_I}\right\} - \sum_{J \neq I} \frac{\partial f(|\mathbf{R}_I - \mathbf{R}_J|)}{\partial \mathbf{R}_I}, \tag{14.28}$$

where the last term is absent if equation (14.27) is used. Pressure and stress are also straightforward since the stress tensor, Eq. (G.4), can be written as the sum of terms with the form

$$\sigma_{\alpha\beta} = -\frac{1}{\Omega}\frac{\partial E_{\text{total}}}{\partial u_{\alpha\beta}} = -\text{Tr}\left\{\hat{\rho}\frac{\partial \hat{H}}{u_{\alpha\beta}}\right\} - \sum_{J \neq I} \frac{\partial f(|\mathbf{R}_I - \mathbf{R}_J|)}{\partial u_{\alpha\beta}}. \tag{14.29}$$

The first term involves the derivative of the matrix elements with distance, and the final term is a sum of two-body contributions, as treated in Section G.2.

Examples of Tight-Binding Models for Total Energies

Perhaps the most useful and widely used tight-binding formulations are for carbon and silicon, for which there are several very successful parameterizations. For carbon, these are extremely useful for the amazing variety of structures found, including graphite, diamond, buckyballs, nanotubes, and amorphous and liquid carbon. For example, the form of Xu et al. [619] was constructed to fit the total energies of C in low-coordination structures, the chain, graphitic, and diamond. The potential has been used for many simulations and agrees well with other calculations, e.g., the bands of nanotubes shown in Fig. 14.7 and simulations of liquid carbon [619]. Other potentials are also successful and widely used, such as [632] and forms given in Section 14.12.

A number of parameterizations, e.g., those in [633–636], have been developed for Si and applied to problems involving defects, diffusion, and many other interesting properties. For example, Fig. 18.7 shows the results of an $O(N)$ molecular dynamics calculation [637, 638] of the structure of complex {311} defects done using the model of Kwon et al. [634], with checks on smaller cells with plane wave density functional theory calculations.

14.12 Transferability: Nonorthogonality and Environment Dependence

There is a basic difficulty in generating tight-binding models that can describe very different structures, in particular, those with different numbers of near neighbors such as open- and close-packed structures [615]. In models that have only two-center matrix elements, the values of the matrix elements must take into account effects of three-center terms. These effects change drastically between structures such as diamond (4 nearest neighbors that lead to $4 \times 3 = 12$ three-center terms) and a close-packed structure (12 nearest-neighbors that lead to $12 \times 11 = 132$ three-center terms). There are two primary approaches toward making tight-binding models that are transferable between such different structures. One is to define *environment-dependent tight-binding matrix elements*, the values of which depend on the presence of other neighbors. The other approach involves nonorthogonal tight-binding, which is more transferable than orthogonal forms. The reasons are brought out in Section 23.4, where it is clear that a small basis of orthogonal functions *must* extend to long range and be environment dependent in order to be accurate; on the other hand, nonorthogonal functions can accurately describe bands even if they are short-range atomic-like functions that are almost environment independent. Indeed, such functions are also the bread and butter of the local orbital methods of Chapter 15.

The different models can be exemplified by the extensive body of work of Cohen, Mehl, and Papaconstantopoulos [631], developed in many subsequent papers, which utilizes nonorthogonal tight-binding with environment-dependent matrix elements. This approach employs shifted eigenvalues ε_i', defined in Eq. (14.26), with the diagonal on-site matrix elements dependent on a sum of densities representing the neighboring atoms. The explicit form suggested [631] for the state at atom I with angular momentum l and spin σ is

$$H_{Il\sigma} = \sum_{n=0}^{3} b_{Il\sigma}^{(n)} \rho_{I\sigma}^{2n/3}, \qquad (14.30)$$

where the $b_{Il\sigma}^{(n)}$ are parameters and $\rho_{I\sigma}$ depends on the surrounding atoms

$$\rho_{I\sigma} = \sum_{J \neq I} \exp(-\lambda_{IJ\sigma}^2 R_{IJ}) \, f\left(\frac{R_{IJ} - R_0}{R_c}\right). \qquad (14.31)$$

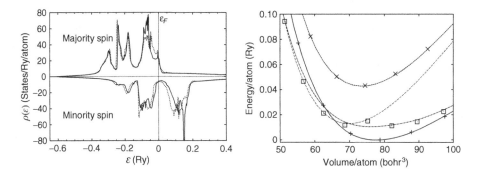

Figure 14.10. Left: density of states of ferromagnetic Fe in the bcc structure with $a = 5.40a_0$, comparing LAPW (solid) and fitted tight-binding (dashed curves). Right: total energy versus volume for Fe in various structures, comparing points from LAPW calculations (+, ferromagnetic bcc; square, ferromagnetic fcc; ×, paramagnetic bcc) and lines from the tight-binding calculations. The left figure and most of the total energies illustrate the quality of the fit. The total energy of the ferromagnetic fcc state was not fitted and the results are a test of transferability. For example, the tight-binding calculations reproduce the delicate collapse of the moment at $\approx 70a_0^3$ where the paramagnetic and ferromagnetic energies cross. From [640].

Here the exponential factor represents a density assigned to a neighboring atom, with the scales set by the parameters $\lambda_{IJ\sigma}$, and f is a cutoff factor taken to be the Fermi function. The spin dependence is needed for magnetic systems. The intersite matrix elements of the hamiltonian and overlap are each parameterized and have the same functional form that can be written

$$K_\gamma(R) = \left(\sum_{n=0}^{3} c_\gamma^{(n)} R^n\right) \exp(-g_\gamma^2 R) \, f\left(\frac{R - R_0}{R_c}\right), \qquad (14.32)$$

where $K_\gamma(R)$ denotes either hamiltonian and overlap matrix elements, and the subscript γ denotes sso, spo, spπ, etc.

This form has been applied to a large number of systems, including carbon [622], silicon [622, 639], and many transition metals. As an example, Fig. 14.10 shows the results for the density of states of ferromagnetic Fe in the bcc structure and the total energies in fcc and bcc structures, both paramagnetic and ferromagnetic. The tight-binding parameters are the same for all structures and are fitted with a total of 106 parameters[5] to the LAPW results at several volumes for paramagnetic and ferromagnetic bcc and paramagnetic fcc Fe. The total energies and moments for ferromagnetic fcc Fe were not fitted but reproduce correctly the energies, including collapse of the moment to form a paramagnetic state at $\approx 70a_0^3$. It is evident that the fits are extremely good and useful for analysis, given that the tight-binding calculations are orders of magnitude faster than the LAPW ones.

---

[5] There are 40 parameters each for intersite hamiltonian and overlaps, and 13 on-site terms (s, p, d_{e_g}, $d_{t_{2g}}$).

SELECT FURTHER READING

Classic paper and books with emphasis on tight-binding theory and methods:

Harrison, W. A., *Electronic Structure and the Properties of Solids* (Dover, New York, 1989).

Harrison, W. A., *Elementary Electronic Structure* (World Publishing, Singapore, 1999).

Slater, J. C. and Koster, G. F., "Simplified LCAO method for the periodic potential problem," *Phys. Rev.* 94:1498–1524, 1954.

Turchi, P. E. A., Gonis, A., and Columbo, L., eds., *Tight-Binding Approach to Computational Materials Science* (Materials Research Society, Warrendale, PA, 1998).

Papaconstantopoulos, D. A., *Handbook of the Band Structure of Elemental Solids: From Z = 1 to Z = 112*, 2nd ed. (Springer, New York, 2015).

Online resources:

Extensive tutorials, databases, and simnulation tools can be found at nanohub.org, and other sources are listed in Appendix R.

Exercises

14.1 See Appendix R for examples of codes that are available (or can be run online) for tight-binding calculations that are both pedagogical and able to treat problems such as semiconductor devices.

14.2 See many excellent problems (and solutions) on tight-binding bands, densities of states, and the meaning of the bands in the book by Mihaly and Martin [598].

14.3 Using translation invariance of the matrix elements, show that matrix elements of the hamiltonian with basis functions $\chi_{m\mathbf{k}}$ and $\chi_{m'\mathbf{k}'}$ are nonzero only for $\mathbf{k} = \mathbf{k}'$, i.e., the Bloch theorem, and derive the form given in Eq. (14.4).

14.4 Derive the factor $A_{m\mathbf{k}}$ in Eq. (14.3) required for the Bloch basis states $\chi_{m\mathbf{k}}(\mathbf{r})$ to be normalized. Show that $A_{m\mathbf{k}} = 1$ if the functions $\chi_m(\mathbf{r} - (\tau_m + \mathbf{T}))$ are orthonormal and that in general $A_{m\mathbf{k}} = (S_{m,m}(\mathbf{k} = 0))^{-\frac{1}{2}}$, where $S(\mathbf{k})$ is defined in Eq. (14.5). This relation is used in Exercise 23.2 in examples of Wannier functions.

14.5 Show that, in general, one has the relation $K_{ll'm} = (-1)^{l+l'} K_{l'lm}$ under interchange on the indices of the K matrix. This follows from a consistent definition of the orbitals.

14.6 Show that for an s band in a line, square lattice, and simple cubic lattice with only nearest neighbor hamiltonian matrix elements, the respective densities of states (DOS) have the forms shown schematically in Fig. 14.4. First determine the form for the DOS for the one-dimensional line analytically. Then use this result along with the fact that the bands, Eq. (14.13), are simply a sum of cosines for orthogonal directions to derive the form of the DOS for the square and simple cubic lattice. Show that the bands are divided into segments of width $4t$ as stated, and show that in three dimensions the DOS is exactly symmetric and flat in the central range.

14.7 Consider an s band in a square lattice with nearest neighbor matrix element t and one electron per cell. Show that the Fermi surface is a square as shown in Fig. 14.3 and there is a divergence in the density of states at the Fermi energy as shown in Fig. 14.4.

14.8   Derive the expression for the tight-binding s band $\varepsilon(\mathbf{k})$ in a simple cubic crystal. Assume the states are orthonormal and have hamiltonian matrix elements t_1, t_2, and t_3 for the first three neighbors. The bands for $t_2 = t_3 = 0$ are an approximation for the s-like conduction bands in CsCl, which has simple cubic structure. Compare with a calculated band structure in the literature.

14.9   Derive the expression for an s band $\varepsilon(\mathbf{k})$ in an fcc crystal with nearest neighbor hamiltonian matrix element t assuming the states are orthonormal. This should be a qualitative approximation for the lowest-conduction band in an fcc metal like Al or an ionic insulator like NaCl, which has the fcc structure. Compare with the nearly-free-electron bands in Fig. 12.1, Fig. 16.6, or calculated band structures in the literature. (Note that there is a relation to the expressions derived in Exercise 14.8 for second neighbors in a simple cubic lattice. Explain the relation in detail.)

14.10 Derive the expression for an s band in an hcp crystal with nearest neighbor hamiltonian matrix element t assuming the states are orthonormal. Assume the c/a ratio is the ideal value. Explicitly evaluate the bands in the direction along the c axis perpendicular to the hexagonal planes. Show that the lower and upper bands touch at the zone boundary,i.e., there is no gap at the zone boundary. Explain why this happens even though there are two atoms per primitive cell.

14.11 Derive expressions for p bands respectively in simple cubic and fcc crystals with nearest neighbor hamiltonian matrix element $t_{pp\sigma}$ and $t_{pp\pi}$. Compare with calculated bands in the literature for the Cl p state in CsCl and NaCl, respectively, to find reasonable values of $t_{pp\pi}$ and $t_{pp\sigma}$.

14.12 There is a close relation of p bands to the equations for phonons as expressed in Exercises 20.3–20.5. As an example, derive the explicit relation of tight-binding equations for p bands in Exercise 14.11 and the phonon dispersion curves in Exercise 20.6 for a nearest neighbor central potential model.

14.13 See Exercise 26.2 for bands in the two-site model in Fig. 26.2.

14.14 Consider a model like that in Exercise 26.2 except that the state on the B atom has p symmetry. For on-site energies $\varepsilon_A > \varepsilon_B$, this is a one-dimensional model for an ionic crystal like NaCl. From the symmetry of the crystal, show that two bands are formed from the s and p_x states decoupled from bands formed by the orthogonal p_y and p_z states. Assume the states are orthonormal and there are only nearest neighbor hamiltonian matrix elements of magnitude t. In terms of $\Delta = \varepsilon_A - \varepsilon_B$ and t, give analytic expressions for s–p_x bands $\varepsilon_i(k)$. Describe any simplifications in the expressions at $k = 0$ and the BZ boundary. Plot the bands in the Brillouin zone for the case $\Delta = 4t$ and show there is qualitative agreement with published bands of NaCl in the (100) direction. What values of Δ and t provide a reasonable description of NaCl bands? Suggest changes that would better describe the bands.

14.15 Show that the model of Cu and O states shown on the left-hand side of Fig. 14.8 leads to effective model nearest neighbor interactions between Cu states as shown on the right-hand side of the figure. Hint: construct a 3×3 matrix and diagonalize to find the highest band that corresponds to the band that crosses the Fermi energy in Fig. 17.8.

14.16 Show that the expression for bands with nonorthogonal basis orbitals, Eq. (14.15), is correct. The bands are no longer symmetric about $\varepsilon = 0$. Why is this? What is the physical interpretation? See the following problem for more general properties.

14.17 This problem is to analyze the general consequences of the overlap term in a nonorthogonal basis. Show that the effect of the overlap can be transformed to an orthogonal form, with the result that the hamiltonian matrix elements have infinite range, decaying exponentially. This is the correct result as shown by the decay of orthonormal Wannier functions in Chapter 23. Thus show that rigorously it can never be fully consistent to assume that the hamiltonian matrix elements are finite range and yet the orbitals are orthogonal. The same conclusion is found in Exercise 23.2.

14.18 Give a simple argument why "cosine" appeared many times in this chapter, whereas "sine" did not appear at all.

14.19 This problem relates to the structure and bands of a plane of graphene. Show that the Brillouin zone has the shape and orientation shown in the two cases in Fig. 14.7; also show that one of the K points is given by $k_x = 4\pi/3a, k_y = 0$ and find the coordinates of all six K points. Show that the bands indeed touch at all six K points.

14.20 Derive Eq. (14.17) for the bands in graphene.

14.21 Show that rolling of a graphene sheet to form $(n, 0)$ and (n, n) tubes leads, respectively, to the structures and Brillouin zones for the "zigzag" and "armchair" tubes that are shown in Fig. 14.7. For the armchair tubes show that the allowed states always include the states at the K point in graphene, so that simple mapping of graphene bands always leads to the prediction of metallic bands. For the zigzag tubes, give the conditions for which the allowed states include the graphene K point.

14.22 Show that the expressions for the force and stress theorems in tight-binding form, Eqs. (14.28) and (14.29), follow immediately from the condition that energy is minimum w.r.t. the coefficients in the wavefunctions.

14.23 Consider a heteropolar diatomic molecule with a total of two electrons. The hamiltonian is approximated by a orthogonal tight-binding model with one state per atom and hamiltonian matrix elements $H_{11} = E_1$, $H_{22} = E_2$, and $H_{12} = H_{21} = t(x)$, where x is the distance between atoms. Find the analytic expression for the ground-state energy E.
(a) Calculate the force on atoms 1 and 2 directly from the derivative of the analytic expression for the energy and also from the force theorem.
(b) Do the same for a generalized force $dE/d\Delta$, where $\Delta = E_1 - E_2$.

14.24 Find expressions for the valence and conduction band eigenvalues in a diamond-structure crystal at $\mathbf{k} = 0$ in terms of the matrix elements of the hamiltonian, the on-site energies E_s and E_p, and the matrix elements $H_{ss\sigma}$, $H_{sp\sigma}$, $H_{pp\sigma}$, and $H_{pp\pi}$. Do this in two steps. First, show that the eigenstates at $\mathbf{k} = 0$ are pure s or pure p. Next, use this fact to find expressions for the four eigenvalues for bonding and antibonding s and p states. Assuming four electrons per atom, identify the valence and conduction states and the gap between filled and empty states at $\mathbf{k} = 0$. Find numerical values for Si using Harrison's "universal" table [354, 615] and compare with bands that can be found in many references. The valence bands should be similar but the conduction bands are quite different.

15

Localized Orbitals: Full Calculations

Summary

As emphasized in the previous chapter, localized functions provide an intuitive description of electronic structure and bonding. This chapter is devoted to quantitative methods in which the wavefunction is expanded as a linear combination of localized atomic(-like) orbitals, such as gaussians, Slater-type orbitals, and numerical radial atomic-like orbitals. Such calculations can be very efficient; they can also be very accurate, as shown by the highly developed codes used in chemistry; and they provide the basis for creation of new methods, such as "order-N" (Chapter 18) and Green's function approaches. There is a cost, however: full self-consistent DFT calculations require specification of the basis, and the price paid for efficiency is loss of generality (in contrast to the "one basis fits all" philosophy of plane wave methods). Since details depend on the basis, we can only describe general principles with limited examples.

It is instructive to note that there are important connections to localized muffin-tin orbitals (MTOs) (Chapter 16), the linear LMTO method (Chapter 17). This has led to an "*ab initio* tight-binding" method (Section 17.7) in which a minimal basis of orthogonal localized orbitals is derived from the Kohn–Sham hamiltonian.

15.1 Solution of Kohn–Sham Equations in Localized Bases

The subject of this chapter is the class of general methods for electronic structure calculations in terms of the localized atom-centered orbitals defined in Section 14.1. The orbitals may literally be atomic orbitals, leading to the LCAO method or various atomic-like orbitals. These are powerful methods widely used in chemistry (see, e.g., [274, 275, 641–643]) and of increasing importance in condensed matter (see, e.g., [643–646]). Unlike the tight-binding methods of the previous chapter, these methods are fully "*ab initio*," i.e., they involve no parameters and solve the full Kohn–Sham or Hartree–Fock equations in a basis of orbitals. Unlike plane waves, however, the orbitals must be chosen for the given system to be accurate and efficient, and there is a problem of "overcompleteness" if one attempts to go to convergence. Nevertheless, there is great experience in constructing appropriate orbitals, so that localized orbitals are often the basis of choice, providing crucial understanding and calculational procedures that can be both fast and accurate with careful choice of orbitals.

In constructing desirable localized basis functions, there are two (often competing) considerations: reduction of the number of basis functions and ease of computation of the needed integrals. The former consideration means each function must be well tailored to the problem, which has led to many choices; only a few examples can be considered here. These competing requirements have led to the two general classes of orbitals discussed in Sections 15.2 and 15.4, which involve, respectively, analytic basis functions and numerical orbitals.

The goal of having a small basis leads to some overall conclusions that can be seen from general principles. The most common approach is to use atom-centered orbitals that are the products of radial functions and spherical harmonics defined in Eq. (14.8). The primary degrees of freedom are captured by a small set of l, m and radial functions, the shape of which must be optimized. It is often advantageous to choose a small set of radial functions that are optimal for a given environment. However, we shall concentrate on more general, flexible methods with a basis of several radial orbitals for each l, m channel.

Basis Functions: Naming Conventions and Examples

The common notation in the field is that multiple radial functions for the same l, m are denoted "multiple zeta," i.e., single-ζ or "SZ," double-ζ or "DZ," triple-ζ or "TZ" for one, two, or three radial functions, etc. The nomenclature arises from the use of ζ to denote the range of the basis functions. There are some general guidelines for the choice of optimal radial basis functions. For example, it is well known that in a molecule or solid, the localized orbitals typically are best described by atomic-like orbitals with shorter range and larger amplitude at the nucleus than in the atom [274, 643]. This is a direct consequence of the fact that the fundamental driving force for the binding of molecules or solids is the lowering of total energy because the electrons can be closer to the nuclei without paying as much cost in kinetic energy, compared to electrons in isolated atoms. Furthermore, the long-range exponential tails of the atomic orbitals are irrelevant or incorrect in regions that overlap other atoms. Thus basis orbitals tend not to be as extended as atomic functions. Different radial functions can be generated in many ways. One of the most elegant uses the ideas of the energy derivative of the wavefunction $\dot{\psi}$ derived in Section 17.2. Using the same principles as invoked in the LMTO approach and in norm-conserving pseudopotentials (Section 11.9), the change in the wavefunction in different environments is described to linear order by a combination of ψ and $\dot{\psi}$. Thus, $\psi, \dot{\psi}, \ddot{\psi}$, etc., form a possible set of localized orthonormal radial functions. On the other hand, it is often essential to include longer-range functions, e.g., to describe the decay of wavefunctions in the vacuum around molecules or at surfaces.

Since the environment of an atom in a molecule or solid is not spherical, in general the basis requires higher angular momenta than the minimal basis in the atom. The first such functions are called "polarization functions," which have angular momentum l^+ one unit larger than the maximum occupied state in the atom. It is pertinent to note that it is *not* appropriate to use the atomic state of angular momentum l^+. Such a state tends to be very diffuse and not relevant to the actual change in the function in the molecule or solid. A much better choice [274, 643, 646] is the actual change in the wavefunction of angular momentum l upon application of a weak electric field; this is a real "polarization function" that is localized and captures the essence of the lowest-order effect of the nonspherical environment. Inclusion of polarization functions in the basis is denoted by "P," e.g., "TZP" for triple-ζ with polarization functions.

The solution of the Kohn–Sham equations has exactly the same form as for the tight-binding equations of Chapter 14 except that the matrix elements must be computed explicitly and the potential must be derived self-consistently. Thus, as in all Kohn–Sham or Hartree–Fock methods, the key problem is to calculate the integrals for the matrix elements of the hamiltonian, the solution of the Poisson equation, and the generation of the potential in the self-consistency cycle. The ease with which one can do these operations is greatly affected by the choice of the basis functions, which has led to a number of methods. Furthermore, it is one of the major reasons for the development of standard sets of basis functions, as given in references such as [274, 275, 641, 642].

15.2 Analytic Basis Functions: Gaussians

By far the most useful and used basis functions for electronic structure calculations of molecules are gaussians multiplied by polynomials, apparently first adopted by Boys [647] and expounded upon in many texts such as [274, 275, 641, 642]. The great virtue is that *all* matrix elements can be computed analytically, greatly simplifying and speeding up calculations.[1]

Gaussians have the property, illustrated in Fig. 15.1, that the product of any two gaussians is a gaussian

$$e^{-\alpha|\mathbf{r}-\mathbf{R}_A|^2}e^{-\beta|\mathbf{r}-\mathbf{R}_B|^2} = K_{AB}e^{-\gamma|\mathbf{r}-\mathbf{R}_C|^2}, \tag{15.2}$$

where (Exercise 15.2)

$$\gamma = \alpha + \beta, \tag{15.3}$$

$$\mathbf{R}_C = \frac{\alpha\mathbf{R}_A + \beta\mathbf{R}_B}{\alpha + \beta}, \tag{15.4}$$

---

[1] The general principle that determines the usefulness of the analytic basis functions is the existence of an "expansion theorem" for the orbital centered on one site in terms of the basis functions on neighboring sites

$$\chi_\alpha(\mathbf{r} - \mathbf{R}) = \sum_{\alpha'} B_{\alpha\alpha'}(\mathbf{R}, \mathbf{R}') \, \chi_{\alpha'}(\mathbf{r} - \mathbf{R}'), \tag{15.1}$$

which greatly facilitates evaluation of the integrals. Examples of functions that possess this property are polynomials multiplied by gaussians ($r^\beta e^{-\alpha r^2}$), Slater-type orbitals ($r^\beta e^{-\alpha r}$), and spherical Bessel, Neumann, and Hankel functions. The advantages of the expansion theorem are emphasized in Chapters 16 and 17, where the expansion formulas for spherical Bessel, Neumann, and Hankel functions are crucial to the formulation of the KKR and (L)MTO methods.

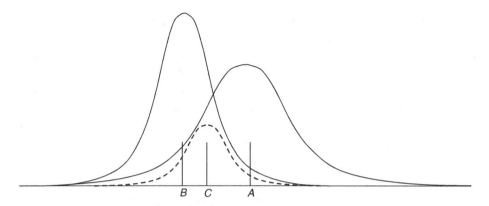

Figure 15.1. The overlap of two gaussians is another gaussian, with center, width, and weight given by Eqs. (15.2)–(15.5). From this basic result, matrix elements of kinetic energy and polynomials can be constructed by simple procedures (see text).

and

$$K_{AB} = \left[\frac{2\alpha\beta}{\pi(\alpha + \beta)} \right]^{3/4} e^{-\frac{\alpha\beta}{\gamma}|\mathbf{R}_A - \mathbf{R}_B|^2}. \tag{15.5}$$

Analytic relations can be found for gaussians multiplied by any polynomial of the radius by differentiating the above formulas using the fact that $(d/dx)e^{x^2} = 2xe^{x^2}$ and so forth up to any power. Similarly, it is straightforward to evaluate the laplacian applied to any gaussian multiplied by a polynomial. Thus for a basis set consisting of gaussians times polynomials (and spherical harmonics) centered at any site, *all multicenter integrals can be evaluated analytically.*

The expressions for the overlap and kinetic energy matrix elements can be easily derived (Exercise 15.3). The charge density $|\psi(\mathbf{r})|^2$, where $\psi(\mathbf{r})$ is a sum of such basis functions, is also readily expressed as a sum of gaussians. Potential matrix elements depend on the form of the potential. Two cases are of particular interest: if the potential is a sum of gaussians, the matrix elements are simply a sum of analytic three-center integrals; in addition, matrix elements of the Coulomb interaction with the nuclei and between the electrons can be computed analytically in terms of "Boys functions." Since these are the only integrals needed in Hartree–Fock calculations, gaussians have long enjoyed their status as the basis of choice. (For density functional theory some of the advantage is lost since the exchange–correlation potential is a nonlinear functional of the density that is not directly expressible as gaussians even if the density is a sum of gaussians.) Detailed expressions for the total energy, etc., are not given here, since they can be found from the expressions in Section 15.5. However, here is one major point. The Hartree–Fock equations can be written directly in terms of four center integrals, since the Coulomb matrix elements involve four orbitals. This is an effective approach for small systems; however, it scales as N_{orbital}^4. For large systems, especially for the Kohn–Sham equations, it is more effective to generate the total potential due to all occupied orbitals and to evaluate the matrix elements using grids [648].

The downside of gaussians is that they are eigenstates of a harmonic oscillator, which has little in common with potentials in a material made of atoms. For this reason there is great use of standardized "Slater-type orbitals" (STOs), which are sums of gaussians with fixed coefficients [274, 275, 641, 642]. An STO retains all the nice features of gaussians while at the same time having a form closer to an atomic-type orbital. In essence, an STO is a radial orbital that is expanded in a convenient basis; however, instead of allowing all the coefficients of the gaussians to vary, standard optimized sets have been generated that can be used to compare calculations with identical bases and to achieve different levels of accuracy with different bases.

15.3 Gaussian Methods: Ground-State and Excitation Energies

Electronic structure calculations using gaussians and STOs are extensively used in calculations for molecules that are far too numerous to attempt to summarize here and they are covered in great detail elsewhere. The methods are so successful that there are many commercially available codes adapted to molecular systems. One of the great advantages of gaussians is that Coulomb integrals can be computed analytically. It is for this reason that gaussians have been the workhorse of Hartree–Fock calculations and they are useful in problems that involve Coulomb integrals. This includes Hartree–Fock, exact exchange (EXX; Section 9.5), hybrid functionals (Section 9.3), and many-body GW calculations (see the companion volume [1]), all of which involve computation of exchange integrals.

Gaussian can also be efficient for periodic systems that are relevant for condensed matter. Many calculations have used the CRYSTAL code [644, 645], for example, using hybrid functionals for materials such as La_2CuO_4 [649] and UO_2 [650], where the correct insulating antiferromagnetic state is found unlike the usual LDAs or GGAs that predict a metallic nonmagnetic state. The GAUSSIAN code has been extended to periodic systems; an example is the bandgaps for crystals using hybrid functionals shown in Fig. 2.24 taken from [242].

The efficiency and flexibility of gaussians is illustrated by application to complex systems, e.g., to electronic bands for the dimerized Ge (1 0 0) surface [651] shown in Fig. 22.4, which compares LDA and GW quasiparticle bands with experimental results. Another example is linear-scaling "order-N" methods have been developed and applied to problems such as the structures of large RNA molecules [652] (see caption of Fig. 18.8).

15.4 Numerical Orbitals

Efficient algorithms using numerical orbitals can be constructed for either all-electron [113, 653, 654] or pseudopotential methods [646, 655]. Construction of the orbitals is very similar in either case. However, evaluation of the matrix elements in a local orbital calculation may be different. Since pseudopotentials and pseudofunctions are smooth, it is possible to carry out many integrals directly on a grid. For all-electron states, on the other hand, it is essential to have a method that accurately integrates the region around each nucleus where the wavefunctions vary rapidly.

Construction of Orbitals

Localized orbitals can be constructed from atomic-like programs with spherically symmetric potentials. It is possible to use the atomic orbitals themselves, but their long-range tails are not desirable. Since they are not really appropriate in a solid or molecule, the tails actually decrease the accuracy of the calculations and at the same time introduce troublesome long-range terms. It is more desirable to define shorter-range, more "compressed" orbitals that are better suited to the final application. The effects are relevant primarily for the valence states since the core states are localized and little affected by boundary conditions.

Many different procedures have been proposed for generation of compressed, short-range orbitals. Each involves modification of the atomic potential so that it is strongly repulsive at large distance. This makes the orbitals confined, which is more realistic and, in addition, has advantages for the calculation since fewer matrix elements need be included. However, such confined orbitals may not be sufficient, especially at surfaces, since they cut off the charge density in the vacuum in an unphysical way. Examples of types of potentials for generating orbitals that have been used for calculations on solids include those shown in Fig. 15.2. The orbitals [620] constructed by the hard-wall confining potential have the advantage that they

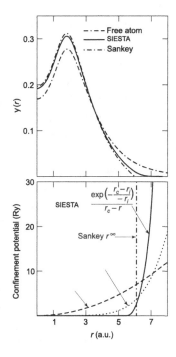

Figure 15.2. Confining potentials and pseudowavefunctions for numerical basis functions of Mg. The orbital is the solution with an atomic potential modified by the addition of a confining potential as shown at the bottom for various choices. The confining functions in the FHI-aims code [113] are similar to those in the SIESTA code [646] except that $1/(r_c - r)$ is replaced by $1/(r_c - r)^2$. Figure from [656].

are strictly localized, but the disadvantage that the second derivative has a divergence that makes a finite contribution to kinetic matrix elements. The other confining potentials are constructed to have chosen degrees of confinement versus smoothness.

Integrals Involving the Orbitals

Many of the needed integrals depend only on the positions of the atoms. Efficiency is not a premium for these terms because they can be calculated in advance and used later. This includes the overlap, kinetic energy matrix element, and nonlocal pseudopotential terms in Eqs. (14.2) and (14.1). The first two are two-center terms that are functions only of the distance R between the centers for the case where the angular momentum axis is along the line joining the centers. Rotation to treat the general case is given in Section 14.2. An effective procedure is to calculate the values at many discrete values of R and interpolate to find the matrix elements during a calculation. Thus we can view these matrix elements as known two-body functions,

$$S_{m,m'}(R) = \int d\mathbf{r}\, \chi_m^*(\mathbf{r})\chi_{m'}(\mathbf{r} - \mathbf{R}), \qquad (15.6)$$

and

$$T_{m,m'}(R) = \int d\mathbf{r}\, \chi_m^*(\mathbf{r})\frac{1}{2}\nabla^2 \chi_{m'}(\mathbf{r} - \mathbf{R}), \qquad (15.7)$$

where m denotes the atom type and orbital on that atom.

Potential terms are not so easy to express since they involve three centers (wavefunctions on two centers and the potential on a third). However, nonlocal pseudopotential terms have special features that can be used to advantage. First, nonlocal terms are fixed for each atom and never change during a calculation. Second, if a separable form (the Kleinman–Bylander form of Section 11.8, the "ultrasoft" pseudopotential of Section 11.11, or the PAW in Section 11.12) is used, then all three-center terms factorize into sums of products of two-center terms. These can be tabulated in advance as a function of distance.

Thus we are left with the problem of treating the matrix elements of the local potentials. This includes the full Kohn–Sham potential or any local parts of the potential. In a pseudopotential method, this can be treated exactly as is done with plane waves – sums on a grid. All operations on the grid, such as finding the density and exchange–correlation functions, can be done exactly as for plane waves. The only differences are that the wavefunctions in the local orbital basis must be transferred to the grid and the integrals are done by summing on the grid points to find the matrix elements

$$V_{m,m'}^{\text{local}}(\mathbf{T}) = \int d\mathbf{r}\, \chi_m^*(\mathbf{r} - \tau_m)V^{\text{local}}(\mathbf{r})\chi_{m'}(\mathbf{r} - (\tau_{m'} + \mathbf{T})). \qquad (15.8)$$

If the wavefunctions are smooth, for example, with pseudopotentials, the integral in Eq. (15.8) can be carried out on a regular grid [646]. In an all-electron method, the integration must be done carefully around each nucleus since the wavefunctions vary rapidly. One approach [657] is to use the "muffin-tin" partitioning of space illustrated in

Fig. 16.1, in which case the integration can be done on radial grids around each nucleus and uniform grids in the interstitial regions very much like the procedures in augmented methods. Another general approach is to break up the integral into overlapping domains using functions $\alpha_i(\mathbf{r})$ that together cover all space. If we define the normalized weight functions $w_i(\mathbf{r}) = \alpha_i(\mathbf{r})/\sum_j \alpha_j(\mathbf{r})$, any integral can be written as

$$\int d\mathbf{r} f(\mathbf{r}) = \sum_i \int d\mathbf{r}\, w_i(\mathbf{r}) f(\mathbf{r}). \tag{15.9}$$

Each integral on the right-hand side can be done on a different grid; in particular, radial grids can be used around each atom to deal with the rapid variations near the nucleus [653, 654, 658].

15.5 Localized Orbitals: Total Energy, Force, and Stress

Any of the expressions for the total energy in Section 7.3 can be used in a local orbital basis. The expressions can be written most compactly in terms of the density matrix $\rho_{mm'}$. If the eigenvectors are $\psi_i(\mathbf{r}) = \sum_m c_{im}\chi_m(\mathbf{r} - \mathbf{R}_m)$, then $\rho_{mm'}$ is given by

$$\rho_{mm'} = \sum_i f(\varepsilon_i)c_{im}^* c_{im'}. \tag{15.10}$$

In the present case, $\chi_m(\mathbf{r} - \mathbf{R}_m)$ are the localized basis functions, where m denotes both orbital α and site I, and $\mathbf{R}_m$ denotes the position of atom $\mathbf{R}_I$ on which orbital m is centered.

All quantities in the total energy expressions can be cast in terms of $\rho_{mm'}$. In particular, the sum of eigenvalues in Eqs. (7.17) and (7.19) is given by

$$E_s = \sum_{i=1}^N \varepsilon_i f(\varepsilon_i) = \sum_{mm'} \rho_{mm'}[H_{KS}]_{mm'}, \tag{15.11}$$

where $[H_{KS}]_{mm'}$ denotes matrix elements of the Kohn–Sham effective hamiltonian, and the spatial electron density is given by

$$n(\mathbf{r}) = \sum_i f(\varepsilon_i)|\psi_i(\mathbf{r})|^2 = \sum_{mm'} \rho_{mm'}\chi_m^*(\mathbf{r} - \mathbf{R}_m)\chi_{m'}(\mathbf{r} - \mathbf{R}_{m'}). \tag{15.12}$$

These are sufficient to determine the energy in any of the various expressions in the Kohn–Sham approach.

It is useful to be more specific because certain forms are advantageous in a localized basis [646, 659]. In particular, calculation of forces, stress, and Coulomb energies can be facilitated by the choice of the energy functional. Calculation of Coulomb terms (including the local pseudopotential term) can be done conveniently by grouping terms due to each nucleus (or ion) I with a compensating localized, spherical charge $n_I(\mathbf{r})$, as is described in Section F.4. Any convenient localized electron density can be used; if the orbitals are constructed from an atomic-like calculation, an obvious choice is the density resulting from these orbitals. The needed expressions for the Coulomb terms are given in Sections F.3 and F.4 in terms of the electron density written as

$$n(\mathbf{r}) \equiv \sum_I n_I(\mathbf{r}) + \delta n(\mathbf{r}), \tag{15.13}$$

which is equivalent to Eq. (F.18).

Inserting the expression for the Coulomb and local potential terms, Eqs. (F.19) and (F.20), into that for total energy Eq. (F.16), a convenient form[2] for the total energy is [646, 659]

$$E_{tot} = \sum_{mm'} \rho_{mm'}\left[T_{mm'} + V_{mm'}^{NL}\right] + \sum_{I<J} U_{IJ}^{NA}(|\mathbf{R}_I - \mathbf{R}_J|) + \sum_I U_I^{NA}$$

$$+ \int d\mathbf{r} V^{NA}(\mathbf{r})\delta n(\mathbf{r}) + \frac{1}{2}\int d\mathbf{r}\delta V_{\text{Hartree}}(\mathbf{r})\delta n(\mathbf{r})$$

$$+ \int d\mathbf{r}\epsilon_{xc}(\mathbf{r})n(\mathbf{r}). \tag{15.14}$$

Here U_I^{NA} denotes the potential energy of the neutral atom, $U_{IJ}^{NA}(|\mathbf{R}_I - \mathbf{R}_J|)$ is the classical interaction of two neutral atom densities, and the two terms involving $\delta n(\mathbf{r})$ are the first- and second-order changes in the potential energy due to the changes in the density in the solid. As discussed in Section F.4, the local part of the pseudopotential is included in U_I^{NA} and $V^{NA}(\mathbf{r})$. Since the exchange–correlation energy is a nonlinear functional of ρ, it cannot be divided in the same way and is left as the usual expression. Similarly, it is more convenient to express the first term that involves the density matrix directly as shown.

Force and Stress

Expression (15.14) is particularly appropriate for calculation of derivatives. The solution of the self-consistent Kohn–Sham equations leads to a minimization of the total energy with respect to the coefficients c_{im} in the wavefunction. Therefore the derivative of E_{tot} with respect to $\rho_{mm'}$ vanishes. Thus the derivatives of the first three terms in Eq. (15.14) can be considered as functions of the distances between the atoms, since $T_{mm'}$ and $V_{mm'}^{NL}$ are functions only of distances, as shown in Section 15.4. It is straightforward [646, 659] to express their contribution to the force on atom I in terms of derivatives that involve its position relative to other atoms $\mathbf{R}_I - \mathbf{R}_J$. The contribution of such two-body terms to the stress can be derived from the analysis of Section G.2. The fourth term is a constant that has no effect.

The fifth term involving V^{NA} contributes a term of exactly the same form

$$-\int d\mathbf{r}\frac{\partial V^{NA}}{\partial \mathbf{R}_I}\delta n(\mathbf{r}), \tag{15.15}$$

---

[2] The term δE_{II} in Eq. (F.16) is omitted here. It should be added to cancel an unphysical term if the extent of the smeared ion cores is allowed to be so large that they overlap.

since the "NA" terms move rigidly with the atom. For the stress this term can be included by scaling the density.

The last three terms all involve n or δn. They contribute to stress as in other formulas involving the density; however, their contributions to the force would vanish if the density had no explicit dependence on the atom positions. Indeed, this is the case for plane waves where the basis is not related to atom positions. However, the local orbitals are displaced with the atom; if the basis is not complete (Exercise 15.4), there are "Pulay corrections" due to the fact that the density changes to first order. These terms can be calculated using the fact that they arise from the change in $n(\mathbf{r})$ *for fixed* $\rho_{\alpha\beta}$. For any functional $F[n]$, the derivative is

$$-\frac{\partial F[n]}{\partial \mathbf{R}_I} = -\int d\mathbf{r} \frac{\delta F[n]}{\delta n(\mathbf{r})} \frac{\partial n(\mathbf{r})}{\partial \mathbf{R}_I}, \tag{15.16}$$

where (here $m \to \alpha, I$ to clarify the role of the position R_I)

$$\frac{\partial n(\mathbf{r})}{\partial \mathbf{R}_I} = \sum_{\alpha,\beta J} \left[\rho_{\alpha,\beta J} \frac{\partial \chi_\alpha^*(\mathbf{r} - \mathbf{R}_I)}{\partial \mathbf{R}_I} \chi_\beta(\mathbf{r} - \mathbf{R}_J) + \text{c.c.} \right]. \tag{15.17}$$

Since the functions χ_α are localized, the sum can be restricted to only include atoms J within some range of I.

15.6 Applications of Numerical Local Orbitals

Numerical local orbitals are used in all-electron [653, 654] and pseudopotential methods [646, 655], and they are convenient for both localized and extended systems. There are several examples in this book.

- Local orbital methods are one of the efficient ways to carry out all-electron calculations, taking into account the localized core states. This is the approach in the FHI-aims codes used for comparison of functionals in Table 2.1 where high accuracy is required.
- Local orbitals are particularly efficient for complicated systems with many atoms per cell or with vacuum regions, where plane waves become expensive to use. Linear-scaling calculations based on numerical orbitals have been developed and applied to problems such as quantum molecular dynamics, structural relaxation, and electronic states. An example using the SIESTA code is the application to a large DNA molecule, as illustrated in Fig. 18.8.
- They are also used in the real-time approach for time-dependent density functional theory for the C_{60} molecule and the optical spectra of LiF crystals.

15.7 Green's Function and Recursion Methods

One of the primary uses of local orbitals is Green's function-type methods that take advantage of the locality. For example, self-consistent calculations of localized defects in semiconductors were calculated using Green's functions to treat the infinite medium around the defect [660, 661]. Similarly, adatoms on metal surfaces have been treated using Green's function gaussian orbital codes [662, 663]. Although Green's function methods are widely used with simpler tight-binding hamiltonians (Chapter 18), these approaches have not been extensively used in fully self-consistent density functional theory calculations because of difficulties in calculation of the hamiltonian. Most self-consistent work has involved "supercell" methods (Section 13.5) and quantum molecular dynamics (Chapter 19) to treat such problems with periodic boundary conditions.

With the reemergence of interest in local nonperiodic methods and the advent of linear-scaling methods, there is now renewed interest in Green's function approaches. Indeed, many approaches described in Chapter 18 are variations of Green's function methods that utilize localized functions.

15.8 Mixed Basis

Mixed-basis methods utilize a combination of localized and delocalized bases, e.g., the appealing choice of gaussians and plane waves [664]. This gives the possibility of two widely used methods, plane waves and gaussians, and any linear combination. The hallmark of a mixed-basis approach is that both bases are used in the same region of space and the equations are expressed in terms of the usual overlap and hamiltonian matrix elements. The motivations have much in common with ultrasoft pseudopotential and projector augmented wave methods, which also include additive localized functions that augment the smooth functions near the nuclei. However, those methods transform the problem so that one needs to solve equations that involve only the smooth plane waves and the localized functions do not appear as explicit basis functions. This is a great advantage that allows much of the additional work to be done once and for all on the atomic reference state, simplifying the final calculation.

A different use of the mixed-basis idea is to utilize plane waves and Gaussians in *different spatial directions* [665]. This can be used to advantage, for example, for surfaces that are periodic in two directions but not in the third. Thus the basis functions become

$$\chi_{\mathbf{k},m,n}(\mathbf{r}) = e^{i(\mathbf{k}+\mathbf{G}_m)\cdot\mathbf{r}} e^{-\alpha|z-z_n|^2}, \tag{15.18}$$

which denotes a Fourier component $\mathbf{k} + \mathbf{G}_m$ in the x, y plane of the surface and multiplied by a gaussian centered at position z_n. A surface or interface can thus be represented by "layer" wavefunctions that are extended in the plane and centered on atomic layers.

SELECT FURTHER READING

Books and papers on localized orbital methods used in condensed matter applications (there is a vast literature for methods used for molecules in the chemistry community):

Delley, B., "From molecules to solids with the DMol3 approach," *J. Chem. Phys.* 113:7756–7764, 2000.

Eschrig, H., *Optimized LCAO Methods* (Springer, Berlin, 1987).

Orlando, V. R., Dovesi, R., Roetti, C. and Saunders, V. R., "*Ab initio* Hartree–Fock calculations for periodic compounds: application to semiconductors," *J. Phys. Condens. Matter* 2:7769, 1990.

Saunders, V. R., Dovesi, R., Roetti, C., Causa, M., Harrison, N. M., Orlando, R., and Zicovich-Wilson, C. M., *CRYSTAL 98 User's Manual* (University of Torino, Torino, 2003). See www.theochem.unito.it/, 2003.

Soler, J. M., Artacho, E., Gale, J., Garcia, A., Junquera, J., Ordejon, P., and Sanchez- Portal, D., "The SIESTA method for *ab intio* order-N materials simulations," *J. Phys.: Condens. Matter* 14:2745–2779, 2002.

Exercises

15.1 There are excellent open-source codes (see Appendix R) freely available online for DFT calculations using gaussian, such as CRYSTAL, and numerical orbitals, such as FHI-aims and SIESTA. These codes can be used for localized systems and crystals. (Plane waves can be used for localized systems also, but that is not appropriate for examples since many plane waves are needed. Real-space grids, wavelets, etc. also require large basis sets.) The codes often have tutorials similar to the exercises and also many examples for materials.

15.2 Show that the product of two gaussians is a gaussian as in Eq. (15.2), and derive the expressions for the coefficients in the product gaussian Eqs. (15.3)–(15.5).

15.3 Find the analytic formula for the kinetic energy matrix element between gaussian basis functions with spreads α and β and separated by displacement **R**.
(a) First consider only simple gaussians with s symmetry and not multiplied by powers of the radius.
(b) Then show that the formulas can be generalized to any l, m and power r^p by taking appropriate derivatives of expressions derived in (*a*). You do not need to work out all the detailed formulas, which can be found in texts.

15.4 Derive Eq. (15.17) using a chain rule and show that the right-hand side vanishes if the basis is complete. Hint: use the completeness relation.

15.5 Construct a simple computer program for a gaussian s band in one dimension. This entails calculating the overlap and hamiltonian matrix elements that are analytic if we assume the potential is also a sum of gaussians centered on each atom. Vary the band shapes from nearly-nearest-neighbor tight-binding-like given in Eq. (14.15) to nearly-free-electron-like.

15.6 Use the results for the eigenvectors from Exercise 15.5 to construct Wannier functions. Construct the atom-centered "maximally projected" form defined in Section 23.2 with the phase (sign) chosen to maximize the function on the central atom.
(a) Show that the function has positive and negative values (a plot is best) and it is longer range than the gaussian basis function.
(b) With a careful fit to the long-range behavior (the log of the absolute value) of the Wannier function, show it decays exponentially as a function of distance as claimed in Section 23.2.

16

Augmented Functions: APW, KKR, MTO

Summary

Augmentation provides a method of constructing a basis that is in some ways the "best of both worlds": the smoothly varying parts of the wavefunctions between the atoms represented by plane waves or other smoothly varying functions and the rapidly varying parts near the nuclei represented as radial functions times spherical harmonics inside a sphere around each nucleus. The solution of the equations becomes a problem of matching the functions at the sphere boundary. The original approach is the augmented plane wave (APW) method of Slater, which leads to equations similar to the pseudopotential and OPW equations but with matrix elements of a more complicated, energy-dependent potential operator. The disadvantage of augmentation is that the matching conditions lead to nonlinear equations, which has led to the now widely used linearized methods described in Chapter 17. The KKR method is a multiple-scattering Green's function approach that yields directly local quantities. The muffin-tin orbital (MTO) approach reformulates the KKR method, leading to physically meaningful descriptions of the electronic bands in terms of a small basis of localized, augmented functions.

16.1 Augmented Plane Waves (APWs) and "Muffin Tins"

The augmented plane wave (APW) method, introduced by Slater [61] in 1937, expands the eigenstates of an independent-particle Schrödinger equation in terms of basis functions, each of which is represented differently in the two characteristic regions illustrated in Fig. 16.1. In the region around each atom the potential is similar to the potential of the atom and the solution for the wavefunction is represented in a form appropriate to the central region of an atom. In the interstitial region between atoms the potential is smooth and the wavefunction is represented in a form appropriate to smooth variations coupling the atomic-like regions.

If the total effective potential is approximated as spherically symmetric $V_{\text{eff}}(\mathbf{r}) \rightarrow V_{\text{eff}}(r)$ within each sphere, and constant $V_{\text{eff}}(\mathbf{r}) \rightarrow V_0$ in the interstitial region, it is termed a "muffin-tin potential." This approximation is very appropriate for many problems and

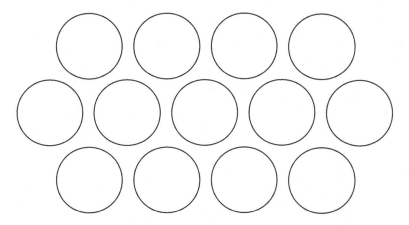

Figure 16.1. The "muffin-tin" division of space into intra-atomic spheres of radius S, and interstitial regions. This is the basis for representing wavefunctions differently in the different regions used in all augmented formulations. The muffin-tin potential approximates the potential in the two regions, but the division into spheres and interstitial regions is more general and can be applied to any potential. (The picturesque name derives from the fact that the figure looks like a pan for cooking muffins.)

allows for dramatic simplifications, since the wavefunctions can be represented in terms of the eigenstates in each region, i.e., spherical harmonics around each atom and plane waves in between. The entire problem is recast into a matching or boundary condition problem. It is most instructive to first discuss the APW approach for such idealized muffin-tin potentials; however, we emphasize that the Schrödinger equation with a general potential (and the full Poisson equation) can be solved using the APW basis. Generalization of the equations to "full-potential" problems is discussed in Chapter 17 after linearization is introduced.

In many ways, the APW approach brings together the "best of both worlds" with the ability to treat highly localized atomic-like states (e.g., core states) using atomic-like spherical functions and delocalized states using delocalized plane waves – the Bloch wavefunctions for the crystal illustrated in Fig. 4.11 are solved separately in the two regions in terms of the APW basis functions (defined in Eq. (16.2) and illustrated in Fig. 16.2). The disadvantage is the difficulty of matching the functions and solving the resulting nonlinear equations in this basis. We will first describe the original nonlinear methods, which illustrate many of the main ideas and the relations to other methods, OPW, pseudopotential, and KKR, etc., through their common features of describing the valence states in terms of the scattering properties of the atoms, which in turn are determined by the phase shifts. A separate chapter (Chapter 17) is devoted to linearized methods because of their conceptual and practical importance in transforming the augmented methods into more useful forms.

Just as in the plane wave (see Eq. (12.11)) or OPW (see Eq. (11.1)) method, each Bloch function $\psi_{i,\mathbf{k}}(\mathbf{r})$ is expanded in a set of basis functions labeled by the reciprocal lattice vectors $\mathbf{G}_m, m = 1, 2, \ldots,$

$$\psi_{i,\mathbf{k}}(\mathbf{r}) = \sum_m c_{i,m}(\mathbf{k}) \chi_{\mathbf{k}+\mathbf{G}_m}^{\mathrm{APW}}(\mathbf{r}). \tag{16.1}$$

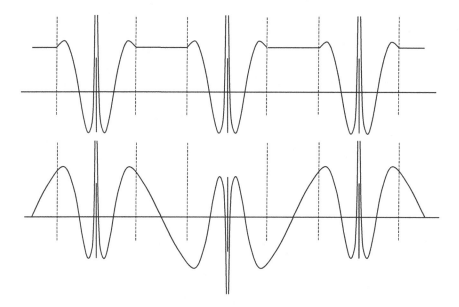

Figure 16.2. Schematic representation of the APW basis functions for $\mathbf{k} = 0$ (top) and the zone boundary (real part shown in bottom panel). The sphere boundaries are represented by the vertical dashed lines. In the interstitial region, each APW is a single plane wave. Inside each sphere the APW is a solution of the radial equation, with the boundary condition that it match the plane wave in value. A single APW has a discontinuous derivative at the boundaries. The solution for the eigenstate minimizes the final discontinuity in the derivative, leading to Bloch states like those illustrated in Fig. 4.11.

However, in the APW method each basis function $\chi_{\mathbf{k}+\mathbf{G}}(\mathbf{r})$ is represented as a single plane wave only in the interstitial region between atoms, and within a sphere of radius S around each atom the function is represented in spherical harmonics:

$$\chi_{\mathbf{k}+\mathbf{G}_m}^{\mathrm{APW}}(\mathbf{r}) = \begin{cases} \exp(i(\mathbf{k}+\mathbf{G}_m)\cdot\mathbf{r}), & r > S, \\ \sum_{Ls} C_{Ls}(\mathbf{k}+\mathbf{G}_m)\psi_{Ls}(\varepsilon,\mathbf{r}), & r < S, \end{cases}$$

where the compact notation for the wavefunction,[1]

$$\psi_{Ls}(\varepsilon,\mathbf{r}) = i^l Y_L(\hat{r})\psi_{ls}(\varepsilon,r), \tag{16.2}$$

is introduced to simplify the notation. The angular momentum is indicated by upper case $L \equiv l, m_l$, and $Y_L(\hat{r}) \equiv Y_{l,m_l}(\theta,\phi)$ are spherical harmonics, with r and $\hat{r}$ referred to an origin τ_s for each atom s in the unit cell. The function $\psi_{ls}(\varepsilon,r)$ is a solution of the radial

---

[1] The factor of i^l is introduced to simplify the coefficients that result when matching plane waves that have the factor i^l; see expansion Eq. (16.4).

Schrödinger equation regular at the origin at energy ε, i.e., it satisfies Eq. (10.12), written here keeping factors of $\hbar$ and m_e [2]

$$\left[\frac{\hbar^2}{2m_e} \left(-\frac{d^2}{dr^2} + \frac{l(l+1)}{r^2} \right) + V_s(r) - \varepsilon \right] r \psi_{ls}(r) = 0. \tag{16.3}$$

It is important here that ε is a variable and need not be an eigenvalue. Schematic forms of two $\chi_{\mathbf{k}+\mathbf{G}_m}^{APW}(\mathbf{r})$ are shown along a line through the atoms in Fig. 16.2.

The coefficients $C_{Ls}(\mathbf{k} + \mathbf{G}_m)$ are obtained by requiring the waves to match at the surface of the muffin-tin spheres, i.e., that the phase shifts match as described in scattering theory, Section J.1. Using the expansion given in Eq. (J.1),

$$e^{i\mathbf{q}\cdot\mathbf{r}} = 4\pi \sum_L i^l \, j_l(qr) \, Y_L^*(\hat{\mathbf{q}}) \, Y_L(\hat{\mathbf{r}}), \tag{16.4}$$

where $j_l(qr)$ are spherical Bessel functions, it follows that χ^{APW} is continuous at the sphere boundary if

$$C_{Ls}(\mathbf{K}_m) = 4\pi \, e^{i\mathbf{K}_m\cdot\tau_s} \, j_l(|\mathbf{K}_m|S_s) \, \frac{Y_L^*(\hat{\mathbf{K}}_m)}{\psi_{ls}(\varepsilon, S_s)}, \tag{16.5}$$

where $\mathbf{K}_m = \mathbf{k} + \mathbf{G}_m$. An APW is, by construction, discontinuous in slope on the muffin-tin boundary (see Fig. 16.2), a fact that must be taken into account when applying the kinetic energy operator.

Within the APW basis, the secular equation can be written

$$\sum_m \left\{ \langle m'|H - \varepsilon_{i,\mathbf{k}}|m\rangle + \langle m'|H^S|m\rangle \right\} c_{i,m}(\mathbf{k}) = 0, \tag{16.6}$$

where

$$\langle m'|H - \varepsilon_n(\mathbf{k})|m\rangle = \int_{cell} d^3r \, \chi_{\mathbf{k}+\mathbf{G}_{m'}}^*(\mathbf{r}) \, [H - \varepsilon_n(\mathbf{k})] \, \chi_{\mathbf{k}+\mathbf{G}_m}(\mathbf{r}), \tag{16.7}$$

and the discontinuity is incorporated into the integral over the sphere surface(s) using Green's identity (Exercise 16.2)

$$\langle m'|H^S|m\rangle = \int_S dS \, \chi_{\mathbf{k}+\mathbf{G}_{m'}}^*(\mathbf{r}) \left[\frac{\partial}{\partial r} \chi_{\mathbf{k}+\mathbf{G}_m}(\mathbf{r}^-) - \frac{\partial}{\partial r} \chi_{\mathbf{k}+\mathbf{G}_m}(\mathbf{r}^+) \right]$$

$$= \int_S dS \chi_{\mathbf{k}+\mathbf{G}_{m'}}^*(\mathbf{r}) \left[\frac{\partial}{\partial r} \ln \chi_{\mathbf{k}+\mathbf{G}_m}(\mathbf{r}^-) - \frac{\partial}{\partial r} \ln \chi_{\mathbf{k}+\mathbf{G}_m}(\mathbf{r}^+) \right] \chi_{\mathbf{k}+\mathbf{G}_m}(\mathbf{r}^-), \tag{16.8}$$

where $+ (-)$ indicates just outside (inside) the sphere.

---

[2] The factor $\hbar^2/2m_e = \frac{1}{2}$ in Hartree atomic units, where $\hbar = m_e = e = 1$ is used in this text. In the present chapter and the next, $\hbar$ and m_e are explicitly indicated where needed to avoid confusion with expressions in the literature, since many authors use "Rydberg atomic units," where $\hbar = 2m_e = e^2/2 = 1$.

One way to proceed is to solve the secular equation (16.7) in terms of matrix elements of the hamiltonian and the overlap matrix, just as for any nonorthogonal orbital. However, one can take advantage of the fact that the basis functions are not fixed but instead are chosen to satisfy the Schrödinger equation inside each muffin-tin sphere at energy ε. Thus the integral Eq. (16.7) is zero inside each sphere and needs to be evaluated only in the interstitial region where the hamiltonian is just the kinetic energy operator and χ is a plane wave. All the information about the way each atom affects the bands is incorporated into the boundary terms, i.e., boundary conditions on the plane waves. This is, of course, not surprising since the wavefunction both inside and outside the spheres must each obey the Schrödinger equation in their respective regions subject to the boundary conditions.[3]

Following this approach, it is straightforward to cast the APW equation (16.6) in a form identical to that in plane wave Fourier methods,

$$\sum_{\mathbf{G}} \left\{ \left[\frac{\hbar^2}{2m_e}(\mathbf{k}+\mathbf{G})^2 - \varepsilon_{i,\mathbf{k}} \right] \delta_{\mathbf{G}'\mathbf{G}} + V_{\mathbf{G}',\mathbf{G}}^{\text{APW}}(\varepsilon_k,\mathbf{k}) \right\} c_{i,\mathbf{G}}(\mathbf{k}) = 0, \qquad (16.9)$$

where the first term is the usual kinetic energy for a plane wave extended throughout the cell including the sphere, the energy is relative to the constant in the muffin-tin potential, and all effects due to the potential in the sphere are collected into an APW "potential" $\hat{V}^{\text{APW}}$, which is an operator that is both nonlocal and energy dependent. The matrix elements of $\hat{V}^{\text{APW}}$ for a sphere at $\tau = 0$ in the unit cell are [157, 666]

$$V_{\mathbf{G}',\mathbf{G}}^{\text{APW}}(\varepsilon_k,\mathbf{k}) = -\frac{4\pi S^2}{\Omega_{\text{cell}}} \left(\frac{\hbar^2}{2m_e}|\mathbf{k}+\mathbf{G}|^2 - \varepsilon_k \right) \frac{j_1\left(|\mathbf{G}-\mathbf{G}'|S\right)}{|\mathbf{G}-\mathbf{G}'|}$$

$$+ \frac{\hbar^2}{2m_e} \frac{4\pi S}{\Omega_{\text{cell}}} \sum_l \left\{ (2l+1) P_l(\cos(\theta_{\mathbf{G}\mathbf{G}'}) j_l\left(|\mathbf{k}+\mathbf{G}'|S\right) j_l\left(|\mathbf{k}+\mathbf{G}|S\right) \right\}$$

$$\times \Delta D_{l,\mathbf{G}}(\varepsilon_k), \qquad (16.10)$$

with $\theta_{\mathbf{G}\mathbf{G}'}$ the angle between vectors $\mathbf{k}+\mathbf{G}$ and $\mathbf{k}+\mathbf{G}'$, and S the sphere radius. For a crystal with more than one sphere centered at positions τ_s, it is simple to show that the potential is a sum of terms with phase factors $\exp(i(\mathbf{G}-\mathbf{G}')\cdot\tau_s)$ just as in the plane wave method (Eq. (12.16) and Exercise 12.2). The first term in the APW potential operator Eq. (16.10) subtracts the kinetic energy for that part of the plane wave inside the spheres (see Loucks [666], p. 32–33), and the last term[4] includes all the effects of the atoms in

---

[3] This is exactly the same condition as used in Chapter 11 where the pseudowavefunctions were shown to equal the true wavefunctions in the outer part of the atom, so long as the eigenvalue was the same and the wavefunction satisfied the boundary conditions at the sphere boundary. Furthermore, the specification of the boundary condition in terms of the logarithmic derivative in Eq. (16.11) is the same as in Eqs. (11.20) or (J.5); here the evaluation is done at the muffin-tin boundary and with the assumption that the total potential has the spherical muffin-tin form.

[4] Another form slightly different and more convenient for computation is given by Loucks [666], p. 37).

terms of the difference of the dimensionless logarithmic derivative $\Delta D_{l,\mathbf{G}}(\varepsilon)$ from that of an empty sphere

$$\Delta D_{l,\mathbf{G}}(\varepsilon) = \left[r\frac{\mathrm{d}}{\mathrm{d}r}\ln\psi_l(\varepsilon,r) - r\frac{\mathrm{d}}{\mathrm{d}r}\ln j_l\,(|\mathbf{k}+\mathbf{G}|r) \right]_{r=S}, \tag{16.11}$$

which follows from Eq. (16.8) for the boundary "kink" term, with the function just inside the sphere given by $\psi_l(\varepsilon,r)$ and the function just outside by the partial wave component of the unscattered plane wave j_l. (The normalization is not needed for the logarithmic derivative.) It is interesting that the "potential" operator involves $\frac{\hbar^2}{2m_e}$; this is because it really is a "matching operator."

16.2 Solving APW Equations: Examples

The APW equations are more difficult than the usual independent-particle equations that are linear in energy ε, such as Eq. (12.9) for plane waves or Eq. (14.7) for localized orbitals, where all the eigenvalues and eigenvectors can be determined at once from a single diagonalization. Instead, the APW equations must be solved separately for each eigenstate as follows:

- Solution of the matrix equation (16.9) has exactly the same form as the usual linear equations, except that the potential operator depends on the logarithmic derivatives in Eq. (16.11) that are functions of the energy $\varepsilon = \varepsilon_{i,k}$, which are not known in advance and are different for each band.
- In order to find the logarithmic derivatives, the radial equations (16.3) for $r\psi_l(\varepsilon,r) \equiv \phi_l(\varepsilon,r)$ must be solved for each band energy $\varepsilon_{i,k}$, individually. However, $\varepsilon_{i,k}$ are established only in conjunction with solution of the plane wave equations (16.9). In general, this requires "root tracing," i.e., searching for the zeros of the determinant on the APW matrix given in Eq. (16.9). This may be done by fixing ε and varying k or vice versa.
- There can be simplification in some cases, e.g., highly localized states, such as core states that are completely contained in the sphere, are fully specified by $\psi_l(\varepsilon,r)$ and there is no $\mathbf{k}$ dependence and it can often be considered to be the same as in an atom. This is termed the "frozen-core approximation."

Illustrative Examples

Two limiting cases illustrate the power and generality of the APW method: the nearly-free-electron case and the opposite limit in which the spherical potential has a strong resonance. It is important that, despite the artificial division of space, the free-electron case is solved trivially. If the potential inside the muffin-tin sphere is set to be the same constant value as in the interstitial, exact solutions inside the spheres are spherical Bessel functions j_l, in which case the *difference* in logarithmic derivatives vanishes $\Delta D_l(\varepsilon_k) = 0$.

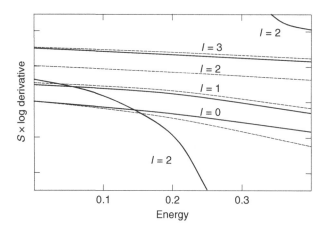

Figure 16.3. Logarithmic derivatives of the radial wavefunctions in Cu, defined in Eq. (16.11) as a function of energy in Hartrees for different angular momenta l, evaluated at a sphere radius $S = 2.415a_0$ appropriate for metallic Cu. For comparison are shown the free-electron curves (Exercise 16.5). The $l = 2$ d channels show strong resonance, leading to narrow bands, whereas the $l = 0, 1, 3$ channels reveal the nearly-free-electron behavior. These are essentially the same as for the Chodorow potential [668] used in the pioneering work of Burdick [669]. From Kubler and Eyert [157], who attribute the figure to Slater and Mattheiss

It follows immediately that the eigenvalues of Eq. (16.9) are just those for free electrons $\varepsilon_k = \frac{1}{2}|\mathbf{k} + \mathbf{G}|^2$. If the phase shift $\Delta D_l(\varepsilon_k) \neq 0$ but is small, then the dispersion ε_k will be only slightly modified. This is the "nearly-free-electron case" and the APW method clarifies an important point: this *does not necessarily mean the potential is small or that the wavefunctions are close to a single plane wave. The difference in phase shift can be small even for strong potentials.* Explicit cases, where actual phase shifts are close to free-electron values, are for Na ([484], p. 242) and many examples in [667]. Logarithmic derivatives for Cu are presented in Fig. 16.3, which shows that the phase shifts for $l = 0, 1, 3$ are very close to the free-electron values (see Exercise 16.5). Thus the APW approach explains the nearly-free-electron character of bands in many materials that are weak scatterers just as well as the OPW or pseudopotential methods.

The opposite limit is a resonance at energy ε_0 where the phase shift becomes large. In an isolated atom, the logarithmic derivative $D(\varepsilon)$ evaluated at large radius diverges at $\varepsilon = \varepsilon_0$, signifying a bound state at that energy. (This is one of the standard methods to find bound states in actual atomic programs.) In a crystal, the fact that the phase shift at radius S changes rapidly with energy means that the Bloch boundary conditions for different $\mathbf{k}$ can be satisfied with only small changes in energy, i.e., a band $\varepsilon(\mathbf{k})$ with only a small dispersion. In the case of Cu, the $l = 2$ logarithmic derivative disperses rapidly corresponding to the d bands in Cu, which are much narrower than the s–p bands. In general, the bands start as parabolic and each resonance introduces a new band, which has the physical interpretation of each atomic state broadening into a band in the crystal.

Calculations of Bands Using the APW Method

The power of the APW approach was first fully realized after the advent of electronic computers, in particular, for the first accurate calculations of bands of transition and rare earth metals. A well-known early example is the band structure of Cu calculated by Burdick [669] in 1963, which are in very good agreement with measurements from angle-resolved photoemission experiments by Thiry et al. [670] in 1979. The logarithmic derivatives in the APW equations were calculated using the Chodorow potential [668], derived in 1939 for the Cu atom, and are essentially the same as shown in Fig. 16.3. The impressive agreement between measured and calculated energies is due to two factors: (1) the potential was modeled as a sum of atomic potentials fitted to the atom and (2) Cu has a filled (closed-shell) d band and a wide s–p band, a case in which independent-particle methods are expected to work well.[5] In general, one should exercise caution in identifying Kohn–Sham eigenvalues as excitation energies and one should not expect such agreement.

Bands for the entire series of 3d transition metals are presented in Fig. 16.4, which shows narrow d bands crossing the wider s–p bands. The 3d bands are much broader than the 4f bands of the lanthanides but narrow enough to indicate that many atomic-like properties carry over to the solid. At the time this work was done in 1964, a central issue was the choice of the potential. Mattheiss used overlapping atomic potentials derived from Hartree–Fock density and Slater's local approximation for exchange [672]. This is very similar to a Kohn–Sham calculation but the exchange differs from the LDA by a factor of 4/3, and correlation is not included. In hindsight, this choice has been found to have some justification (see Fig. 8.6).

The transition metals have partially occupied d bands. This leads to their magnetic behavior and correlations among the d electrons. Nevertheless, independent-particle methods are in many ways adequate to describe the basic properties of these metals [673], e.g., the prediction of magnetism from the Stoner criterion, as illustrated in Fig. 2.8 [104, 157, 673]. At the end of the series is the noble metal Cu in which the d bands are filled but only slightly below the Fermi energy, which leads to its closed-shell, nonmagnetic behavior and its yellow color. The total energies for the series are remarkably well described by the local density approximation, as illustrated in Fig. 2.2.

Looking at the bands in more detail, we see that in all cases the lowest band is minimum at the Γ point ($\mathbf{k} = 0$) in the BZ, with energy defined to be $\varepsilon = 0$; the label Γ_1 designates the same symmetry as an s wave function ($l = 0$). The nearly parabolic ε_k curve is modified because it mixes with narrower bands (examples of a resonance) that start at Γ with labels $\Gamma_{25'}$ and Γ_{12}, which are labels for d ($l = 2$) states in cubic symmetry: $\Gamma_{25'}$ is threefold degenerate and transforms under rotations like xy, yz, and zx, whereas Γ_{12} is twofold degenerate and transforms like $x^2 - y^2$ and $2z^2 - x^2 - y^2$. The bands labeled Δ_5 and Λ_3 are twofold degenerate d states along the lines shown. At higher energy around the Fermi level, the bands are also approximately parabolic; this is the feature that explains why Cu

---

[5] The bands are also expected to be predicted qualitatively by density functional theory in LDA or GGA approximations, except that the energies of the filled d states will be too shallow and the s–p bands too broad in accordance with experience on many systems.

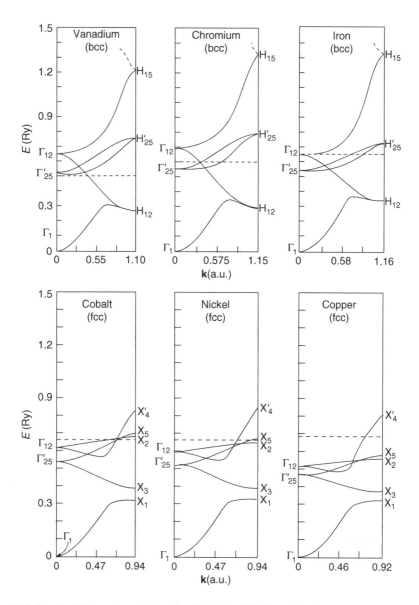

Figure 16.4. Bands of 3d metals calculated by Mattheiss [671] in 1964 using the APW method. The figures show the narrow d bands crossing the wide s band and the progression of band filling across transition series. At the top are elements with body centered cubic (bcc) and at the bottom face centered cubic (fcc) structures. The bands are shown along the lines labeled Δ from Γ–H and Γ–X respectively, in the Brillouin zones shown in Fig. 4.10. The potential was derived in a clever way that preceded DFT as described in the text.

is a good electrical conductor. The states $X_{4'}$ at $\varepsilon \approx 0.80$ Ry and $\Lambda_{2'}$ at $\varepsilon \approx 0.61$ Ry have labels that designate p symmetries ($l = 1$), which is expected for a free-electron band, and the eigenvalues are quite close to free-electron energies at the density of one electron per Cu atom as discussed in Exercise 16.5.

The bands ε_k for the non-d states are close to free-electron bands; however, the wave-functions are far from single plane waves. Although they are close to a single plane wave between the atomic spheres, inside each sphere the wavefunctions have all the oscillations characteristic of atomic 4s and 4p wave functions. The point is that a single plane wave joins smoothly onto the solution of the Schrödinger equation inside the sphere, illustrated in Fig. 4.11, so that the energy is essentially the same as that of the plane wave. In addition, the s–p states hybridize with the d states, which act like a resonance in the scattering of the s electrons from the atoms. This is the basis of the "s–d model" [674], which describes the s–p bands near the energies of the d states and the resulting dispersion in the narrow d bands (Exercise 16.6).

As an example of spin-dependent bands in a ferromagnet, Fig. 16.5 shows the bands for Ni in the fcc structure calculated by Connolly [675]. The solid lines show the majority-spin and the dashed lines the minority-spin bands calculated with the local spin-density formalism of Kohn and Sham. Connolly also found the bands using the Slater local exchange and concluded that it gives much poorer agreement with experiment. The larger exchange causes significant changes in the bands, particularly around the L point.

It should be emphasized, however, that density functional theory approximations often lead to wrong predictions, especially for more strongly correlated electron systems such

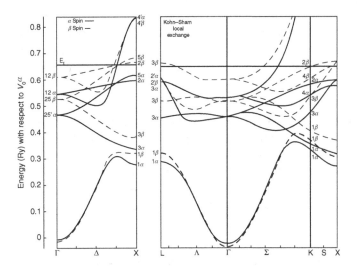

Figure 16.5. Spin-polarized bands of ferromagnetic Ni in the fcc structure [675]. Solid lines indicate the majority-spin and dashed lines the minority-spin bands. The numbers indicate symmetry labels at high-symmetry points. This figure represents the results using the Kohn–Sham local spin density approximation. (The Slater local exchange gives poorer agreement with experiment, especially around the L point, as shown in [675].)

as the rare earths and the transition metal oxides [240]. In a nutshell, methods that start from the homogeneous gas (LDAs and GGAs) tend to predict solutions that are too much like the gas – nonmagnetic and metallic – whereas methods that involve Hartree–Fock-exchange (Hartree–Fock itself and exact exchange tend to predict solutions that are too much like Hartree–Fock – too magnetic and insulating. Methods such as self-interaction correction (SIC) and "DFT+U" can apparently describe aspects of strongly correlated electron systems, but at the cost that the functionals are not universal. Hybrid functionals (Chapter 9) that combine density functionals with a fraction of Hartree–Fock-like exchange have proven to be successful in many cases. The issues related to correlation in 3d and 4f materials is a major topic of [1].

Practical Aspects

How large is the secular equation? The number of plane waves is determined by the fact that they must accurately represent the variations in the wavefunctions in the interstitial region. These are determined by the fact that they must match the solutions inside the spheres, which is incorporated into the logarithmic derivatives in Eq. (16.9). Thus, in general, we expect the number of plane waves (i.e., the number of APWs) to be comparable to the number of plane waves needed in a norm-conserving pseudopotential calculation, so long as the atoms have relatively smooth orbitals and large interstitial regions (like Si). However, unlike norm-conserving pseudopotentials, the number of basis functions does not increase as the states become more localized around the atoms (e.g., 3d states in transition metals or 4f in rare earths) since the augmentation takes care of the regions around the atoms. Indeed, roughly 40 APW basis functions are needed for each atom in the unit cell for transition metals [157]. In this respect, APW methods are more closely related to "ultrasoft pseudopotential" and PAW methods (Chapter 11), which can also describe localized states, since they add localized functions and use plane waves only for the smooth part. In addition, spherical harmonics with high angular momenta are required for accurate description of the bands (see also Section 17.4).

The APW method is not restricted to muffin-tin potentials and it can be extended to general potentials [107]. The APW basis can be defined as before (using an effective muffin-tin potential determined by averaging the full potential) but now one must calculate matrix elements of the full potential. The solution has the same general form as Eqs. (16.9) and (16.10), but it becomes more complicated because the matrix elements are no longer diagonal in angular momentum in the sphere nor in momentum in the interstitial region. The extension becomes much more feasible using linearized methods and further discussion is given in Section 17.10.

16.3 The KKR or Multiple-Scattering Theory (MST) Method

In the words of J. Ziman [676]:

> From mathematical point of view, the most refined method of calculating energy band structures is the subtle procedure invented independently by Korringa [677] and Kohn and

Rostoker [678]. This method is indeed so fundamental that it is to be found in all its essentials in a study by Rayleigh [679] [in 1892] of the propagation of sound waves through an assembly of spheres.

The KKR method, also called "multiple-scattering theory" (MST) or Green's function method, finds the stationary values of the inverse transition matrix T rather than the hamiltonian. This is the method used in the pioneering work of Moruzzi and coworkers [103, 104], highlighted in Section 2.3, that first established the efficacy of density functional theory for calculation of properties of close-packed metals. In addition, KKR is the method of choice for most calculations on liquids, disordered systems, and impurities in various metallic and nonmetallic hosts. Probably the most important features of the KKR or Green's function formulation are (1) it separates the two aspects of the problem, the structure (positions of the atoms) from the scattering (chemical identity of the atoms) and (2) Green's functions provide a natural approach to a localized description of electronic properties that can be adapted to alloys and other disordered systems.

Here we consider only muffin-tin potentials, where each site can be viewed as a spherical scatterer; and electrons propagate between sites with the free propagator or Green's function. This greatly simplifies the equations and was the basis of all KKR calculations until the development of full potential methods [680, 681]. The problem of overlapping potentials is subtle and the reader is referred to papers in the collection in [682] and references given there.

A Green's function G describes propagation of a system from one event to another [683], e.g., $G(\varepsilon, \mathbf{r}, \mathbf{r}')$ describes the system with a particle added at point $\mathbf{r}$ and removed at $\mathbf{r}'$ with energy ε. In terms of a reference Green's function G_0 (for example, the free-particle propagator given in Eq. (16.18) below) and scattering matrix elements t, representing single scattering events from any of the atoms in the system, the full Green's function can be written in schematic form as

$$G = G_0 + G_0 t G_0 + G_0 t G_0 t G_0 + \cdots$$
$$= G_0 + G_0 t G \Rightarrow$$
$$G = (G_0^{-1} - t)^{-1}. \tag{16.12}$$

Similarly, one can sum the series to write G as

$$G = G_0 + G_0 T G_0, \tag{16.13}$$

where T is the full multiple-scattering matrix for the entire system

$$T = t + t G_0 t + t G_0 t G_0 t + \cdots$$
$$= t + t G_0 (t + t G_0 t + \cdots)$$
$$= t + t G_0 T \Rightarrow$$
$$T = (t^{-1} - G_0)^{-1}. \tag{16.14}$$

The stationary states of the system are given by the poles of G or T as functions of ε and hence are obtained from the zeros of the determinant[6]

$$\det(t^{-1} - G_0) = 0. \tag{16.15}$$

For independent-particle electronic structure problems with hamiltonian $\hat{H} = -(\hbar^2/2m_e)\nabla^2 + V_{\text{eff}}(\mathbf{r})$,[7] a convenient starting point is to take G_0 to be the free Green's function. It is useful to first give the well-known solution for the Helmholtz equation $(\nabla^2 + \kappa^2)g(\mathbf{r} - \mathbf{r}') = \delta(\mathbf{r} - \mathbf{r}')$, for which the real solution is

$$g(x) = -\frac{1}{4\pi}\frac{\cos(\kappa x)}{x}, \tag{16.16}$$

where $x = |\mathbf{r} - \mathbf{r}'|$. Thus the Green's function for the Schrödinger equation with $V = 0$ satisfies

$$\left[-\frac{\hbar^2}{2m_e}\nabla_r^2 - (\varepsilon - V_0)\right]G_0(\varepsilon, \mathbf{r} - \mathbf{r}') = \delta(\mathbf{r} - \mathbf{r}'), \tag{16.17}$$

which has the solution

$$G_0(\varepsilon, x) = \frac{2m_e}{\hbar^2}\frac{1}{4\pi}\frac{\cos(\kappa x)}{x}, \tag{16.18}$$

where $(\hbar^2/2m_e)\kappa^2 = \varepsilon - V_0$ and V_0 is the "muffin-tin zero" reference energy. For positive energies ε, $G_0(\varepsilon, x)$ is a slowly decaying oscillatory function; for negative ε, it decays exponentially.

Within the muffin-tin spherical approximation, the scattering amplitude $t(\varepsilon)$ of an electron from each sphere conserves angular momentum $L \equiv \{l, m\}$ referred to the center of that sphere. Because the scattering is unitary and independent of m [684, 685], $t_l(\varepsilon)$ and can be written in terms of the phase shift $\eta_l(\varepsilon)$ (see Section J.8)

$$t_l(\varepsilon) = \frac{i}{2\kappa}(e^{i2\eta_l(\varepsilon)} - 1) = -\frac{1}{\kappa}e^{i\eta_l(\varepsilon)}\sin(\eta_l(\varepsilon)). \tag{16.19}$$

The scattering amplitude plays a key role in many phenomena in physics, such as Friedel oscillations around an impurity, resistivity due to impurities, etc. (see Section J.1). The scattering is represented pictorially in Fig. J.1. In the present case, the great advantage is that the electronic bands and Green's functions can be described by a few phase shifts $\eta_l(\varepsilon)$, typically $l \leq 3$.

The full solution for the multiple-scattering problem for the muffin-tin potential is given by Eq. (16.15), where G_0 depends only on the structure and the energy ε, and t incorporates all the effects of the potential inside each sphere. The expressions needed are for the Green's function $G_0(\varepsilon, |\mathbf{r} - \mathbf{r}'|)$ when $\mathbf{r}$ and $\mathbf{r}'$ are in the same and different spheres. For different spheres, this requires that a spherical wave of angular momenta L about one sphere be

---

[6] As indicated by the form of Eq. (16.13), one must take care to avoid spurious poles appearing in the final solution at the positions of the poles of G_0.

[7] Here $\hbar^2/2m_e$ is explicitly indicated to avoid confusion with references that assume $m_e = 1/2$.

expressed in terms of waves centered at another site, which involves a sum over L' at that sphere. The needed formulas are given in [10, 157, 684], which can be understood using the addition formula for plane waves

$$e^{i\mathbf{k}\cdot(\mathbf{r}-\mathbf{R}_1)} = e^{i\mathbf{k}\cdot(\mathbf{r}-\mathbf{R}_2)}e^{i\mathbf{k}\cdot(\mathbf{R}_2-\mathbf{R}_1)}, \tag{16.20}$$

together with the identity Eq. (16.4). The result for sites $\mathbf{R} \neq \mathbf{R}'$ is given by (see Exercise 16.3)

$$G_0(\varepsilon, |\mathbf{r}-\mathbf{r}'|) = \sum_{L,L'} i^l j_l(\kappa r) Y_L(\hat{\mathbf{r}}) B_{LL'}(-i)^{l'} j_{l'}(\kappa r') Y_{L'}^*(\hat{\mathbf{r}}'), \tag{16.21}$$

where $\mathbf{r}$ and $\mathbf{r}'$ are referred to the centers of their respective spheres, and $B_{LL'}$ denotes the "KKR structure constants"

$$B_{LL'}(\varepsilon, \mathbf{R}-\mathbf{R}') = -4\pi\kappa \sum_{L''} i^{l''} C_{L'L''}^{L} n_{l''}(\kappa|\mathbf{R}-\mathbf{R}'|) Y_{L''}(\widehat{\mathbf{R}-\mathbf{R}'}), \tag{16.22}$$

where the Cs are directly related to the Gaunt coefficients $c^{l''}(l\,m,l'\,m')$ defined in Eq. (K.14),

$$C_{L'L''}^{L} \equiv \int d\Omega Y_L^*(\Omega) Y_{L'}(\Omega) Y_{L''}(\Omega) = \sqrt{\frac{2l''+1}{4\pi}} c^{l''}(l\,m,l'\,m'). \tag{16.23}$$

For the general case where the scattering amplitude $t_l(\varepsilon, \mathbf{R})$ is site dependent, the resulting equation for the Green's function Eq. (16.12) is

$$\left[G_{LL'}(\varepsilon, \mathbf{R}, \mathbf{R}')\right]^{-1} = \left[\left[B_{L,L'}(\varepsilon, \mathbf{R}-\mathbf{R}')\right]^{-1} - t_l(\varepsilon, \mathbf{R})\delta_{\mathbf{R},\mathbf{R}'}\delta_{L,L'}\right], \tag{16.24}$$

and condition Eq. (16.15) for stationary states becomes

$$\det\left[t_l^{-1}(\varepsilon, \mathbf{R})\delta_{\mathbf{R}\mathbf{R}'}\delta_{LL'} - B_{LL'}(\varepsilon, \mathbf{R}-\mathbf{R}')\right] = 0, \tag{16.25}$$

where $\mathbf{R}, \mathbf{R}'$ denote centers of the spheres, and $B_{LL'}(\varepsilon, 0) \equiv 0$ for the same site. As it stands, this is a matrix equation in all the sites and angular momenta – a formal expression valid for crystals, molecules, and disordered solids.

KKR Band Structure Equations

The original equation of Koringa [677] results if we consider a crystal where the scattering is the same at every site, centered on the translations vectors $\mathbf{T}$. Then the determinant equation can be solved separately for each wavevector $\mathbf{k}$. The structure constants can be defined as

$$B_{LL'}(\varepsilon, \mathbf{k}) = \sum_{\mathbf{T}\neq 0} B_{LL'}(\varepsilon, \mathbf{T})\, e^{-i\mathbf{k}\cdot\mathbf{T}}, \tag{16.26}$$

and the bands $\varepsilon = \varepsilon_\mathbf{k}$ are the solution of

$$\det\left[t_l^{-1}(\varepsilon)\delta_{LL'} - B_{LL'}(\varepsilon, \mathbf{k})\right] = 0. \tag{16.27}$$

The well-known form for the KKR equations is found by using expression (16.19) for t (only the real part is needed) in terms of the phase shifts,

$$\sum_{L'} [B_{LL'}(\varepsilon_\mathbf{k}, \mathbf{k}) + \kappa \cot(\eta_l(\varepsilon_\mathbf{k})) \delta_{LL'}] a_{L'}(\mathbf{k}) = 0. \tag{16.28}$$

This can be generalized straightforwardly to more than one atom per cell, $\alpha = 1, \ldots, N$, leading to one band per atom and angular momentum

$$\sum_{L'} \sum_{\beta=1}^{N} [B_{LL'}(\tau_\alpha - \tau_\beta, \varepsilon_\mathbf{k}, \mathbf{k}) + \kappa \cot \eta_{l\beta}(\varepsilon_\mathbf{k}) \delta_{LL'} \delta_{\alpha\beta}] a_{L'\beta}(\mathbf{k}) = 0. \tag{16.29}$$

The dispersion relation $\varepsilon_\mathbf{k}$ can be found from the roots of the determinant of the matrix in square brackets. Often it is most effective to fix the energy and scan the wavevector $\mathbf{k}$ to find the roots, since the phase shifts depend only on energy and the structure constants depend only on $\mathbf{k}$ at a given energy. For example, the Fermi surface can be mapped out conveniently in this way.

The eigenvectors of Eqs. (16.28) or (16.29) determine the wavefunction, since the eigenvectors of the Green's function are the same as those of the hamiltonian. Inside each sphere, the solution is simply a linear combination of the augmentation functions (apart from a normalization factor)

$$\psi_\mathbf{k}(\mathbf{r}) = \sum_{L'} \sum_{\beta=1}^{N} a_{L'\beta}(\mathbf{k}) \psi_{L'\beta}(\mathbf{r}), \tag{16.30}$$

where $\psi_{L'\beta}$ are known since they were used to find the phase shifts and t matrix. Outside the spheres the wavefunction can be found from the Green's function equation (see Exercise 16.8)

$$\psi_\mathbf{k}(\mathbf{r}) = -\int d\mathbf{r}' G_0(\varepsilon_\mathbf{k}^n, |\mathbf{r} - \mathbf{r}'|) V(\mathbf{r}') \psi_\mathbf{k}(\mathbf{r}'), \tag{16.31}$$

which can be evaluated with $\mathbf{r}'$ restricted to the interstitial region with boundary conditions on each sphere or with an integral over all space. The integral can be done in different ways, since the free Green's function G_0 given in Eq. (16.18) is long ranged for energy in the continuum but is exponential for energies below the continuum (see also Section 16.7).

The size of the secular equation depends on the maximum angular momentum needed. In the case of Cu and the elementary transition metals it is sufficient to take $l_{max} = 3$ and therefore the rank of the secular equation is 16. Note that the basis is much smaller than in the APW (or plane wave) method; the number of functions is determined by the principle angular momenta of the atomic states needed and not by an accuracy criterion for representation of the wavefunctions.

The KKR method has provided some of the most influential and insightful examples of electronic structure calculations. For example, Fig. 16.6 shows the bands of Al [558]. This is an ideal case for the muffin-tin approximation and illustrates the simple physics that emerges from the KKR approach. The results are similar to previous OPW [559] and pseudopotential calculations [504], all showing the free-electron character of the valence

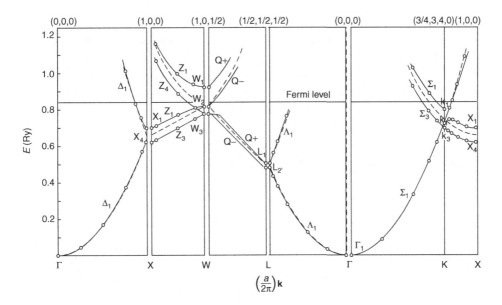

Figure 16.6. Band structure of Al (solid line) calculated by the KKR method [558] compared to free-electron bands (dashed lines). The results can also be easily understood in terms of a weak effective scattering in the plane wave method and the nearly-free-electron approximation, Chapter 12. From [558].

bands. The KKR method conveniently integrates over all plane waves in the analytic Green's function, whereas the plane wave methods make use of the fact that for weak effective scattering only a few plane waves are needed.

KKR is the method used for the first quantitative calculations of the total energy, equilibrium lattice constants, and bulk moduli given in Fig. 2.2; the density of states and Stoner interaction that led to Fig. 2.8; and hosts of other properties, as documented in [104, 673, 682] and many other sources. As a band method, however, it suffers from the same nonlinearity difficulties as the APW method and it is very difficult to extend to a full potential [681]. Therefore, we focus on Green's function approaches where KKR shines.

KKR Green's Function Equations

The power of the KKR approach is most apparent in its formulation as a Green's function method that determines electronic properties directly from $G_{LL'}(\varepsilon)$ in Eq. (16.12). The explicit form in real space for the muffin-tin potential is given by $G_{LL'}(\varepsilon, \mathbf{R}, \mathbf{R}')$ in Eq. (16.24). In a crystal with one atom per cell (the expressions are easily generalized to many atoms per cell) the Green's function is a function only of the relative separation $G_{LL'}(\varepsilon, \mathbf{R} - \mathbf{R}')$. It is most convenient to work with the Fourier transform $G_{LL'}(\varepsilon, \mathbf{k})$, which can be evaluated at each $\mathbf{k}$ separately, as follows from the Bloch theorem. Furthermore, in a crystal, the fact that $G_{LL'}(\varepsilon, \mathbf{k})$ is only a small matrix, of dimension determined by $l_{\max}$, is a great advantage: the inversion of such small matrices is of negligible computational cost and the method can be very efficient depending on the effort required to set up the matrices.

The Green's function provides a spectral representation and many physical properties can be calculated as integrals over energy. The basic relations given in Section D.5 apply to any representation of a Green's function. In particular, the imaginary part of $G(\varepsilon, \mathbf{R})$ provides a *local density of states*, whereas $G(\varepsilon, \mathbf{k})$ provides a "*Bloch spectral representation*" i.e., energy- and wavevector-resolved spectra. For example, the density of states per unit energy ε in the L channel is given by the diagonal part of G with $L = L'$,

$$n_L(\varepsilon, \mathbf{k}) = -\frac{1}{\pi} \mathrm{Im} G_{LL}(\varepsilon + i\delta, \mathbf{k}), \tag{16.32}$$

where δ is a positive infinitesimal. The total density of states at wavevector $\mathbf{k}$ is given by $n(\varepsilon, \mathbf{k}) = \sum_L n_L(\varepsilon, \mathbf{k})$, which is a sum of delta functions of unity weight at the band energies $\varepsilon = \varepsilon_i(\mathbf{k})$.

This Green's function approach provides a convenient of way of calculating the band structure. For example, the Fermi surface can be calculated directly as the locus of states with $\varepsilon_F = \varepsilon_i(\mathbf{k})$ by calculating only the Green's function at ε_F, without calculating the entire band structure. But how does one know ε_F and the potential V_{eff} from which the phase shifts are derived? The Fermi energy can be fixed by a fast procedure for counting the total number of states up to a given energy, which is given by a formula due to Lloyd [686] that effectively evaluates the integral in Eq. (D.30). The potential is fixed by the density, which is considered next.

The density in real space $n(\mathbf{r})$ can be calculated from the projected density at each site $\mathbf{R}$ due to angular momentum component L. The local density of states is

$$n_L(\varepsilon, \mathbf{R}) = -\frac{1}{\pi} \mathrm{Im} G_{LL}(\varepsilon + i\delta, \mathbf{R}), \tag{16.33}$$

and the total density in the sphere at site $\mathbf{R}$ is

$$n_{L, \mathbf{R}} = -\frac{1}{\pi} \int_{\infty}^{E_F} d\varepsilon \, \mathrm{Im} G_{LL}(\varepsilon + i\delta, 0). \tag{16.34}$$

Another quantity that is easily derived is the sum of eigenvalues of occupied states, which is given by

$$\sum_i \varepsilon_i = -\frac{1}{\pi} \sum_L \int_{\infty}^{E_F} d\varepsilon \, \varepsilon \, \mathrm{Im} G_{LL}(\varepsilon + i\delta, 0). \tag{16.35}$$

The last equation represents a way of summing the eigenvalues: each eigenvalue leads to a pole in $G(\varepsilon)$, which gives a contribution of ε to the integral in Eq. (16.35). This provides all the information needed to determine the total energy.

The integrals for total quantities can also be evaluated as contour integrals as shown in Fig. D.1 and given in Eqs. (D.30)–(D.31), which can be evaluated by a discrete sum over points on the contour in the complex plane. Thus one evaluates the Green's function for chosen complex energies z, so that there is no disadvantage due to the nonlinear nature of the secular equations. Furthermore, wherever the contour is far from any pole, the Green's function $G_{LL}(z, \mathbf{R})$ decays exponentially as a function of distance $|\mathbf{R}|$, so that it can be evaluated using only a cluster of atoms. However, in a metal, the contour necessarily approaches the poles at the Fermi energy, and $G(z, \mathbf{R})$ must exhibit long-range oscillatory behavior in real space (Friedel or Ruderman–Kittel oscillations) due to the sharp cutoff in Fourier space at the Fermi surface.

16.4 Alloys and the Coherent Potential Approximation (CPA)

Alloys represent important classes of materials ranging from metallic alloys, where mechanical and magnetic properties can be controlled, to semiconductors where delicate electronic properties are tuned by composition. There are two general types of theoretical approaches: direct calculations on selected supercells and methods that average over disorder. The former approach allows direct studies of effects of short-range order and can be very powerful using clusters and supercell methods (see, e.g., [687] and refences given there). We will concentrate on the coherent potential approximation (CPA), which provides an intuitive, yet accurate, approach when combined with Green's function methods. Such methods are widely applied in crystalline metallic alloys. The formulation that underlies present-day work is due to Soven [688] and Velicky et al. [689], and the earlier work of Lax [690] and Beeby [691].

The general idea of the CPA approach is to formulate an effective (or coherent) potential that, when placed on every site of the alloy lattice, will mimic the electronic properties of the actual alloy. As distinguished from a "virtual crystal approximation" in which the alloy is replaced by an average crystal potential, the coherent potential is derived from averaging the scattering properties of the different atoms embedded in an effective potential as illustrated in Fig. 16.7. Requiring the weighted site average to be the same as the effective potential results in a complex, energy-dependent CPA potential. This is readily treated in terms of Green's functions, in which a complex energy is naturally introduced. An early formulation of the KKR–CPA method, with application to Cu–Ni alloys, is that of Stocks, Temmerman, and Gyorffy [692].

As an example of the "band structure" of alloys calculated in the CPA approximation, Fig. 16.8 shows the Bloch spectral function at five $\mathbf{k}$ points along the Δ direction in a $Cu_{0.77}Ni_{0.23}$ alloy [693]. The peaks (which would be delta functions in a perfect crystal) indicate effective bands that are broadened by scattering due to disorder. The energy- and $\mathbf{k}$-dependent broadening is directly related to scattering rates and lengths, and therefore to transport properties such as resistivity [694].

The KKR–CPA equations can also yield total quantities such as energy, pressure, and magnetization in random substitutional alloys [695]. As an example, a calculation of the total energy and magnetic moments in disordered Fe_xCu_{1-x} alloys [696] finds an abrupt first-order transition from a nonmagnetic Cu-like phase to a magnetic phase with a change in volume. Alloys can also be treated in a response function approach in which the differences are treated in perturbation theory (see, e.g., [697]).

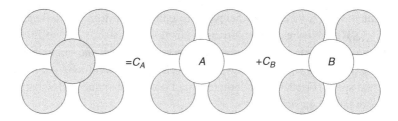

Figure 16.7. Schematic illustration of the averaging over sites in the CPA. The shaded spheres represent an effective average environment and the equation indicates that the average is required to equal the weighted average over sites A and B with concentrations C_A and C_B, each in the same average environment. This leads to the complex CPA potential most readily represented by a Green's function.

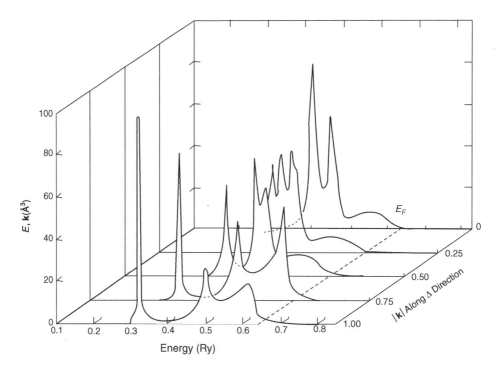

Figure 16.8. Bloch spectral function for a $Cu_{0.77}Ni_{0.23}$ alloy calculated using the KKR–CPA muffin-tin approximation for potential. The figure shows peaks that disperse, revealing the underlying crystal-like bands and broadening that is due to the disorder treated in the coherent potential approximation (CPA). From [693].

16.5 Muffin-Tin Orbitals (MTOs)

Muffin-tin orbitals form a basis of localized augmented orbitals introduced by Andersen [698] in 1971 and subsequently extended into an entire methodology. The goal of the MTO approach is not merely to devise another band structure method but to provide a satisfying interpretation of the electronic structure of materials in terms of a *minimal basis of orbitals*. Like local orbital methods, the electronic states are described in a small number of meaningful orbitals; however, unlike those approaches the minimal basis can be accurate because the MTOs are generated from the Kohn–Sham hamiltonian itself.

This section is devoted to the MTO approach, which sets the stage for the linearized LMTO extension [699, 700] (Chapter 17) that exhibits the real power of the approach. The (L)MTO approach has led to many new concepts and methods, for example, "canonical bands" [699, 700], a new approach to the first-principles tight-binding method [701], and many other features. The (L)MTO methodology has been developed in a way most appropriate for close-packed solids, and the descriptions in the literature are often difficult to penetrate because the basic theory is interwoven with approximations and motivational aspects. The goal of the presentation here and in Section 17.6 is to bring out the simplicity of the (L)MTO approach, the ways in which the concepts enhance our understanding, difficulties in its use in structures that are open or have low symmetry, and the power of the method in actual calculations when used appropriately.

An MTO can be understood in terms of a single atomic sphere with a flat potential in all space outside the sphere, which is the subject of Section J.1 and is closely related to the KKR method. The MTO is defined to be a localized basis function continuous in value and derivative at the sphere boundary. Direct application of the KKR formalism would be to construct an orbital as the energy dependent $\psi_l(\varepsilon, r)$ inside the sphere as in Eq. (J.3), and matching the wavefunction outside the sphere, leading to the form $\propto$ $j_l(\kappa r) - \tan(\eta_l(\varepsilon)) \, n_l(\kappa r)$ outside the sphere, where j_l and η_l are spherical Bessel and Neumann functions. For negative energies, the Neumann functions are replaced by Hankel functions $h_l^{(1)} = j_l + i\eta_l$, which have the asymptotic form $i^{-l} e^{-|\kappa|r}/|\kappa|r$, and the Bessel functions are unbounded. Such orbitals are *not* suitable as basis functions since, at negative energies, they are normalizable only at ε corresponding to eigenvalues where the coefficient of the Bessel function vanishes.

The insight of Andersen [698] was to reformulate the problem defining a new set of functions that depend separately on κ and ε,

$$\text{MTO}_L(\varepsilon, \kappa, \mathbf{r}) = i^l Y_L(\hat{r}) \begin{cases} \psi_l(\varepsilon, r) + \kappa \cot(\eta_l(\varepsilon)) \, j_l(\kappa r), & r < S, \\ \kappa n_l(\kappa r), & r > S, \end{cases}$$

where $Y_L(\hat{r}) \equiv Y_l^m(\hat{r})$ and the factor i^l is a convenient definition. (This is the same as adopted in [157, 673, 699] and it leads to bound-state functions that are real, as shown in Section J.1). The definition in Eq. (16.36) leads to a very simple envelope function outside the sphere with the property that each MTO basis function is well defined, both inside the sphere (since $j_l(\kappa r)$ is regular at the origin) and outside the sphere (since $n_l(\kappa r)$ is regular at ∞). Furthermore, the states are normalizable for all negative energies for any κ. Of course, the χ^{MTO} cannot be eigenstates of a single-muffin-tin potential, but they are basis functions with desirable features for the many-site problem.

The form of Eq. (16.36) contains the seed of an idea that flows through the development of the MTO and LMTO methods: the wavefunction inside the sphere has been modified in a way that takes into account the presence of neighboring atoms to some approximation. The Bessel function $j_l(\kappa r)$ added for $r < S$ is a step toward incorporating into the wavefunction effects due to the neighbors so that a minimal basis of MTO functions χ^{MTO} can accurately describe the system.

The equations for many atoms can be derived using an expansion theorem of the form of Eq. (15.1), which expresses the tail of an MTO extending into another sphere in terms of functions centered on that sphere. Fortunately, there is a well-known expansion analogous to Eq. (16.21),

$$n_L(\kappa, \mathbf{r} - \mathbf{R}) = 4\pi \sum_{L'L''} C^L_{L'L''} n^*_{L''}(\kappa, \mathbf{R} - \mathbf{R}') j_{L'}(\kappa, \mathbf{r} - \mathbf{R}'), \qquad (16.36)$$

where the $C^L_{L'L''}$ are defined by Eq. (16.23). At this point, the MTO basis can be used for calculation of bands by requiring that the total wave function be a solution both inside and outside the spheres, i.e., that the energy and κ be related by $(\hbar^2/2m_e)\kappa^2 = \varepsilon - V_0$. This amounts to a transformation of the KKR method and would lead to nonlinear equations equivalent to Eqs. (16.27) or (16.28).

However, the MTO approach can also be used in a different way. By treating the $\chi_L^{\text{MTO}}(\varepsilon, \kappa, \mathbf{r})$ defined in Eq. (16.36) as functions of ε and κ separately, a judicious fixed choice of κ can be used to define a basis that greatly simplifies the problem and yet is accurate for many problems. This has the advantage that one can define structure constants $S_{L'L}(\mathbf{R})$ or $S_{L'L}(\mathbf{k})$ that depend only on the structure (and the fixed value of κ); in contrast, the KKR "structure constants" are not really constant but are functions of energy ε. This leads to a second hallmark of the (L)MTO approach: developing the method in such a way to take advantage of the fact that an error in the wavefunction leads to higher-order errors in the energy and certain other properties, so that a minimal basis and energy-independent structure constants suffice for accurate calculations.

16.6 Canonical Bands

The simplest version of the MTO equations results if the constant κ is chosen to be $\kappa = 0$, which has been shown to be remarkably accurate for many problems, especially close-packed crystals. The rationale for the freedom to choose κ is that it is finally needed only to represent the variation in the wavefunction in the interstitial between the spheres; if there is only a short distance between the spheres (as in a close-packed solid), the wavefunction will be nearly correct because it has the correct value and slope at the sphere boundary. Many of the applications and much of the motivation for the method [699, 700] is associated with the atomic sphere approximation (ASA) in which the Wigner–Seitz sphere around each atom is replaced by a sphere as shown schematically in Fig. 16.9. It is evident that for close-packed cases the distances between spheres are indeed short; since the spheres overlap, the extrapolation to connect the spheres can be either forward or backward.

For $\kappa = 0$, the wavefunction satisfies the Laplace equation in the interstitial region, i.e., it is equivalent to the electrostatic potential due to a multipole moment. The form

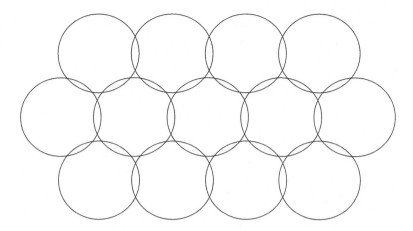

Figure 16.9. Atomic sphere approximation (ASA) in which the muffin-tin spheres are chosen to have the same volume as the Wigner–Seitz cell, which leads to overlapping spheres. The ASA is often a very good approximation for a close-packed solid. Even some open structures (like diamond) can be formed from close-packed spheres (Exercise 16.11) by including "empty spheres" not centered on atoms [702, 703].

can be derived from the previous equations with the $\kappa \to 0$ limit of the Bessel and Neumann functions: inside the sphere $j_l \to (r/S)^l$ with logarithmic derivative $D = l$ and outside, $n_l \to (r/S)^{-l-1}$ with $D = -l - 1$. The MTO in Eq. (16.36) can be written ([157], eq. (1-221); see also [699], eq. (2.2))

$$\chi_L^{\mathrm{MTO}}(\varepsilon, 0, \mathbf{r}) = i^l Y_L(\hat{r}) \psi_l(\varepsilon, S) \begin{cases} \dfrac{\psi_l(\varepsilon, r)}{\psi_l(\varepsilon, S)} - \dfrac{D_l(\varepsilon) + l + 1}{2l + 1} \left(\dfrac{r}{S}\right)^l, & r < S, \\[3mm] + \dfrac{l - D_l(\varepsilon)}{2l + 1} \left(\dfrac{S}{r}\right)^{l+1}, & r > S, \end{cases}$$

where $D_l(\varepsilon)$ the dimensionless logarithmic derivative of $\psi_l(\varepsilon, r)$ evaluated at the boundary $r = S$. This function is continuous and differentiable everywhere (Exercise 16.12). The expansion theorem can be found as the $\kappa \to 0$ limit of Eq. (16.36), which is a well-known multipole expansion,

$$\left[\frac{S}{|\mathbf{r} - \mathbf{R}|}\right]^{l+1} i^l Y_L(\widehat{\mathbf{r} - \mathbf{R}})$$

$$= 4\pi \sum_{L'} \left[\frac{r}{S}\right]^{l'} i^{l'} Y_{L'}(\hat{\mathbf{r}}) \left\{ \frac{(2l'' - 1)!!}{(2l - 1)!!\,(2l' + 1)!!} C_{L'L''}^{L} \left[\frac{S}{|\mathbf{R}|}\right]^{l''+1} i^{-l''} Y_{L''}^*(\hat{\mathbf{R}}) \right\},$$

$$(16.37)$$

where $l'' = l' + l$ and $m'' = m' - m$ and the notation $(\ldots)!!$ denotes $1 \times 3 \times 5 \ldots$.

The essential features of the method are illustrated by a crystal with one atom per cell (extension to more atoms per cell is straightforward). Details of the calculation of the structure constants can be found in [699]; we give only limited results to emphasize that they can be cast in closed form using well-known formulas. The structure factor in $\mathbf{k}$ space is found from the Fourier transform of Eq. (16.37),

$$\sum_{\mathbf{T} \neq 0} e^{i\mathbf{k} \cdot \mathbf{T}} \left[\frac{S}{|\mathbf{r} - \mathbf{T}|} \right]^{l+1} i^l Y_L \widehat{(\mathbf{r} - \mathbf{T})} \equiv \sum_{L'} \frac{-1}{2(2l'+1)} \left[\frac{r}{S} \right]^{l'} i^{l'} Y_{L'}(\hat{\mathbf{r}}) S_{L'L}(\mathbf{k}), \quad (16.38)$$

where the factors have been chosen to make $S_{L'L}(\mathbf{k})$ hermitian [699]. The result is

$$S_{L'L}(\mathbf{k}) = g_{l'm',lm} \sum_{\mathbf{T} \neq 0} e^{i\mathbf{k} \cdot \mathbf{T}} \left[\frac{S}{|\mathbf{T}|} \right]^{l''+1} \left[\sqrt{4\pi} i^{l''} Y_{L''}(\hat{\mathbf{T}}) \right]^*, \quad (16.39)$$

where $g_{l'm',lm}$ can be expressed in terms of Gaunt coefficients [699].

An MTO basis function with wavevector $\mathbf{k}$ is constructed by placing a localized MTO on each lattice site with the Bloch phase factor, e.g., for $\kappa = 0$,

$$\chi_{L,\mathbf{k}}^{\text{MTO}}(\varepsilon, 0, \mathbf{r}) = \sum_{\mathbf{T}} e^{i\mathbf{k} \cdot \mathbf{T}} \chi_L^{\text{MTO}}(\varepsilon, 0, \mathbf{r} - \mathbf{T}). \quad (16.40)$$

The wavefunction in the sphere at the origin is the sum of the "head function" (Eq. (16.37) for $r < S$) in that sphere plus the tails (Eq. (16.37) for $r > S$) from neighboring spheres, and can be written using Eq. (16.38) as

$$\chi_{L,\mathbf{k}}^{\text{MTO}}(\varepsilon, 0, \mathbf{r}) = \psi_l(\varepsilon, r) i^l Y_L(\hat{\mathbf{r}}) - \frac{D_l(\varepsilon) + l + 1}{2l + 1} \psi_l(\varepsilon, S) \left(\frac{r}{S} \right)^l i^l Y_L(\hat{\mathbf{r}})$$

$$+ \frac{l - D_l(\varepsilon)}{2l + 1} \psi_l(\varepsilon, S) \sum_{L'} \left(\frac{r}{S} \right)^{l'} \frac{1}{2(2l'+1)} i^{l'} Y_{L'}(\hat{\mathbf{r}}) S_{LL'}(\mathbf{k}). \quad (16.41)$$

The solution can now be found for an eigenstate as a linear combination of the Bloch MTOs Eq. (16.41),

$$\psi_{\mathbf{k}}(\varepsilon, \mathbf{r}) = \sum_L a_L(\mathbf{k}) \chi_{L,\mathbf{k}}^{\text{MTO}}(\varepsilon, 0, \mathbf{r}). \quad (16.42)$$

Since the first term on the right-hand side of Eq. (16.41) is already a solution inside the atomic sphere, $\psi_{\mathbf{k}}(\varepsilon, \mathbf{r})$ can be an eigenfunction only if the linear combination of the last two terms on the right-hand side of Eq. (16.41) vanishes – called "tail cancellation" for obvious reasons. This condition can be expressed as

$$\sum_L \{ S_{LL'}(\mathbf{k}) - P_l(\varepsilon) \delta_{LL'} \} a_L(\mathbf{k}) = 0, \quad (16.43)$$

where $P_l(\varepsilon)$ is the "potential function"[8]

$$P_l(\varepsilon) = 2(2l + 1) \frac{D_l(\varepsilon) + l + 1}{D_l(\varepsilon) - l}. \quad (16.44)$$

---

[8] We use the symbol P_l since it is the standard term. It should not be confused with a Legendre polynomial.

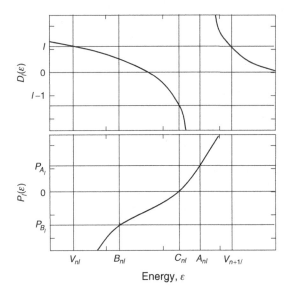

Figure 16.10. Potential function $P_l(\varepsilon)$ (bottom) compared to logarithmic derivative $D_l(\varepsilon)$ (top) versus energy. The functions are related by Eq. (16.44) and the energies $\{A, C, B\}$ denote, respectively, the top, center, and bottom of the nth band formed from states of angular momentum l. The energy V denotes the singularities in $P_l(\varepsilon)$ that separate bands. The key point is that $P_l(\varepsilon)$ is a smooth function for all energies in the band so that it can be parameterized as discussed in the text. (Taken from a similar figure in [699], ch. 2.)

Equation (16.43) is a set of linear, homogeneous equations for the eigenvectors $a_L(\mathbf{k})$ at energies $\varepsilon = \varepsilon_k$ for which the determinant of the coefficient matrix vanishes:

$$\det\left[S_{LL'}(\mathbf{k}) - P_l(\varepsilon)\,\delta_{LL'}\right] = 0. \tag{16.45}$$

This is a KKR-type equation, but here $S_{LL'}(\mathbf{k})$ does not depend on the energy.

The potential function $P_l(\varepsilon)$ contains the same information as the phase shift or the logarithmic derivative $D_l(\varepsilon)$, and the relation between them is illustrated in Fig. 16.10. $P_l(\varepsilon)$ provides a convenient description of the effective potential in Eq. (16.45) because it varies smoothly as a function of energy in the region of the eigenvalues, as opposed to the logarithmic derivative $D_l(\varepsilon)$, which varies strongly and is very nonlinear in the desired energy range. This leads to the simple, but very useful, approximations discussed next.

One of the powerful concepts that arises from Eqs. (16.43) or (16.45) is "canonical bands," which allow one to obtain more insight into the band-structure problem. In essence it is the solution of the problem of states in an atomic sphere (as considered in Section 10.7 but here with nonspherical boundary conditions imposed by the lattice through the structure constants, $S_{LL'}(\mathbf{k})$; see also further discussion in Section 17.2). Since the potential function, $P_l(\varepsilon)$, does not depend on the magnetic quantum number, m, the structure matrix

$$S_{LL'}(\mathbf{k}) \equiv S_{lm,l'm'}(\mathbf{k}) \tag{16.46}$$

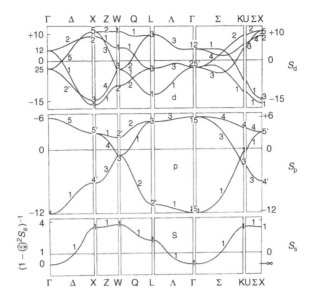

Figure 16.11. Canonical unhybridized bands for an fcc lattice. Comparison with 16.4, and 16.5 shows that this "canonical band" structure has remarkable similarity to the full calculated bands in an elemental fcc crystal. The canonical bands have only one material-dependent factor that scales the overall band width, and even that parameter can be found from simple atomic calculations as discussed in Sections 10.7 and 17.6. Further improvement can be included through information about the potential function as described in the text. Provided by O. K. Andersen; similar to those in [495, 498, 700] and [699]

contains $(2l + 1) \times (2l + 1)$ blocks. If one neglects hybridization, i.e., if one sets the elements of $S_{LL'}(\mathbf{k})$ with $l \neq l'$ equal to zero, the unhybridized bands $[\varepsilon_{li}(\mathbf{k})]$ are simply found as the ith solution of the equation

$$\left| P_l(\varepsilon) - S_{lm,lm'}(\mathbf{k}) \right| = 0. \tag{16.47}$$

This motivates the idea of "canonical bands," which are defined to be the $2l + 1$ eigenvalues $S_{l,i}(\mathbf{k})$ of the block of the structure constant matrix, $S_{lm,lm'}(\mathbf{k})$, for angular momentum l. Since each $S_{l,i}(\mathbf{k})$ depends only on the structure, canonical bands can be defined once and for all for any crystal structure. An example of canonical bands is given in Fig. 16.11 for unhybridized s, p, and d canonical bands for an fcc crystal plotted along symmetry lines in the Brillouin zone [495, 498, 699, 700]. Similar bands for bcc and hcp can be found in [493, 498, 699] and the canonical bands for hcp along $\Gamma - K$ are shown in the left panel of Fig. 17.6. The canonical d densities of states for fcc, bcc, and hcp crystals are given in Fig. 16.12. All the information about the actual material-dependent properties is included in the potential function $P(\varepsilon)$.

The potential function $P(\varepsilon)$ captures information about the bands in a material in terms of a few parameters, all of which can be calculated approximately (often accurately) from

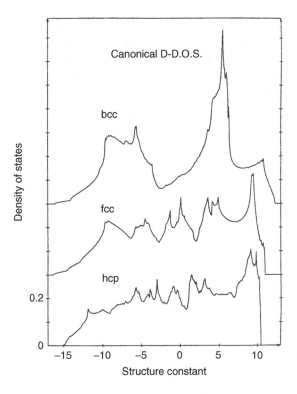

Figure 16.12. Canonical densities of states for unhybridized d bands in the fcc, bcc, and hcp structures. The d states dominate the densities of states for transition metals; the DOS s and p bands are not shown. From [498]. (See also [495, 700], and [699].)

very simple models. A simple three-parameter form that contains the features shown in Fig. 16.10 is

$$P_l(\varepsilon) = \frac{1}{\gamma} \frac{\varepsilon - C_l}{\varepsilon - V_l}, \tag{16.48}$$

which can be inverted to yield[9]

$$\varepsilon(P_l) = C_l + \gamma(C_l - V_l) \frac{P_l}{1 - \gamma P_l} \equiv C_l + \frac{\hbar^2}{2\mu S^2} \frac{P_l}{1 - \gamma_l P_l}, \tag{16.49}$$

where $\hbar^2/2\mu S^2 \equiv \gamma(C_l - V_l)$. This expression has an important physical interpretation with μ an effective mass that sets the scale for the band width. The formulation takes added significance from the fact that μ can be calculated from the wavefunction in the sphere. It is a matter of algebra (Exercise 16.10) to relate μ to the linear energy variation of the

[9] Here we keep the explicit factors of $\hbar$ and m_e to indicate energy units clearly and to avoid confusion with notation in the literature.

logarithmic derivative $D_l(\varepsilon)$ at the band center ($\varepsilon = C_l$, where $D_l(\varepsilon) = -l - 1$), which is then simply related to the value of the wavefunction at the sphere boundary, as given explicitly in Eq. (17.10). This expresses the simple physical fact that the band width is due to coupling between sites, which scales with the value of the wavefunction at the atomic sphere boundary as discussed in Section 10.7. Expressions can also be derived for γ [699].

Combining Eqs. (16.49) with (16.47) leads to the unhybridized band structure

$$\varepsilon_{li}(\mathbf{k}) = C_l + \frac{\hbar^2}{2\mu S^2} \frac{S_{l,i}(\mathbf{k})}{1 - \gamma_l S_{l,i}(\mathbf{k})}. \tag{16.50}$$

This formulation provides a simple intuitive formulation of the energy bands in terms of the "canonical bands" $S_{l,i}(\mathbf{k})$, with center fixed by C_l, the width scaled by an effective mass μ, and a distortion parameter γ that is very similar to the effect of a nonorthogonal basis. If γ is small, the canonical bands illustrated in Fig. 16.11 and the DOS in Fig. 16.12 are a scaled version of the actual bands and DOS of a crystal, thus providing a good starting point for understanding the bands. This can be seen from the remarkable similarity to the calculated bands in Figs. 16.4, and 16.5, and to the tight-binding bands in Fig. 14.9. (Indeed, the real-space interpretation of canonical bands leads to a new formulation of tight-binding described in Section 16.7.) Of course, it is necessary to take into account hybridization to describe the bands fully. This is a notable achievement: *essential features of all five* d *bands are captured by one parameter, the mass in Eq. (16.50).* Furthermore, as we shall see in Section 17.6, the mass can be determined simply from an atomic calculation. In addition, the bands can be improved by including the parameter γ, which distorts the canonical bands as in Eq. (16.50) and which can also be calculated from atomic information. Canonical bands can be used to predict tight-binding parameters, which follows from the structure factors in real space and is discussed in the next section. Finally, many important results for real materials can be found simply using the notions of canonical bands; however, the examples are best deferred to Section 17.8 to include features of the linear LMTO method.

16.7 Localized "Tight-Binding," MTO, and KKR Formulations

The subject of this section is transformations to express the MTO and KKR methods in localized form, with the goal of making possible "first-principles" tight-binding (Chapter 14), localized interpretations, linear scaling methods (Chapter 18), and other developments in electronic structure. The original formulations of the KKR method involve structure constants $B_{LL'}$ in Eq. (16.22) that oscillate and decay slowly as a function of distance $|\mathbf{R} - \mathbf{R}'|$. For positive energies[10] the range is so long that it is not possible to make any simple short-range pictures analogous to the local orbital or tight-binding pictures.

The MTO formalism partly remedies this situation to provide a more localized picture. The distance dependence in Eq. (16.37) illustrates the important features that emerge from

---

[10] If the problem is transformed to complex energies, e.g., in Green's function method, the range can be short. See Section D.5.

the MTO approach: since $\kappa = 0$ has been shown to be a good approximation in many cases and since all the information about interactions between sites is contained in the structure constants, this identity shows the characteristic feature that interactions between orbitals of angular momenta $L = l, m$ and $L' = l', m'$ decrease as a structure factor

$$S_{LL'}(|\mathbf{R}|) \propto \left[\frac{S}{|\mathbf{R}|} \right]^{l+l'+1}. \tag{16.51}$$

For high angular momenta, the sums converge rapidly, which provides a new formulation of tight-binding [615] in which the matrix elements decay as $(1/r)^{l+l'+1}$ and are derived from the original independent-particle Schrödinger equation.

For $l = 0$ and $l = 1$, however, this does not lead to a simple picture because the sums do not converge rapidly – in fact there are singularities in the longest-range terms just as for the Coulomb problem. This is not a pleasant prospect for providing a simple physical picture of electronic states! How can the properties of the MTO basis be interpreted to provide a more satisfying picture? The answer lies in the fact that the long-range terms are Coulomb multipole in nature; the distance dependence has inverse power because it is equivalent to the long-range behavior of electrostatic multipole fields. By a unitary transformation that is equivalent to "screening of the multipoles" one can transform to a fully localized tight-binding form for all angular momenta [701]. There are many ways of screening multipoles, all having the effect shown in Fig. 16.13 that contributions from neighboring sites are added with opposite sign to give net exponential decay of the basis function. An example of a transformation choice is given in [701]. The reason that one can transform to a set of exponentially decaying orbitals is not accidental; this properly can be understood using the same ideas as for the construction of Wannier functions. Since the space spanned by the minimal-basis MTO hamiltonian is a finite set of bands, bounded both above and below in energy, the transformations given in Chapter 23 can be used to construct localized functions that span this finite-basis subspace.

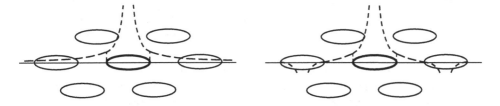

Figure 16.13. Schematic illustration of MTO orbital centered on the site indicated by the dark sphere. Left: a standard MTO. Outside its sphere it decays as a power law with a smooth tail that extends through other spheres. Right: the "screened MTO" that results from linear transformation of the MTOs set of. The essence of the transformation is to add neighboring MTOs with opposite signs as shown; since the tails of the original MTO functions have exactly the same form as the fields due to electronstatic multipoles, the long-range behavior can be "screened" by a linear combination that cancels each multipole field. The transformation to localized functions can also be understood in terms of the construction of Wannier functions; see text.

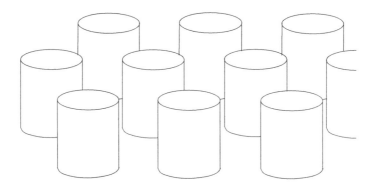

Figure 16.14. Schematic illustration for creation of localized Green's functions G_0 in the KKR method. Because of strong repulsive potentials at every site, G_0 decays exponentially at all energies in the range of the energy bands.

A localized form of KKR also can be generated very straightforwardly, even though the ideas may at first seem counterintuitive. The idea is simply to choose a different reference G_0 instead of the free propagator equation, (16.18), that satisfies Eq. (16.17). If G_0 is chosen to be the solution of the Schrödinger equation for a particle in a set of strongly repulsive potentials, as illustrated in Fig. 16.14, then $G_0(\mathbf{r})$ is localized for all energies of interest [680]. Simply inserting this into the Dyson equation, (16.12), leads to a localized form for any of the KKR expressions in Section 16.3. The greatest advantage is realized in the Green's function formulation, in which the nonlinearity is not a problem and the equations can be made fully localized. "Order-N" methods have been developed using this approach (see, e.g., [704] and Chapter 18).

There are important advantages of using augmented localized orbitals over the standard tight-binding-like local orbital approach. A basis of fixed local orbitals has the inherent difficulty that the tails of orbitals extending into the neighboring atoms are far from the correct solution – e.g., they do not obey the correct cusp conditions at the nucleus – and a sufficient number of orbitals (beyond a minimal basis) must be used to achieve the "tail cancellation" that is built into the KKR and MTO methods. In general, local orbitals are nonorthogonal, whereas the transformed MTO basis can be made nearly orthogonal [701, 705].

On the other hand there are disadvantages in the use of KKR and MTO methods. The KKR formalism is more difficult to apply to general low-symmetry problems where the potential is not of muffin-tin form. The localized MTO form has been developed for close-packed systems and application to open structures requires care and often introduction of empty spheres (see also Section 17.6). Thus these methods have been applied primarily to close-packed metals and high-symmetry ionic crystals but have not been widely applied to molecules, surfaces, and related systems.

16.8 Total Energy, Force, and Pressure in Augmented Methods

Total energies and related quantities are more difficult to calculate than in the pseudopotential method because of the large energies and strong potentials involved. It is especially important to use appropriate functions for the total energy, such as Eq. (7.22), which was derived by Weinert and coworkers [360] explicitly for APW-type methods. Augmented methods have played a key role in total energy calculations since the 1960s when self-consistent calculations became feasible, e.g., for KCl [107], alkali metals [108, 109], and Cu [110]. One of the most complete studies was done by Janak, Moruzzi, and Williams [103, 111], who were pioneers in making Kohn–Sham density functional theory a practical approach to computation of the properties of solids. Their results, shown in Fig. 2.2, were calculated using the KKR method. Many other examples of LAPW calculations are given in Chapters 2 and 17.

Straightforward application of the "force (Hellmann–Feynman) theorem" is fraught with difficulty in any all-electron method. The wavefunctions must be described extremely accurately very near the nucleus in order for the derivative to be accurate, and the wavefunctions must be extremely well converged since the force is not a variational quantity. The problem is in the core electrons. In the atom because of spherical symmetry the force on the nucleus must vanish, which is easy to accomplish since the core states are symmetric. If the nucleus is at a site of low symmetry in a molecule or solid, however, the electric field $\mathbf{E}$ at the nucleus and the net force is nonzero. Even though the core electrons are nearly inert, in fact they polarize slightly and transmit forces to the nucleus. It is only by proper inclusion of the polarized core that one arrives at the correct conclusion that the force due to an electric field on an ion (nucleus plus core) is the "screened" force $\mathbf{F} = Z_{\mathrm{ion}}\mathbf{E}$ instead of the "bare" force $\mathbf{F} = Z_{\mathrm{nucleus}}\mathbf{E}$.

Difficult problems associated with calculation of the force on a nucleus can be avoided by the use of force expression that are alternative to the usual force theorem. As emphasized in Appendix I, difficult core–nucleus terms can be explicitly avoided by displacing rigidly the core around each nucleus long with the nucleus. The resulting expressions then involve additional terms due to displacement of the core. Although they lack the elegant simplicity of the original force theorem, they can be much more intuitive and appropriate for actual calculations. A method for calculation of forces and stresses in APW (and LAPW) approaches has been developed by Soler and Williams [706] and by Yu, Singh, and Krakauer [707]. The general ideas, given in Section I.5, involve finding the force on a sphere in terms of the boundary conditions that transmit forces from the plane waves plus Coulomb forces on the charge in the sphere due to charges outside. The expressions can be found by directly differentiating the explicit APW expressions for the energy.

Within the atomic sphere approximation the pressure can be calculated using the remarkably simple expressions given in Section I.3. Only the wavefunctions at the boundary of the sphere are needed. This can be applied in any of the augmented methods, and examples are given using the LMTO approach in Section 17.8.

SELECT FURTHER READING

Books including augmented methods:

Kübler, J., *Theory of Itinerant Electron Magnetism* (Oxford University Press, Oxford, 2001).

Singh, D. J., *Planewaves, Pseudopotentials, and the APW Method* (Kluwer Academic Publishers, Boston, 1994), and references therein.

Multiple scattering and KKR:

Butler, W. H., Dederichs, P. H., Gonis, A., and Weaver, R. L., *Applications of Multiple Scattering Theory to Material Science* (Materials Reserach Society, Pittsburgh, PA, 1992).

Linear methods and LMTO:

Andersen, O. K., "Linear methods in band theory," *Phys. Rev. B* 12:3060–3083, 1975.

Andersen, O. K. and Jepsen, O., "Explicit, first-principles tight-binding theory," *Physica* 91B:317, 1977.

Skriver, H., *The LMTO Method* (Springer, New York, 1984).

Exercises

16.1 The basic ideas of the APW method can be illustrated by a one-dimensional Schrödinger equation for which the solution is given in Exercise 4.22. In addition, close relations to pseudopotentials, plane wave, KKR, and MTO methods are brought out by comparison with Exercises 11.14, 12.6, 16.7, and 16.13. Consider an array of potentials $V(x)$ spaced by lattice constant a; $V(x)$ is arbitrary except that it is assumed to be like a muffin tin composed of nonoverlapping potentials with $V(x) = 0$ in the interstitial regions. For actual calculations it is useful to treat the case where $V(x)$ is a periodic array of square wells.

   (a) Consider the deep well defined in Exercise 11.14 with width $s = 2a_0$ and depth $-V_0 = -12Ha$. Solve for the two lowest states (analogous to "core" states) using the approximation that they are bound states of an infinite well.

   (b) Construct APW functions that are e^{ikx} outside the well; inside, the APW is a sum of solutions at energy ε (as yet unknown) that matches e^{ikx} at the boundary. Show that the expansion inside the cell analogous to Eq. (16.2), and the plane wave expansion, analogous to Eq. (16.4), are sums only over two terms, sine and cosine, and give the explicit form for the APW.

   (c) Derive the explicit APW hamiltonian for this case. Include the terms from the discontinuity of the derivative. Show that the equation has the simple interpretation of plane waves in the interstitial with boundary conditions due to the well.

   (d) Construct a computer code to solve for the eigenvalues and compare to the results of the general method described in Exercise 4.22.

   (e) Use the computer code also to treat the shallow square well defined in Exercise 12.6 and compare with the results found there using the plane wave method.

   (f) Compare and contrast the APW, plane wave, and the general approach in Exercise 4.22.

16.2 Derive the form for the contribution to the hamiltonian matrix elements from the kink in the wavefunctions given in Eq. (16.8) using Green's identity to transform to a surface integral.

16.3 Derive the identity given in Eqs. (16.21)–(16.23) for the expansion of a spherical wave defined about one center in terms of spherical waves about another center. One procedure is through the use of Eq. (J.1), which is also given in Eq. (16.4).

16.4 Evaluate values for the logarithmic derivatives of the radial wavefunctions for free electrons and compare with the curves shown in Fig. 16.3 for Cu. The expressions follow from Eq. (16.4) (also given in Eq. (J.1)) for zero potential and the functions should be evaluated at the radius $S = 2.415a_0$ appropriate for metallic Cu.

16.5 Show that the nearly parabolic s band for Cu in Fig. 16.4 are well approximated by free-electron values given that Cu has fcc crystal structure with cube edge $a = 6.831a_0$. Show also that the states at the zone boundary labeled would be expected to act like p states ($l = 1$, odd) about each atom. (Quantitative comparisons are given in [157], p. 25).

16.6 As the simplest example of the "s–d" hybridization model, derive the bands for a 2×2 hamiltonian for flat bands crossing a wide band in one dimension: $H_{11}(k) = E_1 + W\cos(2\pi k/a)$, $H_{22}(k) = E_2$, and $H_{12}(k) = H_{21}(k) = \Delta$. Find the minimum gap, and the minimum direct gap in the bands. Show that the bands have a form resembling the bands in a transition metal.

16.7 The KKR method can be illustrated by a one-dimensional Schrödinger equation, for which the solution is given in Exercise 4.22. See [694] for an extended analysis. Close relations to pseudopotentials, plane wave, APW, and MTO methods are brought out by comparison with Exercises 11.6, 12.6, 16.1, and 16.13. As in Exercise 16.1, the KKR approach can be applied to any periodic potential $V(x)$. The KKR solution is then given by Eq. (16.28) with the structure constants defined in Eq. (16.26). (Here we assume $V(x)$ is symmetric in each cell for simplicity. If it is not symmetric there are also cross terms η^{+-}.)
(a) The phase shifts are found from the potential in a single cell. In Exercise 11.6 it is shown that the scattering is described by two phase shifts η^+ and η^-.
(b) In one dimension the structure constants define a 2×2 matrix $B_{L,L'}(\varepsilon, \mathbf{k})$, with $L = +, -$ and $L' = +, -$. Each term is a sum of exponentials that oscillates and does not converge at large distance. Find physically meaning expressions for $B_{L,L'}(\varepsilon, \mathbf{k})$ by adding a damped exponential convergence factor.
(c) Using the relations from Exercise 11.6, show that the KKR equations lead to the same results as the general solution, Eq. (4.49), with $\delta = \eta^+ + \eta^-$ and $|t| = \cos(\eta^+ - \eta^-)$.

16.8 This exercise is to show the relation of the Green's function expression, (16.31), and the Schrödinger equation. This can be done in four steps that reveal subtle features.
(a) Show that application of the free-electron hamiltonian $\hat{H}_0$ to both sides of the equation leads to a Schrödinger-like equation but without the eigenvalue. Hint: use the fact that $\hat{H}_0 G_0 = \delta(|\mathbf{r} - \mathbf{r}'|)$.
(b) Show that this is consistent with the Schrödinger equation using the fact that a constant shift in V has no effect on the wavefunction.
(c) Give an auxiliary equation that allows one to find the eigenvalue.
(d) Finally, give the expression for the full Green's function G analogous to Eq. (16.12) from which one can derive the full spectrum of eigenvalues.

16.9 Show that $\chi_L^{\mathrm{MTO}}(\varepsilon, 0, \mathbf{r})$, defined in Eq. (16.37), is continuous and has continuous derivative (i.e., D is the same inside and outside) at the boundary $r = S$.

16.10 Find the relation of the mass parameter μ to the energy derivative $dD(E)/dE$ evaluated at the band center, assuming P_l has the simple form given in Eq. (16.48).

16.11 The diamond structure can be viewed as a dense-packed structure of touching spheres with some spheres not filled with atoms. Show this explicitly, starting with the crystal structure shown in Fig. 4.7, and insert empty spheres in the holes in the structure.

16.12 Show that Eq. (16.37) indeed leads to a function that is continuous and has a continuous derivative at its boundary.

16.13 The MTO method can be illustrated by a one-dimensional Schrödinger equation. The purpose of this exercise is to show that the solution in Exercises 4.22 and 16.7 can be viewed as "tail cancellation." (An extended analysis can be found in [708].) This reinterpretation of the equations can be cast in terms of the solutions of the single-cell problem given in Exercise 4.22, ψ_l and ψ_r, which correspond to waves incident from the left and from the right; only the part outside the cell is needed. Consider the superposition of waves inside a central cell at $T = 0$ formed by the sum of waves $\psi_l(x)$ and $\psi_r(x)$ from all *other* cells at positions $T \neq 0$ with a phase factor e^{ikT}. Show that the requirement that the sum of waves from all other cells vanishes at any point x in the central cell (i.e., tail cancellation) and leads to the same equations as in Exercises 4.22 and 16.7.

17

Augmented Functions: Linear Methods

Summary

The great disadvantage of augmentation is that the basis functions are energy dependent, so that matching conditions must be satisfied separately for each eigenstate at its (initially unknown) eigenenergy. This leads to nonlinear equations that make such methods much more complicated than the straightforward linear equations for the eigenvalues of the hamiltonian expressed in fixed energy-independent bases such as plane waves, atomic orbitals, gaussians, etc. *Linearization* is achieved by defining augmentation functions as linear combinations of a radial function $\psi(E_\nu, r)$ and its energy derivative $\dot{\psi}(E_\nu, r)$ evaluated at a chosen fixed energy E_ν. In essence, $\psi(E_\nu, r)$ and $\dot{\psi}(E_\nu, r)$ form a basis adapted to a particular system that is suitable for calculation of all states in an energy "window." Any of the augmented methods can be written in linearized form, leading to secular equations like the familiar ones for fixed bases. The simplification has other advantages, e.g., it facilitates construction of full potential methods not feasible in the original nonlinear problem. In addition, the hamiltonian thus defined leads to *linear methods* that take advantage of the fact that the original problem has been reduced to a finite basis. This approach is exemplified in the LMTO method, which defines a minimal basis that both provides physical insight and quantitative tools for interpretation of electronic structure.

17.1 Linearization of Equations and Linear Methods

It should be emphasized from the outset that the terms "nonlinear" and "linear" have *nothing to do with the fundamental linearity of quantum mechanics*. Linearity of the governing differential equation, the Schrödinger equation, is at the heart of the quantum nature of electrons and any nonlinearities would have profoundly undesirable consequences. *Linearization* and *linear methods* have to do with practical matters of solving and interpreting the independent-particle Schrödinger equations.

Formulations in which the wavefunctions are expressed as linear combinations of fixed basis functions, such as plane waves, gaussians, atomic-like orbitals, etc., are manifestly linear. This leads directly to standard linear algebra eigenvalue equations, which is a great advantage in actual calculations. Since the same basis is used for all states, it is simple to express the conditions of superposition, orthogonality, etc., and it is simple to determine many eigenfunctions together in one calculation.

Augmented methods are also linear in the fundamental sense that the wavefunctions can be expressed as linear combinations of basis functions. However, nonlinear equations for the eigenstates arise because the basis is energy dependent. This choice has great advantages, effectively representing electronic wavefunctions both near the nucleus and in the interstitial regions between the atoms. But there is a high price. The matching conditions lead to nonlinear equations due to the intrinsic energy dependence of the phase shifts that determine the scattering from the atoms. This results in greatly increased computational complexity since each eigenstate must be computed separately, as described in Chapter 16.

Linearization of nonlinear equations around selected reference energies allows the construction of operators that act in the same way as ordinary familiar linear operators, while at the same time taking advantage of the desirable attributes of the augmentation and achieving accurate solutions by choice of the energies about which the problem is linearized. The LAPW approach illustrates clearly the advantages of linearization.

Linear methods result from the same process but lead to different formulation of the problem. The resulting hamiltonian matrix is expressed in terms of the wavefunctions and their energy derivatives, which are determined from the original independent-particle Schrödinger equation. Thus this defines a hamiltonian matrix in a reduced space. Working with this derived hamiltonian leads to a class of linear methods that provide physically motivated interpretations of the electronic structure in terms of a minimal basis. This is the hallmark of the LMTO method.

In a nutshell the key idea of linearization is to work with two augmented functions, $\psi_l(r)$ and its energy derivative denoted by $\dot{\psi}_l(r)$, each calculated at the chosen reference energy. These two functions give greater degrees of freedom for the augmentation, which allow the functions to be continuous and to have continuous derivatives at the matching boundaries. However, *the basis does not double in size*: the energy dependence is taken into account to first order by the change of the wavefunction with energy. The wavefunction is correct to first order $\propto (\Delta\varepsilon)$, where $\Delta\varepsilon$ is the difference of the actual energy from the chosen linearization energy; therefore, the energies are correct to $(\Delta\varepsilon)^2$ and variational expressions [699] are correct to $(\Delta\varepsilon)^3$, illustrating the "$2n + 1$" theorem (Section D.6), which is important in actual applications. The methods can be used with any augmentation approach and have led to the widely used LAPW, LMTO, and other methods.

17.2 Energy Derivative of the Wavefunction: ψ and $\dot{\psi}$

In this section we assume that the potential has muffin-tin form, i.e., spherically symmetric within a sphere of radius S about each atom and flat in the interstitial. The equations can be generalized to non-muffin-tin potentials using the same basis functions. Initially, we consider a single spherical potential. The analysis involves radial equations exactly like those for atoms and scattering problems (Section J.1) and the analysis has useful relations to

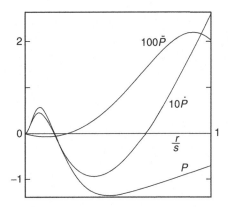

Figure 17.1. Radial d function, $P \equiv S^{1/2}\psi(r)$, and energy derivatives, $\dot P \equiv S^{-2}\partial P/\partial E$, and $\ddot P \equiv S^{-4}\partial^2 P/\partial E^2$, for ytterbium

the derivation of norm-conserving pseudopotentials in Section 11.4 although the application is quite different.

The goal is to sidestep the problems of the nonlinear methods of Chapter 16. Linearized methods achieve this by expanding the solution of the single-sphere Schrödinger equation in terms of $\psi_l(\varepsilon,r)$ belonging to one arbitrarily chosen energy, $\varepsilon = E_v$, i.e., [1]

$$\left(-\frac{\hbar^2}{2m_e}\frac{d^2}{dr^2} + V_{\text{sphere}} - E_v\right) r\psi_l(E_v,r) = 0 \tag{17.1}$$

and its energy derivative

$$\dot\psi(\varepsilon,r) \equiv \frac{\partial}{\partial\varepsilon}\psi(\varepsilon,r)|_{\varepsilon=E_v}. \tag{17.2}$$

If we define the derivative with respect to energy to mean a partial derivative keeping ψ normalized to the same value in the sphere (even though it is not an eigenfunction at an arbitrary E_v), then it is easy to show that ψ and $\dot\psi$ are orthogonal,

$$\langle\psi|\dot\psi\rangle = 0, \tag{17.3}$$

so that the two functions indeed span a larger space. Furthermore, one can readily show that

$$(\hat H - \varepsilon)\dot\psi(\varepsilon,r) = \psi(\varepsilon,r) \tag{17.4}$$

and similar relations in Eq. (17.36) in Exercise 17.1. It is also straightforward to show that each successive energy derivative of the function $\psi(r)$ is given by simple relations like

$$\langle\dot\psi|\dot\psi\rangle = -\frac{1}{3}\frac{\ddot\psi(S)}{\psi(S)}. \tag{17.5}$$

The functions ψ, $\dot\psi$, and $\ddot\psi$ are illustrated in Fig. 17.1 for ytterbium [700], where it is shown that each order of derivative corresponds to a decrease in size by an order of magnitude.

---

[1] The factor $\hbar^2/2m_e$ is included explicitly to avoid confusion with the equations given in other sources.

The augmentation functions as a function of energy can be specified in terms of the dimensionless logarithmic derivative, which is defined as

$$D(\varepsilon) = \left[\frac{r}{\psi(\varepsilon,r)} \frac{d\psi(\varepsilon,r)}{dr} \right]_{r=S}. \tag{17.6}$$

The linear combination of ψ and $\dot{\psi}$ that has logarithmic derivative D is given by

$$\psi(D,r) = \psi(r) + \omega(D) \dot{\psi}(r), \tag{17.7}$$

where $\omega(D)$ has dimensions of energy and is given by

$$\omega(D) = -\frac{\psi(S)}{\dot{\psi}(S)} \frac{D - D(\psi)}{D - D(\dot{\psi})}, \tag{17.8}$$

with $D(\dot{\psi})$ denoting the logarithmic derivative of $\dot{\psi}$. If $\psi(r)$ and $\dot{\psi}(r)$ are calculated at a reference energy E_ν, then Eq. (17.7) is the wavefunction to first order in the energy $E(D) - E_\nu$. It then follows that the variational estimate of the eigenvalue,

$$E(D) = \frac{\langle \psi(D)|\hat{H}|\psi(D)\rangle}{\langle \psi(D)|\psi(D)\rangle} = E_\nu + \frac{\omega(D)}{1 + \omega(D)^2 \langle \dot{\psi}(D)|\dot{\psi}(D)\rangle}, \tag{17.9}$$

is correct to third order and the simpler expression $E_\nu + \omega(D)$ is correct to second order [493].

The logarithmic derivative at the sphere radius S can also be expressed in a Taylor series in $E - E_\nu$. The first term is given by the analysis in Section 11.4, where Eq. (11.28) shows that to first order

$$D(E) - D(E_\nu) = -\frac{m_e}{\hbar^2} \frac{1}{S\psi_l(S)^2}(E - E_\nu), \tag{17.10}$$

where we have substituted $\psi = \phi/r$. (The factor $m_e/\hbar^2 = 1$ in Hartree atomic units and $m_e/\hbar^2 = 1/2$ in Rydberg units.) In deriving Eqs. (17.10) from (11.28), it is assumed that the charge $Q_l(S)$ in the sphere is unity. This is not essential to the logic, but it is convenient and valid in the atomic sphere approximation and is a good approximation in many cases. Higher-order expressions are given in [493] and [699] and related expressions in the pseudopotential literature in [526].

17.3 General Form of Linearized Equations

We are now in a position[2] to define an energy-independent orbital $\chi_j(\mathbf{r})$ everywhere in space for a system of many spheres,

$$\chi_j(\mathbf{r}) = \chi_j^e(\mathbf{r}) + \sum_{L,s} \left[\psi_{l,s}(\mathbf{r} - \tau_s)\Pi_{Lsj} + \dot{\psi}_{l,s}(\mathbf{r} - \tau_s)\Omega_{Lsj} \right] i^l Y_L(\widehat{\mathbf{r} - \tau_s}), \tag{17.11}$$

where $\psi_{l,s}$ and $\dot{\psi}_{l,s}$ are the radial functions in each sphere and Π and Ω are factors to be determined. Between the spheres the function is defined by the envelope function $\chi_j^e(\mathbf{r})$,

---

[2] This section follows the approach of Kübler and V. Eyert [157].

which is a sum of plane waves in the LAPW method. In the LMTO approach, $\chi_j^e(\mathbf{r})$ is a sum of Neumann or Hankel functions as specified in Eq. (16.36) or is proportional to $(r/S)^{-l-1}$ in the $\kappa = 0$ formulation of Eq. (16.37).

The explicit form of the linearized equations depends on the choice of the envelope function, but first we can give the general form. The result is quite remarkable: because of the properties of ψ and $\dot\psi$ expressed in Eqs. (17.3) and (17.4) (see Exercise 17.1), the form of the hamiltonian can be greatly simplified. Furthermore, the "hamiltonian" is expressed in terms of the solution for the wavefunctions; this allows a reinterpretation of the problem as a strictly linear solution of the new hamiltonian expressed as matrix elements in the reduced space of states that span an energy range around the chosen linearization energy.

For a crystal the label s can be restricted to the atoms in one cell, and a basis function with Bloch symmetry can be defined at each $\mathbf{k}$ by a sum over cells $\mathbf{T}$,

$$\psi_{Ls\mathbf{k}}(\mathbf{r}) = \sum_{\mathbf{T}} e^{i\mathbf{k}\cdot\mathbf{T}} \psi_{Ls}(\mathbf{r} - \mathbf{T}), \tag{17.12}$$

and similarly for $\dot\psi_{Ls}$, so that Eq. (17.11) becomes

$$\chi_{j\mathbf{k}}(\mathbf{r}) = \chi_{j\mathbf{k}}^e(\mathbf{r}) + \sum_{L,s} \left[\psi_{Ls\mathbf{k}}(\mathbf{r})\Pi_{Lsj}(\mathbf{k}) + \dot\psi_{Ls\mathbf{k}}(\mathbf{r})\Omega_{Lsj}(\mathbf{k}) \right] i^l Y_L(\widehat{\mathbf{r} - \tau_s}). \tag{17.13}$$

The wavefunction is defined by the coefficients Π_{Lsj} and Ω_{Lsj} that are constructed at each $\mathbf{k}$ so that the basis function $\chi_{j\mathbf{k}}(\mathbf{r})$ satisfies the continuity conditions, with actual equations that depend on the choice of basis (see sections below). The construction of the hamiltonian $H_{ij}(\mathbf{k})$ and overlap matrices $S_{ij}(\mathbf{k})$ can be divided into the envelope part and the interior of the spheres at each $\mathbf{k}$, yielding (see Exercise 17.2)

$$S_{ij}(\mathbf{k}) = \langle i\mathbf{k} | j\mathbf{k} \rangle^e + \sum_{L,s} \left[\Pi_{Lsi}^\dagger(\mathbf{k})\Pi_{Lsj}(\mathbf{k}) + \Omega_{Lsi}^\dagger(\mathbf{k})\langle \dot\psi_{ls} | \dot\psi_{ls} \rangle \Omega_{Lsj}(\mathbf{k}) \right]. \tag{17.14}$$

and

$$H_{ij}(\mathbf{k}) - E_v S_{ij}(\mathbf{k}) = \langle i\mathbf{k} | H - E_v | j\mathbf{k} \rangle^e + \sum_{L,s} \Pi_{Lsi}^\dagger(\mathbf{k})\Omega_{Lsj}(\mathbf{k}). \tag{17.15}$$

The secular equation $\sum_j \left[H_{ij}(\mathbf{k}) - \varepsilon S_{ij}(\mathbf{k}) \right] a_j(\mathbf{k}) = 0$ becomes

$$\sum_j \left[\langle i\mathbf{k} | H - E_v | j\mathbf{k} \rangle^e + V_{ij}(\mathbf{k}) - \varepsilon' S_{ij}(\mathbf{k}) \right] a_j(\mathbf{k}) = 0, \tag{17.16}$$

where $\varepsilon' = \varepsilon - E_v$ is the energy relative to E_v.[3] The potential operator acting inside the spheres is given by

$$V_{ij}(\mathbf{k}) = \frac{1}{2} \sum_{Ls} \left[\Pi_{Lsi}^\dagger(\mathbf{k})\Omega_{Lsj}(\mathbf{k}) + \Pi_{Lsi}(\mathbf{k})\Omega_{Lsj}^\dagger(\mathbf{k}) \right], \tag{17.17}$$

---

[3] For simplicity, a single linearization energy E_v is used here; in general, E_v depends on l and s, leading to expressions that are straightforward but more cumbersome.

which has been made explicitly hermitian [709]. Note that unlike the APW operator V^{APW} in Eq. (16.9), there is no energy dependence in $V_{ij}(\mathbf{k})$. The linear energy dependence is absorbed into the overlap term $\varepsilon' S_{ij}(\mathbf{k})$ in Eq. (17.16).

As promised, the resulting equations are remarkable, with the "hamiltonian" expressed in terms of Π and Ω, i.e., in terms of the wavefunctions ψ and $\dot{\psi}$ calculated in the sphere at the chosen energy E_{ν}. However, this is not the whole story. It would appear that the basis must be doubled in size by adding the function $\dot{\psi}$ along with each ψ; this is exactly what happens in the usual local orbital formulation where one possible way to improve the basis is by adding $\dot{\psi}$ to the set of basis functions. Similarly, the basis is doubled in the related "augmented spherical wave" (ASW) approach [710], which uses functions at nearby energies $\psi(E_{\nu})$ and $\psi(E_{\nu} + \Delta E)$ instead of $\psi(E_{\nu})$ and $\dot{\psi}(E_{\nu})$. However, as we shall see in the following two sections, there is a relation between Π and Ω provided by the boundary conditions. Therefore, it will turn out that *the basis does not double in size*, but nevertheless the wavefunction is correct to linear order in $\varepsilon - E_{\nu}$. Thus errors in the energy are $\propto (\varepsilon - E_{\nu})$,[2] and variational estimates of the energy ([699], Section 3.5) are accurate to $\propto (\varepsilon - E_{\nu})$,[3] an example of the "$2n + 1$" theorem, Section D.6.

17.4 Linearized Augmented Plane Waves (LAPWs)

If we choose a plane wave for the envelope function, we obtain the LAPW method [709] (see also [711–715]). The quantum label j becomes a reciprocal lattice vector $\mathbf{G}_m$ and the form of Eq. (16.2) for an APW can be adapted

$$\chi_{\mathbf{k}+\mathbf{G}_m}^{LAPW}(\mathbf{r}) = \begin{cases} \exp(i(\mathbf{k} + \mathbf{G}_m) \cdot \mathbf{r}), & r > S, \\ \sum_{Ls} C_{Ls}(\mathbf{k} + \mathbf{G}_m)\psi_{Ls}(D_{ls|\mathbf{K}_m|}, \mathbf{r})i^l Y_L(\widehat{\mathbf{r} - \tau_s}), & r < S, \end{cases}$$

where s denotes the site in the unit cell, $L \equiv l, m_l$, $\mathbf{K}_m \equiv \mathbf{k} + \mathbf{G}_m$. The solution inside the sphere of radius S_s is fixed by matching the plane wave, requiring the function to be continuous and have continuous first derivative. This boundary condition leads to $\psi_{ls}(D_{ls|\mathbf{K}_m|}, \mathbf{r})$ as a combination of ψ_{ls} and $\dot{\psi}_{ls}$ as given below. It is this step that includes energy dependence to first order without increasing the size of the basis.

Since the expansion of the plane wave is given by Eq. (J.1), this is accomplished if the logarithmic derivative is the same as for the plane wave,

$$D_{lsK} = \left[x \frac{j_l'(x)}{j_l(x)} \right]_{x=KS_s}, \tag{17.18}$$

which fixes the solution inside the sphere s for a given L and $\mathbf{K}$ to be given by (see Eq. (17.7))

$$\psi_{ls}(D_{lsK}, r) = \psi_{ls}(r) + \omega_{lsK} \; \dot{\psi}_{ls}(r), \tag{17.19}$$

and the total solution in the sphere is given by Eq. (16.5) with

$$j_l(K_m r) \rightarrow \frac{j_l(K_m S_s)}{\psi_{ls}(D_{lsK}, S_s)} \psi_{ls}(D_{lsK}, r)[] * []. \tag{17.20}$$

Thus coefficients Π and Ω are given by

$$\Pi_{LsG_m}(\mathbf{k}) \;=\; 4\pi \; e^{i\mathbf{K}_m \cdot \tau_s} \; \frac{j_l(K_m S_s)}{\psi_{ls}(D_{lsK_m}, S_s)} Y_L(\hat{K}_m) \tag{17.21}$$

and

$$\Omega_{LsG_m}(\mathbf{k}) \;=\; \Pi_{LsG_m}(\mathbf{k}) \; \omega_{lsG_m}, \tag{17.22}$$

where $\mathbf{K}_m = \mathbf{k} + \mathbf{G}_m$ and $K_m = |\mathbf{K}_m|$.

The resulting equations have exactly the same form as the APW equations Eqs. (16.9) and (16.10), with the addition of the overlap term and the simplification that the operator $V_{\mathbf{G'},\mathbf{G}}^{\text{LAPW}}(\mathbf{k})$ is independent of energy. The explicit kinetic energy terms are the same as in Eq. (16.9), which is energy independent. The remaining terms in the secular equation, (17.16), involving Π, Ω, and the overlap can be used conveniently in actual calculations [709] in the form given in Eqs. (17.14)–(17.16) with relations Eqs. (17.21) and (17.22). The expressions can also be transformed into a form for $V_{\mathbf{G'},\mathbf{G}}^{\text{LAPW}}(\mathbf{k})$ that is very similar to the APW expression (16.10), with additional terms but with no energy dependence [157, 673].

Major advantages of the LAPW method are its general applicability for different materials and structures, its high accuracy, and the relative ease with which it can treat a general potential (Section 17.10). Disadvantages are increased difficulty compared to plane wave pseudopotential methods (so that it is more difficult to develop techniques such as Car–Parrinello simulations based on the LAPW method) and the fact that a large basis set is required compared to KKR and LMTO methods (so that it is more difficult to extract the simple physical interpretations than for those methods).

More on the LAPW Basis

How large is the basis required in realistic LAPW calculations? A general idea can be derived from simple reasoning [709]. The number of plane waves, chosen to have wavevector $G < G_{\max}$, is expected to be comparable to pseudopotential calculation for materials without d and f electrons (since the rapidly varying part of the d and f states are taken care of by the radial functions), e.g., ≈ 100 plane waves/atom, typical for high-quality pseudopotential calculations on Si (see also Section 16.2). The size of the basis is somewhat larger than for the APW since each function is continuous in value and slope. The expansion in angular harmonics is then fixed by the requirement that the plane waves continue smoothly into the sphere of radius S: since a Y_{lm} has $2l$ zeros around the sphere, an expansion up to $l_{\max}$ can provide resolution $\approx 2\pi S / l_{\max}$ in real space or a maximum wavevector $\approx l_{\max}/S$. Thus, in order for the angular momentum expansion to match smoothly onto plane waves up to the cutoff $G_{\max}$, one needs $l_{\max} \approx S G_{\max}$, which finally results in $l_{\max} \approx 8$ (see Exercise 17.3) and larger for accurate calculations in complex cases [709, 716].

17.5 Applications of the LAPW Method

The LAPW method, including the full-potential generalization of Section 17.10, is the most accurate and general method for electronic structure at the present time. It is the benchmark for other methods, e.g., in the comparison of many methods in [406]. The calculations can be done for structures of arbitrary symmetry with no bias if the basis is extended to convergence. Extensive tests of convergence are illustrated in Fig. 17.2 taken from the work of Jansen and Freeman [717] in the early development of the full-potential LAPW method. The figure shows the total energy of W in the bcc and fcc crystal structures as a function of volume. The total energy is $\approx -16{,}156$ Ha and the energy is converged to less than 0.001 Ha, including the basis set convergence and integration over the BZ. On the right-hand side of Fig. 17.2 is shown the convergence as a function of plane wave cutoff k_{max} plotted on a logarithmic scale.

However, the generality and accuracy of LAPW comes at a price: there is a large basis set of plane waves and high angular momentum functions, which in turn means that the potentials must be represented accurately (to twice the cutoffs in wavevectors and angular moments used for the wavefunctions) as described in Section 17.10. Other methods are faster, in which case LAPW calculations can serve as a check. Other methods are much more adaptable for generation of new developments that are the subject of Part V, Chapters 18–19. In fact all the developments of quantum molecular dynamics, polarization and localization, excitations, and O (N) methods were stimulated by other approaches and have been adapted to LAPW in only a few cases.

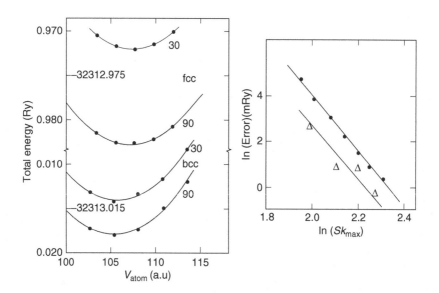

Figure 17.2. Full-potential LAPW calculations of the total energy of W in bcc and fcc structures. Note the break in the vertical scale. The two curves for each structure denote integrations over the irreducible BZ with 30 and 90 points, respectively. The absolute value of the energy is given in the figure on the left. On the right is shown the convergence with plane wave cutoff Sk_{max}, where S is the radius of the muffin-tin sphere. From [717].

Examples of total energies and bands have already been shown in Chapter 2. One is the energy versus displacement for the unstable optic mode that leads to the ferroelectric distortion in $BaTiO_3$ shown in Fig. 2.10. The LAPW results [169] are the standard to which the other calculations are compared for this relatively simple structure. As shown in Fig. 2.10, local orbital pseudopotential methods (and also plane wave calculations) give nearly the same results when done carefully. When using a pseudopotential, it has been found to be essential to treat the Ba semicore states as valence states for accurate calculations. Figure 2.10 also shows the phonons for Mo and Nb and the instability that occurs in Zr. LAPW is the benchmark for the equation of state of Fe at high pressure shown in Fig. 2.13. Another example is the DOS for ferromagnetic Fe shown in Fig. 14.10, where it is compared to tight-binding fit to the LAPW bands.

Perhaps the most important class of application in which the LAPW approach is particularly adapted are compounds involving transition metals and rare earth elements. Understanding many properties of these interesting materials often involves small energy differences due to magnetic order and/or lattice distortions. Linearization simplifies the problem so that one can use full-potential methods with no shape approximations. Since the LAPW approach describes the wavefunctions with unbiased spherical and plane waves, it is often the method of choice.

Perhaps the best example are the bands and total energies for the high-temperature superconductors [716]. For example, the structure of $YBa_2Cu_3O_7$ is shown in Fig. 17.3.

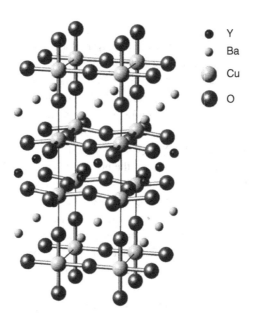

Figure 17.3. Crystal structure of $YBa_2Cu_3O_7$ showing two CuO_2 planes that form a double layer sandwiching the Y atoms, the CuO chain, and the two Ba–O layers per cell. The orthorhombic BZ is shown with the y-direction along the chain axis. Other high-temperature superconductors have related structures all involving CuO_2 planes. Provided by W. Pickett; similar to fig. 6 in [716]

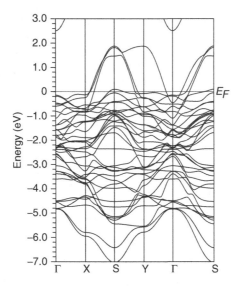

Figure 17.4. Band structure of $YBa_2Cu_3O_7$ computed using the LAPW method [719]. Other calculations [716] with various methods give essentially the same results. The band that protrudes upward from the "spaghetti" of other bands is the antibonding (out-of-phase) Cu–O band that is mainly O 2p in character. Simple counting of electrons (Exercise 17.8) shows that the highest band has one missing electron per Cu, leading to the Fermi level in the band, as shown. Provided by W. Pickett; similar to Fig. 25 in [716]

There are two CuO_2 planes that form a double layer sandwiching the Y atoms, one CuO chain, and two Ba–O layers per cell. The structure must be optimized with respect to all the degrees of freedom and three independent cell parameters. The O atoms in the planes are not exactly in the same plane as the Cu atoms and the "dimpling" has significant effects on the bands. The process of energy minimization with respect to the atomic positions leads to comparison with experiment, including phonon energies that are found to be in very good agreement with experiment, e.g., for $YBa_2Cu_3O_7$ in [718].

Figure 17.4, as an example, shows one of the host of calculations [716] that have led to similar conclusions. The most important conclusion is that the states near the Fermi energy are primarily the one simple band made up of states that involve the Cu $d_{x^2-y^2}$ and O p states in an antibonding combination. The number of electrons is just enough to almost fill the bands, and there is one hole per Cu atom in the band that crosses the Fermi energy (Exercise 17.8). The properties of this band on a square lattice representing one plane have been emphasized in Section 14.9 and a quantitative description of this band, disentangled from the rest, is given later in Figs. 17.8 and 17.9.

Thus the Kohn–Sham equations indicate which states are important at the Fermi energy. Yet there is a fundamental failure of the simplest forms of density functional theory, i.e., the LDA and GGA approximations. Experimentally the CuO systems with one hole per Cu are antiferromagnetic insulators, not metals with half-filled bands. It is far beyond the subject of this book to attempt to summarize all the issues. Let it suffice to say that it appears to be essential to describe both nonlocal exchange (which can be done in Hartree–Fock or exact exchange methods (Chapter 9) and correlation among the electrons in the band near the Fermi energy.

17.6 Linear Muffin-Tin Orbital (LMTO) Method

The LMTO method [699, 700] builds on the properties of muffin-tin orbitals, which have been defined in Section 16.5 in terms of the energy ε and the decay constant κ that characterizes the envelope function. For a fixed value of κ an LMTO basis function inside a sphere is defined to be a linear combination of $\psi(\varepsilon, r)$ and $\dot{\psi}(\varepsilon, r)$ evaluated at the energy $\varepsilon = E_\nu$ as in Eq. (17.11). The differences from an MTO defined in Eq. (16.36) are as follows: (1) inside the "head sphere" in which a given LMTO is centered, it is a linear combination of $\psi_l(E_\nu, r)$ and $\dot{\psi}_l(E_\nu, r)$; and (2) the tail in other spheres is replaced by a combination of $\dot{\psi}_l(E_\nu, r)$. The form of an LMTO can be expressed in a very intuitive and compact form by defining functions J_l and N_l, which play a role analogous to the Bessel and Neumann functions j_l and n_l in Eq. (16.36):

$$\chi_L^{\text{LMTO}}(\varepsilon, \kappa, \mathbf{r}) = i^l Y_L(\hat{\mathbf{r}}) \begin{cases} \psi_l(\varepsilon, r) + \kappa \cot(\eta_l(\varepsilon)) J_l(\kappa r), & r < S, \\ \kappa N_l(\kappa r), & r > S. \end{cases}$$

The form of J_l is fixed by the requirement that the energy derivative of χ_L^{LMTO} vanishes at $\varepsilon = E_\nu$ for $r \leq S$,

$$\frac{d}{d\varepsilon} \chi_L^{\text{LMTO}}(\varepsilon, \kappa, \mathbf{r}) = i^l Y_L(\hat{r}) \left[\dot{\psi}_l(\varepsilon, r) + \kappa \frac{d}{d\varepsilon} \cot(\eta_l(\varepsilon)) J_l(\kappa, r) \right] = 0, \qquad (17.23)$$

which leads to (Exercise 17.4)

$$J_l(\kappa r) = -\frac{\dot{\psi}_l(E_\nu, r)}{\kappa \frac{d}{d\varepsilon} \cot(\eta_l(E_\nu))}, \quad r \leq S. \qquad (17.24)$$

This defines an *energy-independent* LMTO basis function $\chi_L^{\text{LMTO}}(E_\nu, \kappa, \mathbf{r})$ inside the sphere, given by the first line of Eq. (17.23) with $\varepsilon = E_\nu$.

The augmented Neumann functions N_L can be *defined* as the usual n_l in the interstitial, with the tails in other spheres given by the *same expansion* as in Eq. (16.36) with $n_l \rightarrow N_l$ and $j_l \rightarrow J_l$,

$$N_L(\kappa, \mathbf{r} - \mathbf{R}) = 4\pi \sum_{L', L''} C_{LL'L''} n_{L''}^*(\kappa, \mathbf{R} - \mathbf{R}') J_{L'}(\kappa, \mathbf{r} - \mathbf{R}'), \qquad (17.25)$$

where $N_L(\kappa, \mathbf{r}) \equiv i^l Y_L(\hat{\mathbf{r}}) N_l(\kappa r)$, etc. Thus an LMTO is a linear combination of ψ and $\dot{\psi}$ in the central sphere, which continues smoothly into the interstitial region and joins smoothly to $\dot{\psi}$ in each neighboring sphere.

If we chose $\kappa = 0$ for the orbital in the interstitial region, as was done for an MTO in Section 16.6, then the expressions can be simplified in a way analogous to Eq. (16.41).

The wavefunction inside the sphere is chosen to match the solution $\propto (r/S)^{-l-1}$ in the interstitial; this is accomplished for $r < S$ by choosing the radial wavefunction with $D = -l - 1$ as defined in Eq. (17.7), i.e., $\psi_l(D = -l - 1, r) \equiv \psi_{l-}(r)$. In turn this can be expressed in terms of ψ and $\dot{\psi}$ at a chosen reference energy together with ω from Eq. (17.8). The tails from other spheres continued into the central sphere must replace the tail $\propto (r/S)^l$ keeping the same logarithmic derivative, i.e., $(r/S)^l \to \psi_l(D = l, r) \equiv \psi_{l+}(r)$ with the proper normalization. The result is

$$\chi_{L,\mathbf{k}}^{\text{LMTO}}(\mathbf{r}) = \frac{\psi_{L-}(\mathbf{r})}{\psi_{l-}(S)} - \frac{1}{\psi_{l+}(S)} \sum_{L'} \psi_{L'+}(\mathbf{r}) \frac{1}{2(2l'+1)} S_{LL'}(\mathbf{k}). \tag{17.26}$$

This defines an energy-independent LMTO orbital, along with the continuation into the interstitial region. The orbital itself contains effects of the neighbors through the structure constants and through a second effect, the requirement on the logarithmic derivative $D = -l - 1$ in the first term needed to make the wavefunction continuous and have continuous slope. *Thus the orbital contains the tail cancellation to lowest order and the energy dependence to linear order has been incorporated into the definition of the LMTO basis function.*

The LMTO method then finds the final eigenvalues using the LMTO basis and a variational expression with the full hamiltonian. This has many advantages: the energy is thus accurate to second order (and third order using appropriate expressions [699]) and the equations extend directly to full-potential methods. This is analogous to the expression for a single sphere and is accomplished by solving the eigenvalue equation,

$$\det \left| \langle \mathbf{k}L | \hat{H} | \mathbf{k}L' \rangle - \varepsilon \langle \mathbf{k}L | \mathbf{k}L' \rangle \right| = 0, \tag{17.27}$$

by standard methods. It is clear from the form of Eq. (17.26) that the matrix elements of the hamiltonian and overlap will be expressed as a sum of one-, two- and three-center terms, respectively, involving the structure constants to powers 0, 1, and 2. The expressions can be put in a rather compact form after algebraic manipulation, and we will only quote results [157, 699]. Here we consider only a muffin-tin potential, which simplifies the expressions. If we define $\omega_{l-} = \omega_l(-l-1), \omega_{l+} = \omega_l(l), \Delta_l = \omega_{l+} - \omega_{l-},$ and $\tilde{\psi}_l = \psi_{l-}\sqrt{(S/2)}$, then the expression for $\chi_{j\mathbf{k}}(\mathbf{r})$ in Eq. (17.13) can be specified by

$$\Pi_{LL''}(\mathbf{k}) = \tilde{\psi}_l^{-1} \delta_{LL''} + \frac{\tilde{\psi}_{l''}}{\Delta_{l''}} S_{LL''}(\mathbf{k}), \tag{17.28}$$

and

$$\Omega_{LL''}(\mathbf{k}) = \omega_{l''-} \tilde{\psi}_l^{-1} \delta_{LL''} + \frac{\tilde{\psi}_{l''}}{\Delta_{l''}} \omega_{l''+} S_{LL''}(\mathbf{k}). \tag{17.29}$$

The expressions for the matrix elements are, in general, complicated since they involve the interstitial region, but the main points can be seen by considering only the atomic sphere approximation (ASA) as used in Section 16.6 in which the interstitial region is eliminated. Also the equations are simplified if the linearization energy E_ν is set to zero, i.e., the energy

ε is relative to E_v; this is always possible and it is straightforward to allow E_v to depend on l as a diagonal shift for each l. The resulting expressions have simple forms [157, 699]

$$\langle Lk|H|kL'\rangle = \frac{\omega_{l-}}{\tilde{\psi}_l^2}\delta_{LL'} + \left[\frac{\omega_{l+}}{\Delta_l} + \frac{\omega_{l'+}}{\Delta_{l'}}\right] S_{LL'}(\mathbf{k})$$

$$+ \sum_{L''} S_{LL''}(\mathbf{k})\left[\tilde{\psi}_{l''}^2 \frac{\omega_{l''+}}{\Delta_{l''}^2}\right] S_{L''L'}(\mathbf{k}), \qquad (17.30)$$

and

$$\langle kL|kL'\rangle = \left\{(1 + \omega_-^2 \langle\dot{\psi}^2\rangle)/\tilde{\psi}^2\right\}_l \delta_{LL'}$$

$$+ \left\{\{(1 + \omega_+ \omega_- \langle\dot{\psi}^2\rangle)/\Delta\}_l + \{\cdots\}_{l'}\right\} S_{LL'}^k$$

$$+ \sum_{L''} S_{LL''}^k [\tilde{\psi}^2 (1 + \omega_+^2\langle\dot{\psi}^2\rangle)/\Delta^2]_{l''} S_{L''L'}^k. \qquad (17.31)$$

The terms involving $\delta_{LL'}$ are one-center terms (which are diagonal in L for spherical potentials); terms with one factor of $S_{LL''}$ are two center; and those with two factors are three-center terms. The hamiltonian has the interpretation that the on-site terms involve the energy $\omega_{l-} = \omega_l(-l-1)$ of the state with $D = -l - 1$, whereas all terms due to the tails involve the energy $\omega_{l+} = \omega_l(l)$ for the state with $D = l$. Similarly, the overlap terms involve $\langle\dot{\psi}^2\rangle$ and combinations of ω_+ and ω_-.

Thus, within the ASA the LMTO equations have very simple structure, with each term in Eqs. (17.30) and (17.31) readily calculated from the wavefunctions in the atomic sphere. Within this approximation, the method is extremely fast, and the goal has been reached of a minimal basis that is accurate. Only wavefunctions with l corresponding to the actual electronic states involved are needed. This is in contrast to the LAPW method where one needs high l in order to match the spherical and plane waves at the sphere boundary. Furthermore, the interstitial region and a full potential can be included; the same basis is used but the expressions for matrix elements are more cumbersome. The size of the basis is still minimal and the method is very efficient.

There is a price, however, for this speed and efficiency. The interstitial region is not treated accurately since the LMTO basis functions are single inverse powers or Hankel or Neumann functions as in Eq. (17.23). Open structures can be treated only with correction terms or by using "empty spheres." The latter are useful in static, symmetric structures, but the choice of empty spheres is problematic in general cases, especially if the atoms move. Finally, there is no "knob" to turn to achieve full convergence as there is in the LAPW method. Thus the approximations in the LMTO approach are difficult to control and care is needed to ensure robust results.

Improved Description of the Interstitial in LMTO Approaches

One of the greatest problems with the LMTO approach, as presented so far, is the approximate treatment of the interstitial region. The use of a single, energy-independent tail outside each sphere was justified in the atomic sphere approximation (Fig. 16.9) where the distances between spheres is very small (and in the model the interstitial is nonexistent). This approximation fails for open structures where the interstitial region is large, e.g., in the diamond structure, and applications of the LMTO method depend on tricks like the introduction of empty spheres [702, 703]. This can be done for high-symmetry structures, but the method cannot deal with cases like the changing structures that occur in a simulation.

An alternative approach is to generalize the form of the envelope function, generalizing the single Hankel or power law function given in Eqs. (16.37), (16.36), or (17.23). One approach is to work with multiple Hankel functions with different decay constants κ_i that can better describe the interstitial region and yet keep the desirable features of Hankel functions [720, 721]. Using this approach, a full-potential LMTO method has been proposed [722] that combines features of the LMTO, LAPW, and PAW approaches. Like the LAPW it has multiple functions outside the spheres, but many fewer functions. Like the PAW method, the smooth functions are continued inside the sphere where additional functions are included as a form of "additive augmentation."

The form of the basis function proposed is an "augmented smooth Hankel function." In Eq. (17.23), the tail of the LMTO function is a Neumann function, which at negative energy (imaginary κ) becomes a Hankel function, which is the solution of

$$(\nabla^2 + \kappa^2)h_0(\mathbf{r}) = -4\pi\delta(\mathbf{r}).\tag{17.32}$$

This function decays as $i^{-l}e^{-|\kappa|r}/|\kappa|r$ at large r and it diverges at small r as illustrated in Fig. 17.5. The part inside the sphere is not used in the usual LMTO approach and it makes the function unsuitable for continuation in the sphere. Methfessel and van Schilfgaarde [722] instead defined a "smooth Hankel function" that is a solution of

$$(\nabla^2 + \kappa^2)\tilde{h}_0(\mathbf{r}) = -4\pi g(\mathbf{r}).\tag{17.33}$$

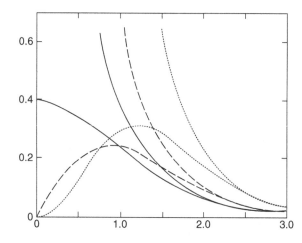

Figure 17.5. Comparison of standard and "smooth" Hankel functions for $l = 0$ (solid lines), $l = 1$ (dashed), and $l = 2$ (dotted) for the case $\kappa = i$ and the smoothing radius $R_{sm} = 1.0$ in the gaussian. From [722].

If $g(\mathbf{r})$ is chosen to be a gaussian, $g(\mathbf{r}) \propto \exp(r^2/R_{sm}^2)$, then $\tilde{h}_0(\mathbf{r})$ is a convolution of a gaussian and a Hankel function. It has the smooth form shown in Fig. 17.5 and has many desirable features of both functions, including analytic formulas for two-center integrals and an expansion theorem. It is proposed that the form of the smooth function near the muffin-tin radius more closely resembles the true function than the Hankel function, and that the sum of a small number of such functions can be a good representation of the wavefunctions in the interstitial region [722].

17.7 Tight-Binding Formulation

It has been pointed out in Section 16.7 that the MTO approach provides a localized basis and tight-binding-type expressions for the Kohn–Sham equations. With a unitary transformation that is equivalent to "screening of electrostatic multipoles," one can transform to a compact short-range form [701]. The transformation applies in exactly the same way in the LMTO approach since it depends only on the form of the envelope function outside the sphere. The matrix elements between different MTOs decrease as $R^{-(l+l'+1)}$, which leads to short-range interactions for large l. Matrix elements for $l + l' = 0$, 1, or 3 can be dealt with by suitable transformations [701].

There are two new features provided by linearization. Most important, the linear equations have the same form as the usual secular equations so that all the apparatus for linear equations can be applied. Second, transformation of the equations leads to very simple expressions for the on-site terms and coupling between sites in terms of ψ and $\dot{\psi}$. The short-range LMTO is the ψ function in one sphere coupled continuously to the tails in neighboring spheres, which are $\dot{\psi}$ functions. This provides an orthonormal minimal basis tight-binding formulation in which there are only two-center terms, with all hamiltonian matrix elements determined from the underlying Kohn–Sham differential equation. The disadvantage is that all the terms are highly environment dependent, i.e., each matrix element depends in detail on the type and position of the neighboring atoms.

This *ab initio* tight-binding method is now widely used for many problems in electronic structure. Because the essential calculations are done in atomic spheres, determination of the matrix elements can be done very efficiently. Combination of the recursion method (Section 18.4) and the tight-binding LMTO [701] provides a powerful method for density-functional calculations for complex systems and topologically disordered matter [724, 725]. For example, in Fig. 18.3 is shown the electronic density of states of liquid Fe and Co determined using tight-binding LMTO and recursion [726]. The calculations were done on 600 atom cells with atomic positions, representing a liquid structure generated by classical Monte Carlo and empirical interatomic potentials. Such approaches have been applied to many problems in alloys, magnetic systems, and other complex structures.

17.8 Applications of the LMTO Method

In its simplest form, the LMTO method can be very effective and informative in addition to providing quantitative results. An example is the calculation of the equation of state, equilibrium volume, and bulk moduli. It is a great advantage to calculate the pressure directly using the formulas valid in the ASA given in Section I.3. The equilibrium volume per atom Ω is the volume for which the pressure $P = 0$, and the bulk modulus is the slope $B = -dP/d\Omega$. The results for 4d and 5d transition metals [498] compare well with the calculations using the KKR method presented in Fig. 2.2. The results are quite impressive and show the way that important properties of solids can be captured in simple calculations with appropriate interpretation.

Band structures of materials with d and f bands often are a jumble of lines like a plate of spaghetti. A helpful way to understand the complicated structure is the progression of energy bands from the simplest unhybridized "canonical" form to the full calculation. Figure 17.6 shows this progression for hcp Os along one line in the BZ from unhybridized canonical bands on the left to fully hybridized relativistic bands on the right.

Although MTOs were originally designed for close-packed metals, the methods can be applied to materials with open structures. By including empty spheres [702, 703] the structure becomes effectively close packed and accurate calculations can be done with only a few basis functions per empty sphere. An example is the calculation of Wannier functions [728] described in Section 23.2 and band offsets of semiconductor structures [729, 730]. As an example of band structure calculations, Fig. 17.7 shows calculations for GaAs LMTO illustrating the large effects of relativity and core relaxation, and spin–orbit interaction [727]. The same work also considered only the scalar relativistic level and showed that

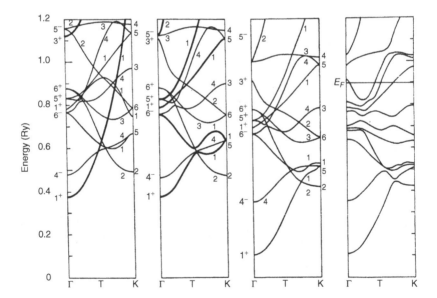

Figure 17.6. Development of the band structure of hcp Os in the LMTO method. From left to right: non-relativistic "canonical" (Section 16.6) bands neglecting hybridization of d and s, p bands (shown dark); including hybridization (with dark lines indicating the most affected bands); relativistic bands without spin orbit; fully relativistic bands. From [498]; original calculations in [723].

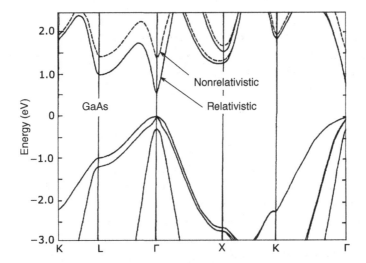

Figure 17.7. Fully relativistic band structure of GaAs (solid lines) calculated using the local density approximation (LDA) and the LMTO method [727]. Dashed line in the conduction band calculated with no relativistic effects. The difference shows the large scalar relativistic effect that lowers the *s* states in the conduction band. The spin–orbit interaction is evident at the top of the valence band. From [727].

the results are the same as using a pseudopotential. In addition it was the first work to show that Ge is a metal in the LDA! (See Fig. 2.23 and related discussion.)

An example that illustrates many different features of the LMTO approach is the work of Duthi and Pettifor [731], which provided a simple explanation for the sequence of structures observed in the series of rare earths elements. The energy differences are very small and the authors used the expressions (I.7) in terms of the difference of the sum of eigenvalues, together with the tight-binding form of LMTO and the recursion method (Section M.5 and [732]). Stabilization results from filling of the d bonding states which is an example of the Friedel argument [733], but it took the combination of ideas in [731] to sort out the way in which bonding varies with structure.

17.9 Beyond Linear Methods: NMTO

Later developments in MTO methods show how approximations that were introduced during development of the LMTO approach can be overcome. The NMTO approach [734, 735] provides a more consistent formalism, treats the interstitial region accurately, and goes beyond the linear approximation.

In the MTO and LMTO approaches, energy-independent orbitals were generated using the approximation of a fixed κ in the envelope function that describes the interstitial region. This breaks the relation of κ and the eigenvalue that causes nonlinearities in the KKR method. However, it also is an approximation that is justified only in close-packed solids. In contrast, the wavefunction inside the sphere is treated more accurately through

linearization. The NMTO method treats the sphere and interstitial equally by working with
MTO-type functions $\psi_L(E_n, \mathbf{r} - \mathbf{R})$ localized around site $\mathbf{R}$ and calculated at fixed energies
E_n both inside the sphere and in the interstitial (assumed to have a flat muffin-tin potential).
The NMTO basis function is then defined to be a linear combination of N, such functions
evaluate at N energies,

$$\chi_{\mathbf{R}L}^{\text{NMTO}}(\varepsilon\mathbf{r}) = \sum_{n=0}^{N} \sum_{\mathbf{R}'L'} \psi_{L'}(E_n, \mathbf{r} - \mathbf{R}') L_{nL'\mathbf{R}',L\mathbf{R}}^{N}(\varepsilon, \mathbf{r}), \qquad (17.34)$$

where $L_n^N(\varepsilon)$ is the transformation matrix that includes the idea of screening (mixing states
on different sites) and a linear combination of states evaluated at N fixed energies.

As it stands, the NMTO function is energy dependent and appears to be merely a way
to expand the basis. However, Andersen and coworkers [734, 735] have shown a way of
generating energy-independent functions $\chi_{\mathbf{R}L}^{\text{NMTO}}(\mathbf{r})$ using a polynomial approximation so
that the Schrödinger equation is solved exactly at the N chosen energies. The ideas are
a generalization of the transformation given in Section 11.9, which were chosen to give
the correct phase shifts at an arbitrary set of energies. The basic ideas can be understood,
following the steps in Exercise 11.12, where the exact transformation, Eq. (11.47), is easily
derived. In the present case, the transformation is more general, mixing states of different
angular momenta on different sites as indicated in Eq. (17.34). The result of the trans-
formation is that each eigenfunction is accurate to order $(\varepsilon - E_0)(\varepsilon - E_1) \cdots (\varepsilon - E_N)$
and the eigenvalue to order $(\varepsilon - E_0)^2(\varepsilon - E_1)^2 \cdots (\varepsilon - E_N)^2$.

As an illustration of the NMTO approach, Fig. 17.8 shows the $d_{x^2-y^2}$ orbital centered
on a Cu atom in $YBa_2Cu_3O_7$. This orbital is not unique; it is chosen to represent the mixed
Cu–O band that crosses the Fermi level, as shown in Fig. 17.4. Note that the state centered

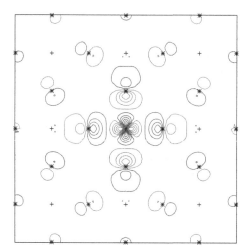

Figure 17.8. Orbital of $d_{x^2-y^2}$ symmetry centered on a Cu atom in $YBa_2Cu_3O_7$ chosen to describe
the actual band crossing the Fermi energy and derived using the NMTO method [735]. The resulting
band derived from this single orbital is shown in Fig. 17.9. From [735].

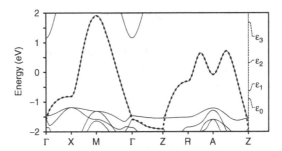

Figure 17.9. Band from the orbital shown in Fig. 17.8 (dark symbols) compared to the full bands (light symbols), which are essentially the same as the bands in Fig. 17.4. Also shown are the energies at which the band is designed to agree. The band is well described even when it has complicated shape and crosses other bands. From [735].

on one Cu atom is extended, with important contributions of neighboring O and Cu sites. The band resulting from that single orbital is shown as dark circles in Fig. 17.9, which can be compared with the states near the Fermi energy in Fig. 17.4. (Also shown are the energies at which the state is required to fit the full-band structure.) The important point is that the procedure leads to an accurate description of the desired band, without the "spaghetti" of other bands. Such a function is derived by focusing on the energy of interest and by "downfolding" the effects of all the other bands by identifying the angular momentum channels of interest in the transformation, Eq. (17.34).

Although it is beyond the scope of this book, we can draw two important conclusions about the promise of the NMTO approach. First, for MTO-type methods, it appears to remove the limitation to close-packed structures, and, second, it allows accurate solution for general structures. This means that the MTO approach can provide a "first-principles tight-binding approach" (see Sections 16.7 and 17.6) applicable to general structures of crystals and molecules. Furthermore, if calculations can be done efficiently, then NMTO calculations can provide forces and can be used in molecular dynamics. Taking a broader perspective, the NMTO approach is a promising addition to all-band structure methods, potentially providing new approaches beyond the present linearized methods.

17.10 Full Potential in Augmented Methods

One of the most important outcomes of linearization is the development of full-potential augmented methods, e.g., for LAPW [709, 713–715] and LMTO [736] methods. Although actual implementations may be cumbersome and cannot be described here, the basic ideas can be stated very simply. Since the linearized methods have been derived in terms of matrix elements of the hamiltonian in a fixed basis, one simply needs to calculate matrix elements of the full nonspherical potential ΔV in the sphere and the full spatially varying potential in the interstitial. The basis functions are still the same APW, PAW, or LMTO functions χ_L, which are derived from a spherical approximation to the full potential. However, the spheres merely denote convenient boundaries defining the regions where the basis functions and the potential have different representations. In principle, there are no approximations on the wavefunctions or the potential except for truncations at some l_{max} and G_{max}. If the basis is carried to convergence inside and outside the spheres, the accuracy is, in principle, limited only by the linearization.

Inside each sphere the potential is expanded in spherical harmonics,

$$V(r,\theta,\phi) = \sum_L V_l(r)i^l Y_{lm}(\theta,\phi), \tag{17.35}$$

so that matrix elements $\langle L|V|L'\rangle$ can be calculated in terms of radial integrals. Similarly, the interstitial matrix elements are no longer diagonal in plane waves, but they can be found straightforwardly by integrating in real space. In the PAW method and the multiple-κ LMTO method, the smooth functions continue into the sphere and it is convenient also to define the potential as a smooth part everywhere plus a sharply varying part restricted to spheres. In that case, the matrix elements of the smooth part can be calculated by FFT methods just as is done in pseudopotential methods (Section 13.1).

Of course, in the self-consistent calculation one also needs to calculate the potential arising from the density. This necessitates a procedure in which the Poisson equation is solved taking into account the sharply varying charge density inside the spheres. This is always possible since the field inside can be expanded in spherical harmonics and outside the spheres can be represented by smooth functions plus multipole fields due to the charge inside the spheres. Perhaps the simplest approach is to define both the density and the potential as smooth functions everywhere, with sharply varying components restricted to spheres [542, 709].

There is a quantitative difference between the LAPW and LMTO approaches in the requirements on the full potential. Since the minimal basis LMTO only involves functions with l_{max} given by the actual angular momenta of the primary states making up the band (e.g., $l = 2$ for transition metals), only angular momenta up to $2l_{max}$ are relevant. However, for the LAPW methods, much higher angular momenta in the wavefunctions (typically $l_{max} \approx 8$–12 for accurate calculations) are required to satisfy the continuity conditions accurately. In principle, very large values of l_{max} are needed for the potential, and in practice accurate numerical convergence can be reached with $l_{max} \approx 8$–12. The difference results from the fact that the LAPW basis is much larger; in order to represent the interstitial region accurately many plane waves are needed, which leads to the need for high angular momenta in order to maintain the continuity requirements (see Exercises 17.5–17.7).

SELECT FURTHER READING

The references at the end of Chapter 16 also treat the lineraized methods.

Andersen, O. K. and Jepsen, O., "Explicit, first-principles tight-binding theory," *Physica* 91B: 317, 1977.

Blaha, P., Schwarz, K., Sorantin, P., and Trickey, S.B., "Full-potential, linearized augmented plane wave programs for crystalline systems," *Computer Phys. Commun.* 59(2): 399, 1990.

Exercises

17.1 Derive Eq. (17.4) from the definition of $\dot{\psi}$. In addition, show the more general relation

$$(\hat{H} - \varepsilon)\psi^{(n)}(\varepsilon, r) = n\psi^{(n+1)}(\varepsilon, r), \qquad (17.36)$$

where n is the order of the derivative. Hint: use the normalization condition.

17.2 Carry out the manipulations to show that the hamiltonian and overlap matrix elements can be cast in the linearized energy-independent form of Eqs. (17.14) to (17.17). Thus the matrix elements are expressed in terms of Π and Ω, which are functions of the wavefunctions ψ and $\dot{\psi}$ calculated in the sphere at the chosen energy E_ν.

17.3 Derive the result that $l_{max} \approx 8$ in LAPW calculations. Consider a simple cubic crystal with one atom/cell with the volume of the atomic sphere $\approx 1/2$ the volume of the unit cell. The order of magnitude of ≈ 100 planes waves is reasonable since it corresponds to a resolution of $\approx 100^{1/3}$ points in each direction. If the plane waves are in a sphere of radius G_{max}, find G_{max} in terms of the lattice constant a. This is sufficient to find an estimate of l_{max} using the arguments in the text. If the number of plane waves were increased to 1,000, what would be the corresponding l_{max}?

17.4 The condition Eq. (17.23) requires that the LMTO be independent of the energy to first order and is the key step that defines an LMTO orbital; this removes the rather arbitrary form of the MTO and leads to the expression in terms of $\dot{\psi}$. Show that this condition leads to the expression, (17.24), for the J function proportional to $\dot{\psi}$ inside the sphere.

17.5 If the augmented wavefunction (LAPW or LMTO) is expanded in Y_{lm} up to l_{max}, what is the corresponding $l_{max}^{density}$ needed in an exact expansion for the charge density for the given wavefunction? Give reasons why it may not be essential to have $l_{max}^{density}$ this large in an actual calculation.

17.6 If the density is expanded in Y_{lm} up to $l_{max}^{density}$, what is l_{max} for the Hartree potential? For V_{xc}?

17.7 What is the maximum angular momentum l_{max}^{pot} of the potential Eq. (17.35) needed for exact evaluation of matrix elements $\langle L|V|L'\rangle$ if the wavefunction is expended up to l_{max}? Just as in Exercise 17.5, give reasons why smaller values of l_{max}^{pot} may be acceptable.

17.8 Consider the compound $YBa_2Cu_3O_7$. Determine the number of electrons that would be required to fill the oxygen states to make a closed-shell ionic compound. Show that for $YBa_2Cu_3O_7$ there is one too few electrons per Cu atom. Thus, this material corresponds to one missing electron (i.e., one hole per Cu).

18

Locality and Linear-Scaling O(N) Methods

Nearsightedness

W. Kohn

Throwing out k-space

V. Heine

Summary

The concept of localization can be imbedded directly into the methods of electronic structure to create new algorithms that take advantage of locality or "nearsightedness" as coined by W. Kohn. As opposed to the textbook starting point for describing crystals in terms of extended Bloch eigenstates, many physical properties can be calculated from the *density matrix* $\rho(\mathbf{r}, \mathbf{r}')$, which is exponentially localized in an insulator or a metal at finite T. For large systems, this fact can be used to make "order-N" or O(N) methods where the computational time scales linearly in the size of the system. There are two aspects of the problem: building the hamiltonian and solving the equations. Solving the problem presents more fundamental issues, but constructing the hamiltonian is essential for useful methods. In this chapter are representative O(N) approaches that either treat $\rho(\mathbf{r}, \mathbf{r}')$ directly or work in terms of Wannier-like localized orbitals.

18.1 What Is the Problem?

Every textbook on solid state physics begins with the symmetry of crystals. The entire subject of electronic structure is cast in the framework of the eigenstates of the hamiltonian classified in k space by the Bloch theorem. So far this volume is no exception. However, the real goal is to understand the properties of materials from the fundamental theory of the electrons and this is not always the best approach, neither for understanding nor for calculations.

What does one do when there is no periodicity, e.g., an amorphous solid or a liquid or a large molecule? One approach is to use periodic boundary conditions on artificial "supercells" chosen to be large enough that the effects of the boundary conditions are small or can be removed from the calculation by an analytic extrapolation procedure. This approach can be very effective and is widely used, especially with efficient plane wave methods (Chapter 13) and molecular dynamics simulations (Chapter 19).

The subject of this chapter is an alternative approach built upon the general principle of *locality*, i.e., that properties at one point can be considered independent of what happens at distant points. If the theory is cast in a way that takes advantage of the locality, this can lead to algorithms that are "linear scaling" ("order-N" or $O(N)$), i.e., the computational time and computer memory needed is proportional to N as the number of particles N is increased to a large number. $O(N)$ methods emerge naturally in classical mechanics. If there are only short-range interactions, the forces on each particle depend on only a small number of neighboring particles. One step in a molecular dynamics calculation can be done updating the position of each particle in time $\propto N$.[1] The same conclusion holds even if there are long-range Coulomb forces, since there are various methods to sum the long-range forces in time $\propto N$ [737].

The problem is that quantum mechanics is inherently *not* local: eigenstates in extended systems, in general, are extended and the solutions of the wave equation depend on boundary conditions. Indistinguishability of identical particles requires that the wavefunctions obey the symmetry or antisymmetry conditions among all particles, whether they are nearby or far away. Quantum entanglement of pairs or groups of particles can occur at a large distance.[2] In practical applications with independent-particle methods, the ground state of N electrons is an antisymmetric combination of N eigenstates, each of which is, in general, extended through a volume also proportional to N. Working in terms of the independent-particle eigenstates leads to scaling at least $\propto N^2$, and specific methods are often worse. Full matrix diagonalization scales as N_{basis}^3, where the number of basis functions $N_{\text{basis}} \propto N$. The widely used Car–Parrinello-type algorithms in a plane wave basis scales as $N^2 N_{\text{basis}}$, which is much better since $N \ll N_{\text{basis}}$. Solutions of correlated many-body problems, in general, scale much worse, growing exponentially in N for exact solutions and as high powers for practical configuration interaction calculations [738].

Despite the inherent nonlocality in quantum mechanics, many important properties can be found without calculating the eigenstates, using information that is only "local" (as defined in the following section). For example, the density and total energy are integrated quantities that are invariant to unitary transformations of the states, and these quantities are sufficient to determine the stable ground state and the force on every nucleus. In this chapter we discuss algorithms for calculation of such quantities with computational time $\propto N$. It is not possible to cover all methods; here we discuss selected examples that illustrate the developments and several of the available codes are listed in Appendix R. More extensive

---

[1] It is a more difficult question if all desired properties can be found in time proportional to the total size of the system; for example, there could be slow relaxation modes that become increasingly difficult to determine as the system size increases. We will not deal with such issues here.

[2] Topological insulators have introduced a new aspect of the problem: entanglement of the electron wavefunctions over macroscopic distances. One manifestation is surface states that can exist only because of properties of the bulk. This is not a short-range effect and it is closely related to the conclusion that it is not possible to construct localized Wannier functions in a system with nontrivial topology of the electronic structure (see Section 25.8). These are properties that can be considered separately from the issues considered in this chapter.

exposition of linear-scaling methods and codes can be found in the review by Bowler and Miyazaki [739] and developments related to chemistry and biology are the topics of [740] and [741].

18.2 Locality in Many-Body Quantum Systems

The term "nearsightedness" has been coined by Walter Kohn [742] for such integrated quantities that can be calculated at one point **r** in terms only of information at points **r'** in a neighborhood of **r**. This embodies the ideas developed by Friedel [733] and Heine and coworkers [499] on locality and other work such as the 1964 paper by Kohn, "The nature of the insulating state," in which the key idea is the localization of electronic states in insulators. Nearsightedness is a property of a many-body system of particles: the density of an individual eigenstate at any point is dependent on the boundary conditions and the potential at all other points; however, for systems of many particles, the net effect is reduced due to interference between the different independent-particle eigenstates (i.e., in the sum in Eqs. (18.1) and (18.2) below). In insulators and metals at nonzero temperature, the one-electron density matrix decays exponentially, and in metals at $T = 0$ as a power law ($1/R^3$ in three dimensions), with Gibbs oscillations due to the sharp cutoff at the Fermi surface. Interactions introduce correlations with the longest range being van de Waals type that decay as $1/R^6$ in the energy and $1/R^3$ in the wavefunction [738]. We will use the term "local" in this sense to mean "dependent on only distant regions to an extent which decays rapidly": exponential decay is sufficient to ensure convergent algorithms; the power law decay in metals at $T = 0$ is problematic [743] but a possible approach is to use the fact that the states near the Fermi energy vary smoothly with energy [744].

All the linear-scaling O(N)methods discussed here take advantage of the decay of the density matrix with distance, which is zero or can be truncated at some point, as depicted in Fig. 18.1. One of the great advantages of the density matrix formalism is that it is a general approach applicable at finite temperature, where all correlations become shorter range. Therefore, in general, the range of the density matrix is decreased and O(N) algorithms should become more feasible and efficient. In addition, the possibility of continuous variation of fractional occupation of states is of great advantage in problematic cases where occupation must be fractional by symmetry or jumps discontinuously at $T = 0$. This problem is well known in standard electronic structure algorithms and is discussed in Chapter 7, where a fictitious temperature is often added to improve calculations by smoothing details of the state distribution.

For noninteracting particles, the density matrix can be written (see Section 3.6) as

$$\hat{\rho} = \sum_{i=1}^{M} |\psi_i\rangle f_i \langle\psi_i| \text{ or } \rho(\mathbf{r},\mathbf{r}') = \sum_{i=1}^{M} \psi_i^*(\mathbf{r}) f_i \psi_i(\mathbf{r}'), \tag{18.1}$$

where $f_i = 1/(1 + \exp(\beta(\epsilon_i - \mu))$ is the Fermi function and $\beta = 1/k_B T$. For $T \neq 0$, the number of states M must be greater than the number of electrons N, which is related to the Fermi energy μ by $N = \sum_{i=1}^{M} f_i$. At $T = 0$ this becomes

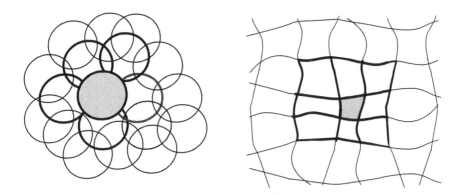

Figure 18.1. Schematic diagram of treating a quantum system in terms of overlapping regions, often in terms of functions centered on atoms (left) or with strictly nonoverlapping regions as shown at the right for an irregular disordered system. A typical region at the center interacts only with neighboring regions (dark lines), which generates a sparse hamiltonian. Different methods use such divisions to create linear-scaling O(N) methods.

$$\hat{\rho} = \sum_{i=1}^{N} |\psi_i\rangle\langle\psi_i| \text{ or } \rho(\mathbf{r}, \mathbf{r}') = \sum_{i=1}^{N} \psi_i^*(\mathbf{r})\psi_i(\mathbf{r}'). \tag{18.2}$$

Even if the independent-particle eigenfunctions are extended, the density matrix is exponentially localized and vanishes at large relative separation $|\mathbf{r} - \mathbf{r}'|$ in all cases for $T \neq 0$ (see discussion following Eq. (5.10)) and at $T = 0$ in an insulator.

The sum of independent-particle energies E_s, written in terms of eigenstates, is given by

$$E_s = \sum_{i=1}^{M} \frac{1}{1 + \exp(\beta(\epsilon_i - \mu))} \epsilon_i, \tag{18.3}$$

where the sum is over all eigenvalues. (This is also sufficient for Kohn–Sham theory (Section 7.3) where the total energy can always be found from E_s plus terms that involve only the density $n(\mathbf{r}) = \rho(\mathbf{r}, \mathbf{r})$.) In general, one can rewrite the sum over all eigenvectors as a trace, so that energy can be written

$$E_s = \text{Tr}\left\{ \frac{1}{1 + \exp(\beta(\hat{H} - \mu))} \hat{H} \right\} = \text{Tr}\{\hat{\rho}\hat{H}\}. \tag{18.4}$$

Similarly, the grand potential that describes the energy for different numbers of particles is given by

$$\Omega_s = \text{Tr}\left\{ \frac{1}{1 + \exp(\beta(\hat{H} - \mu))} (\hat{H} - \mu) \right\} = \text{Tr}\{\hat{\rho}(\hat{H} - \mu)\}. \tag{18.5}$$

The difficulty in using the Fermi function is the exponentiation of the operators that are nondiagonal, if one wishes to avoid diagonalization, e.g., in an O(N) scheme.

At $T = 0$ in an insulator, the expressions are closely related to the construction of Wannier functions. The N eigenstates can be transformed to N localized orthogonal Wannier-like functions w_i (Chapter 23)

$$\hat{\rho} = \sum_{i=1}^{N} |w_i\rangle\langle w_i|, \quad T = 0. \tag{18.6}$$

It may also be advantageous [745, 746] to work in terms of nonorthogonal functions $\tilde{w}_i$ which are more localized, so that

$$\hat{\rho} = \sum_{i=1,j}^{N} |\tilde{w}_i\rangle S_{ij}^{-}\langle \tilde{w}_j|, \quad T = 0, \tag{18.7}$$

where S^- is the inverse of the overlap matrix.[3] For $T \neq 0$ the form of the decay can be derived for model problems, such as the free-electron gas discussed in Section 5.2.

For all cases, the challenge is to determine efficient, robust ways to find the density matrix $\hat{\rho}$ or the generalized Wannier functions w_i or $\tilde{w}_i$. In the latter case, the functions are not unique, which leads to possible advantages that can accrue by using particular choices and possible problems due to approximations that violate the invariance of the functionals.

There are two aspects to the creation of linear-scaling methods, both relying upon sparsity of the hamiltonian and overlap matrices, assumed to have nonzero elements only for a finite range as shown schematically in Fig. 18.1:

- Building the hamiltonian, i.e., generating the nonzero matrix elements in a sparse form that is linear in the size of the system. For many approaches, this is the rate-limiting step for sizes up to hundreds of atoms, and therefore it is relevant even if the solution is done with traditional $O(N^2)$ or $O(N^3)$ methods.
- Solving the equations. This is the more fundamental aspect that is necessary for O(N) scaling in the large N limit. We will consider approaches that treat $\rho(\mathbf{r}, \mathbf{r}')$ (or a Green's function) or work in terms of Wannier-like localized orbitals. The key division is between "nonvariational methods," which truncate well-known expansions, and "variational methods," which work with variational functionals.

18.3 Building the Hamiltonian

The hamiltonian can be constructed in a sparse form in real space in any case in which the basis functions are localized to regions much smaller than the system size. The most obvious basis for such an approach are local atomic-like orbitals, as in Chapter 14, which are illustrated at the left in Fig. 18.1. The tight-binding model approach is ideal for this purpose since the matrix elements of $\hat{H}$ and the overlap matrix $\hat{S}$ are defined to be short range. In the full local orbital method, matrix elements of the hamiltonian vanish beyond some distance

---

[3] Here we use the notation of [746], which generalizes the definition of S^{-1} as shown later in Eq. (18.28).

only if the orbitals are strictly localized. This can be accomplished in a basis of numerical orbitals purposefully chosen to be strictly localized as described in Section 15.4. This is the approach in the SIESTA code used for the calculation shown in Fig. 18.8 and in the FHI-aims all-electron code. In general, analytic bases such as gaussians decay exponentially but never vanish entirely, so they must be used with care.

The real space methods in Sections 12.8 and 12.9 naturally suggest ways to use localized functions. The CONQUEST and ONETEP codes use grids with representations in localized regions done respectively with Blip (b-splines) and with FFT restricted to a box. The construction of discontinuous functions illustrated in Fig. 12.6 generates nonoverlapping basis function like those indicated at the right in Fig. 18.1. Finite elements and multiresolution methods involve functions that are overlapping but are strictly localized as described in Section 12.8. For example, the BigDFT code is based on Daubechies wavelets, which have been used for calculations on large systems [588].

The augmented approaches can also be cast in localized forms. LMTOs can be transformed to an orthogonal tight-binding form, in which the hamiltonian has a power law decay (Section 17.6) that is much shorter range than in the original method. In Green's function methods, such as KKR, G is generated in terms of G_0, which is very long range for positive energies, but decays exponentially for negative energies. This has been used to construct an $O(N)$ KKR method [704] in which G_0 is the numerically calculated Green's function for an electron in an array of repulsive centers, as described in Section 16.7. In these methods one must invert a matrix labeled by the orbital quantum numbers at the atomic sites, and linear scaling is accomplished in constructing the hamiltonian or Green's function matrix including only neighbors in a "local interaction zone." The approach, termed "locally self-consistent multiple scattering" (LSMS), involves a calculation on a finite cell around each site; an alternative approach is to use Lanczos or recursion methods to solve the multiple scattering problem around each site [747].

It is, however, *not essential* that the basis be localized. One of the original ideas is due to Galli and Parrinello [748], who combined the plane wave Car–Parrinello algorithm (Chapter 19) with transformations of the wavefunctions to a localized form as in Eq. (18.5). Physical properties are unchanged due to the invariance of the trace. By constraining orbitals to be localized to regions, they showed that one can construct an algorithm that automatically generates localized functions and a sparse hamiltonian, even though the plane wave basis is not localized.

18.4 Solution of Equations: Nonvariational Methods

Green's Functions, Recursion, and Moments

The original ideas for electronic structure methods that take advantage of the locality grew out of Green's function approaches, using the facts that the density matrix and the sum of eigenvalues are directly expressible in terms of integrals over the Green's function. The basic relations are given in Section D.5 and in Eqs. (16.33)–(16.35), which we rewrite here in slightly different notation. If the basis states are denoted χ_m (here assumed to be orthonormal for simplicity), the local density of states projected on state m is given by

$$n_m(\varepsilon, \mathbf{R}) = -\frac{1}{\pi} \mathrm{Im} G_{m,m}(\varepsilon + i\delta). \tag{18.8}$$

For example, m might denote a site and basis orbital on that site. The sum of eigenvalues of occupied states projected on state m can be found from the relation

$$\sum_i \varepsilon_i |\langle i | \chi_m \rangle|^2 = -\frac{1}{\pi} \int_{-\infty}^{E_F} d\varepsilon \, \varepsilon \, \mathrm{Im} G_{m,m}(\varepsilon + i\delta, 0), \tag{18.9}$$

and the total sum of eigenvalues is given by

$$\sum_i \varepsilon_i = -\frac{1}{\pi} \sum_m \int_{-\infty}^{E_F} d\varepsilon \, \varepsilon \, \mathrm{Im} G_{m,m}(\varepsilon + i\delta, 0). \tag{18.10}$$

The left-hand sides of Eqs. (18.9)–(18.10) are in the standard eigenstate form, whereas the right-hand sides are in the form of Green's functions that can be evaluated locally. This provides all the information needed to determine the total energy and related quantities from integrals over the Green's functions.

Elegant methods have been devised to calculate the Green's functions as local quantities. The basic idea can be illustrated by Fig. 18.1 in which we desire to determine the Green's function $G_{0,0}(\varepsilon + i\delta)$ for the orbital 0 shown as dark gray. This can be accomplished by repeated applications of the hamiltonian, called recursion [732], which has close relations to the Lanczos algorithm (Section M.5). The ideas are used in tight-binding approaches (summarized, e.g., in [749]), KKR Green's function methods (see Section 16.3 for the basic ideas and references such as [704] and [747] for O(N) algorithm developments), and LMTO Green's function methods that use recursion (see, e.g., [725]). The diagonal elements of the Green's function can be used to find the charge density and the sum of single-particle energies; however, there are difficulties in finding the off-diagonal elements of the Green's function needed for forces. This problem is addressed by "bond-order" potential methods [750–752].

The basic idea of the recursion method [732, 753] is to use the Lanczos algorithm (Section M.5) as a method to construct a Green's function, using the properties of a tridiagonal matrix. The Lanczos recursion relation, Eq. (M.9), for the hamiltonian applied to a sequence of vectors,

$$\psi_{n+1} = C_{n+1}[\hat{H}\psi_n - H_{nn}\psi_n - H_{nn-1}\psi_{n-1}], \tag{18.11}$$

generates the set of vectors ψ_n and the coefficients in the tridiagonal matrix Eq. (M.10), $\alpha_n = H_{nn}$ and $\beta_n = H_{n-1n}$. The normalization constant is readily shown (Exercise 18.1) to be $C_{n+1} = 1/\beta_n$, $n \geq 1$. If the starting vector ψ_0 is a basis state localized at a site,[4] and the hamiltonian is short range, then the algorithm generates a sequence of states in "shells"

---

[4] Here 0 denotes the starting state, which can be any state in the basis, e.g., any of the localized atomic-like states on site i.

around the central site. The diagonal part of the Green's function for state 0 is given by the continued fraction [732, 749, 753]

$$G_{0,0}(z) = \cfrac{1}{z - \alpha_0 - \cfrac{\beta_1^2}{z - \alpha_1 - \cfrac{\beta_2^2}{z - \alpha_2 - \cfrac{\beta_3^2}{\ddots}}}} \qquad (18.12)$$

where z is the complex energy.

The properties of the continued fraction and a proper termination are the key features of the recursion method, and an introductory discussion can be found in [754]. If the fraction is terminated at level N, then the density of states consists of N delta functions that is useful only for integrations over $G(z)$. If a constant imaginary energy δ is used as a terminator, this is equivalent to a lorentzian broadening of each delta function. However, there are other approaches that are just as simple but much more elegant and physical. One follows from the observation [754] that the coefficients α_n and β_n tend to converge quickly to asymptotic values that can be denoted α_∞ and β_∞. If one sets the coefficients $\alpha_n = \alpha_\infty$ and $\beta_n = \beta_\infty$ for all $n > N$, then one can evaluate the remainder of the fraction analytically. This leads to replacement of the β_{N+1}^2 coefficient by $t(z)$ given by

$$t(z) = \cfrac{1}{z - \alpha_{N+1} - \cfrac{\beta_{N+2}^2}{z - \alpha_{N+2} - \cfrac{\beta_{N+3}^2}{z - \ldots}}} = \frac{1}{z - \alpha_\infty - \beta_\infty^2 t(z)}, \qquad (18.13)$$

which has the solution [754] (Exercise 18.3)

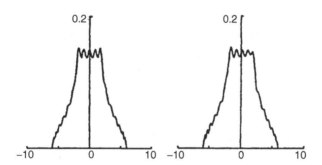

Figure 18.2. Density of states (DOS) for an s band in a simple cubic lattice with nearest-neighbor interactions. The recursion is up to $N = 20$ levels on cells of size $(21)^3$ with open boundary conditions (left) and periodic boundary conditions (right). Compared to the schematic DOS in Fig. 14.4 recursion yields correct features, although there are oscillations. The similarity of the two results demonstrates the insensitivity of local properties to the boundary conditions. (The small asymmetry results from odd-length paths to its image in an adjacent cell.) From [754].

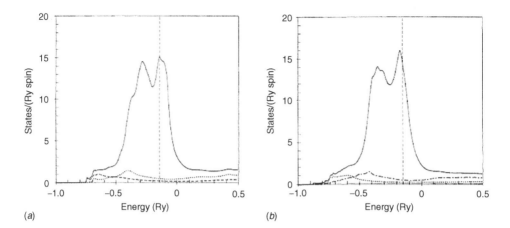

Figure 18.3. Electronic density of states (DOS) for liquid Fe (a) and Co (b) calculated with the tight-binding LMTO method (Section 17.6) and recursion [726]. The liquid was simulated by 600 atom cells generated by classical Monte Carlo and empirical interatomic potentials. The density is similar to the crystal (see, for example, the canonical DOS in Fig. 16.12) broadened by the disorder. From [726].

$$t(z) = \frac{1}{2\beta_\infty^2} \left\{ (z - \alpha_\infty) - \left[(z - \alpha_\infty)^2 - 4\beta_\infty^2 \right]^{1/2} \right\}. \tag{18.14}$$

This has the appealing form of a terminator that is an analytic square root function, which is real outside the range $\alpha_\infty \pm 2|\beta_\infty|$ and has a branch cut corresponding to a band width $4|\beta_\infty|$. The ideal choice is that which yields the correct band width. In actual practice it is important not to choose $|\beta_\infty|$ too small in which case there are spurious delta functions in the range between the real band width and the approximate interval $\alpha_\infty \pm 2|\beta_\infty|$. If the range is overestimated, there is broadening but no serious problems.

An example of the use of the recursion method is shown in Fig. 18.2 for the density of states (DOS) for an s band in a simple cubic lattice with nearest-neighbor interactions. The bands are given analytically in Section 14.5 and the DOS is shown schematically in Fig. 14.4. In comparison, the DOS calculated with the recursion method up to $N = 20$ levels shows correct features, but with added oscillations. Other types of terminators can improve the convergence to the exact result [732]. The two results shown in Fig. 18.2 are for open and periodic boundary conditions, showing the basic point of the insensitivity of local properties to the boundary conditions.

The recursion method is very powerful and general. It is not limited to tight-binding and it can be applied to *any* hermitian operator to generate a continued fraction form for the Green's function. This is a stable method to generate the density of states so long as care is taken in constructing a proper terminator for the continued fraction [732, 749, 753]. An example is shown in Fig. 18.3, which shows the electronic density of states of liquid Fe and Co determined using the tight-binding LMTO method (Section 17.6) and recursion [726]. The actual calculations were done on 600 atom cells with atomic positions generated by classical Monte Carlo methods with empirical interatomic potentials. This illustrates a powerful combination of the recursion approach with a basic electronic structure method that is now widely used for many problems in complicated structures and disordered systems.

Determination of the moments of the density of states is also a powerful tool to extract information in an O(N) fashion. Moments of the local DOS for basis function m,

$$\mu_m^{(n)} \equiv \langle m|[\hat{H}]^n|m\rangle, \tag{18.15}$$

in principle contain all the information about the local DOS, and thus about all local properties. The basic ideas are described in Section L.7, where the two issues, generating the moments efficiently and inverting the moment information to reconstruct the DOS are emphasized.

There are a number of ways to generate the moments; one approach, which is very stable, is to construct the moments in terms of the tridiagonal matrix elements generated by the Lanczos algorithm. For the state labeled 0, the moments can be written [749]

$$\mu_0^{(0)} = 1,$$
$$\mu_0^{(1)} = \alpha_0,$$
$$\mu_0^{(2)} = \alpha_0^2 + \beta_1^2, \tag{18.16}$$
$$\mu_0^{(3)} = \alpha_0^3 + 2\alpha_0\beta_1^2 + \alpha_1\beta_1^2,$$
$$\mu_0^{(4)} = \alpha_0^4 + 3\alpha_0^2\beta_1^2 + 2\alpha_0\alpha_1\beta_1^2 + \alpha_1^2\beta_1^2 + \beta_1^2\beta_2^2 + \beta_1^4,$$

and so forth.

Inversion of the moments to find the DOS is the well-known, difficult, "classical moment problem" [822] discussed in Section L.7. As an example of the construction of the density of states using moments for an actual problem done in O(N) fashion, Fig. 18.4 shows the calculated [755] density of states (DOS) of the series of fullerenes (carbon clusters with flat graphene-like facets and 12 pentagons at the vertices, calculated by an O(N) method [756] as described in Section 18.5) showing the progression from discrete spectra small molecules to the continuous spectrum of graphene with critical points of a two-dimensional crystal. The results for the large fullerenes up to C_{3840} demonstrate the exquisite details that can be resolved with $\approx$150 moments using the maximum entropy method [757] discussed in Section L.7. Similarly, the phonon spectrum can be calculated in an O(N) manner [758].

Bond Order and Forces in Recursion

A key problem with the recursion method is the difficulty in computing off-diagonal matrix elements of G that are essential for forces. The expression in terms of the density matrix is given in Eq. (14.28); omitting the simple two-body term, the contribution from the sum of eigenvalues is given by

$$\mathbf{F}_I = -\text{Tr}\left\{\hat{\rho}\frac{\partial\hat{H}}{\partial\mathbf{R}_I}\right\} = -\sum_{m,m'}\rho_{m,m'}\frac{\partial H_{m,m'}}{\partial\mathbf{R}_I}. \tag{18.17}$$

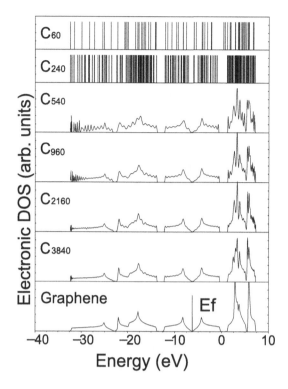

Figure 18.4. Density of states (DOS) for electrons in the fullerenes with size from C_{60} to C_{3840} [755] calculated using tight-binding model of [619]. The structures were also calculated by an O(N) method [756] as described in Section 18.5. The smaller molecules and graphene were computed using diagonalization methods and the DOS for the larger fullerenes was calculated by the method of moments as implemented in [757] (see Section L.7). The figure illustrates the evolution of the DOS from delta functions of the C_{60} molecule to the spectrum of graphene shown in the bottom panel, which has the critical point features of a two-dimensional DOS, as shown in Fig. 14.4. From [755].

General expressions in terms of Green's functions with localized bases are given by Feibelman [663]. The generic problem is the calculation of 'bond order,'' which is the off-diagonal components of the density matrix $\rho_{m,m'}$ that correspond to bonding and are given by integrals over $G_{m,m'}(z)$ analogous to Eqs. (18.8)–(18.10). There has been considerable work to derive efficient recursion-type expressions for the bond order [750–752]. The basic idea is that off-diagonal terms $G_{m,m'}(z)$ can be calculated using recursion with a starting vector

$$\psi_0 = \frac{1}{\sqrt{2}}[\chi_m + e^{i\theta}\chi_{m'}], \tag{18.18}$$

and computing $G_{m,m'}(z)$ from

$$G_{m,m'}(z) = \frac{\partial G_{0,0}^{\lambda}(z)}{\partial \lambda}, \tag{18.19}$$

with $\lambda = \cos(\theta)$. Further generalizations of this idea have been derived as described in [752] with a summary in [749].

"Divide and Conquer" or "Fragment" Method

One of the first $O(N)$ methods is based directly on the argument that the interior of a large region depends only weakly on the boundary conditions. The procedure termed "divide and conquer" [759] or a fragmentation method (see [740] for a review) is to divide a large system into small subsystems each of size N_{small}, for example the central orbital (gray) plus the set of orbitals shown as heavy circles in the left side of Fig. 18.1. For each of these systems one can solve for the electronic eigenstates using ordinary N^3 methods. For each small system one must add buffer regions of size N_{buffer} (the outer orbitals in Fig. 18.1) large enough so that the density and energy in the original small subsystem converges and is independent of the buffer termination. The solution for the density and other properties is then kept only for the interior of each small region. The division of the system into nonoverlapping regions shown in the right side of Fig. 18.1 also involves using a buffer to calculate the functions in a region, as explained in Fig. 12.6. Using traditional methods, the cost is of order $(N_{small} + N_{buffer})^3$ for each subsystem, which may be prohibitive, especially for three-dimensional systems where N_{buffer} may need to be very large. However, the calculations for each subsystem can be done in parallel and the overall method is $O(N)$ for large enough systems. This approach is particularly applicable for long linear molecules with large energy gaps (i.e., small localization lengths (see Chapter 24)) that are important in biochemistry (see, e.g., [741] which makes case for large-scale DFT calculations).

Polynomial Expansion of the Density Matrix

One class of methods uses the form in Eq. (18.4) directly and expands the Fermi function in Eq. (18.4) in powers of the hamiltonian. This is the approach used, e.g., in [760] and [761]. The basic requirement is for the expansion to have the same properties as the Fermi function, i.e., that all states with eigenvalues far above the Fermi energy μ have vanishing weight, all those well below μ have unity weight, and the variation near μ is reproduced accurately. This is difficult to accomplish in an expansion; however, if the eigenvalues are limited to the range $[E_{min}, E_{max}]$, then the expansion can be done efficiently using Chebyshev polynomials T_n defined in Section K.5. The advantage of the Chebyshev polynomials is that they are orthogonal and fit the function over the entire range $[-1, +1]$ (see Section K.5) and they can be generated recursively using the relation (K.19). Let us define $\Delta E = E_{max} - E_{min}$, the scaled hamiltonian operator $\tilde{H} = (\hat{H} - \mu \hat{I})/\Delta E$, and the scaled temperature $\tilde{\beta} = \beta \Delta E$. Then we can express the expansion of the Fermi operator as

$$\hat{F}[\hat{H}] = \frac{1}{1 + e^{\beta(\hat{H}-\mu)}} \rightarrow \frac{c_0}{2}\hat{I} + \sum_{j=1}^{M_p} c_j T_j(\tilde{H}). \tag{18.20}$$

The highest power needed in the expansion depends on the ratio of the largest energy in the spectrum to the smallest energy resolution required. The higher the temperature the lower the power needed, since the Fermi function is smoother. For a metal, T must be of the order of the actual temperature (or at least smaller than the energy scale on any variations in the states near μ). For insulators, a larger effective T can often be chosen so long as states below (above) the gap are essentially filled (empty). For hamiltonians that are bounded (such as in tight-binding), the ratio E_{max}/T can be estimated and powers ≈ 10 to 100 are needed for realistic cases.

The key idea is that all operations can be done by repeated applications of $\hat{H}$ to a basis function, which amounts to repeated multiplication of a matrix times a vector. Each of the basis functions shown in Fig. 18.1 is treated one at a time.[5] In the general case, this procedure scales as N_{basis}^2, since it involves multiplication of a vector by the matrix. However, if $\hat{H}$ is sparse, only a few matrix elements are nonzero (e.g., the nonzero elements in a tight-binding hamiltonian that involve only a few neighbors) and the multiplication, times one localized basis function, is independent of the size of the system. Furthermore, if we invoke the localization property of the density matrix, all matrix operations can be made sparse, so that the calculation scales as order $N_{basis} \propto N$. This method has two great advantages: (1) the computation scales linearly with the number of other basis functions included in the maximum range, compared to cubic scaling in the "divide and conquer" and the variational function methods and (2) the algorithm is perfectly parallel. Major disadvantages are (1) that the results of the expansion are not variational since the truncation errors can be of either sign and (2) the fact that an independent calculation must be done for each orbital means that information is discarded, in comparison to the variational methods that exchange information between the subparts of the system.

The Chebyshev expansion method can be a very efficient procedure if the basis set is small, e.g., in tight-binding models, where M is only a factor of order 2 larger than N. However, it fails for unbounded spectra, because the polynomial expansion must be taken to higher and higher orders as the range of the spectrum is increased. For a typical plane wave calculation, high powers would be required since the energy range is large and $N_{basis} \gg N$.

Inverse Power Expansion of the Density Matrix

There are several approaches that work with an unbounded spectrum employing operators that properly converge at high energies. One is the inverse power method, which is in essence a Green's function approach, Section D.5, and is closely related to the recursion method, Section M.5. In this approach, one expands the Fermi function as follows:

$$\hat{F}[\hat{H}] = \frac{1}{1 + \exp[\beta(\hat{H} - \mu)]} \rightarrow \sum_{i=1}^{M_{pole}} \frac{w_i}{\hat{H} - z_i}. \qquad (18.21)$$

Using the well-known relations for contour integration, the sum over poles on the real axis can be converted into an integral in the complex plane enclosing the poles [306, 704, 763]

---

[5] The Chebyshev polynomial expansion can also be used with random vectors to generate a statistical estimate of extensive quantities, such as the total energy [757, 762], "maximum entropy," or related methods. This is useful for extremely large matrices where it is not feasible to take the trace.

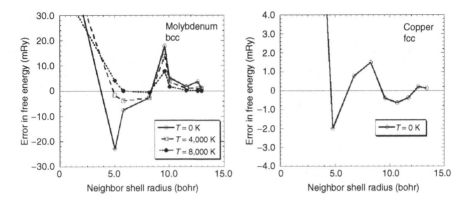

Figure 18.5. Total energy of Cu and Mo calculated using multiple-scattering theory with localized regions, plotted as a function of the radius of the localization region [704]. These are, in fact, hard cases, since the density matrix is most delocalized in perfect crystals, leading to the sharp variations shown. Provided by Y. Wang; similar to [704]

as discussed in Section D.5 and shown schematically in Fig. D.1. For each of the terms with z_i in the complex plane, the inverse operator $\hat{G}_i(z_i)$ can be found in a basis by solving linear equations $(\hat{H} - z_i)\hat{G}_i(z_i) = I$. In terms of basis functions in real space, the operators $\hat{G}_i(z_i)$ are more localized for large complex z_i, which is advantageous for calculations (Exercise 18.5). Their maximum range is where the contour crosses the real axis. In order to describe the contour integral accurately in an insulator one needs the number of poles $M_{pole} \propto (\mu - E_{min})/E_{gap}$; in a metal there is no gap and an accurate evaluation of the integral near the axis requires poles with a more dense spacing $\propto T$. An advantage of this approach is that, unlike the power expansion, it always converges independent of the high-energy spectrum. Furthermore, it corresponds to the physical picture that high-energy processes are more localized, and low-energy ones more delocalized. An example of multiple-scattering Green's function calculations [704] is shown in Fig. 18.5.

Exponential Operators

Perhaps the most fundamental approach of all is to work directly with the time-dependent Schrödinger equation. The density matrix for quantum statistics is $\exp(-\beta\hat{H})$, which is equivalent to the imaginary time propagator. Thus essentially the same techniques can be used as in the real-time methods of Section 21.6. As $\beta \to \infty$, the operation of $\exp(-\beta\hat{H})$ on any wavefunction Ψ projects out of the ground state provided it is not orthogonal to Ψ. The expressions can be evaluated using the fact that $\exp(-\beta\hat{H}) = (\exp(-\delta\tau\hat{H}))^n$, where $\delta\tau = \beta/n$ represents a temperature that is higher by a factor n. In the high-temperature, short-time regime, the operations can be simplified using the Suzuki–Trotter decomposition, as in Eq. (21.18) rewritten here,

$$\exp[-\delta\tau(T + V)] \simeq \exp\left(-\frac{1}{2}\delta\tau V\right) \exp(-\delta\tau T) \exp\left(-\frac{1}{2}\delta\tau V\right), \qquad (18.22)$$

which is factored into exponentials of kinetic and potential terms. One approach for the kinetic term is to use an implicit method that solves linear equations (see Section M.10 and [764]). One can also use FFTs to transform from real space (where V is diagonal) to reciprocal space (where T is diagonal) as in Section M.11. The former is widely used for quantum dots [764] and the latter has been used in simulations, e.g., of hydrogen fluid at high temperature and pressure [765].

18.5 Variational Density Matrix Methods

Two properties must be satisfied[6] by the density matrix $\hat{\rho}$ at $T = 0$:

- "Idempotency," which literally means $\hat{\rho}^2 = \hat{\rho}$, which is equivalent to requiring all eigenvalues of $\hat{\rho}$ to be 1 or 0.
- The eigenvectors of the density matrix with eigenvalue 1 are the occupied eigenvectors of the hamiltonian.

Li et al. [767] showed how to use a minimization method to drive the density matrix to its proper $T = 0$ form. See also Exercise 18.7 for further details. The starting point is the "McWeeny purification [275]" idea: if $\tilde{\rho}_{ij}$ is an approximate trial density matrix with eigenvalues between 0 and 1, then the matrix $3\tilde{\rho}_{ij}^2 - 2\tilde{\rho}_{ij}^3$ is always an improved approximation to the density matrix with eigenvalues closer to 0 or 1. This is illustrated in Fig. 18.6 (left panel), which shows the function $y = 3x^2 - 2x^3$. It is easy to see that for $x < 1/2$, $y < x$, i.e., the occupation is closer to zero, whereas for $x > 1/2$, $y > x$, i.e., the occupation is closer to one. However, if one iterates the matrix using the purification equation alone, there is no reason for the eigenvectors to satisfy the second requirement, i.e., to correspond to the lowest-energy states. In order to make a functional that when minimized yields the proper idempotent density matrix that also minimizes the total energy, one can modify the usual expression, (18.5), for the grand potential at $T = 0$ to use the "purified" form,

$$\Omega_s = \mathrm{Tr}\hat{\rho}(\hat{H} - \mu) \rightarrow \mathrm{Tr}(3\hat{\rho}^2 - 2\hat{\rho}^3)(\hat{H} - \mu). \tag{18.23}$$

Since the functional is minimum for the true density matrix, the energy given by Eq. (18.23) is variational.

The functional can be minimized by iteration using the gradients

$$\frac{\partial \Omega_s}{\partial \hat{\rho}} = 3[\hat{\rho}(\hat{H} - \mu) + (\hat{H} - \mu)\hat{\rho}] - 2[\hat{\rho}^2(\hat{H} - \mu) + \hat{\rho}(\hat{H} - \mu)\hat{\rho} + (\hat{H} - \mu)\hat{\rho}^2], \tag{18.24}$$

which denotes the matrix expression for the derivative with respect to each of the elements of the density matrix $\hat{\rho}$. So long as the density matrixis never allowed to go into the

---

[6] The density matrix minimization approach can also be extended to finite temperature. Corkill and Ho [766] have used the Mermin functional to allow a continuous variation of the occupation instead of the strict idempotency requirements of the original method.

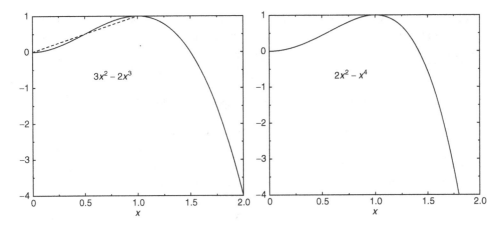

Figure 18.6. Functional forms for the cubic "McWeeny" purification algorithm [767] for the density matrix (left) and the quadratic Mauri–Ordejon–Kim functional [768–770] for the wavefunctions (right). "Purification" of the density matrix results because the function $3x^2 - 2x^3$ is always closer to 0 or 1 than the input value x, as shown by the dashed line ($= x$). The form $2x^2 - x^4$ applied to the localized wavefunctions is minimized for occupied functions normalized to 1, empty functions to 0.

unphysical region where the eigenvalues are <0 or >1, then the algorithm is stable and the gradients always point toward a lower energy and an improved density matrix. Any matrix satisfying the physical conditions can be used as a starting point. A possible choice is $\rho_{ij} = \delta_{ij}(N_{\text{elec}}/N_{\text{basis}})$, and more optimal choices can be found for any particular problem. The method can be extended to nonorthogonal bases [771]. A principle difficulty with this method is that it requires explicit operations of multiplying matrices that are of the size of the basis N_{basis}. Thus it is appropriate for small bases such as in tight-binding, but not for large bases, such as plane waves where $N_{\text{basis}} \gg N_{\text{elec}}$ (but see Section 18.8 for alternative methods).

An example of the density matrix purification algorithm of [767] is the study of selected large icosahedral fullerenes (up to C_{8640}) for which the structures were optimized [756] using the tight-binding potential of Xu et al. [619]. The calculations indicate a progression from near-spherical shape for smaller fullerenes to a faceted (polyhedral) geometrical shape made up of graphite-like faces, sharply curved edges, and 12 pentagons at the vertices. The same conclusions were reached independently [772] using the linear-scaling Wannier function approach (Section 18.6). These structures up to C_{3840} were used in the calculation of the DOS in Fig. 18.4.

The density matrix methods [767] can also be employed for MD calculations using the force theorem for the forces in terms of the density matrix and the derivative of the hamiltonian, as given in Section 14.11. The same tight-binding potential [619] as employed for the giant fullerenes has been used in simulations of liquid and amorphous carbon. The calculated radial density distribution $g(r)$ of liquid carbon at ordinary pressure agrees well with plane wave Car–Parrinello calculations and have been done with both usual matrix diagonalization [619] and linear-scaling density matrix methods [773]. In addition, the results are essentially the same as found in [760], which used the Fermi projection operator approach of Section 18.4. The combination of tight-binding and linear-scaling methods has allowed calculations on larger sizes and longer times than is feasible using other methods.

18.6 Variational (Generalized) Wannier Function Methods

A different approach is to work with the localized wavefunctions in Eq. (18.2) rather than with the density matrix itself. However, in searching for the Wannier-like functions it is not convenient to require the constraint of orthonormality implicit in Eq. (18.2). One possibility is to work with nonorthogonal functions and use the correct general expression

$$E_{\text{total}} = \text{Tr} \hat{\rho} \hat{H} = \sum_{i,j=1}^{N} S_{ij}^{-1} H_{ji}, \tag{18.25}$$

where S is the overlap matrix. Here matrices are defined by $H_{ji} = \langle \tilde{w}_j | \hat{H} | \tilde{w}_i \rangle$ (or with $\tilde{w} \to w$ for orthogonal functions), etc. However, the entire problem can be rewritten in a more advantageous form through the invention of a new class of functionals [768, 769], the simplest of which is

$$\tilde{E}_{\text{total}} = \sum_{i=1}^{N} H_{ii} - \sum_{i,j=1}^{N} (S_{ij} - \delta_{ij}) H_{ji} = \sum_{i,j=1}^{N} (2\delta_{ij} - S_{ij}) H_{ji}, \tag{18.26}$$

where the terms $S_{ij} - \delta_{ij}$ are like Lagrange multipliers that replace the constraint of orthonormality. The special property of $\tilde{E}$ is that $\tilde{E}_{\text{total}} \geq E_{\text{total}}$ for all wavefunctions whether or not they are normalized or orthogonal. Since $\tilde{E}_{\text{total}} = E_{\text{total}}$ for the orthonormal functions, it follows that one can minimize $\tilde{E}_{\text{total}}$ with respect to the wavefunctions with no constraints, leading to orthonormal functions at the minimum.

The behavior of $\tilde{E}_{\text{total}}$ considered as a functional of the wavefunctions can be seen by expressing Eq. (18.26) in terms of the eigenvectors. (Added discussion can be found in Exercise 18.8.) If the coefficient of the jth eigenvector is c_j, then the contribution to $\tilde{E}$ is $\epsilon_j(2c_j^2 - c_j^4)$. Figure 18.6 (right panel) plots the function $y = 2x^2 - x^4$, which shows the consequences graphically. For eigenvalues that are negative there is an absolute minimum at $|c_j| = 1$, i.e., a properly normalized function. The same holds for the general case where we have many states, and minimization leads to an orthonormal set of functions that spans the space. For positive eigenvalues there is a local minimum at $c_j = 0$, but also a runaway solution in the unphysical $|c_j| > 1$ that must be avoided.

The functional has been extended by Kim et al. [770] to include more states than electrons and to minimize the grand potential $\Omega_s = \text{Tr}\{\hat{\rho}(\hat{H} - \mu)\}$, which can be written in matrix form analogous to Eq. (18.26) (Exercise 18.9)

$$\tilde{E}_s' = \text{Tr}\{(2 - S)(H - \mu I)\}, \tag{18.27}$$

where S and H denote matrices S_{ij} and H_{ij} and I is the unit matrix. From the above reasoning it follows that Eq. (23.27) is minimized by filling all states below the Fermi

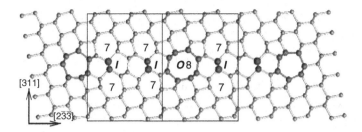

Figure 18.7. Projection on the $\{0\bar{1}1\}$ plane of a rodlike $\{311\}$ defect in silicon containing four interstitial chains I. The structure of $\{311\}$ defects commonly observed in ion-implanted silicon is characterized by random combinations of $/I\,I\,/$ and $/I\,O\,/$ units indicated by the boxes. The calculations showed that the extended defects are much lower in energy than isolated interstitial. Figure provided by J. Kim

energy, and leaving empty all those above. (It is not difficult to restrict the variations to avoid the runaway solution.) One can work with a limited number of states (somewhat larger than the number of electrons) and all operations scale as the number of these states. Thus this functional achieves the additional desired feature that the incorporation of extra states makes it easier to reach the minimum, and one can avoid being trapped in the wrong states at level crossings, etc.

An example of calculations using the Kim et al. functional is calculation to determine the structure of extended defects commonly observed in ion-implanted silicon [637, 638]. The structure shown in Fig. 18.7 was determined using by $O(N)$ molecular dynamics tight-binding calculations, which allowed simulations to be run long enough for the atoms to form rodlike structures. After determination of the extended structure, DFT calculations were used to establish that the extended defects are significantly lower in energy (≈ 2 eV per interstitial) than for isolated interstitials in the bulk.

Nonorthogonal Orbitals

Generalizing the functional to non-orthogonal localized orbitals $\tilde{w}_i$, as in Eq. (18.7), has the advantage that the $\tilde{w}_i$ can be shorter range and more transferable than orthogonal Wannier functions. The latter property is illustrated in Section 23.4; it means that good guesses can be made for the orbitals initially and as the atoms move in a simulation. The difficulty is twofold: finding an efficient way to construct the inverse and dealing with the singular S matrix that results if one allows extra orbitals that have zero norm as in the Kim functional, Eq. (18.27). In the latter case, the size of S is the number of orbitals, but the rank (see below) is the number of electrons, i.e., $N = \text{rank}\{S\}$.

An elegant formulation has been given in [746] and [774], building upon earlier work [745, 775], and using the same ideas as for generating nonorthogonal functions in Section 23.4. The first step is to define the inverse S^- of a singular matrix S by the relation [746]

$$SS^-S = S, \tag{18.28}$$

so that $S^- = S^{-1}$ if S is nonsingular. Next, a functional can be defined that accomplishes the inverse in a way similar to the "Hotelling" method [776]

$$\text{Tr}\{BS^-\} = \min\text{Tr}\{B(2X - XSX)\}, \tag{18.29}$$

where B is any negative definite matrix and the minimization is for all possible hermitian matrices X (Exercise 18.6). Putting this together with the generalization of expression (18.25) for the energy

$$E_{\text{total}} = \text{Tr}\hat{\rho}\hat{H} = \sum_{i,j=1}^{N} S_{ij}^- H_{ji} \tag{18.30}$$

leads to the functional

$$\tilde{E}'_{\text{total}}(N) \rightarrow \text{Tr}[2X - XSX][H - \eta S] + \eta N, \tag{18.31}$$

where the constant η is added to shift the eigenvalues to negative energies. Finally, even the constraint on the rank of S can be removed by defining the functional in terms of the Fermi energy μ as

$$\tilde{E}'_{\text{total}}(N) \rightarrow \text{Tr}[2X - XSX][H - \eta S] + (\eta - \mu)\text{rank}(S) + \mu N, \tag{18.32}$$

where the same trick can be used for rank(S)

$$\text{rank}(S) = \text{Tr}[SS^-] = -\min\{\text{Tr}[(-S)(2X - XSX)]\}. \tag{18.33}$$

The energy $\tilde{E}'_{\text{total}}(N)$ is the minimum for all nonorthogonal wavefunctions $\tilde{w}_i$ that define the S and H matrices and all hermitian matrices X.

If the orbitals are confined, then the minimum of functionals Eq. (18.31) or (18.32) is for nonorthogonal orbitals, since they are more compact than orthogonal ones. Furthermore, one can constrain the orbitals further to require maximal localization, in which case the functionals still give the same energy, and the orbitals are more physical and intuitive, as discussed in Section 23.4.

Combining Minimization and Projection

There are advantages to both the projection methods and the variational methods. An example of a calculation that combines these methods is the calculation of Wannier functions in disordered systems [777–779]. The variational approaches using Wannier functions suffer from the problem that they scale as M^3, where M is the size of the localization region. Furthermore, it has been found in practice [769] that convergence is slow due to the fact that the energy function is only weakly dependent on the tails of the function. On the other hand, the projection method scales as M and can be used to improve the functions in the tails. Stephan and coworkers [777, 778] combined the methods to (1) project Wannier-like functions in a large region; (2) find the largest coefficients, which leads to the best "self-adaptive" functions instead of imposing an arbitrary cutoff; and (3) use minimization method functions to improve the final functions. An example of the density of a bonding-type Wannier function calculated for a model of amorphous Si containing 4,096 atoms. The result is plotted on a logarithmic scale extending over 20 orders of magnitude in shown a beautiful color figure in [779].

18.7 Linear-Scaling Self-Consistent Density Functional Calculations

Full self-consistent density functional theory calculations can be cast in an $O(N)$ linear-scaling form, since the charge density can be computed in $O(N)$ fashion and thus one can build the hamiltonian (Section 18.3) with effort that scales as $O(N)$. The difference from the tight-binding calculation is that one must explicitly represent the orbitals and one must deal with self-consistency. Because of the difficulties in carrying out such full calculations on very large systems, much less work has been done than with the simpler tight-binding methods. One such approach is illustrated in Fig. 18.5 using the KKR multiple-scattering approach [704]. Methods using wavelets have been used on many systems, such as MD simulations on clusters of water molecules, large molecules, etc. [588, 592].

Here we give results only for one of the earliest calculations that shows the possibilities, a calculation [780] for a complete turn of a selected DNA molecule is shown in Fig. 18.8.

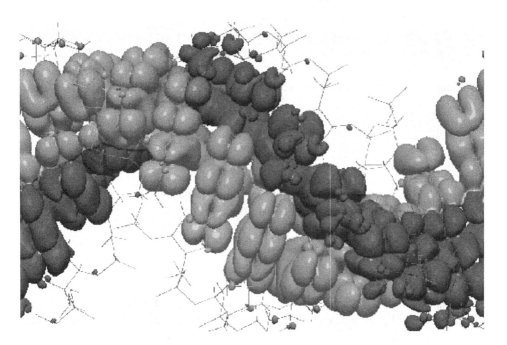

Figure 18.8. Electron density of selected eigenstates in a DNA molecule calculated one complete turn using the self-consistent local orbital SIESTA code and a GGA functional. The density contour shown is $5 \times 10^{-4} a_0^{-3}$. The lighter shaded region is the sum of densities of the 11 highest occupied (HOMO) states, and the darker region represents the 11 lowest unoccupied (LUMO) states. Calculations of comparable complexity have been done using gaussian bases and linear-scaling density matrix methods, e.g., for a fragment of an RNA molecule with 1,026 atoms [652]. Figure provided by E. Artacho; similar to results in [780]

The density, potential, and thermal simulations of the atoms were calculated using O(N) procedures in the SIESTA code [646]. For a given structure, the resulting potential was then used in an ordinary N^3 diagonalization (One could also use a more efficient inverse iteration procedure such as the RMM-DIIS method; see Section 18.9 and Appendix M.) to find selected eigenstates, in particular the fundamental gap between the lowest unoccupied (LUMO) and highest occupied (HOMO) orbitals, which are shown in Fig. 18.8. Further information is given in [780], which investigated the effects of disorder (a mutation) upon localization of the states and electrical conductivity. Calculations on similar size systems, including a fragment of an RNA molecules with 1,026 atoms, have been done using linear-scaling gaussian density matrix methods [652]. See, for example, [739–741] for reviews of DFT and quantum chemistry wavefunction methods and applications.

18.8 Factorized Density Matrix for Large Basis Sets

Large basis sets such as plane waves and grids are desirable in order to have robust methods that always converge to the correct answer. For such approaches, however, straightforward application of linear-scaling approaches may not be feasible. In particular, the density matrix formalism leads to unwieldy expressions since it requires matrices of size $N_\text{basis} \times N_\text{basis}$. Even if the matrix is sparse, it still becomes very large in the limit of finely spaced grids. How can it be feasible to use density matrix methods with such bases? The answer is remarkably simple: only a limited number of orbitals are needed, but each orbital needs to have many degrees of freedom. This is the basis for replacing Eq. (18.23) with a factorized form [781]

$$\rho(\mathbf{r}, \mathbf{r}') = \sum_{ij} \Phi_i^*(\mathbf{r}) K_{ij} \Phi_j(\mathbf{r}'), \qquad (18.34)$$

where $\Phi_j(\mathbf{r}) = \sum_\nu c_j^\nu \phi_\nu(\mathbf{r})$ denotes an orbital expanded in a (large) basis set $\phi_\nu(\mathbf{r})$ and K_{ij} is a matrix of a size comparable to that needed in a tight-binding calculation. Thus the "purified" density matrix in the reduced space of the orbitals can be written

$$K = 3LOL - 2LOLOL, \qquad (18.35)$$

where L is a trial density matrix and O is the overlap matrix of the orbitals. In this form, one minimizes the functional with respect to L_{ij} and the coefficients c_j^ν. Different choices lead to different trade-offs between the number of orbitals involved in the matrix operations in Eq. (18.35) and the number of basis functions needed for each orbital in Eq. (18.34). Clearly, the general nonorthogonal formulation in Eqs. (18.31) or (18.32) can be even more advantageous as a general approach to making the orbitals as confined as possible, and therefore as optimal as possible for representation in a large basis.

The same idea can, of course, be applied to the Wannier-function-type methods. In fact, the local orbital approach is already in this form, since a limited number of localized Wannier functions are each expanded in a basis of atomic-like orbitals. One can also expand each localized function in a representation on a grid, as is done, e.g., in [782], where nonorthogonal functions are used in a method similar to that proposed by Stechel and coworkers [745, 775]. An expansion in overlapping spherical waves has been proposed by Haynes and Payne [783].

18.9 Combining the Methods

Most of the $O(N)$ methods described in this chapter are based on the localization of the density matrix in space: they capture the physics and perform well for wide-bandgap insulators, highly disordered materials, and metals at (very) high temperature. But they fail on all counts whenever the localization length is large, e.g., in a good metal at ordinary temperatures. Such states can be isolated by methods that separate the physics by localization in energy, i.e., spectral methods (Appendix M), rather than by localization in space. How can one combine the methods to take advantage of the properties of each? There can be many approaches, but all have the general feature that $O(N)$ methods can be used for solution at one level, e.g., a metallic system at very high temperature T, and the spectral method for the difference between the high- and low-T solutions that depends only on states near the Fermi energy.

Spectral methods described in Appendix M are designed to find selected states efficiently. If the hamiltonian operator can be cast in sparse form (i.e., the hamiltonian is localized in real space or one can use transforms as in Section M.11) each state can be found with effort $\propto N_{basis}$ or $\propto N_{basis} \ln N_{basis}$. Of course, the usual approach in which all states are desired requires effort that scales as $\propto N \times N_{basis} \propto N^2$ or $\propto N^2 \ln N_{basis}$. The effort needed to treat only the states near the Fermi energy is $\propto N^2 \times k_B T/\mu$, where T is the needed smearing for an efficient $O(N)$ scheme and the Fermi energy μ represents the characteristic total energy range of the electronic states. Although the effort scales as $\propto N^2$, there is an approach [744] that takes advantage of the fact that states vary smoothly near the Fermi energy and create a linear-scaling spectral resolution approach that is applicable to metals.

SELECT FURTHER READING

Reviews:

Bowler, D. R. and Miyazaki, T., "O(N) methods in electronic structure calculations," *Rep. Prog. Phys.*, 75:036503, 2012. A review with extensive description of methods references to available codes.

Cole, D. J. and Hine, N. M. H., "Applications of large-scale density functional theory in biology" *J. Phys. Condens. Matter* 28:393001, 2016. Makes case for large-scale DFT and refers to various O(N) methods.

Gordon, M. S., Fedorov, D. G., Pruitt, S. G., and Slipchenko, L. V., "Fragmentation methods: A route to accurate calculations on large systems," *Chem. Rev.* 112:632, 2012. A review of methods used in chemistry.

Goedecker, S., "Linear scaling electronic structure methods," *Rev. Mod. Phys.* 71:1085–1123, 1999. Compares many methods.

Arias, T. A., "Multiresolution analysis of electronic structure: Semicardinal and wavelet bases," *Rev. Mod. Phys.* 71:267–311, 1999. Provides a more mathematical exposition of the analysis.

Method using Daubechies wavelets:

Mohr, S., Ratcliff, L. E., Boulanger, P., Genovese, L., Caliste, D., Deutsch, T., and Goedecker, S., "Daubechies wavelets for linear scaling density functional theory," *J. Chem. Phys.* 140:204110, 2014.

The recursion method:

Haydock, R., in *Recursion Method and Its Applications*, edited by D. G. Pettifor and D. L. Weaire (Springer-Verlag, Berlin, 1985).

Discussion of locality:

Fulde, P., *Electron Correlation in Molecules and Solids,* 2nd ed. (Springer-Verlag, Berlin, 1993). Discusses localization in a many-body context.

Exercises

18.1 Derive the result stated in the text that the normalization constant in Eq. (18.11) is given by $C_{n+1} = 1/\beta_n$, $n \geq 1$. Show this by directly calculating the normalization of ψ_{n+1} assuming ψ_n is normalized.

18.2 Derive the continued fraction representation of Eq. (18.12) using the Lanczos algorithm for the coefficients. It follows that the spectrum of eigenvalues is given by the poles of the continued fraction (i.e., the zeros of the denominator) in Eq. (18.12). (Thus the spectrum of eigenvalues is given either by the continued fraction form or by the zeros of the polynomial of the previous problem.)

18.3 Derive the form of the terminator given in Eqs. (18.13) and (18.14). Show that imaginary part is nonzero in the band range indicated, so that no poles and only continuous DOS can result in this range. In fact, there is another solution with a plus sign in the square root in Eq. (18.14); show that this is not allowed since $t(z)$ must vanish for $|z| \to \infty$.

18.4 Show that the square root form for terminator Eq. (18.14) satisfies Kramers–Kronig relations, Eq. (D.18), as it must if $G(z)$ is a physically meaningful Green's function.

18.5 Show that the Green's function $G(z) = 1/(\hat{H} - z)$ becomes more localized for large z for the case where $\hat{H}$ is a short-range operator in real space. This is the essence of localization in both the recursion and Fermi function expansion methods. Hint: first consider the $z \to \infty$ limit and then terms in powers of $\hat{H}/z$.

18.6 Derive Eq. (18.29) by showing that the variation around $X = S^-$ is quadratic and always positive for matrices B that are negative definite.

18.7 This exercise is to derive the form of the "purification" functional, Eq. (18.24), that leads to idempotency of the density matrix.
(a) The first step is to demonstrate that the function $y = 3x^2 - 2x^3$ has the form shown in the left panel of Fig. 18.6 and that the result y is always closer to 0 or 1 than the input x.
(b) Next, generalize this to a matrix equation for any symmetric matrix leading to eigenvalues closer to 0 or 1.
(c) Finally, show that the functional Eq. (23.23) minimized using the gradients Eq. (23.24) leads to the desired result.

18.8 This exercise is designed to provide simple examples of the properties of the unconstrained
functional, Eq. (18.26).

(a) Consider a diagonal 2×2 hamiltonian with $H_{11} = \epsilon_1 < 0$, $H_{22} = \epsilon_2 > 0$, and $H_{12} = H_{21} = 0$ in an orthonormal basis, ψ_1 and ψ_2. Show that minimization of the functional leads to the ground state ψ_1 properly normalized.

(b) Now consider the same basis as in part (a) but with a hamiltonian that is not diagonal: $H_{11} = H_{22} = 0$ and $H_{12} = H_{21} = t$. Show that in this case the functional also leads to the properly normalized ground state $\psi = \frac{1}{\sqrt{2}}(\psi_1 + \psi_2)$.

18.9 Show that the functional, Eq. (18.27), has the property that it leads to orthonormal eigenvectors for states below the Fermi energy μ and projects to zero the amplitude of any states with eigenvalue above the Fermi energy.

PART V

FROM ELECTRONIC STRUCTURE TO PROPERTIES OF MATTER

19

Quantum Molecular Dynamics (QMD)

It always seems impossible until it is done.

Nelson Mandela

Summary

Of all the developments for efficient, quantitative computation of the properties of materials from density functional theory, one stands out: quantum molecular dynamics (QMD) simulations pioneered by Car and Parrinello in 1985 [85]. This work and subsequent developments have led to a revolution in the capabilities of theory to treat real, complex molecules, solids, and liquids including thermal motion (molecular dynamics), with the forces derived from the electrons treated by (quantum) density functional methods. Altogether, four advances created a new approach to electronic structure. These comprise

- optimization methods (instead of variational equations),
- equations of motion (instead of matrix diagonalization),
- fast Fourier transforms (FFTs) (instead of matrix operations), and
- a trace of occupied subspace (instead of eigenvector operations).

Car and Parrinello combined these features into one unified algorithm for electronic states, self-consistency, and nuclear movement. The advances of Car and Parrinello have led to an explosion of alternative approaches that utilize the force theorem, together with efficient iterative methods described in Appendix M. These are referred to as Born–Oppenheimer molecular dynamics and are described in the present chapter as well as the Car–Parrinello method per se.

19.1 Molecular Dynamics (MD): Forces from the Electrons

The basic equations for the motion of classical objects are Newton's equations. For a set of nuclei treated as classical masses with an interaction energy $E[\{\mathbf{R}_I\}]$ dependent on the positions of the particles $\{\mathbf{R}_I\}$, the equations of motion are

$$M_I \ddot{\mathbf{R}}_I = -\frac{\partial E}{\partial \mathbf{R}_I} = \mathbf{F}_I[\{\mathbf{R}_J\}]. \tag{19.1}$$

Such equations can be solved analytically only in the small-amplitude harmonic approximation. In general, the solution is done by numerical simulations using discrete time steps based on discrete equations such as the Verlet algorithm, the properties of which are well established (see, e.g., [382, 784]). At each time step t the position of each nucleus is advanced to the next time step $t + \Delta t$ depending on the forces due to the other nuclei at the present time step:

$$\mathbf{R}_I(t + \Delta t) = 2\mathbf{R}_I(t) + \mathbf{R}_I(t - \Delta t) + \frac{(\Delta t)^2}{M_I} \mathbf{F}_I[\{\mathbf{R}_J(t)\}], \tag{19.2}$$

where the first two terms are just the law of inertia. The key property of the Verlet algorithm, well established in classical simulations, is that *the errors do not accumulate*. Despite the fact that the equations are only approximate for any finite Δt, the energy is conserved and the simulations are stable for long runs.

Of course, the forces on the nuclei are determined by the electrons in addition to direct forces between the nuclei. In the past this has been done by effective potentials (such as the Lennard–Jones potential) that incorporate effects of the electrons. These are adequate for many cases like rare gas atoms, but it is clear that one must go beyond such simple pair potentials for real problems of interest in materials. One approach is to use empirical models that attempt to include additional effects and, usually, are parameterized. Advances in electronic structure calculations have made molecular dynamics (MD) simulations possible with forces derived directly from the electrons with no parameters. Such simulations are often termed *ab initio* or "first principles," but here we shall use the nomenclature "quantum MD (QMD)" simulations. Within the Born–Oppenheimer (adiabatic) approximation[1] the electrons stay in their instantaneous ground state as the nuclei move. Thus the correct forces on the nuclei are given by the force (Hellmann–Feynman) theorem, Eq. (3.18), which is practical to implement in pseudopotential density functional theory as shown by many examples in previous chapters. We repeat here the basic formulas from Chapter 7 in slightly different notation. Within the Kohn–Sham approach to density functional theory, the total energy of the system of ions and electrons is given by

$$E[\{\psi_i\}, \{\mathbf{R}_I\}] = 2\sum_{i=1}^{N} \int \psi_i^*(\mathbf{r}) \left(-\frac{1}{2}\nabla^2\right) \psi_i(\mathbf{r}) d\mathbf{r} + U[n] + E_{II}[\{\mathbf{R}_I\}], \tag{19.3}$$

$$U[n] = \int d\mathbf{r} V_{\text{ext}}(\mathbf{r}) n(\mathbf{r}) + \frac{1}{2} \int \int d\mathbf{r} d\mathbf{r}' \frac{n(\mathbf{r})n(\mathbf{r}')}{|\mathbf{r} - \mathbf{r}'|} + E_{\text{xc}}[n], \tag{19.4}$$

$$n(\mathbf{r}) = 2\sum_{i=1}^{N} |\psi_i(\mathbf{r})|^2, \tag{19.5}$$

$$\mathbf{F}_I = -\frac{\partial E}{\partial \mathbf{R}_I}, \tag{19.6}$$

---

[1] This is an excellent approximation for many properties such as phonon energies, as discussed in Chapter 3 and Appendix C.

where ψ_i are the one-electron states, $\mathbf{R}_I$ are the positions of the ions, E_{II} is the ion–ion interaction, $n(\mathbf{r})$ is the electronic charge density, $V_{\text{ext}}(\mathbf{r})$ is the electron–ion interaction, $E_{\text{xc}}[n]$ is the exchange–correlation energy, and $\mathbf{F}_I$ is the force given by the force theorem expressed in Eqs. (3.18), (13.3), and other forms.

The key problem, however, is that the calculations must be *very efficient*. The development of new efficient algorithms by Car and Parrinello [85] and others [385, 785] has led to the explosion of many forms of QMD simulations since 1985. *The bottom line is that different approaches can each be used to great advantage; each works well if used with care, and each has particular advantages that can be utilized in individual situations.*

Examples of use of the Born–Oppenheimer and Car–Parrinello methods are illustrated in the examples in Sections 2.10, 2.11, and 19.6.

19.2 Born-Oppenheimer Molecular Dynamics

It is appropriate to first consider the types of methods that can be called "Born–Oppenheimer molecular dynamics" because they are the straightforward application of the Born–Oppenheimer approximation that the electrons remain in their ground state as the atoms move. The key point is that the problem is separated into two parts: calculation of accurate forces at each step and moving the atoms. The step of calculating forces is the same as used for relaxation of atomic position to find the minimum energy structures and the step of moving the atoms uses standard methods of molecular dynamics summarized in Eqs. (19.1) and (19.2). Contemporaneous with the Car–Parrinello work, progress using the Born–Oppenheimer was underway using self-consistent density functional theory methods (e.g., [375]) and simpler tight-binding-type methods (e.g., [786]). There has been great progress in creating efficient iterative algorithms described in Appendix M so that there are now various approaches to efficiently calculate the forces.

These methods are now the most widely used for quantum MD (QMD) simulations (often termed *ab initio* or "first principles"). Since they only require methods to calculate forces and iterative algorithms described elsewhere in this book, what else is there to specify in a calculation?

- The time step, Δt in Eq. (19.2), is the parameter that governs the accuracy and efficiency of the molecular dynamics. A balance must be chosen with steps long enough to make the calculation feasible and short enough to be accurate. The choice is governed by the nuclear dynamics, i.e., it is the same as in an ordinary classical MD calculation. This is longer (by about an order of magnitude) than the step in the Car–Parrinello unified algorithm, which is governed by the electronic time scale, so that the atoms move further in one step.

- At each step, the electrons must be solved accurately. Calculation of forces requires conditions for self-consistency to be much more rigorous than for energy, which is at a variational minimum for electrons in the ground state. The requirements are much more stringent than in the Car–Parrinello method. This requires more cycles of self-consistency at each MD step, roughly an order of magnitude more calculations per MD step than in the Car–Parrinello method. Thus, to a first approximation, the two approaches require similar amounts of computation.

- Different iterative methods (Appendix M) can be chosen to find the eigenvalues and eigenvectors of all the occupied states, or only the occupied subspace that spans the eigenvectors. The latter is in general faster; the former has the advantage that the eigenvalues can be used to calculate the Fermi function needed to treat metals.
- Since the computation needed to reach self-consistency is such an important factor in the iterative methods, there can be significant advantages with algorithms designed to give a better starting guess for the potential and wavefunctions at each MD step and faster convergence to self-consistency. However, one must be careful since using information from a previous step may lead to a bias in the forces that leads to a drift in the total energy.

19.3 Car–Parrinello Unified Algorithm for Electrons and Ions

The essential feature of the Car–Parrinello approach takes advantage of the fact that the total energy of the system of interacting ions and electrons is a function of *both* the classical variables $\{\mathbf{R}_I\}$ for the ions *and* the quantum variables $\{\psi_i\}$ for the electrons. Instead of considering the motion of the nuclei and the solution of the equations for the electrons at fixed $\{\mathbf{R}_I\}$ as separate problems (an approach that is inherent in the flowcharts describing the usual approach in Fig. 7.2 and Fig. M.1), the Car–Parrinello approach considers these as *one unified problem*. Within the Born–Oppenheimer (adiabatic) approximation, the problem becomes one of minimizing the energy of the electrons and solving for the motion of the nuclei simultaneously. This applies to relaxation of the nuclei to find stable structures as well as to thermal simulations of solids and liquids using MD methods. In one stroke, calculation of the ground-state electronic structure and simulation of material phenomena has been unified.[2]

In the Car–Parrinello approach, the total Kohn–Sham energy is the potential energy as a function of the positions of the nuclei. Molecular dynamics for the nuclei using forces from this energy is the defining criterion for all forms of so-called *ab initio* MD using density functionals. *The special feature of the Car–Parrinello algorithm is that it also solves the quantum electronic problem using MD.* This is accomplished by adding a *fictitious* kinetic energy for the electronic states, which leads to a fictitious lagrangian for both nuclei and electrons [85][3]

$$\mathcal{L} = \sum_{i=1}^{N} \frac{1}{2}(2\mu) \int d\mathbf{r} |\dot{\psi}_i(\mathbf{r})|^2 + \sum_I \frac{1}{2} M_I \dot{\mathbf{R}}_I^2 - E[\psi_i, \mathbf{R}_I]$$

$$+ \sum_{ij} \Lambda_{ij} \left[\int d\mathbf{r} \psi_i^*(\mathbf{r}) \psi_j(\mathbf{r}) - \delta_{ij} \right]. \tag{19.7}$$

---

[2] It is essential to emphasize that the Car–Parrinello algorithm does *not* treat the real dynamics of electrons, which requires a time-dependent Schrödinger equation, (21.3). The algorithm is designed to find the ground state (adiabatic or Born–Oppenheimer) solution for the electrons as the nuclei move.

[3] Note the similarity to the lagrangian in Eq. (M.14), except that here the signature ingredient of the Car–Parrinello method, the "fictitious electronic mass," is added. Such fictitious lagrangians are also used in other quantum field theories [787].

The final term in Eq. (19.7) is essential for orthonormality of the electronic states. This lagrangian leads to MD equations for *both* classical ionic degrees of freedom $\{\mathbf{R}_I\}$ *and* electronic degrees of freedom, expressed as independent-particle Kohn–Sham orbitals $\psi_i(\mathbf{r})$. The resulting equations of motion are

$$\mu\ddot{\psi}_i(\mathbf{r},t) = -\frac{\delta E}{\delta\psi_i^*(\mathbf{r})} + \sum_k \Lambda_{ik}\psi_k(\mathbf{r},t)$$

$$= -H\psi_i(\mathbf{r},t) + \sum_k \Lambda_{ik}\psi_k(\mathbf{r},t), \tag{19.8}$$

$$M_I\ddot{\mathbf{R}}_I = \mathbf{F}_I = -\frac{\partial E}{\partial\mathbf{R}_I}. \tag{19.9}$$

The equations of motion, Eq. (18.8) and Eq. (19.9), are just Newtonian equations for acceleration in terms of forces, subject to the constraint of orthogonality in the case of electrons. The masses of the ions are their physical masses, and the "mass" for the electrons is chosen for optimal convergence of the solution to the true adiabatic solution. Thus the equations can be solved by the well-known Verlet algorithm with the constraints handled using standard methods for holonomic constraints [788]. This can be achieved by solving the equations for Λ_{ik} at each time step so that ψ_i are exactly orthonormal using an iterative method called SHAKE [788, 789]. The resulting discrete equations for time, $t^n = n\delta t$, are

$$\psi_i^{n+1}(\mathbf{r}) = 2\psi_i^n(\mathbf{r}) - \psi_i^{n-1}(\mathbf{r}) - \frac{(\Delta t)^2}{\mu}\left[\hat{H}\psi_i^n(\mathbf{r}) - \sum_k \Lambda_{ik}\psi_k^n(\mathbf{r},t)\right],$$

$$\mathbf{R}_I^{n+1} = 2\mathbf{R}_I^{n+1} - \mathbf{R}_I^{n-1} + \frac{(\Delta t)^2}{M_I}\mathbf{F}_I. \tag{19.10}$$

Note the similarity to those equations for minimization of electronic energy, e.g., Eq. (M.15). The most time-consuming operation (applying the hamiltonian to a trial vector) is exactly the same in *all* the iterative methods; the only difference is the way in which the wavefunctions are updated as a function of time t or step n.

The Stationary Solution

The meaning of the equations can be clarified by considering a stationary solution of the equations, which we now show is equivalent to the usual Kohn–Sham variational equations. At steady state, all time derivatives vanish and Eq. (19.9) leads to

$$H\psi_i(\mathbf{r},t) = \sum_k \Lambda_{ik}\psi_k(\mathbf{r},t), \tag{19.11}$$

which is the usual solution with Λ_{ik} the matrix of Lagrange multipliers. Taking the matrix elements, Eq. (19.11) shows that Λ is the transpose of H ($\Lambda_{ik} = H_{ki}$), where H is the usual Kohn–Sham hamiltonian. Diagonalizing Λ leads to the eigenvalues of the Kohn–Sham equations. Furthermore, this is a self-consistent solution since we have minimized the full Kohn–Sham energy, Eq. (19.3). Thus the solution is stationary if, and

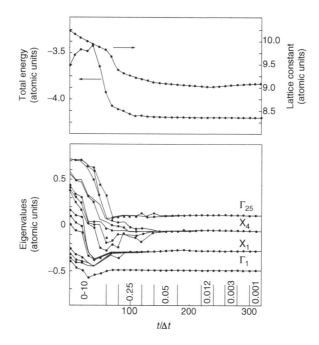

Figure 19.1. Eigenvalues at $\mathbf{k} = 0$ for crystalline Si calculated by quenching the "fictitious kinetic energy" in the lagrangian to reach the steady state [85].

only if, the Kohn–Sham energy is at a variational minimum (or saddle point). In fact, cooling the system down by reducing the kinetic energy is termed *dynamical simulated annealing*, which is a way to find the minimum of the nonlinear self-consistent Kohn–Sham equations. This is illustrated in the original paper of Car and Parrinello; their results copied here in Fig. 19.1 show the eigenvalues reaching the values that would also be found in a self-consistent calculation.

It is instructive to apply the Car–Parrinello algorithm to simplified problems. Examples are described in detail in Exercises 19.2 and 19.3 that illustrate the simplest two-state problem; finding the eigenstates of a simple problem by quenching (the analog of the original calculation of Car and Parrinello in Fig. 19.1), and the equations of motion using the fictitious lagrangian.

Nuclear Dynamics

The real power of the Car–Parrinello method is found in simulations of the coupled motion of nuclei and electrons. This leads to the ability to include the real dynamics of the nuclei in *ab initio* electronic structure algorithms, treating, e.g., thermal motion, liquids, thermal phase transitions, etc. Also, by quenching, one can search for stable structures. Examples are given later in Section 19.6.

It should be emphasized that the *fictitious kinetic energy* has nothing to do with the real quantum mechanical energy of the electrons and the electron dynamics that result from this lagrangian are also fictitious; *it does not represent the real excitations of the electron system*. The purpose of this fictitious kinetic energy is to allow the ground state of the electrons to move efficiently through the space of basis functions, always staying close to the true ground state. This is in fact realized in many cases but is also problematic in other cases as discussed below.

Difficulties in the Car–Parrinello Unified Algorithm

There are three primary disadvantages in using the Car–Parrinello approach.

First, any effects of the fictitious lagrangian must be examined and reckoned with if they are problematic. The method works well for systems with an energy gap (for all steps in the simulation). The characteristic frequency of the fictitious oscillations of the electron degrees of freedom are $\propto E_{gap}/\mu$ (Exercise 19.2) and, if all such frequencies are much greater than typical nuclear vibration frequencies, then the electrons follow the nuclei adiabatically as they should. This is tested by checking the conservation of proper energy (not including the fictitious kinetic energy). Simple examples and discussion are given in [787, 790, 791], and the exercises at the end of this chapter. Even in the best cases, however, one still needs to choose the mass so that the adiabatic condition is satisfied to within acceptable accuracy. There has been controversy on this point [792], but a joint study of Car, Parrinello, and Payne [793] concluded that, with care, problems can be avoided.

Second, the time step Δt must be short. It is governed by the fictitious electronic degrees of freedom and must be chosen smaller than in typical simulations for ions alone. A typical "mass" for the electrons is $\mu = 400 m_e$ (as used for carbon [794]). In general, the value depends on the basis functions, and an issue arises in the plane wave method where Eq. (19.14) below reveals a problem for high Fourier components of the wavefunction. Since the diagonal part of the hamiltonian $H(\mathbf{G}, \mathbf{G}) \propto |\mathbf{G}|^2$ for large $|\mathbf{G}|$, a coefficient $c_i^n(\mathbf{G})$ is multiplied by $((\Delta t)^2/\mu)|\mathbf{G}|^2$. This endangers one of the desirable properties of plane waves: that the cutoff can be increased indefinitely to achieve convergence. It has been proposed to integrate the high Fourier components over the time step interval (since they obey a very simple harmonic oscillator equation) instead of taking the linear variation [785]. Another approach is to take different masses for different Fourier components [795, 796].

Finally, it was recognized from the beginning that problems occur with level crossing, where the gap vanishes, and in metals. This leads to unphysical transfer energy to the fictitious degrees of freedom (Exercise 19.2). The problem has been sidestepped by use of "thermostats" that pump energy into the ion system and remove it from the fictitious kinetic energy of the electron system. This has been used, e.g., in calculations of metallic carbon at high pressure [797].[4] However, problems with metals simulations have led to the widespread use of alternative approaches such as Born–Oppenheimer molecular dynamics (Section 19.2).

---

[4] It is also possible to treat the occupations as dynamical variables in a way related to the ensemble density functional theory method [370], which can potentially allow the Car–Parrinello unified algorithm to apply directly to metals.

19.4 Expressions for Plane Waves

The Car–Parrinello equations can be made more transparent by choosing an explicit basis. The equations have exactly the same form for any orthonormal basis and we choose plane waves as the best example. For simplicity of notation we consider Bloch states only at the center of the Brillouin zone, $\mathbf{k} = 0$, in which case the Bloch functions can be written

$$u_i(\mathbf{r}) = \sum_{\mathbf{G}} c_i(\mathbf{G}) \frac{1}{\sqrt{\Omega}} \exp(i\mathbf{G} \cdot \mathbf{r}), \qquad (19.12)$$

where Ω is the volume of the unit cell. Since each band holds one electron per cell (of a given spin) the $c_i(\mathbf{G})$ are orthonormal

$$\sum_{\mathbf{G}} c_i^*(\mathbf{G}) c_j(\mathbf{G}) = \delta_{ij}. \qquad (19.13)$$

The discrete time step equation corresponding to Eq. (19.10) becomes

$$c_i^{n+1}(\mathbf{G}) = 2c_i^n(\mathbf{G}) - c_i^{n-1}(\mathbf{G}) - \frac{(\Delta t)^2}{\mu} \left[\sum_{\mathbf{G}'} H(\mathbf{G}, \mathbf{G}') c_i^n(\mathbf{G}') - \sum_k \Lambda_{ik} c_k^n(\mathbf{G}) \right], \qquad (19.14)$$

where δt denotes the time step. The equation for Λ_{ik} is derived by assuming that $c_i^n(\mathbf{G})$ and $c_i^{n-1}(\mathbf{G})$ are each orthonormal and imposing the condition that $c_i^{n+1}(\mathbf{G})$ is also orthonormal. The complete solution is then found by updating the electron density at each iteration, finding the new Kohn–Sham effective potential, and, if desired, moving the atoms according to Eq. (19.2) using the force theorem. The procedure then starts over with a new iteration. Thus all the operations have been combined into one unified algorithm.

The algorithm as presented is still too slow to be useful because of the matrix multiplication in Eq. (19.14), for which the number of operations scales as the square of the number of plane waves N_{PW}^2. To circumvent this bottleneck, Car and Parrinello used fast Fourier transforms (FFTs) to reduce the scaling to $N_{\text{PW}} \log N_{\text{PW}}$. The ideas have very general applicability and are described in Sections 12.7 and M.11, where the algorithms are summarized in Figs. 12.4 and M.2. The key steps are the operation $\hat{H}\psi$ and calculation of the density. The kinetic energy operation is a diagonal matrix in Fourier space, whereas multiplication by V is simple in real space where V is diagonal. By the use of the FFT, the operations can be carried out, respectively, in Fourier and real spaces, and the results collected in either space. The limiting factor is the FFT, which scales as $N_{\text{PW}} \log N_{\text{PW}}$. The sequence of steps is described in Fig. M.2.

Finally, at every step the energy and force on each nucleus can be calculated using the force theorem expressed in plane waves, Eq. (13.3). A variant of this form, however, may be more convenient for simulations with large cells. As explained in Section F.3 and [748], the force on an ion due other ions (the Ewald term) can be combined with the local pseudopotential term, leading to a combined expression in reciprocal space plus correction terms expressed as short-range forces between ions, Eq. (F.17). The last are easily included in a standard MD simulation.

The method can be extended to "ultrasoft" pseudopotentials [603] and to the PAW method [542, 544], which is very useful for simulations with atoms that require high plane wave cutoffs using norm-conserving pseudopotentials, e.g., transition metals. The basic idea is that in any such approach the same general formulation can be used for updating the plane wave coefficients, calculating forces, etc. The difference is that there are additional terms rigidly attached to the nuclei that must be added in the expressions [542, 544, 603].

19.5 Non-self-consistent QMD Methods

Much simpler (and faster computationally) simulation methods can be devised if there is no requirement for the full self-consistent Kohn–Sham equations to be solved. The simplest approach uses the empirical tight-binding method (Chapter 14) in which the hamiltonian is given strictly in terms of matrix elements that are simple functions of the positions of the atoms. Since the basis set is also small (several orbitals per atom) it may be more efficient simply to diagonalize the matrices rather than use an iterative method. Then eigenvectors, energies, and forces can be calculated for all positions of the atoms, usually much faster than a typical self-consistent plane wave algorithm. This approach [786] was developed simultaneously with the Car–Parrinello work and still enjoys widespread use because of its speed and simplicity.

Another approach is to solve the electronic problem within a basis using an approximate non-self-consistent form of the hamiltonian. Such methods are *ab initio* since they are not parameterized and the approximation forms have been used effectively for many problems with a total potential that is a sum of atomic-like potentials [630, 632]. Together with the explicit functional for the energy in terms of the input density, Eq. (7.22), and the usual expressions for the forces, this enables a complete, albeit approximate, DFT QMD algorithm. Self-consistency can be added in limited ways that still preserve efficiency, as done, e.g., in [798].

19.6 Examples of Simulations

Because of advances in QMD simulations, it is now possible to determine equilibrium thermodynamic phases and dynamics of the nuclei as a function of temperature and pressure. Three examples in Section 2.10 show the power of the methods to treat important problems.

- One is the challenge to understand the composition and dynamics of the Earth, including the mantle made of minerals and core made of iron with other elements in solid and liquid phases, at pressures and temperatures that vary greatly. First-principles calculations of equations of state and QMD can provide crucial information, and an example is simulations of liquid iron under conditions like those in the core, which are illustrated in Fig. 2.13.
- The other is liquid water and aqueous solutions, which are very difficult problems because of the hydrogen bonding that leads to a plethora of structures and intricate correlations in

the liquid. In this case there are vast numbers of experiments and measured properties; however, there are still many aspects not understood. There is no sense in which we can address the issues in any depth. The brief description in Section 2.10 is meant to convey the enormous progress made to even start attacking such problems, and the fact that density functional methods are just now being developed to the point where it may be possible to make definitive predictions.

• QMD methods provide a general approach to calculations of reaction paths and catalysis. This is far too great a subject to attempt to cover in this book. The example of the Ziegler–Natta reaction shown in Fig. 2.16 illustrates the ways that QMD can provide insight into the nature of atomic-scale reactions and clarify proposed mechanisms [201, 202]. However, a word of caution is in order: reaction barriers are particularly sensitive to electronic correlations and functionals for exchange and correlation often are simply not accurate enough for many problems.

Carbon at High Temperature and Pressure

Carbon is of great interest in many fields of science, with technical, geological, and astrophysical implications, as indicated by the conditions expected inside the Earth and the gaseous planets. Previous to the advent of molecular dynamics based on density functional theory there were wildly divergent proposals for the phase diagram [802]. The graphite–diamond boundary is known from thermodynamic arguments [802], but there were many remaining questions. Does the melting temperature of diamond increase with pressure as in most materials or does it decrease as in Si and Ge? If it increases is there a maximum melting temperature, i.e., a pressure where the slope of the melting curve changes sign? Is liquid C a metal or an insulator? Since the electrons are treated with quantum Kohn–Sham theory as the atoms move, the simulations also yield information on the nature of liquid carbon and can answer such questions.

There are three separate calculations that all agree on the general conclusions and they provide instructive examples of the use of Born–Oppenheimer and Car–Parrinello methods and three different approaches to calculation of the melting temperature. The first QMD calculations for carbon [797, 803] used the LDA approximation and the Car–Parrinello unified algorithm for the simulations of the liquid. The result was a prediction that the melting temperature of diamond increases with pressure (opposite to Si and Ge), which has been confirmed by experiments [802, 804]. The melting curve was *not* directly observed in the simulation, but it was inferred from the Claussius–Clapyron equation that relates the slope dP_{melt}/dT_{melt} to the change in specific volume at the transition. At low P the liquid is less dense than diamond [797, 803] but for $P > \approx 500$ GPa [794], the nature of molten carbon changes to a higher coordination (>4) dense phase, so that the slope of melting curve must change to be like that known to occur in molten Si and Ge at $P = 0$. This method was used to infer a maximum in the melting temperature of ≈ 8000K at ≈ 500 GPa. At still higher pressure, static total energy calculations [118, 119] have found transitions to dense tetrahedrally coordinated structures (BC8, ST12, etc.) and to

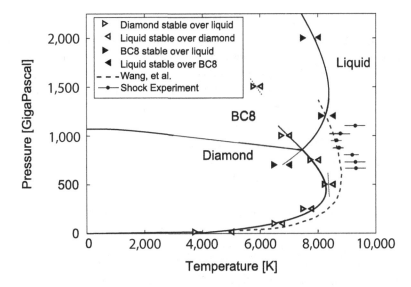

Figure 19.2. Calculated phase diagram of carbon as a function of pressure and temperature omitting the graphite stability region for simplicity. Results indicated by triangles were found using the two-phase method where there is an interface between solid and liquid in the simulation cell [799]. The dashed line was determined by separate calculations of solid and liquid free energies [800]. The light dashed line is the slope of the transition curve calculated independently using the Clausius–Clapeyron equation. Experimental data from [801] is shown as dots with error bars. Adapted from figure provided by A. Correa

the simple cubic metallic phase at $P \approx 30$ Mbar. This last prediction was confirmed by the QMD simulations, where the phase boundary could be determined directly by melting and solidification as the temperature is raised or lowered [794].

Two more recent calculations determine the melting curve directly in two very different ways. Each uses the PBE approximation for the density functional. The work of Correa et al. [799] uses the Born–Oppenheimer approach and a simulation cell that contains both solid and liquid phases. The procedure is to observe whether the solid or the liquid tends to grow as the simulation proceeds. The results are shown in Fig. 19.2 where the melting of the diamond and BC8 phases is determined. The triangle symbols indicate the change from favoring the solid to favoring the liquid. They pin down the transition in a narrow range, and the simulations can detect the melt line for diamond and BC8 even in regions where they are metastable, as shown by the continuation of the curves beyond the triple point. The solid–solid transition is not observed directly. The transition pressure from diamond to the more dense BC8 phase is known at $T = 0$ from total energy calculations, and this work determined the triple point where diamond, BC8, and the liquid converge. The diamond/BC8 transition line shown in the figure is an interpolation between the two end points.

The work of Wang et al. [800] used the Car–Parrinello method and determined the melting temperature by computing the free energies of the solid and liquid. Calculation of the entropy was done by a two-step process following the same method as used previously for Si [805], which involves a two-stage approach with a gradual adiabatic switching of the Kohn–Sham hamiltonian to an intermediate reference system (in this case a model interatomic potential) and then to a system (an Einstein crystal for the solid and an ideal gas for the liquid) with known entropy (see [800] for details). The BC8 phase was not considered. The results shown in Fig. 19.2 are similar to the those found by Correa et al. [799], but the melting temperature is somewhat higher at all pressures considered, and there is a smaller decrease in the melting temperature at high pressure.

After the theoretical work, there have been tour de force experiments [801] in which high pressure and temperature were generated by laser-induced shock waves. In general, as the shock energy increases, the temperature and pressure increases following a Hugoniot equation of state. This is observed for higher temperatures beyond the range in the figure, but there is a region in which an anomalous temperature pressure relation is shown. This indicates a change of phase with a large latent heat and was interpreted as the melting curve. The results are close to the calculations of Wang et al. [800]. Given the difficulty of the calculations (and even more difficulty of the experiments!) the agreement with both calculations is impressive and it shows the value of calculations that can be done for systems in regimes beyond any pervious measurements.

A great advantage of QMD methods is that both electronic and thermal nuclear properties are accessible in the same calculation. One can determine properties like whether liquid carbon is insulating (diamond like) or metallic. The time-averaged electronic density of states does not answer the question. As shown on the left-hand side of Fig. 19.3, the

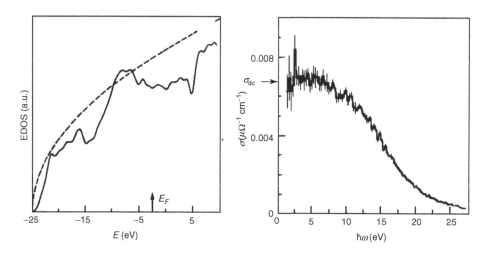

Figure 19.3. Electronic properties of liquid carbon at $P \approx 0$ and $T = 5{,}000$ K. Left: time-averaged density of states, which is close to the free-electron parabola (dashed line). Right: conductivity $\sigma(\omega)$ calculated for a given ionic configuration using Eq. (E.11) and averaged over configurations. The form of $\sigma(\omega)$ is similar to the Drude form, expected for metal, and the dc conductivity can be estimated from the extrapolation $\omega \rightarrow 0$. From [797].

average density of states is almost free-electron-like. On the other hand, there is very strong scattering of the electrons by the ions, which might lead to localization. Theory can avoid semantics and directly calculate the conductivity $\sigma(\omega)$ given by Eq. (E.11) and the well-known Eq. (21.9) in terms of momentum matrix elements. This yields $\sigma(\omega)$ at any configuration of the nuclei and the final results are found by averaging over configurations in the MD simulation. The result shown in Fig. 19.3 is a conductivity at $T = 5,000$ K that has typical Drude form [280, 285] with a very short mean free path of the order of the interatomic spacing. This is borne out by later experiments [801] that observed metallic-like reflectivity of the liquid for temperatures above 10,000 K.

Structures of Defects, Surfaces, Clusters

An ever-expanding area of research involves nanoclusters, where theory has much to add since information about such structures is difficult to determine experimentally. Examples of Si clusters shown in Fig. 2.20 have all been determined by relaxation or MD. An instructive example is Si_{13}, where there is competition between a symmetric structure with 12 outer atoms surrounding a central atom [224] and a low-symmetry structure found by quenching finite temperature MD simulations [223]. In fact, quantum Monte Carlo calculations [225, 226] confirm that the low-symmetry structure is lower in energy.

As an example involving magnetism, Fig. 19.4 shows predicted structures of Fe_3 and Fe_5 molecules [357]. This work used ultrasoft pseudopotentials and plane waves, and, most importantly, used noncollinear formalism (Section 8.3) to predict the spin density shown in the figure. Such molecules can be treated with other methods that have also found noncollinear spin density. Furthermore, noncollinear formalism is essential for treating bulk magnetism at finite temperature [355, 356].

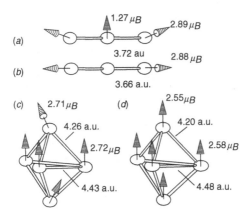

Figure 19.4. Equilibrium structures and magnetic moments of small Fe clusters as calculated [357] using plane waves and ultrasoft pseudopotentials. Of particular note are the predicted noncollinear spin states. Such calculations with the noncolinear spin formalism are needed for treating magnetism at finite temperature [355, 356]. From [357].

SELECT FURTHER READING

Original work:

Car, R. and Parrinello, M., "Unified approach for molecular dynamics and density functional theory," *Phys. Rev. Lett.* 55:2471–2474, 1985.

Books:

Allen, M. and Tildesley, D., *Computer Simulation of Liquids* (Oxford University Press, Oxford, U.K., 1989).

Marx, D. and Hutter, J., Ab Initio *Molecular Dynamics: Basic Theory and Advanced Methods* (Cambridge University Press, Cambridge, U.K., 2009).

Thijssen, J. M. *Computational Physics* (Cambridge University Press, Cambridge, U.K., 2000).

Reviews and tutorial articles:

Remler, D. K. and Madden, P. A., "Molecular dynamics without effective potentials via the Car–Parrinello approach," *Mol. Phys.* 70:921, 1990.

Pastore, G. Smargiassi, E., and Buda, F., "Theory of *ab initio* molecular-dynamics calculations," *Phys. Rev. A* 44:6334, 1991.

Payne, M. C., Teter, M. P., Allan, D. C., Arias, T. A., and Joannopoulos, J. D., "Iterative minimization techniques for *ab initio* total-energy calculations: molecular dynamics and conjugate gradients," *Rev. Mod. Phys.* 64:1045–1097, 1992.

Special issue, "Techiniques for simulations," *Comput. Mater. Sci.* 12, 1998.

Exercises

19.1   In the text it was stated that the "SHAKE" algorithm [788, 789] maintains constraints in a holonomic manner, i.e., with no energy loss. An alternative might be the Gram–Schmidt procedure, in which one updates the wavefunctions with $\hat{H}\psi_i$ and then orthonormalizes starting with the lowest state.

(a) Show that this will cause energy loss. Hint: one way is to consider the two-state problem in Exercise 19.2. Treat the wavefunctions explicitly and show that there is a difference from the equations given below in which the constraint is imposed analytically.

(b) Read the references for SHAKE [788, 789] and summarize how it works.

19.2   Car–Parrinello-type simulation for one electron in a two-state problem is the simplest case and is considered in the tutorial-type paper by Pastore, Smargiassi, and Buda [791]. In this case, the wavefunction can always be written as a linear combination of any two orthonormal states ϕ_1 and ϕ_2,

$$\psi = \cos\left(\frac{\theta}{2}\right)\phi_1 + \sin\left(\frac{\theta}{2}\right)\phi_2.$$

With this definition orthogonality and normalization are explicitly included and we can consider θ to be the variable in the fictitious lagrangian (written for simplicity in the case where ϕ_1 and ϕ_2 are eigenvectors):

$$L = \mu\left|\frac{d\theta}{dt}\right|^2 + \epsilon_1\cos^2 + \epsilon_2\sin^2$$

Solving the Lagrange equations gives

$$\mu \frac{d^2\theta(t)}{dt^2} = (\epsilon_2 - \epsilon_1) \sin(\theta(t) - \theta_0),$$

which is the equation for a pendulum. For small deviations $\theta - \theta_0$, the solution is simple harmonic oscillations of frequency $\omega_e^2 = \Delta E/\mu$. Thus so long as the oscillations are small, the electronic degrees of freedom act like simple oscillators.

Pastore et al. [791] have analyzed the two-state model and large cell calculations to identify the key features, as illustrated in the figures from their paper. If μ is chosen so that the fictitious electronic frequencies are well above all lattice frequencies and motions are small, then there is only slow energy transfer and the Car–Parrinello method works well. This can be done in an insulator. But level crossing, metals, etc. give interesting difficulties.

The exercise is to analyze the algorithm for three cases in which the system is driven by an external perturbation of frequency ω_0:

(a) For the case of small amplitudes and μ chosen so that $\omega_e \gg \omega_0$, show that the electrons respond almost instantaneously adiabatically following the driving field.

(b) For the more difficult case with ω_e of order ω_0, show that the electrons couple strongly with large nonlinear oscillations. (Note: this fictitious dynamics is *not* the correct quantum dynamics.)

(c) For the case where there is a level crossing and ΔE changes sign, show that the electrons can undergo real transitions. (See note in (b).)

19.3 Project for simulation of quantum systems with Car–Parrinello methods. The purpose of this problem set is to write programs and carry out calculations in simple cases for the Car–Parrinello method for simulation of quantum systems by molecular dynamics techniques. Ignore the spin of the electron, which only adds a factor of 2 in paramagnetic cases with even numbers of electrons per cell.

(a) For the case of an "empty lattice" where the potential energy is a constant set equal to zero, write down the Car–Parrinello equations of motion for the electrons. Work in atomic units.

(i) Set up the problem on a one-dimensional lattice, where the wavefunctions are required to be periodic with length L. Write a program that iterates the Verlet equation for a single wavefunction expressed in terms of Fourier coefficients up to $M \times (2\pi/L)$.

(ii) Choose $L = 10$ a.u., $\mu = 300$ a.u., and $M = 16$, which are reasonable numbers for solids. Start with a wavefunction having random coefficients, velocities zero, and iterate the equations. Choose a time step and show that the fictitious energy is conserved for your chosen time step. Show that you can carry out the exercise equivalent to the original calculation of Car and Parrinello in Fig. 19.1. Extract energy from the system by rescaling the velocities at each step. Show that the system approaches the correct ground state with energy zero. Make a graph of the energy versus time analogous to Fig. 19.1.

(iii) Now consider several states. Add the orthogonalization constraints, and find the ground state for two, three, and four filled states. Verify that you find the correct lowest states for a line with periodic boundary conditions.

Make a graph of the total energy and fictitious kinetic energy as a function of time. Show the variation in total energy on a fine scale to verify that it is well conserved.

(b) Now add a potential $V(x) = A \sin(2\pi x/L)$. Use an FFT to transform the wavefunction to real space, multiply by the potential, and the inverse FFT to transform back to Fourier space.

(i) For two electrons per cell (up and down) one has a filled band with a gap to the next band. Find the ground wavefunction and electron density for a value of $A = 1$ Hartree, a reasonable number for a solid. (All results can be verified by using the plane wave methods and diagonalization as described in Chapter 12.)

(c) Consider a system with the electrons coupled to slow classical degrees of freedom; let A be coupled to an oscillator, $A = A_0 + A_1 x$, and the energy of the oscillator $E = 0.5 M \omega_0^2 x^2$. Choose values typical for ions and phonon frequencies (Chapter 20).

(i) Choose a fictitious mass μ so that all the electronic frequencies are much greater than ω_0. See Exercise 19.2.

(ii) Start the system at $x = 0$, which is not the minimum, and let it evolve. Does the oscillator go through several periods before significant energy is transferred to the electron state? Plot the total energy of the system and the fictitious kinetic energy as a function of time. Show that the total energy is accurately conserved, and the fictitious kinetic energy is much less than the oscillator kinetic energy for several cycles.

(iii) The oscillator should oscillate around the minimum. Check, by calculating the total energy by the quenching method, for fixed x, for several values of x near the minimum. Is the minimum in energy found this way close to the minimum found from the oscillations of the dynamic system?

20

Response Functions: Phonons and Magnons

Summary

Many properties of materials – mechanical, electrostatic, magnetic, thermal, etc. – are determined by the variations of the total energy around the equilibrium configuration, defined by formulas such as Eqs. (2.1)–(2.6). Experimentally, vast amounts of information about materials are garnered from studies of vibration spectra, magnetic excitations, and other responses to experimental probes. This chapter is devoted to the role of electronic structure in providing *ab initio* predictions and understanding of such properties, through the total energy and force methods described in previous chapters, as well as advances in efficient methods that are now widely used for calculation of response functions. Through these developments, calculation of full phonon dispersion curves, dielectric functions, infrared activity, Raman scattering intensities, magnons, anharmonic energies to all orders, phase transitions, and many other properties have been brought into the fold of practical electronic structure theory.

20.1 Lattice Dynamics from Electronic Structure Theory

Total energy and forces are sufficient to treat a vast array of problems including stability of structures, phase transitions, surfaces and interfaces, spin polarization, elastic constants, *ab initio* molecular dynamics, etc. One can also use such direct methods to calculate all the derivatives of the energy with respect to perturbations, by carrying out full self-consistent calculations for various values of the perturbation, and extracting derivatives from finite difference formulas. For problems like structural phase transitions, the methods are very useful and successful as illustrated for the ferroelectric $BaTi_2O_3$ and the instability in Nb shown in Fig. 2.10.

The first part of this chapter is devoted to ways to use such direct calculations to determine other properties. The approach is often termed the "frozen phonon" method since all calculations are done for static structures. A great advantage of this approach is that it requires only the calculation of energy and forces available in essentially any density functional calculation. The disadvantage is that it often requires calculations on supercells, which are more computation intensive.[1]

Is it possible to calculate the derivatives directly instead of extracting derivatives from energy differences and forces? The answer is, of course, yes, since it is just a matter of well-known perturbation theory and response functions, for which the general theory is summarized in Appendix D. The subject of the rest of the chapter is developments that allow the perturbation expressions to be rewritten in ways that are much more efficient for actual calculations. In contrast to the frozen phonon methods, the perturbation theory approach makes it possible to calculate phonon frequencies at any $\mathbf{k}$ from a much smaller calculation based on one unit cell.

The starting point is the same in the previous chapter. If the nuclei are treated classically, their dynamics is determined by Newton's equation

$$M_I \frac{\partial^2 \mathbf{R}_I}{\partial t^2} = \mathbf{F}_I(\mathbf{R}) = -\frac{\partial}{\partial \mathbf{R}_I} E(\mathbf{R}), \qquad (20.1)$$

where $\mathbf{R}_I$ and M_I are the coordinate and mass of nucleus I, and $\mathbf{R} \equiv \{\mathbf{R}_I\}$ indicates the set of all the nuclear coordinates. Since nuclear motion is slow compared to typical electron frequencies, it is usually an excellent approximation to neglect any dependence of electronic energies on the velocities of nuclei, i.e., the *adiabatic or Born and Oppenheimer (BO) approximation* [90] (see Appendix C). All effects of the electrons on the dynamics of the nuclei are completely determined by the forces $\mathbf{F}_I(\mathbf{R})$, which are the sum of Coulomb forces between the nuclei and the forces due to the electrons that are determined by the electronic structure.

In the previous chapter the dynamics was determined by molecular dynamics. However, for stable solids at moderate temperature, it is much more useful and informative to cast the expressions in terms of an expansion of the energy $E(\mathbf{R})$ in powers of displacements and external perturbations, as in Eqs. (2.1)–(2.6). Equilibrium positions $\{\mathbf{R}_I^0\} = \mathbf{R}^0$ are determined by the zero-force condition on each nucleus,

$$\mathbf{F}_I(\mathbf{R}^0) = 0. \qquad (20.2)$$

Quantum zero-point motion, thermal vibrations, and response to perturbations are described by higher powers of displacements,

$$C_{I,\alpha;J,\beta} = \frac{\partial^2 E(\mathbf{R})}{\partial \mathbf{R}_{I,\alpha} \partial \mathbf{R}_{J,\beta}}, \quad C_{I,\alpha;J,\beta;K,\gamma} = \frac{\partial^3 E(\mathbf{R})}{\partial \mathbf{R}_{I,\alpha} \partial \mathbf{R}_{J,\beta} \partial \mathbf{R}_{K,\gamma}}, \ldots, \qquad (20.3)$$

where Greek subscripts $\alpha, \beta, \ldots$, indicate cartesian components.

Within the *harmonic approximation* [91], the vibrational modes at frequency ω are described by displacements

$$\mathbf{u}_I(t) = \mathbf{R}_I(t) - \mathbf{R}_I^0 \equiv \mathbf{u}_I e^{i\omega t}, \qquad (20.4)$$

---

[1] In fact, the *ab initio* molecular dynamics method called "Born–Oppenheimer" in the previous chapter use this approach; at each step the forces are calculated with the electrons in their ground state for the positions of the nuclei at that step.

so that Eq. (20.1) becomes for each I

$$-\omega^2 M_I u_{I\alpha} = -\sum_{J\beta} C_{I,\alpha; J,\beta} u_{J\beta}. \qquad (20.5)$$

The full solution for all vibrational states is the set of independent oscillators, each with vibrational frequency ω, determined by the classical equation

$$\det \left| \frac{1}{\sqrt{M_I M_J}} C_{I,\alpha; J,\beta} - \omega^2 \right| = 0, \qquad (20.6)$$

where the dependence on the masses M_I, M_J has been cast in a symmetric form.

For a crystal, the atomic displacement eigenvectors obey the Bloch theorem, Eqs. (4.32) and (4.33), i.e., the vibrations are classified by $\mathbf{k}$ with the displacements $\mathbf{u}_s(\mathbf{T_n}) \equiv \mathbf{R}_s(\mathbf{T_n}) - \mathbf{R}_s^0(\mathbf{T_n})$ of atom $s = 1, S$ in cell $\mathbf{T_n}$ given by

$$\mathbf{u}_{s,\mathbf{T_n}} = e^{i\mathbf{k}\cdot\mathbf{T_n}} \mathbf{u}_s(\mathbf{k}). \qquad (20.7)$$

Inserting this into Eq. (20.6) leads to decoupling of the equations at different $\mathbf{k}$ (just as for electrons – Exercise 20.2), with frequencies $\omega_{i\mathbf{k}}, i = 1, 3S$ called dispersion curves that are solutions of the $3S \times 3S$ determinant equation

$$\det \left| \frac{1}{\sqrt{M_s M_{s'}}} C_{s,\alpha; s',\alpha'}(\mathbf{k}) - \omega_{i\mathbf{k}}^2 \right| = 0, \qquad (20.8)$$

where the reduced force constant matrix for wavevector $\mathbf{k}$ is given by

$$C_{s,\alpha; s',\alpha'}(\mathbf{k}) = \sum_{\mathbf{T_n}} e^{i\mathbf{k}\cdot\mathbf{T_n}} \frac{\partial^2 E(\mathbf{R})}{\partial \mathbf{R}_{s,\alpha}(0) \partial \mathbf{R}_{s',\alpha'}(\mathbf{T_n})} = \frac{\partial^2 E(\mathbf{R})}{\partial \mathbf{u}_{s,\alpha}(\mathbf{k}) \partial \mathbf{u}_{s',\alpha'}(\mathbf{k})}. \qquad (20.9)$$

Since the vibrations are independent, quantization is easily included as usual for harmonic oscillators: phonons are the quantized states of each oscillator with energy $\hbar\omega_{i\mathbf{k}}$.

A useful analogy can be made between phonons and electrons described in a tight-binding model. Since the nuclei have three spatial degrees of freedom, the equation of motion, Eq. (20.8), has exactly the same form as Eq. (14.7) for the case with only three states of p symmetry for the electrons. The set of exercises comprising Exercises 20.1–20.7 is designed to show the relationships, derive phonon dispersion curves in simple cases, and explicitly transform a computational code for the tight-binding algorithm into a code for phonon frequencies with force constants described by models analogous to the parameterized models used in tight-binding methods.

Examples of dispersion curves are given in Figs. 2.11, 20.1, and 20.3. There are three acoustic modes with $\omega \to 0$ for $\mathbf{k} \to 0$ and the other $3S - 3$ modes are classified as optic. In insulators, there may be nonanalytic behavior with different limits for longitudinal and transverse modes, illustrated in Figs. 20.1 and 2.11.

The framework is set for derivation of the lattice dynamical properties from electronic structure so long as attention is paid to certain features:

- Careful treatment of the long-range effects due to macroscopic electric fields in insulators
- Formulation of the theory of elasticity in ways that facilitate calculations.

The key aspects of dealing with macroscopic electric fields are treated in Section E.6. In particular, "Born effective charges" $Z^*_{I,\alpha\beta}$ are defined in Eq. (E.20) in terms of the polarization per unit displacement *in the absence of a macroscopic electric field*. Similarly, proper piezoelectric constants $e_{\alpha,\beta\gamma}$ are defined in terms of polarization per unit strain in the absence of a macroscopic electric field. Fortunately, the theory and practical expressions for polarization are well established due to the advances described in Chapter 24. The result is that $C_{s,\alpha;s',\alpha'}(\mathbf{k})$ can be divided into terms that are analytic, as in the small $\mathbf{k}$ limit, plus nonanalytic contributions that can be treated exactly in terms of $Z^*_{I,\alpha\beta}$ and $e_{\alpha,\beta\gamma}$. These considerations are required in any approach that aspires to be a fundamental theory of lattice properties.

The basic definition of stress is given in Eq. (G.4) and the elastic constants, defined in Eq. (G.5), are the subject of many volumes [285, 300, 806, 807]. The important point to emphasize here is that both the electrons and the nuclei contribute directly to the stress, for which there are rigorous formulations (Appendix G). The expressions use the (generalized) force theorem, which depends on the requirement that all internal degrees of freedom be at their minimum energy values. This includes the electron wavefunctions and the nuclear positions, which are described by "internal strains" that are determined by the zero-force condition, Eq. (G.13). In many simple cases, e.g., in Bravais lattices, the force is zero by symmetry for zero internal strain; however, in general, the positions of the nuclei in a strained crystal are not fixed by symmetry, and any fundamental calculations of elastic properties must find the internal strains from the theory.

20.2 The Direct Approach: "Frozen Phonons," Magnons

The most direct approach is simply to calculate the total energy and/or forces and stresses as a function of the position of the nuclei, i.e., "frozen phonons," using any of the expressions valid for the electronic structure. Then the relevant quantities are defined by numerical derivatives for the force $\mathbf{F}$ and polarization $\mathbf{P}$ as a function of displacement $\mathbf{R}$

$$C_{I,\alpha;J,\beta} \approx -\frac{\Delta F_{I,\alpha}}{\Delta R_{J,\beta}}, \quad Z^*_{I,\alpha\beta}|e| \approx \left.\frac{\Delta P_\alpha}{\Delta R_{I,\beta}}\right|_{\mathbf{E}_{\text{mac}}}, \tag{20.10}$$

and for stress σ and polarization $\mathbf{P}$ as a function of strain ϵ,

$$C_{\alpha\beta;\gamma\delta} \approx -\frac{\Delta\sigma_{\alpha\beta}}{\Delta u_{\alpha\beta}}, \quad e_{\alpha\beta\gamma} \approx \left.\frac{\Delta P_\alpha}{\Delta\epsilon_{\alpha\beta}}\right|_{\mathbf{E}_{\text{mac}}}. \tag{20.11}$$

Such calculations are widely used since they require no additional computational algorithms; furthermore, this is the method of choice in cases where there are large displacements since the energy is automatically found to all orders. The direct approach played a critical role in early work, for example [171] and [808], where it was shown that phonons in materials (other than the sp-bonded metals) could be derived from the electronic structure.

An example of the energy versus displacements of atoms is shown in the left side of Fig. 2.10 for displacement of Ti atoms in BaTiO$_3$. The origin corresponds to atoms in centrosymmetric positions in perovskite structure shown in Fig. 4.8. If the energy had increased with displacement the structure would be stable and the curvature would determine the frequency of an optic mode. In this case there is an instability and at the lowest-energy state it has a finite displacement. One could describe this in terms of perturbation theory with fourth (and higher)-order anharmonic terms; however, it is convenient to calculate the energy directly for finite displacement. Then it is just an example of relaxation to the find the stable structure, a feature in all present-day codes. As pointed out in Section 2.9 the polarization can be calculated using Berry phases (Chapter 24), which is explicitly constructed as a method to calculate polarization directly to all orders.

Supercells: Phonons and Instabilities

One can also treat displacements that increase the size of the cell, i.e., supercells illustrated in Fig. 13.1. Transition metals are an excellent example where electronic structure calculations of "frozen phonon" energies can provide much information on the total bonding and the states near the Fermi energy that couple strongly to the phonons. The phonon energies for many transition metals have been shown to be well described, e.g., in [809] and [170], by calculations at wavevectors $\mathbf{k}$ along high-symmetry directions. For example, there is an interesting anomaly in the longitudinal frequency for $\mathbf{k} = \left(\frac{2}{3}, \frac{2}{3}, \frac{2}{3}\right)$ in the bcc structure crystals Mo and Nb. This is a precursor to the phase transition that actually occurs in Zr, which has bcc structure only above 1,100 K [157, 170]. The left side of Fig. 2.10 shows the calculated energy versus displacement for this phonon for Mo, Nb, and Zr. The inset shows the displacements that correspond to the "ω phase" with each third plane undisplaced, and the other two planes displaced to form a more dense bilayer. The LDA calculations agree well with the phonon frequencies in Mo and Nb and with the observed low-temperature structure of all three elements. The transition to bcc at high temperature is believed to be an effect of entropy, since it is well known that many metals have bcc structure at high temperature. A simple explanation is provided by the general arguments of Heine and Samson [810] that such a superlattice can occur for a 1/3 filled d band, which is the case for Zr. The electronic structure calculations provide a more complete picture, showing that this is not a delicate Fermi surface effect but is a combination of effects of s–p states and directional (covalent) d bonding involving states in a range around the Fermi energy [170].

Supercells and Dispersion Curves

It is also possible [608–610] to derive full-dispersion curves from the direct force calculations using supercells as illustrated in Fig. 13.1 for a zinc blende crystal. Any given phonon corresponds to a displacement of planes of atoms perpendicular to the wavevector $\mathbf{k}$, as shown on the left-hand side of the figure. All the information needed in Eq. (20.10) *for all phonons with a given direction* $\hat{\mathbf{k}}$ can be derived by displacing each inequivalent plane of atoms and calculating the force of all the atoms [608]. One must do a separate calculation for each inequivalent plane and each inequivalent displacement (four in the case shown in Fig. 13.1, for longitudinal and transverse displacement of Ga and As). If the size of

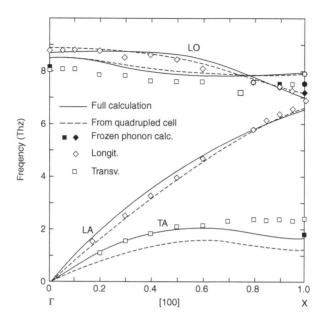

Figure 20.1. Dispersion curves for phonons in GaAs in the [1 0 0] direction calculated using the frozen phonon method. The supercell used for the calculation is a longer version of the cell shown in Fig. 13.1 with displacements as indicated that are either transverse or longitudinal. The same methods are used for interfaces, sublattices, and surfaces treated by making periodic cells, which illustrates the versatility of supercells for many problems. From [608].

the supercell exceeds twice the range of the forces, then all terms can be identified with no ambiguities. The results [608] for GaAs are shown in Fig. 20.1, compared with experiment. These are early calculations that used a semiempirical local pseudopotential. It is satisfying that better agreement with experiment has been found in more recent calculations [181, 811] shown in Fig. 2.11, which use *ab initio* norm-conserving potentials and the response function method (Section 20.3). The inverse dielectric constant ϵ^{-1} and the effective charges Z_I^* can also be calculated from the change in the potentials due to an induced dipole layer [608].

The practical advantage of this approach is immediately apparent from the fact that exactly the same methods can be used to calculate the properties of superlattices and interfaces. As shown in the middle figure of Fig. 13.1, a superlattice can be created by the theoretical alchemy of replacing Ga with Al in part of the supercell. Furthermore, the same methods apply directly to surfaces where part of the supercell is vacuum. This is illustrated on the right-hand side figure of Fig. 13.1, where the surface may undergo massive reconstruction beyond any perturbation expansion.

Another instructive example of the use of supercells is the calculation [611] of the inverse dielectric constant $\epsilon^{-1}(\mathbf{k})$ by imposing a periodic electrostatic potential of wavevector $\mathbf{k}$, i.e., a "frozen field." If the atoms are held fixed, the resulting potential leads to the static electronic inverse dielectric constant $\epsilon_0^{-1}(\mathbf{k})$; if the atoms are allowed to displace so that

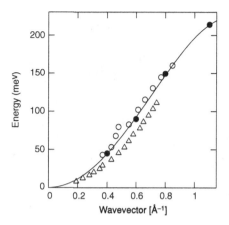

Figure 20.2. Dispersion curves for magnons in Fe: open circles are experimental points, compared to the dark circles that are theoretical results [161] calculated using the Berry phase approach [160]. Triangles show magnon energies for an alloy with 12% Si. From [161].

the forces are zero, one finds the result including the lattice contribution $\epsilon_\infty^{-1}(\mathbf{k})$. In each case, the $\mathbf{k} \to 0$ limit can be found by extrapolation from a few values of $\mathbf{k}$ and using the fact that $\epsilon_\infty^{-1}(\mathbf{k}) \propto k^2$ at small $\mathbf{k}$.

Spin excitations, or magnons, can be treated in the same way by calculating the energy of "frozen magnons." But how does one freeze the magnetization? Unlike phonons, magnetization is a continuous function of position that must be calculated. An elegant way of doing this is to generalize the Berry phase idea, Section 24.3, for electric polarization to the magnetic case of Niu and Kleinman [160]. An example of results [161] from this method for Fe, calculated using plane waves with a pseudopotential, are shown in Fig. 20.2: this shows excellent agreement with experiment. In this case, a *supercell is not needed*: because the magnon is a spiral excitation, the excitation obeys a generalized Bloch theorem [159] in which each cell is rotated equally with respect to its neighbors. See also Fig. 20.4.

20.3 Phonons and Density Response Functions

Historically,[2] the first approach for calculation of phonons was based on response functions: since all harmonic force constants, elastic constants, etc. involve only second derivatives of the energy, they can be derived using second-order perturbation theory. Furthermore, this builds on the fact that in many cases one can calculate small changes more accurately than the total energy itself. This section is devoted to the formulation, which is directly useful for simple cases and leads up to the efficient methods of Section 20.4.

The general expressions for the response to an external perturbation $V_{\text{ext}}(\mathbf{r})$ that varies with parameters λ_i (where λ denotes the position of an atom, the strain, etc.) are

---

[2] See references 2–13 in [180].

$$\frac{\partial E}{\partial \lambda_i} = \frac{\partial E_{II}}{\partial \lambda_i} + \int \frac{\partial V_{ext}(\mathbf{r})}{\partial \lambda_i} n(\mathbf{r}) d\mathbf{r}, \tag{20.12}$$

$$\frac{\partial^2 E}{\partial \lambda_i \partial \lambda_j} = \frac{\partial^2 E_{II}}{\partial \lambda_i \partial \lambda_j} + \int \frac{\partial^2 V_{ext}(\mathbf{r})}{\partial \lambda_i \partial \lambda_j} n(\mathbf{r}) d\mathbf{r} + \int \frac{\partial n(\mathbf{r})}{\partial \lambda_i} \frac{\partial V_{ext}(\mathbf{r})}{\partial \lambda_j} d\mathbf{r},$$

plus higher-order terms. The first equation is just the force (Hellmann–Feynman) theorem, which involves only the external potential and the unperturbed density. The first two terms in the second equation involve only the unperturbed density; however, the last term requires knowledge of $\partial n(\mathbf{r})/\partial \lambda_i$. Using the chain rule, the last term can be written in symmetric form [180]

$$\int \frac{\partial V_{ext}(\mathbf{r}')}{\partial \lambda_i} \frac{\partial n(\mathbf{r})}{\partial V_{ext}(\mathbf{r}')} \frac{\partial V_{ext}(\mathbf{r})}{\partial \lambda_j} d\mathbf{r} d\mathbf{r}' = \int \frac{\partial V_{ext}(\mathbf{r}')}{\partial \lambda_i} \chi(\mathbf{r}, \mathbf{r}') \frac{\partial V_{ext}(\mathbf{r})}{\partial \lambda_j} d\mathbf{r} d\mathbf{r}', \tag{20.13}$$

where χ is the density response function, Eq. (D.9). The expressions may be written in either $\mathbf{r}$ or $\mathbf{q}$ space as in Eq. (D.10), and may be expressed in terms of inverse dielectric function using the relation $\epsilon^{-1} = 1 + V_C \chi$, given explicitly in Eq. (E.15).

Following this approach, expressions for any second derivative of the energy can be found in two steps: using a relation like Eq. (D.14),

$$\chi = \chi^0 [1 - \chi^0 K]^{-1}, \text{ or } \chi^{-1} = [\chi^0]^{-1} - K, \tag{20.14}$$

that expresses χ in terms of the kernel K in Eq. (D.13) and the noninteracting response function χ^0. In turn, χ^0 is given by Eqs. (D.6)–(D.8), which are derived from standard expressions of perturbation theory in terms of sums over the entire spectrum of eigenstates. The form that is most useful for comparison with Green's function methods is Eq. (D.5), repeated here,

$$\Delta n(\mathbf{r}) = 2 \sum_{i=1}^{N} \sum_{j=N+1}^{\infty} \psi_i^*(\mathbf{r}) \psi_j(\mathbf{r}) \frac{\langle \psi_j | \Delta V_{KS} | \psi_i \rangle}{\varepsilon_i - \varepsilon_j}. \tag{20.15}$$

The formulation of response functions in Eqs. (20.14) and (20.15) is extremely fruitful in cases where the response function can be approximated by a simple form. For example, calculations of phonon dispersion curves in sp-bonded metals are well described by χ from the homogeneous electron gas, i.e., the static Lindhard dielectric function $\epsilon^{-1}(|\mathbf{q}|)$, Eq. (5.38), which is an analytic, scalar function of the one-dimensional magnitude $|\mathbf{q}|$. The foundation for the theory is the nearly-free-electron approximation with the ions represented by weak pseudopotentials [506, 508].[3]

However, the approach is very difficult for accurate calculations on general materials. There are two primary problems: The quantities involved are six-dimensional functions $\epsilon^{-1}(\mathbf{q}, \mathbf{q}')$, which become large matrices $\epsilon_{GG'}^{-1}(\mathbf{k})$ in crystals. Furthermore, calculation of *each of the entries in the matrix* involves a sum over the BZ of the filled and empty bands

---

[3] The formulas in terms of density response functions do not apply directly to nonlocal pseudopotentials, but appropriate modifications can be made [812].

(as in Eqs. (D.6)–(D.8)) up to some energy sufficient for convergence. This is a very difficult task that we will not belabor; the next section describes a much more efficient approach.

20.4 Green's Function Formulation

An alternative, much more effective, approach is "density-functional perturbation theory" (DFPT) [181, 811, 813, 814], which builds upon earlier classic works [52, 815]. Instead of calculation of the inverse dielectric function $\epsilon_{GG'}^{-1}(\mathbf{k})$, which gives the response to *all possible perturbations*, DFPT is designed to calculate the needed response to a particular perturbation. Instead of the standard perturbation theory sums over empty states, the expressions are transformed into forms that involve only the occupied states, which can be calculated using efficient electronic structure methods. This leads to two related types of expressions: (1) self-consistent equations for the response function in terms of the change in the wavefunctions to a given order, or (2) variational expressions in which the calculation of a response at any given order of perturbation is cast as a problem of minimizing a functional defined at that order. The theory can be applied to any order (Sections D.1 and D.6) but the main ideas can be seen in the lowest-order linear response.

The formulation can be understood by first returning to the fundamentals of perturbation theory. In terms of the wavefunctions, the first-order change in density is

$$\Delta n(\mathbf{r}) = 2\text{Re} \sum_{i=1}^{N} \psi_i^*(\mathbf{r}) \Delta \psi_i(\mathbf{r}), \qquad (20.16)$$

where $\Delta \psi_i(\mathbf{r})$ is given by first-order perturbation theory as

$$(H_{\text{KS}} - \varepsilon_i)|\Delta \psi_i\rangle = -(\Delta V_{\text{KS}} - \Delta \varepsilon_i)|\psi_i\rangle. \qquad (20.17)$$

Here H_{KS} is the unperturbed Kohn–Sham hamiltonian, $\Delta \varepsilon_i = \langle \psi_i | \Delta V_{\text{KS}} | \psi_i \rangle$ is the first-order variation of the KS eigenvalue, ε_i, and the change in the effective potential is given by

$$\Delta V_{\text{KS}}(\mathbf{r}) = \Delta V_{\text{ext}}(\mathbf{r}) + e^2 \int d\mathbf{r}' \frac{\Delta n(\mathbf{r}')}{|\mathbf{r} - \mathbf{r}'|} + \int d\mathbf{r}' \frac{dV_{\text{xc}}(\mathbf{r})}{dn(\mathbf{r}')} \Delta n(\mathbf{r}')$$

$$\equiv \Delta V_{\text{ext}}(\mathbf{r}) + \int d\mathbf{r}' K(\mathbf{r}, \mathbf{r}') \Delta n(\mathbf{r}'). \qquad (20.18)$$

The kernel $K(\mathbf{r}, \mathbf{r}')$ is pervasive in the response function formalism and is discussed further in Chapter 9 and Appendix D (where K is given in $\mathbf{k}$ space in Eq. (D.13)). It incorporates the effects of Coulomb interaction and exchange and correlation to linear order through $f_{\text{xc}}(\mathbf{r}, \mathbf{r}') = dV_{\text{xc}}(\mathbf{r})/dn(\mathbf{r}')$.

The standard perturbation theory approach is to expand Eq. (20.17) in eigenfunctions of the zero-order Schrödinger equation, which leads to Expression (20.15). Although this approach is not efficient because it requires knowledge of the full spectrum of the zero-order hamiltonian and sums over many unoccupied states, nevertheless, the expressions are useful for demonstration of important properties. In particular, Eq. (20.15) shows that

$\Delta n(\mathbf{r})$ involves only mixing of the unoccupied ($j > N$) space into the space of the occupied ($i \leq N$) states because contributions from the occupied states cancel in pairs in the sum.

It is much more effective for actual calculations to view the set of equations, Eqs. (20.16)–(20.18), as a self-consistent set of equations for Δn and ΔV_{KS} to linear order in ΔV_{ext}. It might appear that there is a problem since the left-hand side of Eq. (20.17) is singular because the operator has a zero eigenvalue, for which the eigenvector is ψ_i. However, as shown in Eq. (20.15), the response of the system to an external perturbation only depends on the component of the perturbation that couples the manifold occupied states with the empty states. The desired correction to the occupied orbitals can be obtained from Eq. (20.17) by projecting the right-hand side onto the empty-state manifold,

$$(H_{KS} - \varepsilon_i)|\Delta \psi_i\rangle = -\hat{P}_{empty}\Delta V_{KS}|\psi_i\rangle, \tag{20.19}$$

where the projection operator is given by (see also Eq. (23.18))

$$\hat{P}_{occ} = \sum_{i=1}^{N} |\psi_i\rangle\langle\psi_i|; \quad \hat{P}_{empty} = 1 - \hat{P}_{occ}. \tag{20.20}$$

In practice, if the linear system is solved by the conjugate-gradient or any other iterative method in which orthogonality to the occupied-state manifold is enforced during iteration, there is no problem with the zero eigenvalue.

The basic algorithm for DFPT consists of solving the set of linear equation (20.19) for $\Delta \psi_i$ given the definition in Eq. (20.20) and expression (20.18) for ΔV_{KS} in terms of Δn, which is given by Eq. (20.16). Since Δn is a function of the set of occupied $\Delta \psi_i$, this forms a self-consistent set of equations. Any of the efficient iterative methods (Appendix M) developed for electronic structure problems can be applied to reach the solution by iteration. This is a more efficient approach than the standard "textbook" approach outlined following Eq. (20.13): the equivalent of the matrix inverse in Eq. (19.14) is accomplished by the self-consistent solution for $\Delta \psi_i$ and ΔV_{KS}, and the sum over excited states is accomplished by mixing of the unoccupied space into the occupied space using Eq. (20.19).

20.5 Variational Expressions

Variational expressions in perturbation theory have a long history, for example the ingenious use of the variational principle for accurate solution of the two-electron problem by Hylleraas [52] in the 1930s. The ideas have been brought to the fore by Gonze [816, 817] (see also [181]), who has derived expressions equivalent to the Green's function formulas of Sections 20.4 and 20.6. The variational perspective provides an alternative approach for solution of electronic structure problems and is one that involves minimization rather than solution of linear equations.

The basic idea of variational expressions in perturbation theory is very simple. If a system has internal degrees of freedom, which are free to adjust (within any constraints that they must obey), then the static response to an external perturbation requires that all internal degrees of freedom adjust to minimize the energy. Exercise 20.5 gives a simple example of a system made up of two harmonic springs with an internal degree of freedom. In electronic structure the perturbation is a change in the external potential ΔV_{ext}, and the internal degrees of freedom are the density $n(\mathbf{r})$ or the wavefunctions ψ_i. These can be viewed as independent variables, i.e., one can define a functional $E^{(m)}[n]$ or $E^{(m)}[\psi_i]$ valid at a chosen order m of perturbation theory, and the correct solution is found by minimizing the functional. Just as for the variational principle that leads to the Schrödinger or Kohn–Sham equations, the energy, ψ_i, and $n(\mathbf{r})$ are determined by minimizing the energy.

The variational principle in perturbation theory can be derived directly from the same variational principle that leads to the Schrödinger or Kohn–Sham equations, but the new point is that it is applied only to a given order. For example, to second order in the changes $\Delta V_{ext} = V_{ext} - V_{ext}^0$ and $\Delta n = n - n^0$, the Kohn–Sham energy functional, Eqs. (7.5) or (7.20), can be written in a form similar to Eq. (7.21) with the addition of terms involving ΔV_{ext},[4]

$$E^{(2)}[V_{ext}, n] = E[V_{ext}^0, n^0] + \int d\mathbf{r}\, n^0(d\mathbf{r})\Delta V_{ext}(\mathbf{r})$$

$$+ \frac{1}{2}\int d\mathbf{r}d\mathbf{r}'\left[\frac{\delta^2 E}{\delta V_{ext}(\mathbf{r})\Delta n(\mathbf{r}')}\right]\Delta V_{ext}(\mathbf{r})\Delta n(\mathbf{r}') \qquad (20.21)$$

$$+ \frac{1}{2}\int d\mathbf{r}d\mathbf{r}'\left[\frac{\delta^2 E}{\delta n(\mathbf{r})\delta n(\mathbf{r}')}\right]\Delta n(\mathbf{r})\Delta n(\mathbf{r}'),$$

where derivatives are evaluated at the minimum energy solution with V_{ext}^0 and n^0. There is no term involving ΔV_{ext}^2 since the functional is linear in V_{ext}. The first intergral in Eq. (20.21) is the force theorem. The middle line is linear in Δn. The last line is quadratic in Δn and is always positive since the functional is minimum at V_{ext}^0 and n^0. Minimization of the functional is, in principle, merely a matter of minimizing a quadratic form, no harder than the harmonic springs in Exercise 20.5. However, practical solutions must be done in terms of the wavefunctions ψ_i since the functionals of density are unknown.

In terms of the orbitals, Expression (20.21) becomes [181] (omitting terms that are zero for $\Delta V_{ext} = 0$)

$$E^{(2)}[V_{ext}, \psi_i] = E[V_{ext}^0, \psi_i^0]$$

$$+ \frac{1}{2}\sum_i \int d\mathbf{r}d\mathbf{r}'\left[\frac{\delta^2 E}{\delta V_{ext}(\mathbf{r})\delta \psi_i(\mathbf{r}')}\right]\Delta V_{ext}(\mathbf{r})\Delta \psi_i(\mathbf{r}') \qquad (20.22)$$

$$+ \frac{1}{2}\sum_{ij} \int d\mathbf{r}d\mathbf{r}'\left[\frac{\delta^2 E}{\delta \psi_i(\mathbf{r})\delta \psi_j(\mathbf{r}')}\right]\Delta \psi_i(\mathbf{r})\Delta \psi_j(\mathbf{r}'),$$

---

[4] Spin indices are omitted for simplicity and the subscript "KS" is omitted because the expressions apply more generally.

where

$$\left[\frac{\delta^2 E}{\delta V_{\text{ext}}(\mathbf{r})\delta\psi_i(\mathbf{r}')}\right] = \psi_i^0(\mathbf{r})\delta(\mathbf{r} - \mathbf{r}'), \tag{20.23}$$

$$\left[\frac{\delta^2 E}{\delta\psi_i(\mathbf{r})\delta\psi_j(\mathbf{r}')}\right] = H_{\text{KS}}^0(\mathbf{r}, \mathbf{r}') + K(\mathbf{r}, \mathbf{r}')\Delta\psi_i(\mathbf{r})\Delta\psi_j(\mathbf{r}')\Delta\psi_i(\mathbf{r})\Delta\psi_j(\mathbf{r}'), \tag{20.24}$$

with K defined in Eq. (20.18); see also Eq. (D.13).

The solution can be found directly by minimizing Eq. (20.22) with respect to $\Delta\psi_i$ subject to the orthonormality constraint

$$\langle\psi_i^0 + \Delta\psi_i|\psi_j^0 + \Delta\psi_j\rangle = \delta_{ij}. \tag{20.25}$$

The steepest descent directions for $\Delta\psi_i$ are found by writing out the equations, which also show equivalence to the Green's function method [181]. Minimizing $E^{(2)}[V_{\text{ext}}, \psi_i]$ in Eq. (20.22) with condition Eq. (20.25) leads to

$$H_{\text{KS}}^0\Delta\psi_i - \sum_j \Lambda_{ij}\Delta\psi_j = -(\Delta V_{\text{eff}} - \varepsilon_i)\psi_i + \sum_j \Lambda_{ij}\Delta\psi_j, \tag{20.26}$$

where ΔV_{eff} is the change in total effective potential given to the second order by Eq. (20.18). Taking matrix elements with respect to the orbitals, one recovers (Exercise 20.10) the form given in Eq. (20.19).

20.6 Periodic Perturbations and Phonon Dispersion Curves

The DFPT equations have a marvelous simplification for the case of a crystal with a perturbation at a given wavevector, e.g., a phonon of wavevector $\mathbf{k}_p$, with displacements given by Eq. (20.7). To linear order, the change in density, external potential, and Kohn–Sham potential all have Fourier components only for wavevectors $\mathbf{k}_p + \mathbf{G}$, where $\mathbf{G}$ is any reciprocal lattice vector. The expressions can be written

$$\Delta V_{\text{ext}}(\mathbf{r}) = \Delta v_{\text{ext}}^{\mathbf{k}_p}(\mathbf{r})e^{i\mathbf{k}_p\cdot\mathbf{r}} = \sum_{\mathbf{T}} \frac{V_s[\mathbf{r} - \mathbf{R}_s(\mathbf{T})]}{\partial \mathbf{R}_s(\mathbf{T})}e^{-i\mathbf{k}_p\cdot(\mathbf{r}-\mathbf{R}_s(\mathbf{T}))}\mathbf{u}_s(\mathbf{k}_p)e^{i\mathbf{k}_p\cdot\mathbf{r}}, \tag{20.27}$$

$$\Delta V_{\text{KS}}(\mathbf{r}) = \Delta v_{\text{KS}}^{\mathbf{k}_p}(\mathbf{r})e^{i\mathbf{k}_p\cdot\mathbf{r}}, \tag{20.28}$$

$$\Delta n(\mathbf{r}) = \Delta n^{\mathbf{k}_p}(\mathbf{r})e^{i\mathbf{k}_p\cdot\mathbf{r}}. \tag{}$$

The wavefunction for an electron at wavevector $\mathbf{k}_e$ is modified to linear order *only by mixing of states with wavevector* $\mathbf{k}_e + \mathbf{k}_p$, so that Eq. (20.19) becomes

$$\left(H_{\text{KS}}^{\mathbf{k}_e} - \varepsilon_i^{\mathbf{k}_e}\right)|\Delta\psi_i^{\mathbf{k}_e+\mathbf{k}_p}\rangle = -\left[1 - \hat{P}_{\text{occ}}^{\mathbf{k}_e+\mathbf{k}_p}\right]\Delta V_{\text{KS}}^{\mathbf{k}_p}|\psi_i^{\mathbf{k}_e}\rangle. \tag{20.29}$$

To linear order the density is given by

$$\Delta n^{\mathbf{k}_p}(\mathbf{r}) = 2\sum_{\mathbf{k}_e, i} u_{\mathbf{k}_e, i}^*(\mathbf{r})\Delta u_{\mathbf{k}_e+\mathbf{k}_p, i}(\mathbf{r}), \tag{20.30}$$

where u denotes the periodic part of the Bloch function, and the Kohn–Sham potential is

$$\Delta v_{KS}^{\mathbf{k}_p}(\mathbf{r}) = \Delta v_{ext}^{\mathbf{k}_p}(\mathbf{r}) + \int d\mathbf{r}' \left[\frac{1}{|\mathbf{r} - \mathbf{r}'|} + f_{xc}(\mathbf{r}, \mathbf{r}') \right] \Delta n^{\mathbf{k}_p}(\mathbf{r}). \tag{20.31}$$

The DFPT algorithm for the calculation of the response to any periodic external perturbation $\Delta v_{ext}^{\mathbf{k}_p}(\mathbf{r})$ is the solution of the set of equations Eqs. (20.29)–(20.31). Note that the calculation involves only *pairs* of wavevectors, $\mathbf{k}_e$ and $\mathbf{k}_e + \mathbf{k}_p$ for the electrons, in the linear equation (20.29), and a sum over all $\mathbf{k}_e$ and filled states i for the self-consistency. The calculations can be done using the same fast Fourier transform (FFT) techniques that are standard in efficient plane wave methods (Chapter 13 and Appendix M): Eqs. (20.29) and (20.31) can be solved partly in $\mathbf{r}$ space and partly in $\mathbf{k}$ space, and Eq. (20.30) is most efficiently done in $\mathbf{r}$ space, with transformations done by FFTs.

Figure 2.11 shows the results of DFPT calculations for the phonon dispersion curves of GaAs [182], done using the local approximation (LDA). Similar results are found for other semiconductors and it is clear that agreement with experiment is nearly perfect. Calculations have been done for many other materials, and an example is the set of results presented in Fig. 20.3 for the phonon dispersion curves of a set of metals. From top to bottom, these represent increasing electron–phonon interactions and increasing complexity of the Fermi surface. The LDA is essentially perfect for Al (as expected!), but the GGA provides a significant improvement in Nb where there is a soft phonon (see discussion in Section 20.2 and calculation for Nb in Fig. 2.10). Similar results are found for many materials using many methods, finding agreement with experimental frequencies within $\approx 5\%$ is typical.

20.7 Dielectric Response Functions, Effective Charges

Electric fields present a special problem due to long-range Coulomb interaction. This arises in any property in which electric fields are intrinsically involved, e.g., dielectric functions, effective charges, and piezoelectric constants. One approach is to formulate the theory of response functions at finite wavelengths and take limits analytically [180, 819]. Is it possible to generate an efficient Green's function approach that involves only the infinite wavelength $\mathbf{q} = 0$ limit? The answer is yes, but only with careful analysis. The problem is that the limit corresponds to a homogeneous electric field with potential $V_{ext}(\mathbf{r}) = \mathbf{E} \cdot \mathbf{r}$, which leads to an ill-defined hamiltonian in an extended system; see Chapter 24. The saving grace is that *perturbation theory in the electric field* involves matrix elements of the form of Eq. (20.15) or Eq. (20.19), i.e., only *off-diagonal* matrix elements of the perturbing potential between eigenfunctions of the unperturbed hamiltonian. These matrix elements are well defined even for a macroscopic electric field, which can be seen by rewriting them in terms of the commutator between $\mathbf{r}$ and the unperturbed hamiltonian,

$$\langle \psi_i | \mathbf{r} | \psi_j \rangle = \frac{\langle \psi_i | [H, \mathbf{r}] | \psi_j \rangle}{\epsilon_i - \epsilon_j}, \quad i \neq j. \tag{20.32}$$

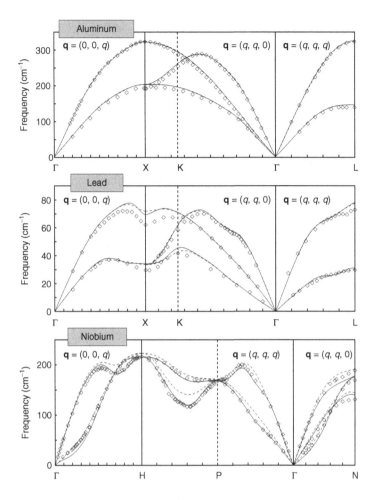

Figure 20.3. Phonon dispersion curves calculated for the metals Al, Pb, and Nb compared with experiment [818]. The agreement using the LDA is excellent for Al and progressively worse for the other metals, which are further from the homogeneous gas, where GGA functionals improve the agreement. The dips in the phonon dispersion curves for Pb and Nb indicate strong electron–phonon interactions and sensitivity to Fermi surface features that are important for superconductivity in these elements.

If the total potential acting on the electrons is local, the commutator is simply proportional to the momentum operator,

$$[H, \mathbf{r}] = -\frac{\hbar^2}{m_e}\frac{\partial}{\partial \mathbf{r}} = i\frac{\hbar}{m_e}\mathbf{p}. \tag{20.33}$$

For nonlocal potentials, the commutator involves an explicit contribution from the potential [547, 548] as defined in Eq. (11.72). Littlewood [820] has used the momentum form for the matrix elements to derive expressions for the dielectric functions of crystals in terms of the periodic Bloch functions, and the explicit iterative Green's function algorithm corresponding to Eqs. (20.16)–(20.20) is given in [181].

An example of the application of the ideas is the calculation of effective charges Z^* in ionic insulators. Linear response calculations have been done for many materials, e.g., the ferroelectric perovskite materials with formula unit ABO_3. Anomalously large effective charges of the B atoms and the O atoms moving along the line joining them have been found and interpreted as resulting from covalency [821, 822]. Results are essentially the same as those given in table 24.1 from [823] using the Berry phase method.

20.8 Electron–Phonon Interactions and Superconductivity

Electron–phonon coupling plays a crucial role in the theory of transport and superconductivity in solids, and there are excellent sources showing the relation of the microscopic interactions to the phenomena [824–826]. In particular, the basic quantity in the Eliashberg [827] equations for phonon-mediated superconductivity is $\alpha^2 F(\omega)$, where $F(\omega)$ is a phonon density of states and α denotes an average over all phonons of energy ω.

The quantities that can be derived from the underlying electronic structure are the electron bands and density of states, the phonon dispersion curves and density of states, and electron–phonon coupling (Appendix C). The matrix element for scattering an electron from state $i\mathbf{k}$ to $j\mathbf{k} + \mathbf{q}$ while emitting or absorbing a phonon $\nu\mathbf{q}$ with frequency ω is given by [825]

$$g_{i\mathbf{k};\, j\mathbf{k}+\mathbf{q}}(\nu) = \frac{1}{\sqrt{2M\omega_{\nu\mathbf{q}}}} \langle i\mathbf{k}|\Delta V_{\nu\mathbf{q}}|j\mathbf{k}+\mathbf{q}\rangle, \tag{20.34}$$

where M is the (mode-dependent) reduced mass and $1/\sqrt{(2M\omega_{\nu\mathbf{q}})}$ is the zero-point phonon amplitude.

For scattering at the Fermi surface, one can define the dimensionless coupling to the phonon branch ν, where $\nu = 1, \ldots, 3S$, with S the number of atoms per cell, by [825]

$$\lambda_\nu = \frac{2}{N(0)} \sum_{\mathbf{q}} \frac{1}{\omega_{\nu\mathbf{q}}} \sum_{ijk} \left| g_{i\mathbf{k};\, j\mathbf{k}+\mathbf{q}}(\nu) \right|^2 \delta(\varepsilon_{i\mathbf{k}}) \delta(\varepsilon_{j\mathbf{k}+\mathbf{q}} - \varepsilon_{i\mathbf{k}} - \omega_{\nu\mathbf{q}}), \tag{20.35}$$

where $N(0)$ is the electronic density of states per spin at the Fermi energy and $\omega_{\nu\mathbf{q}}$ is the energy of phonon ν with the wavevector $\mathbf{q}$. The energy of electron band i with the wavevector $\mathbf{k}$ is $\varepsilon_{i\mathbf{k}}$ and $g_{i\mathbf{k};\, j\mathbf{k}+\mathbf{q}}(\nu)$ is the matrix element between the states $i\mathbf{k}$ and $j\mathbf{k} + \mathbf{q}$ due to the induced potential when phonon $\nu\mathbf{q}$ is excited. The delta functions restrict the electron scattering to the Fermi surface.

A key quantity in the theory is the induced potential ΔV per unit displacement needed in the matrix element Eq. (20.34). To linear order it is given by

$$\Delta V_{\nu\mathbf{q}} = \frac{\partial V}{\partial u_{\nu\mathbf{q}}} = \sum_{I\alpha} X_{\nu\mathbf{q}}(I\alpha) \frac{\partial V}{\partial u_{I\alpha}}, \tag{20.36}$$

where $u_{\nu\mathbf{q}}$ is the phonon normal coordinate and $X_{\nu\mathbf{q}}(I\alpha)$ is the eigenvector of the dynamical matrix, Eq. (20.8), expressed in terms of displacements $u_{I\alpha}$ of the nucleus I in the α direction. Since the phonon is low frequency, the potential $\Delta V_{\nu\mathbf{q}}$ is a "screened potential," i.e., the electrons react to contribute to the effective potential.

There are four general methods for calculation of the matrix elements with the properly screened $\Delta V_{\nu\mathbf{q}}$:

- Displacement of rigid atomic-like potentials, e.g., muffin-tin potentials [828]. The justification is that the coupling is primarily local [829] and the potential is very much like a sum of atomic potentials, which has been shown to be appropriate for transition metals, even with displaced atoms. The same ideas apply for displacement of any effective total potential, such as empirical pseudopotentials.
- Calculations using a general expression for the screening, which can be conveniently expressed in Fourier space as

$$\frac{\partial V(\mathbf{q}+\mathbf{G})}{\partial u_{\nu\mathbf{q}}} = \sum_{\mathbf{G}'} \epsilon^{-1}(\mathbf{q}+\mathbf{G},\mathbf{q}+\mathbf{G}')\frac{\partial V_{\text{ion}}(\mathbf{q}+\mathbf{G}')}{\partial u_{\nu\mathbf{q}}}, \qquad (20.37)$$

following the definition of ϵ^{-1}, given, e.g., above Eq. (20.14). This is particularly convenient when one can use a simple model for ϵ^{-1}, e.g., the Lindhard function, Eq. (5.38), which is appropriate for simple metals. In general, however, it is more efficient to use the Green's function approach.
- Methods using direct calculation of the linear order screened potential [181, 830]. This can be done conveniently using the Green's function technique described in Sections 20.4 and 20.6. As is clear from Eq. (20.37), one of the by-products of the calculation of phonons is the screened $\Delta V_{\nu\mathbf{q}}$ itself. This needs to be done for all relevant phonons needed for the integral over the Fermi surface.
- Self-consistent "frozen phonon" calculations [718, 831]. As described in Section 20.2, the calculations involve V directly to all orders in the displacement. The linear change in potential can be extracted as the linear fit to calculations with finite displacements of the atoms. Alternatively, one can find the mixing of wavefunctions from which the matrix elements can be found. The "frozen phonons" can be determined on a grid of $\mathbf{k}$ points and interpolated to points on the Fermi surface.

The full calculation requires a double sum over the Brillouin zone and determination of the matrix elements g at each pair of points and for each phonon ν.

20.9 Magnons and Spin Response Functions

Response functions for spin excitations are the fundamental quantities measured in spin-dependent neutron scattering from solids just as lattice dynamical response functions are the fundamental quantities measured in lattice dynamics experiments. Since spin dynamics is strongly affected by magnetic order and excitations tend to be damped by coupling to

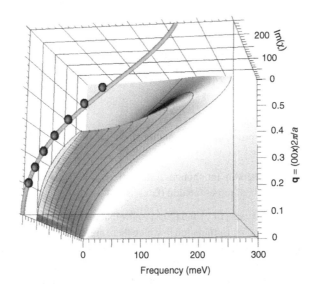

Figure 20.4. Spin response spectral functions $\text{Im}\chi\,(\mathbf{q},\omega)$ for Fe. The line shows the dispersion corresponding to the maximum in $\text{Im}\chi$ compared with experimental values (points). Compare also with Fig. 20.2. Note that the excitations broaden at higher frequency and momentum. From [833].

the electron degrees of freedom, the form of the spectral function $\text{Im}\chi\,(\mathbf{q},\omega)$ is often of importance. For many years the basic formalism has been known and calculations have been done with standard Green's function techniques and realistic band structures of metals, e.g., in the work of Cooke [832] and many more recent calculations. The method is based on a simple extension of the random-phase approximation of the relevant Green's function equation and an interpolation formalism for treating wavefunctions and matrix elements.

There is a Green's function method that generalizes the approach used for phonons to treat spin response functions [833]. The formulas are closely related and involve spin density instead of number density. An example of results for Fe using the LMTO approach is shown in Fig. 20.4. Comparison with Fig. 20.2 shows good agreement with calculations done using the Berry phase method and the plane wave pseudopotential method. In addition, however, the results in Fig. 20.4 show the full spectral function, with its broadening increasing rapidly for wavevectors near the zone boundary, in qualitative agreement with experiment. Such work has been extended to alloys using the KKR–CPA method and for the treatment of disorder within linear response theory [697].

SELECT FURTHER READING

Classic text for lattice vibrations.

> Born, M. and Huang, K., *Dynamical Theory of Crystal Lattices* (Oxford University Press, Oxford, 1954).

Theory of phonons in terms of the electronic resonse functions.

> Pick, R., Cohen, M. H., and Martin, R. M., "Microscopic theory of force constants in the adiabatic approximation," *Phys. Rev. B* 1:910–920, 1970.

Review of density-functional perturbation theory and Green's Function methods.

Baroni, S., de Gironcoli, S., Dal Corso, A., and Giannozzi, P., "Phonons and related crystal properties from density-functional perturbation theory," *Rev. Mod. Phys.* 73:515–562, 2000.

Exercises

20.1 See many excellent problems (and solutions) on phonons in the book by Mihaly and Martin [598].

20.2 Show that the Bloch theorem for phonons follows from exactly the same logic as for electrons treated in the local orbital representation (Exercise 14.3).

20.3 By comparing Expressions (14.7) and (20.8), show that the equations for phonons are exactly the same as a tight-binding formulation in which there are three states of p symmetry for each atom, corresponding to the three degrees of freedom for the atomic displacements. Show explicitly the correspondence of the terms of the two problems, especially the fact that the eigenvalue in the electron problem corresponds to the *square* of the frequency in the phonon problem.

20.4 This exercise is to explain a salient difference in the tight-binding electron and the phonon problems. The force constant has the property of *translational invariance*; show that the fact that the total energy does not change if all the atoms are displaced uniformly leads to the relation $\sum_J C_{I,\alpha;J,\beta} = 0$. This condition can be used to fix the self-term $C_{I,\alpha;I,\beta}$ so that, unlike the electron tight-binding problem, the on-site term is not an independent variable. In addition, show that this means that there are three zero frequency phonon modes at $\mathbf{k} = 0$.

20.5 The simplest model for phonons is the central force model in which the energy is a function only of the distance between the nearest neighbors. Find expressions for a force constant $C_{I,\alpha;J,\beta}$ using the definition as a second derivative of the energy expressed as a sum of pair terms $E = \sum_{I<J} E_{IJ}(|\mathbf{R}_I - \mathbf{R}_J|)$. Show that the resulting expressions are equivalent to a tight-binding problem of electron p states, Exercise 14.11, with the $t_{pp\pi}$ matrix elements equal to zero.

20.6 Find expressions for phonon dispersion curves respectively in elemental simple cubic and fcc crystals using the simplest model for phonons, a nearest-neighbor central potential model with energy given by $E = \frac{1}{2} \sum_{I<J} K |\mathbf{R}_I - \mathbf{R}_J|^2$, where J is restricted to nearest neighbor. Show the relation to tight-binding equations for p bands in Exercise 14.11 as explicit examples of the relationship given in Exercises 20.3–20.5.

(a) Show that there are two dispersion curves that have zero frequency for all k in simple cubic but not in fcc crystals. Explain why this instability occurs in a simple cubic crystal in a central potential model.

(b) There is a corresponding result in the tight-binding model for p bands in a simple cubic crystal if the only nonzero matrix element is $t_{pp\sigma}$ between nearest neighbors. Show that in this case there are two bands with no dispersion.

20.7 Computational exercise: using the properties established in Exercises 20.3–20.5, construct a computer program (or use a tight-binding code that is available) to evaluate phonon frequencies in a model analogous to a tight-binding model for electrons.

20.8 Why is there no term involving $\partial^2 n/(\partial\lambda_i\partial\lambda_j)$ in the expression for $\partial^2 E/(\partial\lambda_i\partial\lambda_j)$ in Eq. (20.12)?

20.9 Derive Eq. (20.15), which is the same as from the basic perturbation expressions (D.4) and (20.16). In particular, show that all contributions involving i and j, both occupied, vanish in the expression for the change in the density.

20.10 Show that Eq. (20.19) follows from Eq. (20.26) by taking matrix elements of the equation.

20.11 This is an example of the variational principle in perturbation theory. Consider a system composed of three points x_0, x_1, x_2 in a line connected by two springs. The energy is $E = \frac{1}{2}k_1(x_1 - x_0)^2 + \frac{1}{2}k_2(x_2 - x_1)^2$. Suppose forces $f = f_0 = -f_2$ are applied to the two ends.
(a) Identify the functional $F[f, x_1]$ valid for all f and x_1.
(b) If the middle position x_1 is free to move, calculate the change in position $x_1 - x_0$ and total length $x_2 - x_0$ as a function of f.
(c) See Exercise D.7 for extension to nonlinear springs.

21

Excitation Spectra and Optical Properties

Look, Simba, everything the light touches is our kingdom.

King Mufasa in *The Lion King*

Summary

Excitation spectra reveal the properties of matter in terms of the response to time- or frequency-dependent perturbations. Particularly important examples are the dielectric function $\epsilon(\omega)$ and the inverse function $\epsilon^{-1}(\omega)$ in solids and the polarizability $\alpha(\omega)$ in molecules. The basic formulas relating the response to the electronic structure are rooted in perturbation theory and response functions in Appendixes D and E. This chapter is devoted to time-dependent density functional theory (TDDFT), which provides an exact framework in principle. In practice, approximate forms are very successful and TDDFT is now a tool used expansively for molecules and nanosize clusters. There are special issues in solids that provide insights into the effects of interactions.

21.1 Overview

As emphasized in Sections 2.14 and 2.15, two types of excitations are of primary importance for electronic structure: those in which an electron is added or removed from the system, and those in which the number of electrons remains fixed. In independent-particle theories, such as the Hartree–Fock or Kohn–Sham approaches, the addition and removal energies are the eigenstates of the independent-particle hamiltonian, which form bands $\varepsilon_{i\mathbf{k}}$ in a crystal. As discussed in Section 7.7, the Kohn–Sham equations are meant to determine the total energy and it is a misuse of the original Kohn–Sham approach to identify the eigenvalues with true electron removal or addition energies. Even if the theory is on a more firm basis in the generalized Kohn–Sham picture (Section 9.2), the eigenvalues are still just a band structure that cannot describe the broadening of the spectra and satellites that occur in real systems due to interactions, as illustrated in Fig. 2.22. Many examples of bands are given in other chapters and will not be covered further here.

A new set of concepts emerges for excitations in which the number of electrons does not change. An example is optical spectra, which are among the most important properties of materials and one of the most important experimental tools for studying the electronic structure of materials. There are also other techniques such as inelastic scattering of charged particles and X-rays [834]. In the independent-particle approximation the spectrum is only a convolution of one-particle processes in which an electron is removed from an occupied state (creating a hole) and added to an empty state with no interactions. However, in the real world, the added electron interacts with the hole, and the interaction is screening by the other electrons. Since the Kohn–Sham method cannot describe aspects of the one-particle spectra for the electron and hole individually, it may seem surprising that there could be ways to use the Kohn–Sham approach that are better for problems in which both are present and there are even more effects of interaction.

The topic of this chapter is time-dependent density functional theory (TDFT), the extension of the Kohn–Sham theory that provides a way to calculate spectra, in principle exactly. The ideas are similar to the time-dependent Hartree–Fock approximation, which has a long history [329, 835], and were perhaps first used by Ando et al. [836] for model problems in semiconductors and by Zangwill and Soven [245] for atoms. It is only since the 1980s with the work of Runge and Gross [246] that it has been put on a firm footing analogous to the Kohn–Sham formulation for the ground state. TDDFT has been extremely successful and is the method of choice for many problems, especially in the chemistry community. It is widely applied to clusters and nanosystems (Section 21.7) and there are more recent developments that make it more practical for solids (Section 21.8). At the end of this chapter are listed a few of the many books and reviews that provide penetrating analyses and practical information.

21.2 Time-Dependent Density Functional Theory (TDDFT)

The derivation of timedependent density functional theory (TDDFT) is fundamentally different from that for DFT, yet there is much in common. Like the static version, the equations are remarkably simple; the difficult issues of treating exchange and correlation are buried in a functional that is unknown, but feasible approximations turn out to be very successful. For a time-dependent problem, the equations are not derived by minimizing an energy, like the original static Kohn–Sham theory; instead, the time-dependent Kohn–Sham equations can be derived from the stationarity principle for the action A [246],

$$\frac{\delta A}{\delta n(\mathbf{r}, t)} = 0, \tag{21.1}$$

where

$$A = \int_{t_0}^{t_1} dt \, \langle \Psi(t) | \left[i \frac{d}{dt} - \hat{H}(t) \right] | \Psi(t) \rangle. \tag{21.2}$$

If one adds the Kohn–Sham idea of replacing the density with the density of independent particles, this leads to time-dependent Kohn–Sham density functional theory (TDDFT), in which there is a time-dependent Schrödinger-like equation

$$i\hbar \frac{d\psi_i(t)}{dt} = \hat{H}_{\text{eff}}(t)\psi_i(t), \tag{21.3}$$

with an effective hamiltonian that depends on time t

$$\hat{H}_{\text{eff}}(t) = -\frac{1}{2}\nabla^2 + V_{\text{ext}}(\mathbf{r},t) + \int \frac{n(\mathbf{r}',t)}{|\mathbf{r}-\mathbf{r}'|}d\mathbf{r}' + V_{\text{xc}}[n](\mathbf{r},t), \qquad (21.4)$$

where $V_{\text{xc}}[n](\mathbf{r},t)$ is a function of $\mathbf{r}$ and t and a *functional* of density $n(\mathbf{r}',t')$.

The original derivation [246] was in terms of the current density $\mathbf{j}(\mathbf{r},t)$; however, in a finite system it can be transformed to a theory strictly in terms of the density using the continuity equation $\nabla \cdot \mathbf{j}(\mathbf{r},t) = -\dot{n}(\mathbf{r},t)$. In the applications later in this chapter, we will first consider (Section 21.7) finite systems such as molecules and nanoscale clusters using Eqs. (21.3) and (21.4) as they are written. Afterward we consider (Section 21.8) electromagnetic waves in crystals where it is advantageous to use the formulation in terms of the current and the vector potential $\mathbf{A}(\mathbf{r},t)$, but the approach will still be called TDDFT, which is traditional in the literature.

The seemingly simple equations of TDDFT conceal subtleties that we will not deal with, but which should be pointed out; the issues are discussed in reviews such as [395, 427, 837, 838], which give references to the literature. In the formally exact theory, $V_{\text{xc}}[n](\mathbf{r},t)$ is a functional of $n(\mathbf{r}',t')$ for *all earlier times* $t' \leq t$. This leads to issues related to causality that have been controversial but been resolved [839]. Most applications get around the difficulties by making the *adiabatic approximation* in which $V_{\text{xc}}[n(t)](\mathbf{r})$ depends only on the density at the same time, e.g., in the adiabatic LDA (ALDA), it is simply $V_{\text{xc}}(\mathbf{r},t) = V_{\text{xc}}(n(\mathbf{r},t))$. However, it is clear that the adiabatic approximation is not always sufficient. For example, in a system driven near a resonance, the functional should take into account the particular states involved in the transition. The effects are not easily captured in terms of the density alone; however, orbital-dependent functionals naturally incorporate at least some aspects of spatial and dynamical effects and they play an essential role in the theory and applications.

One of the main applications of TDDFT is to calculate optical spectra, which is emphasized in the rest of this chapter. But it should be kept in mind that TDDFT is a very general theory that applies for various perturbations. Inelastic scattering of charged particles is described in terms of the inverse dielectric function $\epsilon^{-1}(\mathbf{q},\mathbf{q}',\omega)$ defined in Section E.4. This is discussed more fully in [1], Section 14.3. It is the inverse function that is most convenient to derive the expression for the macroscopic dielectric constant including the microscopic internal fields as expressed in Eq. (E.17). This is also the response function needed to describe energy loss for high-energy charged particles and stopping power of energetic ions (see, e.g., [840]). Inelastic X-ray scattering is another active area of experiments that can be addressed by TDDFT (see, e.g., the review [834]).

21.3 Dielectric Response for Noninteracting Particles

Before proceeding, it is useful to first define the dielectric function and give the expressions in the independent-particle approximation. The forms are illuminating and examples of the independent-particle spectra are shown later for comparison with the TDDFT spectra. For a finite system, such as a cluster, with size much less than the wavelength of light, the expressions in terms of a scalar potential and the position vectors $\mathbf{r}$ is a convenient approach. The external perturbation[1] can be written in terms of the applied electric field $\mathbf{E}_{\text{ext}}(t)$ acting upon an electron at point $\mathbf{r}$,

$$\Delta\hat{H}(t) = \Delta V_{\text{ext}}(\mathbf{r}, t) = -e\mathbf{E}_{\text{ext}}(t) \cdot \mathbf{r}, \tag{21.5}$$

instead of Eq. (21.8), and the response is the induced dipole moment per unit volume given by

$$\Delta\mathbf{d}(t) = \frac{-e}{\Omega} \int_{\text{all space}} d\mathbf{r} \, \Delta n(\mathbf{r}, t) \, \mathbf{r}. \tag{21.6}$$

This is the approach used directly in the real-time method. In the frequency domain the expression for linear response is the polarizability $\alpha(\omega) = d(\omega)/E(\omega)$ (omitting vector indices for simplicity), which involves matrix elements of the position operator $\mathbf{r}$,

$$\alpha(\omega) = e^2 \sum_{ij} (f_i - f_j) \frac{\langle \psi_i | \mathbf{r} | \psi_j \rangle \langle \psi_j | \mathbf{r} | \psi_i \rangle}{\varepsilon_i - \varepsilon_j + \omega + i\eta}. \tag{21.7}$$

It is not apparent but Exercise 9.8 is to show that α has units of volume. In TDDFT these expressions are modified due to the induced variation of the Kohn–Sham potential. The general form of the expressions is discussed below with references to papers where detailed expressions can be found.

In a crystal the formulation in terms of electric fields, dipoles, and matrix elements of the position $\mathbf{r}$ are ill-defined and they have led to enormous confusion over years.[2] Those problems are avoided if the macroscopic field is treated in terms of the vector potential $\mathbf{A}(t)$ instead of the scalar potential V. Then the perturbation can be included in the hamiltonian as given in Eq. (E.19),

$$\Delta\hat{H}(t) = \frac{1}{2m_e} \sum_i \left\{ \left[\hat{\mathbf{p}}_i - \frac{e}{c}\mathbf{A}(t) \right]^2 - \hat{\mathbf{p}}_i^2 \right\}, \tag{21.8}$$

where $\mathbf{E}(t) = -\frac{1}{c}\frac{d\mathbf{A}}{dt}$ and $\mathbf{E}(\omega) = -\frac{i\omega}{c}\mathbf{A}(\omega)$. The desired response is the macroscopic average current density $\mathbf{j} = -e\langle\mathbf{v}\rangle$, and since $\mathbf{p} = m\mathbf{v} - \frac{e}{c}\mathbf{A}$, it follows that $\mathbf{j} = \frac{-e}{m}\langle\mathbf{p} + \frac{e}{c}\mathbf{A}\rangle$, and the relation of $\mathbf{j}(\omega)$ to $\mathbf{E}(\omega)$ determines the conductivity $\sigma(\omega)$ and $\epsilon(\omega)$ (see Section E.2). Using the general form of a response function Eq. (D.19) one finds the expression

$$\epsilon_{\alpha\beta}(\omega) = \delta_{\alpha\beta} - \frac{e^2}{m_e\Omega}\frac{1}{\omega^2}\sum_i \left[f_i\delta_{\alpha\beta} + \sum_j \frac{f_i - f_j}{\hbar m_e} \frac{\langle\psi_i|p_\alpha|\psi_j\rangle\langle\psi_j|p_\beta|\psi_i\rangle}{\varepsilon_i - \varepsilon_j + \omega + i\eta} \right], \tag{21.9}$$

---

[1] Here the electron charge $-e$ is explicitly included to avoid confusion, since the standard definition of $\mathbf{E}$ is the field acting on a positive charge.

[2] The issues are recounted in Chapter 24 and the resolution is part of the saga leading to Berry phases and topology. However, those developments are not needed here.

where the f_i are occupation numbers and $\eta > 0$ is a small damping factor. (The first term in the square brackets comes from the contribution of $\mathbf{A}$ to the current operator (Exercise 21.1) and the second is the response function χ^0 for noninteracting particles.)

This expression shows the basic reason that measurements of optical spectra are such powerful tools for studies of electronic properties of crystals. If the $\mathbf{p}$ matrix elements do not vary rapidly as a function of energy for transitions between each pair of bands of electronic states, the imaginary part of $\epsilon(\omega)$ (or the real part of $\sigma(\omega)$) directly reveals singularities in the density of states for optical transitions. In the noninteracting approximation, this is a joint density of states for transitions between pairs of filled and empty bands weighted by the matrix elements. Examples of $\epsilon(\omega)$ calculated in the independent-particle approximation are given in Figs. 2.25 and 21.5; see [561] and [565] for many other examples.

21.4 Time-Dependent DFT and Linear Response

In this chapter are three methods for calculations using time-dependent DFT. The first is based on well-established perturbation theory and response functions. It has a great advantage that it is close in spirit to many-body perturbation theory (MBPT), in particular the Bethe–Salpeter equation (BSE) and implementations that are described in [1]. The relation of TDDFT and MBPT is the topic of the review by Onida et al. [841]. One of the fortunate consequences is that codes developed for calculation of spectra using the BSE can be readily adapted to TDDFT, including hybrid functionals that involve the nonlocal Hartree–Fock exchange. A disadvantage of this approach is that it is formulated in terms of transitions to empty states, i.e., all combinations of pairs of occupied and empty states (notice the number of indices on the kernel $K_{ij\sigma,i'j'\sigma'}$ and the expression for each term in Eq. (21.14), which leads to large matrices especially for cases where there is a large basis such as plane waves or grids.

The general form of response functions in self-consistent field theories is given in Appendix D in terms of the noninteracting response functions χ^0 and the interaction kernel K. The dynamical density response function is given by Eq. (D.23), which can also be written

$$\chi(\omega) = \chi^0(\omega)[1 - \chi^0(\omega)K(\omega)]^{-1} \quad \text{or} \quad \chi(\omega) = \chi^0(\omega)[1 + \chi(\omega)K(\omega)] \qquad (21.10)$$

where K is the Fourier transform of the space- and frequency-dependent kernel given in Eq. (D.22). To put flesh on these bare bones, we can give useful explicit expressions following [447]. Expanding the expressions in terms of the time-independent Kohn–Sham orbitals,[3] the matrix elements of the effective potential are given by $\delta[V_{\text{eff}}]_{ij}^\sigma \equiv \langle \psi_i^\sigma | \delta V_{\text{eff}}(\mathbf{r}, \omega) | \psi_j^\sigma \rangle$, and the density can be written in terms of the density matrix ρ_{ij}^σ using

$$\delta n^\sigma(\mathbf{r}, \omega) = \sum_{ij} \psi_i^\sigma(\mathbf{r}) \rho_{ij}^\sigma \psi_j^\sigma(\mathbf{r}). \qquad (21.11)$$

---

[3] The expressions in terms of the wavefunctions or density matrix are general and apply with nonlocal potential operators, such the exchange term, in Hartree–Fock and hybrid functionals.

The response of the system can be expressed in terms of the density matrix and one can define a noninteracting χ^0 given by [447]

$$\chi^0_{ij\sigma,i'j'\sigma'} = \frac{\delta\rho^\sigma_{ij}}{\delta[V_{\text{eff}}]^{\sigma'}_{i'j'}} = \delta_{ii'}\delta_{jj'}\delta_{\sigma\sigma'}\frac{f^\sigma_i - f^\sigma_j}{\omega - (\varepsilon^\sigma_i - \varepsilon^\sigma_j)}. \tag{21.12}$$

The interacting response function can be derived from the relation $\delta V_{\text{eff}} = \delta V_{\text{ext}} + \delta V_H + \delta V_{\text{xc}} \equiv \delta V_{\text{ext}} + \delta V_{H\text{xc}}$, which to linear order is given by $\delta V_{\text{eff}} = \delta V_{\text{ext}} + K\delta n$. The full expression can be written in the form of Eq. (21.10),

$$\chi_{ij\sigma,i'j'\sigma'} = \frac{\delta\rho^\sigma_{ij}}{\delta[V_{\text{ext}}]^{\sigma'}_{i'j'}} = \chi^0_{ij\sigma,i'j'\sigma'}\left[\delta_{ii'}\delta_{jj'}\delta_{\sigma\sigma'} + \sum_{i''j''\sigma''}\chi_{i''j''\sigma'',i'j'\sigma'}\,K_{ij\sigma,i''j''\sigma''}\right], \tag{21.13}$$

where K is the array of matrix elements of the interaction terms

$$K_{ij\sigma,i'j'\sigma'} = \frac{\delta[V_{H\text{xc}}]^\sigma_{ij}}{\delta\rho^{\sigma'}_{i'j'}} \tag{21.14}$$

$$= \int d\mathbf{r}\int d\mathbf{r}'\psi^\sigma_i{}^*(\mathbf{r})\psi^\sigma_j(\mathbf{r})\left[\frac{\delta_{\sigma\sigma'}}{|\mathbf{r}-\mathbf{r}'|} + f^{\sigma\sigma'}_{\text{xc}}(\mathbf{r},\mathbf{r}',\omega)\right]\psi^{\sigma'}_{i'}{}^*(\mathbf{r}')\psi^{\sigma'}_{j'}(\mathbf{r}'),$$

where f_{xc} is the second derivative of $E_{\text{xc}}[n]$ given explicitly in Eqs. (D.21) and (D.22). The frequency dependence of f_{xc} presents difficult issues and essentially all work to date uses the adiabatic approximation where $E_{\text{xc}}[n]$ and f_{xc} are the same as in the static Kohn–Sham method.

With some algebra (Exercise 21.2), the solution of the equation can be cast in the form of an eigenvalue equation often called the Casida equation [447]

$$\left[\omega^2_{ij\sigma}\delta_{ii'}\delta_{jj'}\delta_{\sigma\sigma'} + 2\sqrt{f_{ij\sigma}\omega_{ij\sigma}}\,K_{ij\sigma,i'j'\sigma'}\sqrt{f_{i'j'\sigma'}\omega_{i'j'\sigma'}}\right]F_n = \Omega^2_n F_n, \tag{21.15}$$

where $f_{ij\sigma} = f^\sigma_i - f^\sigma_j$ and $\omega_{ij\sigma} = \varepsilon^\sigma_i - \varepsilon^\sigma_j$. The TDDFT problem thus becomes a matrix problem with the basis of *pairs of Kohn–Sham states* $ij\sigma$. If there are N_{occ} filled orbitals and one includes N_{empty} empty orbitals of each spin, then the size of the matrices is $N_{\text{pair}} \times N_{\text{pair}}$, where $N_{\text{pair}} = 2N_{\text{occ}} \times N_{\text{empty}}$. These are very general equations for the density matrix, which can be utilized in the problem of polarizability of a molecule or cluster in terms of matrix elements of the position vector $\langle\psi_i|\mathbf{r}|\psi_j\rangle$ and the density matrix. An efficient approach is to use the eigenvectors F_n of the Casida equation (21.15), which is explained in detail in references such as [575].

21.5 Time-Dependent Density-Functional Perturbation Theory

Dynamical response functions can also be formulated in ways closely related to the iterative or variational Green's function methods of Section 20.4 called density functional perturbation theory (DFPT). The general idea is that the response functions developed in the previous section can describe the response to *all possible perturbations*. However, what we need in any particular case is the response to the particular perturbation. In the static case, DFPT was developed as a framework in which the desired result to a given order in perturbation theory can be found by iterative solution of a self-consistent set of equations involving only the occupied subspace.

The corresponding formulation has been developed for dynamic perturbations in two papers by Walker et al. [842] and Rocca et al. [843], and there is an overview in a paper [844] that describes the resulting computer code turboTDDFT that is publicly available. In static DFPT, the basic algorithm is iteratively solving the set of equations (20.19) for the change in the occupied set of wavefunctions projected onto the empty space, i.e., orthogonal to the occupied space. The corresponding problem in the time-dependent case is cast in terms of a Liouvillian superoperator that acts on pairs of functions. Like the static form, it is transformed into a set of equations that do not require calculation of the empty states. The calculation can be much more efficient than the standard perturbation theory, especially for large systems or large basis sets, such as plane waves or real-space grids. With efficient adaptations of the Lanczos method, the entire spectrum of a molecule or extended system can be computed with computational effort comparable to that required for a ground-state calculation.

21.6 Explicit Real-Time Calculations

An alternative approach is to propagate the system in time. The evolution of each one-particle state $\psi_i(t)$ is given by a Schrödinger-like equation (21.3) with an effective hamiltonian that depends on time. If we adopt the adiabatic approximation, as is done in the other methods, then $V_{xc}(\mathbf{r}, t)$ depends on only the density at the same time for each step. Many of the basic ideas originated in nuclear physics [845] and an excellent exposition can be found in the text by Koonin and Meredith [485].

The main difference from the other approaches in that the evolution is not limited to linear response (or nonlinear response, which is more and more difficult for higher orders). Instead it treats the effect nonperturbatively and can be for nonlinear effects, including extreme conditions created by laser pulses. (See, e.g., [395] and references therein.) A great advantage is that, like the time-dependent DFPT methods, there is no need to explicitly calculate excited states. One needs only to propagate N wavefunctions for N particles. A convenient choice is for the initial wavefunctions $\psi_i(t = 0)$ to be the filled ground-state functions before the perturbation is turned on. When the external field is turned on, this is not the ground state; the wavefunctions are not eigenstates and they evolve in time mixing in components of the states that were empty of the initial system. The algorithm for real-time calculations is very straightforward and it automatically includes the effects of f_{xc} as the hamiltonian is updated.

The fact that the one does not have to know f_{xc} is also a disadvantage; because the effects are hidden, it is more difficult to understand the physics and one has lost the connection to the many-body pertubation theory. Another drawback is the fact that the time step has to be very small in order to capture the details of the response of the electrons.

The key requirement for the iterations in time is that the propagation is unitary, which is essential for particle number conservation. If the functions $\psi_i(t)$ are expanded in a fixed time-independent basis

$$\psi_i(t) = \sum_\alpha c_{i,\alpha}(t)\chi_\alpha, \qquad (21.16)$$

then the iteration from $c_{i,\alpha}^n$ at time t^n to $c_{i,\alpha'}^{n+1}$ at time $t^{n+1} = t^n + \delta t$ is given by

$$c_{i,\alpha}^{n+1} = \sum_{\alpha'} [e^{-i\hat{H}\delta t}]_{\alpha,\alpha'} c_{i,\alpha'}^n, \qquad (21.17)$$

where $\hat{H}$ is a matrix in the basis α, α'. The hamiltonian $\hat{H}$ depends on the density $n(\mathbf{r}, t)$, but this is not written explicitly to keep the notation simple. Because the complex exponential operator has unity modulus, Eq. (21.17) is a unitary transformation. The size of the time step δt is limited by condition that $\hat{H}(t)$ can be considered constant over the interval δt; nevertheless, one can still choose the time at which $\hat{H}(t)$ is evaluated. Since $\hat{H}$ must be updated as a function of the time-dependent density, this can have important consequences for efficiency. Two examples are the predictor–corrector method [846] and the "railway curve interpolation" [847].

Here we describe four types of approaches used in actual calculations:

- Explicit operations with the exponential. In general it is not possible to perform the exponentiation of an operator exactly, and one must bring the operators to diagonal form. In that case, $\exp(A) = \sum_i |\zeta_i\rangle \exp(A_{ii})\langle\zeta_i|$, $A_{ii} = \langle\zeta_i|A|\zeta_i\rangle$, with ζ_i an eigenvector of A, $A_{ii} = \langle\zeta_i|A|\zeta_i\rangle$, and $\exp(A_{ii})$ the ordinary exponential of a scalar (Exercise 21.3). This can be done separately for the potential (V) and kinetic (T) operators, which are diagonal, respectively, in real and reciprocal space. However, the hamiltonian involves both operators, which cannot in general be separated since they do not commute. Nevertheless, one can use a Suzuki–Trotter expansion, of which a simple example is

$$\exp[-i(T + V)\delta t] \simeq \exp\left(-i\frac{1}{2}V\Delta t\right)\exp(-iT\Delta t)\exp\left(-i\frac{1}{2}V\Delta t\right), \qquad (21.18)$$

plus corrections $\propto \delta t^2$ (Exercise 21.4). This approach is well suited for plane wave or real-space methods, where efficient fast Fourier transform algorithms provide an *exact* transformation between *finite* plane wave expansions and real-space grids. The transformations are exactly the same as used in ground-state plane codes and described in Section M.11, especially Fig. M.2.
- Expansion of the exponential. The simplest approach is to expand the exponential in Eq. (21.17) in powers of the hamiltonian, which leads to

$$c^{n+1} = \left[1 - i\hat{H}\delta t - \frac{1}{2}\hat{H}^2\delta t^2 + \cdots\right]c^n. \qquad (21.19)$$

The expansion can easily be carried to high orders and the calculation done by iterative applications of $\hat{H}$. Just as for other iterative methods (Appendix M), the operations can be done efficiently so long as the hamiltonian is sparse, e.g., with localized states, or using transformations such the FFT to make all operations sparse (Section M.11, especially Fig. M.2). Although the operations are not manifestly unitary, they can be used for practical calculations [846] with small δt.

- Unitary expansion of the exponential. The expansion of the exponential can be done in an alternative form using the Crank–Nicholson operator [485],

$$c^{n+1} = \frac{1 - i\hat{H}\frac{\delta t}{2} + \cdots}{1 + i\hat{H}\frac{\delta t}{2} + \cdots} \, c^n. \tag{21.20}$$

This method is unitary, strictly preserving the orthonormality of the states for an arbitrary δt. For time-independent hamiltonians, it is also explicitly time-reversal invariant, and exactly conserves energy. In practice, with a suitable choice of δt, energy is satisfactorily conserved even when the hamiltonian changes with time. The disadvantage is that it involves an inverse operator, which requires a matrix inversion or solution of linear equations [485].

- Expansion in Chebyshev polynomials. An alternative to iterative approaches is to expand in Chebyshev orthogonal polynomials [848] that provide a *global* fit to the propagation over the entire time range,

$$e^{-iHt} \simeq e^{-iE_{av}t} \sum_{n=0}^{N} a_n \left(\frac{\Delta E t}{2} \right) T_n(H_{norm}), \tag{21.21}$$

where T_n are the Chebyshev polynomials (Section K.5) and the expansion coefficients $a_n(x)$ can be shown to be analogous to Bessel functions of the first kind of order n. Here $H_{norm} \equiv (2H - E_{av})/\Delta E$ is a normalized hamiltonian, where $E_{av} = (E_{max} + E_{min})/2$ and $\Delta E = E_{max} - E_{min}$, with E_{max} and E_{min} the maximum and minimum eigenvalues of H. The Chebyshev polynomials are chosen because their error decreases exponentially when N is large enough, due to the uniform character of the Chebyshev expansion [848].

21.7 Optical Properties of Molecules and Clusters

TDDFT has become a standard tool in chemistry to characterize spectra of molecules. There are vast numbers of calculations described in references such as [447, 838, 849], and comparisons of functionals, for example in [850]. Here we focus on methods and applications on clusters and large molecules (exemplified by C_{60}) that are close to work in condensed matter. Many calculations have been done using the adiabatic LDA and GGA, which improve agreement with experiment significantly compared to the noninteracting approximation.

For a finite system with size much less than the wavelength of light, such as molecules and clusters, the relevant response function is the polarizability $\alpha(\omega) = d(\omega)/E(\omega)$ where d is the induced dipole defined as a function of time in Eq. (21.6) or the Fourier transform to frequency after Eq. (21.6). The spectra are often presented as the imaginary part of the polarizability $\alpha(\omega)$ or as the dipole strength function[4]

$$S(\omega) = \frac{2m}{\pi e^2 \hbar} \omega \, \mathrm{Im}\alpha(\omega). \tag{21.22}$$

The strength function is proportional to the experimentally measured photoabsorption cross-section, and it satisfies the f sum rule,

$$\hbar \int_0^\infty d\omega S(\omega) = \sum_i f_i = N_e, \tag{21.23}$$

where f_i are the oscillator strengths (see Exercise 21.5).

Metal clusters are an interesting class of systems that can be varied from atomic to macroscopic size [222]. The simplest approximation is spherical jellium ignoring the atomic structure, which leads to the correct general features of the optical spectra dominated by a plasmon-like peak.[5] The question for quantitative calculations is the extent to which the real atomic structure matters. An example of one of the first quantitative calculations on metal clusters is shown in Fig. 21.1. The structure was determined by a ground-state LDA calculation and the spectrum by time-dependent DFT using the LDA (TDLDA) and a plane wave basis. There is a two-peak structure very different from the jellium model, with the main peak shifted and in better agreement with experiment. The shift in the main peak from the noninteracting particle response, Eq. (21.9), shows the effects of the Coulomb interaction in this confined geometry. Many calculations for metallic clusters have been reported (e.g., [851]), including quantum molecular dynamics (QMD) simulations [852] that account for thermal broadening due to dynamical motion of the nuclei.

Semiconductor clusters, or "quantum dots," terminated by H or other elements, are of great interest because of their enhanced optical properties (see, e.g., [854]). Figure 21.2 shows TDLDA spectra for selected structures compared to independent-particle spectra calculated using a real-space finite difference method as described in Section 12.8 [575]. For the smallest system (the SiH_4 molecule), TDLDA results in a small shift to lower energy due to the fact that the electron and hole are confined to a small volume reducing the energy due to Coulomb attraction. For large clusters, still much smaller than the wavelength of light, the spectra shift to higher energy and have a plasma-like resonance that couples to light because of finite size effects.

The lowest-energy excitations and an effective gap corresponding to the onset of strong absorption are shown in Fig. 21.3 for H-terminated Si clusters as a function of diameter. Experimental results are shown for molecules and for large clusters. The gap increases with decreasing cluster size due to quantum confinement effects, and it is evident that the theory explains the trends in general agreement with experiment. For example, the gap of

---

[4] Here m, e, and $\hbar$ are written out explicitly to enable the conversion to usual units.

[5] In a bulk solid, the plasma peak [280] is due to a longitudinal density response dominated by a peak in the inverse function $\mathrm{Im}\,\epsilon^{-1}(\omega)$ for $\omega \approx \omega_p$; however, in a small confined system, the distinction between longitudinal and transverse is lost and there is a peak in the absorption of light at $\omega \approx \omega_p$.

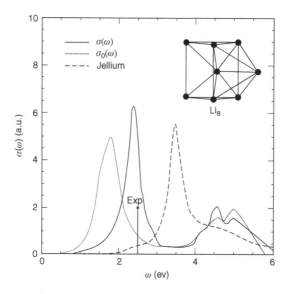

Figure 21.1. Example of optical spectrum of a metal cluster (Li_8) calculated [853] using TDDFT with the adiabatic local density approximation (TDLDA). The inset shows the structure of the cluster determined by the usual ground-state density functional theory. The resulting spectrum is significantly changed from the spherical jellium model, in better agreement with experiment (arrow), and from the noninteracting approximation (dotted line labeled σ_0). From [853].

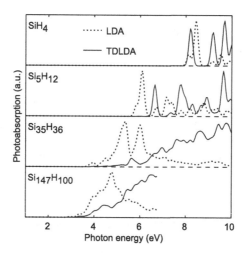

Figure 21.2. Optical spectra of selected hydrogen-terminated Si systems, from the molecule SiH_4 to the $Si_{147}H_{100}$ cluster. The solid lines are spectra from TDLDA calculations, which are compared to uncorrected independent-particle spectra (dotted lines) that follows from Eq. (21.9) applied to a finite system. For SiH_4, TDLDA results in a small shift to lower energy; whereas for the larger clusters, the intensity of the optical absorption shifts to higher energy due to Coulomb interactions and finite size effects (see text). Provided by I. Vasiliev

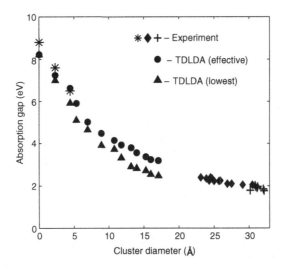

Figure 21.3. Optical properties of Si nanoclusters predicted using TDLDA compared to experiment. The graph shows results for the lowest gap and for a higher energy, designating a threshold in the optical strength that corresponds to experimental assignments of an effective gap [575]. Provided by I. Vasiliev

Si nanoclusters is increased above the bulk gap, and it has been found that Si clusters can become efficient emitters of blue light [855].

Real-Time Calculations for Molecules and Clusters

In the real time approach for a small system, the dipole moment $\mathbf{d}(t)$ is calculated as the response to an applied field, and the polarizability $\alpha(\omega) = d(\omega)/E(\omega)$ is found by a Fourier transform. A convenient procedure is to start with the equilibrium ground-state wavefunctions $\psi_i^{\bar{E}}$ of the system in a constant applied electric field $\bar{E}$, i.e., with an added potential $\Delta V(\mathbf{r}) = -e\bar{E}x$. At time $t = 0$, the field $\bar{E}$ is suddenly removed and for $t > 0$ the system evolves as given by the time-dependent Schrödinger equation (21.3) with initial independent-particle states $\psi_i^{\bar{E}}$ at $t = 0$ and the usual Kohn–Sham hamiltonian, which is a function of time for $t > 0$ since the density $n(\mathbf{r}, t)$ is a function of time, even though there is no applied field. The electric field $E(\omega)$ is the Fourier transform of a step function $E(t) = \bar{E}\Theta(-t)$, so that $E(\omega) = \bar{E}/(i\omega)$ and $\mathrm{Im}\alpha(\omega) = \omega\mathrm{Re}\alpha(\omega)/\bar{E}$. Finally, $d(\omega)$ can be calculated from $d(t) = e\int d\mathbf{r}n(\mathbf{r},t)x = e\sum_i\langle\psi_i(t)|x|\psi_i(t)\rangle$ and Fourier transforming to give

$$d(\omega) = \int_0^{\infty} dt e^{i\omega t - \delta t} d(t), \tag{21.24}$$

where the factor $e^{-\delta t}$ is a damping factor introduced for convergence at large times.

There are various methods for real time calculations. One of the first is that of Yabana and Bertsch [846], who used a real-space grid and a hamiltonian with a finite difference

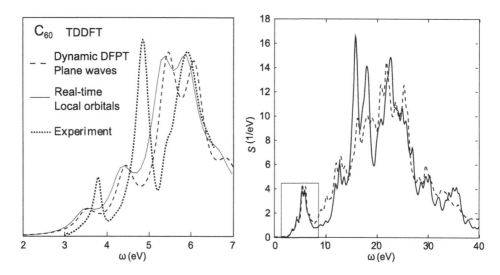

Figure 21.4. Spectrum of the dipole strength function $S(\omega)$ for C_{60} calculated by two different methods. The solid lines were calculated using a real-time method and local orbitals [856], and the dashed lines with the dynamic density functional perturbation theory method (DFPT) and a plane wave basis [843]. At the right are the spectra for a large energy range; the agreement is good, given that the local orbital basis is limited and not expected to accurately represent high energy states, and plane waves are affected by the size of the simulation box. At the left is the comparison with experiment in the low-energy range. The calculations agree well as expected in the low-energy range. However, the energies of the transitions are somewhat too low, and the strengths of the first two peaks are much too low, compared to experiment [857]. The spectra at the left are normalized to have the same integrated weight in the low-energy range for comparison with the experiment. Provided by D. Rocca and A. Tsolakidis

expression for the kinetic energy. The real-time method can also be implemented [856] for localized bases where it is difficult to exponentiate the hamiltonian operator directly. Instead, it is convenient to use the Crank–Nicholson form in Eq. (21.20). This is explicitly unitary and inversion of the matrix $1 + i\hat{H}\frac{\delta t}{2} + \cdots$ can be done since a local orbital basis can be very small and since there can be efficient inversion methods for matrices close to the unit matrix.

Figure 21.4 shows the spectra for a C_{60} molecule calculated by the real-time local orbital method [856] and the dynamic density functional perturbation theory (DFPT) method described in Section 21.5 [843] with a very large basis of plane waves. The real-time simulation was done with time step of 2.128×10^{-17} s and the total time of 130 fs, and both methods are expected to be accurate in the low-energy range. As shown at the left in Fig. 21.4, the two methods agree well and agree with experiment except for the peak at ≈ 5 eV. The two calculations used LDA and PBE functionals respectively, but they should be comparable since the spectra of the two functionals are comparable.

The lowest-energy excitations in the calculated spectra are too low and are too weak compared to the experiment. A discrepancy in the energies ≈ 0.4 eV is also found in a large survey of other systems [857, 858]. Presumably the difference from experiment is due to the approximations for the functional – the local or semilocal form and the adiabatic approximation. Possibly this is related to the failures of such functionals for excitons in solids, as described in the following section; in any case, it shows that care must be taken in the interpretation of the calculations.

21.8 Optical Properties of Crystals

Crystals and other macroscopic systems with size much greater than the wavelength of light are in the opposite regime from clusters. The optical properties are governed by the dielectric function $\epsilon(\omega)$ instead of polarizability $\alpha(\omega)$, and similarly for nonlinear response functions. In that case the appropriate formulation is in terms of the macroscopic currents $\mathbf{j}$ and the vector potential $\mathbf{A}$ instead of charge n and the scalar potential V. This is what is done in the response function for independent particles in Eq. (21.9) and the goal is to take into account effects of exchange and correlation using TDDFT.[6]

Even though there is a vast amount of work on molecules and clusters, TDDFT has been used much less to calculate optical properties of crystals. There are difficulties in solids that can be understood on physical grounds. One of the great problems with LDA and GGA functionals is the large error in the bandgap. One might hope that TDDFT would be a way to alleviate this problem, but, in fact, the gap in TDDFT is the same as the difference in Kohn–Sham eigenvalues. This can be seen immediately from the form of the response functions, e.g., in the left side of Eq. (21.10). In a crystal there is a continuum of excitations and the absorption is the imaginary part of a response function. Since the independent-particle response function $\chi^0(\omega)$ has an imaginary part across the entire band, it follows that the full function $\chi(\omega)$ also has an imaginary part fo the same range. Thus even though the shape of the spectrum may change, the band edges do not.

Another problem has been the subject of much work for years: the difficulty is describing excitons, which are bound states below the continuum. Excitons are very important in the spectra of many solids, as illusioned by the spectrum for LiF in Figs. 2.26 and 21.5. Since the band edges do not change in TDDFT, the only way a bound state can emerge from the theory is for there to be a pole in the response function. From the expression in the left part of Eq. (21.10), this can happen if the denominator in the response function χ goes to zero, which means that the Kernel K has to be sufficiently large and attractive. However, we can see from Eq. (21.14) that this can happen only if the term f_{xc} is sufficiently large that it overcomes the $1/|\mathbf{r} - \mathbf{r}'|$ Coulomb term. It is a stringent requirement for a functional of only the density to capture this behavior and, in general, it means that $f_{xc}^{\sigma\sigma'}(\mathbf{r},\mathbf{r}',\omega)$ must be a very nonlocal function. Local and semilocal functionals are not enough; they simply fail to capture the physics. There have been various suggestions to create functionals f_{xc} that

---

[6] The method is still called TDDFT even through it involves the macroscopic $\mathbf{A}$ and $\mathbf{j}$. There are changes in the periodic charge density in each cell induced by the macroscopic fields, and it is more useful to treat them in terms of potentials and densities. The induced fields are called local field corrections. They can be dealt with as described in Section E.4 and they are included in the TDDFT response function.

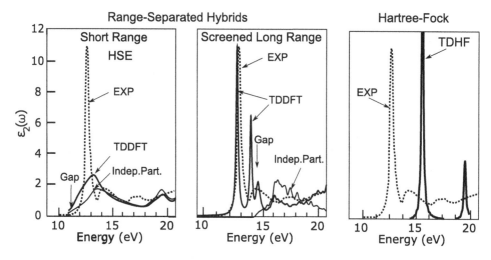

Figure 21.5. Optical spectra of LiF calculated in the independent-particle approximation and using time-dependent DFT (TDDFT) for two functionals and time-dependent Hartree–Fock. The left panel illustrates the fact that a short range functional (in this case HSE) does not produce bound states below the band gap in either the IPA or TDDFT. The right figure illustrates the fact that the gap in Hartree–Fock is much too large (off the scale and not shown) and time-dependent Hartree–Fock (TDHF) leads to bound excitons that are at energy too high and have oscillator strength too large. In the middle is the results for a range-separated hybrid functional with a long-range part (screened Hartree–Fock), which has an exciton series below the gap. The middle and right panels are calculated using real-time methods; the spectrum of the hybrid functional is the Fourier transforms of the data shown in the lower part of Fig. 21.6. Provided by C. D. Pemmaraju with the curves for HSE provided by J. Paier

have the requisite properties. The history and the physics issues are discussed, for example, in the reviews by Onida et al. [841] and Maitra [395].

Both problems are addressed by hybrid functionals. There is considerable improvement in the bandgaps for many materials, as shown in Fig. 2.24. Regarding the ability to describe excitons, there is a qualitative effect. If the exchange has a long-range form such as the Coulomb interaction in Hartree–Fock, it leads to an attractive term that can lead to a pole in the response function, i.e., an exciton that is a bound state in the gap. A range-separated hybrid functional like HSE that has only short-range exchange may improve bandgaps but only functionals with long-range Hartree–Fock-like exchange can describe excitons.

It is ironic that for more than 80 years (at least since the work of Bardeen in 1936 [254]) condensed matter theory has striven to avoid the Hartree–Fock singularity at the Fermi surface of metals discussed in Chapter 5. A great accomplishment of the RPA invented by Pines and others in the 1950s was to show that summation of selected diagrams leads to short-range exchange and correlation. (See the list of classic books at the end of this chapter and extensive discussion in [1].) Now we find that long-range Hartree–Fock-like effects must be included to solve another problem in condensed matter. A working strategy is to create functionals that have long-range exchange scaled by $1/\epsilon$ so that it vanishes in a metal.

This can be justified in the approach of range-separated functions: just as keeping only a short range was justified in part by arguments based on RPA, so also can keeping some amount of the long-range exchange be justified in insulators. Such functions are described in Section 9.3 and two examples are the "tuned" [859] and DDH functionals [244], each supported by arguments to make functionals with no parameters.

Optical Spectrum of LiF

The spectrum of LiF is a quintessential example of an exciton. As shown in Fig. 2.26, the spectrum is completely dominated by the exciton feature at ≈ 12.6 eV that looks nothing like an independent-particle spectrum. Since the gap is known experimentally to be around 14.2 eV, the peak is well below the continuum with a binding energy more than 1 eV, which indicates strongly bound localized state referred to as a Frenkel exciton.

Here we illustrate the spectra calculated in two ways. The spectrum shown in Fig. 2.26 was calculated using the response function method based on the Casida equations in Section 21.4 [253]. The functional used in the calculation is one called "optimally tuned" with screened exchange [859], a range-separated hybrid functional in which the range is proposed to be an appropriate variable that determines the character of the nonlocal exchange with the added feature of the fraction of long-range exchange scaled by $1/\epsilon$. In a finite system like a molecule, the range is determined by a condition on the asymptotic form of the wavefunctions [440]. However, this does not work in a solid and the range was adjusted to give the correct experimental gap. With that adjustment, the spectrum including the exciton feature is in excellent agreement with experiment. Any similar functional would give such exciton features, but the gap and the exciton energy might be further from the experimental values.

The results of a real-time calculation for the same problem are shown in the middle panel of Fig. 21.5. The functional was chosen to be very close to the one used in Fig. 2.26 so they could be compared, and the calculations were done using numerical local orbitals as defined in SIESTA (see Section 15.4). The spectra calculated by the two methods are very close except for two peaks above the main exciton, which are more clearly resolved in Fig. 21.5. The independent-particle approximation is also shown, which makes it clear that the sharp peaks are bound states below the continuum. The other panels of Fig. 21.5 are chosen to illustrate the limits of no long-range to full long-range exchange. At the left is the spectrum calculated with the short-range HSE06 functional from [252] that totally misses the exciton peaks. At the right is the Hartree–Fock spectrum, where the gap is so large that the continuum is off the scale, and there are exciton peaks with oscillator strength and binding energy that are much too large. The middle panel shows that TDDFT with this hybrid functional captures features of short-range exchange and correlation that go beyond Hartree–Fock, and also features that are like a screened version of Hartree–Fock.

Real-Time Calculations for Macroscopic Systems

The real-time calculations for clusters in Section 21.7 were in terms of the dipole moment induced by an electric field. Yabana and Bertsch [860] proposed an analogous real-time

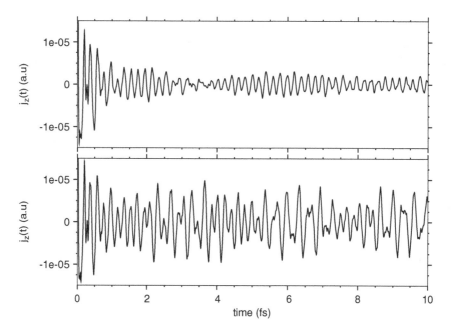

Figure 21.6. Real-time current $j(t)$ that is the response to a delta function electric field (a step in the vector potential $\mathbf{A}(t)$ at time $t = 0$). At the top is $j(t)$ that decays quickly as a function of time, which leads to a smooth spectrum like that for the short-range functional shown at the left in Fig. 21.5. At the bottom is $j(t)$ that leads to a spectrum shown in the middle of Fig. 21.5, where the sharp peaks result from oscillations in the current that last for a long time with little decay. Figure provided by C. D. Pemmaraju

approach for the dielectric function in solids in terms of the vector potential and currents that avoids the fundamental difficulties of attempting to define dipoles in macroscopic systems.[7] In practice the calculations involve a procedure analogous to the calculations for a cluster. The system is perturbed by a step in the vector potential $\mathbf{A}(t)$ at time t_0 (a delta function in the macroscopic electric field at time t_0 since $\mathbf{E}(t) = d\mathbf{A}(t)/dt$), and the calculation of the resulting macroscopic average of the current $\mathbf{j}(t - t_0)$ as a function of time. Since the effect of vector potential in the hamiltonian, $\mathbf{p} \rightarrow \mathbf{p} - (e/c)\mathbf{A}$, a step in $\mathbf{A}$ is equivalent to a shift of the momenta. In the calculation this amounts to modifying the ground-state wavefunction by replacing each Bloch function $u_{\mathbf{k}}$ by the function $u_{\mathbf{k}+\delta\mathbf{k}}$, which creates a state that is not an eigenstate. The system is propagated forward in time with the periodic potential $V(\mathbf{G})$ updated as the density $n(\mathbf{G})$ changes, and the current is calculated as a function of time. The spectrum as a function of frequency is calculated using a Fourier transform, the same as for the time-dependent moment in finite systems.

---

[7] Of course, the general ideas have been known for a long time in many contexts. A closely related method was proposed by Krieger and Iafrate [861], who considered the time evolution of electrons in a homogeneous electric field and a fixed crystal potential. The new feature in TDDFT is that the effective hamiltonian varies during the evolution.

Two examples are shown in Fig. 21.6. The bottom curve is the real-time data that leads to the TDDFT spectra in the middle panel Fig. 21.5. Sharp peaks in the spectrum are due to the oscillations in the current that last for a long time as shown in the figure. The top curve is for a case where the current decays rapidly as a function of time, which leads to a smooth spectrum with few sharp features like that shown at the left in Fig. 21.5.

21.9 Beyond the Adiabatic Approximation

A fully satisfactory time-dependent theory must go beyond the adiabatic approximation [862]. In the time-dependent Kohn–Sham approach with a functional of the density, it must be nonlocal in time, i.e., a function of time differences, in order to take into account the way that the transitions to excited states affect the functional. The conceptual issues are discussed in references such as the reviews by Maitra [395] and Burke [247] and other sources listed at the end of the end of this chapter. There is, however, an alternative approach along the lines of the generalized Kohn–Sham (GKS) theory in Section 9.2. A functional in terms of the wavefunctions automatically creates a dependence on the excitations. The importance of long-range exchange that can be included in functionals of the wavefunctions was stressed in Section 21.8, where it makes possible the description of excitons. This corresponds to a pole in the response function, which is an extreme frequency dependence. Methods like "exact exchange" (EXX) [863] and hybrid functionals that include exchange to varying extents are known to improve many aspects of description of spectra. There are connections to other problems such as current functionals [342, 343], and this is an area that deserves attention in the future.

SELECT FURTHER READING

The basic theory of dielectric functions and optical properties is in the texts listed at the end of Chapter 2.

Original paper:
Runge, E. and Gross, E. K. U., "Density-functional theory for time-dependent systems," *Phys. Rev. Lett.* 52:997–1000, 1984.

Reviews:
Maitra, N. T., "Perspective: Fundamental aspects of time-dependent density functional theory," *J. Chem. Phys.* 144:220901, 2016. Provides insights into many aspects of TDDFT along with many references.
Onida, G. Reining, L., and Rubio, A., "Electronic excitations: Density-functional versus many-body Green's-function approaches," *Rev. Mod. Phys.* 74:601, 2002. Describes relation of TDDFT and Green's function methods. See also the companion book [1].

Books that provide in-depth exposition of theory and applications:
Ferre, N., Filatov, M. and Huix-Rotllant, M., editors, *Density-Functional Methods for Excited States* (Springer International, Switzerland, 2016).
Marques, M. A. L., et al., editors, *Fundamentals of Time-Dependent Density Functional Theory*, Lecture Notes in Physics 837 (Springer, Berlin, 2012).

Marques, M. A. L., et al., editors, *Time-Dependent Density Functional Theory*, Lecture Notes in Physics 706 (Spinger, Berlin, 2016). See also earlier volumes in the series.

Ullrich, C., *Time-Dependent Density-Functional Theory: Concepts and Applications* (Oxford University Press, Oxford, 2012).

Exercises

21.1 Derive expression (21.9) for the dielectric function for noninteracting particles. Show that the first term in brackets comes from the A^2 term as stated following Eq. (21.9).

21.2 Derive the matrix equation (21.15) for the eigenvalues Ω_n of the density response from the preceding equations. Although there are many indices, this is a straightforward problem of matrix manipulation.

21.3 The general approach for exponentials of operators is described in the text preceding Eq. (21.18); it is also used in the rotation operators in Appendix N. Show that for any operator A, $\exp(A) = \sum_i |\zeta_i\rangle \exp(A_i)\langle\zeta_i|$, where A_i and ζ_i are eigenvalues and eigenvectors of A. Hint: use the power series expansion of the exponential and show the equivalence of the two sides of the equation at every order.

21.4 Show that Eq. (21.18) has error of order $\propto \delta t^2$. Would the error be of the same order if the potential part were not symmetric?

21.5 Derive the f sum rule for the strength function $S(\omega)$ in Eq. (21.23). The proof is analogous to Exercise E.2.

21.6 Describe qualitatively how the top and bottom parts of Fig. 21.6 lead to broad and sharp spectra, respectively.

22

Surfaces, Interfaces, and Lower-Dimensional Systems

The papers by Kronig and Penney, Tamm and Shockley belong to that category
of classical papers which are always quoted, but never read.

Maria Steślicka [864]

Summary

This chapter is devoted to a diverse set of problems with the common feature that
they are lower-dimensional systems. Each type of system is a large field in itself
and we can consider only very limited examples chosen to fit with other topics in this
book. Electron bands at surfaces are represented by semiconductors and the Shockley
midgap surface state on gold, which actually has topological character! Interfaces
are exemplified by semiconductor heterostructures (possibly the best understood of
all interfaces) and oxide interfaces where rapid advances have created news fields of
research. The unique properties of two-dimensional layers are illustrated by MoS_2
(and graphene, which is in other chapters), made even more interesting by the
possibilities for stacked "van der Waals" structures as illustrated in Section 2.12.
The example of one-dimensional systems is nanotubes, which are also the topic of
Section 14.8.

22.1 Overview

Surfaces and interfaces are major areas of research in materials science, physics, chemistry,
and other fields. Even small parts of these fields are subjects of entire journals and books.
It is also a large field for electronic structure theory and calculations that cannot be covered
here. Nevertheless, we can identify a set of problems that are very general and elucidate
some interesting aspects of electronic structure. The topics in this chapter are phenomena
in a diverse set of systems with the common feature that the electronic states have one- or
two-dimensional character (or zero dimensions in the case of the surface [end] of a one-
dimensional system). Some examples are materials that are in fact two-dimensional; others
are surfaces or interfaces for three-dimensional materials, where there may be electronic
states bound to the surface or interface.

Sections 22.2–22.3 set up the problems of idealized flat surfaces and interfaces, and
highlight the distinction between states that are caused by the potential at the surface or

interface and other types of states that are intrinsically connected to the nature of the bulk states. The original work identifying such a bulk-boundary relation was by Shockley in 1939, which is emphasized in this chapter because it is a valuable precedent for topological insulators in Chapters 25–28. Prototypical examples of surface states are considered in Sections 22.4 and 22.5.

Interfaces are illustrated by two types of systems. Almost certainly the most nearly ideal interfaces with the cleanest examples of electronic states are semiconductors (Section 22.6) which have been perfected over the years and are essential for present-day technology. There has been an enormous interest in oxide interfaces (Section 22.7), since the discovery in 2004 [214] of two-dimensional metallic systems at the interface between two insulating oxides. These hold great potential, but they are much less well understood than the semiconductors. The goal is to identify characteristic features and the role of electronic structure in unraveling the complexities and revealing interesting phenomena.

Two dimensions was largely a theoretical field for many years, but this has changed with the advent of graphene as a extraordinary freestanding material that can be made in the laboratory and engineered into many forms. Materials like transition metal pnictides and chalcogenides can be grown and stacked in an endless set of possible structures. An example that was instrumental is creating the interest in these materials is MoS_2, discussed in Section 22.8.

22.2 Potential at a Surface or Interface

A schematic illustration of a, the potential for a crystal surface, is shown in Fig. 22.1, which is the same as that for the bulk crystal in Fig. 4.11 but terminated by a surface with a modified potential that increases up to a constant value in the vacuum. In a realistic system with Coulomb interactions the average potential inside the crystal relative to the vacuum determines the work function ϕ, defined to be the minimum energy required to take an electron from the solid to the vacuum, which is indicated in the figure where the line is

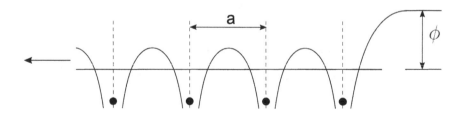

Figure 22.1. Schematic graph of a potential for a finite crystal large enough to have a bulk region like Fig. 4.11 extending to the left and a surface where then potential rises to the vacuum value. *The most fundamental point is that the absolute energy (relative to the vacuum) of the electronic states inside the crystal depends on the state of the surface and is not an intrinsic property of the crystal.* The work function ϕ is defined to be the potential and the Fermi energy in a metal indicated by the horizontal line. In a semiconductor or insulator it depends on the definition and may be controlled by extrinsic defects. The same principles apply to interfaces.

the energy of the highest occupied state. One can see immediately an important general fact: *the energy of the eigenstates for electrons in a crystal relative to vacuum is not an intrinsic property of the crystal*. The potential relative to the vacuum is determined by the details of the surface and is shifted by an average effect due to the surface dipole layer as shown in Section F.5. This is in contrast to a neutral atom or small molecule where the potential is defined to be zero in the vacuum. In a solid, however, there is a bulk region and a surface region as indicated in the figure.[1] In the bulk region the potential is an intrinsic quantity except that it can be shifted by adding a constant with no effect on the electronic states in the bulk except to shift the energies rigidly. Thus the band structure calculations discussed in the rest of this book are defined relative to some reference. In a metal the convenient reference is the Fermi energy, and the work function ϕ defines the relation to the vacuum. In an insulator there are additional effects. The position of the Fermi energy may be determined by extrinsic effects such as doping, which also can cause band bending, which is essential to take into account in semiconductor devices. Nevertheless, all the issues can be taken into account as long as we specify the problem carefully.

The same considerations apply for interfaces and the relative energies of states in the two materials must be defined. In this case also there are simple conclusions for metals. In equilibrium the Fermi energy must be the same everywhere and charge flows from one region to the other to shift the potentials so that the Fermi energies of the bulk of the two materials are the same. In insulators and semiconductors the problem must be specified, for example, in the band offsets considered in Sections 22.6 and 22.7.

In model problems the issues are swept up into the specification of the energies. For example, in tight-binding models one must specify the relative energies of the states at each site in the bulk crystals, which may be different at the surface or interface. Work functions are taken into account by the values of the energies relative to vacuum, and band offsets, by the relative value for the two materials.

22.3 Surface States: Tamm and Shockley

Tamm States

States bound to surfaces of crystals are often called Tamm states[2] after the work of Igor Tamm [865] in 1932. This was a year after Kronig and Penny proposed their famous model of delta functions [866], which are used in textbooks as a solvable model for extended Bloch states in a crystal. However, Kronig and Penny also considered a finite crystal with a step function potential up to the vacuum level. They calculated amplitudes for reflection and

---

[1] As a reminder that the energies inside the crystal relative to vacuum are not an intrinsic property of the bulk crystal, recall that an important practical way to lower the work function of electron emitters is by coating the surface with a material that tends to donate electrons and have net positive charge. See Exercise 22.1 for an estimate of the size of the effect.

[2] A special issue of *Progress in Surface* listed in "Select Further Reading" at the end if this chapter includes papers that describe Tamm's work and discuss the similarities and distinctions between Tamm and Shockley states.

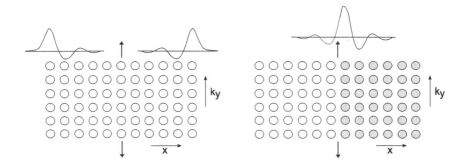

Figure 22.2. Left: illustration of a crystal with a surface perpendicular to x, considered as finite size in the x direction but periodic parallel to the surface in the y direction. A surface band can be labeled by momentum k_y and it has finite extent into the crystal and the vacuum, as illustrated by the curves for the left and right surfaces. In general the form and extent of the wavefunction in the x depends on the momentum k_y. Right: the same considerations apply at an interface where there can be a state that decays into the materials on either side of the interface. A striking example is the interface between two insulators that has a metallic interface if the states are partially filled, which happens in the oxides that are described in Section 22.7.

transmission of electrons from a surface and showed there is total reflection for energies in the bandgaps and partial transmission otherwise.[3] Tamm recognized that for the finite crystal there also may be states in the gaps between the bands that decay into the vacuum and into the crystal as shown schematically in Fig. 22.2.

One way to understand the formation of surface or interface states is by the fact that the potential near the surface is different from the bulk, in general expected to be higher than in the bulk as indicated in Fig. 22.1. The nature of the surface states can be illustrated by a one-dimensional crystal with a single s band where the density of states is given in Fig. 14.4. Because of the logarithmic divergences at the band edges the change in the potential at the boundary always creates a bound state at the surface, either above the band edge for a repulsive potential or below the band edge for an attractive potential (see Exercise 22.2). Since there is one state per cell of the crystal, the formation of the bound state means there is one less state in the continuum.

There is a long history of controversy over whether or not there is a distinction between Tamm and Shockley states. The states derived by Tamm were in gaps between bands like those found by Shockley. However, there is an important distinction. What Shockley discovered was a transition in the bulk band structure that is a precursor of modern-day topological insulators. The surface state is a consequence but not the essential point. For that reason, Shockley's work is considered in some detail and it is the basis for many of the concepts and models in Chapters 25–28.

---

[3] Electron scattering from a crystal had been observed by Davisson and Germer published in 1927 [867] and this provided a simplified description although they did not address the Bragg scattering from a three-dimensional crystal.

Shockley Transition in the Bulk Band Structure and Surface States

In a landmark paper in 1939 [42], Shockley[4] considered the lowest two bands of a one-dimensional crystal and derived two important results: a transition in the character of the bulk bands and the conditions for which the electronic structure of the bulk crystal leads to a surface state in the gap. He considered a finite crystal having a potential that is periodic inside the bulk and at the surface approaches the value in the vacuum monotonically like that shown Fig. 22.1; the bulk potential can have any form except that it was assumed to have inversion symmetry. He used the method of matching the eigenfunctions at the cell boundaries, which leads to transcendental equations for the band edges of the lowest two bands, which are Bloch states formed from linear combinations of even and odd in one unit cell. The transition at which the nature of the electronic states change occurs when the bulk bands exchange character as a function of the crystal momentum k, which may be called an "inverted band structure."

The problem can be illustrated by a two-band s–p model in one dimension, which is treated in Sections 26.2 and 26.3. (The model and discussion are postponed to Chapter 26 because it is intertwined with models for topological insulators.) As can be seen in Fig. 26.3, the s–p model captures the essence of a variation from atomic-like (s at lower energy than p) to covalent (bonding and antibonding). Figure 26.4 shows two examples of bands for where there is a center of inversion, in which case the bands can be classified by their parity at the two high-symmetry points in the Brillouin zone, Γ and X. At all other points in the Brillouin zone the states are mixed and they repel to form a gap. Thus one can always follow each band from $k = 0$ to $k = \pi/a$ and determine whether each band has the same parity or they have exchanged parity. The transition between these two possibilities can occur only if the gap goes to zero. The figure also show the states of a finite crystal with two ends which illustrates Shockley's proof that a midgap surface state occurs in the regime where the bands exchange parity.

The conclusions can be illustrated in the way done by Shockley and shown in Fig. 1.1. For a potential like that illustrated in Fig. 22.1 if the cells are far apart the states are atomic-like with an even s state lowest and an odd p state next. As atoms are brought together to form a crystal, the overlap increases and the bands broaden into bands and at some point bonds are formed between the atoms. This is a transition from one insulating state to another, atomic-like to covalent bonding. This is the essence of the problem and transition always occurs at a point where the gap vanishes, so long as there is inversion symmetry.

Shockley surface states are affected by surface conditions, as pointed out in the original paper. They actually correspond to dangling bonds (see Section 22.5) and in a semiconductor like Si or Ge they are expected to be eliminated by passification by bonding with adatoms, reconstruction of the surface, etc. But Shockley argued they would survive in metals, which has been borne out by experiments on many metals such as gold that is the example in Section 22.4.

---

[4] As noted on page 5, this is the same Shockley who shared the Nobel prize, along with Bardeen and Brattain, for development of the transistor. The first transistors worked by controlling the conductivity in the surface region and the properties of surface states was crucial.

Surface Bands in Two and Three Dimensions

The one-dimensional analysis also is relevant for higher dimensions. As illustrated in Fig. 22.2 the momentum parallel to the surface k_y is conserved. For each k_y the states of the finite crystal can be considered as a one-dimensional problem with a hamiltonian that depends on k_y. Various examples in two dimensions can be found in Chapter 27. The example of a graphene ribbon is discussed in some detail in Section 14.7 where it is shown that for a zig-zag edge, there is a Shockley transition as a function of momentum of the edge state, going from a regime with normal bands and no states in the gap to an inverted band structure an edge state, and there is a zero gap at the transition point. Even though that chapter is focused on effects of spin–orbit interaction, there are cases where the spin–orbit interaction is set to zero. For example, a Shockley surface state in a metal is illustrated by the model with an s and two p states on a square lattice. The leftmost panel in Fig. 27.6 is the bulk band structure for parameters chosen to have overlap of the s and p and it is metallic. The upper left panel in Fig. 27.6 shows the states for a strip of finite width. In parts of the surface (edge) Brillouin zone there are gaps in the projected bulk states and there are states in the gap that persists in the entire region where there is a gap. If there is spin–orbit interaction, gaps open all across the Brillouin zone; it is not an insulator because the bands overlap, but nevertheless the bands are split apart. This also happens in real materials such as gold, as described in the following section.

22.4 Shockley States on Metals: Gold (111) Surface

The current usage of the term "Shockley state" often means a surface state on a metal, which can be illustrated by the (111) surface of gold that has been studied for decades. As illustrated in Fig. 22.3 there is a region of the surface Brillouin zone where there is a gap and a surface state in the gap. The state has been observed in photoemission and the left side of the figure from [868] shows bands dispersing up to the Fermi energy. The reason there are two bands is a relativistic effect that couples spin and momentum. As explained in Appendix O and shown in the top part of Fig. O.1 there is a linear dispersion around $\mathbf{k} = 0$ opposite for the two spin states, which happens on any surface due to the lack of inversion symmetry at the surface (it also must happen for interface states) called the Rashba effect [870, 871]. In a heavy element like Au with a large spin–orbit interaction it is greatly amplified by the nature of the states. It may seem surprising that there is an effect in this case since the conduction band of gold is made of 6s atomic-like states; however, there is a simple explanation due to the mixing of p and higher angular momentum states due to the asymmetry of a surface state, as shown in Fig. O.2.

More recent photoemission experiments have observed the surface state below the Fermi energy and also used a two-photon pump–probe method that makes it possible to observe the states above the Fermi energy [869]. The results are shown in the middle of Fig. 22.3 reveal that the spin-split states continue to disperse upward until they merge into the bulk band continuum, with one state merging with the lower band below the gap and the other with the band above the gap. The figure shows the data superposed on theoretical results

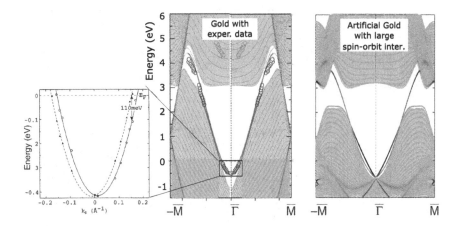

Figure 22.3. Shockley surface states of gold. Left: the dispersion of states below the Fermi energy from [868], which shows the effect of spin–orbit interaction. Middle: data (circles) for ARPES and two-photon ARPES [869] with DFT calculations (see text). Right: bulk and surface bands found theoretically if the spin–orbit interaction increased by 350%. From [869].

of a DFT calculation using the VASP code for a 35-atom-thick slab that is large enough to identify bulk and surface bands. At the right is shown theoretical results found in [869] if the spin–orbit interaction is artificially increased by 350%; this makes the dispersion of the surface bands more visible. If it had one more electron, gold would be a topological insulator!

The results for the bands of gold can be understood by considering a simple model for s and p bands on a square lattice worked out in Section 27.5. If there is no spin–orbit interaction the bands shown at the upper left in Fig. 27.6 show the classic behavior of a metal with a Shockley state in the gap. As the spin–orbit interaction is increased the other panels in the figure show the bands that are split and disperse until they merge in the continuum with one joining the lower and the other joining the upper band. Finally the lower right shows the insulator for large spin–orbit interaction and the discussion in the chapter gives the reasoning that it is a topological insulator with surface bands that must cross the gap from valence to conduction bands.

22.5 Surface States on Semiconductors

Surfaces of covalently bonded materials can exhibit an array of surface reconstructions. For C, Si, and Ge calculated absolute surface energies for various surfaces of C, Si, and Ge are reported in [872], which gives many earlier references. A well-known example of surface reconstruction is the (1 0 0) surface of Si and Ge. As indicated in Fig. 2.17, each surface atom has broken bonds and the surface reconstructs with atoms dimerizing to form a new bond so that each surface atom has three bonds. This leaves one electron per surface atom: for a symmetric unbuckled dimer, the two extra electrons can make a π bond, whereas if the dimer buckles so that the two atoms have inequivalent positions, the two electrons can form a "lone pair" in the lower-energy state. The energy differences are small and

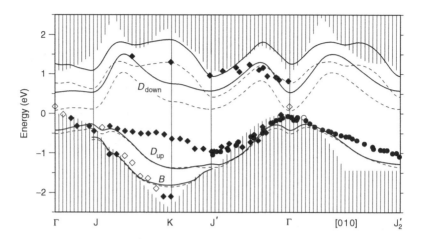

Figure 22.4. Bands for the (1 0 0) surface of Ge [651], which has the buckled dimer structure shown in Fig. 2.17. The bulk states are denoted by shaded areas, and surface states in the gap by lines. Bands denoted D and B are surface states associated with the surface "dangling" bonds and with the "back" bond between the first and second layers. The dashed lines are for the LDA (showing the zero-gap problem) and the solid lines are calculated in the GW approximation. Dots indicate experimental results referenced in [651]. From [651].

calculations using DFT, e.g., [203], and quantum Monte Carlo calculations [873] find the lowest-energy state to be the buckled dimer.

Such calculations also predict the electronic structure that can be measured experimentally. For example, surface bands for Ge [651] are shown in Fig. 22.4, comparing GW quasiparticle and LDA bands with experiment. That work uses gaussians, but essentially the same results are found with plane waves. The results illustrate the famous problem of a zero bandgap for Ge for the Kohn–Sham bands in the LDA shown in Fig. 2.23 and the consequent errors in the surface bands. For the bulk crystal the GW quasiparticle bands are close to experiment, and Fig. 22.4 shows that the results for the surface bands are also generally in agreement with experiment.

This example reveals a crucial aspect of Shockley surface states. The dangling bond for the unreconstructed surface is an example of surface states that appear in the covalent regime as found in the original Shockley work. However, dangling bonds can be removed by passivation, i.e., adatoms such at hydrogen or oxygen, reconstruction, etc., which shifts the energies. The two surface bands for Ge in Fig. 22.4 illustrate this and with large enough effects the surface band may be shifted out of the gap. A qualitative difference in the topological insulators in Chapters 27 and 28 is that there are always bands that cross the gap so long as there is time reversal symmetry.

22.6 Interfaces: Semiconductors

The interfaces that join bulk materials are among the major problems of materials science. This section addresses a small part of the field with examples that illustrate basic

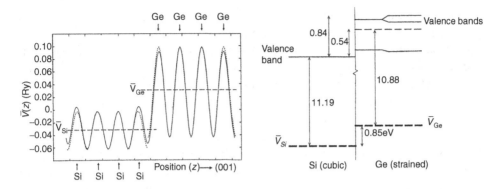

Figure 22.5. Average potential and band alignment at a strained layer interface Ge grown on a Si substrate [874]. The left panel shows the average potential offset determined by the interface calculation. The right panel shows the alignment of the valence bands relative to the average potential, which is determined by a bulk calculation for each material, cubic Si and strained Ge. The top of the valence band (omitting spin–orbit splitting) is threefold degenerate in cubic Si but split in the strained Ge. The energies of the conduction bands were fixed by adding the experimental bandgaps.

principles and interesting electronic phenomena that can be studied by experiment and theoretical calculations.

Perhaps the most perfect interfaces are those grown with atomic precision in semiconductors. Fourfold coordination in the diamond and zinc blende structures can be maintained across the interface with essentially no defects. The change is just that due to chemical identity of the atoms and relaxations of the structure. Therefore this is a case where detailed comparison can be made between theory and experiment. Here we consider simplest cases, which are called nonpolar, where there is no issue regarding buildup of charge at the interface. The interface is nonpolar if the two materials have the same valence, namely two group IV elements like Si/Ge, III–V compounds like GaAs/AlAs, or II–VI compounds like HgTe/CdTe. For a (1 1 0) interface there is no net polarity and interfaces are nonpolar even for cases like Ge/GaAs or GaAs/ZnSe. These are the ones most commonly grown experimentally because of the instabilities in nonpolar interfaces discussed below.

Extensive calculations have been done for many combinations of semiconductors. See, for example, the review [211]. A representative example [874] for Si/Ge is shown in Fig. 22.5. In this case, there is a uniaxial strain induced by a lattice-matching condition in addition to the change in chemical identity. The parallel lattice constant is fixed by the substrate, creating a uniaxial strain in the thin layer. The figure illustrates the results for strained Ge grown on a Si substrate. The left panel shows the planar averaged potential,[5] which quickly reaches the periodic bulk potential, with an offset determined by the interface calculation. The relative alignment of the Si and Ge bands (the band "offset") is fixed by

---

[5] The potential shown in Fig. 22.5 is the local pseudopotential; if done correctly, the final results are invariant to choice of pseudopotential or all-electron potential since the alignment is due to an intrinsic part plus terms that depend only on the interface dipole (Appendix F).

referring the bands in the bulk of each material to the average potentials indicated. The strain in the Ge layers can be divided into a dilation, which shifts the energies of the bands, and a uniaxial strain, which splits the top of the valence bands, as shown on the right-hand side of the figure. Similarly, growth on a Ge substrate leads to strained Si. Intermediate cases and alloys can be treated by interpolation.

In more recent work, band offsets have been calculated with various functions and many-body GW calculations. An example is [875] where there are comparisons of a representative GGA (PBE), hybrid functional (PBE0) compared to GW calculations and experiment. The general conclusion is that the interface dipole is well accounted for with semilocal density functionals. This greatly simplifies the problem because the interface calculation can be done with the GGA including relation of the atomic positions at the interface. With the assumption that the dipole is the same for each functional, all other effects are intrinsic effects in the bulk materials, i.e., the positions of the bands relative to the average electrostatic potential just as was done in the Si/Ge example in Fig. 22.5. For a given functional, this can be determined by one calculation for each material and the results apply to any interface. The hybrid functional and GW methods lead to consistent valence-band offsets in close agreement with experiment, slightly improving upon semilocal functionals. However, the offset in the conduction bands is greatly affected by the choice of functionals, which is primarily due to change in the bandgap.

What about polar interfaces? An example might be Ge/GaAs in the (100) direction where there is abrupt change from Ge to Ga (or As) at one plane with no defects. For an ideal interface, simple counting rules [205, 876] lead to the conclusion that there would be 1/2 electron per cell missing from the valence band for a Ge–Ga interface (or 1/2 electron per cell added to the conduction band for a Ge–As interface). These are the same issues as confronted in the oxide interfaces considered next.

22.7 Interfaces: Oxides

An entire field of research on oxide interfaces was stimulated by the discovery of a two-dimensional electron gas (2DEG) at the $SrTiO_3$/$LaAlO_3$ interface (STO/LAO) by Ohtomo and Hwang [214]. The carrier densities are very high and the systems exhibit a range of phenomena including superconductivity and magnetism [877, 878]. This is made possible by fabrication of heterostructures of oxides using methods that allow unprecedented control over layer thickness, such as pulsed layer deposition and molecular beam epitaxy.

It is remarkable that it is possible to create a high-density metal at an interface between two insulators, and there is debate over the origin of the carriers. Are they the predicted carriers that must exist at an ideal polar interface with no defects? Or are they due to defects? Is there a combination of both effects or some other mechanism? A summary of the mechanisms can be found in [879] and aspects are discussed briefly in this section because they raise basic issues.

Figure 22.6 illustrates the prototypical example of LAO/STO. At the left is a schematic illustration of an ideal interface between a TiO_2 layer and a LaO layer. (See the structure of a ABO_3 provskite crystal in Fig. 4.8, which can be viewed as layers of AO and BO_2

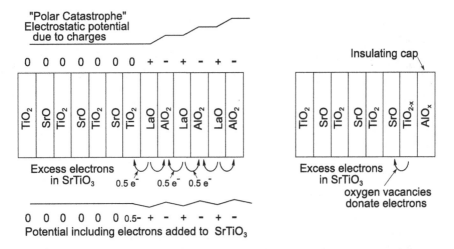

Figure 22.6. Left: an ideal interface of $SrTiO_3/LaAlO_3$ (STO/LAO). The planes of SrO and TiO_2 are neutral, and the formal charges of LaO and AlO_2 are positive and negative as shown. The "polarization catastrophe" denotes the fact that the potential for an electron would increase indefinitely as shown at the top if these are the only charges. (See Exercise 22.4.) Such a system can be stable for a thin layer of a few cells of $LaAlO_3$, but a system with many layers can be stable only if there are compensating electrons or negatively charged defects in the interface region. At the bottom are two ways to envision the effect. By adding 1/2 negative change per cell it is straightforward to show (Exercise 22.4) that the catastrophe is avoided. A graphic way to picture the origin of the 1/2 charge is to create the LaO^+ and AlO_2^- layers by transfer of 1/2 electron to the left and the right, which leaves 1/2 extra electron per cell at the interface. Other possibilities are charged defects as discussed in the text. Right: $SrTiO_3$ capped by an AlO_x layer that creates a partially filled surface band. A common feature of the interface and the capped system is that the interface bands are primarily in Ti d states that form bands as shown in Fig. 22.7.

stacked along the (0 0 1) direction perpendicular to the layers.) Consideration of interfaces like this is what led to the notion of a "polar catastrophe" [876], and various ways in which it is avoided are closely related to the origin of the carriers. The catastrophe is the potential shown at the top, which grows indefinitely. This is the potential for a negative charge and can be understood as the field in series of capacitors all with electric field in the same direction. To avoid such effects there must be compensating charges near the interface. (See Exercise 22.4.) As shown at the bottom of Fig. 22.6, this is one of the ways to show that for the ideal interface there must be 1/2 an electron per cell in the region near the interface. The excess electrons are mainly in the $SrTiO_3$ since the band alignment causes the conduction band of STO to be lower that in LAO [880]. This is a definitive result for an ideal interface between two thick pieces of STO and LAO, and it is one of the proposals for the origin of the 2DEG.

There are, however, other possibilities. The compensating negative changes could simply be replacement of 1/2 the La atoms in the first layer by Ti atoms, and many other possible rearrangements of the charges, and/or charged defects. These are possible ways the electron density in the band can be reduced to values that are observed. In addition, in the real

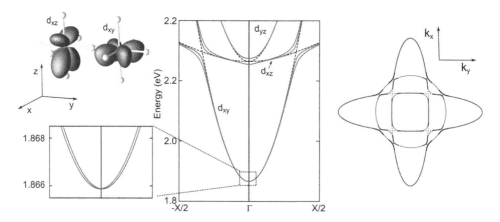

Figure 22.7. Bands for interfaces or surfaces of $SrTiO_3$ (STO). At the upper left are illustrations of the Ti d states that comprise the conduction bands: d_{xy}, d_{xz} and d_{yz} (the same as d_{xz} rotated by $\pi/2$ and not shown). The orbitals are distorted with unequal upper and lower parts, which happens at an interface or surface. As discussed in the text, these form three bands that are mixed strongly only at the avoided crossing points indicated by the dashed lines. The inset at the left shown the tiny Rashba effect, much smaller than in gold in Fig. 22.3, and at the right is a schematic Fermi surface showing the three bands and small regions of avoided crossing.

systems, there are only a few layers of LAO so that the surface may be a source or sink of electrons. Much work has been done to understand the effect of the surface and certainly it can donate or accept carriers [881].

A simpler system has been made that captures the essence of electrons added to STO. At the right in Fig. 22.6 is shown a surface of STO with added aluminum that forms a copping layer, The result of adding Al is formation of an insulating Al oxide layer by extracting oxygen atoms from the STO. Each oxygen vacancy frees two electrons that can contribute to a surface band. By this technique a 2DEG can be realized at the surface where it has been probed by photoemission [882], which finds bands like those shown in Fig. 22.7.

Even though the mechanisms cannot be sorted out here, there are common features that can be brought out. The main examples known so far involve added electrons in STO, where the nature of the conduction bands has a simple interpretation. The conduction band of bulk STO is formed from Ti d states as has been understood since the work of Mattheiss in the 1970s [883]. The 5d states are split into e_g and t_{1g} with the triplet t_{1g} lowest. The states can be labeled d_{xy}, d_{xz}, d_{yz} as illustrated in the upper left of Fig. 22.7. (The d_{yz} orbital is not shown but is the same as d_{xz} rotated $\pi/2$ around the z-axis.) The resulting band structure has been determined by many people, and the figure show representative results taken from [884]. It is useful to first consider three separate bands and afterward put in the coupling between the bands and effects of spin–orbit interaction. A d_{xy} state has large overlap with d_{xy} on neighboring atoms in the x and y directions in the plane of the interface and hence large dispersion and a wide band in both x and y directions as indicated in the figure. However, the bands formed from d_{xz} and d_{yz} have anisotropic dispersion. The d_{xz} forms a wide band in the x direction since there is large overlap with neighbors in that direction, but the band width is much smaller in the y direction since the extent of the d_{xz} state is smaller in the y direction. The bands in the y direction are not shown, but the effect can be seen in the figure for the d_{yz} band, which is very narrow in the x direction. For a band localized to an interface or surface perpendicular to the (0 0 1) direction the d_{xz} and d_{yz} are degenerate at the Γ point, but the d_{xy} band is split off as indicated. The Fermi energy is determined by the density of carriers and an example is shown at the right in Fig. 22.7 where there are three sheets of the Fermi surface, one isotropic and two anisotropic.

Finally, coupling between the bands, spin–orbit interaction and the Rashba effect (see Section O.2 and the discussion for gold bands in Section 22.4) can be taken into account which leads to the avoided crossings indicated. The usual Rashba effect is very small, barely visible on the scale of a millivolt as shown in the inset. (Compare with the similar effect at a gold surface in Fig. 22.3, which is much larger.) However, the Rashba effect and the consequences of spin–orbit interaction are greatly enhanced in the small region with avoided crossing [884].

22.8 Layer Materials

There are a fabulous variety of layer materials that exhibit a host of fascinating phenomena. An almost unlimited set of materials, called van der Waals heterostructures [219, 220], can be made by stacking layers such as graphene and a variety of transition metal pnictides and chalcogenides. An example is the electron microscope image in Fig. 2.19 of stacked single layers of graphene, MoS_2 and WSe_2, which illustrates the high quality with which such materials can be grown. Certainly graphene is the most studied material, experimentally and theoretically; and it figures prominently in many places in this book, including in Section 27.8 where it is an example for Shockley edge states and is one of the prototypical models for topological insulators.

As representative of other layered materials, we consider MoS_2, which has been of great interest since the experimental realization [217, 885] of the large increase in optical absorption and photoluminescence in a single layer compared to the bulk (see [886] for a review). This is a case that brings out the predictive power of theoretical calculations. Years before the experiments, it was predicted that the bandgap would change from indirect in the bulk to direct in a single layer [216]. The calculated bands are shown in Fig. 22.8 for the crystal and the monolayer, which shows that there is a very large shift in the states near the Γ point, which are the states responsible for the indirect gap in the crystal. The calculations show that these states have a large contribution of p_z sulfur character whose overlap between layers changes as a function of distance, whereas the zone boundary points are mainly Mo d states, which are tightly bound in a layer and changed little in going from the monolayer to the crystal. The results in Fig. 22.8 were found using a GGA functional where the gaps are too small; but the conclusions about the changes in the gaps from bulk to monolayer have been confirmed by later work using the HSE functional [218] and the GW method [887], which also yield band gaps closer to experiment.

A more recent work [888] has calculated the band structure for 51 related sulfides, selenides, and tellurides, comparing LDA and GW results for the bandgaps. In addition, this

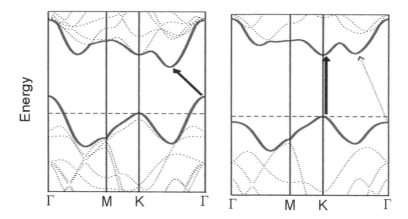

Figure 22.8. Calculated bands for bulk and single-layer MoS_2 showing the transition from the indirect gap in the bulk crystal (left) to the direct gap in the single layer (right). Taken from a figure [885], essentially the same as the bands in [216].

work reported absolute positions of the bands relative to vacuum. Unlike three-dimensional solids, this is a well-defined property just as for a molecule. One can argue that in the weakly bonded layers systems the bands should line up accordingly. The placement of the absolute energies of the highest occupied energy (top of the valence band) is about 1 eV lower in the LDA than in the GW calculation, except for a few cases including MoS_2, $MoSe_2$, CrS_2, and $CrSe_2$ in the 2H structure, where the LDA and GW are very close. Thus the lineups of bands for pairs of layers is similar except for cases involving those four compounds paired with one of the others.

22.9 One-Dimensional Systems

As an example of plane wave calculations for a one-dimensional system, Fig. 22.9 from [620] shows the bands for a small-diameter carbon nanotube. Nanotubes are described in more detail in Section 14.8 where they are considered as an excellent example where tight-binding is the natural description. Yet methods that use general basis functions like plane waves may provide new results and insights. The example in Fig. 22.9 is for a tube that would be an insulator if it were simply "rolled graphite" and, indeed, that is the result of the simplest tight-binding models. However, for small tubes there can be large changes. The band labeled (a) was discovered in plane wave calculations [620] to be pushed down to cross the Fermi energy and create a metallic state. This is due to curvature and resulting mixing of states, which is shown on the right-hand side of the figure by the charge density for this band. The fact that the density is higher on the outside than the inside shows that the wavefunction is not simply derived from the π state of graphite, which would yield equal density inside and out. Although the same effect can be captured using other methods [621], a great virtue of plane waves is finding states that were not expected based on usual bonding pictures.

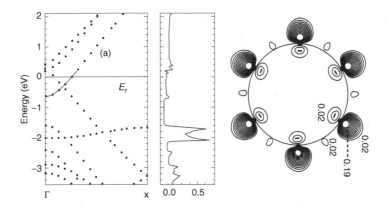

Figure 22.9. Electronic states of small nanotubes are significantly modified from simple "rolled graphite." Left: bands for a (6, 0) "zigzag" tube (see Section 14.8) calculated with plane waves [620]. The fact that graphene-like bands are strongly mixed causes one band to be lowered below the Fermi level to make the tube metallic. Right: density of the lowered band, which is primarily on the outside, not symmetric as would be the case if it were radial graphene-like p states.

SELECT FURTHER READING

Book and collection with historical overview and more recent work on surfaces:

A special issue of *Progress in Surface Science* (vol. 42, no. 1, 1993) is a tribute to Igor Tamm, with many papers devoted to the history of surface states, the similarities and distinctions between Tamm and Shockley states, and more recent developments.

Davison, S. G. and Steslicka, M., *Basic Theory of Surface States* (Clarendon Press, Oxford), 1992. Provides information on surfaces and theories in acessible form, with historical references along with many examples.

Reviews on oxide interfaces:

Bjaalie, L., Himmetoglu, B., Weston, L., Janotti, A., and Van de Walle, C. G., "Oxide interfaces for novel electronic applications," *New J. Phys.* 16:025005, 2014.

Hwang, H. Y., et al., "Emergent phenomena at oxide interfaces," *Nat. Mater.* 111:103, 2012.

Zubko, P., et al., "Interface physics in complex oxide heterostructures," *Annu. Rev. Condens. Matter Phys.* 2:141–165, 2011.

Review on Van der Waals heterostructures:

Geim, A. K. and Grigorieva, I. V., "Perspective: Van der Waals heterostructures," *Nature,* 499:419, 2013.

Exercises

22.1 The work function ϕ in Section 22.2 involves intrinsic properties of the material plus surface effects that are equivalent to the change in potential of a capacitor as described in Section F.5. An important effect is the reduction of the work function of a metal by adding a layer of atoms that donate electrons (become positively charged). The magnitude can be estimated by simple models. Consider a metal with a layer of adatoms in a square array separated by 4 a_0 positive charge $|e|$ at a height of 4 a_0 above the surface layer of atoms (the compensating negative charge is assumed to be on the surface layer of the metal). Find the change in the work function in eV.

22.2 Show that there is always a surface (end) state due to the form of the density of states in a one-dimensional system shown in Fig. 14.4. This is an example of a state caused by a potential as described by Tamm. (Hint: use a Green's function formulation. Since the density of states diverges at the edge of the band any localized effect, such as the potential at the surface, always leads to a bound state either below the band for attractive potentials or above the band for repulsive potentials.)

22.3 See Exercise 26.7 for an analytic solution for a Shockley midgap surface state wavefunction as it decays into the crystal for a two-site model.

22.4 Show that the potential has the "polar catastrophe" form shown at the top of Fig. 22.6 for the sequences of + and − charged planes. Show that if there is added a layer with 1/2 negative change per cell, the resulting potential does not increase indefinitely.

22.5 Describe qualitatively the bands shown in Fig. 22.7 in terms of the d states shown in the figure. Explain why the three bands would be independent for a layer of Ti atoms in vacuum, but they are mixed and repel one another because it is a surface layer. Explain also that spin–orbit interaction splits the states at Γ and mixes the states at other $\mathbf{k}$ points.

23

Wannier Functions

Summary

Wannier functions are enjoying a renaissance as practical tools with deep conceptual significance in electronic structure theory. There is a long history building upon the description of properties of crystals in terms of localized Wannier functions, even though their properties were not well understood and they are inherently nonunique. The focus of this chapter is the construction of functions that provide a natural description of electronic properties and are very useful, for example, in the intuitive understanding of bonds in molecules and solids, reduction of large problems to a system of only a few relevant bands with no loss of information, and in "order-N" methods (Chapter 18) that require localized functions. Maximally localized Wannier functions defined in Section 23.3 are especially useful and they provide a bridge to the role of Berry phases and such important phenomena as electric polarization and topological insulators, the topics of Chapters 24–28. In this chapter it is assumed that Wannier functions exist, and it is comforting that topological arguments provide the long-sought proof for when it is possible to construct localized Wannier functions and when it is not.

23.1 Definition and Properties

Wannier functions are orthonormal localized functions that span the same space as the eigenstates of a band or a group of bands. They were first derived by Wannier in 1937 [889] and there are extensive reviews by Wannier [348], Blount [890], Nenciu [349], and Marzari et al. [891]. In this chapter we focus on the role of Wannier functions in practical calculations and in understanding the electronic properties of materials. However, it is very useful to point out that Wannier functions also play a central role in the theoretical advances that are providing new paradigms for condensed matter physics in terms of Berry phases and topology, which are the topics of Chapters 24–28.[1]

---

[1] As far as the present chapter is concerned, the essential assumption is that it is possible to find localized Wannier functions. The proof in one dimension was given years ago by Kohn [892], but the conditions for existence of

The eigenstates of electrons in a crystal are extended throughout the crystal with each state having the same magnitude in each unit cell. This has been shown in the independent-particle approximation (Section 4.3) using the fact that the hamiltonian $\hat{H}$ in Eq. (4.22) commutes with the translations operations $\hat{T}_{\mathbf{n}}$ in Eq. (4.23). Thus eigenstates of hamiltonian $\hat{H}$ are also eigenstates of $\hat{T}_{\mathbf{n}}$, leading to the Bloch theorem, Eqs. (4.33) or (12.11),

$$\psi_{i\mathbf{k}}(\mathbf{r}) = e^{i\mathbf{k}\cdot\mathbf{r}}u_{i\mathbf{k}}(\mathbf{r}), \tag{23.1}$$

which here is taken to be normalized in one cell. However, Eq. (23.1) does not uniquely specify the functions since overall phase of each eigenstate is arbitrary. If we are given a set of functions $\tilde{\psi}_{i\mathbf{k}}(\mathbf{r})$, e.g., the output of a diagonalization routine, we can choose functions $\psi_{i\mathbf{k}}(\mathbf{r})$ related by a "gauge transformation"

$$\psi_{i\mathbf{k}}(\mathbf{r}) = e^{i\alpha_i(\mathbf{k})}\tilde{\psi}_{i\mathbf{k}}(\mathbf{r}) \tag{23.2}$$

The transformation is a $\mathbf{k}$-dependent change of the phase of each wavefunction, which leaves all physically meaningful quantities unchanged, but it can be chosen so that the functions $\psi_{i\mathbf{k}}(\mathbf{r})$ obey desired conditions.[2]

Wannier functions are the Fourier transforms of the $\mathbf{k}$-dependence of the Bloch eigenstates. In this chapter we assume that there exists some set of functions denoted $\psi_{i\mathbf{k}}(\mathbf{r})$ that are smooth functions of $\mathbf{k}$.[3] It is also convenient to require that the Bloch functions $\psi_{i\mathbf{k}}(\mathbf{r})$ are periodic in reciprocal space, i.e., a "periodic gauge" (See also Eq. (25.2) where this is also used in the Berry phase formulation.) where they satisfy the condition

$$\psi_{i\mathbf{k}}(\mathbf{r}) = \psi_{i\mathbf{k}+\mathbf{G}}(\mathbf{r}) \tag{23.3}$$

for all reciprocal lattice vectors $\mathbf{G}$. For a single band i separated from other bands, the Wannier function associated with the cell $\mathbf{n} = (n_1, n_2, \ldots)$ at position $\mathbf{T}_{\mathbf{n}}$ is defined by

$$w_{i\mathbf{n}}(\mathbf{r}) = \frac{\Omega_{\text{cell}}}{(2\pi)^3} \int_{\text{BZ}} d\mathbf{k} e^{-i\mathbf{k}\cdot\mathbf{T}_{\mathbf{n}}} \psi_{i\mathbf{k}}(\mathbf{r}). \tag{23.4}$$

The notation $w_{i\mathbf{n}}$ for the function associated with cell n is analogous to the Bloch function $\psi_{i\mathbf{k}}$ for momentum $\mathbf{k}$. The functional form can be defined as $w_i(\mathbf{r})$, and for other cells it is the same function translated by $\mathbf{T}_m$ $w_{i\mathbf{n}}(\mathbf{r}) = w_i(\mathbf{r} - \mathbf{T}_m)$, as shown schematically in Fig. 23.1. Conversely, Eq. (23.4) leads to

$$\psi_{i\mathbf{k}}(\mathbf{r}) = \sum_m e^{-i\mathbf{k}\cdot\mathbf{T}_m} w_i(\mathbf{r} - \mathbf{T}_m). \tag{23.5}$$

---

localized functions in higher dimensions have been derived only recently [893]. The proof uses the topological properties summarized in Section 25.8 to show that it is possible to construct exponentially localized Wannier functions for crystals in which the electronic structure has "trivial topology." This justifies the assumption of localized Wannier functions for large classes of crystals, like those considered in this chapter. But one must be aware that there are other cases with nontrivial topology where it is not possible to construct localized functions.

[2] The properties of Wannier functions are closely related to Berry phases where it is also useful to choose a smooth gauge. The analysis in Eqs. (P.7)–(P.9) is the proof that Berry phases are gauge invariant.

[3] It is natural to expect that this is required for a Fourier transformation to lead to localized Wannier functions. In fact, the derivation of the conditions under which such a set of smooth functions can be found is a topological property as shown in Section 25.8.

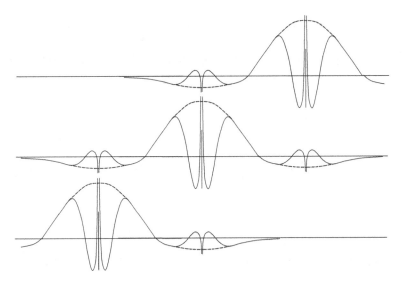

Figure 23.1. Schematic example of Wannier functions that correspond to the Bloch functions in Fig. 4.11. Each line shows a function centered on a different site. These are for a band made of 3s atomic-like orbitals and the smooth functions denote the smooth part of the wavefunction as illustrated in Fig. 11.2.

The Bloch functions defined by Eq. (23.5) clearly obey the periodic gauge requirement (Eq. (23.3)) and in practice this is a convenient way to generate Bloch functions that are smooth functions of $\mathbf{k}$, provided it is possible to find localized Wannier functions (see footnote 1).

It is most useful to work with the periodic parts of the Bloch functions so that Eq. (23.4) can be written

$$w_{i\mathbf{n}}(\mathbf{r}) = \frac{\Omega_{\text{cell}}}{(2\pi)^3} \int_{\text{BZ}} d\mathbf{k} e^{i\mathbf{k}\cdot(\mathbf{r}-\mathbf{T_n})} u_{i\mathbf{k}}(\mathbf{r}). \qquad (23.6)$$

Wannier functions, labeled $i = 1, 2, \ldots$, can be defined for a set of bands separated by an energy gap from other bands, as in insulator. In general, the functions are defined as a linear combination of the Bloch functions of different bands, so that the definition is an extension of Eq. (23.2). Each Wannier function is given by Eq. (23.6) with

$$u_{i\mathbf{k}} = \sum_j U_{ji}^{\mathbf{k}} u_{j\mathbf{k}}^{(0)}, \qquad (23.7)$$

where $U_{ji}^{\mathbf{k}}$ is a $\mathbf{k}$-dependent unitary transformation. For example, in the diamond or zinc blende semiconductors, four occupied bands together are needed to form Wannier functions with sp^3 character, which have a simple interpretation in terms of chemical bonding.

It is straightforward to show (Exercise 23.3) that the Wannier functions are orthonormal

$$\langle i\mathbf{m}|j\mathbf{m}'\rangle = \int d\mathbf{r} w_i^*(\mathbf{r} - \mathbf{T_m}) w_j(\mathbf{r} - \mathbf{T_{m'}}) = \delta_{ij}\delta_{\mathbf{mm'}}, \qquad (23.8)$$

using Eq. (23.4) and the fact that the eigenfunctions $\psi_{ik}(\mathbf{r})$ are orthonormal. The integral in Eq. (23.8) is over all space to account for the tails of the Wannier functions.

Expressions for moments of the Wannier functions can be derived by noting that

$$\langle u_{ik}|u_{j\mathbf{k}+\mathbf{q}}\rangle = \langle \psi_{ik}|e^{-i\mathbf{q}\cdot\mathbf{r}}|\psi_{j\mathbf{k}+\mathbf{q}}\rangle = \sum_{\mathbf{n}} e^{-i\mathbf{k}\cdot\mathbf{T_n}}\langle i\mathbf{n}|e^{-i\mathbf{q}\cdot\mathbf{r}}|0j\rangle, \qquad (23.9)$$

and expanding in powers of $\mathbf{q}$. For example, the first moment is the expectation value of the position operator $\hat{\mathbf{r}}$, which can be expressed using notation analogous to Eq. (23.8) as

$$\langle i\mathbf{n}|\hat{\mathbf{r}}|0j\rangle = i\frac{\Omega}{(2\pi)^3}\int d\mathbf{k}e^{-i\mathbf{k}\cdot\mathbf{T_n}}\langle u_{ik}|\nabla_{\mathbf{k}}|u_{j\mathbf{k}}\rangle. \qquad (23.10)$$

Conversely

$$\langle u_{ik}|\nabla_{\mathbf{k}}|u_{j\mathbf{k}}\rangle = -i\sum_{m} e^{-i\mathbf{k}\cdot\mathbf{T_n}}\langle i\mathbf{n}|\hat{\mathbf{r}}|0j\rangle, \qquad (23.11)$$

where it is understood that $\nabla_{\mathbf{k}}$ acts only to the right. Of particular importance is the first moment, which is called a "Wannier center," which is given by

$$\langle 0i|\hat{\mathbf{r}}|0i\rangle = i\frac{\Omega}{(2\pi)^3}\int d\mathbf{k}\langle u_{ik}|\nabla_{\mathbf{k}}|u_{ik}\rangle, \qquad (23.12)$$

for cell 0 and is simply shifted by the lattice vector $\mathbf{T_n}$ for other cells. Expansion to second order in $\mathbf{q}$ leads to (see Exercise 23.4)

$$\langle i\mathbf{n}|\hat{\mathbf{r}}^2|0j\rangle = -\frac{\Omega}{(2\pi)^3}\int d\mathbf{k}e^{-i\mathbf{k}\cdot\mathbf{T_n}}\langle u_{ik}|\nabla_{\mathbf{k}}^2|u_{j\mathbf{k}}\rangle. \qquad (23.13)$$

Even though Eqs. (23.10) and (23.12) may appear to be an innocuous integral over the Brillouin zone, this is where Berry phases and topology come into play, as discussed below and in Chapter 25.

A drawback of the Wannier representation is that the functions are not uniquely defined, i.e., there are many choices with different shapes and ranges. This can be seen from Eq. (23.4) together with Eqs. (23.2) or (23.7): the Wannier function changes because variations in $\alpha_i(\mathbf{k})$ or $U_{ji}^{\mathbf{k}}$ change the *relative* phases and amplitudes of Bloch functions for band i as a function of $\mathbf{k}$.

In addition, for decades there was no proof that exponentially localized Wannier exist except for the proof by Kohn [892, 894]. The proof has finally been provided by topological analysis, which shows that it is possible for band structures that have trivial topology, and it is not possible to construct localized Wannier functions that have the crystal symmetry for cases with nontrivial topology as discussed in Section 25.8.

Nevertheless, there is an important property of Wannier functions that is unique: it is invariant to the choice of gauge, i.e., it is the same for all choices of $\alpha_i(\mathbf{k})$ and $U_{ji}^{\mathbf{k}}$.[4] The first moment of a Wannier function $\langle 0i|\hat{\mathbf{r}}|0i\rangle$ is the mean position, termed a Wannier center,

---

[4] There are some restrictions on the allowed choices of $\alpha_i(\mathbf{k})$ and $U_{ji}^{\mathbf{k}}$ that they must be continuous and obey the boundary conditions. See also Section P.2.

is given in Eq. (23.12). Blount [890] has shown that the *sum of the centers of the Wannier functions in a cell is invariant*, and a more recent proof based on Berry phases is derived in Chapter 24. Of course, Blount and others understood that a rigid shift of all the Wannier functions by a lattice constant must have no effect except possibly an overall phase. It is now understood that this can be incorporated in the theory in terms of Berry phases that are invariant except for additions of multiples of 2π, with profound implications in the quantum mechanical theory of polarization and topological insulators, which are the topics of Chapters 24 and 25.

23.2 Maximally Projected Wannier Functions

The term "maximally projected Wannier functions" is introduced here to describe a simple, intuitive approach for construction of Wannier functions sufficient to choose the phases of the Bloch functions so that the Wannier function is maximum at a chosen point. The simplest example has been analyzed by Kohn [892, 894]: for a crystal with one atom per cell and a single band derived from s-symmetry orbitals, the Wannier function $w_i(\mathbf{r})$ on site $\mathbf{T}_m = 0$ can be chosen to be the sum of Bloch functions $\psi_\mathbf{k}(\mathbf{r})$ with phases such that $\psi_\mathbf{k}(0)$ is real and positive for each $\mathbf{k}$. The Wannier function thus defined by Eq. (23.6) is maximal on site 0 and decays as a function of distance from 0. In the case of a one-dimensional crystal, it has been proven [892, 894] that the decay is exponential and that this is the *only* exponentially decaying Wannier function that is real and symmetric about the origin.

This approach can be extended to more general cases by requiring that the phase of the Bloch function be chosen to have maximum overlap with a chosen localized function, i.e., maximum projection of the function. An example is a p-symmetry atom-centered Wannier function chosen to have maximal overlap with a p atomic-like state on an atom in the cell at the origin. Maximum projection on any orbital in the basis is easy to accomplish in localized basis representations, simply by choosing the phase of each Bloch function so that the amplitude is real and positive for the given orbital. In a plane wave calculation, for example, it means taking a projection much like the projectors for separable pseudopotentials (Section 11.8). For bond-centered functions, one can require maximal overlap with a localized bonding-like function.

Stated in this broad way, the construction leads to a transformation of the electronic structure problem into a new basis of localized orthonormal orbitals. This is the basis of the formulations of Bullett [895] and Anderson [896], which provide a fundamental way of deriving generalized Hubbard-type models [460, 461]. This approach is used, for example, to calculate model parameters for orbitals centered on Cu and O in CuO_2 materials [897] and for orbitals that span a space of d and s symmetry functions in Cu metal [898].

A construction often used in "order-N" calculations is to find functions localized to a sphere within some radius around a given site. This can be interpreted as "maximal overlap" with a function that is unity inside the sphere and zero outside, usually applied with the boundary condition that the function vanish at the sphere boundary.

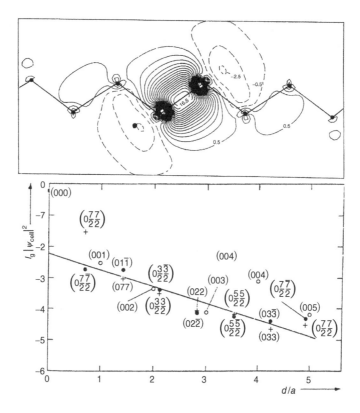

Figure 23.2. Top: bond-centered Wannier function for Si calculated [728] by requiring the phases of the Bloch functions to have real, positive amplitude at one of the four equivalent band centers. Note the similarity to Fig. 23.3. Bottom: the decay of the Wannier function in various directions, plotted on a log scale.

Bond-Centered Wannier Function in Silicon

The construction of bond-centered Wannier functions in diamond structure crystals is discussed by Kohn [894], and careful numerical calculations have been done by Satpathy and Pawlowska [728] for Si – the standard test case, of course. The calculations used the LMTO method (Chapter 17), in which the orbitals are described in terms of functions centered on the atoms (and on empty spheres). A bond-centered Wannier function is generated simply by choosing the phases of the Bloch functions to be positive on one of the four bond centers in a unit cell. The function can then be plotted in real space as shown in Fig. 23.2; note the striking resemblance to the "maximally localized" function shown below on the left-hand side of Fig. 23.3.

Satpathy and Pawlowska [728] also showed numerically that the bond-centered function decays exponentially, as shown in Fig. 23.2. This is perhaps the first such accurate numerical test of the exponential decay in solids like Si, which has since been found in many other calculations. Presumably the reason for the similarity of the Wannier functions for Si in Figs. 23.2 and 23.3 is related to Kohn's proof in one dimension that the function is uniquely fixed by the requirements that the function be real, symmetric, and exponentially decaying, bolstered by the knowledge that such exponentially decaying functions exist in systems like Si with trivial topology of the electronic structure.

23.3 Maximally Localized Wannier Functions

Finding highly localized Wannier functions (or transforms of Wannier functions) with desired properties is a venerable subject chemistry (see, e.g., the paper of Boys [899] and many references cited in the review [891], where they are called "localized molecular orbitals"). Such functions are useful in constructing efficient methods (see Chapter 18 on "order-N" algorithms) and in providing insight through simple descriptions of the electronic states using a small number of functions.

Although there are many possible ways to define "maximally localized,"[5] one stands out: minimization of the mean square spread Ω defined by

$$\Omega = \sum_{i=1}^{N_{bands}} \left[\langle r^2 \rangle_i - \langle \mathbf{r} \rangle_i^2 \right], \qquad (23.14)$$

where $\langle \cdots \rangle_i$ means the expectation value over the ith Wannier function in the unit cell (whose total number N_{bands} equals the number of bands considered). As shown by Marzari and Vanderbilt [900], this definition leads to an elegant formulation, in which a part of the spread, Eq. (23.14), can be identified as an invariant (Eq. (23.16) below). Furthermore, this invariant part leads to a physical measure of localization as shown by Souza et al. [901] (see Section 24.6).

Because the Wannier functions are not unique, the Ω defined in Eq. (23.14) is not invariant under gauge transformations of the Wannier functions [900]. Nevertheless, Marzari and Vanderbilt were able to decompose Ω into a sum of two positive terms: a gauge-invariant part Ω_I, plus a gauge-dependent term $\widetilde{\Omega}$:

$$\Omega = \Omega_I + \widetilde{\Omega}, \qquad (23.15)$$

$$\Omega_I = \sum_{i=1}^{N_{bands}} \left[\langle r^2 \rangle_i - \sum_{\mathbf{T}j} |\langle \mathbf{T}j|\hat{\mathbf{r}}|0i\rangle|^2 \right], \qquad (23.16)$$

$$\widetilde{\Omega} = \sum_{i=1}^{N_{bands}} \sum_{\mathbf{T}j \neq 0i} |\langle \mathbf{T}j|\hat{\mathbf{r}}|0i\rangle|^2. \qquad (23.17)$$

Clearly, the second term $\widetilde{\Omega}$ is always positive. The clever part of the division in Eq. (23.15), however, is that Ω_I is *both invariant and always positive*. Furthermore, it has a simple interpretation that may be seen by identifying the projection operator $\hat{P}$ onto the space spanned by the N_{bands} bands,[6]

---

[5] For example, a widely used criterion is to maximize the self-Coulomb interaction.
[6] This is the same as defined in Eq. (20.20), except that here the sum need not be over all occupied states.

$$\hat{P} = \sum_{i=1}^{N_{\text{bands}}} \sum_{\mathbf{T}} |\mathbf{T}i\rangle\langle\mathbf{T}i| = \sum_{i=1}^{N_{\text{bands}}} \sum_{\mathbf{k}} |\psi_{i\mathbf{k}}\rangle\langle\psi_{i\mathbf{k}}|, \tag{23.18}$$

and $\hat{Q} = 1 - \hat{P}$ defined to be the projection onto all other bands. Writing out Eq. (23.16) leads to the simple expression (here α denotes the vectors index for $\mathbf{r}$)

$$\Omega_I = \sum_{i=1}^{N_{\text{bands}}} \sum_{\alpha=1}^{3} \langle 0i|\hat{\mathbf{r}}_\alpha \hat{Q} \hat{\mathbf{r}}_\alpha |0i\rangle, \tag{23.19}$$

which is manifestly positive (Exercise 23.5). The presence of the $\hat{Q}$ projection operator leads to an informative interpretation of Eq. (23.19) as the quantum fluctuations of the position operator from the space spanned by the Wannier functions into the space of the other bands. This can also be viewed as a consequence of the fact that the position operator does not commute with $\hat{P}$ or $\hat{Q}$ (Exercise 23.8), so that expression (23.19) is *not* simply the mean square width of the Wannier function. Instead Ω_I is an invariant, as is apparent in the explicit expressions in $\mathbf{k}$ space given below.[7] Furthermore, the fact that Eq. (23.19) represents fluctuations leads to the physical interpretation of Ω_I is brought out in Section 24.6.

Practical Expressions in k Space

Expressions for Ω_I and $\tilde{\Omega}$ in terms of the Bloch states can be derived by substituting the definitions of the Wannier functions, Eq. (23.6), into Eqs. (23.16) and (23.17). It is an advantage for practical calculations to write the formulas in terms of discrete sums instead of integrals using Eq. (12.14). If one uses a finite difference approximation for the derivatives w.r.t $\mathbf{k}$ in Eqs. (23.10) and (23.13), one finds [900]

$$\langle \mathbf{r} \rangle_j = \frac{i}{N_k} \sum_{\mathbf{kb}} w_{\mathbf{b}} \mathbf{b} \left[\langle u_{j\mathbf{k}}|u_{j\mathbf{k}+\mathbf{b}}\rangle - 1 \right], \tag{23.20}$$

and

$$\langle r^2 \rangle_j = \frac{1}{N_k} \sum_{\mathbf{kb}} w_{\mathbf{b}} \left[2 - \text{Re}\langle u_{j\mathbf{k}}|u_{j\mathbf{k}+\mathbf{b}}\rangle \right], \tag{23.21}$$

where $\mathbf{b}$ denotes the vectors connecting the points $\mathbf{k}$ to neighboring points $\mathbf{k} + \mathbf{b}$, and $w_{\mathbf{b}}$ denotes the weights in the finite difference formula.

Although these formulas reduce to the integral in the limit $\mathbf{b} \to 0$, they are not acceptable because they violate the fundamental requirement of translation invariance for any finite $\mathbf{b}$. If one makes the substitution $\psi_{i\mathbf{k}}(\mathbf{r}) \to e^{-i\mathbf{k}\cdot\mathbf{T}_m}\psi_{i\mathbf{k}}(\mathbf{r})$, the expectation values should change by a translation,

---

[7] An interesting feature is that Ω_I can be expressed in terms of a "metric" that defines the "quantum distance" along a given path in the Brillouin zone [900]. This "distance" quantifies the change of character of the occupied states $u_{n\mathbf{k}}$ as one traverses the path, leading to the heuristic interpretation of Ω_I as representing a measure of the dispersion throughout the Brillouin zone [900].

$$\langle \mathbf{r} \rangle_j \rightarrow \langle \mathbf{r} \rangle_j + \mathbf{T}_m,$$
$$\langle r^2 \rangle_j \rightarrow \langle r^2 \rangle_j + 2\langle \mathbf{r} \rangle_j \cdot \mathbf{T}_m + T_m^2, \tag{23.22}$$

so that Ω is unchanged. These properties are not obeyed by Eqs. (23.20) or (23.21).

Acceptable expressions can be found [900] that have the same limit for $\mathbf{b} \rightarrow 0$ yet satisfy Eq. (23.22). Functions with the desired character are complex log functions that have a Taylor series expansion, $\ln(1 + ix) \rightarrow ix - x^2 + \cdots$ for small x (similar to Eqs. (23.20) and (23.21) for x real), but are periodic functions for large $\mathrm{Re}\{x\}$. If we define $\langle u_{i\mathbf{k}} | u_{j\mathbf{k}+\mathbf{b}} \rangle \equiv M_{ij}(\mathbf{k}, \mathbf{b})$, Eqs. (23.20) and (23.21) can be replaced by[8] (Exercise 23.6)

$$\langle \mathbf{r} \rangle_j = \frac{i}{N_k} \sum_{\mathbf{kb}} w_\mathbf{b} \, \mathbf{b} \, \mathrm{Im} \, \ln M_{jj}(\mathbf{k}, \mathbf{b}), \tag{23.23}$$

and

$$\langle r^2 \rangle_j = \frac{1}{N_k} \sum_{\mathbf{kb}} w_\mathbf{b} \left\{ 1 - |M_{jj}(\mathbf{k}, \mathbf{b})|^2 + [\mathrm{Im} \ln M_{jj}(\mathbf{k}, \mathbf{b})]^2 \right\}. \tag{23.24}$$

The invariant part can be found in a way similar to Eq. (23.24) with the result

$$\Omega_I = \frac{1}{N_k} \sum_{\mathbf{kb}} w_\mathbf{b} \left[N_{\mathrm{bands}} - \sum_{ij}^{N_{\mathrm{bands}}} |M_{ij}(\mathbf{k}, \mathbf{b})|^2 \right], \tag{23.25}$$

which is positive (Exercise 23.7). The meaning of this term and closely related expressions are given in Section 24.6.

In one dimension it is possible to choose Wannier functions so that $\widetilde{\Omega} = 0$, i.e., the minimum possible spread. However, in general, it is not possible for $\widetilde{\Omega}$ to vanish in higher dimensions. This follows (Exercise 23.8) from the expression for $\widetilde{\Omega}$, given later in Eq. (23.30), and the fact that the *projected* operators $\{\hat{P}\hat{x}\hat{P}, \hat{P}\hat{y}\hat{P}, \hat{P}\hat{z}\hat{P}\}$ do not commute, i.e., the matrices representing the matrix elements $\langle \mathbf{T}i | \hat{x} | \mathbf{T}'j \rangle$ do not commute.

Minimization by Steepest Descent

Finding Wannier functions that are maximally localized can be accomplished by minimizing the spread, Eq. (23.24), as a function of the Bloch functions. (This means minimizing $\widetilde{\Omega}$ since Ω_I is invariant.) For a given set of Bloch functions $u_{j\mathbf{k}}^{(0)}$, one can consider all possible unitary transformations given by Eq. (23.7), which can be written in the form

$$\mathbf{M}(\mathbf{k}, \mathbf{b}) = [\mathbf{U}^\mathbf{k}]^\dagger \mathbf{M}^{(0)}(\mathbf{k}, \mathbf{b}) \mathbf{U}^{\mathbf{k}+\mathbf{b}}, \tag{23.26}$$

where $\mathbf{M}$ and $\mathbf{U}$ are understood to be matrices in the band indices. To minimize, one can vary $\mathbf{U}^\mathbf{k}$, which is done most conveniently by defining

$$\mathbf{U}^\mathbf{k} = e^{\mathbf{W}^\mathbf{k}}, \tag{23.27}$$

---

[8] These forms are not unique and alternatives are pointed out in [900]. See Eq. (24.13) for the corresponding expressions derived using Berry phases.

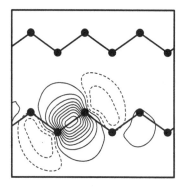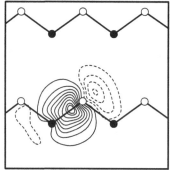

Figure 23.3. "Maximally localized" Wannier functions for Si (left) and GaAs (right) from [900]. Each figure shows one of the four equivalent functions found for the four occupied valence bands. Provided by N. Marzari

where $\mathbf{W^k}$ is an antihermitian matrix (Exercise 23.15). The solution can be found by the method of steepest descent (Appendix L). The gradient can be found by considering infinitesimal changes, $\mathbf{U^k} \rightarrow \mathbf{U^k}(1 + \delta \mathbf{W^k})$. Expressions for the gradient,

$$\frac{\delta \Omega}{\delta \mathbf{W^k}} = \mathbf{G^k}, \tag{23.28}$$

in $\mathbf{k}$ space are given in [900]; we will give equivalent expressions in (23.30) and (23.31) that bring out the physical meaning. Choosing $\delta \mathbf{W^k} = \epsilon \delta \mathbf{G^k}$ along the steepest decent direction leads to a useful minimization algorithm, which corresponds to updating the $\mathbf{M}$ matrices at each step n

$$\mathbf{M^{(n+1)}(k, b)} = e^{-\delta \mathbf{W^k}} \mathbf{M^{(n)}(k, b)} e^{\delta \mathbf{W^{k+b}}}, \tag{23.29}$$

where the exponentiation can be done by diagonalizing $\delta \mathbf{W}$.

Examples of Wannier functions calculated by this "maximal localization" prescription are shown in Fig. 23.3 for Si and GaAs [900]. These functions are derived by considering the full set of four occupied valence bands, leading to four equivalent bonding-like orbitals. For GaAs there is an alternative possibility: since the four valence bands consist of one well-separated lowest band plus three mixed bands at higher energy, Wannier functions can be derived separately for the two classes of bands. The result is one function that is primarily s-like centered on the As atom and three functions primarily p-like on the As atom [900]. However, these Wannier functions do not lead to the maximum overall localization, so that the bonding orbitals appear to provide the most natural picture of local chemical bonding.

Wannier Functions in Disordered Systems

Up to now the derivations have focused entirely on crystals and have used Bloch functions. How can one find useful maximally localized functions for noncrystalline systems, such as molecules or disordered materials? This is particularly important for interpretation purposes and for calculations of electric polarization in large Car–Parrinello-type simulations (Chapter 19), where often calculations are done only for periodic boundary conditions, i.e., for $\mathbf{k} = 0$. Many properties, such as the total dipole moment of the sum of Wannier functions, are invariant (Section 24.4), so any approach that finds accurate Wannier functions is sufficient. For other properties, it is desirable to derive maximally localized functions.

The most direct approach is to construct "maximally projected" functions (Section 23.2) that are chosen to have weight at a center or maximum overlap with a chosen function. A closely related procedure is actually used in the "order-N" linear-scaling methods (see Section 18.6) that explicitly construct Wannier-like functions constrained to be localized to a given region [777]. These methods provide an alternative approach for direct construction of Wannier functions without ever constructing eigenstates.

It is also useful to construct "maximally localized" functions. For example, they are directly useful in the concept of localization (Section 24.6). The functions can be derived by minimizing $\widetilde{\Omega}$ given by Eq. (23.17). It follows from the definitions (as shown in Appendix A of [900] and further elucidated in [902] and [903]), that Eq. (23.17) can be written as

$$\widetilde{\Omega} = \text{Tr}[\hat{X}'^{\,2} + \hat{Y}'^{\,2} + \hat{Z}'^{\,2}], \qquad (23.30)$$

where $X_{ij} = \langle 0i | \hat{x} | 0j \rangle$, $[X_D]_{ij} = X_{ii} \delta_{ij}$, and $X'_{ij} = X_{ij} - [X_D]_{ij}$, with corresponding expressions for $\hat{Y}$ and $\hat{Z}$. For an infinitesimal unitary transformation $|i\rangle \rightarrow |i\rangle + \sum_j \delta W_{ji} |j\rangle$, the gradient of Eq. (23.30) can be written as (see Exercise 23.16) $\delta \Omega = 2\text{Tr}[\hat{X}' \delta \hat{X} + \hat{Y}' \delta \hat{Y} + \hat{Z}' \delta \hat{Z}]$, where $\delta \hat{X} = [\hat{X}, \delta \hat{W}]$, etc. Finally, one finds $\delta \Omega = \text{Tr}[\delta \hat{W} \hat{G}]$, where

$$\frac{\delta \widetilde{\Omega}}{\delta \hat{W}} = \hat{G} = 2\{[\hat{X}', \hat{X}_D] + [\hat{Y}', \hat{Y}_D] + [\hat{Z}', \hat{Z}_D]\}. \qquad (23.31)$$

These forms are the most compact expressions for $\widetilde{\Omega}$ and its gradient. They are directly useful in real-space calculations in terms of Wannier functions $w_i(\mathbf{r})$; the corresponding forms in $\mathbf{k}$ space [900] can be derived using the transformations of Section 23.1 and used in Eq. (23.28).

Examples of Wannier functions calculated at steps in a quantum molecular dynamics simulation of water under high-pressure, high-temperature conditions are shown in Fig. 2.14. Three "snapshots" during a simulation show a sequence that involves a proton transfer and the associated transfer of a Wannier function (only this one of all the Wannier functions is shown) to form H^+ and $(H_3O)^-$.

23.4 Nonorthogonal Localized Functions

One can also define a set of nonorthogonal localized orbitals $\tilde{w}_i$ that span the same space as the Wannier functions w_i and which can be advantageous for practical applications and for intuitive understanding. Just as for Wannier functions, one must choose some criterion for "maximal localization" to fix the $\tilde{w}_i$. The work of Liu et al. [774] is particularly illuminating since it uses the same mean square radius criterion as in Eq. (23.14). It provides a practical approach for calculating the functions directly related to optimizing functionals in O(N) methods (Section 18.6).

The transformation to nonorthogonal $\tilde{w}_i$ can be defined by

$$\tilde{w}_i = \sum_{j=1}^{N_{\text{bands}}} A_{ij} w_j, \tag{23.32}$$

where A is a non-singular matrix that must satisfy

$$\sum_{i=1}^{N_{\text{bands}}} (A_{ij})^2 = 1, \tag{23.33}$$

since the $\tilde{w}_i$ are defined to be normalized. The mean square spread, Eq. (23.14), generalizes to [774]

$$\Omega[A] = \sum_{i=1}^{N_{\text{bands}}} \left[\langle \tilde{w}_i | r^2 | \tilde{w}_i \rangle - \langle \tilde{w}_i | \mathbf{r} | \tilde{w}_i \rangle^2 \right], \tag{23.34}$$

which is to be minimized as a function of the matrix A subject to two conditions: A is nonsingular and satisfies Eq. (23.33). It is simple to enforce the latter condition; however, it is not so simple to search only in the space of nonsingular matrices. It is shown in [774] that one can use the fact that a nonsingular matrix must have full rank, i.e., $\text{rank}(A) = N$. Using $\text{rank}(A) = \text{rank}(A^\dagger A)$ and the variational principle, Eq. (18.33), developed for minimizing the energy functional [746] with $S \to A^\dagger A$, the result is

$$\text{rank}(A) = -\min\{\text{Tr}[(-A^\dagger A)(2X - XA^\dagger AX)]\}, \tag{23.35}$$

which is minimized for all hermitian matrices X. Defining a constraint functional $\Omega_a[A, X] = (N - \text{Tr}((A^\dagger A)(2X - XA^\dagger AX)))^2$, it follows that maximally localized nonorthogonal orbitals $\tilde{w}_i$ can be found by minimizing $\Omega[A] + C_a \min\{\Omega_a[A, X]\}$, for all matrices A that satisfy Eq. (23.33). Here C_a is an adjustable positive constant and the second term ensures that the final transformation matrix A is nonsingular (Exercise 23.17).

An example of maximally localized orbitals for a benzene molecule is shown in Fig. 23.4, where we see that the nonorthogonal orbitals are much more localized and much easier to interpret as simple bonding orbitals than the corresponding orthogonal orbitals. The short range of the nonorthogonal orbitals can be used in calculations to reduce the cost, for example, in O(N) methods as discussed in Section 18.6.

23.5 Wannier Functions for Entangled Bands

The subject of this section is construction of Wannier-type functions that describe bands in some energy range *even though they are not isolated and are "entangled" with other bands*. Strictly speaking, Wannier functions as defined in Section 23.1 will not be useful; if the bands cannot be disentangled then there will be nonanalytic properties resulting

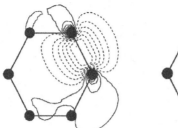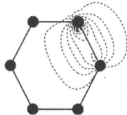

Figure 23.4. Comparison of orthogonal and nonorthogonal maximally localized orbitals for C–C
σ bonds (left) and C–H bonds (right) in benzene C_6H_6. The nonorthogonal orbitals are more
localized and more transferrable than the orthonormal ones. In order to be orthonormal they must
have tails that extend into neighboring atoms, and the tails must change when transferred to another
system with different neighbors. From [774].

from mixing with other bands in the integrals over the Brillouin zone. However, one can
define useful functions that have real-space properties like Wannier functions and form an
orthonormal, localized basis for a subspace of bands that span a desired range of energies.

There are two basic approaches for construction of functions that span a desired subspace.
One approach is to identify the type of orbitals involved and to generate a reduced set of
localized functions that describes the energy bands over a given range. Outside that range,
the full band structure is, of course, not reproduced: the reduced set of bands has an upper
and a lower bound, i.e., they form a set of isolated bands in the reduced space. This is in
essence the idea of "maximally projected" functions in Section 23.2, but now constrained
only to match the bands over some range. An example of such an approach is the "downfolding" method [735, 905], the results of which are illustrated in Figs. 17.8 and 17.9. The
single orbital centered on a Cu atom is sufficient to accurately describe the main band that
crosses the Fermi energy without explicitly including the rest of the "spaghetti" of bands.

The second approach [898, 904] generalizes the idea of "maximally localized" Wannier
functions (Section 23.3) to maximize the overlap with Bloch functions *only over an energy
window*. This also generates a finite subspace of bands that describes the actual bands only
within the chosen range. Of course, the functions are not unique since there are many
choices for the energy range and weighting functions. Two calculations for Cu, done using
pseudopotentials and plane waves [904] and the LMTO method [898], give very similar
results for the desired bands, but with different localized functions. For example, maximally
localized functions constructed from 6d and s bands taken together are each centered in
interstitial positions near the Cu atom [904]. The bands for the six-dimensional subspace of
orbitals are given in Fig. 23.5, which shows that the band structure is accurately represented
for energies extended to well above the Fermi energy, even though the higher bands are
missing. The lower panel of the figure shows a different decomposition with the subspace
decomposed into 5d orbitals (which have the expected form of atom-centered d orbitals
in a cubic symmetry crystal) plus the complement that is an optimal s-symmetry orbital.
Similar results for the bands are found using the LMTO method [898] where the authors
also showed that the functions decay exponentially (or at least as a very high power).

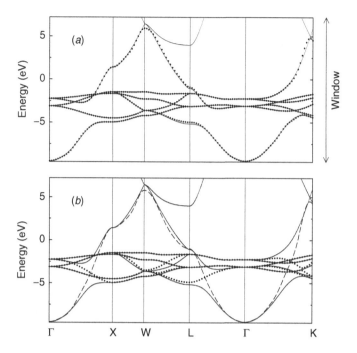

Figure 23.5. Bands of Cu produced by maximally localized Wannier-like functions [904]. Top panel: functions that span the six-dimensional subspace for the five d states and one s state compared to the full band structure. Similar results for the bands are found in [898]. The bands are accurately reproduced up to well above the Fermi energy, even though the higher bands are missing. The lower panel shows results if the subspace is decomposed into the five d orbitals (chosen as maximally localized with a narrow energy window around the primarily d bands) plus the complement that is the s orbital. From [904].

23.6 Hybrid Wannier Functions

The derivations in the previous section can be modified to define functions that are "hybrid" Wannier and Bloch functions.[9] In dimensions greater than one, these are localized along only one direction but extended in the other perpendicular directions. The crystal can be considered to be composed of planes separated by the distance L_{cell}, the length of the cell in the $\hat{x}$ direction, and X_n denotes translations that are integral multiples of L_{cell}. If x denotes the position and $\mathbf{k}$ is resolved into components $k, \mathbf{k}_\perp$, where k is along the $\hat{x}$ direction and $\mathbf{k}_\perp$ along the perpendicular directions, a hybrid function can be defined by the one-dimensional integral over a cell,

$$w_{in,\mathbf{k}_\perp}(\mathbf{r}) = \frac{L_{cell}}{2\pi} \int dk e^{ik(x-X_n)} u_{ik,\mathbf{k}_\perp}(\mathbf{r}), \qquad (23.36)$$

---

[9] It would be more accurate and informative to call these hybrid Wannier–Bloch functions, but we choose the simpler notation hybrid Wannier functions.

which is a function of the momentum $\mathbf{k}_\perp$ along the perpendicular directions. A hybrid is a function of position $\mathbf{r}$ like the usual Wannier and Bloch functions, but it is expressed as a function that is localized in one direction and extended in the other directions with momentum $\mathbf{k}_\perp$. Hybrid functions can be used like one-dimensional Wannier functions of a single variable x with $\mathbf{k}_\perp$ considered as a parameter. For example, an important property is the Wannier center in the x direction for function i in the one-dimensional cell 0 for each $\mathbf{k}_\perp$ given by

$$\langle 0i|x|0j\rangle_{\mathbf{k}_\perp} = i\frac{L_{\text{cell}}}{2\pi}\int dk\langle u_{ik,\mathbf{k}_\perp}|\partial_k|u_{ik,\mathbf{k}_\perp}\rangle, \tag{23.37}$$

for each $\mathbf{k}_\perp$.

Hybrid functions are especially useful for problems that involve surfaces or interfaces that are periodic in a plane (or a line for the edge of a two-dimensional system) but there may be bands of states that are localized in the direction parallel to the surface normal. The centers of the hybrid functions play a central role in understanding topological insulators as described in Chapters 25–28.

23.7 Applications

There are a host of uses of Wannier functions or Wannier-like localized functions. Here we list a few examples and refer to the review [891] for a host of other examples and references.

One major use is to construct Bloch functions with desired properties using the reverse transformation in Eq. (23.5). For example, the fine details of band structures can be worked out using Bloch functions that vary smoothly with $\mathbf{k}$, called Wannier interpolation. This is much more efficient and instructive than a brute force method of calculation on a very fine grid in $\mathbf{k}$. A relatively crude interpolation may not provide the most accurate energies, but the fact that the Bloch functions vary smoothly can reveal the details of bands near critical points and avoid crossings with very small gaps. Great accuracy in the energies and the fine details can be achieved by high-quality interpolation. An example is the details of the band structure of Fe including spin–orbit interaction where there are many tiny features that have large effects in transport [906]. The details are especially important for phenomena due to Weyl nodes (Section 28.4), both for quantitative accuracy and for qualitative understanding. A much more extensive discussion of applications is in [891].

Wannier functions appear often in the following chapters on polarization and topology. The theoretical expressions often assume smooth Bloch functions to derive the elegant formulations in terms of Berry phases. In turn, the results are often best expressed in terms of Wannier functions, which provide intuitive, instructive interpretations.

Another category of applications is large-scale calculations where localized functions present great advantages. Examples are cases like calculation of the band structure and conductance of complex nanostructures [907]. One of the methods in Section 13.4 to treat nonlocal exchange in hybrid functionals in plane wave calculations uses a transformation to Wannier functions. Even if they are not used in the calculation, they are very useful in analysis, for example, the snapshots of the Wannier function for an electron associated with a proton that is transferred in a simulation of water in Fig. 2.14. Linear-scaling $O(N)$ methods in Chapter 18 are based on the fact that the hamiltonian is sparse in localized functions. There are many choices and in general it is best *not* to use orthogonal Wannier functions, which are longer range and less transferable than nonorthogonal functions. However, generalized Wannier functions (see Section 18.6) are an effective approach.

Wannier functions are also a means to develop models that go beyond the purview of standard Kohn–Sham or other independent-particle calculations. Essentially all models used in interacting many-body problems are based on localized orthogonal basis functions, for example, the Hubbard model. In order to apply the models to actual problems in materials the parameters must be gotten from some other method, and more and more, they are derived from Wannier functions from Kohn–Sham calculations. This is often the approach to derive the interaction U in the DFT+U methods in Section 9.6, which is also the basis for the parameters in the models used in dynamical mean field theory (see [1]). Despite the inherent nonuniqueness, there are arguments supporting the choices and often maximally localized functions are used. The issues and applications are a major topic of [1].

SELECT FURTHER READING

Basic theory of Wannier functions:

Ashcroft, N. W. and Mermin, N. D., *Solid State Physics* (W. B. Saunders Company, Philadelphia, 1976).

Weinreich, G., *Solids: Elementary Theory for Advanced Students* (John Wiley & Sons, New York, 1965).

In-depth presentation including topology and relation to polarization:

Vanderbilt, D. H., *Berry Phases in Electronic Structure Theory* (Cambridge University Press, Cambridge, 2018).

Review of maximally localized functions with many references to older work:

Marzari, N., Mostofi, A. A., Yates, J. R., Souza, I., and Vanderbilt, D. H., "Maximally localized Wannier functions: Theory and application," *Rev. Mod. Phys.* 84:1419–1475, 2012.

Exercises

23.1 This exercise is to construct a localized Wannier function for the s bands described in Section 14.5 and Exercises 14.6 and 14.7. The hamiltonian has only nearest-neighbor matrix elements t and the basis is assumed to be orthogonal. For all cases (line, square, and simple cubic lattices), show that one can choose the periodic part of the Bloch functions $u_{i\mathbf{k}}(\mathbf{r})$ to be real, in which case they are independent of $\mathbf{k}$. Next, show from the definition, Eq. (23.6), that this choice leads to the most localized possible Wannier function, which is identical to the basis function.

23.2 This exercise is to analyze Wannier functions for s bands as described in Exercise 23.1, except that the basis is nonorthogonal with nearest-neighbor overlap s. Show that one can choose the periodic part of the Bloch functions $u_{i\mathbf{k}}(\mathbf{r})$ to be real, and find the $\mathbf{k}$ dependence of $u_{i\mathbf{k}}(\mathbf{r})$ as a function of s. (Hint: for nonorthogonal functions, the normalization coefficient given in Exercise 14.4 is $\mathbf{k}$ dependent, which is relevant for constructing the Wannier function using Eq. (23.6).) Show that the resulting Wannier function has infinite range; even though it decays rapidly, its amplitude does not vanish at any finite distance.

23.3 Derive Eq. (23.8) using definition Eq. (23.6) and properties of the eigenfunctions.

23.4 Show that Expression (23.13) to second order in $\mathbf{q}$ follows in analogy to the expansion that leads to Eq. (23.9).

23.5 Show that Ω_I is always positive by noting that $\hat{Q} = \hat{Q}^2$ so that Eq. (23.19) can be written as the sum of expectation values of squares of operators.

23.6 Show that Eqs. (23.23) and (23.24) have the same limit for $\mathbf{b} \to 0$ as Eqs. (23.20) and (23.21), and that they obey the translation invariance conditions, Eq. (23.22). Show further that this means that Ω is unchanged.

23.7 Show that Ω_I in Eq. (23.25) is positive using the definition of the $\mathbf{M}$ matrices and the fact that the overlap of Bloch functions at different $\mathbf{k}$ points must be less than unity.

23.8 Explain why it is not possible to make $\widetilde{\Omega}$ vanish in higher dimensions.
(a) First show that the *projected* operators $\{\hat{P}\hat{x}\hat{P}, \hat{P}\hat{y}\hat{P}, \hat{P}\hat{z}\hat{P}\}$ do not commute. Show this is equivalent to the statement that $\hat{x}$ and $\hat{P}$ do not commute. Then show that $\hat{x}$ and $\hat{P}$ do not commute.
(b) Use the fact that noncommuting operators cannot be simultaneously diagonalized to complete the demonstration.

23.9 Demonstrate that it is possible to find functions with $\widetilde{\Omega} = 0$ in one dimension by explicitly minimizing $\widetilde{\Omega}$ for a one-band, nearest-neighbor tight-binding model with overlap (see definitions in Section 14.5):

$$H_{i,i\pm1} = t; \quad S_{i,i\pm1} = s, \tag{23.38}$$

where $S_{i,i} = 1$.
(a) First consider the case with $s = 0$: show that in this artificial model the minimum spread is the spread of the basis function. However, one can also choose more delocalized states, e.g., the eigenstates.
(b) For $s \neq 0$, find the minimum spread Ω_I as a function of t. Show it is greater than in part (a). For explicit evaluation of the spread Ω_I, use Eq. (21.23) with the eigenvectors given by analytic solution of the Schrödinger equation and the sum over k done approximately on a regular grid of values in one dimension.

23.10 See exercises in Chapter 26 for solutions for the two-site model in Fig. 26.2, which is also used in Exercises 23.11 and 23.12.

23.11 This exercise is to construct maximally localized Wannier functions for the one-dimensional ionic dimer model in Fig. 26.2 using the fact that the gauge-dependent term in it can be made to vanish.
(a) Let $t_1 = t_2$ so that each atom is at a center of symmetry. Show that the maximally localized Wannier function for the lower band is centered on the atom with lower energy ε_A or ε_B, and the function for the upper band is centered on the atom with higher

energy. (Hint: if there is a center of inversion the periodic part of the Bloch functions can be made real.)

(b) Similarly, there are two centers of inversion if $\varepsilon_A = \varepsilon_B$ and $t_1 \neq t_2$. Show that in this case the Wannier functions are centered respectively on the strong and the weak bonds between the atoms.

(c) In each of the cases above, calculate the maximally localized Wannier function as a sum of localized basis functions. The eigenfunctions can be calculated analytically and the Wannier functions constructed using the definition in Eq. (21.3) and approximating the integral by a sum over a regular grid of k points. (This can be done with a small computer code. Note that the grid spacing must be small for a small gap between the bands.)

23.12 Using the model of Exercise 23.11 and the methods described in Section 23.3, construct a computer code to calculate the centers of the Wannier functions in a general case, $\varepsilon_A \neq \varepsilon_B$ and $t_1 \neq t_2$. This can be used to find polarization and effective charges as described in Exercises 24.8 and 24.9.

23.13 Construct the maximally localized Wannier function for the lowest band in the one-dimensional continuum model of Exercise 12.5. Show that the function is centered at the minimum of the potential. Calculate the functions using the analytic expressions for the Bloch functions and the same approach as in Exercise 23.11, part (c).

23.14 See Exercise 15.6 for a project to construct Wannier functions in one dimension.

23.15 Show that $\mathbf{U}^{\mathbf{k}}$, defined in Eq. (23.27), is unitary if $\mathbf{W}^{\mathbf{k}}$ is antihermitian, i.e., $W_{ij} = -W_{ji}^*$.

23.16 Show that the gradient, Eq. (23.31), follows from the definitions. To do this, verify the operator commutation relations, and note that $\mathrm{Tr}[\hat{X}'\hat{X}_D] = 0$, etc.

23.17 Show that the minimization of the functional $\Omega[A] + C_a \min\{\Omega_a[A, X]\}$ leads to the desired solution of a nonsingular transformation to nonorthogonal orbitals. Hint: use the conditions stated following Eq. (18.33) and the relations given in Eqs. (23.32)–(23.35).

24

Polarization, Localization, and Berry Phases

The only thing certain is change.
Variation of saying attributed to Heraclitus, a Greek philosopher

Summary

Electric polarization is one of the basic quantities in physics, essential to the theory of dielectrics, effective charges in lattice dynamics, piezoelectricity, ferroelectricity, and other phenomena. However, descriptions in widely used texts are often based on oversimplified models that are misleading or incorrect. The basic problem is that the expression for a dipole moment is ill-defined in an extended system, and there is no unique way to find the moment as a sum of dipoles by "cutting" the charge density into finite regions. However, polarization can be expressed in terms of the quantum mechanical wavefunction of the electrons, with an elegant formulation in terms of a Berry phase and alternative expressions using Wannier functions. The other essential property of insulators is "localization" of the electrons. Although the concept of localization is well known, theoretical advances in recent years have provided insights and quantitative measures. This chapter is closely related to Chapter 23 on Wannier functions, in particular to the gauge-invariant center of mass and contribution to the spread of Wannier functions, and to Chapters 25–28 on the topology of the wavefunctions.

24.1 Overview

The theory of electrodynamics of matter [480, 908] (see Appendix E) is cast in terms of electric fields $\mathbf{E}(\mathbf{r}, t)$ and currents $\mathbf{j}(\mathbf{r}', t')$. (Here we ignore magnetic fields.) In metals, there are *real currents* and, in the static limit, electrons flow to screen all macroscopic electric fields. Thus, the description of the metal divides cleanly into two parts: the bulk, which is completely unaffected by static external fields, and surface regions, where there is an accumulation of charge $\delta n(\mathbf{r})$ that adjusts to bring the surfaces to an equipotential. The surface thus determines the *absolute value* of the potential in the interior relative to vacuum (see Section 13.5), but this has no affect on any physical properties intrinsic to the bulk interior of the metal.

The fundamental definition of an insulator, on the other hand, is that it can support a static electric field. In insulators, charge cannot flow over macroscopic distances so long as the field is not too large, but there can be time-dependent currents termed *polarization currents*. The physical, measurable quantities are the electric field $\mathbf{E}(\mathbf{r}, t)$, the density $n(\mathbf{r}, t)$ and current $\mathbf{j}(\mathbf{r}, t)$, and the polarization is defined by the relations

$$\nabla \cdot \mathbf{P}(\mathbf{r}, t) = -\delta n(\mathbf{r}, t), \tag{24.1}$$

or, using the conservation condition $\nabla \cdot \mathbf{j}(\mathbf{r}, t) = -\mathrm{d}n(\mathbf{r}, t)/\mathrm{d}t$,

$$\frac{\mathrm{d}\mathbf{P}(\mathbf{r}, t)}{\mathrm{d}t} = \mathbf{j}(\mathbf{r}, t) + \nabla \times \mathbf{M}(\mathbf{r}, t), \tag{24.2}$$

where $\mathbf{M}(\mathbf{r}, t)$ is an arbitrary vector field. (See also Appendix E, especially Eq. (E.5).) The theory of dielectrics [480, 908] is based on the existence of local constitutive relations of the macroscopic average $\mathbf{P}(t)$ to the time-dependent macroscopic electric field, atomic displacements, strain, etc. (see also Chapter 20).

The first part of this chapter addresses the problem of the definition of polarization in condensed matter. This is treated in some detail because there has been great confusion for many years and the resolution of the issues was instrumental in a revolution in condensed matter theory that encompasses topological insulators (Chapters 25–28) and other phenomena. There is no issue with the calculation of the induced polarization $\mathbf{P}(t)$ due to an applied field $\mathbf{E}(t')$, which can be formulated in perturbation theory to any order. The issue is the way to determine the static macroscopic polarization $\mathbf{P}$ (the average value of $\mathbf{P}(\mathbf{r})$) in the absence of an applied electric field as an *intrinsic* property of the bulk of an insulating crystal with no dependence on surface termination. The question for electronic structure theory is: *Can one determine the macroscopic polarization $\mathbf{P}$ in terms of the intrinsic bulk ground-state wavefunction?* This is the fundamental problem if we want to find proper theoretical expressions for the polarization in a ferroelectric or pyroelectric material. Expressions for energy, force, magnetization,[1] and stress have been given in previous chapters; electric polarization completes the set of properties needed to specify the macroscopic state of insulators.

Traditional texts on solid state physics published before 1990, such as Ashcroft and Mermin [280] and Kittel [285], electrodynamics [480, 908], other texts including the Feynman lectures [909] and the classic references for ferroelectrics [910] are little help. In a typical text, ionic crystals are represented by point charge models and polarization is considered only in models, such as the Clausius–Mossotti model of a solid as a collection of polarizable units, each with a dipole moment. However, the electron density $n(\mathbf{r})$ is a continuous function of $\mathbf{r}$ and there is no way of finding a unique value of $\mathbf{P}$ as a sum of dipole moments of units derived by "cutting" the density into parts [911]. Attempts to make such identifications have led to much confusion and claims that properties such as

---

[1] Only spin was treated explicitly; orbital magnetization presents difficult issues that require special care.

piezoelectric constants are not true bulk properties (see [912] for a review). The polarization of ferroelectrics or pyroelectrics is even more problematic [910].

The resolution to these issues is an elegant – yet practical – quantum mechanical formulation in terms of Berry phases. This places polarization firmly in the body of electronic structure theory and condensed matter physics. As shown in Section 24.2, the key steps are to relate *changes* in the polarization to integrals over currents flowing through the interior of the body. This is the basis for the new developments (Section 24.3) that express the integrated current as a geometric Berry phase involving integrals over the *phases of the electronic wavefunctions*. This was realized by King-Smith and Vanderbilt [175], who built upon the earlier work of Thouless and coworkers [913–915]. These developments are included in more recent texts (e.g., by Marder [300], Cohen and Louie [916], and Kaxiras and Joannopoulos [917]) and an in-depth exposition is in the book by Vanderbilt on *Berry Phases in Electronic Structure Theory* [918]. See also reviews such as [176, 919].

The formulation of polarization in terms of phases of the wavefunctions has led to a reexamination of density functional theory, since the density is independent of the phases [350]. The issues are very subtle and can only be summarized here. In the absence of a macroscopic electric field, the bulk polarization is, in principle, a *functional* of the bulk density in the spirit of the original Hohenberg–Kohn theorem since the many-body wavefunction is a functional of the density. But if there is a macroscopic electric field, the state of the bulk is *not determined by the bulk density alone* [920] but can be written in terms of a "density polarization theory" ([351, 352] and references cited there). *In addition, the Kohn–Sham equations are meant to describe only the density and it is an approximation to calculate the polarization in terms of the independent-particle wavefunctions.*

The properties of an insulator are fundamentally related to *localization* [921], and it has been shown [901, 922–924] how to express polarization and localization in a unified way in terms of the ground-state wavefunction. A summary of this work is given in Section 24.6, including useful explicit formulas for the localization length in an insulator, which are experimentally measurable [901, 924] and which reduce to the invariant part of the spread of the Wannier functions, Eq. (23.19), in the independent-particle approximation.

The formulation of polarization and localization in terms of Berry phases is closely related to topological insulators; in fact, the "quantum of polarization" in Section 24.3 is the topological invariant discovered by Thouless [914], which leads to electron pumps and is closely related to the quantum Hall effect and topological insulators. Discussion of the relationships are postponed to Chapter 25 in order to separate the different aspects of the theory. The present chapter is devoted to the quantitative, fractional part of the Berry phases that is the basis for the theory of intrinsic bulk polarization. Qualitative, topological properties are described in Chapters 25 and 26.

24.2 Polarization: The Fundamental Difficulty

In a finite system, as shown schematically in Fig. 24.1, there is no problem in defining the average value of the polarization **P**. From Eq. (24.1) one can integrate by parts and use

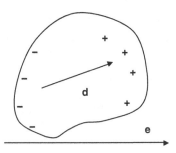

Figure 24.1. Illustration of finite system for which the total dipole moment is well defined. However, the total dipole čannot be used to find the bulk polarization in the large system limit because there is a surface contribution that does not vanish. A bulk theory must be cast solely in terms of bulk quantities.

the requirement that $\mathbf{P}(\mathbf{r}) = 0$ outside the body (see Exercise 24.1) to express $\mathbf{P}$ in terms of the total dipole moment $\mathbf{d}$,

$$\mathbf{P} \equiv \frac{\mathbf{d}}{\Omega} = \frac{1}{\Omega} \int_{\text{allspace}} d\mathbf{r} \, n(\mathbf{r}) \, \mathbf{r}. \tag{24.3}$$

This integral is well defined since the density vanishes outside the finite system and there is no difficulty from the factor $\mathbf{r}$. This expression has the desired properties: in particular, a change in polarization $\Delta\mathbf{P} = \mathbf{P}^{(1)} - \mathbf{P}^{(0)}$ is given strictly in terms of the density difference $\Delta n = n^{(1)} - n^{(0)}$ independent of the path along which the density changes in going from the starting point 0 to the end point 1.

Before tackling the problem of the intrinsic bulk macroscopic polarization in an extended system, we first define the meaning of "intrinsic bulk." A well-defined thermodynamic reference state of a bulk material can be specified by requiring the macroscopic electric field $\mathbf{E}_{\text{mac}}$ to vanish. This is the *only* well-defined reference state [280, 285, 300, 600] since it is only in this case that the bulk is not influenced by *extrinsic* charges at long distance. With this requirement, the electrons in a crystal are in a periodic potential, i.e., the polarization is calculated for a periodic system. Of course, this is not the whole story: there are real physical effects due to long-range electric fields, which is the subject of the dielectric theory of insulators [480, 908]. The complete description requires a full quantum mechanical theory involving both the thermodynamic reference state with $\mathbf{E}_{\text{mac}} \equiv \mathbf{0}$, which is the topic of this chapter, and changes caused by applying macroscopic electric fields, which can be treated by perturbation theory (Appendix E and Chapter 20).

The next step is to realize that in an extended system, any interpretation of polarization based on Eq. (24.3) suffers from a fatal difficulty originating in the factor of the position vector $\mathbf{r}$, which is unbounded. This is illustrated for a periodic array of point charges in Fig. 24.2. Suppose we consider a large finite system of charges that repeat the pattern shown. Depending on the termination of the charges there can be a surface contribution due to factor $\mathbf{r}$ that remains even in the infinite system limit. If we attempt to define polarization as the dipole moment divided by the volume of each unit cell, then the choices shown

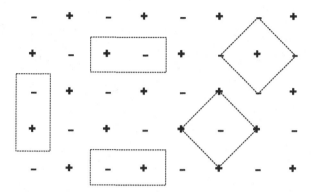

Figure 24.2. Point charge model of an ionic crystal. The dipole is obviously not unique since the cells shown all have different moments. However, a change in moment is the same for all cells so long as charges do not cross boundaries. The cells at the right illustrate cells centered on the two inequivalent sites with inversion symmetry.

in Fig. 24.2 illustrate three choices all with different moments. Another approach is the Clausius–Mossotti-type models where the material is assumed to be a set of localized "molecule-like" densities, each of which has a moment and is polarizable (see [280], chapter 27, and [285], chapter 13). However, this is not justified because the density is continuous and there is not a unique way to divide it into pieces.[2] Finally, all expressions for the polarization **P** involve derivatives, as in Eqs. (24.1) and (24.2). We should not expect that the macroscopic average **P** is uniquely determined. Instead we will find expressions for a change in polarization Δ**P**, and clarify that this is also what is relevant for actual measurable quantities.

A proper definition of the polarization of a macroscopic system requires that the expression in terms of the *ill-defined* **r** operator be replaced by a different form. This can be done using the relation to the current in Eq. (24.2), which is a local relation, unlike expressions such as Eq. (24.3). Averaging over space, the change in the macroscopic average polarization can be found from the time integral of the macroscopic average current $\mathbf{j}_{mac}(t')$

$$\mathbf{P}(t) - \mathbf{P}(t_0) = \int_{t_0}^{t} dt' \frac{d\mathbf{P}(t')}{dt'} = \int_{t_0}^{t} dt' \, \mathbf{j}_{mac}(t'). \tag{24.4}$$

In an insulator, **j** is termed a polarization current and this expression provides a well-defined procedure for calculating the changes in polarization in terms of the current inside the bulk.

To clarify that the integrated current is the desired quantity, we must carefully specify the experimentally measurable quantity that can be identified as a physical polarization. As shown schematically in Fig. 24.3, the basic experimental measurement is a *current that flows through an external circuit under the conditions that the internal macroscopic*

---

[2] It turns out that well-defined expressions for the polarization can be derived in terms the densities of localized, overlapping Wannier functions even though the functions are not individually unique (see Section 24.4).

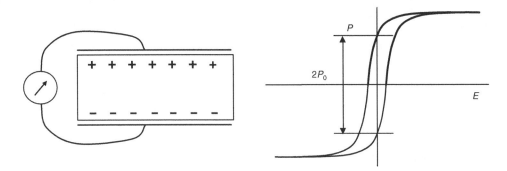

Figure 24.3. Left: schematic illustration for measurement of a change in polarization $\Delta\mathbf{P}$. In order to keep the two surfaces at the same potential (i.e., zero macroscopic field), the integrated current that flows in the external circuit exactly balances any change in surface charge. Since the surface charge is given by (see Eq. (24.1)) $\Delta n_{\text{surface}} = -\int_{\text{surface}} \nabla \cdot \Delta\mathbf{P}(\mathbf{r}) = \Delta\mathbf{P}_{\text{bulk}}$, this is a direct measure of $\Delta\mathbf{P}$. Right: schematic hysteresis loop for a ferroelectric showing that a change in polarization is the quantity actually measured to determine the remnant permanent polarization P_0 for zero macroscopic electric field E.

field vanishes. This makes it clear that *changes in polarization* $\Delta\mathbf{P}$ are the quantities measured directly. Since the current is physically measurable this expression eliminates the difficulties that occur if one tries to extend Eq. (24.3) to an infinite system. As examples of physical measurements, piezoelectric constants are changes $\Delta\mathbf{P}$ caused by a strain ϵ_{ij} measured under conditions where there is no flow of internal free charges that could "short out" the external circuit [925].[3] Ferroelectrics are materials whose state can be switched by an external field, so that the magnitude of the intrinsic polarization $|\mathbf{P}|$ can be found as one-half the measured change $\Delta\mathbf{P}$ between states of opposite remnant polarization.

The static theory can be formulated as the limit of a slow *adiabatic* evolution of the system with "time" replaced by a parameter λ that characterizes the evolution (e.g., λ could represent a particular pattern of displacement of the nuclei),

$$\Delta\mathbf{P} = \int_0^1 d\lambda \, \frac{d\mathbf{P}(\lambda)}{d\lambda} \equiv \int_0^1 d\lambda \, \mathbf{j}_{\text{mac}}(\lambda), \qquad (24.5)$$

where the macroscopic electric field is required to vanish for all λ. Thus a *change* in polarization $\Delta\mathbf{P}$ can be determined strictly from the polarization current that flows *through* the bulk [911, 928–930]. In a crystal the polarization current $\mathbf{j}(\mathbf{r})$ is a periodic function with an average value that is independent of the choice of the cell.

There is still one feature missing. Up to now we have only used definitions and classical theory of dielectrics. The expression, (24.5), for the change in polarization is correct

---

[3] There is a distinction between "proper" and "improper" piezoelectricity, the latter being the change of moment when a material with a permanent moment is rotated. Only the former is a real response of the material [926, 927] as is clear in Fig. 24.3 where obviously nothing happens if the sample and electrodes are rotated. Vanderbilt [927] has shown that the Berry phase expressions below give the proper terms since the $\mathbf{G}$ vectors rotate with the crystal.

but not sufficient. What is lacking is a proof that the value of the polarization is independent of the path in the integral and a useful method for calculation for a solid. Thus as it stands, Eq. (24.5) is not acceptable as a definition of an intrinsic bulk property. Quantum mechanics including the phases of the wavefunctions provides the needed proof and a practical method for calculation: this is the subject of the next section.

24.3 Geometric Berry Phase Theory of Polarization

This is a critical juncture in the organization of this book. The resolution of the problem of polarization required a new approach that played an important role in a revolution in the theory of condensed matter.[4] The extraordinary phenomena are properties of the wavefunction that are described by the Berry phase and related properties. Because the Berry phase is useful in many areas of physics, it is described in general terms in Appendix P and expressions needed for calculations in crystals are derived in Section 25.4. In an abstract form the Berry phase may seem esoteric, but it can be understood starting only from the principle of superposition in quantum mechanics, as described in only a few pages in Appendix P. In this section we will see that the problem of polarization is an excellent example of the practical consequences for actual measurable physical quantities. Furthermore, it is the same theoretical formulation that is the key to understanding topological properties that are the topic of Chapters 25–28. and many other fascinating properties of condensed matter!

The breakthrough that provided a proper quantum mechanical theory of polarization was due to King-Smith and Vanderbilt [175]. The first step is the expression for the integrand in Eq. (24.5). In an independent-particle system[5] described by a single-particle potential, e.g., the Kohn–Sham potential $V_{KS}^\lambda(\mathbf{r})$ as function of the parameter λ, $\partial \mathbf{P}/\partial \lambda \equiv \mathbf{j}_{mac}(\lambda)$ can be expressed using perturbation theory [929],

$$\frac{\partial \mathbf{P}}{\partial \lambda} = -i\frac{e\hbar}{\Omega m_e} \sum_{\mathbf{k}} \sum_{i}^{occ} \sum_{j}^{empty} \frac{\langle \psi_{\mathbf{k}i}^\lambda | \, \hat{\mathbf{p}} \, | \psi_{\mathbf{k}j}^\lambda \rangle \langle \psi_{\mathbf{k}j}^\lambda | \, \partial V_{KS}^\lambda/\partial \lambda \, | \psi_{\mathbf{k}i}^\lambda \rangle}{(\varepsilon_{\mathbf{k}i}^\lambda - \varepsilon_{\mathbf{k}j}^\lambda)^2} + c.c., \qquad (24.6)$$

where the current is given terms of momentum matrix elements that are well defined in the infinite system, and the sum over i, j is assumed to include a sum over the two spin states. This expression can be cast in a form involving only the occupied states following the approach of Thouless and coworkers [256, 913], which is essentially the same as used in Section 20.4 to cast expressions for phonons in terms of the occupied states. In terms of the $\mathbf{k}$-dependent hamiltonian $\hat{H}(\mathbf{k}, \lambda)$, whose eigenfunctions are the strictly periodic part of the Bloch functions $u_{\mathbf{k}i}^\lambda(\mathbf{r})$ as expressed in Eq. (4.37), the needed relations are (Exercise 24.2)

$$\langle \psi_{\mathbf{k}i}^\lambda | \hat{\mathbf{p}} | \psi_{\mathbf{k}j}^\lambda \rangle = \frac{m_e}{\hbar} \langle u_{\mathbf{k}i}^\lambda | [\partial/\partial \mathbf{k}, \hat{H}(\mathbf{k}, \lambda)] | u_{\mathbf{k}j}^\lambda \rangle \qquad (24.7)$$

---

[4] See, for example, the book by Vanderbilt [918] and reviews by Resta [176, 919].
[5] See [931] for Berry phase formulation for many-body systems.

and

$$\langle \psi^\lambda_{\mathbf{k}i} | \partial V^\lambda_{\mathrm{KS}}/\partial\lambda | \psi^\lambda_{\mathbf{k}j} \rangle = \langle u^\lambda_{\mathbf{k}i} | [\partial/\partial\lambda, \hat{H}(\mathbf{k},\lambda)] | u^\lambda_{\mathbf{k}j} \rangle. \tag{24.8}$$

Substituting in Eq. (24.6) and using completeness relations (Exercise 24.3) leads directly to the result [175, 931] for the electronic contribution to $\Delta\mathbf{P}$

$$\Delta\mathbf{P}_\alpha = -|e|\frac{2}{(2\pi)^3}\mathrm{Im}\int_{\mathrm{BZ}}d\mathbf{k}\int_0^1 d\lambda \sum_i^{\mathrm{occ}}\left\langle \frac{\partial u^\lambda_{\mathbf{k}i}}{\partial k_\alpha}\bigg|\frac{\partial u^\lambda_{\mathbf{k}i}}{\partial\lambda}\right\rangle. \tag{24.9}$$

The problem of polarization involves the integrated current along the α direction due to the variation of the crystal potential as a function of λ. Thus it is most useful to express Eq. (24.9) integrals over λ and $k \equiv k_\alpha$ in the direction parallel to the polarization axis, averaged over $\mathbf{k}_\perp$ in the two perpendicular directions, with a factor of 2 for spin,

$$\Delta\mathbf{P}_\alpha = \frac{2}{(2\pi)^2}\int_{\mathrm{2DBZ}}d\mathbf{k}_\perp \Delta\mathbf{P}_\alpha(\mathbf{k}_\perp), \tag{24.10}$$

$$\Delta\mathbf{P}_\alpha(\mathbf{k}_\perp) = -|e|\frac{1}{(2\pi)}\int_0^1 d\lambda\,\mathrm{Im}\int_{\mathrm{1DBZ}}dk\sum_i^{\mathrm{occ}}\left\langle \frac{\partial u^\lambda_{\mathbf{k}i}}{\partial k_\alpha}\bigg|\frac{\partial u^\lambda_{\mathbf{k}i}}{\partial\lambda}\right\rangle. \tag{24.11}$$

By defining a reduced dimensionless vector $(a/2\pi)k$, the right-hand side of Eq. (24.9) is readily shown to be proportional to $[ea/\mathrm{volume}]$ the proper dimensions for polarization. The integral Eq. (24.11) in one space dimension is $|e|$ times a dimensionless factor, which is the quantity that must be determined.

At this point we have derived expressions for $\Delta\mathbf{P}$, but this does not solve the original problem! It is not clear that the results are intrinsic properties of the system. The expression is an integral over λ and there is no proof that the integrals are independent of the way the λ is varied or the way the wavefunctions are defined (a gauge transformation; see Eqs. (23.2) and (P.7)). A satisfactory theory must the express the change in polarization only in terms of the states of the system at the starting $\lambda = 0$ and ending $\lambda = 1$ points.

The proof is provided by the theory of the Berry phase, which is discussed in Section 25.4. The key point is that the Eq. (24.11) is a two-dimensional integral over an area which can be depicted as either the rectangle or the cylinder in Fig. 25.1 with axes k and the parameter λ. The integrand is a sum of Berry curvatures $\Omega_1(k,\lambda)$ for each occupied band i, which are expressed in terms of the Bloch functions as in Eq. (25.8) with $\mathbf{k}$ replaced by (k,λ). The integral is a Berry flux in Eq. (25.9) that is gauge invariant, i.e., it is uniquely defined, which provides the desired proof that the result is an intrinsic property of the system.

However, this is still not what we want. A satisfactory theory for an intrinsic quantity should involve only the states of the system at $\lambda = 0$ and $\lambda = 1$, without knowing anything about the integrand for intermediate λ. The Berry theory provides the desired expression, as worked out in Section 25.5. The integral over the area is equal to the line integral around the boundary, as shown in the left side of Fig. 25.1. In the case of a crystal the integrals over the vertical segments at the two sides on the Brillouin zone cancel and the result is the sum of the segments at the top and bottom. Since the one-dimensional Brillouin zone can also be represented by a circle, the area can be represented as the cylinder as in the

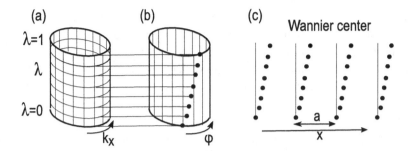

Figure 24.4. Calculation of polarization $\Delta \mathbf{P}$ using the Berry phase formula, Eq. (24.12) or the discrete version in Eq. (24.13). (a) circles formed by one-dimensional k from 0 to 2π for each value of λ. (b) Change in Berry phase ϕ for each λ. (c) Displacement of center of Wannier function proportional to ϕ (Eq. (24.14)). Equation (24.9) requires an integral of λ. The two-point expression (24.12) requires only the Berry phase at $\lambda = 0$ and $\lambda = 1$ but it is modulo 2π. If λ is a periodic function use of Eq. (24.9) can describe a quantized charge pump as depicted in Fig. 25.2.

middle figure in Fig. 25.1. This is the representation used in Fig. 24.4 to illustrate the calculation of the polarization, where the line segments are the circles at the top and bottom. Since the direction of k is opposite on the two segments, the result is the difference between the Berry phases for $\lambda = 1$ and $\lambda = 0$, with the recognition that there can be added factors of 2π.

The result is the "two-point" formula [175, 176] that the change in polarization $\Delta \mathbf{P}_\alpha (\mathbf{k}_\perp)$ is the charge $|e|$ times the difference in Berry phases at $\lambda = 1$ and $\lambda = 0$,[6]

$$\Delta \mathbf{P}_\alpha (\mathbf{k}_\perp) = -i \frac{|e|}{(2\pi)} \int_{\text{1DBZ}} dk \sum_i^{\text{occ}} \left[\langle u_{\mathbf{k}i}^{\lambda=1} | \partial_{k_\alpha} u_{\mathbf{k}i}^{\lambda=1} \rangle - \langle u_{\mathbf{k}i}^{\lambda=0} | \partial_{k_\alpha} u_{\mathbf{k}i}^{\lambda=0} \rangle \right]$$
$$+ (\text{integer}) \times |e|, \tag{24.12}$$

where the last term in Eq. (24.12) represents multiples of the "quantum of polarization" that originate in the fact that the line integral determines the Berry phase modulo 2π. Interestingly, the factors of 2π can be interpreted as transport of an integer number of electrons across the entire crystal, leaving the bulk invariant. In a finite crystal this is closely related to the surface states of topological insulators and the quantum Hall effect as brought out in Chapter 25. This is the part of the transport that was emphasized by Thouless et al., for the quantum Hall effect [913] and quantized charge transport in an insulator [914, 915] in the case where the hamiltonian is changed along a closed path returning to the same point, i.e., $\hat{H}(\mathbf{k}, \lambda = 1) = \hat{H}(\mathbf{k}, \lambda = 0)$. In contrast, changes in the Berry phase by fractions of 2π correspond to polarization, which is an intrinsic property of the bulk crystal.

Note that the geometric phase is not equal to zero or π only if the periodic functions $u_{\mathbf{k}i}^\lambda$ are complex; this occurs if there is no center of inversion, which is of course exactly the condition under which there may be a nonzero polarization. Hence, *the change in macroscopic polarization between two different insulating states can be regarded as*

---

[6] Spin is included in the sum over i. Some authors include a factor of 2 for spin, assuming no spin dependence.

a measure of the phase difference between the initial and final wavefunctions. In all mean-field approaches, this means Slater determinants of single-body functions $u_{\mathbf{k}i}^\lambda$, but the formula generalizes directly to correlated many-body wavefunctions [931].

For actual calculations in crystals, it is convenient to express the polarization terms of the Bloch functions calculated on a grid in the Brillouin zone, rather than the derivatives required in Eq. (24.12). The grid can be constructed with lines of J points in the α direction along which the derivative is to be calculated. The Bloch functions ψ_i at the two sides on the BZ ($\mathbf{k}_0$ and $\mathbf{k}_J$) are required to be the same, i.e., the periodic gauge. The practical expression is the discrete from in Eq. (25.4). (See also Eq. (23.23).) For a system with multiple bands it takes the form

$$
-i \int_{\mathrm{BZ}} d\mathbf{k}_\alpha \sum_i^{\mathrm{occ}} \left[\langle u_{\mathbf{k}i}^\lambda | \partial_{k_\alpha} u_{\mathbf{k}i}^\lambda \rangle \right] \to \mathrm{Im} \left[\ln \Pi_{j=0}^{J-1} \det(\langle u_{\mathbf{k}_j i}^\lambda | u_{\mathbf{k}_{j+1} i'}^\lambda \rangle) \right], \tag{24.13}
$$

where the determinant is that of the $N \times N$ matrix formed by allowing i and i' to range over all occupied states. Note that the overall phase of each $u_{\mathbf{k}_j i}^\lambda$ cancels in the sum since each function appears in a bra and a ket in the product. This is the expression that is now widely used for calculations of polarization, as exemplified below.

24.4 Relation to Centers of Wannier Functions

The "two-point" expression for a change in polarization, Eq. (24.12), can be expressed in terms of the centers of Wannier functions defined in Chapter 23. Using the relation Eq. (23.10), the result is

$$
\Delta P_\alpha = \frac{-|e|}{\Omega} \sum_i^{\mathrm{occ}} \left[\langle 0i | \hat{\mathbf{r}} | 0i \rangle^{\lambda=1} - \langle 0i | \hat{\mathbf{r}} | 0i \rangle^{\lambda=0} \right]. \tag{24.14}
$$

This has the simple interpretation that the change in polarization is the same as if the electrons were localized at points corresponding to the centers of the Wannier functions. In general, the center of each function is *not* unique, but the sum of the moments of all the functions is unique as shown by Blount [890] and follows from the gauge invariance of the Berry phase.

Thus the derivation in terms of Wannier functions provides a rigorous basis for the simplified models of charges in crystals. For example, the model in terms of point charges can be identified with the centers of the Wannier functions; each charge is not unique, but any change in polarization is well defined, modulo the "quantum of polarization." Furthermore, the "quantum" has the simple interpretation that the electrons can be shifted by a translation between equivalent Wannier functions, which of course is a symmetry operation of the infinite crystal. Similarly, the Clausius–Mossotti model becomes a rigorous theory of polarization in insulators (at least in the independent-particle approximation) if the polarizable units are taken to be overlapping Wannier functions. Even though the way the system is divided into units is not unique, the total polarization is well defined.

The relation of the Berry phase and Wannier centers also provides one of the most useful and informative ways to calculate topological invariants as explained in Section 25.5 and used in Chapters 26–28.

24.5 Calculation of Polarization in Crystals

As examples of the calculation of polarization it is appropriate to consider effective charges, given by Eq. (E.20), because the results can be compared with experiments that can measure the charges accurately in terms of the splittings of longitudinal and transverse modes. For example, several groups [821–823] have derived the anomalous effective charges in perovskites, such as $BaTiO_3$, which have ferroelectric transitions. In such materials there are several infrared (IR) active modes and it is not possible to determine directly from the experiment the individual effective charges of the atoms because all the IR modes interact with one another and are mixed. The strengths of the modes can be decomposed into contributions from different atoms only by using information on the lattice dynamics from a theoretical model. In contrast, the *ab initio* calculations determine the atomic effective charges and the vibrational modes; thus the effective charges for the eigenmodes can be directly predicted and compared with experiment. The prediction is that there is a great mixing of the IR modes, so that the lowest transverce optic (TO) mode is most closely associated with the highest longitudinal optic (LO) mode, giving a very large effective charge for that mode, i.e., the phonon that softens at the ferroelectric phase transition. The anomalously large effective charges of the B atoms and the O atoms moving along the line toward the B atoms are interpreted as resulting from covalency. Selected results are shown in Table 24.1, taken from [823]; essentially the same results have been found using linear response methods [821, 822]; however, the Berry phase approach has the advantage that it applies directly to the finite polarizations that develop in the ferroelectric states.

Table 24.1. Born effective charges for the atoms and mode effective charges for the IR active modes in ABO_3 perovskites. The two charges $Z_1^*(O)$ and $Z_2^*(O$ for O atoms are for displacements in the plane formed by the O atom and its 4B neighbors, and in the perpendicular direction. Taken from Zhong et al. [823].

Type	$BaTiO_3$	$PbTiO_3$	$NaNbO_3$
Z*(A)	2.75	3.90	1.13
Z*(B)	7.16	7.06	9.11
$Z_1^*(O)$	−5.69	−5.83	−7.01
$Z_2^*(O)$	−2.11	−2.56	−1.61
|Z*(TO1)|	8.95	7.58	6.95
|Z*(TO2)|	1.69	4.23	2.32
|Z*(TO3)|	1.37	3.21	5.21

Examples of calculations of linear (Gonze et al. [932]) and nonlinear susceptibilities (Dal Corso and Mauri [933]) have been done using transformations between the Wannier and Bloch orbitals and the "$2n + 1$ theorem" (Section D.6).

Spontaneous Polarization

Spontaneous polarization occurs in any insulating crystal that lacks a center of inversion. In ferroelectrics, the value P_0 can be measured as indicated in Fig. 24.3 because the direction of the polarization can be reversed. Values of the ferroelectric remnant polarization have been calculated for a limited number of ferroelectric materials, with values in general agreement with measured ones [176]. For example, the calculated residual polarization [821] of $KNbO_3$ is $\Delta \mathbf{P} = 0.35$ C/m^2 compared with the measured value [934] of 0.37 C/m^2, although it is very difficult to find the intrinsic moment experimentally [910].

In other crystals, such as wurtzite structure, there is a net asymmetry with the positive direction of the c-axis inequivalent from the negative direction. The two directions are detected experimentally by the fact that their surfaces are inequivalent. It is not so easy to reverse this axis since it requires breaking and remaking of all the bonds in the crystal. Such crystals are pyroelectrics [807] because the change in the polarization with temperature can be measured. How can the absolute value of the polarization be determined? The absolute value can be expressed in terms of quantities that can be defined theoretically: the difference $\Delta \mathbf{P}$ given the Berry phase formula between the acual crystal and a crystal in which $\mathbf{P} = 0$ by symmetry. This can be calculated by constructing a path between the two crystals by "theoretical alchemy," in which the charge on the nuclei is varied, or by large displacements of atoms to change the crystal structure, so long as no gap goes to zero. Both of these possibilities are straightforward in calculations. Although there is no corresponding direct experiment, there are experimental consequences that occur at an interface between regions with different polarization, so that $\nabla \cdot \mathbf{P}$ leads to a net charge.

24.6 Localization: A Rigorous Measure

An insulator is distinguished from a conductor at zero temperature by its vanishing dc conductivity and its ability to sustain a macroscopic polarization, with and without an applied electric field [480, 908]. The theory of polarization, presented thus far, has shown the fundamental relation of the latter property to the ground-state wavefunction for the electrons. Regarding the former property, the classic paper "Theory of the insulating state" by W. Kohn [921] has clarified that the many-body system of electrons in an insulator is "localized" in contrast to the delocalized state in a metal. However, a rigorous quantitative measure of the degree of localization was worked out only later. Such a measure is provided by the theory of polarization: not only the average value, but also the fluctuations of the polarization [901, 922, 935].

The relation between polarization and localization was established by Kudinov [936], who proposed to measure the degree of localization in terms of the *mean square quantum fluctuation of the ground-state polarization.* Kudinov considered the quantum fluctuation of the net dipole moment, $\langle \Delta \hat{\mathbf{d}}^2 \rangle = \langle \hat{d}^2 \rangle - \langle \hat{\mathbf{d}} \rangle^2$ in a large, but finite, volume Ω. Using the zero-temperature limit of the fluctuation–dissipation theorem [278, 937–940], the mean square fluctuation is related to the linear response function by

$$\frac{\langle \Delta \hat{d}_\alpha^2 \rangle}{V} = \frac{\hbar}{\pi} \int_0^\infty d\omega \frac{1}{\omega} \mathrm{Re}\sigma_{\alpha\alpha}(\omega) = \frac{\hbar}{\pi} \frac{1}{4\pi} \int_0^\infty d\omega \mathrm{Im}\epsilon_{\alpha\alpha}(\omega). \tag{24.15}$$

For a metal with $\sigma(\omega) \neq 0$ for $\omega \to 0$, the right-hand side diverges, i.e., the mean square fluctuation of the dipole moment diverges in the large Ω limit. However, for an insulator $\sigma(\omega = 0)$ is finite and Kudinov proposed that the integral has a well-defined limit for large volume. Since the dipole is a charge times a displacement, $\hat{\mathbf{d}} = -e\hat{\mathbf{X}}$, where $\hat{\mathbf{X}} = \sum_i^N \hat{\mathbf{x}}_i$ is the center of the mass position operator of N electrons in the volume, Relation Eq. (24.15) can be used to define a mean square displacement of the electrons, and thus a localization length ξ. Souza et al. [901] have shown that the arguments carry over to the infinite system with proper interpretation of the position operator $\mathbf{X}$ consistent with the polarization operator,[7] with the result for the length in the α direction

$$\xi_\alpha^2 = \lim_{N\to\infty} \frac{1}{N} \left[\langle \hat{X}_\alpha^2 \rangle - \langle \hat{X}_\alpha \rangle^2 \right]$$

$$= \lim_{N\to\infty} \frac{\Omega^2}{e^2 N} \left[\langle \hat{P}_\alpha^2 \rangle - \langle \hat{P}_\alpha \rangle^2 \right]. \tag{24.16}$$

In terms of measurable conductivity, ξ_α^2 is given by

$$\xi_\alpha^2 = \frac{\hbar}{\pi e^2 n} \int_0^\infty d\omega \frac{1}{\omega} \mathrm{Re}\sigma_{\alpha\alpha}(\omega). \tag{24.17}$$

Bounds can be placed on the length [901, 924]. Using the inequality

$$\int_0^\infty d\omega \frac{1}{\omega} \mathrm{Re}\sigma_{\alpha\alpha}(\omega) \leq \frac{1}{E_{\mathrm{gap}}^{\mathrm{min}}} \int_0^\infty d\omega \mathrm{Re}\sigma_{\alpha\alpha}(\omega), \tag{24.18}$$

where $E_{\mathrm{gap}}^{\mathrm{min}}$ is the minimum direct gap and the sum rule, Eq. (E.13), leads to an upper bound

$$\xi_\alpha^2 \leq \frac{\hbar^2}{2m_e E_{\mathrm{gap}}^{\mathrm{min}}}. \tag{24.19}$$

On the other hand, arguments based on standard perturbation formulas for the static susceptibility $\chi = (\epsilon - 1)/4\pi$ lead to a lower bound [924]

$$\xi_\alpha^2 \geq \frac{E_{\mathrm{gap}}^{\mathrm{min}}}{2n} \chi, \tag{24.20}$$

which also illustrates that ξ diverges in a metal where χ necessarily diverges. The inequalities can be derived, as described in more detail in Exercise 24.5, in terms of integrals

---

[7] That is, the expectation value $\langle \mathbf{X} \rangle$ is equivalent [901] to the Berry phase expressions given in Section 24.3.

of the real part of the conductivity $\sigma'(\omega)$ and the average gap defined by Penn [941] in terms of the electron density and the polarizability.

The localization length also relates to important theoretical quantities, establishing rigorous relations with experimental measurables and providing tests for approximate theories. The demonstrations have been done in several ways [901, 942] (see a review in [935]). For particles that are independent except that they are indistinguishable, there is a general relation, Eq. (3.55), between the correlation function and the density matrix, repeated here for fermions

$$\Delta n_{\mathrm{ip}}(\mathbf{x}; \mathbf{x}') = -|\hat{\rho}_\sigma(\mathbf{x}, \mathbf{x}')|^2. \tag{24.21}$$

As shown by Sgiarovello et al. [942], transformations of the density matrix lead to the relation [901]

$$\sum_{\alpha=1}^3 \xi_\alpha^2 = \frac{\Omega_I}{m_e}, \tag{24.22}$$

in terms of the invariant part of the spread of the Wannier functions, Eq. (23.16), thus giving a physical meaning to the invariant spread.

The localization length can be determined directly from the ground-state wavefunction using the expressions in Section 23.3 for Eq. (23.16) or alternative forms given in [942]. Quantitative calculations for many semiconductors have been carried out [942] where they satisfy the bounds in Eq. (24.19).

24.7 The Thouless Quantized Particle Pump

If the change of the hamiltonian described by the variation of the parameter λ is so large that the Berry phase approaches 2π then the polarization corresponds to shifting all the electrons by a lattice constant. A pump is a system in which λ is a periodic function so that each cycle pumps the electrons by one lattice constant. This small observation is part of a great story that has revolutionized the theory of condensed matter. It is described in the larger context in Chapter 25 and worked out explicitly in Chapter 26, but we can appreciate a number of aspects at this point. In the pump, the hamiltonian varies in a cyclic fashion; for a bulk crystal there is no difference between the initial and final states of the crystal for a cycle and, since the Berry phase is defined only modulo 2π, one cannot distinguish any multiple of 2π. However, the fact that charge has been displaced through the crystal can be detected by keeping track of the Berry phase as λ varies; so long as λ varies continuously, one can uniquely specify how much charge has been transferred to the right or left in each cycle. For a finite crystal the consequence is that there is a net transfer of electrons through the insulating crystal from one side to the other. This is what has been shown by Thouless in his 1983 paper [256] to signify "'quantization of particle transport" in a one-dimensional crystal using the same mathematical formulation as done the year before in the famous TKNN paper [89] that derived the quantization of Hall conductance in a two-dimensional crystal. The quantized value is a Chern number, which is a topological invariant as discussed in Section P.4 and Chapters 25 and 27. The derivations are described in more detail in Chapter 25.

24.8 Polarization Lattice

The fact that polarization is defined as a change of the electron system as the hamiltonian is varied leads us to a new concept: a "polarization lattice." Even if a bulk periodic crystal is the same before and after the hamiltonian is varied continuously to displace the electrons by a multiple of the lattice constant, nevertheless the polarization differs by a multiple of the lattice constant. This defines a lattice of values of the polarization.[8]

How can this be reconciled with the idea that polarization is an intrinsic property of a crystal? The first part of this chapter went to great lengths to show that the quantity calculated by the Berry phase was experimentally measurable and well defined! The resolution is hidden in the "quantum of polarization" that was glossed over as just a matter of 2πs. The assumption is well justified for real materials with nonzero polarization like ferroelectrics and pyroelectrics where the magnitude of the polarization corresponds to displacement of the electrons by much less than a lattice constant so that the change can always be defined in terms of the change from any one of periodic positions of the polarization lattice, whichever one corresponds to the intial state. Even though this may resolve one problem it leaves two others.

The formulation of polarization as a quantity defined on a lattice implies that there must be such a lattice even for a crystal with center of inversion: the polarization has the same value – zero – at each point on the lattice. This might seem totally trivial but there are two nontrivial consequences. One is due to the fact that if a crystal has a center of inversion, then it also has another center of inversion that is not equivalent. Examples are the one-dimensional crystal in Fig. 4.11 where there is a center of inversion at the atom sites and between the atoms, and in an ionic crystal like Fig. 24.2 where there is a center of inversion at each of the two ions in the unit cell. The change in polarization is well defined only if it is stipulated that the change is relative to one or the other center. As far as a measurement of an intrinsic polarization is concerned it makes no difference so long the change is small enough that it is always around one center. However, a calculation must determine the value relative to a center and it may not be obvious which site is the appropriate choice.

The other problem is that there may be a transition between states that are centered on different sites. This is a remarkable story that is at the heart of topological properties of the electronic structure. It appears in Section 22.3 as the Shockley transition between two insulating states in the bulk, which leads to surface states in certain cases so long as symmetry is maintained. The transition is expressed in terms of the Berry phase. An example is shown in Fig. 26.7, where the two-site model has four points that can be centers of inversion. The relation to the Berry phase is shown at the bottom of the figure, with $\phi = 0, \pi/2, \pi, 3\pi/2$ for the four sites in the polarization lattice. Winding numbers

---

[8] A detailed discussion can be found in [918].

in Section 26.4 are changes of ϕ by 2π; a continuous variation defines the topological quantized particle transport and it is a step along the path to topological insulators as explained in Chapters 25–28.

SELECT FURTHER READING

Recent book with a very readable, in-depth exposition:

Vanderbilt, D. H., *Berry Phases in Electronic Structure Theory* (Cambridge University Press, Cambridge, 2018).

Reviews with the basic theory of insulators and polarization:

Resta, R., "The insulating state of matter: A geometric approach" *Eur. Phys. J. B* 79:121–137, 2011.

Resta, R., "Macroscopic polarization in crystalline dielectrics: The geometric phase approach," *Rev. Mod. Phys.* 66:899–915, 1994.

Exercises

24.1 Verify that the well-known expression for a dipole moment, Eq. (24.3), follows from the definition of the polarization field, Eq. (24.1), with the boundary condition given.

24.2 Show that Expressions (24.7) and (24.8) for the expectation values in terms of the commutators follow from the definition of $\hat{H}(\mathbf{k}, \lambda)$.

24.3 Show that Eq. (24.9) follows from the previous equations as stated in the text. Hint: use completeness relations to eliminate excited states.

24.4 Show that the dipole moment averaged over all possible cells vanishes in any crystal.

24.5 Define a "localization gap" E_L by turning Eq. (24.19) into an equality: $\xi^2 \equiv \hbar^2/(2m_e E_L)$. (a) Using the f sum rule and Eq. (24.17), show that E_L can be expressed as the first inverse moment of the optical conductivity distribution:

$$E_L^{-1} = \frac{1}{\hbar} \frac{\int \omega^{-1}\sigma'(\omega)d\omega}{\int \omega^0 \sigma'(\omega)d\omega}.$$

(b) Use the f sum rule and the Kramers–Krönig expression for $\epsilon(0)$ to show that the Penn gap [941] E_{Penn} defined via the relation $\epsilon(0) = 1 + (\hbar\omega_p/E_{Penn})$, where ω_p is the plasma frequency, can be expressed as the second inverse moment:

$$E_{Penn}^{-2} = \frac{1}{\hbar^2} \frac{\int \omega'^{-2}\sigma'(\omega)d\omega}{\int \omega^0 \sigma'(\omega)d\omega}.$$

(c) Using the results of (a) and (b), show that inequalities Eqs. (24.19) and (24.20) can be recast in a compact form as follows:

$$E_{Penn}^2 \geq E_L E_{gap}^{min} \geq (E_{gap}^{min})^2.$$

24.6 Find a reasonable estimate and upper and lower bounds for $\sum_{\alpha=1}^{3} \xi_\alpha^2$ using gaps and dielectric constants of typical semiconductors and the expressions given in Exercise 24.5. The lowest direct gaps can be taken from [942] or other sources and values of the dielectric functions can be found in texts such as [280, 285, 300], e.g., $\epsilon \approx 12$ in Si.

24.7 It is also instructive to calculate values of the average "Penn gap" [941], which is defined in Exercise 24.5. The Penn gap is an estimate of the average gap in the optical spectrum and is directly related to the inequalities in Exercises 24.5 and 24.6. As an example, find the gap in Si with $\epsilon \approx 12$ and compare with the minimum direct gap. Find values for other semiconductors as well, using standard references or [941].

24.8 Construct a small computer code to calculate the electronic contribution to the polarization from the Berry phase expressions given in Section 24.3 for the one-dimensional ionic dimer model of Fig. 26.2 in a general case, $\varepsilon_A \neq \varepsilon_B$ and $t_1 \neq t_2$. Compare with the calculations of the centers of the Wannier function found in Exercise 23.12.

24.9 The effective charge, Eq. (E.20), is defined by the change in the polarization induced by displacement on an atom. An important part of the charge is the "dynamical" electronic contribution that results from changes in the electronic wavefunctions in addition to rigid displacements. A simple model for this is given by the one-dimensional ionic dimer model of Fig. 26.2. Consider $\varepsilon_A \neq \varepsilon_B$ and let $t_1 = t + \delta t$ and $t_2 = t - \delta t$. A change in δt causes a change in polarization in addition to any change due to rigid displacement of ionic charges. For a given $\Delta \varepsilon \equiv \varepsilon_A - \varepsilon_B$ calculate $\delta \mathbf{P}/\delta t$ for small δt using computer codes for the Berry phase (Exercise 24.8) or the centers of the Wannier functions (Exercise 23.12). Show that the contribution to the effective charge can be large and have either sign (depending on the variation of t with displacement), which can explain large "anomalous effective charges" as described in Section 24.5.

24.10 Consider the one-dimensional continuum model of Exercise 12.5 for which Wannier functions are found in Exercise 23.13. The polarization is zero since the crystal has a center of inversion and the eigenfunctions can be chosen to be real. If the entire crystal is shifted rigidly a distance Δx, so that $V(x) \to V_0 \cos(2\pi(x - \Delta x)/a)$, the origin is not at the center of inversion and the eigenfunctions are not real. Using the Berry phase expressions in Section 24.3, show that the change in the electronic contribution to the polarization is $\Delta P = -2|e|\Delta x/a$. The interpretation of this simple result is that the electrons simply move rigidly with the potential. Give the reasons that this shift does not actually lead to a net polarization since in a real crystal this electronic term is exactly canceled by the contribution of the positive nuclei, which shift rigidly.

PART VI

ELECTRONIC STRUCTURE AND TOPOLOGY

25

Topology of the Electronic Structure of a Crystal: Introduction

Imagination is more important than knowledge.

Albert Einstein

Summary

The topic of this chapter is an introduction to the topology of the electronic structure as a function of **k** in the Brillouin zone. Each step in the derivations can be understood in terms of (1) the Bloch functions using only basics of band theory and (2) the Berry phase, which can be understood using only the superposition principle of quantum mechanics. In this chapter are all the equations needed to use the Berry phase in the relevant problems in crystals, but the reader should also read Appendix P, where the Berry phase can be understood in a few pages. The result is an intuitive picture that provides a practical way to calculate properties such as polarization in Chapter 24 in terms of the Berry phase and topology in terms of winding of the Berry phase in two dimensions, where the integer winding number is called a Chern number. This leads up to the following chapters with the topological Thouless pump in Chapter 26, Chern insulator and topological insulator in two dimensions in Chapter 27, and topological insulators in three dimensions in Chapter 27, which can be understood in terms of the two-dimensional classifications.

25.1 Introduction

It is remarkable that, after so many years since the advent of quantum mechanics, there are still new discoveries of qualitatively new phenomena, even for systems where we have exact solutions and in principle everything is known. Topological insulators is such a case.[1]

---

[1] See, for example, the reviews by Hasan and Kane [943] and by Qi and Zhang [944]. Pedagogical presentation of Berry phases and topological properties in electronic structure can be found in the book by Vanderbilt [918] and the basic theory in the lecture notes by Asboth et al. [945]. Many aspects are treated more thoroughly in the book by Bernevig [946], and an extensive review of quantitative calculations for materials is given by Bansil, Lin, and Das [947].

Even though the equations for electrons in the independent-particle approximation can be found in textbooks and they can be solved in great detail using computers, qualitatively new effects were discovered only a few years ago. The extraordinary properties of topological insulators are caused by the spin–orbit interaction together with the behavior of the Bloch states as a function of momentum **k** in the Brillouin zone, which can be classified by their topology.[2] This is a property of the bulk crystal, but it leads to observable consequences for the surface: surface bands in which particles move in only one direction (opposite directions for the two spin states) displaying phenomena analogous to the quantum Hall effect but without the need for a magnetic field.

Just as amazing is that the underlying theory can be readily understood by anyone with only a familiarity with basic quantum mechanics and the Bloch theorem at the level of an undergraduate or beginning graduate-level textbook – if only pointed in the right direction. Topological insulators can be understood in terms of 2×2 matrices and the essential features can be illustrated by simple models for a crystal in which there are two bands that can be related to one another in two distinct ways. For a finite crystal the consequence is that the nature of the bulk bands determines whether or not there *must* be surface bands in the bulk band gap that can conduct with no resistance. The problems connect directly to other important areas of the theory of solids, including the atomic-covalent Shockley transition illustrated in the first figure of this book in Chapter 1, surface physics in Chapter 22, Wannier functions in Chapter 23, polarization in ferroelectric materials in Chapter 24, and other important areas.

The mathematical formulation that provides an elegant description of all these phenomena is based on the Berry phase and Chern number. The essential aspects for electronic states in crystals are recounted in Section 25.4; however, a reader interested in understanding the basic concepts and profound physics should turn to Appendix P, which is a short introduction to Berry phases and the way the theory also leads to Chern numbers that are the topological invariants. Even if these are unfamiliar, they can be understood in only a few pages starting only from the superposition principle of quantum mechanics. Two examples in Appendix P bring out the physics – the Bohm–Aharonov effect in Section P.6, and the Dirac magnetic monopole in Section P.7. Apparently the monopole does not exists in nature, but an analogue is found in topological insulators!

There is a precedent, the quantum Hall effect (see Appendix Q), which occurs only in the presence of a strong magnetic field. The famous TKNN paper [89] was a hallmark in condensed matter physics because it showed that the precise quantization is a topological property of the electronic states in a strong magnetic field. The role of topology in quantum

---

[2] There are many types of systems with nontrivial topology. Here the term topological insulator denotes a system with time-reversal symmetry where spin plays an essential role. There are also crystalline topological insulators that are protected by a spatial symmetry [948].

mechanics is one of the great stories of physics,[3] and the discovery of topological insulators has brought topology squarely into electronic structure, the topic of this book.

25.2 Topology of What?

Topology refers to the classification of shapes. A famous example is the difference between a sphere and a torus. A torus has a hole and no matter how much the torus is twisted and distorted it still has one and only one hole unless there is a "violent event" when two surfaces touch to fill in the hole or open a new hole. All such shapes with one hole are considered to have the same topology. ("A topologist is someone who cannot tell the difference between a doughnut and a coffee cup.") Similarly the sphere can be distorted to make any shape that does not have a hole and all such shapes with no holes have the same topology. The points at which the topology changes form one type to another occurs where surfaces touch, e.g., where two parts of the surface of a sphere are brought together to initiate a hole and form a torus. Even though the field of topology is far beyond the scope of this book, the role of topology in the problems considered here can be understood starting from the basic principles of quantum mechanics as bought out in Appendix P.

The relevant topology for our purposes is *not* the shape in real space. The topological transitions considered here do not involve any change of symmetry of the crystal and the structure does not change at the transition. It is *not* the topology of the Brillouin zone, which does not change at the transition. Instead, the topology is defined by the way the eigenvectors change as a function of $\mathbf{k}$ in the Brillouin zone. The theory is cast in terms of the periodic parts of the Bloch functions $u_{\mathbf{k}}$ or equivalently in terms of the $\mathbf{k}$-dependent hamiltonian $H(\mathbf{k})$.[4] Of course, the fundamental hamiltonian in terms of the kinetic energy operator and the Coulomb interaction is the same for all materials. What changes is the way the eigenstates are divided into occupied and unoccupied states, which can be formulated in

---

[3] The Nobel Prize in Physics 2016 was to David J. Thouless, F. Duncan, M. Haldane, and J. Michael Kosterlitz "for theoretical discoveries of topological phase transitions and topological phases of matter" (see the Nobel prize lecture [949]). Their work in the 1980s was the forerunner of the theoretical discovery and present-day understanding of topological insulators. An insightful book by Thouless is entitled *Topological Quantum Numbers in Nonrelativistic Physics* [950]. There are connections to other fields of physics. For example, the book *The Phases of Quantum Chromodynamics: From Confinement to Extreme Environments* by Kogut and Stephanov [951] points out that "Domain-wall methods to describe light fermions were known in statistical physics for over 60 years [referring to the 1939 paper by Shockley] but were rediscovered in lattice gauge theory in the 1990's," and the relevance to quantum field theory and topological insulators is pointed out in a paper by Wilczek [952] entitled "Particle physics and condensed matter: the saga continues."

[4] Essentially all work in this field is formulated in terms of independent-particle theories. From the outset this book and its companion [1] have attempted to always be clear on what aspects of such theories are justified. The arguments are based on the continuity principle, stated in Chapter 1, that certain properties of a system of independent particles remain the same if interactions are increased continuously. A quintessential example is topology, which does not change continuously. The topology must remain the same so long as the ground state of the interacting system evolves continuously and the system remains an insulator.

terms of the matrix elements of $H(\mathbf{k})$. The problems considered here are the topology of the occupied states of $H(\mathbf{k})$ as a function of $\mathbf{k}$ in two- or three-space dimensions, or, in the case of one space dimension, the topology of the occupied states of $H(k, \lambda)$ as a function of k and a parameter λ that can be varied continuously so that (k, λ) can be viewed as coordinates in a two-dimensional space. The variation of the eigenvalues and eigenfunctions depends on the details, but no matter how complicated the problem, a system can be classified by a topological invariant, and the only way the topology can change is if a bandgap goes to zero.

Why is a change in topology associated with a gap going to zero? At the points in the BZ where the gap between filled and empty states vanishes, there are eigenstates of $H(\mathbf{k})$ that become degenerate so that the filled and empty states can exchange eigenfunctions, thus changing the way the set of filled eigenfunctions connect in the BZ. (The same reasoning applies in the space defined by k and λ.) In an insulator there is a gap between the filled and the empty bands and all systems that can be transformed into one another by continuous variation of the hamiltonian, without a gap going to zero, are considered to have the same topology. There can be transitions between insulating states with different topologies only if a gap vanishes.

25.3 Bulk-Boundary Correspondence

The most directly observable consequence is surface states in the gap, which are also the phenomena of most practical importance that is driving much of the interest in topological insulators. Topology is a global property of the electronic structure of the bulk, but it is not easy to find a measurement that determines the topology directly. Since topology of bulk states is the fundamental property but surface states are the measurable properties, it is important to establish the relationship that is called "bulk-boundary correspondence." In addition, it is always useful to be able to describe physical effects in more than one way and it may be most illuminating to use the nature of the surface states in some cases and the topology of the bulk bands in others.

The role of bulk-boundary correspondence has been recognized since the 1970s, e.g., in the work of Jackiw and Rebbi [953], who gave a simple, compelling argument for an interface where the hamiltonian slowly interpolates between the two insulating states with different topologies. At some point the energy gap has to vanish because otherwise it is impossible for the topological invariant to change. Thus, there must be low-energy electronic states bound to the interface region, and these states form bands that propagate along the interface. The relation of topology and gapless states appears in many contexts in physics, such as the gradual interpolation between regions described by Dirac hamiltonians with positive and negative masses,[5] which can be solved analytically [953] as described in the first part of the review [943]. The most prominent example is the edge state in the

---

[5] In this context, "mass" corresponds to "gap" in condensed matter (not curvature of bands) and the model corresponds to positive and negative (or inverted) gaps.

quantum Hall effect (Appendix Q); other cases are solitons in one dimension at a boundary in the Su, Schrieffer, and Heeger model for polyacetalene [954], and Majorana modes at the surface of superconductors (Section 25.10), all of which are closely related to the Shockley transition in the bulk (see Section 26.2) and surface states (Section 22.3).

Despite the importance of the surface states, it is important to keep in mind that the topology is a property of the states as a function of **k** in an infinite periodic crystal with no surfaces. It is the topology arguments that are the most basic because they guarantee that the results are robust, i.e., the qualitative conclusions will not change due to changes in the hamiltonian, so long as the topology does not change and the system remains an insulator with a gap in the bulk.

25.4 Berry Phase and Topology for Bloch States in the Brillouin Zone

The Berry phase and related quantities are defined in Appendix P for the wavefunction in a general problem. In this section the derivations are summarized in a form appropriate for application to crystals; however, the reader is encouraged to also turn to Appendix P, where the general theory can be understood within only a few pages. The entire formulation follows from one observation: if a wavefunction is varied as a function of some parameter along a path until it returns to the same point, the parameter has the same value as at the start, but the wavefunction may have acquired a phase. The relative phase is physically meaningful and is defined to be the Berry phase ϕ, which is restricted to the range 0 to 2π, plus integral multiples of 2π.

From this information alone, we can anticipate that it may be possible to relate a quantitative property to a Berry phase, since it can take a continuous range of values, whereas the integers that specify the multiples of 2π (called Chern numbers denoted by C) form a discrete set that may describe a qualitative property. Indeed, we will see that the same theory can describe a phenomenon such as polarization in terms of a Berry phase and topological classification of the electronic structure in terms of Chern numbers.

It is valuable to derive the expressions in both a discrete form, which is the most useful for actual calculations and a continuous form that brings out the mathematical relations. For a discrete set of points with wavefunction u_j, the relative phase between neighboring points can be expressed as $\Delta\phi_{j,j+1} = \text{Im}\ln\langle u_j | u_{j+1}\rangle$ using the definition of the complex logarithm as the phase of the argument in the range 0 to 2π (see Eq. (P.1)). The total phase around a loop (a closed path) can be written as in Eq. (P.2) which is repeated here,

$$\phi = -\text{Im}\ln[\langle u_0|u_1\rangle\langle u_1|u_2\rangle\langle u_2|\ldots|u_{N-2}\rangle\langle\langle u_{N-2}|u_{N-1}\rangle\langle u_{N-1}|u_0\rangle], \qquad (25.1)$$

where a closed loop is indicated by the choice that the function u_0 at the end (the point after $N-1$) is identical to the starting point 0; thus Eq. (25.1) is the phase acquired along the path from 0 to $N-1$. If the path is specified by a continuous variable, the continuum expressions for the Berry phase can be found by taking a limit as worked out in Section P.2 and given below in the notation appropriate for a crystal.

Expressions for the Berry Phase and Berry Connection in a Crystal

Here we specialize the notation to a Bloch state $u_{\mathbf{k}}(\mathbf{r})$, which is the ground state of a hamiltonian $H(\mathbf{k})$ as a function of $\mathbf{k}$ in a crystal. In some cases it is important to also take into account the variation of $u_{\mathbf{k}}(\mathbf{r})$ as a function of other parameters in the hamiltonian denoted by λ. The formulation is a special case of the general derivations in Appendix P, except for one important difference. For integrals over the Brillouin zone the path does not return to the starting point as required in Eq. (25.1). Instead, it starts at one point $\mathbf{k}$ and extends across the Brillouin zone to a point $\mathbf{k} + \mathbf{G}$, where $\mathbf{G}$ is a reciprocal lattice vector, and the relation of the eigenvectors at $\mathbf{k}$ and $\mathbf{k} + \mathbf{G}$ must be specified.[6] A convenient choice is the periodic gauge where the wavefunction is required to be periodic (the usual choice, e.g., in Eq. (23.3)),

$$\psi_{\mathbf{k}}(\mathbf{r}) = \psi_{\mathbf{k}+\mathbf{G}}(\mathbf{r}). \tag{25.2}$$

This means that the periodic part of the Bloch function $u_{\mathbf{k}}$ given by $\psi_{\mathbf{k}}(\mathbf{r}) = e^{i\mathbf{k}\cdot\mathbf{r}}u_{\mathbf{k}}(\mathbf{r})$ obeys the condition

$$u_{\mathbf{k}+\mathbf{G}}(\mathbf{r}) = e^{-i\mathbf{G}\cdot\mathbf{r}}u_{\mathbf{k}}(\mathbf{r}), \tag{25.3}$$

so that the relative phase includes a factor $e^{-i\mathbf{G}\cdot\mathbf{r}}$ as $\mathbf{k}$ varies across the Brillouin zone.[7] Thus a discrete expression for a Berry phase can be written as a slightly modified form of Eq. (25.1),

$$\phi = -\mathrm{Im}\ln[\langle u_{\mathbf{k}_0}|u_{\mathbf{k}_1}\rangle\rangle\langle u_{\mathbf{k}_1}|u_{\mathbf{k}_2}\rangle\ldots\langle u_{\mathbf{k}_{N-1}}|e^{-i\mathbf{G}\cdot\mathbf{r}}|u_{\mathbf{k}_0}\rangle], \tag{25.4}$$

which defines ϕ in the range to 0 to 2π. This is the form used in actual calculations for single occupied band, and the extension to multiple bands is the form in Eq. (24.13), which involves a determinant of the $N \times N$ matrix for N occupied bands.

The differential form can be derived if $u_{\mathbf{k}}$ is a smooth, differentiable function of $\mathbf{k}$. The expression is found by expanding the logarithm as is done in Eq. (P.4), which leads to

$$\phi = -\mathrm{Im}\oint dk\langle u_k|\partial_k u_k\rangle = \oint dk\mathcal{A}(k), \tag{25.5}$$

where we have used the fact that $\langle u_k|\partial_k u_k\rangle$ is purely imaginary (see Exercise P.1). If the integral is from $\mathbf{k}$ to a different point $\mathbf{k} + \mathbf{G}$, the phase of $u_{\mathbf{k}}(\mathbf{r})$ must be chosen as a function of $\mathbf{k}$ that satisfies the periodic gauge condition Eq. (25.3). In the last expression $\mathcal{A}(k) = -\mathrm{Im}\langle u_k|\partial_k u_k\rangle$ is the Berry connection, which is not gauge invariant. However, the integral ϕ is gauge invariant modulo 2π, which is proven in Eqs. (P.7)–(P.9). We will refer to ϕ as a Berry phase with the understanding that it is the gauge-invariant Berry phase plus or minus factors of 2π.

---

[6] Here and in the following chapters, the term "Brillouin zone" denotes a primitive cell of the reciprocal lattice, which is most conveniently chosen to be a parallelepiped formed by the reciprocal lattice vectors. In some contexts (see Chapter 4) the term is defined more narrowly to be the Wigner–Seitz cell of the reciprocal lattice, which is not convenient for the derivations of topological properties.

[7] It is essential that the theory involves $u_{\mathbf{k}}$, not $\psi_{\mathbf{k}}$, because all functions $u_{\mathbf{k}}$ obey the same periodic boundary conditions, which is required for matrix elements like $\langle u_j|u_{j+1}\rangle$ to be nonzero.

Berry Curvature and Flux

In two dimensions the Berry connection has two components that can be written $\mathcal{A} = (\mathcal{A}_x, \mathcal{A}_y)$ where

$$\mathcal{A}_\mu(\mathbf{k}) = -\mathrm{Im}\langle u_\mathbf{k} | \partial_\mu u_\mathbf{k} \rangle, \quad \mu = x, y, \tag{25.6}$$

where $\partial_\mu = \partial_{k_\mu}$, and the Berry phase for any closed loop in the plane is the line integral

$$\phi = \oint \mathcal{A}(\mathbf{k}) \cdot d\mathbf{k} \tag{25.7}$$

or the discrete form in Eq. (25.4).

As discussed in Sections P.3 and P.4 the topological properties can be defined in terms of the curvature of the surface defined by the two-component Berry connection $\mathcal{A}(\mathbf{k})$. (The result is really only Stokes's theorem, but the derivation shows the physical basis.) The curvature is defined by considering the Berry phase for a loop that encloses a small area (a patch); it is gauge invariant for a smooth function since there can be no added factors of 2π since the change must be a small quantity. It is straightforward to show that in the limit of a small patch the result is (see Section P.3 and Exercise 25.2)

$$\Omega(\mathbf{k}) = \partial_x \mathcal{A}_y - \partial_y \mathcal{A}_x = -2\mathrm{Im}\langle \partial_x u(\mathbf{k}) | \partial_y u(\mathbf{k}) \rangle, \tag{25.8}$$

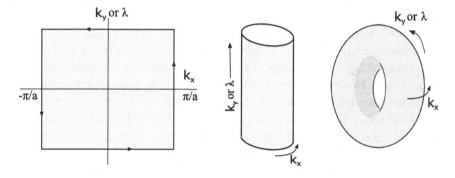

Figure 25.1. Three representations of an area in two dimensions for a crystal hamiltonian $H(k_x, k_y)$ or a one-dimensional crystal with a hamiltonian $H(k_x, \lambda)$. The Berry phase ϕ is defined for the path that encloses the area (Eq. (25.7)); the flux Φ is the integral of the Berry curvature over the area (Eq. (25.9)). Left: the boundary of the rectangle is a closed path $[k_x = -\pi/a \rightarrow \pi/a,$ $\lambda = \lambda_i \rightarrow \lambda_f, k_x = \pi/a \rightarrow -\pi/a, \lambda = \lambda_f \rightarrow \lambda_i]$, or with $\lambda \rightarrow k_y$. With the periodic boundary condition, $\psi_{k_x = -\pi/a} = \psi_{k_x = \pi/a}$, the left and right segments cancel, leaving only the top and bottom segments $\phi(\lambda_f) - \phi(\lambda_i)$. This applies for any range of λ or k_y. Middle: the cylinder is the same, taking into account the periodicity in k_x, with only the top and bottom circles $\phi(\lambda_f) - \phi(\lambda_i)$. Right: the torus for the Brillouin zone of a two-dimensional crystal or for λ that is periodic. In Figs. 24.4 and 25.2, the representation as a cylinder (middle) is used because it applies for polarization and it is easier to depict winding on the cylinder than on the torus.

which is the curvature of the surface.[8] The last form can be derived using the condition that the wavefunction is normalized as shown in Exercise 25.2. The integral over the entire surface can be determined by covering the surface with small patches, where the Berry phases for the loop around each patch is exactly canceled by the neighboring patches, except for patches at the boundaries. In the limit of small patches the integral, which is called the flux Φ, is equal to the phase calculated for the loop around the entire area,

$$\Phi = \int_S \mathbf{\Omega}(\mathbf{x}) \cdot d\mathbf{S}$$

$$= \oint \mathcal{A} \cdot d\mathbf{k} \mod(2\pi), \tag{25.9}$$

where $\mathbf{S}$ is the area vector directed along the normal to the surface. The first line is the flux, which is equal to the Berry phase modulo integral factors of 2π (the second line). It is the integral multiples of 2π that characterize the topology, the next topic.

25.5 Berry Flux and Chern Numbers: Winding of the Berry Phase

What happens if the surface is closed, i.e., there is no edge, a torus, like a donut or a coffee cup? It would appear from Eq. (25.9) that the flux Φ must be zero. However, that is not always true. A nonzero value is called a Chern number (see Section P.4), which is defined for a closed surface as an integer $C = \Phi/2\pi$. The Chern number can take any integer value, which is called Z, the set of integers. If there is a smooth gauge defined everywhere, the derivation of the curvature in Eq. (25.9) shows that C must be zero. However, there are cases where it is not possible to find a gauge that is smooth everywhere, in which case C may be nonzero. An example is the wavefunction for a spin on the surface of a sphere that encloses a source term analogous to a magnetic monopole, as discussed in Section P.7.

If one has an analytic formulation of the Berry curvature, the Chern number can be calculated directly. However, in a band structure calculation for a material, it is not obvious how to get the information. Fortunately, there is straightforward method that can be used to calculate the polarization using the Berry phase (see Section 24.3) and also to determine the integer Chern number by calculating the winding of the Berry phase. It is a practical approach and it provides insight into the meaning of the topological invariants.

The method is a straightforward implementation of the derivation in the previous section. The Berry curvature was defined in terms of patches, which is a gauge-invariant function in the limit of small patches that cover the area. This can be implemented as by covering the area of the cylinder with small squares as shown at the left in Fig. 25.2. The Berry phase can be calculated for each line and it can be plotted as an angle on the cylinder shown in the middle of the figure. By this process the integral over the area can be calculated from the Berry phase calculated as a function of k_y or λ with no uncertainty of factors of 2π.

---

[8] Note that $\Omega(\mathbf{k})$ in Eq. (25.8) is defined to be $\Omega(\mathbf{k})_{xy}$, and we should take care of the order and $\Omega(\mathbf{k})_{xy} = -\Omega(\mathbf{k})_{yx}$.

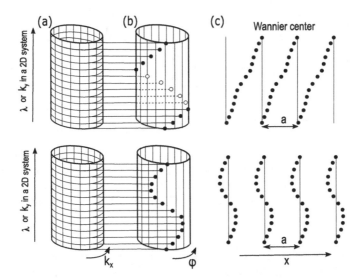

Figure 25.2. Schematic figure for calculation of the Chern number C for a two-dimensional problem described by a hamiltonian $H(k_x)$ that is a function of k_y, or a one-dimensional problem with a hamiltonian $H(k_x)$ that is a function of a parameter λ that varies periodically analogous to k_y. The left side marked (a) is the Brillouin zone drawn as a cylinder with periodic boundary conditions. The middle (b) shows the Berry phase ϕ as a function of k_y or λ also plotted as points on a cylinder since ϕ is a phase angle. At the right (c) is the shift of the centers of the Wannier functions, which are proportional to ϕ. The lower part of the figure illustrates a case where ϕ does not wind ($C = 0$) and the Wannier centers return to the original position. The top part represents a case where ϕ winds around the cylinder once to approach 2π ($C = 1$), and the electrons have shifted by a lattice constant. This illustrates a quantized charge pump (Chapter 26) or a Chern insulator (Chapter 27), which exhibits a quantum Hall effect.

If the surface is closed, i.e., k_y extending across the Brillouin zone for a two-dimensional crystal or a periodic variation of λ, the result is the Berry flux, which is just the winding number, ie., the number of times the Berry phase winds around the cylinder. Thus we have calculated the Chern number as the winding number! Exercise 25.3 is to show that winding once is analogous to the Chern number for a Dirac monopole; the cylinder shown at the top of Fig. 25.2 contains a monopole (a source of Berry flux)!

There is also a very useful interpretation of the winding in terms of Wannier functions. As shown in Section 24.4, the Wannier functions shift by an amount that is proportional to the change in the Berry phase ϕ, which is indicated in the right side of Fig. 25.2. A change of the Berry phase by 2π corresponds to shift of a hybrid Wannier function from one plane of cells to the next plane along the direction defined by derivatives of the Bloch functions, the x direction in this case. This provides a physical picture of transfer of particles through an insulator in a pump (Section 26.6), which leads to the transfer of electrons to edge and surface states that are the hallmark of quantum Hall systems and topological insulators. These are the edge states that move in only one direction in Fig. 25.3. Calculation of the flow of Wannier functions is useful for practical calculations of the topological indices

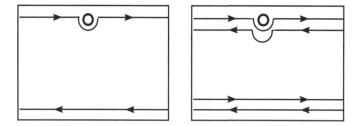

Figure 25.3. Left: edge currents for the quantum Hall effect (QHE) as depicted in Fig. Q.1 or for a Chern insulator (see Section 27.2). The current is quantized and not affected by imperfections as illustrated for the top line. Right: the two oppositely directed currents at the edge of a topological insulator, called the quantum spin Hall effect (QSHE) even though spin is not actually conserved, as described in Chapter 27.

as discussed in Section 27.4. This approach provides an intuitive picture of the charge flow and it is a practical method to calculate the Chern number as shown by Soluyanov and Vanderbilt [955] and by Yu et al. [956]. An automated method has been developed by Gresch et al. [957].

25.6 Time-Reversal Symmetry and Topology of the Electronic System

Time-reversal symmetry plays a special role in the possible topologies for electrons in crystals that can be understood in terms of the curvature $\Omega(\mathbf{k})$.

- If the crystal has time-reversal symmetry (TRS) then $\Omega(\mathbf{k}) = -\Omega(-\mathbf{k})$, since $\langle u_k|\partial_k u_k\rangle$ is odd in $\mathbf{k}$ (but see the next point below). Thus the integral over the BZ vanishes, which means that the Chern number $C = 0$. This seems to rule out any possibility for topological classification in any system with TRS. Indeed, the original classification of the topology of an electronic system in the famous TKNN paper was for the quantum Hall effect, which occurs only in a system with a magnetic field.
- However, there is a way around this roadblock. Time reversal changes $\mathbf{k}$ to $-\mathbf{k}$ and reverses the spin direction, and Kramers theorem requires that the two states be degenerate. Even in a system with TRS, spin–orbit interaction can lead to effective magnetic fields acting oppositely on opposite spin states. This is the reason that spin–orbit interaction is essential for topological insulators in the absence of magnetic fields, and the reason that the surface bands in the gap of a topological insulator always come in pairs with opposite momenta and spin state.
- In the problem of polarization in Chapter 24 there is no magnetic field and the hamiltonian is manifestly invariant under time reversal. It might seem that there could be no effect related to topology and no way there could be an analogy with systems with broken TRS. However, the way out of this dilemma is that the correct formulation of polarization in Section 24.3 is a integrated current, i.e., a directed flow of current due to variation of the potential as a function of λ that plays the role of "time." If the change in λ is reversed

then so also the change in direction of polarization is reversed. Similarly, in a pump (Section 26.6) the hamiltonian is varied in a periodic cycle and the direction the particles move is determined by the sense of direction of the cycle, i.e., an externally imposed effect with a direction analogous to time.

- If the crystal has inversion symmetry then $\Omega(\mathbf{k}) = \Omega(-\mathbf{k})$. Thus if the crystal has both time-reversal and inversion symmetries, $\Omega(\mathbf{k}) = 0$ at all $\mathbf{k}$. Nevertheless, even in these materials Berry phases have roles in understanding certain properties, for example, the relation of the values 0 and $\pm\pi$ of the Berry phase and two centers of inversion in Section 26.4. These are also the sites for the polarization lattice in Section 24.8.

25.7 Surface States and the Relation to the Quantum Hall Effect

The principle of "bulk-boundary correspondence" guarantees that there will be surface states in the gap if the Chern number is nonzero, and there are additional possibilities when spin is included as discussed in Chapter 27. There are different physical consequences of the bulk-boundary correspondence depending on the nature of the hamiltonian. The topic of Chapter 26 is a pump, where conservation of particle number relates the transfer of particles between the boundaries and the cyclic variation of the hamiltonian in the bulk. The direction of the cycle determines the direction of the transfer, and the transfer can take place only if there is a surface state that crosses the gap in order for the surface to accept or give up a particle (see Chapter 26).

For a two-dimensional crystal in Chapter 27, the surface states form bands that propagate in only one direction and extend across the gap from the valence to conduction bands. However, we have not derived the most remarkable feature of these bands, that electrons can flow with no resistance even if there is disorder. For this we appeal to the understanding of the quantum Hall effect, which is described briefly in Appendix Q. The states at the boundary are conducting states that move in only one direction, and adding (subtracting) a particle leads to a net increase (decrease) of the current by a quantized amount, as described in Section Q.2. The current is not affected by impurities or imperfections in the surface region because the particles cannot scatter backward; so long as the effect is not so large that the gap is destroyed, the current goes around a defect as indicated in the left side of Fig. 25.3. The same happens a Chern insulator in Section 27.2 because the hamiltonian for the two dimensions naturally incorporates the corresponding effect as if it is in a large magnetic field.

The right side of Fig. 25.3 is for a topological insulator described in Chapter 27. There are two opposite currents of coupled spin and momentum due to the spin–orbit interaction, which are analogous to the edge currents in a Chern insulator. However, a topological insulator is not the same as two copies of a Chern insulator, which would be pairs of integers for the set of independent right and left circulating modes (C_1, C_2), which is a $Z \times Z$ classification. As discussed in Section 27.4, the topology of a topological insulator can be classified by one of two intergers, 0 or 1, which is called a Z_2 topological classification. There is a bulk-boundary correspondence and counter propagating edge currents, so long as the system has time-reversal symmetry in both the bulk and the surface. This is called a quantum spin Hall effect (QSHE) even though spin is not conserved; a more proper description is helical currents with opposite spin-momentum character.

25.8 Wannier Functions and Topology

Displacements of the centers of Wannier functions are used in many places in Chapters 24–28 to show the consequences of the topology of the electronic structure and as a way to detect nontrivial topology. This is due to the identification of the displacement of the centers as a Berry phase, e.g., in Section 24.4. The question arises: Is there a consequence of the topological properties of the electronic structure for Wannier functions themselves? The answer leads to the need for careful formulation of the issues and a result that is a major theoretical accomplishment.

First, the analysis of topological properties of the electronic structure actually involves "hybrid Wannier functions," which are defined in Section 23.6 as functions that are confined in one direction, but extended in the other direction(s), and they can be expressed as in Eq. (23.36): $w_{in,\mathbf{k}_\perp}(\mathbf{r})$, where n is a band index, i the position in one direction and $\mathbf{k}_\perp$ along the perpendicular directions. It has been established by Kohn [892] and others that in one dimension it is possible to construct exponentially localized Wannier functions so long as there is an insulating gap. The analysis of topology of the electronic structure in terms of hybrid functions is on solid grounds since it depended only on the functions being localized at each value of $\mathbf{k}_\perp$.

However, it has been a major theoretical issue whether or not it is possible to find Wannier functions that are localized in all directions in two and three dimensions. See, for example, the review [891]. It is intuitively clear that a necessary requirement is that it is possible to find Bloch functions that are smooth functions of $\mathbf{k}_\perp$ in order to Fourier transform to Wannier functions that are localized along all directions. This is a global property of the Bloch functions at all $\mathbf{k}$ in the bulk Brillouin zone independent of the directions called $\mathbf{k}_{||}$ and $\mathbf{k}_\perp$. One can suspect that there may be a difficulty for topological insulators since the behavior of the hybrid Wannier centers as a function of $\mathbf{k}_\perp$ along particular directions is a signature of the topology, as discussed in Section 25.5. Indeed, even before the work of Berry it was shown that obstacles to construction of localized Wannier functions have topological origins (see references in [893]). It was shown in [893] that that localized functions cannot be constructed if the Chern number is nonzero, and that a trivial topology (all Chern numbers zero) is a sufficient condition. This is the long-sought proof that it is always possible to construct Wannier functions that are exponentially localized in all directions if and only if the occupied electronic structure has trivial topology as a function of $\mathbf{k}$.[9]

---

[9] There are additional considerations in topological insulators that have time-reversal symmetry and involve spin–orbit interaction, where it is possible to construct localized functions if they are allowed to break the time-reversal symmetry. See, for example, [955].

25.9 Topological Quantum Chemistry

Perhaps the simplest way to explain the essential difference between crystals that have nontrivial topological character and those that do not is the way that the electronic system behaves as atoms are brought together to form a crystal, a characterization that has been termed topological quantum chemistry [958]. The term "chemistry" denotes formation of bonds and "topological quantum" denotes a purely quantum effect that is a global property of the electronic system, not the local properties that are the province of the field commonly known as "quantum chemistry." One way to describe the distinction is the consequence of an obstruction to the existence of localized Wannier functions. The central idea is that, if such localized functions exist, they span the occupied space; if the solid is pulled apart to approach the atomic limit and no gap goes to zero in the process, then the Wannier functions continuously approach disconnected localized functions. Thus the system has the same properties as weakly coupled atoms with trivial topology. Conversely, if it is not possible to approach a well-defined atomic-like limit with localized functions without a gap going to zero, then it is also not possible to construct such functions in the solid, which characterizes a system with nontrivial topology.

Up to this point the topological character has been described in terms of the behavior of the Bloch states as a function of $\mathbf{k}$ in the Brillouin zone. The classification in terms of Wannier functions is a real-space picture and the transformation between the two pictures is the essence of the Wannier function transformation. Difficulties have been recognized in the past, but to the knowledge of the author, it is only with the topological arguments that the connection is firmly established.

An example is the Shockley atomic-covalent transition (see Sections 26.2, 22.4, and 26.4) in which a gap vanishes at the transition. However, it is a distinct transition only if there is a center of inversion. In this sense it is a crystalline topological insulator protected by inversion symmetry [948]. A topological insulator, defined as a system that is protected only by time-reversal symmetry, is one that cannot be pulled apart to disconnected atoms without a gap vanishing so long as time-reversal symmetry is preserved. This applies no matter what is the spatial symmetry and it can occur in two or higher dimensions.

25.10 Majorana Modes

There is much work on Majorana fermions at surfaces of superconductors, which are possible topological modes for quantum computing. Superconductivity is outside the scope of this book, but it is interesting to point out that Majorana modes are closely related to Shockley states. See, for example, the review by Beenakker [959], who points out that "Majorana bound states can be understood as superconducting counterparts of the Shockley states from surface physics. The closing and reopening of a band gap in a chain of atoms leaves behind a pair of states in the gap, bound to the end points of the chain. Shockley states are unprotected and can be pushed out of the band gap by local perturbations. In contrast, in a superconductor, particle-hole symmetry requires the spectrum to be symmetric in energy, so an isolated bound state is constrained to lie at $E = 0$ and cannot be removed by any local perturbation."

SELECT FURTHER READING

See the review articles listed in footnote 1.

Books with pedagogical overview of Berry phases and topology in condensed matter:

Asboth, J. K., Oroszlany, L., and Palyi, A., *A Short Course on Topological Insulators: Band Structure and Edge States in One and Two Dimensions* (Springer, Heidelburg, 2016).

Bernevig, A., *Topological Insulators and Topological Superconductors* (Princeton University Press, Princeton, NJ, 2013).

Vanderbilt, D. H., *Berry Phases in Electronic Structure Theory* (Cambridge University Press, Cambridge, 2018).

Exercises

25.1 Apply same arguments as in Exercise P.1 to show that $\langle u_k | \partial_k u_k \rangle$ is purely imaginary.

25.2 See Exercise P.2 for the general form of Berry curvature. Apply the arguments to derive the first equality in expression for $\Omega(\mathbf{k})$ in Eq. (25.8) from the previous expressions the Berry phases. Discuss how added factors of 2π might occur in crystals. Show that the second equality follows from the normalization condition for the Bloch functions.

25.3 In Section P.7 it is pointed out that one way to treat the monopole is to define two different smooth gauges for the northern and southern hemispheres and "glue" them together, which is allowed because the Berry phase is defined modulo 2π. The integer factor of 2π is the Chern number. The winding around the cylinder results from a smooth variation across the Brillouin zone. Discuss the way this is analogous when we "glue" together the two ends of the cylinder to describe the behavior of the Berry phase on the torus that is the Brillouin zone (or the torus defined by a cyclic variable λ).

26

Two-Band Models: Berry Phase, Winding, and Topology

To turn, turn will be our delight,
Till by turning, turning we come round right

Elder Joseph, "Simple Gifts"

Summary

The first part of this chapter sets up the problem of two bands, which is formulated in terms of Pauli matrices. This is the simplest problem in which there is a bandgap and it turns out that any topological insulators can be characterized by a two-band model. Examples are the two-site and $s - p_x$ models, which provide physical interpretation. The rest of the chapter is devoted to the steps that lead up to the first example of topology – the Thouless electron pump. First is the Shockley transition that can be characterized by a winding number for the hamiltonian $H(k)$ in which the Berry phase can take two values. Next is the definition of a two-dimensional problem with $H(k, \lambda)$ in which λ varies in a cycle. This is the step beyond the work of Shockley and it leads to a new classification of the two-dimensional problem by a global quantity: winding of the Berry phase and the Chern number, which is a topological invariant. The topology of the pump carries over to the Chern insulator in Chapter 27, which is a step on the path to a topological insulator.

26.1 General Formulation for Two Bands

Two bands are the minimum that can describe a system with a band gap and nontrivial topological structure, which originates in the way the occupied and empty states exchange character (or not) as a function of momentum **k**. It turns out that is also sufficient: *all* topological insulators with time-reversal symmetry can be described by 2×2 Dirac hamiltonians.[1] For realistic, quantitative calculations one needs many bands, and even in model problems it is often convenient to use more bands to capture physical effects such as

---

[1] This is shown in [960–964] and described in the reviews [943] and [944]. A table for the 10 symmetry classes of single-particle hamiltonians (in terms of time-reversal, particle-hole, and sublattice symmetries) in one, two, and three dimensions is given in [961].

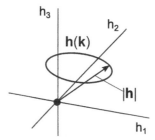

Figure 26.1. The three-dimensional space defined by the coefficients h_1, h_2, h_3 of the Pauli matrices are expressed as vector $\mathbf{h}$. The curve depicts the hamiltonian $\mathbf{h}(\mathbf{k})$ for $\mathbf{k}$ on a closed line in $\mathbf{k}$-space. The eigenvalues are $\pm|\mathbf{h}|$ where $|\mathbf{h}|$ is the distance to the origin, and the gap vanishes only if the curve passes through the origin where $|\mathbf{h}| = 0$.

spin–orbit interaction, which is essential for topological insulators. Nevertheless, even in those cases, the topological properties are captured in terms of only two bands, as shown in examples in Chapters 27 and 28.

For two bands the 2×2 hamiltonian can always be written in terms of the Pauli matrices. The two-band problem as a function of momentum $\mathbf{k}$ is equivalent to the two-level spin 1/2 problem in a field that depends on a parameter that varies continuously. In some cases we want to consider both band and spin and it is helpful to distinguish the ways the matrices are used even though the algebra in terms of the matrices is the same. Here we denote them with the symbol τ to emphasize that they are the matrices in the space of the two bands:

$$\tau_1 = \begin{bmatrix} 0 & 1 \\ 1 & 0 \end{bmatrix}, \quad \tau_2 = \begin{bmatrix} 0 & -i \\ i & 0 \end{bmatrix}, \quad \tau_3 = \begin{bmatrix} 1 & 0 \\ 0 & -1 \end{bmatrix}, \quad \text{and} \quad \tau_0 = \begin{bmatrix} 1 & 0 \\ 0 & 1 \end{bmatrix}. \tag{26.1}$$

Any 2×2 hermitian matrix can be expressed as a linear combination of these four matrices, e.g., a hamiltonian can be written

$$H(\mathbf{k}, \lambda) = \sum_{i=1}^{3} h_i(\mathbf{k}, \lambda)\tau_i + h_0(\mathbf{k}, \lambda)\tau_0, \tag{26.2}$$

where λ represents one or more parameters that can be varied to change the hamiltonian continuously and each lower-case $h_i(\mathbf{k}, \lambda)$ and $h_0(\mathbf{k}, \lambda)$ is a scalar function of $\mathbf{k}$ and λ. Here $H(\mathbf{k}, \lambda)$ is the $\mathbf{k}$-dependent hamiltonian whose eigenfunctions are the periodic part of the Bloch functions and $\mathbf{k}$ can be considered as a parameter that varies over the Brillouin zone.

The last term in Eq. (26.2) can be ignored for many purposes because it is a shift in the energy that is the same for the two bands, which does not affect the gaps or the eigenfunctions. This is a quantitative effect that must be taken into account for realistic problems; however, qualitative properties such as the existence of a gap, exchange of character of the bands, the winding number, and topology do not change if the hamiltonian is changed continuously, so long as the system remains insulating everywhere along the path.

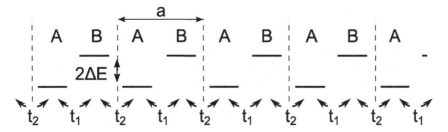

Figure 26.2. The two-site model of a chain of sites labeled A and B with alternating energies $\pm \Delta E$ and hopping matrix elements $\bar{t} \pm \delta t$, which can represent molecules on a lattice if δt is large or an ionic crystal for large ΔE.

If we ignore the last term involving τ_0, $H(\mathbf{k}, \lambda)$ can be expressed as a vector in a three-dimensional space

$$\mathbf{h}(k, \lambda) = [h_1(\mathbf{k}, \lambda), \ h_2(\mathbf{k}, \lambda), \ h_3(\mathbf{k}, \lambda)], \tag{26.3}$$

as indicated in Fig. 26.1, no matter what the dimension of the actual space is or the number of parameters in the hamiltonian. The eigenvalues are simply $\pm |h(\mathbf{k}, \lambda)|$ and there is always an energy gap unless all three (h_1, h_2, h_3) are zero. Rotations and inversions leave the energies invariant but express the eigenfunctions differently, which provides the mapping between different problems.

26.2 Two-Band Models in One-Space Dimension

This rest of this chapter is devoted to systems in one-space dimension but it will become clear how it leads up to two- and three-dimensional systems, which are the topics of the following chapters. The interesting cases are where the hamiltonian $H(k, \lambda)$ can describe different regimes as parameters denoted by λ are varied. This is the one-dimensional problem used by Shockley [42] in 1939 to identify a transition in the bulk that leads to a

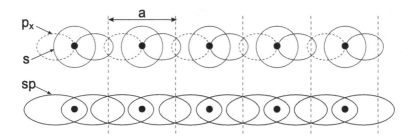

Figure 26.3. Two-band models derived from a crystal of atoms with s (even) and p (odd) states per cell. At the bottom the states are portrayed in a basis of sp hybrid functions $(s \pm p)/\sqrt{2}$. This is a model for weakly coupled atoms if the hopping is much less than the $s - p$ energy difference, or for a covalently bonded system in the hopping dominates. In the bulk crystal the eigenvalues are invariant to the choice of the basis but the eigenvectors are different and there are consequences for the surface.

surface state (now called a Shockley state; see Chapter 22). It is also a model for "quantized particle transport" (Section 26.6), which was shown to be a consequence of the topological character by Thouless [256] in 1983 before the work of Berry. We illustrate the problem with tight-binding models, but it is completely general; as pointed out by Shockley the lowest two bands of any one-dimensional crystal can be described by states that are even and odd. The models are simple enough to be solved analytically and they illustrate surface states in Chapter 22 and polarization in Chapter 24. The topological analysis for the quantized pump carries over directly to the Chern insulator in Chapter 27, and it is only a few steps derive the properties of topological insulators in Chapters 27 and 28.

Two-Site Model

The simplest model for a lattice of diatomic molecules or an ionic crystal has two sites per cell, each with one state, as depicted in Fig. 26.2, and where hopping t is only between nearest-neighbor A and B sites. This is the Mele–Rice model [965] or the Su–Schrieffer–Heeger (SSH) model [954] if $\Delta E = 0$, which are widely used in the literature. The hamiltonian can be expressed as

$$H(k) = \begin{bmatrix} \Delta E & t(k) \\ t^*(k) & -\Delta E \end{bmatrix}, \tag{26.4}$$

where the on-site energies $\pm \Delta E$ are chosen so the average energy is zero. There is no k-dependence of the diagonal terms since there is no $A - A$ or $B - B$ hopping. There are various ways to express $t(k)$. If the unit cell is defined as shown in Fig. 26.2 and we choose to assign the k-dependence to the intercell terms t_2, then $t(k) = t_1 + t_2 exp(-ika)$ where a is the lattice constant. The eigenvalues can be expressed as $\varepsilon(k) = \pm \sqrt{(\Delta E)^2 + |t(k)|^2}$, with $|t(k)|^2 = t_1^2 + t_2^2 + 2t_1 t_2 \cos(ka)$ (see Exercise 26.1.). In a bulk crystal there is no difference in the eigenvalues if the cell is defined differently so that t_2 is the intracell term and t_1 intercell, but the eigenvectors change and there are consequences for surface states.

Two-State sp Model

A model with even and odd states is equivalent to a basis of s and p states illustrated in Fig. 26.3. If there is only nearest neighbor hopping, the 2×2 hamiltonian $H(k)$ can be expressed as

$$H(k) = \begin{bmatrix} \varepsilon_p + 2t_{pp} \cos(ka) & -2it_{sp} \sin(ka) \\ 2it_{sp} \sin(ka) & \varepsilon_s + 2t_{ss} \cos(ka) \end{bmatrix}, \tag{26.5}$$

where ε_s and ε_p denote the on-site energies and the nearest-neighbor hopping matrix elements are denoted t_{ss}, t_{pp}, and t_{sp}. If the difference $\varepsilon_p - \varepsilon_s$ is larger than the band widths, i.e., the atomic-like regime, the lowest-energy band has mainly s character and the higher energy state is p-like. The covalent regime corresponds to a large band width due to the hopping matrix elements so that bands have bonding and antibonding character. The transition between these two regimes is marked by a gap going to zero and a qualitative change of the nature of the bands, called the "Shockley transition," the classic paradigm for formation of bands shown in the first figure of this book Fig. 1.1.

Relation of the Models

It is useful to use the two-site model in some cases and the two-band $s - p$ model in other cases to represent different physical problems. However, they are related by a simple transformation that is brought out in the representation in terms of Pauli matrices. Each model can be written in terms of a general form or as a vector in the three-dimensional space

$$\mathbf{h}(k) = [u + v\cos(k), \quad w\sin(k), \quad \Delta + d\cos(k)]. \tag{26.6}$$

The $s - p$ model in Eq. (26.5) and Fig. 26.3 can be expressed as

$$\mathbf{h}(k) = [u, \quad w\sin(k), \quad \Delta + d\cos(k)], \tag{26.7}$$

where $\Delta = \Delta_{sp} = (\varepsilon_p - \varepsilon_s)/2$, $d = t_{ss} - t_{pp}$, and $w = 2t_{sp}$. In this model there is no component in the h_1 direction: $v = 0$ by symmetry since the p states are odd. (The term u is zero if there is a center of inversion at the atom site. It is included here because it is important in Section 26.5.) The analysis applies just as well to the two-site model in Eq. (26.4) and Fig. 26.2, which can be expressed as

$$\mathbf{h}(k) = [u + v\cos(k), \quad w\sin(k), \quad \Delta], \tag{26.8}$$

with $u = t_1$, $v = w = t_2$ and $\Delta = \Delta E$. In this model $d = 0$ because there is no A–A and B–B hopping. The mapping between these models is accomplished simply by rotating around the h_2 axis (see Fig. 26.1), which exchanges h_1 and h_3, and renaming the variables, i.e., u, Δ and d in Eq. (26.7) renamed as Δ, u, and v in Eq. (26.8).

Even though the two models are related by transformations, it is often more useful to work with one or the other. The $s - p$ model carries over to the models in two and three dimensions that are most useful for including spin–orbit interaction in Chapters 27 and 28, but the two-site model is often easier to use. An example is the pump in Section 26.6.

26.3 Shockley Transition in the Bulk Band Structure and Surface States

For a two-band model the 2×2 equations are readily solved and the properties can be illustrated by either the two-site or the $s - p$ model. Here we consider the $s - p$ model with hamiltonian given in Eq. (26.5) and Fig. 26.4 shows the bands for two cases (see Exercise 26.4). At the left is shown an example where each band has the same parity at the zone center and the zone boundary. In this case the tight-binding parameters are on-site energies $\varepsilon_s = -1.0$, $\varepsilon_p = 1.0$, and the nearest-neighbor intersite matrix elements, $t_{ss} = -t_{pp} = 0.4$, and $t_{sp} = 0.10$ in arbitrary units. The right side shows what happens when the band widths are increased with $t_{ss} = -t_{pp} = 0.75$, keeping other parameters the same. Now the lowest-energy state at the zone center is odd (p) but at the zone boundary the even (s) state is lowest, i.e., an inverted band structure. At other points in the Brillouin zone the states are mixed by the t_{sp} matrix element, which leads to the gap between the bands.

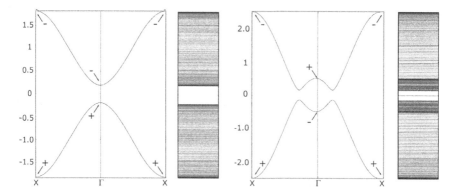

Figure 26.4. Band structures for a one-dimensional crystal and a finite chain with 99 sites for two choices of parameters in the model with s and p_x states. Bands for the periodic crystal are plotted for $k = 0$ (Γ) to the BZ boundary X, and the states of the finite chain are shown as lines with no dispersion. At Γ and X the bands have definite parity (+ or −) as shown. Left: example where the bands do not exchange and there is no state in the gap. Right: example where the bands exchange character as a function of k and there are surface states in the gap at each end of the crystal.

Before doing any calculations for a crystal with a surface, we can understand the emergence of a midgap surface state at the transition from atomic-like states to covalent bonding. In the atomic-like regime the hopping terms are small compared to the $s - p$ separation ε_s and ε_p and the bands separated by a gap with the lowest-energy band formed primarily from atomic s states filled with two electrons. On the other hand, as the atoms are brought together bonds are formed between the atoms; in the bulk each atom contributes one electron to each of two covalent $s - p$ bonds on the left and right sides. However, an atom at the surface has only one neighbor, so that one electron is left in a half-filled "dangling bond" state like those pictured at the ends in the bottom part of Fig. 26.3.

Turning to quantitative calculations, we can solve the problem for an infinite periodic crystal and for a finite crystal that is large enough to identify bulk states in the interior and surface states localized at each end, if they exist. The states of a finite chain are discrete energies that are shown in the bar graphs in Fig. 26.4. The calculations were done with a crystal of 99 sites that has two equivalent ends. It is instructive to first do the calculation for an artificial supercell of 100 atoms and construct the finite system by removing one site. For the case where the bands are well separated in normal order with the same parity, there is no state in the gap and the results are hardly changed when one site is removed. For the case where the bands exchange parity as a function of k in the perfect crystal, a state appears in the gap at each end of the finite crystal. These are the Shockley states that are localized at the ends of the chain and they are exactly at midgap in this case since the bands are symmetric.

The Two-Site Model and Analytic Solution for the Surface State

The two-site model should be viewed as another representation of the Shockley problem. This model is simpler (only two parameters) and it is the generic interpretation two-band systems used in the literature, e.g., this is the model proposed later by Su, Schrieffer, and Heeger (SSH) [954] as a model for one-dimensional polyacetylene. The generalization to two inequivalent sites is shown schematically in Fig. 26.2 and is often referred to as the Rice–Mele model [966]. Since we ultimately want to derive topological properties that are independent of details, the conclusions will be the same for all models so long as the hamiltonians can be connected smoothly without a gap going to zero. In this case the transition occurs at the point where $\Delta E = 0$ and $t_1 = t_2$. A state in the gap occurs if the crystal ends on a weak bond.[2] For this model there is an analytic solution (see Exercise 26.7) for the surface state wavefunction as it decays into the crystal [967] and [960].

26.4 Winding of the Hamiltonian in One Dimension: Berry Phase and the Shockley Transition

The topic of this section and Sections 26.5 and 26.6 is the properties of electronic structure in one space dimension, which provides valuable insights and useful steps toward topological insulators.[3] Here we assume there is a center of inversion and the consequences of breaking inversion symmetry is considered in the following section. One can interpret the two bands as either the two-site model with $\Delta = 0$ in Eq. (26.8) or the $s - p$ model in Eq. (26.7) with $u = 0$. The conclusions are the same for the winding of the Berry phase since the only difference is a rotation in the space of the Pauli matrices. Here we choose the orientation corresponding to the $s - p$ model because it provides the most physical interpretation of the Shockley transition and surface states in 22.3 and it is the model that carries over directly to the topological insulator in Chapter 27.

In the $s - p$ model with a center of inversion ($u = 0$ in Eq. (26.7)) $\mathbf{h}$ is restricted to the $h_1 = 0$ plane and the variation of $\mathbf{h}(k)$ for k varying from 0 to 2π is shown as ovals in the $h_2 - h_3$ plane in Fig. 26.5, for fixed values of d and w and two choices of Δ, one where $\mathbf{h}(k)$ winds around the origin and one where $\mathbf{h}(k)$ does not wind. If Δ is varied continuously between these two cases there must be a transition where the curve touches the origin and the gap vanishes. This is the point where the bands touch in the Brillouin zone, which can occur only at $k = 0$ or $k = \pm\pi$, the bands exchange eigenvectors, i.e., a qualitative change in the electronic structure that was pointed out by Shockley [42] and Zak [968].

The consequence of winding (or not) can be related to a Berry phase (see Section 25.4), which is the phase acquired by the ground-state wavefunction as k goes across the BZ, i.e., the hamiltonian goes around the oval in Fig. 26.5. In this case, the crystal has inversion symmetry so that the wavefunction can be chosen to be real and the only possibilities are a Berry phase with values 0 or π. (See Exercise 26.6.) For the case where the circle does not wind, the hamiltonian vector returns to the starting point with an angle that increases and

---

[2] The termination for the $s - p$ model is easy to understand; however, termination on an A or B site is arbitrary and one must be careful to interpret the results correctly. An example is the pump in Section 26.6.

[3] The discussion follows the lecture notes by Asboth et al. [945], where the discussion is cast in terms of the two-site model. See also [967].

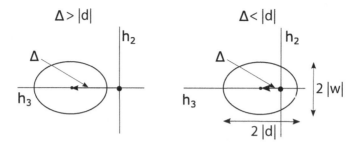

Figure 26.5. The hamiltonian in Eq. (26.7) with $u = 0$ in the space of the $h_2 - h_3$ plane of the space of the Pauli matrices. Each oval depicts $\mathbf{h}(k)$ with energy $\pm|\mathbf{h}|$, i.e., the gap is $2|\mathbf{h}|$. In the example at the right $\mathbf{h}(k)$ winds once around the origin, which also means that the bands interchange as a function of k. At the left is a case where there is no winding and bands that do not interchange.

decreases with no net change, and thus the Berry phase is zero. If the hamiltonian vector winds around the origin, the angle changes by 2π so that the Berry phase changes by π, i.e., a change of sign as k goes across the BZ.[4]

There are two important corollaries. One is that the two distinct insulating states found by Shockley[5] (see Section 22.3) are characterized by winding numbers or equivalently by discrete values of the Berry phase. This is a property of the electronic structure of the bulk crystal and it is a rigorous distinction so long as there is inversion symmetry. The transition corresponds to an atomic-covalent transition as indicated in Fig. 26.3. For a finite crystal, the transition may also lead to a surface state in the gap, called a Shockley state (see Sections 1.2 and 22.3). Whether or not the state is in the gap depends on the details of the surface; nevertheless, it is a stepping stone toward topological insulators where the bulk is characterized by topological invariants and the surface band is guaranteed to cross the gap.

The second corollary is that the transition corresponds to a shift of the center mass of the electrons by half the lattice constant, which can be understood as a shift between two centers of inversion: if there is inversion around the center of a cell, there is also another center of inversion at the cell boundary.[6] This was recognized by Zak [968], who defined "band centers," and it is an example of the relation of the centers of Wannier functions to the Berry phase ϕ, where $\phi = \pi$ corresponds to a shift by 1/2 a lattice constant. This is an example of the polarization lattice in Section 24.8. It is also just what is expected for

---

[4] This can be seen as a special case of the fact that the Berry phase around a loop is 1/2 the solid angle enclosed as deduced in Section P.7: the circle representing k is like the equator in the sphere Fig. P.4, and the Berry phase is 1/2 the solid angle of the upper or lower hemispheres, which is π. The ground-state eigenvector of the two-band problem as a function of k is equivalent to the variation of the lowest-energy state of a spin-1/2 where the spin is carried around a circle in the presence of a magnetic field that causes the spin to always be directed outward (or inward). For a spin-1/2 there is a change of sign as it completes a full circle, which corresponds to the sign change in the ground state eigenvector of the two-band problem as k varies across the Brillouin zone.

[5] The conclusions apply to any one-dimensional problem and are not limited to the nearest-neighbor expressions used here; this is also apparent from Shockley's original derivation that did not invoke a tight-binding model.

[6] This also holds for crystals in higher dimensions and the fact that there can be multiple inversion sites is intimately related to the concept of a polarization lattice. See Section 24.8.

the atomic-covalent picture of the transition in which the bonding shifts from centered on atoms to centered on bonds, as indicated in Section 22.3.

Finally, this is an example of a Z_2 classification, where Z_2 denotes the set of integers modulo 2, which divide into two distinct classes. In this case, the phase ϕ can take values $N\pi$ where N is even or odd. The Berry phase is defined to be in the range 0 to 2π, with 2π equivalent to 0, so that there are only two possibilities, 0 or 1. Note that this is just one of many possible Z_2 classifications and the physical meaning is *not* the same as for topological insulators that is derived in Section 27.4. This is an example of a crystalline topological insulator and there are also other classifications given in [948, 969, 970] and other references.

26.5 Winding of the Berry Phase in Two Dimensions: Chern Numbers and Topological Transitions

How can there be a transformation between the two different winding numbers for the hamiltonian shown in Fig. 26.5 without a gap vanishing? The answer is that the space must be expanded by moving into the third dimension, in this case by changing the hamiltonian to have a component in the h_1 direction. A schematic figure for such a cyclic variation of the hamiltonian is shown in Fig. 26.6 plotted as vectors **h** in three dimensional space of Pauli matrices.[7] At the left the dark ovals are the same as in Fig. 26.5 plotted here in the $h_1 = 0$ plane, with light ovals that indicate a sequence of hamiltonians with a transition where the gap vanishes. At the right is a path in the three-dimensional space with $h_1 \neq 0$ that connects the same hamiltonians in the $h_1 = 0$ plane without ever having a zero gap, i.e., without touching the origin. This figure represents the class of problems where there is a directed path in which the Berry phase changes by $\pm\pi$ (the part above the $h_1 = 0$ plane) and then continues below the $h_1 = 0$ plane until the change is $\pm 2\pi$, i.e., the Berry phase winds once around the origin.

In the $s - p$ model the displacement above or below the $h_1 = 0$ plane means $u \neq 0$ in Eq. (26.7), which corresponds to a real term in the off-diagonal matrix elements that can occur only if the center of inversion is broken. Then the system is polarized either one direction or the opposite depending on the sign of u, and it is possible to shift the center of the Wannier functions continuously from 0 to either $+1/2$ or $-1/2$, or equivalently the Berry phase varies continuously from 0 to $\pm\pi$. It is also possible to continue varying until the shift is by a full lattice constant; this corresponds to a case where the hamiltonian varies in a cyclic manner so that it is the same after each cycle, but the electrons have shifted by a lattice constant during each cycle. This is an example of winding like that illustrated in the upper part of Fig. 25.2, where the Berry phase changes by 2π during a cycle, and the Wannier functions shift. The consequence is that this an electron pump discussed further in Section 26.6 (see also Section 24.7).

---

[7] Figure 26.6 is *not* the same as the torus in Fig. 25.1, where the surface represents the two-dimensional space (k, λ) or (k_x, k_y). Instead it is the hamiltonian expressed in terms of (h_1, h_2, h_3) where each h_i varies as a function of (k, λ) or (k_x, k_y).

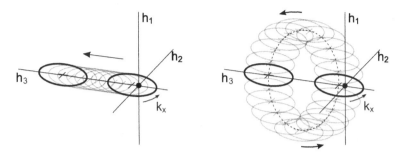

Figure 26.6. Schematic illustrations of a sequence of hamiltonians represented as vectors $\mathbf{h} = [h_1, h_2, h_3]$ in a three-dimensional space of the Pauli matrices as indicated in Fig. 26.1. Left: the same information as in Fig. 26.5 plotted in the $h_1 = 0$ plane, with the light ovals indicating a continuous variation between the case that winds and the one that does not. Right: a hamiltonian that changes in a cycle that includes the two dark ovals in the left figure and returns to the starting hamiltonian. The gap is never zero and the components h_1 and h_3 vary together to form a closed loop shown as the dashed circle. The figure shows an example where the variation of the hamiltonian generates a surface that encloses the origin; the Berry phase winds once around the origin so that the Chern number is one and there is nontrivial topology (like a surface that encloses a monopole in Fig. P.4). See also Fig. 25.2 for an alternative view of winding of the Berry phase. A hamiltonian that does not enclose the origin would have trivial topology. Application to a pump is in Section 26.6 and to a Chern or topological insulator is in Chapter 27. The figure and description follow [945].

Another example is a two-dimensional Chern insulator (see Section 27.2) that is a forerunner of topological insulators. The pump may seem contrived and only a theoretical device. A Chern insulator is different: there is no need for breaking a symmetry and no need for an externally driven parameter in the hamiltonian; the effect occurs as a function of the momentum k_y and it is intimately related to breaking of time-reversal symmetry. The properties carry over to topological insulators with some very important differences.

Topological Classification

The right side of Figure 26.6 indicates a topological property by the fact that the hamiltonian vector h either encloses the origin (the singular point where the gap vanishes) or it does not. In the example shown, $\mathbf{h}(k, \lambda)$ has nontrivial topology as a function of k and λ, analogous to the surface that encloses a monopole in Fig. P.4 or the winding on the surface depicted in Fig. 25.2, where the Chern number is one. If the parameters are changed so that the surface does not enclose the origin, this corresponds to bands that do not exchange character and a hamiltonian with trivial topology. A transition between states with different topology can occur only if a gap vanishes. It also follows that the Berry phase winds as one proceeds around the path defined by the variation in the $h_1 - h_3$ plane in Fig. 26.6 if and only if the surface traced out by $\mathbf{h}$ encloses the origin. This is the relation to the topological winding number, called a Chern number, as discussed in Section 25.5.

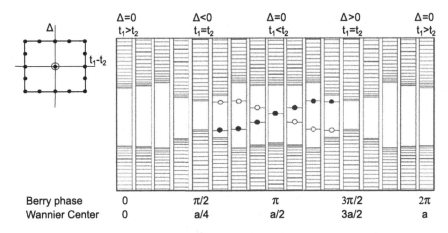

Figure 26.7. States of the finite chain of 99 cells in a small energy range around the gap for a set of calculations for the hamiltonian in Eq. (26.4), with $t_2 = 1$ and $t_1 - t_2$ and Δ in the range ± 0.1 indicated by the dots in the rectangle at the left, which encircle the point where the gap vanishes. There are four points with inversion symmetry: the first panel is for a case where there are well-separated bands (the atomic-like regime of the $s - p$ model) and the center panel in the covalent regime where the bands exchange character as a function of k and there is a state in the gap. The last panel is the same as the first at the end of a cycle of the hamiltonian. The surface states that cross the gap are highlighted by open and solid dots for the left and right ends, respectively.

26.6 The Thouless Quantized Particle Pump

The topological character of the cyclic change of the hamiltonian was derived by Thouless [256] in 1983 (before the work of Berry) using the same principles as he used in the TKNN paper [89] for the quantum Hall effect (QHE) summarized in Appendix Q. In the QHE the number of edge modes is a topological invariant, the Chern number that can take any integer value, denoted by Z. The same logic leads to the Z classification for the number of particles transported through the one-dimensional crystal that can be any integer.

An example of a pump is illustrated in Fig. 26.7 for a cyclic variation of the parameters in the two-site model with the hamiltonian Eq. (26.4) (or Eq. (26.8)). (Exercise 26.9 is the same for the $s - p$ model in Eq. (26.5) (or Eq. (26.7)).) The values of $t_1 - t_2$ and Δ are indicated by the dots in the rectangle at the left, which are discrete steps in a square cycle. The essential point is that $t_1 - t_2$ and Δ each change sign in a way that encircles the point $t_1 = t_2, \Delta = 0$ where the gap vanishes, which has the same qualitative behavior as the cycle depicted in Fig. 26.6. The pump action can be detected by calculation of the Berry phase or Wannier centers for the periodic crystal or a calculation for a large finite system, where there must be states at the ends of the crystal that shift to remove a state at one end and add it at the other.

First, consider a periodic crystal. There are four special points where there is inversion symmetry,[8] and either h_1 or h_3 vanishes, which happens for $t_1 = t_2$ and $\Delta = 0$, respectively,

---

[8] This is an example of a polarization lattice in Section 24.8.

as indicated at the top of Fig. 26.7. In this case the centers of the Wannier functions can be determined by inspection: progressing from an A–B bond center, to a B site, to a B–A bond center, an A site, and the A–B bond center in the next cell. The change in the position of the Wannier center is indicated in the bottom line of the figure. The change in Berry phase is proportional to the Wannier center, varying from 0 to 2π, so that it winds as depicted in the top part of Fig. 25.2.

For the finite chain, the energies of the states at the ends depends upon the termination. Since the topological character is independent of termination, we can consider any one case and it is instructive to see how the conclusions apply for other choices. The results in Fig. 26.7 are found for the case where the finite crystal is an integer number of A–B cells, terminating on A sites at one end and B sites at the other end, with the site energies the same as in the bulk. The energies for the finite system in a range around the gap are shown in Fig. 26.7. The left side of the figure with $\Delta = 0$, $t_1 > t_2$ corresponds to ending on a strong bond, so that there is no surface state. In the middle $\Delta = 0$, $t_1 < t_2$, the crystal ends on weak bonds and there is a surface state. These are an example of the Shockley analysis without and with a surface state, where the Berry phase ϕ is 0 and π respectively, as shown in Fig. 26.5. The energies of the states for the entire cycle show the way that the surface state disappears into the bulk continuum, which is essential for a pump to transport a particle through the crystal as the energy increases on one end and decreases on the other with the crystal returning to the same state as at the start of the cycle.

What happens with other terminations? It is not hard to see that if the crystal is an integer number of B–A cells, the role of t_1 and t_2 are exchanged. The bulk is unchanged but now the crystal is terminated by a weak bond for $t_1 > t_2$ and the end state occurs in a different part of the cycle. The essential point is that there still is a state that transverses from the valence to the conduction band at one end and the opposite at the other end. If the surface is perturbed by changes of parameters, the energies are modified but there still must be such states that transverse the gap and join the bulk continuum that is unaffected by surface conditions.

A similar calculation for a cycle that does not encircle the zero-gap point could have three possibilities: no surface state over the range of parameters; a state that is fully inside the gap for the entire cycle; or a state that emerges from one band (upper or lower) and rejoins the same band during the cycle. In all those cases it is not hard to see that a band in the gap can be created or removed by modifying the surface.

Continuous Variation of the Hamiltonian

At this point it is most useful to adopt a general notation and a simple choice for a smooth variation is a circle in which h_1 and h_3 vary as sine and cosine of a variable λ, which varies from 0 to 2π. In the two-site model, this can be expressed as

$$\mathbf{h} = [v + D\cos(\lambda) + v\cos(k), \quad w\sin(k), \quad D\sin(\lambda)], \tag{26.9}$$

where $v = w = t_2$, $t_1 - t_2 = D\cos(\lambda)$ and $\Delta = D\sin(\lambda)$. In the $s - p$ model a corresponding form is

$$\mathbf{h} = [D\sin(\lambda), \quad w\sin(k), \quad \bar{\Delta} + D\cos(\lambda) + d\cos(k)], \tag{26.10}$$

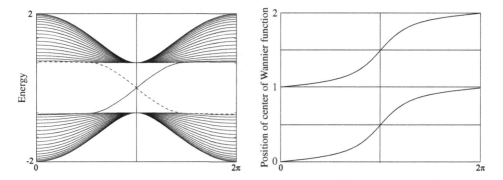

Figure 26.8. Left: states of the finite chain of 19 cells for a cycle of a pump defined in Eqs. (26.9) or 26.10 for the generic choice with all parameters set equal to unity: $D = v = w = 1$ or $\bar{\Delta} = D = d = w = 1$. The dark solid and dashed lines show the surface states that cross the gap in an upward direction at one end and a downward direction at the other end, which pumps an electron through the crystal. Right: displacement of the centers of the Wannier functions for the perfect crystal. Provided by J. Shallcrass Susinos and J. Junquera

where $s - p$ energy difference is $\bar{\Delta} + D \cos(\lambda)$ and there is an on-site sp term that breaks inversion symmetry and varies as $D \sin(\lambda)$. Notice that k and λ have comparable roles, but there is a very important difference: the factor of i difference between the Pauli matrices τ_1 and τ_2. The form of the $s - p$ model in Eq. (26.10) carries over directly to the two-dimensional Chern insulator in Eq. (27.2).

Figure 26.8 illustrates the surface state and the pumping action in a finite crystal of 19 cells for $D = d = w = 1$, which is chosen so that the coefficients of the terms involving k and λ are the same, which is useful in relating to the examples in Chapter 27. It is also convenient to set $\bar{\Delta} = 1$; the qualitative behavior is the same for any D large enough that the cycle includes hamiltonians that span a range from winding to nonwinding. The two surface states shown in the figure are at opposite ends of the crystal and they cross the gap in opposite directions as the parameter λ is varied from 0 to 2π, just as in Fig. 26.7. At the right in Fig. 26.8 is the shift of the Wannier centers, which corresponds to transfer of all the electrons by one cell, so that in a finite crystal the net effect is to transfer an electron from one end to the other. The direction of the electron transfer and the sense of circulation in Fig. 26.6 is defined by the phase of the sine and cosine terms.

26.7 Graphene Nanoribbons and the Two-Site Model

It is not easy to find an actual example of a one-dimensional system that can be varied to make a pump. Is there a clear example of the bulk and surface states in a system with the even/odd classification in Section 26.3?

Graphene can be used to form an amazing variety of systems that are almost ideal realizations of model problems. In Fig. 2.21 is shown one such system, a ribbon of graphene with alternating widths grown on an Ag substrate using a "bottom-up" procedure that

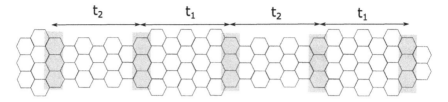

Figure 26.9. Schematic diagram of the nanoribbon in Fig. 2.21, which consists of segments with width 7 and 9 lines of carbon atoms. The shaded areas represent the midgap interface states due to the difference between the 7- and 9-atom-wide segments, as discussed in the text. These states form a one-dimensional two-site problem with alternating hopping matrix elements t_1 and t_2 that appears to be just like the two-site model in Fig. 26.2 with $\Delta = 0$. However, there is a difference because this is a band in the gap formed by the nanoribbons. The properties of the states are determined by the winding numbers for all the bands up to the gap. The theory is worked out in [236] with the result that is opposite to what would happen if one considers only the two states per cell.

involves patterning with precursor molecules. The figure shows the end of a long ribbon, which can be used to observe the effect that there are two possible states, one with a midgap state at the end and one that does not. This is an experimental realization of a one-dimensional two-site model like that depicted in Fig. 26.2 with $\Delta = 0$. In fact, it illustrates the model in more than one way, and the results reveal a feature that is not contained in the one-dimensional models as presented in Sections 26.2 and 26.3.

The description of nanoribbons and the properties of various repeated patterns have been worked out in a paper by Cao et al. [236]. Here the goal is to explain some of the key points that illuminate the transition between two states of one-dimensional systems, and the detailed proofs are left to exercises and references to [236]. Figure 26.9 is a schematic picture of a ribbon with repeated sections 7 and 9 atoms wide, like the experimental structure. The ribbon is oriented so the edge has the "armchair" shape shown in Fig. 14.7, where it describes the end of a graphene nanotube.[9] The first step is Exercise 26.10 to show that a long narrow ribbon is an insulator with no edge bands. The existence of a gap is essential for all that follows, which involves midgap states. As we have seen in Section 26.4, the bands in one dimension can be classified by winding numbers that are 0 or 1; the nanoribbon is not just a single chain but nevertheless the reasoning applies and Exercise 26.11 is to work out the winding numbers, following [236]. The 7-atom and 9-atom ribbons have different winding numbers, which means there is a bound state at each junction as shown in Fig. 26.9. Finally, we have arrived at a two-site $t_1 - t_2$ model that is apparently like the one pictured in Fig. 26.2 with $\Delta = 0$. It follows that there are two possibilities, one with an end state and one with no end state. However, this is not simply a two-site problem in a vacuum; it is a two-site band in a gap formed by ribbons that have different winding numbers! To determine the winding index one has to count all the bands up to the gap. As shown in [236], the end state occurs for the case where the ribbon terminates on a strong bond, the opposite of the conclusion for a simple two-site model.

[9] The relation of ribbons and nanotubes may be useful. A nanotube that is short but has large circumference is a ribbon!

SELECT FURTHER READING

See list at end of Chapter 25. Much of this chapter is patterned after the lectures of Asboth, Oroszlany, and Palyi referred to there.

Exercises

26.1 Derive the expression for the hamiltonian for the two-site model Fig. 26.2 in Eq. (26.4) and the eigenvalues $\varepsilon(k)$ given after Eq. (26.4). Show that the results are the same if the cell is chosen with B sites on the left and A on the right.

26.2 Consider the one-dimensional tight-binding model with two atoms per cell labeled A and B. If the basis is one s state on each atom, the model can be depicted pictorially by Fig. 26.2 where $\varepsilon_A, \varepsilon_B$ are the on-site energies and t_1, t_2 the hopping matrix elements. By varying the parameters, this model describes a symmetric ionic crystal ($\varepsilon_A \neq \varepsilon_B$, $t_1 = t_2$), a molecular elemental crystal ($\varepsilon_A = \varepsilon_B$, $t_1 \neq t_2$), and any ionic/molecular combination. Derive the bands as a function of the parameters and show that there is a gap between the two bands for all cases except the one-atom/cell limit where $\varepsilon_A = \varepsilon_B$, $t_1 = t_2$. See Exercise 26.1 for a related exercise and Exercises 23.11, 23.12, 24.8, and 24.9 for examples of Wannier functions, polarization, and effective charges using this model.

26.3 Show the transformation of the calculations in Exercise 26.2 to the $s - p$ model in Fig. 26.3 using the relations given in Section 26.2.

26.4 Derive the equations for of the dispersion curves for the one-dimensional lattice illustrated in Fig. 26.4 and show that the bands have the normal and inverted order respectively in the two cases.

26.5 See Exercise 14.14 for a related problem of an $A–B$ model with p states on the B sites.

26.6 Show that the Berry phase takes takes the value π (mod(2 π)) for the case where the oval encloses the origin. Explain how this relates to the fact that the wavefunction for a fermion changes sign for a 2π rotation. See also Exercise 28.7 for Weyl points.

26.7 Derive the analytic solution for the surface state wavefunction as it decays into the crystal for the two-site model. Hint: first show that the state has nonzero amplitude only on A sites if the crystal is terminated on an A site (and similarly for B sites). The solution can be found in [967].

26.8 Use the results of Exercise 26.7 to explain why the two end states in Fig. 26.4 are degenerate even though the calculation is for a finite-size crystal.

26.9 In the text is worked out the pump using the two-site model (Eq. (26.8)) and a cycle of parameters as shown in Fig. 26.7. Find an equivalent cycle using the $s - p$ model Eq. (26.7) and show that there are equivalent results for the pump.

26.10 Show that a narrow armchair ribbon is an insulator with no edge bands. It is useful to check Exercise 27.9, which shows the gap vanishes in the limit of a wide ribbon, but there are no edge bands.

26.11 The counting of the winding numbers is done in the paper of Cao et al. [236]. The exercise is to go through the arguments in that paper to find the winding numbers as a function of width, and the show that the 7 and 9 atom ribbons are different.

26.12 Show that a junction between two ribbons with different winding numbers has an interface
state. The simplest example is the two-site model in Fig. 26.2 with $\Delta = 0$, which is the
Su-Schrieffer–Heeger (SSH) model [954]. The famous result is that there is a bound state
an interface between two sections, one with $t_1 > t_2$ and the other with $t_1 < t_2$.

26.13 This exercise is at two levels. One is to explain the qualitative behavior of the pump shown in
Figs. 26.7 and Fig. 26.8. Explain why a cycle that does not enclose the zero point in the upper
left figure in Fig. 26.7 is not a pump. The second level is to do a calculation to derive the
results. This can be done using available tight-binding codes for a large finite system. Verify
that a pump transfers a state from one end to the other.

27

Topological Insulators I: Two Dimensions

Summary

Topological insulators are crystals with time-reversal symmetry in which there is nontrivial topology of the electronic structure due to the spin–orbit interaction; there are many examples in nature and there has been an explosion of activity to discover new systems and utilize their fascinating phenomena, often called the quantum spin Hall effect. This chapter is devoted to two-dimensional systems that are the primary examples of topological insulators and are also the basis for three-dimensional systems in the following chapter. The topological properties are illustrated using sp^2 models that are generalizations of the one-dimensional models used to describe surfaces in Chapter 22; they are the natural basis to take into account spin–orbit interaction and they embody the essential features of Hg/CdTe quantum wells (Section 27.7), which were the first experimental realization of a topological insulator. It is useful to first consider Chern insulators, which can be understood using only the material in Chapter 25, and then proceed to topological insulators and derive the Z_2 topological classification. The topology and edge states are illustrated by a square lattice with spin–orbit interaction increasing from zero to larger than the band width (Section 27.5) and a progression from chains to planes where there is a topological transition (Section 27.6). The two-site model provides a natural way to understand graphene as a topological insulator (Section 27.8).

The culmination of this part of the book is the theory of topological insulators that have generated great interest because of their extraordinary properties.[1] As discussed in Chapter 25, the role of topology in condensed matter had already been established in the 1980's for two-dimensional systems that exhibit the quantum Hall effect (QHE), where there are surface states in which electrons move in only one direction with zero resistance; however, this occurs only in a strong magnetic field. Haldane proposed a model for a

---

[1] In-depth presentation of the topics in this chapter and the next can be found in reviews by Hasan and Kane [943] and by Qi and Zhang [944], the book by Vanderbilt [918], and the basic theory in the lecture notes by Asboth et al. [945].

honeycomb lattice that involves complex hopping matrix elements that can lead to a QHE even in the absence of a net magnetic field, but it still has broken time-reversal symmetry. It was only in 2005–2006 that it was shown by two different lines of reasoning, by Kane and Mele [86, 87] and by Bernevig and Zhang [88], that spin–orbit interaction could lead to closely related effects in systems with time-reversal symmetry, now called topological insulators.[2]

The previous chapters provided an introduction to topological properties of electronic states in crystals. The topic of this chapter is the special properties of topological insulators. The goal is to understand the topological properties based only upon knowledge of band structure at the level of introductory texts on solid state physics, such as Kittel or Ashcroft and Mermin, illustrated by simple, pedagogical models. The natural formulation in two dimensions is an sp^2 model in Section 27.1, which leads directly to a two-band model for a Chern insulator (Section 27.2). Togther with the spin–orbit interaction, it is then only a few steps to the topological insulator in Section 27.4. Illustrative examples are given in Section 27.5, which is the model used to describe the first experimental example of a topological insulator, Hg/CdTe quantum well structures in Section 27.7, and in Section 27.6, which illustrates the evolution from chains to planes with a topological transition at a point where the gap closes. Graphene is the topic of Section 27.8, where it is an example of a different way to generalize the one-dimensional two-site model in Section 26.2.

27.1 Two Dimensions: sp^2 Models

In previous chapters the $s - p$ tight-binding model in one dimension has been used in many ways. It is a model two-band system in Section 26.2, with the s and p states depicted in Fig. 26.3 and the 2×2 hamiltonian is given in Eq. (26.5). There are two regimes depending on whether the bands cross or not, which is perhaps the first example where there are two distinct insulating states and a transition between them that can occur only if a gap vanishes. The mathematical formulations are readily understood simply as a transition from atomic to covalent character with a dangling-bond surface state, as brought out in Section 22.3. The goal here is to understand topological insulators in a way that brings out the beauty in the rigorous, concise framework characterized by topology, and yet understandable in terms of pictures there are almost as simple as dangling bonds.

The natural extension to two dimensions is a two-dimensional lattice and two-dimensional p_x and p_y states, which leads to a three-band model sp^2 (s, p_x, p_y) (six bands including spin) as shown in Fig. 27.1. It can also defined in terms of the complex p states $(p_x \pm ip_y)/\sqrt{2}$ with orbital angular momentum $L = 1$ and $m = \pm 1$, as illustrated in Fig. 27.2. It is a simple, intuitive model that can capture the essential features of real materials made of atoms where the primary spin–orbit interaction is centered on atoms and

---

[2] The term "topological insulator" has come to be used in different ways. Here we use it to denote systems with time-reversal symmetry that can be realized in systems with spin–orbit interaction. Other cases where topology plays an essential role are indicated explicitly, for example, crystalline topological insulators, Chern insulators, and quantum Hall systems.

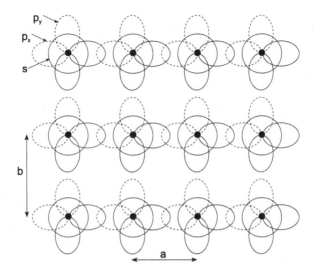

Figure 27.1. Generalization of the sp model one dimension shown in Fig. 26.3 to a two-dimensional lattice with three bands: one s and two p states per site. The figure depicts chains with lattice constant a separated by distance b, which can vary from $b \gg a$ (isolated chains) to $b = a$ (square lattice). The two p states can be considered as p_x and p_y as shown here or as states with angular momentum $L = 1$ and $m = \pm 1$ as shown in Fig. 27.2.

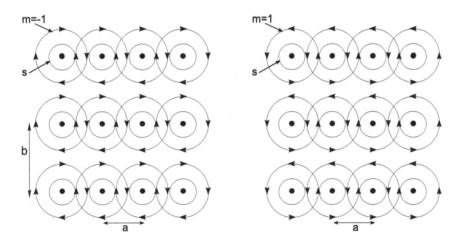

Figure 27.2. The three-band model in Fig. 27.1 can be cast in terms of p states $(p_x \pm i p_y)/\sqrt{2}$ angular momentum $m = \pm 1$. In a Chern insulator time-reversal symmetry is broken and one of the $m = \pm 1$ states is preferred. Together with the s state, this forms the two-band problem that is a precursor to topological insulators. The two figures provide graphic images of currents circulating around the boundary in opposite directions: A Chern insulator with only one p orbital $m = 1$ or $m = -1$ has an edge current like the quantum Hall effect in Fig. Q.2. A topological insulator with time-reversal symmetry has currents propagating in both directions as explained in the text.

most readily represented in terms of degenerate atomic-like states. We will consider the progression from one to two dimensions, i.e., chains to planes, and Figs. 27.1 and 27.2 are drawn to depict chains with lattice constant a, separated by a distance b, which can span the range from isolated chains with $b \gg a$ to a square lattice with $b = a$.

27.2 Chern Insulator and Anomalous Quantum Hall Effect

There is a case where one can readily show that the problem reduces from six to two bands, which is the most concise way to derive the topological properties (see Chapter 25). At this point we ignore spin–orbit interaction and the bands can be classified as purely ↑ or ↓ spin. In a magnetic insulator (which breaks time-reversal symmetry) one of the spin states and one of the p states with angular momentum ($m = \pm 1$) are favored. Similarly the lowest energy s state is nondegenerate and has the same spin. If the splitting of the spin and angular momentum states is large enough, the lowest bands are the eigenstates of a 2×2 problem that is called a Chern insulator,[3] and it is very instructive to consider this as a precursor of topological insulators.

Consider a two-dimensional lattice in Fig. 27.2 with an s and two p bands with $m = 1$, $(p_x + ip_y)/\sqrt{2}$ states per site. At this point, it is sufficient to assume it is a square lattice with on-site energies and nearest-neighbor interactions t_{ss}, t_{pp}, and t_{sp} that are the same in the x and y directions. (See Section 27.6 for cases where x and y are not equivalent.) The hamiltonian is the extension the one-dimensional form given in Eq. (26.5) or the form with electron-hole symmetry in Eq. (26.7), which can be expressed as (in units where the lattice constant $a = 1$)

$$H(k_x, k_y) = \begin{bmatrix} \Delta_{sp} + V(\cos(k_x) + \cos(k_y)) & v(i\sin(k_x) + \sin(k_y)) \\ v(-i\sin(k_x) + \sin(k_y)) & -\Delta_{sp} - V(\cos(k_x) + \cos(k_y)) \end{bmatrix},$$
(27.1)

and the vector is the space of Pauli matrices can be written

$$\mathbf{h} = [v\sin(k_y), \ v\sin(k_x), \ \Delta_{sp} + V(\cos(k_x) + \cos(k_y))], \tag{27.2}$$

where $V = t_{pp} - t_{ss}$ and $v = \sqrt{2}t_{sp}$.

The crucial feature that distinguishes this hamiltonian from one that has time-reversal symmetry is the factor of i difference in the sp matrix elements in the x and y directions, $v(i\sin(k_x) + \sin(k_y))$, which results from the form of the basis state $\propto p_x + ip_y$. The consequence is expressed in terms of the Pauli matrices in Eq. (27.2) (where the first two terms are the coefficients of the real τ_1 and imaginary τ_2) in a way that is qualitatively different from a system with time-reversal symmetry.[4]

The topology of this hamiltonian is illustrated in the right side of Fig. 26.6 where the ovals represent the variation with k_x for a given value of k_y. The points $k_y = 0$ and

---

[3] A Chern insulator is interesting in itself. Since spin plays no essential role, it can be thought of as a spinless problem. This is what was studied in the TKNN paper that demonstrated the topological character of the quantum Hall effect. However, it is not common and, to the knowledge of the author, there is no known natural crystal that is a Chern insulator in the absence of an applied magnetic field.

[4] Equation (27.2) is the model of Qi, Wu, and Zhang [971] for a Chern insulator, which is the stepping stone toward the model for a topological insulator.

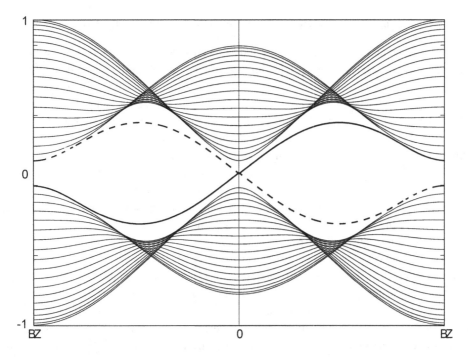

Figure 27.3. Energy bands for a Chern insulator strip that is 19 atoms wide with the hamiltonian in Eq. (27.1) and parameters as described in the text. The two curves in the gap are edge states on the two sides of the strip; electrons can flow in the edge states without resistance in analogy to the edge states of the quantum Hall effect depicted in Fig. 25.3. As discussed in text, the hamiltonian for this problem can also represent an electron pump with the momentum k_y replaced by a variable parameter λ. Note the similarity to the bands of a topological insulator in Fig. 27.7. The flow of the hybrid Wannier functions as a function of momentum is not shown, but it is qualitatively the same as in Fig. 26.8.

$k_y = \pi$ correspond to the cases with $h_1 = 0$ (a center of symmetry in the models) and the progression around the circle in the direction of the arrows indicate the variation of the hamiltonian as a function of k_y. Figure 26.6 illustrates an example of nontrivial topology of the band structure, which occurs if the parameters in Eq. (27.1) are chosen so that the periodic variation includes a point where **h** winds around the origin and one that does not, i.e., the surface encloses the origin in the three-dimensional space of the Pauli matrices. The topology of the hamiltonian is described by the Chern number, which is a topological invariant, as described in Sections 25.4 and P.4 and, as shown by Thouless and coworkers, the Chern number equals the number of states on an edge, which is also equivalent to the winding of the Berry phase, as worked out in Section 25.5.

An example of the bands is shown in Fig. 27.3 for a square lattice made into a strip of 19 atoms in width. The hamiltonian is given by Eq. (27.1) with $\Delta_{sp} = 1.0$, $t_{ss} = 0.25$, $t_{pp} = -0.5$, and $t_{sp} = 0.25$. The two curves that cross the gap are the two surface bands, one on each side of the strip, that propagate in only one direction like a quantum Hall state as depicted in Fig. 25.3. In this case there is one state on each edge that corresponds to a Chern number equal to one. The bands have qualitatively the same form as in Fig. 26.8 and are similar to each of the time-reversed bands of the topological insulator in Fig. 27.7.

Relation to the Electron Pump and Shift of the Wannier Centers

A two-band model for a Chern insulator and an electron pump can each be expressed with the same form for the hamiltonian, but with different interpretations for the terms. In a Chern insulator the coefficient of τ_1 in Eq. (27.2) arises naturally in an sp^2 model for a system with inversion symmetry but with broken time-reversal symmetry and it is a periodic function of the momentum k_y with period 2π. A pump can also be represented by Eq. (27.2) and Fig. 26.6 if k_y is replaced by a parameter λ that varies from 0 to 2π. Viewed as a one-dimension crystal with hamiltonian that depends on a parameter λ, the interpretation is that the coefficient of τ_1 is a term that appears if inversion symmetry is broken and there are nonzero matrix elements between s and p states on the same site that vary as $\sin(\lambda)$. On the diagonal is an added term that varies as $\cos(\lambda)$. Depending on the phase relation of the sin and cos terms, the electrons are pumped one way or the other by one lattice constant for each cycle of the pump. This is the aspect of the pump that is analogous to broken time-reversal symmetry in the Chern insulator.

In the pump, the natural interpretation is the shifts of the centers of the Wannier functions that lead to transfer of an integer number of electrons from one end to the other through the one-dimensional crystal even though it is an insulator, so long as the hamiltonian as a function of λ varies in particular ways. As described in Section 25.5, this provides a convenient method to determine the Chern number from the changes of the Berry phase by integer multiples of 2π.

The Anomalous Quantum Hall Effect

The same analysis applies to a Chern insulator, but in this case Wannier centers shift as a function of the momentum k_y parallel to the edge and the transfer of electrons is from surface (edge) bands on one side to bands on the other side. This sets up the situation that leads to the anomalous quantum Hall effect illustrated by the left side of Fig. 25.3. As explained in Section Q.2, adding (subtracting) an electron from a band that propagates in only one direction increases (decreases) the current along the edge by a quantized amount. The fact that there must be surface bands that propagate in only one direction is determined by the topology of the bulk electronic structure; however, the filling of the surface bands can be controlled by external potentials that set the Fermi level. In this case, the electric field across the strip leads to current along the strip, i.e., a Hall-like effect. The reason is that the voltage difference increases the number of electrons on one side and decreases the number on the other side. Since they flow in opposite directions on the two sides, the two contributions add to produce a net change in the current that is quantized exactly like the quantum Hall effect.

27.3 Spin–Orbit Interaction and the Diagonal Approximation

Up to this point the examples of topological insulators have involved broken time reversal symmetry: the quantum Hall effect in a large magnetic field (see Appendix Q) and a Chern insulator that has ferromagnetic orbital currents and an anomalous quantum Hall effect.

The understanding of the Berry curvature and Chern numbers in Section 25.4 might appear to rule out the possibility that there could a nontrivial topology if there is time-reversal symmetry. However, there is an exception to this line of reasoning that lay unnoticed until the discovery of topological insulators. In a system with time-reversal symmetry, all states come in pairs with reversed velocity and spin (the Kramers theorem) so that there can be effects that have the properties of a system with broken time-reversal symmetry so long as there are pairs of states that act in diametrically opposed ways. This is what is accomplished by the spin–orbit interaction.

The spin–orbit interaction is a consequence of special relativity that can be interpreted as an effective magnetic field (opposite for the two spin states), and it is this property that is the basis for topological insulators that have time-reversal symmetry. The general form of the spin–orbit interaction is derived in Appendix O and the expression needed for p states in tight-binding models are given in Section 14.3, where the magnitude of the effect is determined by the parameter ζ. The only things one needs to do are to double the size of the matrix and add H_{SO} in the p components of the hamiltonian, as shown in Eq. (14.11). This describes a 6×6 hamiltonian that involves all three p bands, and it is important to keep the p_z components in order to describe the full spin–orbit interaction, even if the p_z states are only a small component of the relevant low-energy bands.

Diagonal Approximation and Two-Dimensional Systems

The diagonal approximation is simply to ignore the off-diagonal terms $H_{SO}(\downarrow, \uparrow)$ and $H_{SO}(\uparrow, \downarrow)$ and we can see immediately what happens: there are two separate systems, $\uparrow$ and $\downarrow$, which each act as if it is in a magnetic field that is opposite for the two spins. It is not a Zeeeman term that splits $\uparrow$ and $\downarrow$; it is a velocity-dependent term in the spatial part of the hamiltonian that leads to circulating currents, opposite for the two spins. Thus the system is two copies of a Chern insulator for $\uparrow$ and $\downarrow$ spins and, in this approximation, the two circulating currents are independent and can be classified by two Chern numbers $(C_\uparrow, C_\uparrow)$ each of which can be any integer.

The diagonal approximation is especially important for two-dimensional systems, as can be seen in Eq. (14.12) for the matrix elements of the spin–orbit interaction for p states in three dimensions. If the p_z state is ignored to create a strictly two-dimensional system with p_x and p_y states, there are no off-diagonal terms and the diagonal approximation is exact. Then we are only concerned with the spatial part of the hamiltonian, which is the same for $\uparrow$ and $\downarrow$ spins. There are two possibilities: If the band structure has a normal form with separated bands that do not cross, the system has trivial topology and there are no edge states. If the bands cross, it has nontrivial topology and there are edge currents, opposite for spins. It follows that in a strictly two-dimensional system, where both space and spin are restricted to two dimensions, there is quantum Hall effect for each spin with equal and opposite currents for $\uparrow$ and $\downarrow$ spins on each edge.

There is, however, a hole in this argument: real systems are not strictly two-dimensional. The spin should be treated in three dimensions and even though the bands may have two-dimensional quantum numbers. The wavefunctions are not confined to a plane with zero thickness and there is some three-dimensional character. In the present models this means that the p_z state has a nonzero effect even if it is at high energy. There are consequences for the topology that are discussed next.

27.4 Topological Insulators and the Z_2 Topological Invariant

The previous sections have dealt with two systems that display properties analogous to the quantum Hall effect: a Chern insulator, which has broken time-reversal symmetry, and an approximation to a system with time-reversal symmetry in which spin is conserved ($\uparrow$ and $\downarrow$ along some axis), which is equivalent to two independent copies of a Chern insulator. The topic of this section is the qualitative change that leads to a new topological classification of the electronic structure when the spin–orbit interaction is treated properly. The results follow from the requirement that, in a system with time-reversal symmetry, there are always degenerate pairs of states with reversed spin and momentum (Kramers degeneracy).

There are two complementary ways to understand the effect by considering the topology of the bulk system or by considering the edge states. It is useful to see the effect in both ways, which provides insight and brings out the bulk-boundary correspondence. In each way of approaching the problem, the analysis is based on the same fundamental property: the two spin systems are not independent. At general point $\mathbf{k}$ in the Brillouin zone, the bands cannot cross.[5] The only points where the states must be degenerate are momenta $\mathbf{k}$ that are invariant under time reversal, called time-reversal invariant momenta (TRIM) denoted $\mathbf{K}_a$, which are shown in Fig. 27.4 for the bulk crystal and the edge.[6]

Z_2 Topology of the Bulk Band Structure

In a two-dimensional crystal a BZ can always be defined to be a parallelogram; for our purposes it can be represented by a square. (It is important to realize the analysis applies to any crystal and is *not* restricted to a square lattice.) In Fig. 27.4 is shown 1/4 the BZ, where the TRIM are at the corners $(0,0)$, $(0, \pm \pi)$, and (π,π), which are designated $\mathbf{K}_a$, $a = 1,4$. If the crystal has a center of inversion there is a simple procedure that uses the fact that all Bloch states at the four TRIM $\mathbf{K}_a$ are invariant under inversion as well as time reversal. From this Fu and Kane [972] showed that the Z_2 index $\nu = 0$ or 1 is given by

$$(-1)^{\nu} = \prod_{a=1}^{4} \prod_{i=1}^{N/2} \zeta_{ai}, \tag{27.3}$$

where $\zeta_{ai} = \pm 1$ is the parity of band i at $\mathbf{K}_a$. It is important to note that the parity at any one TRIM is *not* uniquely defined. In general the character of the Bloch functions can be

---

[5] Exceptions are semimetals with Weyl points; however, they can occur only in three or higher dimensions (Section 28.4).

[6] There may be also other degeneracies depending on the crystal symmetry that must be considered in each case individually.

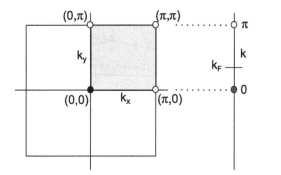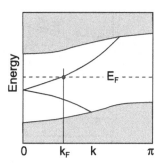

Figure 27.4. The time-reversal invariant momenta (TRIM) $\mathbf{K}_a$ for a lattice in two dimensions and the projection onto a one-dimensional edge. At the left the shaded area is 1/4 the BZ for the crystal and the single filled circle denotes an example of the Z_2 odd phase where the product of signs is -1 (see Eqs. (27.3) or (27.4)). One-half the edge BZ is indicated in the middle with a possible position of the Fermi point k_F. Schematic surface bands, shown at the right, are filled to the Fermi energy corresponding to k_F.

varied by choice of the unit cell and center of inversion, called a choice of gauge, but the product in Eq. (27.3) is unique, i.e., gauge invariant.

If there is no center of inversion, the problem is more complicated; however, an expression analogous to Eq. (27.3) has been derived in [972] (Exercise 27.1) in terms of the Bloch functions at the TRIM $\mathbf{K}_a$:

$$(-1)^{\nu} = \prod_{a=1}^{4} \delta_a, \quad \text{where} \quad \delta_a \frac{\sqrt{\det[W(\mathbf{K}_a)]}}{\text{Pf}[W(\mathbf{K}_a)]} = \pm 1, \tag{27.4}$$

where Pf denotes a Paffian which is defined so that its square is the determinant of the skew-symmetric matrix (a matrix with $A_{ij} = -A_{ji}$, see Exercise 27.1) $W_{mn}(\mathbf{K}_a) = \langle u_{m-\mathbf{K}_a} | \mathcal{T} | u_{n\mathbf{K}_a} \rangle$ where $\mathcal{T}$ is the time-reversal operator. The δ_a depend on the choice of gauge but the product in Eq. (27.4) is gauge with a consistent choice of the square root [972].

There is another approach that has already been explained in Chapter 25 because it is so useful and instructive: the flow of the Wannier functions, which is equivalent to the winding of the Berry phase in Section 25.5 and illustrated in Fig. 25.2. The application in topological insulators is described clearly by Vanderbilt [918], who refers to the original works [955, 956] and an automated procedure [957]. It provides a physical picture and is closely related to the analysis in terms of surface states; however, it only involves calculations for the bulk crystal. It can be also applied readily in three dimensions, which is illustrated for a real material in Fig. 28.5.

Nature of the Edge States

Edge states are the primary manifestation of the topology that can be measured experimentally. Correspondingly, they are a way to determine the topological character theoretically. Calculation of surface states requires greater effort than for the bulk crystal, but once

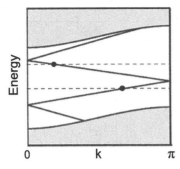 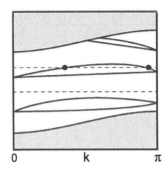

Figure 27.5. Schematic illustration of the bulk and surface bands of a topological insulator in one half the Brillouin zone. As shown in Fig. 27.4 there are two TRIM, the zone center $k = 0$ and boundary $k = \pi$, where the surface bands must be degenerate. At the left is a case with an odd number of surface bands crossing at any energy, e.g., the two dashed lines, and at the right is an example with an even number of bands. This defines two classes (even and odd) called Z_2. An even number can always be eliminated by surface effects, but an odd number signifies topologically protected metallic surface bands that must cross the gap from the bulk valence to conduction band edges. Based on a figure in [943].

it is done one only needs to count the number of edge states. Using the bulk-boundary correspondence one can work backward to determine the topology in the bulk. The one-dimensional edge states are labeled by the momentum k in the surface BZ, $-\pi$ to π in dimensionless units, which applies to any crystal. The surface BZ and the two TRIM (at the zone center $k = 0$ and boundary $k = \pi$) are shown in the middle in Fig. 27.4, where only 1/2 the BZ is indicated. At the right in Fig. 27.4 is a schematic figure of surface bands filled to the Fermi energy in the gap. At the TRIM, the states are required to be degenerate by Kramers theorem, but at any other point in the Brillouin zone, the bands cannot cross, which is illustrated in Fig. 27.5. There are two possibilities illustrated by the two parts of the figure. At any energy in the gap, e.g., the two dashed lines, there is either an odd or an even number of bands in 1/2 the BZ, shown at the left and right, respectively. Note that the criterion is *not* in terms of the number of states in the gap at the TRIM; at those points, just like any other k-point, the bands may be shifted out of the gap by a surface effect. There may be many bands in the gap, but all cases can be classified as even or odd, which is the set of integers modulo 2 denoted by Z_2.

It is useful to verify that any case can be reduced to even or odd by considering various ways that the bands can be modified by surface conditions. The object of Exercise 27.2 is to show that in any case with an even number of bands all surface bands can be removed from the gap by a strong enough surface potential, i.e., all cases are topologically equivalent to zero. Similarly, any odd number of bands is equivalent to one. Also the exercise is to connect these conclusions to the fact that the Shockley states for a one-dimensional chain can be pushed out of the gap, unless there is a symmetry requirement such as electron-hole symmetry, which is called chiral symmetry.

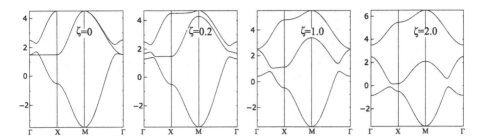

Figure 27.6. Band structures for the sp^2 model with a square lattice. Each graph plots the bands in the two-dimensional Brillouin zone $\Gamma - X - M - \Gamma$, where Γ denotes the zone center $(0,0)$, X is a zone face $(\pi/a, 0)$, and M a zone corner $(\pi/a, \pi/a)$. In all cases the Slater–Koster parameters are $\varepsilon_s = -0.5$, $\varepsilon_p = 3.0$, $t_{ss} = 0.75$, $t_{pp} = -0.75$, and $t_{sp} = 0.20$, chosen so there is band crossing, and the zero of energy is in the middle of the gap for $\zeta = 2.0$. The panels show the results with no spin–orbit interaction at the left, and three values of the spin–orbit parameter $\zeta = 0.2$, 1.0, and 2.0 in the other three panels. Note that there is no gap if $\zeta = 0$ and there is a complete gap between the lowest bands, even for small ζ.

27.5 Example of a Topological Insulator on a Square Lattice

In this section we consider a square lattice with spin–orbit interaction that increases from zero to a magnitude that is larger than the band widths due to dispersion. The calculations include an s and three p states per site and the full spin–orbit interaction in three dimensions, which is the crux of the arguments that lead to the Z_2 topological invariant. The p_z state is set to high (but finite) energy so that effectively this becomes a two-dimensional, three-band model sp^2, but without the diagonal approximation. In all the examples the parameters are chosen so the bands at Γ are inverted, i.e., the s is above the p state, and the character reverses at the zone boundary points. We shall see that in this case the topology for the lowest two bands remains the same for all nonzero values of the spin–orbit interaction, even if the quantitative results depend on the details. A topological transition occurs if the parameters are varied so that the bands do not cross, which is illustrated for the upper bands in Fig. 27.7.

Figure 27.6 shows the bands along high-symmetry lines calculated using the sp^2 model for a two-dimensional crystal with a square lattice and parameters given in the figure caption. At the left are the bands with no spin–orbit interaction. Since the p states are degenerate at the zone center (Exercise 27.3) and some points on the zone boundary, there is no gap. The other panels in Fig. 27.6 show the effect of the spin–orbit interaction for three values of $\zeta = 0.2$, 1.0 and 2.0. At $\mathbf{k} = 0$ the p states are split into $J = 1/2$ and $J = 3/2$ states. There are gaps separating the three bands in all directions and a gap over the entire Brillouin zone between the lowest and the middle band even for small spin–orbit interaction. For $\zeta = 1.0$ the upper bands just touch, and for larger ζ there is a gap between the upper bands over the entire BZ. The topology of the band structure can be found by the condition in Eq. (27.3) in terms of the parities of the bands and Exercise 27.4 is to show that in all these cases the system has nontrivial Z_2-odd topology.

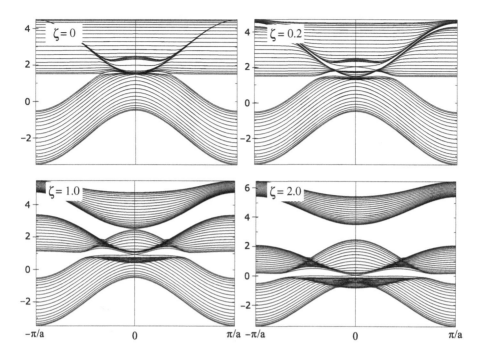

Figure 27.7. Results of a calculation for a strip 19 atoms wide for the same values of the spin–orbit interaction and the parameters as in the bulk calculations in Fig. 27.6. For $\zeta = 0$, the states in the gap are traditionally known as Shockley states. For each case with nonzero spin–orbit interaction, there are gaps between the bands with surface states that have linear dispersion at $k = 0$, and the bands in the lower gap are qualitatively the same for a Chern insulator in Fig. 27.3. A topological transition occurs in the upper bands at $\zeta = 1.0$ as explained in the text. The bands in the lower right figure are shown on an enlarged scale in Fig. 27.8.

Figure 27.7 shows the bands for a strip that is periodic in the y direction and 19 atoms wide in the x direction with two edges that are abruptly terminated like the end in the one-dimensional case in Sections 22.3 and 26.4. In the upper left panel of Fig. 27.7 are the bands for the slab if there is no spin–orbit interaction. In this case, there are parts of the surface BZ with a gap and parts where there is no gap. As a function of momentum parallel to the surface k, states in the gaps form bands as k is varied until it reaches the place where there is no gap. These are traditionally called Shockley states as discussed in Section 22.3. Note that the dispersion is quadratic around $k = 0$, as it must be for any problem where there are no velocity-dependent terms.

The other panels of Fig. 27.7 show the effects of spin–orbit interaction, which leads to gaps and additional surface bands. The energy range near the gap for the lowest two bands for $\zeta = 2.0$ are also presented on an expanded scale in Fig. 27.8, which clearly shows the way the surface states cross the gap and join the bulk states. At the TRIM $k = 0$ there must two degenerate states, which must continue in some way away from $k = 0$. Due to the spin–orbit effect (see Fig. O.1), the bands must have linear dispersion near $k = 0$ that

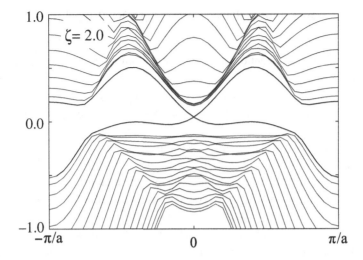

Figure 27.8. Electronic bands for a strip with $\zeta = 2.0$ on an expanded scale. Each edge has two bands with opposite spin shown as heavier lines, which join the bulk bands at point in the Brillouin zone as can be seen since the spacing between states is the same as for the other bulk bands. Note the similarity to the Chern insulator in Fig. 27.3 where there is one band on each edge.

is opposite for the two time-reversed spin–momentum states. One band goes upward and the other downward. If there are two electrons per cell that fill the lower band, the system is an insulator in the bulk and the surface states cross the gap from the occupied to the empty bands. This a topological insulator for all values of the spin–orbit interaction, and the surface state is an example of the bulk-boundary correspondence.

Figure 27.7 also illustrates a topological transition in the upper bands. For small values of the spin–orbit interaction ζ, there are surface bands that cross the upper gap. As ζ is increased, the upper band moves to higher energy and at $\zeta = 1.0$, the middle and upper bands just touch. This is a topological transition where the gap between the upper bands is zero at the zone center, as shown in the lower left panel. For larger ζ there is a gap as shown at the lower right; the highest band is well separated and does not cross the other bands at any k. Thus, the gap to the highest band is not inverted, the bands are in the normal band order, and there is no surface state between these bands.

This example brings out a crucial aspect of the topological characterization of the band structures of crystals: the classification depends upon which bands are included. If we count only the lowest band as filled, the system has nontrivial topology with surface states crossing the gap to the higher bands. However, if we consider the lowest and middle bands as filled, the electronic structure counting these bands as a whole has trivial topology for $\zeta > 1$. As shown in the figure, there are still surface bands between the lower bands, which can potentially be observed in an experiment such as photoemission, but no surface band crossing the gap to the highest band, which is consistent with trivial topology for the lowest and middle bands treated together.

27.6 From Chains to Planes: Example of a Topological Transition

The relation to one-dimensional problems is brought out if we consider a system of chains with a fixed spin–orbit interaction and vary the coupling between the chains until they become a square lattice. This corresponds to chains separated by a distance b, as depicted in Fig. 27.1 or 27.2, where b is varied from $b \gg a$ to $b = a$ for a square lattice. The spin–orbit interaction is set to $\zeta = 2.0$, which corresponds to the lower right panel of Fig. 27.7, and the calculations include the full spin–orbit interaction with no diagonal approximation. For this large spin–orbit interaction the upper spin–orbit split band is at high energy and we have effectively a two-band problem as discussed in the previous section.

The bands for a strip with a progression of increasing interactions between the chains is shown in Fig. 27.9. The on-site energies and the Slater–Koster hopping parameters in the x direction are the same as for the square lattice in Section 27.5, but those involving s and p_y in the y direction are reduced by a factor that varies from 0 to 1. At the left is the isolated chain with a state in the gap that is qualitatively the same as the Shockley surface state in Fig. 26.4. As interchain hopping is increased, the states broaden to become two bands in the gap with spin–orbit splitting except at the momenta with time-reversal symmetry (TRIM) $k = 0$ and $k = \pi/a$, where they are degenerate by time-reversal symmetry. The gap in the bulk bands

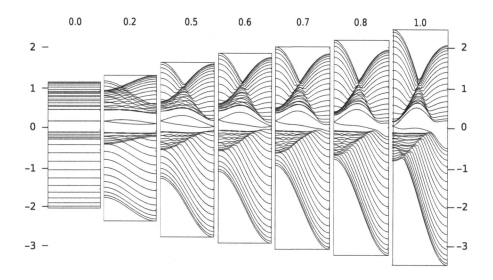

Figure 27.9. The bands of a 19-atom strip for different values of the coupling between chains. The spin–orbit interaction is fixed at $\zeta = 2.0$, and the ratio of interchain to intrachain varies from 0 to 1, where 0 denotes isolated chains and 1 denotes the square lattice, the same as in the lower right panel of Fig. 27.7. Only 1/2 the surface BZ is shown for each case; the other half is determined by time-reversal symmetry. We see that for small coupling illustrated by the ratios 0.2 and 0.5, the Shockley state forms a surface band with spin–splitting that is fully in the gap. At ≈ 0.6 the gap at the zone boundary vanishes (see Exercise 27.5), and for larger ratios the gap opens again to form a topological insulator.

decreases as the interchain hopping matrix elements increase due to the competition with the on-site spin–orbit interaction. The figure shows that in this case the gap goes to zero between the cases marked 0.6 and 0.7. This marks the topological transition in the bulk.

At the point in the surface Brillouin zone where the projected bulk bands have zero gap, the bulk and surface states of the slab just touch. Exercise 27.5 is to find the point where the gap vanishes due to competition between spin–orbit interaction and hopping. Beyond this point the bands have different character as a function of k in the surface BZ. In part of the surface BZ near $k = 0$ the bands remain inverted, but they are in normal order at larger k so that there is no surface state. Since one spin state merges into the upper band and the other merges into the lower band in half of the surface Brillouin zone. The surface bands in the gap are not protected at any one k: their energy is affected by surface conditions and they can be shifted out of the gap. However, for some range of k, there must be bands that cross the gap. This follows from the fact that the surface bands must be degenerate at $k = 0$ and one must vary continuously to join the upper band and one varies to join the lower band. For each spin the bands cross the Fermi energy an odd number of times as depicted in the right side of Fig. 27.5. This is the consequence of the bulk topological transition for the boundary states.

27.7 Hg/CdTe Quantum Well Structures

The first experimental example of a topological insulator was a quantum well of HgTe sandwiched between CdTe barriers. This forms a two-dimensional system like other quantum wells in semiconductors; in this case the special feature is that the band order is determined by the competition of spin–orbit interaction and the confinement in the quantum well. The prediction [973], and experimental observation [974], is that, in the regime where the bands have inverted order, there must be edge states that exhibit the quantum spin Hall effect.

For the bulk crystals of both materials the states near the Fermi energy are at the Γ point: CdTe has a large gap with normal order; however, HgTe has inverted order and zero gap. The essential features of the band structure of CdTe and HgTe have been known since the 1960s and it is sufficient to use experimental information. For our purposes, however, it is useful to consider this example as a test of the theoretical methods and their ability to predict topology of band structures. Figure 27.10 shows the results of calculations using two functionals. The results using the PBE generalized gradient functional have the well-known problem that the gaps are too low; in these cases the relevant gap is the difference between the states with s and p symmetry at $\mathbf{k} = 0$. For HgTe the gap is negative and "too low" means "too negative," i.e., a negative gap that is too large in magnitude. As shown in the upper right part of Fig. 27.10, this means the bands have the wrong character. The HSE functional in general is much more accurate as shown in Fig. 2.24. For HgTe it results in a band structure with the correct result that the s state is between the spin–orbit split p states.

For quantum wells the relevant states are determined by the energies at Γ and the masses (curvature) of the bands as indicated in Fig. 27.11. Since (HgCd)Te materials are of interest as narrow-gap semiconductors, there are well-established models, such as the Kane model

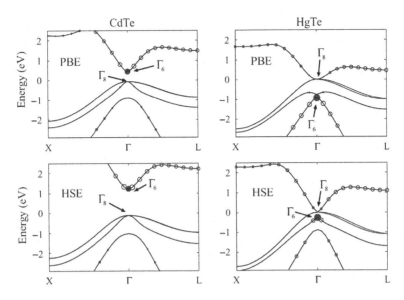

Figure 27.10. Band structure of CdTe and HgTe calculated using the generalized gradient approximation (GGA) functional PBE and the hybrid functional HSE. As usual, the gaps are too small using the local or GGA approximation, and much better results are found with hybrid functionals. The circles correspond to the weight of the s-like projection, and the filled circle is the purely s Γ_6 state. The essential point for a topological insulator is that in HgTe the gap is inverted, i.e., $\varepsilon(\Gamma_6) - \varepsilon(\Gamma_8) < 0$, but by an amount smaller than the spin–orbit splitting $\varepsilon(\Gamma_8) - \varepsilon(\Gamma_7)$. Using HSE the normal gap in CdTe and the inverted gap and relation to the spin–orbit splitting are both given correctly. Taken from [975].

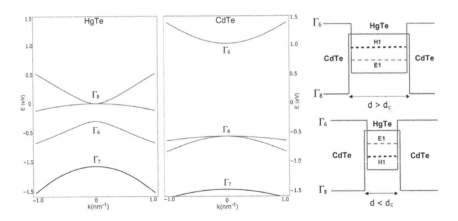

Figure 27.11. Band structure near Γ for HgTe, which inverted order of the bands, with the s (Γ_6) energy lower than the degenerate spin–orbit coupled p state (Γ_8), and for CdTe where the bands are in normal order. At the right are the energy levels for a quantum well where the order changes as a function of well-thickness d. Taken from figure in [973].

[976] with realistic parameters (see [973] supplementary material). In bulk HgTe the states labeled Γ_8 are fourfold degenerate corresponding the $J = 3/2, m_J = \pm 1/2, \pm 3/2$ and the Γ_6 have degeneracy 2, corresponding to $J = 1/2, m_J = \pm 1/2$. In the quantum well the states have quantum numbers $\mathbf{k}$ in two dimensions and a discrete label due to confinement in the perpendicular direction. The Γ_8 states are split so that finally there are two lowest-energy bands (four bands counting spin) at Γ in the two-dimensional Brillouin zone that are labeled as $E1$ ($m_J = \pm 1/2$ states that are linear combinations of Γ_6 and Γ_8) and $H1$ ($\Gamma_8, m_J = \pm 3/2$). A topological transition occurs at the critical well width d_c where the $E1$ and $H1$ states cross at Γ and there is a zero gap. The result is that in the regime of large d the bands are inverted. Thus the role of the quantum well is to confine the states in the HgTe to make a two-dimensional system with well-defined $E1$ and $H1$ states, but it should not be so thin that the band inversion is lost.

It is instructive to notice that the only information used to determine the topology is the nature of the bands near Γ, which can be described in a continuum model that does not depend on the lattice symmetry. In [973] it is shown that the bands are conveniently described in terms of a square lattice like that in Fig. 27.6, which has nothing the do with the actual diamond structure lattice. The point is that topological properties are robust and the only essential aspects of the bands is the inverted order at Γ and the fact that there is normal order in the rest of the Brillouin zone. The interpretation of the model for the HgTe quantum well is that $E1$ corresponds to an s, $\pm 1/2$ state and $H1$ corresponds to the two-dimensional spin–orbit coupled $(1/\sqrt{2})(p_x \pm ip_y)$, $\pm 1/2$. In the diagonal approximation for the spin–orbit interaction, the notation $\pm 1/2$ denotes the two spin states, and the system is properly termed a quantum spin Hall insulator, where the two spins make two copies of the quantum Hall effect. The notation is still used even if spin is not conserved since Kramers theorem guarantees that there are two oppositely directed bands at each edge with oppositive spin states, so long as the system has time-reversal symmetry.

27.8 Graphene and the Two-Site Model

The topic of this section is graphene, the other system for which the quantum spin Hall effect was first proposed. The story of the topological character of the bands starts with the observation by Haldane [977] in the 1980s that a gap can be opened in two ways. One way is to make the two atoms in the unit cell different, like in BN, which is an insulator. Haldane proposed another avenue for opening the gap: a magnetic field that averages to zero inside the unit cell and does not break any spatial symmetry of the lattice. He showed that this leads to a quantum Hall state that is distinct from a normal insulator (Fig. 14.5).

The major step to create a topological insulator with time-reversal symmetry was made in 2005 by Kane and Mele [87], who recognized that the spin–orbit interaction leads to a Haldane-like state for each spin, which they called the quantum spin Hall effect. If the spin is approximated as $\uparrow$ or $\downarrow$ (the diagonal spin approximation in Section 27.3) this leads to the Haldane hamiltonian with effective magnetic fields opposite for the two spins. In a separate paper [86] they showed that if the spin is teated properly, it leads to the Z_2 classification that sets a topological insulator apart from a quantum Hall system. Their example was graphene, but the argument applies to any two-dimensional system, as explained in Section 27.4.

The approach of Kane and Mele was to treat the spin–orbit interaction as a small perturbation that leads to a 4-band model (2 sites and 2 spins) per cell, which reduces to a 2×2 hamiltonian using time-reversal symmetry. This is an elegant approach in which the effective hamiltonian has complex hopping terms like the Haldane model. Here we describe an alternative approach that treats the sp^3 system directly in a straightforward application of the approach in the previous sections. This is a 16-band model (2 sites, 4 states per site, and 2 spins), which is not limited to small perturbations and not hard to treat using standard band structure methods. The results for bulk graphene are described in Section 14.7. The advantage is that the method applies directly to other problems where the effects are large, like the models in Sections 27.5 and 27.6 and bismuth on a honeycomb lattice in the following section; similar models apply to materials like Bi_2Se_3 in the following chapter. However, for graphene the spin–orbit interaction is an extremely small effect and the results must be the same as in the approach of Kane and Mele. Exercise 27.6 is to derive the perturbation expressions for effect of spin–orbit interaction, and to show that in the approximation where spin is conserved (↑ or ↓), spin–orbit interaction leads to a hamiltonian with complex second-neighbor hopping exactly like the Haldane model with opposite magnetic field for the two spins.

The π Bands of Graphene

Graphene is a single layer of carbon atoms in a two-dimensional honeycomb structure with a hexagonal Bravais lattice and two atoms per cell shown in Fig. 4.5. The Brillouin zone is shown in Fig. 27.12 with the corners labeled K and K', where each point K has a corresponding point with opposite momentum. The key feature is the π bands with Dirac points at K and K'. In the absence of spin–orbit interaction the valence and conduction

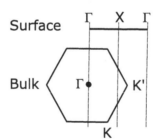

Figure 27.12. The Brillouin zone (BZ) for graphene and the surface BZ for a zigzag strip. The bands for the bulk crystals are shown in Fig. 14.5 where K and K' are the points where the gap in the π bands vanishes with Dirac linear dispersion. The BZ for the surface of a zigzag strip (the real-space structure is shown in Fig. 27.13) is shown centered on the zone boundary labeled X, and the dashed lines indicate the projection of the bulk BZ onto the surface. Bands for a strip are shown in Fig. 27.14, where we see that a surface state exists in the range between the projections of the K and K' points.

bands touch with linear dispersion and zero gap, which is readily understood using the simple tight-binding model described in Section 14.7 for which the bands are given by Eq. (14.19). The left panel of Fig. 14.5 shows the bands with the Dirac point for a calculation that includes only the π bands with hopping set to $V_{pp\pi} = -1.0$.

As explained in Sections 14.7 and 27.3, there would be no effect of spin–orbit interaction if the bands were strictly described by only the p_z states. There is a weak effect, however, because the spin–orbit interaction mixes the p_z with the other p states. The resulting bands with a gap are shown in right-hand part of Fig. 14.5 where the spin–orbit interaction is chosen to be much larger than the actual value for carbon to make the effect more visible. Note that a point K' is related to K by time-reversal symmetry such that it has opposite momentum and opposite spin. The eigenvalues are the same as at K but, since the spins are opposite, this is referred to as a negative gap.

Surface (Edge) States of a Graphene Strip

A graphene strip (also called a ribbon) is long in one direction with a width that is finite, but wide enough that there is a bulk-like region and a surface (edge) region on each side. Here we consider a zigzag strip that has the structure shown in Fig. 27.13. The bands $\varepsilon(k_\parallel)$ as a function of $k_\parallel$ parallel to the edge in the surface BZ are shown in Fig. 27.14.

With No Spin–Orbit Interaction. If there is no spin–orbit interaction there are Dirac points with zero gap at **K** and **K'**, which lead to the zero gap points in the projected bulk states, as shown in the left side of Fig. 27.14. There is a surface state that extends over part of the surface BZ, ending at the zero-gap points. It is straightforward to show (see Exercise 27.8) that this is exactly what is expected for a Shockley state (see Section 22.3) with zero gap at the transition point where the surface state ends. This can be understood as illustrated in the right side Fig. 27.13, which shows that the strip can be considered as double

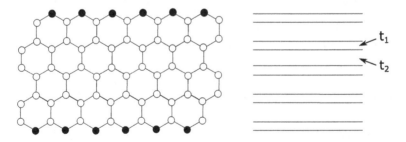

Figure 27.13. Left: graphene zigzag strip atomic structure where the black circles represent edge atoms and the strip is extended in the horizontal direction. Right: the one-dimensional representation of pairs of lines of atoms, which is sufficient to describe the eigenstates for the strip as a function of $k_\parallel$ parallel to the edge in the surface BZ. As discussed in the text this is equivalent to a two-site model shown in Fig. 26.2 with hopping matrix elements that vary with $k_\parallel$. This is sufficient to show that there is a transition as a function of $k_\parallel$ (the Shockley transition in Section 22.3) at the points where the gap vanishes. An edge state occurs only in one regime, which is the state at zero energy shown in the left side of Fig. 27.14.

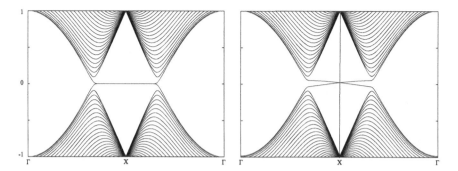

Figure 27.14. Electronic bands for a graphene strip 49 cells wide with the same parameters as for the bulk bands shown in Fig. 14.5. At the left is shown the Shockley surface state if there is no spin–orbit interaction. At the right are the surface bands of the topological insulator with addition of spin–orbit interaction that is much larger than the actual value in graphene to make the effects visible.

lines of atoms parallel to the edge with alternating distances. As described in the figure caption, this defines a one-dimensional problem that is equivalent to the two-site model in Fig. 26.2 with $\Delta E = 0$ and alternating hopping t_1 and t_2 that are functions of k. At $k = 0$, the edge is terminated by strongly bonded double lines ($|t_1| > |t_2|$). This is the normal order and there is no surface state. However, for k at the BZ boundary, the coupling t_1 within the double lines vanishes since the two neighbors are out of phase, whereas the hopping t_2 between the double lines is unchanged. Thus $t_1 = 0$, the double layer is decoupled, and the strip ends with a nonbonded (dangling bond) edge that forms an edge band in the gap.[7] The edge band must end at a point where the projected bulk gap vanishes. Thus we have derived the fact that the surface state must exist at an abruptly terminated zigzag edge using the arguments in Section 22.3. (The same analysis shows that there is not a surface band on the armchair strip as shown in Exercise 27.9 and in agreement with [978].)

With Spin–Orbit Interaction. The effect of spin–orbit interaction is to open a gap at the K and K' points. As discussed before, the gaps have opposite signs for each spin so that the surface band for each spin goes from the valence state at K (K') to the conduction band at K' (K) points. The results are very similar to those shown in Fig. 2 of the paper by Kane and Mele [87], which were calculated using the Haldane model with complex matrix elements that mimic the effects of spin–orbit interaction.

Calculations with realistic parameters for graphene have the same characteristic properties near the Dirac points. In addition, there are ordinary surface bands on the edge associated with the different bands. A realistic value of the spin–orbit interaction gives a very small gap that would not be visible on the scale of the bands in Figs. 14.5 and 27.14.

---

[7] A small second-neighbor hopping leads to dispersion of the surface band, but the gap still vanishes at the same points and there is still a transition between normal and inverted bands.

27.9 Honeycomb Lattice Model with Large Spin–Orbit Interaction

The model of graphene is a honeycomb lattice with the p_z states forming a π band at the Fermi energy. The other p_x and p_y states are tied up in strong covalent bonding with bands far below (bonding) and above (antibonding) the Fermi energy. They play little role in states at the Fermi energy, and the spin–orbit interaction is very small because it is a second-order effect due to virtual mixing of the p_z with the other p orbitals, and it can be understood in terms of the Haldane model.

There is a another possibility that is diametrically opposite: for a honeycomb lattice that is stabilized by bonding to a substrate, the p_z orbitals are tied up in covalent bonds with bonding and antibonding states that make filled and empty bands well below and above the Fermi energy. If the p_x and p_y states are partially filled and more weakly bound, they can be near the Fermi energy to form a metallic state or an insulator with a small gap. More important for topological insulators, the spin–orbit interaction is a first-order effect where linear combinations of the p_x and p_y states form $J = 3/2, m_J = \pm 3/2$ and $J = 1/2$, $m_J = \mp 1/2$ states, the states described in Section 27.1 and illustrated in Fig. 27.1. An example of a system like this with large spin–orbit interaction is a single layer of bismuth atoms in a honeycomb structure on a substrate, which has been grown experimentally [979]. See Exercise 27.7 for an estimate of the effects.

SELECT FURTHER READING

Books by Vanderbilt and Bernevig listed at end of Chapter 25.

Two original papers on prototypical topological insulators in two dimensions:

Kane, C. L. and Mele, E. J., "Z_2 topological order and the quantum spin Hall effect," *Phys. Rev. Letters* 14:146802, 2005.

Bernevig, A. Hughes, T. L., and Zhang, S.-C., "Quantum spin Hall effect and topological phase transition in HgTe quantum wells," *Science* 314:1757–1761, 2006.

Reviews of topological insulators in two and three dimensions:

Bansil, A., Lin, H. and Das, T., "Colloquium: Topological band theory," *Rev. Mod. Phys.* 88:021004, 2016.

Hasan, M. Z. and Kane, C. L., "Colloquium: Topological insulators," *Rev. Mod. Phys.* 82:3045–3067, 2010.

Qi, X.-L. and Zhang, S.-C., "Topological insulators and superconductors," *Rev. Mod. Phys.* 83:1057–1110, 2011.

Exercises

27.1 Show that the matrix $W_{mn}(\mathbf{K}_a) = \langle u_{m-\mathbf{K}_a}|\mathcal{T})|u_{n\mathbf{K}_a}\rangle$ defined after Eq. (27.4) is skew symmetric (a matrix with $A_{ij} = -A_{ji}$). Check the original paper [972] and summarize the derivation of the formula involving Pfafians.

27.2 Work out explicitly for yourself the conclusion that all cases reduce to even or odd and the relation to the Shockley problem stated in Section 27.4.

27.3 Show that for zero spin–orbit interaction the bands for the square lattice have the general form shown in the left figure in Fig. 27.6 with degeneracies at high symmetry points as shown.

27.4 Show that the bands do not cross at general **k**-points and the parities of the bands at the TRIM lead to a Z_2 odd state for any nonzero spin–orbit interaction. Give an example of a gauge transformation that can change the parity at any one point, but the product remains the same. (Hint: there is more than one center of inversion.) Show also that the definition in terms of Pfaffians gives the same result.

27.5 Find the value of the ratio that leads to a zero gap for the model used in Fig. 27.9. This can be done with a relatively simple calculation for the bulk bands at the zone boundary. Does the result agree with the transition point in Fig. 27.9?

27.6 Starting with the full sp^3 model on a honeycomb lattice, show that in a model appropriate for graphene with only p_z states near the Fermi energy and nearest-neighbor hopping, derive the perturbation expressions for the effects of spin–orbit interaction. Show that in the approximation that spin is conserved ($\uparrow$ or $\downarrow$), this leads to an effective hamiltonian with complex second-neighbor hopping. Hint: even if there is no direct hopping to second neighbors, show that there is hopping to a second neighbor due to spin–orbit interaction on the atom that is in between – a second-order effect. Estimate the size of the interaction using the spin–orbit interaction for the atom, which can be found in the literature, and show that the final effect is indeed extremely small.

27.7 Show that in the bismuthene system described in Section 27.9 the spin–orbit interaction is first-order effect, i.e., it does not require other bands. Calculate the eigenvalues at high-symmetry points for a nearest-neighbor hopping model and estimate the size of the effects. Compare with the observations in [979].

27.8 Show that a wide strip of graphene with a zigzag edge has an edge state like that shown in left side of Fig. 27.14. The answer is given in the text and this exercise is to verify the statements.

27.9 An armchair edge is one that ends on pairs of atoms. It is the same as the edges on the nanoribbon in Fig. 26.9 for the case where the ribbon is very wide. (It is also the same as an end of a nanotube in Fig. 14.7.) Show that the projected bulk density of states has only one point in the surface BZ where the gap vanishes, and there is no surface state. See also Exercise 26.10 for a narrow nanoribbon.

28

Topological Insulators II: Three Dimensions

Summary

In three dimensions there are new topological classifications and phenomena that do not occur in lower dimensions. Topological insulators can be divided into ones called "weak," which are fragile and susceptible to disorder, and others called "strong," which are robust and have two-dimensional states on all surfaces with Fermi surfaces, Dirac cones, spin textures, and other phenomena. Sections 28.2 and 28.3 describe a simple model and quantitative theory for Bi_2Se_3. Sections 28.4 and 28.5 are devoted to another type of system with interesting topology that can occur only in three (and higher) dimensions: Weyl semimetals where the bands touch at isolated points that have topological character. Such systems have remarkable properties such as surface states that are not possible in an isolated two-dimensional system.

28.1 Weak and Strong Topological Insulators in Three Dimensions: Four Topological Invariants

In the previous chapter, the Z_2 even and odd classification of topological insulators in two dimensions was derived using the principle that bands do not cross except at special points in the Brillouin zone called time-reversal invariant momenta (TRIMs). The principle applies in three dimensions; however, there are more possibilities for Z_2-odd states to occur in three dimensions, leading to different topological classifications. Since the surface of a three-dimensional crystal is a two-dimensional system, there is a greater range of possibilities for the surface states.

One of the ways to approach the relationship of two and three dimensions is to start with the fact that the structure of a crystal in three dimensions can always be viewed as a one-dimensional array of two-dimensional planes of nuclei. One possibility is that the electronic structure can also be described in this way to create a topological insulator in three dimensions analogous to stacks of planes each with Z_2-odd topology. However, there is reason to be cautious. For example, consider two Z_2-odd planes brought together gradually. This is still a two-dimensional system with the same TRIMs. Two decoupled planes would have separate circulating currents. If they are weakly coupled there will still be a gap in the bulk; however, the currents can interact, and the arguments of the previous chapter immediately lead to the conclusion that the bilayer should be Z_2-even with trivial topology. (See Exercise 28.1 for steps in the reasoning.) Thus, even without detailed knowledge of the topology, we can expect at least four possibilities for bringing together planes: one where the planes are Z_2-even and the resulting system is Z_2-even with trivial topology, and three ways that Z_2-odd planes can be brought together, one for each dimension. In the latter case, it seems clear that the properties of such systems may be problematic in some sense.

However, there may be other possibilities in three dimensions. It may be better to view the problem in the reverse direction: pulling the crystal apart. The previous examples can be described as classes of three-dimensional crystals that can be pulled apart to form isolated planes while remaining insulators at every point along the path, with four possibilities for the resulting planes. If there are cases where it is not possible to pull the crystal apart without a gap going to zero, they are fundamentally different and must have different topology of the electronic structure.

To put the arguments on a firm footing we follow the approach of Fu, Kane, and Mele [980] illustrated in Fig. 28.1. In the figures the Brillouin zone is represented as a cube, but it is important to emphasize that the analysis applies to any Bravais lattice since the Brillouin zone can be chosen to be a parallelepiped and all arguments carry over to that case. This

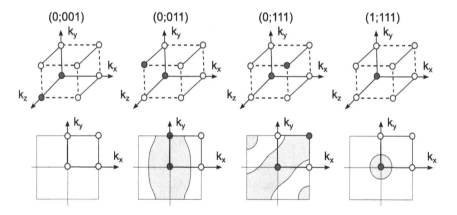

Figure 28.1. Schematic figures of the four phases of Z_2 invariants characterizing topological insulators in three dimensions, which is a variation a figure in [980]. At the top are figures with the eight time-reversal invariant momentum (TRIM) points in 1/8 of the BZ, where open and closed (dark) circles denote positive and negative signs. As explained in the text, the individual signs are not unique (subject to choice of a gauge) but the patterns are gauge invariant. In each case the TRIM in the $k_z = 0$ plane are chosen to be the same as in Fig. 27.4 for 1/4 the BZ in two dimensions (an example of the Z_2-odd case with one black circle). The change in the signs at the four TRIM with $k_z = \pi$ define the possible phases in three dimensions. The TRIM for the (0 0 1) surface are shown in the bottom set of figures along with possible examples of Fermi surfaces. The nature of the surface states for the first three "weak" cases is delicate, but the single Dirac cone for the "strong" case is robust.

is the analogue of Fig. 27.4 for two dimensions, but now there are more possibilities that are labeled by the integers at the top of Fig. 28.1. The top line of Fig. 28.1 illustrates the eight TRIM is 1/8 the Brillouin zone in three dimensions, with different possibilities for arrangement of the signs (the δ factors defined in Eq. (27.4) or parity in Eq. (27.3)). The examples shown in the figure are not unique; the individual signs are gauge dependent, but the final result of Z_2 odd or even is gauge invariant. This is the analogue of Fig. 27.4 for two dimensions but now there are more possibilities.

The example at the left labeled $(0; 001)$ represents a crystal with Z_2-odd planes stacked in the z direction, and the figure at the bottom indicates that it is insulating on the $(0\ 0\ 1)$ surface just like a single plane, which has circulating edge currents. For the other faces the edge currents become dispersive as a function of k_z so that those surfaces can be metallic with a Fermi surface. However, this a a fragile state. Doubling of the periodicity destroys the topological protection in analogy to the bilayer. Disorder can couple different momenta and lead to resistance in the surface currents. For that reason it has been termed "weak," which is denoted by the first index 0. The other two cases, $(0; 011)$ and $(0; 111)$, can be considered to be stacks along other directions, which completes the picture of three cases that are equivalent to stacked two-dimensional Z_2-odd layers. (See Exercise 28.2.)

The fourth example in Fig. 28.1 labeled $(1; 111)$ is different. So long as time-reversal symmetry is maintained such a system cannot be pulled apart to make planes without a gap going to zero at some point. This is indicated by the fact that there is only one point with odd parity or δ. In the figure this has been chosen to be at the origin (the Γ point), which leads to a Dirac point at Γ and a Fermi surface centered on Γ. This is the choice that applies for Bi_2Se_3 with the Brillouin zone and bulk bands in Fig. 28.4, which has a small region of overlap around the Γ point. This is one way to depict the $(1; 111)$ classification; in [980] is a different choice centered on the corner of the Brillouin zone.

The Z_2 phase $(1; 111)$ is termed "strong" because it is robust, with surface bands in the gap on all surfaces, which are protected by time-reversal symmetry. The fact that there is only one TRIM with negative sign means that the electronic structure is entangled in all directions and the system cannot be pulled apart in any way without a gap vanishing, so long as time-reversal symmetry is maintained. The surface of a weak topological insulator has an even number of Dirac cones. However, a strong topological insulator has an odd number of Dirac cones, e.g., the single cone observed on Bi_2Se_3. To appreciate how striking this is, recall that spin is counted. For example, a system with no spin–orbit interaction has degenerate $\uparrow$ and $\downarrow$ bands, so that it always has an even number of cones. In any isolated two-dimensional system even with spin–orbit interaction there are always two states related by time-reversal symmetry. In general, two states are required by the famous "fermion doubling theorem" [981]. The loophole that allows an odd number for a surface of a strong topological insulator is that the general theorem is for the entire system counting all surfaces; the partner of a single Dirac cone of one surface is on the opposite surface! Two-dimensional systems with an odd number of cones can exist *only on a surface of a strong topological insulator.* There are many other fascinating phenomena not feasible to cover here and the reader is directed to the exposition in the book by Vanderbilt [918].

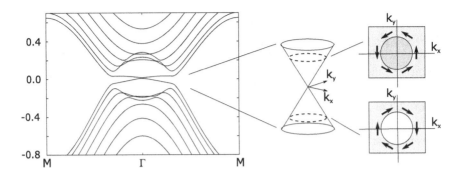

Figure 28.2. Electronic bands for a model similar to one for Bi_2Se_3. At the left is surface state and the projection of the bulk bands on the (0 0 1) surface where M denoted to corner of the 2D Brillouin zone. At the center is a schematic illustration of the Dirac cone at the Γ point, and at the right are typical Fermi surfaces above and below the point. Darker shading indicates increasing occupation of the surface band.

28.2 Tight-Binding Example in 3D

In this section we consider an sp^3 model with parameters chosen to have a strong topological insulator with one Dirac cone, which is analogous to Bi_2Se_3 in Fig. 28.3. All three p states are included, which is essential to include the spin–orbit interaction properly, but the bands near the Fermi energy are composed primarily of an even and an odd atomic-like states.[1] Figure 28.2 shows the resulting surface band and projected bulk bands for the (0 0 1) surface around the Dirac point at the zone center in the surface Brillouin zone. The middle and right parts of the figure are schematic pictures of the Dirac cone for the two-dimensional surface and typical Fermi surfaces for two energies above and below the Dirac point. Note that the spin, indicated in the figures at the right, has a chiral behavior, with $(\mathbf{k}, \uparrow)$ and $(-\mathbf{k}, \downarrow)$ related by time-reversal symmetry. These are single bands for one spin-momentum state; the corresponding band with the opposite chirality is on the opposite (0 0 −1) surface.

Diamond Structure and the Two-Site Model

In the previous chapter graphene was the example of a topological insulator that exemplifies the two-site model. The model has four bands (two sites and two spins) and complex hopping parameters that can be considered to be a consequence of spin–orbit interaction. This reduces to a 2×2 hamiltonian using time-reversal symmetry. The analogous model in three dimensions is the diamond structure, which has two sites per cell. The model is given

---

[1] The model is a three-dimensional tetragonal lattice, with the same parameters as used in the square lattice in Section 27.5 with spin–orbit interaction $\zeta = 0.5$, and with the p_z state added at lower energy than the p_x and p_y. A $t_{pp\pi}$ term is included (essential for dispersion of the p_z state in the (x, y) plane), and hopping between planes scaled to be a factor of 5 smaller than the hopping inside each plane to mimic a layered system like Bi_2Se_3.

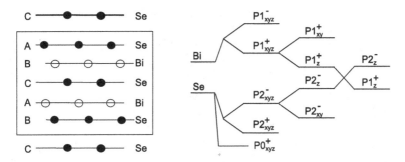

Figure 28.3. Left: structure for Bi_2Se_3, and Sb_2Se_3 showing the 2 Bi and 3 Se atomic layers per cell. The letters A, B, C denote the positions in a triangular lattice as shown in Fig. 4.9. Note that adjacent cells are displaced so that the translation vector is not orthogonal to the layers. Right: evolution of the bands near the Fermi energy from p states of Bi and Se at the left; bonding between the Bi and Se states; including the crystal fields and hopping that split the p_z and the $p_{x,y}$ levels; and including spin–orbit interaction that inverts the band order. The energies at the last two steps correspond to the states at the Γ-point without and with spin–orbit interaction in Fig. 28.4. Based on figures in [257].

in [980] and is discussed in the book by Vanderbilt [918], where it is the example worked out for trivial, weak, and strong phases.

28.3 Normal and Topological Insulators in Three Dimensions: Sb_2Se_3 and Bi_2Se_3

The prototypical example of a topological insulator in three dimensions is Bi_2Se_3. The surface of a three-dimensional crystal is two-dimensional and a signature of a topological insulator is a surface band with a linear dispersion that forms a single Dirac cone, which is protected by time-reversal symmetry. The theory and experimental observations have been done by two groups working independently, with results that are very similar and support one another. The topological nature of the bulk band structure has been worked out by H. Zhang et al. [257] and Xia et al. [982] and ARPES experiments by Xia et al. [982] and Chen et al. [255, 982]. Here we follow theoretical work in [257] and the experimental results from [255], which are compared in quantitative detail in Fig. 2.27. The experimental data are the basis for the figure on the cover of this book. A corresponding normal insulator Sb_2Se_3 completes the picture since it is similar in essentially every aspect except that the bands are not inverted. The analysis for Bi_2Se_3 and contrast with Sb_2Se_3 are described in some detail because it provides an especially clear example of both theory and experiment.

The crystal structure of Bi_2Se_3 is shown in Fig. 28.3, which is the same for Sb_2Se_3 with Bi replace by Sb. Each primitive cell contains one formula unit (2 Bi and 3 Se) arranged to form a crystal with five layers of atoms as shown in the figure, with the layers arranged in sites A, B, and C in the triangular lattice like the close-packed structures illustrated in Fig. 4.9. The five-layer cells are connected by weak Se–Se bonds to form a layer crystal analogous to the transition metal sulfides an selenides in Fig. 4.6. The stacking of the five-layer cells is displaced horizontally as shown in Fig. 28.3 so that the translation vector is not

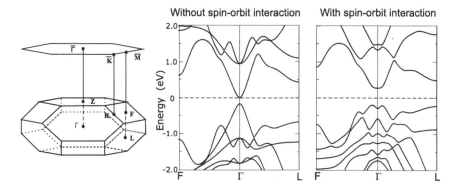

Figure 28.4. Brillouin zone for Sb_2Se_3 and Bi_2Se_3, and bulk bands for Bi_2Se_3: left without spin–orbit interaction and right with spin–orbit interaction. Note that the bands are inverted at the Γ point as indicated in the right side of Fig. 28.3. Taken from figures in [257].

along the direction perpendicular to the layers, which is a complication in the analysis but not a fundamental problem. The overall crystal structure has threefold rotational symmetry about the axis perpendicular to the layers and the Brillouin zone is shown in Fig. 28.4.

The calculated band structure is shown at the right in Fig. 28.4 without and with spin–orbit interaction. With spin–orbit interaction the sum of the parities of the bands up to the gap shows that it is a topological insulator because the bands are in normal order throughout the Brillouin zone except at Γ. It is a strong topological insulator consistent with the fact that there is only one Dirac cone. However, it is instructive to dissect the bands to understand the origin of the effects.[2] There are only two bands (four including spin) close to the Fermi energy. The states at Γ can be classified by parity and the two bands have opposite parity. Since there is time-reversal symmetry all solutions come in pairs and we can already see that a 2×2 hamiltonian suffices to describe the relevant bands as far as topology is concerned. As a first step in analysis the bands structures in Fig. 28.4 show that the order of the bands changes when the spin–orbit interaction is turned off. This is the hallmark of a topological transition, not just this particular example.

To go further in understanding the original of the bands, the right side of Fig. 28.3 shows to way a model can be used. The entire manifold of many bands near the Fermi energy can be described in terms of Bi and Se p states. If one completely turns off hopping, the Bi states are above the Se states. Including hopping pushes the Se states down and mixes in Bi character forming bands. Figure 28.3 is a schematic illustration of the states at Γ, which can be labeled by parity. The progression can be described in steps, the first ignoring the difference between p_z and $p_{x,y}$ and the second step taking into account the difference in the energies and hopping matrix elements. The result is two bands (four including spin) with opposite parity near the Fermi energy, as shown. The last step is to include spin–orbit interaction, which leads to an insulating gap with inverted bands at Γ like the calculated bands shown in Fig. 28.4.

---

[2] In earlier work [983] the gap and bands were shown to be given well by DFT calculations using the PBE functional, in contrast to many other examples where the gap is too small.

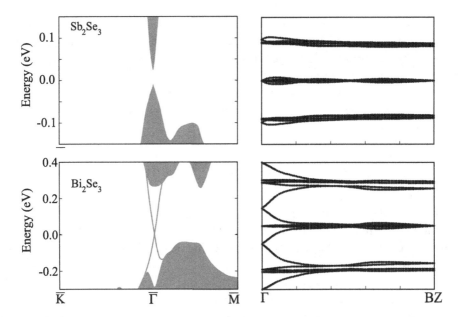

Figure 28.5. Composite figure of surface bands and Wannier centers for Sb_2Se_3 and Bi_2Se_3. Only Bi_2Se_3 has a topologically required surface state, which is also shown in Fig. 2.27 and compared quantitatively with an experimental spectrum. Provided by Haijun Zhang and Rui Yu; the left side is adapted from figures in [257].

The surface bands and projected bulk density of states for Sb_2Se_3 and Bi_2Se_3 are shown in the left side of Fig. 28.5. The figures show a narrow energy range and the surface bands for Bi_2Se_3, which are absent in Sb_2Se_3. The same information for Bi_2Se_3 is plotted in Fig. 2.27 where it is compared quantitatively with the experimental measurements in [255], which were published over a year later. The agreement is impressive.

The topological character can be established in terms of the parities of the bands at the TRIM points, which is done in [257]. It can also be determined by the flow of the centers of hybrid Wannier functions as described in Section 25.4. Since the functions are extended in two directions they can labeled by momenta in those direction and the flow can be calculated as a function of the in-plane momentum. In this case there are many Wannier functions per cell and there are complications because the translation vectors are not orthogonal; nevertheless, the main conclusions can be appreciated independent of the details. The right side of Fig. 28.5 shows the Wannier centers for Sb_2Se_3 and Bi_2Se_3. In both cases the functions rearrange within each layer, and only in Bi_2Se_3 near the Γ point is there a function that transfers from one cell to the next.

28.4 Weyl and Dirac Semimetals

It has been known since the work of Herring in the 1930s [984] that there may be degeneracies not required by symmetry. Herring called them "accidental degeneracies," meaning that they are not due to the symmetry of the crystal, but he realized that they are robust, i.e., the degeneracy is not lifted by small perturbations. The point in **k**-space where such a degeneracy occurs may change, but a degeneracy remains for a range of hamiltonians under appropriate conditions.

Such degeneracy points are now called Weyl nodes because the hamiltonian that describes the bands near the degeneracy point has the same form as the one introduced in 1929 by Weyl [985],

$$H(\mathbf{k}) = \chi\, c\, \mathbf{p} \cdot \boldsymbol{\sigma}, \qquad (28.1)$$

where c is the speed of light, $\mathbf{p}$ is the momentum and $\chi = \pm 1$ is the chirality, which denotes the handedness, i.e., whether the particle has spin parallel or antiparallel to the momentum. No such elementary particles with only one chirality have been discovered but they can be found in the electronic structure of crystals.[3] There are many ways that Weyl nodes can occur. Here the goal is to illustrate the remarkable consequences of the topology properties and we consider only the simplest case.

Before proceeding it is important to emphasize that the touching points of the bands (called Dirac and Weyl nodes) occur in the bulk band structure. This should not be confused with the previous sections that dealt with topological insulators, which are insulators in the bulk but have surface bands that touch at points (also called Dirac nodes) and cones as illustrated in Fig. 28.2. Because the semimetal character occurs in the bulk, the topology has experimentally measurable consequences for transport in the bulk. See, for example, the sources in the footnote in the previous paragraph for bulk phenomena, which are not covered here. There are also consequences for surface states, one of which is described in Section 28.5 as an illustration of the fascinating phenomena that result from nontrivial topology of the electronic structure in Weyl semimetals.

A zero gap at a general **k** point occurs only in three or higher dimensions. The minimum hamiltonian for which a gap can be defined involves two bands and a 2×2 hamiltonian, which can always be expressed as a linear combination of Pauli matrices, τ_i in Eq. (26.1). Here the τ matrices can represent space, spin, or a combination, and we consider only variations of the hamiltonian as a function of momentum **k**. Then Eq. (26.2) can be written as

$$H(\mathbf{k}) = \sum_{i=1}^{3} h_i(\mathbf{k})\tau_i, \qquad (28.2)$$

where the term involving τ_0 is omitted since it only shifts the bands and is not relevant for determining the gap. The gap is give by $\sum |h_i|^2 = 0$ so that a zero gap occurs only if all three coefficients $h_i(\mathbf{k})$ vanish. It is only in three or more dimensions that there are enough

---

[3] Resources for Weyl semimetals include the review by Armitage et al. [986] and the book by Vanderbilt [918]. For reviews that focus more on materials, see Hasan et al. [987] and Yan and Felser [988]. For a review of band theory and calculations, see Bansil, Lin, and Das [947]. A pedagogical overview is in the viewpoint "Where the Weyl things are" by Vishwanath [989]. The role of Weyl nodes in condensed matter is brought out in the book by Volovik [990].

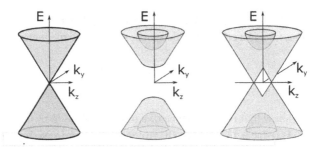

Figure 28.6. Schematic illustrations of a massless Dirac cone (left) and two possibilities when a mass and a Zeeman field along the z direction are added as in Eqs. (28.3) or (28.4): a magnetic semiconductor (center) or a system with two Weyl nodes (right).

degrees of freedom to make all three $h_i(\mathbf{k})$ zero at a general point in the Brillouin zone (see Exercise 28.3). Near the degeneracy point the bands disperse linearly to form two cones (up and down) as illustrated at the left in Fig. 28.6. Note that it is not guaranteed that a zero gap can be found at a generic $\mathbf{k}$-point if there are conditions on the hamiltonian; an example is in the next paragraph.

It is instructive to first define a Dirac semimetal where there is inversion and time-reversal symmetry. In general, near a point where bands touch we have a four-band problem (two spin and two band labels) like the Dirac equation for particles and antiparticles with spin. A Dirac semimetal occurs when the band touch with zero gap, corresponding to a massless Dirac equation. This is illustrated in the left side of Fig. 28.6 with a cone shaded in gray to emphasize that it is different from the cone in Fig. 28.2.[4] For four bands and a hamiltonian that obeys time-reversal symmetry, a zero gap can occur only at the TRIM points. An example is a transition point between a normal and a topological insulator with time-reversal symmetry. Any perturbation that preserves the symmetry opens a gap.

However, perturbations that break symmetries can lead to other possibilities. The simplest case a topological insulator near the transition point with an added Zeeman field. This is a perturbation that splits the bands up and down and it is not hard to imagine that there might be a way to have a zero gap for some $\mathbf{k}$. In the presence of a Zeeman field, the hamiltonian for the four bands can be expressed as (see [986])

$$h(k) = v\tau_x(\boldsymbol{\sigma} \cdot \mathbf{k}) + m\tau_z + b\sigma_z, \tag{28.3}$$

or

$$h(k) = \begin{bmatrix} m\sigma_0 + b\sigma_z & v\boldsymbol{\sigma} \cdot \mathbf{k} \\ v\boldsymbol{\sigma} \cdot \mathbf{k} & -m\sigma_0 + b\sigma_z \end{bmatrix}, \tag{28.4}$$

---

[4] For the surface, the cone is the dispersion of the two-dimensional bands. Here it is the bands of the three-dimensional bulk crystal plotted as a function of k_y and k_z for a value of k_x that passes through the degeneracy point. This is the edge of the continuum in the bulk bands with a gap that increases in all directions.

where τ_i and σ_i are the 2×2 Pauli matrices in the band and spin components, σ_0 is the unit matrix, m is the mass,[5] v is a characteristic speed, and b is the magnetic Zeeman field. The eigenvalues are

$$\varepsilon(k) = \pm\sqrt{m^2 + b^2 + v^2|\mathbf{k}|^2 \pm 2b\sqrt{v^2 k_z^2 + m^2}}. \tag{28.5}$$

Exercise 28.4 is to show that in the absence of a magnetic field ($b = 0$), the hamiltonian in Eqs. (28.3) or (28.4) reduces to the Dirac hamiltonian, and to derive Eq. (28.5).

There is a transition between two regimes at the point $m = b$ where it is apparent from Eq. (28.4) that two eigenvalues are zero for $\mathbf{k} = 0$. For $m > b$ the system is a magnetic semiconductor exemplified by the middle figure in Fig. 28.6. For $b > m$ the solution has two Weyl nodes at $\mathbf{k}_0$ and $-\mathbf{k}_0$ along the k_z axis (the axis along the direction of the magnetic field). Each of the two cones is 1/2 a Dirac cone, i.e., a two-band system with a hamiltonian analogous to Eq. (28.1) (see Exercise 28.6), which describes particles with only one chirality, the particles proposed by Weyl.

Topology enters the picture when we realize that each Weyl node is a source or sink of Berry flux, i.e., a sphere around one of the nodes has Chern number ± 1.[6] The proof is the same as the derivation of the Chern number for the sphere around a magnetic monopole in Section P.7 (see Exercise 28.7). There it was shown that wavefunction for the low-energy state of spin parallel to a radial magnetic field has a nonzero Chern number. The same holds for the high-energy state with antiparallel spin with the opposite Chern number. These correspond to the Berry flux of $\pm 2\pi$ for the upper and lowers cones.

Since this example has inversion symmetry it follows that Weyl nodes come in pairs at $\pm\mathbf{k}$. Furthermore, if we consider positive energies, a source of Berry flux at one Weyl node at $\mathbf{k}$ is balanced by a sink at $-\mathbf{k}$. The same happens at negative energy. For other problems, Weyl nodes can occur at different energies and momenta, but there is always the requirement that the sum of Berry flux (sum of Chern numbers) must be zero. This is because the net flux out of the Brillouin zone must be zero.

There are many other possibilities for Weyl nodes, which lead to interesting phenomena that cannot be covered here. Weyl nodes can also be created by broken inversion symmetry. Examples are TaAs and related materials, where two research teams independently reported the experimental discovery of Weyl nodes and Fermi arcs (see Section 28.5) in TaAs by Lv et al. [991] and Xu et al. [992]. (The "Viewpoint" by Vishwanath [989] is an introduction to those works.) There are 24 nodes in the TaAs class of materials [986, 993], so the analysis is considerably more complicated than the magnetic Weyl semimetal with two nodes considered here.

28.5 Fermi Arcs

One of the most remarkable consequences of Weyl nodes is surface states that form Fermi arcs, which are lines in the two-dimensional surface Brillouin zone where there is a surface

[5] Recall that mass in the Dirac equation corresponds to 1/2 the gap in the condensed matter context.
[6] There are other problems that higher-order nodes with Chern numbers that are larger in magnitude.

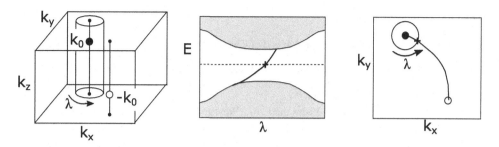

Figure 28.7. Left: the Brillouin zone of a semimetal with two Weyl nodes at $\mathbf{k}_0$ and $-\mathbf{k}_0$. The cylinder indicates a surface that encloses one Weyl node projected on a (0 0 1) surface of the crystal. At all other points in the bulk Brillouin zone there is a gap, and there is a surface state that crosses the gap as discussed in the text. Middle: the points along the cylinder are denoted by λ and the cross indicates the point where the surface state is at the Fermi energy. Right: the crossing point varies as the radius of the cylinder is increased and the locus of points forms a line in the surface Brillouin zone. The line has the form of 1/2 a Fermi surface (a line in two dimensions), which is called a Fermi arc.

state with energy equal to the Fermi energy.[7] Consider the example shown in Fig. 28.6 where there are two nodes at the same energy.[8] If the Fermi energy is at the energy of the Weyl nodes ($E_F = 0$), the bulk Fermi surface is just the two points and the Fermi energy is in a gap in the rest of the bulk Brillouin zone. However, due to the topological character of the Weyl nodes, there is a surface band that crosses from gap. (There is only one surface band for a 2×2 band problem.) The line in the surface Brillouin zone where the surface band is at the Fermi energy can be determined by the procedure indicated in Fig. 28.7. At the left is the bulk Brillouin zone with the two Weyl nodes and the cylinder that encloses one node indicates the projection onto the (0 0 1) surface. In the middle figure the cross indicates the point along the cylinder (parameterized by λ) where the surface state has energy zero. As the radius the cylinder is increased, the point $\mathbf{k}$ at which the surface band has energy zero defines a line that extends until it reaches the other Weyl node, where it terminates.

To appreciate how strange this is, in an ordinary two-dimensional system the Fermi surface is a line that is closed to form a loop. Here the Fermi line extends from one node in the surface Brillouin zone to another and terminates instead of closing. Where is the other line that forms closed Fermi loop? It is on the opposite surface of the crystal! This is analogous to the various cases for topological insulators in the previous sections and chapters where the partner is on the opposite surface, but it is even more striking since it is half a line. For further discussion and analysis for more complicated problems, see the references in footnote 3.

---

[7] Here we follow the analysis in [918].

[8] In general the nodes may be at different energies and there may be more than two nodes, so that the situation is more complicated.

SELECT FURTHER READING

Books by Vanderbilt and Bernevig listed at end of Chapter 25.

Reviews by Hasan and Kane, Qi and Zhang, and Bansil et al. listed at end of Chapter 27.

Reviews of Weyl and related semimetals:

Armitage, N. P. and Mele, E. J., and Vishwanath, A., "Weyl and Dirac semimetals in three-dimensional solids," *Rev. Mod. Phys.* 90:015001, 2018. Provides pedagogical theory and examples.

Yan, B. and Felser, C., "Topological materials: Weyl semimetals," *Annu. Rev. Condens. Matter Phys.*, 8:337–354, 2017. Primarily focused on materials.

Exercises

28.1 Consider a bilayer formed by bringing together two planes each with Z_2-odd. For each plane there are two oppositely circulating currents. Give the arguments that with weak coupling no gap should vanish in the bulk, and the counting rules for even and odd number of crossings in Section 27.4 imply that this is a Z_2-even case. Sketch how this could happen by considering anticrossing points between weakly coupled bands.

28.2 Show that the cases marked $(0; 011)$ and $(0; 111)$ in Fig. 28.1 correspond to stacks of planes along other directions in the crystal.

28.3 Describe explicitly how one can ensure the vanishing of an energy gap for some range of hamiltonians in three dimensions but not two. Hint: for **k** in three dimensions, the condition $h_1(\mathbf{k}) = 0$ can occur for a two-dimensional surface. Continue the argument to reach the conclusion.

28.4 Show that in the absence of a magnetic field ($b = 0$) the expression (28.4) reduces to a Dirac-like equation with the speed of light c replaced by v. The addition of a Zeeman term $b\sigma_z$ completes the derivation of Eq. (28.4). Also derive Eq. (28.5).

28.5 Solve for the energies for three cases: (1) $m = b = 0$; (2) $m = 1$, $b = 0.5$; and (3) $m = 0.5$, $b = 1.0$. Show that these parameters lead to the results with the form shown in Fig. 28.6. (These are three of the cases that are shown in fig. 2 of [986])

28.6 Show that near the zero-gap nodes, the hamiltonian has the form of the Weyl hamiltonian in Eq. (28.1).

28.7 Go through the arguments to convince yourself that this problem is indeed the same as the problem for spins. See also Exercise 26.6 for the one-dimensional two-band problem.

PART VII

APPENDICES

Appendix A

Functional Equations

Summary

A *functional* $F[f]$ is a mapping of an entire function f onto a value. In electronic structure, functionals play a central role, not only in density functional theory but also in the formulation of most of the theoretical methods as functionals of the underlying variables, in particular the wavefunctions. This appendix deals with the general formulation and derivation of variational equations from the functionals.

A.1 Basic Definitions and Variational Equations

The difference between a *function* $f(x)$ and a *functional* $F[f]$ is that a function is defined to be a mapping of a variable x to a result (a number) $f(x)$; whereas a functional is a mapping of an entire function f to a resulting number $F[f]$. The functional $F[f]$, denoted by square brackets, depends on the function f over its range of definition $f(x)$ in terms of its argument x. Here we a describe some basic properties related to the functionals and their use in density functional theory; a more complete description can be found in [273], appendix A. A review of functional derivatives or the "calculus of variations" can be found in [261, 994].

To illustrate functionals $F[f]$ we first consider two simple examples:

- A definite integral of $w(x)f(x)$, where $w(x)$ is some fixed weighting function,

$$I_w[f] = \int_{x_{\min}}^{x_{\max}} w(x)f(x)\mathrm{d}x \tag{A.1}$$

- The integral of $(f(x))^\alpha$, where α is an arbitrary power:

$$I_\alpha[f] = \int_{x_{\min}}^{x_{\max}} (f(x))^\alpha \mathrm{d}x \tag{A.2}$$

A functional derivative is defined by a variation of the functional

$$\delta F[f] = F[f + \delta f] - F[f] = \int_{x_{\min}}^{x_{\max}} \frac{\delta F}{\delta f(x)} \delta f(x) \mathrm{d}x, \tag{A.3}$$

where the quantity $\delta F / \delta f(x)$ is the functional derivative of F with respect to variation of $f(x)$ at the point x. In Eq. (A.1), the fact that the functional is linear in $f(x)$ leads to a simple result for the functional derivative:

$$\frac{\delta I_w}{\delta f(x)} = w(x). \tag{A.4}$$

The variational derivation of the many-body Schrödinger equation in (3.10) and (3.12) is an example of this simple form.

The second example of a nonlinear functional is of the form needed to minimize the Thomas–Fermi expression, (6.4). From definition Eq. (A.3) one can also show (Exercise A.1) that

$$\frac{\delta I_\alpha}{\delta f(x)} = \alpha (f(x))^{\alpha - 1}, \tag{A.5}$$

following the same rules as normal differentiation. In general, however, the functional derivative at point x depends also on the function $f(x)$ at all other points. Clearly, the definition can be extended to many variables and functions $F[f_1, f_2, \dots]$.

A.2 Functionals in Density Functional Theory Including Gradients

In Kohn–Sham density functional theory, the potential, Eq. (7.13), is a sum of functional derivatives. The external term has the linear form of Eq. (A.1); the Hartree term is also simple since it is bilinear; and $V_{\mathrm{xc}}^\sigma(\mathbf{r})$ is found by varying the more complex functional having the form

$$E_{\mathrm{xc}}[n] = \int n(\mathbf{r}) \epsilon_{\mathrm{xc}}(n(\mathbf{r}), |\nabla n(\mathbf{r})|) \mathrm{d}\mathbf{r}. \tag{A.6}$$

Variations of the gradient terms can be illustrated by the general form

$$I[n] = \int g(f(\mathbf{r}), |\nabla f(\mathbf{r})|) \mathrm{d}\mathbf{r}, \tag{A.7}$$

so that varying the function f leads to

$$\delta I[g, f] = \int \left[\frac{\delta g}{\delta f} \delta f(\mathbf{r}) + \frac{\delta g}{\delta |\nabla f|} \delta |\nabla f(\mathbf{r})| \right] \mathrm{d}\mathbf{r}. \tag{A.8}$$

Now using

$$\delta |\nabla f(\mathbf{r})| = \delta \nabla f(\mathbf{r}) \cdot \frac{\nabla f(\mathbf{r})}{|\nabla f(\mathbf{r})|} = \frac{\nabla f(\mathbf{r})}{|\nabla f(\mathbf{r})|} \cdot \nabla[\delta f(\mathbf{r})] \tag{A.9}$$

and integrating by parts, one finds a standard form of variations of gradients:

$$\delta I[g, f] = \int \left\{ \frac{\delta g}{\delta f} - \nabla \cdot \left[\frac{\delta g}{\delta |\nabla \mathbf{f}|} \frac{\nabla \mathbf{f}(\mathbf{r})}{|\nabla \mathbf{f}(\mathbf{r})|} \right] \right\} \delta f(\mathbf{r}) d\mathbf{r}. \qquad (A.10)$$

This form is used in Section 8.6, where other forms for the functional derivative are also given that may be advantageous in actual calculations.

SELECT FURTHER READING

A compact description can be found in:

Parr, R. G., and Yang, W. *Density-Functional Theory of Atoms and Molecules* (Oxford University Press, New York, 1989), app. A.

Basic theory of functionals can be found in:

Evans, G. C., *Functionals and Their Applications* (Dover, New York, 1964).

Matthews, J. and Walker, R. L., *Mathmatical Methods of Physics* (W. A. Benjamin Inc., New York, 1964), ch. 12.

Exercises

A.1 Show that Eq. (A.5) follows from Eq. (A.3), and that application of the expression to the Thomas–Fermi approximation leads to Expression (6.4).

A.2 Derive the form of the variational expression in (A.10) involving the gradient terms.

Appendix B
LSDA and GGA Functionals

Summary

In this appendix are given representative forms for the exchange–correlation energy and potential in the LSDA and GGA approximations. The forms given here are chosen because they are widely used and are relatively simple. Actual programs that provide energies and potentials for these and other forms can be found online.

B.1 Local Spin Density Approximation (LSDA)

The local density approximation is based on the exact expressions for the exchange energy, Eq. (5.15), and various approximations and fitting to numerical correlation energies for the homogeneous gas. Comparison of the forms is shown in Fig. 5.4. The first functions were the Wigner interpolation formula, Eq. (5.22), and the Hedin–Lundqvist [241] form; the latter is derived from many-body perturbation theory and is given below. As described in Chapter 5, the quantum Monte Carlo (QMC) calculations of Ceperley and Alder [311], and more recent work [315, 316, 318], provide essentially exact results for unpolarized and fully polarized cases. These results have been fitted to analytic forms for $\epsilon_c(r_s)$, where r_s is given by Eq. (5.1), leading to two widely used functionals due to Perdew and Zunger (PZ) [313] and Vosko, Wilkes, and Nusiar (VWN) [312], which are very similar quantitatively. Both functionals assume an interpolation form for fractional spin polarization, and Ortiz and Balone [315] report that their QMC calculations at intermediate polarization are somewhat better described by the VWN form. In all cases, the correlation potential is given by

$$V_c(r_s) = \epsilon_c(r_s) - \frac{r_s}{3} \frac{d\epsilon_c(r_s)}{dr_s}. \tag{B.1}$$

Here are listed selected forms for the unpolarized case (complete expressions can be found in [372, 413, 420]):

1. Hedin–Lundqvist (HL) [241]

$$\epsilon_c^{HL}(r_s) = -\frac{C}{2} \left[(1 + x^3) \ln\left(1 + \frac{1}{x}\right) + \frac{x}{2} - x^2 - \frac{1}{3} \right] [] * [], \tag{B.2}$$

where $A = 21$, $C = 0.045$, and $x = r_s/A$. The correlation potential is

$$V_c^{HL}(r_s) = -\frac{Ce^2}{2} \ln\left(1 + \frac{1}{x}\right). \tag{B.3}$$

2. Perdew–Zunger (PZ) [313]

$$\epsilon_c^{PZ}(r_s) = -0.0480 + 0.0311 \ln(r_s) - 0.0116 r_s + 0.0020 r_s \ln(r_s), r_s < 1$$

$$= -0.1423/(1 + 1.0529\sqrt{r_s} + 0.3334 r_s), r_s > 1. \tag{B.4}$$

The expression [313] for V_c^{PZ} is not given here since it is lengthy, but it is straightforward. For fractional spin polarization, the interpolation for $\epsilon_c^{PZ}(r_s)$ is assumed to have the same function form as for exchange, Eq. (5.17), with f given by Eq. (5.18).

3. Vosko–Wilkes–Nusiar (VWN) [312]

$$\epsilon_c^{VWN}(r_s) = \frac{Ae^2}{2}\left[\ln\left[\frac{y^2}{Y(y)}\right] + \frac{2b}{Q} \tan^{-1}\left(\frac{Q}{2y+b}\right)\right.$$

$$\left. - \frac{by_0}{Y(y_0)}\left\{\ln\left[\frac{(y-y_0)^2}{Y(y)}\right] + \frac{2(b+2y_0)}{Q}\tan^{-1}\left(\frac{Q}{2y+b}\right)\right\}\right] \tag{B.5}$$

Here $y = r_s^{1/2}$, $Y(y) = y^2 + by + c$, $Q = (4c - b^2)^{1/2}$, $y_0 = -0.10498$, $b = 3.72744$, $c = 12.93532$, and $A = 0.0621814$. The corresponding potential can be obtained from Eq. (B.1) with [372]:

$$r_s\frac{d\epsilon_c^{VWN}(r_s)}{dr_s} = A\frac{e^2}{2}\frac{c(y-y_0) - by_0 y}{(y-y_0)(y^2 + by + c)}. \tag{B.6}$$

B.2 Generalized-Gradient Approximation (GGAs)

There are many different forms for gradient approximations; however, it is beyond the scope of the present work to give the formulas for even the most widely used forms. The reader is referred to papers and books listed as "Select Further Reading."

B.3 GGAs: Explicit PBE Form

The PBE form is probably the simplest GGA functional. Hence we give it as an explicit example. The reader is referred to other sources such as the paper on "Comparison Shopping for a Gradient-Corrected Density Functional" by Perdew and Burke [409]. The PBE functional [412] for exchange is given by a simple form for the enhancement factor F_x defined in Section 8.5. The form is chosen with $F_x(0) = 1$ (so that the local approximation is recovered) and $F_x \to$ *constant* at large s,

$$F_x(s) = 1 + \kappa - \kappa/(1 + \mu s^2/\kappa), \tag{B.7}$$

where $\kappa = 0.804$ is chosen to satisfy the Lieb–Oxford bound. The value of $\mu = 0.21951$ is chosen to recover the linear response form of the local approximation – i.e., it is chosen to cancel the term from the correlation. This may seem strange, but it is done to agree better with quantum Monte Carlo calculations. This choice violates the known expansion at low s given in Eq. (8.12), with the rationale of better fitting the entire functional.

The form for correlation is expressed as the local correlation plus an additive term both of which depend on the gradients and the spin polarization. The form chosen to satisfy several conditions is [412]

$$E_c^{GGA-PBE}[n^\uparrow, n^\downarrow] = \int d^3r \, n \left[\epsilon_c^{hom}(r_s, \zeta) + H(r_s, \zeta, t) \right], \tag{B.8}$$

where $\zeta = (n^\uparrow - n^\downarrow)/n$ is the spin polarization, r_s is the local value of the density parameter, and t is a dimensionless gradient $t = |\nabla n|/(2\phi k_{TF} n)$. Here $\phi = ((1 + \zeta)^{2/3} + (1 - \zeta)^{2/3})/2$ and t is scaled by the screening wavevector k_{TF} rather than k_F. The final form is

$$H = \frac{e^2}{a_0} \gamma \phi^3 \ln \left(1 + \frac{\beta}{\gamma} t^2 \frac{1 + At^2}{1 + At^2 + A^2 t^4} \right), \tag{B.9}$$

where the factor e^2/a_0, with a_0 the Bohr radius, is unity in atomic units. The function A is given by

$$A = \frac{\beta}{\gamma} \left[\exp \left(\frac{-\epsilon_c^{hom}}{\gamma \phi^3 \frac{e^2}{a_0}} \right) - 1 \right]^{-1}. \tag{B.10}$$

SELECT FURTHER READING

See list at the end of Chapter 8

Appendix C

Adiabatic Approximation

Summary

The only small parameter in the electronic structure problem is the inverse nuclear mass $1/M$ – i.e., the nuclear kinetic energy terms. The adiabatic or Born–Oppenheimer approximation is a systematic expansion in the small parameter that is fundamental to all electronic structure theory. It comes to the fore in the theory of phonons, electron–phonon interactions, and superconductivity (Chapter 20).

C.1 General Formulation

The fundamental hamiltonian for a system of nuclei and electrons, Eq. (3.1), can be written

$$\hat{H} = \hat{T}_N + \hat{T}_e + \hat{U}, \tag{C.1}$$

where U contains all the potential interaction terms involving the set of all-electron coordinates $\{\mathbf{r}\}$ (which includes spin) and the set of all nuclear coordinates $\{\mathbf{R}\}$. Since the only small term is the kinetic energy operator of the nuclei $\hat{T}_N$, we treat it as a perturbation upon the hamiltonian, Eq. (3.2), for nuclei fixed in their instantaneous positions. The first step is to define the eigenvalues and wavefunctions $E_i(\{\mathbf{R}\})$ and $\Psi_i(\{\mathbf{r}\} : \{\mathbf{R}\})$ for the electrons, which depend on the nuclear positions $\{\mathbf{R}\}$ as parameters. This is the same as Eq. (3.13) except that the positions of the nuclei are indicated explicitly, and $i = 0, 1, \ldots,$ denotes the complete set of states at each $\{\mathbf{R}\}$.

The full solutions for the coupled system of nuclei and electrons[1]

$$\hat{H}\Psi_s(\{\mathbf{r}, \mathbf{R}\}) = E_s \Psi_s(\{\mathbf{r}, \mathbf{R}\}), \tag{C.2}$$

where $s = 1, 2, 3, \ldots,$ labels the states of the coupled system, can be written in terms of $\Psi_i(\{\mathbf{r}\} : \{\mathbf{R}\})$,

$$\Psi_s(\{\mathbf{r}, \mathbf{R}\}) = \sum_i \chi_{si}(\{\mathbf{R}\})\Psi_i(\{\mathbf{r}\} : \{\mathbf{R}\}), \tag{C.3}$$

since $\Psi_i(\{\mathbf{r}\} : \{\mathbf{R}\})$ defines a complete set of states for the electrons at each $\{\mathbf{R}\}$.

---

[1] Adapted from notes of K. Kunc and the author.

The states of the coupled electron–nuclear system are now specified by $\chi_{si}(\{\mathbf{R}\})$, which are functions of the nuclear coordinates and are the coefficients of the electronic states Ψ_{mi}. In order to find the equations for $\chi_{si}(\{\mathbf{R}\})$, insert Expression (C.3) into (C.2), multiply the expression on the left by $\Psi_i(\{\mathbf{r},\mathbf{R}\})$, and integrate over electron variables $\{\mathbf{r}\}$ to find the equation

$$[T_N + E_i(\{\mathbf{R}\}) - E_s]\,\chi_{si}(\{\mathbf{R}\}) = -\sum_{i'} C_{ii'}\chi_{si'}(\{\mathbf{R}\}), \qquad (C.4)$$

where $T_N = -\frac{1}{2}(\sum_J \nabla_J^2/M_J)$ and the matrix elements are given by $C_{ii'} = A_{ii'} + B_{ii'}$, with[2]

$$A_{ii'}(\{\mathbf{R}\}) = \sum_J \frac{m_e}{M_J}\langle \Psi_i(\{\mathbf{r}\}:\{\mathbf{R}\})|\nabla_J|\Psi_{i'}(\{\mathbf{r}\}:\{\mathbf{R}\})\rangle\nabla_J, \qquad (C.5)$$

$$B_{ii'}(\{\mathbf{R}\}) = \sum_J \frac{m_e}{2M_J}\langle \Psi_i(\{\mathbf{r}\}:\{\mathbf{R}\})|\nabla_J^2|\Psi_{i'}(\{\mathbf{r}\}:\{\mathbf{R}\})\rangle. \qquad (C.6)$$

Here $\langle \Psi_i(\{\mathbf{r}\}:\{\mathbf{R}\})|\mathcal{O}|\Psi_{i'}(\{\mathbf{r}\}:\{\mathbf{R}\})\rangle$ means integrations over only the electronic variables $\{\mathbf{r}\}$ for any operator $\mathcal{O}$.

The adiabatic or Born–Oppenheimer approximation [90] is to ignore the off-diagonal $C_{ii'}$ terms – i.e., the electrons are assumed to remain in a given state m as the nuclei move. Although the electron wavefunction $\Psi_i(\{\mathbf{r}\}:\{\mathbf{R}\})$ and the energy of state m change, the electrons do not change state and no energy is transferred between the degrees of freedom described by the equation for the nuclear variables $\{\mathbf{R}\}$ and *excitations* of the electrons, which occurs only if there is a change of state $i \to i'$. The diagonal terms can be treated easily. First, it is simple to show (see Exercise C.1) that $A_{ii} = 0$ simply from the requirement that Ψ be normalized. The term $B_{ii}(\{\mathbf{R}\})$ can be grouped with $E_i(\{\mathbf{R}\})$ to determine a modified potential function for the nuclei $U_i(\{\mathbf{R}\}) = E_i(\{\mathbf{R}\}) + B_{ii}(\{\mathbf{R}\})$. Thus, in the adiabatic approximation, the nuclear motion is described by a purely nuclear equation for each electronic state i

$$\left[-\sum_J \frac{m_e}{2M_J}\nabla_J^2 + U_i(\{\mathbf{R}\}) - E_{ni}\right]\chi_{ni}(\{\mathbf{R}\}) = 0, \qquad (C.7)$$

where $n = 1, 2, 3, \ldots$, labels the nuclear states. Within the adiabatic approximation, the full set of states $s = 0, 1, \ldots$, is a product of nuclear and electronic states.

Equation (C.7) with the neglect at the B_{ii} term is the basis of the "frozen phonon" or perturbation methods for calculation of phonon energies in the adiabatic approximation (Chapter 20). So long as we can justify neglecting the off-diagonal terms that couple different electron states, we can solve the nuclear motion problem, Eq. (C.7), given the function $U_i(\{\mathbf{R}\})$ for the particular electronic state i that evolves adiabatically with nuclear motion. (The term B_{ii} is typically very small due to the large nuclear mass.) In general, this is an excellent approximation except for cases where there is degeneracy or near degeneracy

---

[2] Here the mass of the electron m_e is written explicitly to emphasize that it is the ratio m_e/M_J that is essential in the adiabatic approximation.

of the electronic states. If there is a gap in the electronic excitation spectrum much larger than typical energies for nuclear motion, then the nuclear excitations are well determined by the adiabatic terms. Special care must be taken for cases such as transition states in molecules where electronic states become degenerate or in metals where the lack of an energy gap leads to qualitative effects.

C.2 Electron-Phonon Interactions

Electron–phonon interactions result from the off-diagonal matrix elements $C_{ii'}$ that describe transitions between different electronic states due to the velocities of the nuclei. The dominant terms are given in Eq. (C.5), which involves a gradient of the electron wavefunctions with respect to the nuclear positions and the gradient operator acting on the phonon wavefunction χ. The combination of these operators leads to an electronic transition between states i and i' coupled with emission or absorption of one phonon.

The steps involved in writing the formal expressions are to express the nuclear kinetic operator ∇_J in Eq. (C.5) in terms of phonon creation and annihilation operators [306] and to write out the perturbation expression for the matrix element. The latter step can be accomplished by noting that the variation in the electron function due to the displacement of nucleus J is caused by the change in the potential V due to the displacement. To linear order the relation is

$$\langle \Psi_i(\{\mathbf{r}\} : \{\mathbf{R}\}) | \nabla_J | \Psi_{i'}(\{\mathbf{r}\} : \{\mathbf{R}\}) \rangle = \frac{\langle \Psi_i(\{\mathbf{r}\} : \{\mathbf{R}\}) | \nabla_J V | \Psi_{i'}(\{\mathbf{r}\} : \{\mathbf{R}\}) \rangle}{E_{i'}(\{\mathbf{R}\}) - E_i(\{\mathbf{R}\})}. \tag{C.8}$$

This leads to the form of the electron–phonon matrix elements in Section 20.8 and treated in references such as [825].

SELECT FURTHER READING

Born, M. and Huang, K., *Dynamical Theory of Crystal Lattices* (Oxford University Press, Oxford, 1954).

Ziman, J. M., *Principles of the Theory of Solids* (Cambridge University Press, Cambridge, 1989).

Exercises

C.1 Show that the requirement that Ψ be normalized is sufficient to prove $A_{ii} = 0$. Hint: use the fact that any derivative of $\langle \Psi | \Psi \rangle$ must vanish.

C.2 Derive the equation for nuclear motion, Eqs. (C.7), from (C.4) using the assumption of the adiabatic approximation as described before Eq. (C.7).

C.3 For small nuclear displacements about their equilibrium positions, show that Eq. (C.7) leads to harmonic oscillator equations.

C.4 For a simple diatomic molecule treated in the harmonic approximation, show that Eq. (C.7) leads to the well-known result that the ground-state energy of the nuclear–electron system is $E_{min} + \frac{1}{2}\hbar\omega$, where ω is the harmonic oscillator frequency.

Appendix D

Perturbation Theory, Response Functions, and Green's Functions

Summary

Perturbation theory, response functions, and Green's functions are the bread and butter of theoretical physics. The general form of the equations for perturbation theory are summarized in Section D.1 and the "$2n + 1$" theorem is derived in the last section Section D.6. Response functions are used in many contexts and are the connection to important experimental measurements; this appendix is devoted to characteristic forms and properties of response functions, sum rules, and Kramers–Kronig relations. The most important example is the dielectric function described in Appendix E. Useful expressions are given for self-consistent field methods, which lead to "RPA" and other formulas needed in Chapters 5, 20, and 21. The basic aspects of Green's functions relevant for this book are given in Section D.5.

D.1 Perturbation Theory

Perturbation theory describes the properties of a system with hamiltonian $\hat{H}^0 + \lambda \Delta \hat{H}$ as a systematic expansion in powers of the perturbation, which is conveniently done by organizing terms in powers of λ. The first-order expressions depend only on the unperturbed wavefunctions and $\Delta \hat{H}$ to first-order and have already been given as the force or "generalized force" in Section 3.3. To higher order one must determine the variation in the wavefunction. The general form valid in a many-body system can be written in terms of a sum over the excited states of the unperturbed hamiltonian [10, 995, 996],

$$\Delta \Psi_i(\{\mathbf{r}_i\}) = \sum_{j \neq i} \Psi_j(\{\mathbf{r}_i\}) \frac{\langle \Psi_j | \Delta \hat{H} | \Psi_i \rangle}{E_i - E_j}. \tag{D.1}$$

The change in the expectation value of an operator $\hat{O}$ in the perturbed ground state can be cast in the form

$$\Delta \langle \hat{O} \rangle = \sum_{j \neq i} \langle \Delta \Psi_j | \hat{O} | \Psi_i \rangle + \text{c.c.} = \sum_{j \neq i} \frac{\langle \Psi_i | \hat{O} | \Psi_j \rangle \langle \Psi_j | \Delta \hat{H} | \Psi_i \rangle}{E_i - E_j} + \text{c.c.}, \tag{D.2}$$

which can readily be generalized to finite T. An advantage of writing the general many-body expression is that it shows immediately that the perturbation of the many-body ground state Ψ_0 involves only the excited states, an aspect that has to be demonstrated in the simpler independent-particle methods.

In an independent-particle approximation the states are determined by the hamiltonian $\hat{H}_{\text{eff}}$ in the effective Schrödinger equation Eq. (3.36). The change in the individual independent-particle orbitals, $\Delta\psi_i(\mathbf{r})$ to first order in perturbation theory can be written in terms of a sum over the spectrum of the unperturbed hamiltonian $\hat{H}_{\text{eff}}^0$ as [10, 995, 996],

$$\Delta\psi_i(\mathbf{r}) = \sum_{j \neq i} \psi_j(\mathbf{r}) \frac{\langle \psi_j | \Delta\hat{H}_{\text{eff}} | \psi_i \rangle}{\varepsilon_i - \varepsilon_j}, \tag{D.3}$$

where the sum is over all the states of the system, occupied and empty, with the exception of the state being considered. Similarly, the change in the expectation value of an operator $\hat{O}$ in the perturbed ground state to lowest order in $\Delta\hat{H}_{\text{eff}}$ can be written

$$\Delta\langle \hat{O} \rangle = \sum_{i=1}^{\text{occ}} \langle \psi_i + \delta\psi_i | \hat{O} | \psi_i + \delta\psi_i \rangle$$

$$= \sum_{i=1}^{\text{occ}} \sum_{j}^{\text{empty}} \frac{\langle \psi_i | \hat{O} | \psi_j \rangle \langle \psi_j | \Delta\hat{H}_{\text{eff}} | \psi_i \rangle}{\varepsilon_i - \varepsilon_j} + \text{c.c.} \tag{D.4}$$

In Eq. (D.4) the sum over j is restricted to conduction states only, which follows from the fact that the contributions of pairs of occupied states i, j and j, i cancel in Eq. (D.4) (Exercise 3.21). Expressions (D.3) and (D.4) are the basic equations upon which is built the theory of response functions and methods for calculating static (Chapter 20) and dynamic responses (Chapter 21) in materials.

D.2 Static Response Functions

Static response functions play two important roles in electronic structure. One is the calculation of quantities that directly relate to experiments, namely the actual response of the electrons to static perturbations such as strain or applied electric fields, and the response at low frequencies that can be considered "adiabatic" (Appendix C) that governs lattice dynamics, etc. This is the subject of Chapter 20. The other role is the development of methods in the theory of electronic structure to derive improved solutions utilizing perturbation expansions around more approximate solutions. This is the basis of the analysis in Chapter 7.

The basic equations follow from perturbation theory – in particular, Eq. (D.4). The most relevant quantity for static perturbations is the density, for which Eq. (D.4) becomes

$$\Delta n(\mathbf{r}) = \sum_{i=1}^{\text{occ}} \sum_{j}^{\text{empty}} \psi_i^*(\mathbf{r}) \psi_j(\mathbf{r}) \frac{\langle \psi_j | \Delta V_{\text{eff}} | \psi_i \rangle}{\varepsilon_i - \varepsilon_j} + \text{c.c.} \tag{D.5}$$

The response to a variation of the *total potential* $V_{eff}(\mathbf{r})$ at point $\mathbf{r} = \mathbf{r}'$ (see Appendix A for definition of functional derivatives) defines the density response function

$$\chi_n^0(\mathbf{r}, \mathbf{r}') = \frac{\delta n(\mathbf{r})}{\delta V_{eff}(\mathbf{r}')} = 2 \sum_{i=1}^{occ} \sum_{j}^{empty} \frac{\psi_i^*(\mathbf{r})\psi_j(\mathbf{r})\psi_j^*(\mathbf{r}')\psi_i(\mathbf{r}')}{\varepsilon_i - \varepsilon_j}, \tag{D.6}$$

which is symmetric in $\mathbf{r}$ and $\mathbf{r}'$ since it is the response of $n(\mathbf{r})$ to a perturbation $V_{eff}(\mathbf{r}')n(\mathbf{r}') \propto n(\mathbf{r}')$. Equation (D.6) may also be written in a convenient form

$$\chi_n^0(\mathbf{r}, \mathbf{r}') = \sum_{i=1}^{occ} \psi_i^*(\mathbf{r})G_0^i(\mathbf{r}, \mathbf{r}')\psi_i(\mathbf{r}'), \quad G_0^i(\mathbf{r}, \mathbf{r}') = \sum_{j \neq i}^{\infty} \frac{\psi_j(\mathbf{r})\psi_j^*(\mathbf{r}')}{\varepsilon_i - \varepsilon_j}, \tag{D.7}$$

where G_0^i is an independent-particle Green's function (Section D.5).

The Fourier transform of $\chi_n^0(\mathbf{r}, \mathbf{r}')$ is the response to particular Fourier components, which is often the most useful form. If we define $\Delta V_{eff}(\mathbf{r}) = \Delta V_{eff}e^{i\mathbf{q}\cdot\mathbf{r}}$ and $n(\mathbf{q}') = \int d\mathbf{r}n(\mathbf{r})e^{i\mathbf{q}'\cdot\mathbf{r}}$ in Eq. (D.5), then one finds (Exercise D.1)

$$\chi_n^0(\mathbf{q}, \mathbf{q}') = \frac{\delta n(\mathbf{q}')}{\delta V_{eff}(\mathbf{q})} = 2 \sum_{i=1}^{occ} \sum_{j}^{empty} \frac{M_{ij}^*(\mathbf{q})M_{ij}(\mathbf{q}')}{\varepsilon_i - \varepsilon_j}, \tag{D.8}$$

where $M_{ij}(\mathbf{q}) = \langle \psi_i | e^{i\mathbf{q}\cdot\mathbf{r}} | \psi_j \rangle$. This can be a great simplification, for example, in a homogeneous system, $\chi_n^0(\mathbf{q}, \mathbf{q}') \neq 0$ only for $\mathbf{q} = \mathbf{q}'$ (Chapter 5), or in crystals (Chapters 20 and 21).

The response function χ^0 plays many important roles in electronic structure theory. The simplest is in approximations in which the electrons are considered totally noninteracting; then, $\Delta V_{eff} = \Delta V_{ext}$ and χ^0 represents the response to an external perturbation. However, in an effective mean-field theory, like the Hartree–Fock or Kohn–Sham theories of Chapters 7–9, the internal fields also vary and the effective hamiltonian must be found in a self-consistent procedure. This leads to the following section, in which χ^0 still plays a crucial role.

D.3 Response Functions in Self-Consistent Field Theories

In a self-consistent field theory, the total effective field depends on the internal variables – e.g., in the Kohn–Sham approach, $V_{eff} = V_{ext} + V_{int}[n]$. Since the electrons act as independent particles in the potential V_{eff}, χ_n^0 is still given by Eqs. (D.6)–(D.8). However, the relation to the external field is changed. To linear order, the response to an external field is given by

$$\chi = \frac{\delta n}{\delta V_{ext}}, \tag{D.9}$$

which is shorthand for the functional form that can be written in $\mathbf{r}$ space or $\mathbf{q}$ space,

$$\chi(\mathbf{r}, \mathbf{r}') = \frac{\delta n(\mathbf{r})}{\delta V_{ext}(\mathbf{r}')} \quad \text{or} \quad \chi(\mathbf{q}, \mathbf{q}') = \frac{\delta n(\mathbf{q})}{\delta V_{ext}(\mathbf{q}')}. \tag{D.10}$$

Similarly, the linear response of the spin density $m = n^\uparrow - n^\downarrow$ to an external Zeeman field $\Delta \hat{H} = V_{\text{ext}}^m$ has the same form

$$\chi = \frac{\delta m}{\delta V_{\text{ext}}^m}, \tag{D.11}$$

so that the analysis applies to both total density and spin density.

The response function can be written (omitting indices for simplicity)

$$\chi = \frac{\delta n}{\delta V_{\text{eff}}} \frac{\delta V_{\text{eff}}}{\delta V_{\text{ext}}} = \chi^0 \left[1 + \frac{\delta V_{\text{int}}}{\delta n} \frac{\delta n}{\delta V_{\text{ext}}} \right] = \chi^0 [1 + K\chi], \tag{D.12}$$

where the kernel K is given in $\mathbf{r}$ space in Eq. (7.25) or in $\mathbf{q}$ space as

$$K(\mathbf{q}, \mathbf{q}') = \frac{\delta V_{\text{int}}(\mathbf{q})}{\delta n(\mathbf{q}')} = \frac{4\pi}{q^2} \delta_{\mathbf{q}, \mathbf{q}'} + \frac{\delta^2 E_{\text{xc}}[n]}{\delta n(\mathbf{q}) \delta n(\mathbf{q}')} \equiv V_C(q) \delta_{\mathbf{q}, \mathbf{q}'} + f_{\text{xc}}(\mathbf{q}, \mathbf{q}'). \tag{D.13}$$

Solving Eq. (D.12) (Exercise D.2), leads to the ubiquitous form [306, 997, 998]

$$\chi = \chi^0 [1 - \chi^0 K]^{-1} \quad \text{or} \quad \chi^{-1} = [\chi^0]^{-1} - K, \tag{D.14}$$

which appears in many contexts. The approximation $f_{\text{xc}} = 0$ is the famous "random phase approximation" (RPA) [297] for the Coulomb interaction; many approximations for f_{xc} have been introduced and any of the exchange–correlation functionals imply a form for f_{xc}. The density response function $\chi(\mathbf{r}, \mathbf{r}')$ or $\chi(\mathbf{q}, \mathbf{q}')$ is central in the theory of phonons (Chapter 20), dielectric response in Appendix E, and other response functions. The extension to dynamical response leads to the theory for much of our understanding of electronic excitations, Chapter 21. For spin response, the Coulomb term V_C is absent and the kernel f_{xc}^m leads to the Stoner response function, Eq. (2.3), and the RPA expressions for magnons.

The classic approach for finding χ is to calculate χ^0 from Eqs. (D.6)–(D.8) and solve the inverse matrix equation (D.14). Despite the elegant simplicity of the equations, the solution can be a laborious procedure except in the simplest cases. An equally elegant approach much more suited for calculations in real electronic structure problems is described in Chapter 20.

D.4 Dynamic Response and Kramers–Kronig Relations

Harmonic Oscillator

The basic ideas of linear response can be appreciated starting with the simple classical-driven harmonic oscillator as described eloquently by P. C. Martin [940]. The equation for the displacement x is

$$M \frac{d^2 x(t)}{dt^2} = -K x(t) - \Gamma \frac{dx(t)}{dt} + F(t), \tag{D.15}$$

where $F(t)$ is the driving force and Γ is a damping term. If the natural oscillation frequency is denoted $\omega_0 = \sqrt{K/M}$, the response to a force $F(t) = F(\omega)e^{-i\omega t}$ with frequency ω is

$$\chi(\omega) \equiv \frac{x(\omega)}{F(\omega)} = \frac{1}{M} \frac{1}{\omega_0^2 - \omega^2 - i\omega\Gamma/M}. \tag{D.16}$$

Note that for real ω the imaginary part of $\chi(\omega)$ is positive since $\Gamma > 0$ corresponds to energy loss. Furthermore, as a function of complex ω, $\chi(\omega)$ is analytic in the upper half-plane, $\Im\omega > 0$; all poles in the response function $\chi(\omega)$ are in the lower half-plane. This leads to the causal structure of $\chi(\omega)$ that implies the Kramers–Kronig relations below (Exercise D.4).

Frequency–Dependent Damping

The well-known form for the harmonic oscillator response with a constant Γ suffers from a fatal problem: a constant Γ violates mathematical constraints on the moments of $\chi(\omega)$, and it violates physical reasoning since loss mechanisms vary with frequency. If one introduces a more realistic $\Gamma(\omega)$, there is a simple rule: it is also a response function and must obey the laws of causality – i.e., $\Gamma(\omega)$ must also be a causal function that obeys Kramers–Kronig relations. For example, it might be modeled by a form like Eq. (D.16),

$$\Gamma(\omega) = \frac{1}{\omega_1^2 - \omega^2 - i\omega\gamma_1}, \tag{D.17}$$

and so forth. Clearly, this can continue, leading to a continued fraction that is an example of the general memory function formulation of Mori [999].

Kramers–Kronig Relations

Because the response functions represent the causal response of the system to external perturbations, they must obey analytic properties illustrated for the harmonic oscillator in Eq. (D.16). That is, the response function $\chi(\omega)$ continued into the complex plane is analytic for all $\Im\omega > 0$ in the upper half-plane and has poles only in the lower half-plane. By contour integrations in the complex plane [297, 300] (Exercise D.5), one can than derive the Kramers–Kronig relations that allow one to derive the real and imaginary parts from one another in terms of principle value integrals:

$$\mathrm{Re}\chi(\omega) = -\frac{1}{\pi} \int_{-\infty}^{\infty} d\omega' \frac{\mathrm{Im}\chi(\omega')}{\omega - \omega'},$$

$$\mathrm{Im}\chi(\omega) = \frac{1}{\pi} \int_{-\infty}^{\infty} d\omega' \frac{\mathrm{Re}\chi(\omega')}{\omega - \omega'}. \tag{D.18}$$

Dynamic Response of a Quantum System

The response to a time-dependent perturbation is given by Eq. (3.6), which is conveniently solved for a periodic perturbation $\propto e^{-i\omega t}$. The analysis is given in original references

[1000–1003] and in many texts [260, 280, 297, 300, 306, 1004], leading to the Kubo–Greenwood formula. A general response function in the noninteracting approximation[1] can be written as a complex function, with a small imaginary damping factor $\eta > 0$,

$$\chi^0_{a,b}(\omega) = 2 \sum_{i=1}^{\text{occ}} \sum_{j}^{\text{empty}} \frac{[M^a_{ij}]^* M^b_{ij}}{\varepsilon_i - \varepsilon_j + \omega + i\eta}, \qquad (D.19)$$

where the $M^a_{ij} = \langle \psi_i | \hat{O}^a | \psi_j \rangle$ and M^b_{ij} are matrix elements of appropriate operators – e.g., the Fourier components defined following Eq. (D.8) or the momentum matrix elements in the expression for the dielectric function in Eq. (21.9). The real and imaginary parts can be written explicitly as

$$\text{Re}\chi^0(\omega)_{a,b} = \sum_{i=1}^{\text{occ}} \sum_{j}^{\text{empty}} \frac{[M^a_{ij}]^* M^b_{ij}}{(\varepsilon_i - \varepsilon_j)^2 - \omega^2},$$

$$\text{Im}\chi^0(\omega)_{a,b} = \sum_{i=1}^{\text{occ}} \sum_{j}^{\text{empty}} |[M^a_{ij}]^* M^b_{ij} \delta(\varepsilon_j - \varepsilon_i - \omega). \qquad (D.20)$$

An important result from Eq. (D.20) is that the imaginary part of the response function $\chi^0(\omega)$ is just a joint density of states (Section 4.7) as a function of $\omega = \varepsilon_j - \varepsilon_i$, weighted by the matrix elements.

Dynamical Response in Self-Consistent Field Theories

The generalization of the independent-particle expressions to self-consistent field approaches is straightforward using the expressions derived in Section D.3. The only change is that the effective field is itself time or frequency dependent, $V_{\text{eff}} \to V_{\text{eff}}(t)$ or $V_{\text{eff}}(\omega)$. Within the linear response regime, the relevant quantity is the kernel K given in $\mathbf{r}$ space in Eq. (7.25) or in $\mathbf{q}$ space by Eq. (D.13), generalized to include time dependence. The explicit expression in $\mathbf{q}$ space is

$$K(\mathbf{q}, \mathbf{q}', t - t') = \frac{\delta V_{\text{int}}(\mathbf{q}, t)}{\delta n(\mathbf{q}', t')}$$

$$= \frac{4\pi}{q^2} \delta_{\mathbf{q},\mathbf{q}'} \delta(t - t') + \frac{\delta^2 E_{\text{xc}}[n]}{\delta n(\mathbf{q}, t) \delta n(\mathbf{q}', t')}, \qquad (D.21)$$

where the Coulomb interaction is taken to be instantaneous and we have used the fact that K can only depend on a time difference. Fourier transforming leads to the form

$$K(\mathbf{q}, \mathbf{q}', \omega) = V_C(q) \delta_{\mathbf{q},\mathbf{q}'} + f_{\text{xc}}(\mathbf{q}, \mathbf{q}', \omega) \qquad (D.22)$$

---

[1] The full many-body expressions can also formally be written in exactly the same form with $M^a_{ij} = \langle \Psi_i | \hat{O}^a | \Psi_j \rangle$ and $\varepsilon_i \to E_i$, which shows that properties such as the Kramers–Kronig relations apply in general and are not restricted to independent-particle approximations.

and a similar expression in $\mathbf{r}$ space. Thus the dynamical generalization of Eq. (D.14) can be written in compact form as

$$\chi(\omega) = \chi^0(\omega)[1 - \chi^0(\omega)K(\omega)]^{-1}. \tag{D.23}$$

Note that K itself is a response function, so that it also must have the analytical properties required by causality, it must vanish at high frequency, etc. Specific expressions that illustrate how to use this general expression are given in Section 21.4.

D.5 Green's Functions

Green's functions are widely used in theoretical physics [306, 683, 1004]. For independent-particle hamiltonians, the most important Green's function is the spectral function in terms of the time-independent eigenstates of the hamiltonian

$$G(z, \mathbf{r}, \mathbf{r}') = \sum_i \frac{\psi_i(\mathbf{r})\psi_i(\mathbf{r}')}{z - \varepsilon_i}, \tag{D.24}$$

where z is a complex variable. This may be written in a more general form in terms of any complete set of basis states $\chi_\alpha(\mathbf{r})$,

$$G(z, \mathbf{r}, \mathbf{r}') = \sum_{\alpha, \beta} \chi_\alpha(\mathbf{r}) \left[\frac{1}{z - \hat{H}} \right]_{\alpha, \beta} \chi_\beta(\mathbf{r}'), \tag{D.25}$$

or

$$G_{\alpha, \beta}(z) = \left[z - \hat{H} \right]^{-1}_{\alpha, \beta}. \tag{D.26}$$

The density of states per unit energy projected on the basis function α is given by

$$n_\alpha(\varepsilon) = -\frac{1}{\pi} \mathrm{Im} G_{\alpha, \alpha}(z = \varepsilon + i\delta), \tag{D.27}$$

where δ is a positive infinitesimal, and the total density of states is given by

$$n(\varepsilon) = -\frac{1}{\pi} \mathrm{Im} \mathrm{Tr} G(z = \varepsilon + i\delta). \tag{D.28}$$

Total integrated quantities at $T = 0$ can be derived by contour integrations in the complex z plane as illustrated in Fig. D.1. Integration of $G(z)$ around each pole in a counterclockwise direction gives $2\pi i$. The contour C can be any closed line that encircles the poles, so that the density matrix is given by

$$\rho(\mathbf{r}, \mathbf{r}') = \frac{1}{2\pi i} \int_C dz \, G(z, \mathbf{r}, \mathbf{r}'); \tag{D.29}$$

the density by $n(\mathbf{r}) = \rho(\mathbf{r}, \mathbf{r})$; the total number of electrons by

$$N = \int_{-\infty}^{E_F} d\varepsilon \, n(\varepsilon) = \frac{1}{2\pi i} \int_C dz \, \mathrm{Tr} G(z); \tag{D.30}$$

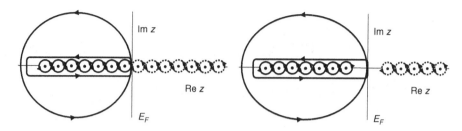

Figure D.1. Contours for line integration over the spectral function to derive integrated quantities. The contours shown enclose all poles below the Fermi energy. (Dotted contours indicate empty states not included.) The integral of the trace $\mathrm{Tr}(G(z)) = \sum_\alpha G_{\alpha,\alpha}(z)$ is the total number of particles. The sum of independent-particle energies is $\mathrm{Tr}\hat{H}[G(z)]$, etc. The left-hand figure indicates a metal where the contour necessarily passes arbitrarily close to a pole. The right-hand figure indicates an insulator where the contour passes through a gap. Whenever z is far from any pole, G decays as a function of distance and therefore can be considered localized.

and sum of occupied eigenvalues by

$$\sum_{i \ occ} \varepsilon_i = \int_{-\infty}^{E_F} d\varepsilon \ \varepsilon \ n(\varepsilon) = \frac{1}{2\pi i} \int_C dz \ z \ \mathrm{Tr}G(z). \tag{D.31}$$

Since the total energy in the Kohn–Sham method can be derived from the sum of eigenvalues and the density, it follows that all quantities related to total energy can be derived from the independent-particle Kohn–Sham Green's function. Expressions for energies and forces are given in Chapter 18.

D.6 The "2n + 1 Theorem"

The "$2n + 1$ theorem" states that knowledge of the wavefunction *to all orders 0 through n determines the energy to order $2n + 1$*. Perhaps the first example of this theorem was by Hylleraas [52] in 1930 in a study of two-electron systems, where he showed that the first-order derivative of an eigenfunction with respect to a perturbation is sufficient to find the second and third derivatives of the energy. In the same paper, Hylleraas observed that there is an expression for the second derivative that is variational (minimal) with respect to errors in $d\psi/d\lambda$ (see Section 20.5). In the intervening years there have been many works proving the full "$2n + 1$ theorem," which has been extended to density functional theory and other functionals obeying minimum energy principles [181, 813, 816].

It is instructive to write down an example of the third-order energy to see the relation to variational principles following the approach in [813]. The principles can be illustrated for a single state, and the derivation is readily extended to many states [813]. If $\hat{H}$ is expanded in powers of λ, $\hat{H} = \hat{H}^{(0)} + \lambda\hat{H}^{(1)} + \lambda^2\hat{H}^{(2)}$, and similarly for ψ, and the eigenvalue ε, then the Schrödinger equation $(\hat{H} - \varepsilon)\psi = 0$ to order m can written

$$\sum_{k=0}^{m} (\hat{H} - \varepsilon)^{(m-k)} \psi^{(k)} = 0, \tag{D.32}$$

with the constraint

$$\sum_{j=0}^{m} \langle \psi^{(j)} | \psi^{(m-j)} \rangle = 0, \, m \neq 0. \tag{D.33}$$

Here are collected all terms of order λ^m and then λ is set to 1. Taking matrix elements of Eq. (D.32) leads to

$$\sum_{j=0}^{m} \sum_{k=0}^{m} \Theta(m - j - k) \langle \psi^{(j)} | (\hat{H} - \varepsilon)^{(m-j-k)} | \psi^{(k)} \rangle = 0, \tag{D.34}$$

where $\Theta(p) = 1, \, p \geq 0; 0, \, p < 0$.

The desired expressions can be derived by applying the condition that Eq. (D.34) be variational with respect to $\psi^{(k)}$ at each order $k = 0, \dots, m$. This is facilitated by writing Eq. (D.34) in the form of an array, illustrated here for $m = 3$:

$$si^{(3)} | \bar{H}^{(0)} | \psi^{(0)} \rangle$$

$$si^{(2)} | \bar{H}^{(1)} | \psi^{(0)} \rangle + \langle \psi^{(2)} | \bar{H}^{(0)} | \psi^{(1)} \rangle$$

$$si^{(1)} | \bar{H}^{(2)} | \psi^{(0)} \rangle + \langle \psi^{(1)} | \bar{H}^{(1)} | \psi^{(1)} \rangle + \langle \psi^{(1)} | \bar{H}^{(0)} | \psi^{(2)} \rangle$$

$$si^{(0)} | \bar{H}^{(3)} | \psi^{(0)} \rangle + \langle \psi^{(0)} | \bar{H}^{(2)} | \psi^{(1)} \rangle + \langle \psi^{(0)} | \bar{H}^{(1)} | \psi^{(2)} \rangle + \langle \psi^{(0)} | \bar{H}^{(0)} | \psi^{(3)} \rangle. \tag{D.35}$$

Variation of each $|\psi^{(k)}\rangle$ ($\langle \psi^{(k)}|$) in turn means that the sum of elements in each row (column) of Eq. (D.35) vanishes. From this it follows that one can eliminate the higher order $\psi^{(k)}, \, k = 2, 3$, with the result (Exercise D.6)

$$\varepsilon^{(3)} = \langle \psi^{(0)} | \hat{H}^{(3)} | \psi^{(0)} \rangle + \langle \psi^{(1)} | \hat{H}^{(2)} | \psi^{(0)} \rangle + \text{c.c.}$$

$$+ \langle \psi^{(1)} | \hat{H}^{(1)} - \varepsilon^{(1)} | \psi^{(1)} \rangle. \tag{D.36}$$

Such expressions are used in electronic structure theory to derive accurate energies from approximate wavefunctions – e.g., in certain expressions in the linearized methods of Chapter 17.

SELECT FURTHER READING

Doniach, S. and Sondheimer, E. H., *Green's Functions for Solid State Physicists* (W. A. Benjamin, Reading, MA, 1974), reprinted in Frontiers in Physics Series, no. 44.

Fetter, A. L. and Walecka, J. D., *Quantum Theory of Many-Particle Systems* (McGraw-Hill, New York, 1971).

Mahan, G. D., *Many-Particle Physics*, 3rd ed. (Kluwer Academic/Plenum Publishers, New York, 2000).

Martin, P. C., *Measurement and Correlation Functions* (Gordon and Breach, New York, 1968).

Pines, D., *Elementary Excitations in Solids* (Wiley, New York, 1964).

Exercises

D.1 Derive the general form of the density response function χ_n in Fourier space, Eq. (D.8). This applies to any function, periodic or nonperiodic.

D.2 Derive the second form given in Eq. (D.14) from the first expression. Hint: move all terms involving χ to the left-hand side, solve for χ in terms of χ^0 and K, and invert both sides of the equation.

D.3 See Exercise 7.20 for the way in which the response function can be used to analyze the form of the energy functionals near the minimum.

D.4 Show that the response of a harmonic oscillator, Eq. (D.16), obeys the KK relations. Hint: The key point is the sign of the damping term that corresponds to energy loss – i.e., $\Gamma > 0$. See Exercise D.5 for an explanation.

D.5 Derive the KK relations, Eq. (D.18), from the analytic properties of the response functions. Causality requires that all poles as a function of complex frequency z be in the lower plane $\Im z < 0$. Hint: An integral along the real axis can be closed in the upper plane with a contour that is at $|z| \to \infty$. Since the contour encloses no poles, the line integral vanishes; also the integral at infinity vanishes. The integral along the axis can be broken into the principal value parts and the residue parts leading to Eq. (D.18). See [297, 300].

D.6 Derive the formula Eq. (D.36) for energy to third order from the preceding equations.

D.7 Exercise 20.11 considers the variational principle in perturbation theory applied to a system composed of two springs. Let each spring have a nonlinear term $\frac{1}{2}\gamma_1(x_1 - x_0)^3$ and similarly for spring 2. Find an explicit expression for the change in energy to third order due to the applied force.

Appendix E

Dielectric Functions and Optical Properties

Summary

Dielectric functions are the most important response functions in condensed matter physics: photons are perhaps the most important probe in experimental studies of matter; electrical conductivity and optical properties are among the most important phenomena in technological applications as well as everyday life. Dielectric functions can be defined in terms of currents and fields, which is most appropriate for conductivity and optical response, or in terms of densities and scalar potentials, which is most appropriate for static problems. The needed expressions follow from Maxwell's equations; however, care must be taken in defining the polarization in extended matter, which is treated here and in Chapter 24. This appendix provides the phenomenological definitions; the role of electronic structure is to provide the fundamental foundations in terms of the underlying quantum theory of the electrons, which is the subject of Chapters 20, 21, and 24.

E.1 Electromagnetic Waves in Matter

Maxwell's equations for electromagnetic fields interacting with particles having charge Q ($Q = -e$ for electrons) and number density n

$$\nabla \cdot \mathbf{E} = 4\pi Q n, \qquad \nabla \times \mathbf{E}(t) = -\frac{1}{c}\frac{d\mathbf{B}}{dt},$$

$$\nabla \cdot \mathbf{B} = 0, \qquad \nabla \times \mathbf{B}(t) = \frac{4\pi}{c}\mathbf{j} + \frac{1}{c}\frac{d\mathbf{E}}{dt}, \tag{E.1}$$

are the fundamental equations that describe the interactions of particles in matter. The arguments $\mathbf{r}, t$ have been omitted for simplicity and $\mathbf{j}$ is the charge current density that satisfies the continuity equation

$$\nabla \cdot \mathbf{j} = -Q\,\frac{dn}{dt}. \tag{E.2}$$

The basic equations of electronic structure – in particular, the hamiltonian, Eq. (3.1) – are based on Eq. (E.1) in the nonrelativistic limit where the speed of light $c \to \infty$ in which case it is sufficient to take $\mathbf{B} = 0$ and work with the scalar potential V, which satisfies the Poisson equation

$$\nabla^2 V = -4\pi Q n , \quad \text{with} \quad \mathbf{E} = -\nabla V. \tag{E.3}$$

However, in order to describe important physical phenomena, such as the propagation of electromagnetic waves in matter and the response to external fields, it is essential to return to the full equations in Eq. (E.1). Here we summarize[1] the phenomenological theory of matter interacting with external time-dependent fields, defining the appropriate quantities carefully to set the stage for proper derivation from electronic structure theory (see especially Chapters 21 and 24).[2]

Two steps are crucial for defining the structure of the theory:

- In order to derive properties of matter under the influence of external fields, the charges and currents in Maxwell's equations must be divided into "internal" and "external,"

$$n = n_{\text{int}} + n_{\text{ext}}; \quad \mathbf{j} = \mathbf{j}_{\text{int}} + \mathbf{j}_{\text{ext}}. \tag{E.4}$$

Although such a division can be made for any perturbation, electromagnetic interactions are of special importance because the long-range interactions lead to effects that extend over macroscopic distances into the interior of bodies.

- The polarization $\mathbf{P}$ is defined by

$$\nabla \cdot \mathbf{P}(\mathbf{r}, t) = -Q n_{\text{int}}(\mathbf{r}, t). \tag{E.5}$$

Together with (E.2) this leads to the relation

$$\frac{d\mathbf{P}(\mathbf{r}, t)}{dt} = \mathbf{j}(\mathbf{r}, t) + \nabla \times \mathbf{M}(\mathbf{r}, t), \tag{E.6}$$

where $\nabla \cdot (\nabla \times \mathbf{M}(\mathbf{r}, t)) = 0$ for a vector field $\mathbf{M}$. *Note that each equation leaves the macroscopic average value of* $\mathbf{P}$ *defined only to within an additive constant.* This is readily remedied in a finite system where $\mathbf{P}$ is given by Eq. (24.3) but is an issue in extended matter that has been fully resolved by using quantum mechanics and Berry phases, as summarized in Chapter 24.

In terms of the displacement field $\mathbf{D} = \mathbf{E} + 4\pi \mathbf{P}$, Maxwell's equations can be written in the form

$$\nabla \cdot \mathbf{D} = 4\pi Q n_{\text{ext}}, \qquad \nabla \times \mathbf{E}(t) = -\frac{1}{c}\frac{d\mathbf{B}}{dt},$$

$$\nabla \cdot \mathbf{B} = 0, \qquad \nabla \times \mathbf{B}(t) = \frac{4\pi}{c}\mathbf{j}_{\text{ext}} + \frac{1}{c}\frac{d\mathbf{D}}{dt}. \tag{E.7}$$

---

[1] Following the clear presentation of [300], Section 20.2.
[2] The derivations are strictly applicable only to materials with trivial topology. Materials like Chern insulators and quantum Hall systems are insulators in the bulk but have conducting surface states; there are fascinating consequences for the electromagnetic properties as discussed in Chapters 25–28 and Appendix Q.

The advantage of this form is that all source terms are "external." In the interior of a sample, n_{ext} and j_{ext} vanish even though they can lead to fields inside the sample. As shown by Eqs. (E.1) and (E.7), $\mathbf{E}$ is the *total field* in the material, whereas $\mathbf{D}$ is the field due only to external sources. Thus the value of $\mathbf{D}$ at any point is independent of the material and is the same as if the material were absent.

E.2 Conductivity and Dielectric Tensors

Solution of the equations requires the material relation of j_{int} or n_{int} to the total fields $\mathbf{E}$ and $\mathbf{B}$. To linear order, the most general relation is

$$\mathbf{j}_{int}(\mathbf{r}, t) = \int d\mathbf{r}' \int^t dt' \sigma(\mathbf{r}, \mathbf{r}', t - t') \mathbf{E}(\mathbf{r}', t'), \tag{E.8}$$

where $\sigma(\mathbf{r}, \mathbf{r}', t - t')$ is the microscopic conductivity tensor. For a perturbation with time dependence $\propto \exp(i\omega t)$, Eq. (E.8) becomes

$$\mathbf{j}_{int}(\mathbf{r}, \omega) = \int d\mathbf{r}' \sigma(\mathbf{r}, \mathbf{r}', \omega) \mathbf{E}(\mathbf{r}', \omega), \tag{E.9}$$

which implies

$$\mathbf{D}(\mathbf{r}, \omega) = \int d\mathbf{r}' \epsilon(\mathbf{r}, \mathbf{r}', \omega) \cdot \mathbf{E}(\mathbf{r}', \omega) \text{ or } \mathbf{E}(\mathbf{r}, \omega) = \int d\mathbf{r}' \epsilon^{-1}(\mathbf{r}, \mathbf{r}', \omega) \mathbf{D}(\mathbf{r}', \omega), \tag{E.10}$$

where

$$\epsilon(\mathbf{r}, \mathbf{r}', \omega) = \mathbf{1}\, \delta(\mathbf{r} - \mathbf{r}') + \frac{4\pi i}{\omega} \sigma(\mathbf{r}, \mathbf{r}', \omega). \tag{E.11}$$

Note that ϵ and σ are the response to the *total field* $\mathbf{E}$, whereas ϵ^{-1} is the response to an *external field*. Interestingly, $\sigma(\omega)$, $\epsilon(\omega) - 1$, and $\epsilon^{-1}(\omega) - 1$ are all response functions and each satisfies Kramers–Kronig relations, Eq. (D.18).

The macroscopic average functions $\bar{\epsilon}(\omega)$ or $\bar{\sigma}(\omega)$ are directly measured by the index of refraction for photons and response to macroscopic electric fields – e.g., conductivity and dielectric response where the measured voltage is the line integral of the internal electric field $\mathbf{E}$. On the other hand, the scattering of charged particles directly measures $\epsilon^{-1}(\mathbf{q}, \omega)$ ([297] p. 126), where $\mathbf{q}$ and ω are the momentum and energy transfers.

E.3 The f Sum Rule

The dielectric functions satisfy the well-known "f sum rule," for which Seitz [40] attributes the original derivation to Wigner [1005] and Kramers [1006]. A simple way to derive the sum rule ([297], p. 136) is to note that in the $\omega \to \infty$ limit, the electrons act as free particles, from which it follows that (Exercise E.1)

$$\epsilon_{\alpha\beta}(\omega) \to \delta_{\alpha\beta} \left[1 - \frac{\omega_p^2}{\omega^2} \right], \tag{E.12}$$

where ω_p is the plasma frequency $\omega_p^2 = 4\pi(NQ^2/\Omega m_e)$, with N/Ω the average density. (As a check note that this is the first term in square brackets in Eq. (21.9).) Combining this with the Kramers–Kronig relations, Eq. (D.18) leads to (Exercise E.2)

$$\int_0^\infty d\omega\, \omega\, \mathrm{Im}\epsilon_{\alpha\beta}(\omega) = \frac{\pi}{2}\omega_p^2\delta_{\alpha\beta}, \quad \text{or} \quad \int_0^\infty d\omega\, \omega\, \mathrm{Re}\sigma_{\alpha\beta}(\omega) = \frac{\pi}{2}\frac{Q^2}{m_e}\frac{N}{\Omega}\delta_{\alpha\beta}. \tag{E.13}$$

A similar sum rule is satisfied by $\epsilon_{\alpha\beta}^{-1}(\omega)$. Finally, all the versions of the f sum rule apply to the exact many-body response as well as to the simple noninteracting approximation, because the sum rule depends only on the Kramers–Kronig relations and the high $\omega \to \infty$ limit, in which the electrons always act as uncorrelated free particles.

E.4 Scalar Longitudinal Dielectric Functions

The dielectric relations, Eq. (E.10), can also be written in terms of scalar potentials. This is sufficient for static problems and is convenient for many uses, especially applications in density functional theory, which is cast in terms of potentials and densities. This is called "longitudinal" because it only applies to electric fields that can be derived from a potential $\mathbf{E}(\mathbf{r}) = -\nabla V(\mathbf{r})$. Thus the electric field in Fourier space $\mathbf{E}(\mathbf{q}) = i\mathbf{q}V(\mathbf{q})$ is longitudinal – i.e., parallel to $\mathbf{q}$. Combining Eqs. (E.3), (E.4), and (E.7), it follows that [180]

$$\epsilon^{-1}(\mathbf{q},\mathbf{q}',\omega) = \frac{\delta V_{\text{total}}^C(\mathbf{q},\omega)}{\delta V_{\text{ext}}(\mathbf{q}',\omega)} \quad \text{or} \quad \epsilon(\mathbf{q},\mathbf{q}',\omega) = \frac{\delta V_{\text{ext}}(\mathbf{q},\omega)}{\delta V_{\text{total}}^C(\mathbf{q}',\omega)}, \tag{E.14}$$

where the total Coulomb potential is denoted V_{total}^C – i.e., the potential acting on an infinitesimal test charge, which does not include the effective exchange–correlation potential V_{xc} that acts on an electron.

Expressions for ϵ and ϵ^{-1} in terms of electronic states can be derived from the general formulas for response functions χ^0 (Eqs. (D.19) and (D.20)) and χ (Eq. (D.23)). In particular, it follows that (Exercise E.3)

$$\epsilon^{-1}(\mathbf{q},\mathbf{q}',\omega) = \delta(\mathbf{q} - \mathbf{q}') + V_C(q)\chi(\mathbf{q},\mathbf{q}',\omega), \tag{E.15}$$

where $V_C(q) = 4\pi e^2/q^2$ is independent of ω (the same as in Eq. (D.13) and we have set $Q = -e$). For a theory in which the electrons interact via an effective field, as in the Kohn–Sham approach, χ is most readily calculated using the expression (D.14)

$$\epsilon^{-1} = 1 + \frac{V_C\chi^0}{1 - (V_C + f_{xc})\chi^0} = \frac{1 - f_{xc}\chi^0}{1 - (V_C + f_{xc})\chi^0}. \tag{E.16}$$

The equation appears simple because the arguments have been omitted, but actual evaluation can be tedious since products such as $f_{xc}\chi^0$ stand for convolutions over all internal wavevectors and frequencies. The inverse dielectric functions is the response to an external field and it describes the energy loss for high energy charged particles – e.g., in electron energy loss spectroscopy. (This is discussed in [1], section 14.3.)

The simplest case is the electron gas (Chapter 5), where χ is nonzero only for $\mathbf{q} = \mathbf{q}'$ and the expressions can be evaluated analytically. The Lindhard expressions, (5.38), for $\chi^0(q,\omega)$ are given in Section 5.5, from which can be derived all the other response functions. In a crystal, the wavevectors can always be written as $\mathbf{q} = \mathbf{k} + \mathbf{G}$ and $\mathbf{q}' = \mathbf{k} + \mathbf{G}'$, where $\mathbf{k}$ is restricted to the first Brillouin zone, so that $\epsilon(\mathbf{k} + \mathbf{G}, \mathbf{k} + \mathbf{G}', \omega)$ is a matrix $\epsilon_{\mathbf{GG}'}(\mathbf{k}, \omega)$ and, similarly, for the inverse matrix, $\epsilon_{\mathbf{GG}'}^{-1}(\mathbf{k}, \omega)$. Optical phenomena involve long wavelengths, $\mathbf{G} = 0$ and $\mathbf{G}' = 0$, and are described by the macroscopic dielectric function $\epsilon(\mathbf{k}, \omega)$, defined by the ratio of internal to external macroscopic fields ($\mathbf{G} = \mathbf{G}' = 0$). Since there are no applied fields at short wavelengths, this corresponds to inverting the matrix for keeping the short wavelength ($\mathbf{G}' \neq 0$) *external* fields fixed. However, there are changes in the short wavelength internal fields called "local field corrections." It follows that [180, 1003, 1007] (Exercise E.4)

$$\epsilon(\mathbf{k}, \omega) = \frac{\delta V_{\text{ext}}(\mathbf{k}, \omega)}{\delta V_{\text{total}}^{C}(\mathbf{k}, \omega)} = \frac{1}{\epsilon_{00}^{-1}(\mathbf{k}, \omega)}. \tag{E.17}$$

Note that the independent-particle approximation corresponds to the use of the response function χ^0 in Eq. (E.16) with $f_{\text{xc}} = 0$, which means that the electrons are independent except that the potential takes into account the average Coulomb (Hartree) potential.

Finally, the full dielectric tensor can be recovered considering different directions $\hat{\mathbf{k}}$ using the fact [180] that for long wavelengths, the scalar dielectric function is related to the dielectric tensor by [180]

$$\epsilon(\mathbf{k}, \omega) = \lim_{|\mathbf{k}| \to 0} \hat{\mathbf{k}}_\alpha \epsilon_{\alpha\beta}(\mathbf{k}, \omega) \hat{\mathbf{k}}_\beta. \tag{E.18}$$

In a cubic crystal $\epsilon_{\alpha\beta} = \epsilon \delta_{\alpha\beta}$, but in general Eq. (E.18) depends on the direction in which the limit is taken.

E.5 Tensor Transverse Dielectric Functions

The general cases of time-dependent electric and magnetic fields can conveniently be treated by calculation of the current response to the vector potential $\mathbf{A}$. The perturbation can be written in terms of $\mathbf{A}$ as

$$\Delta \hat{H}(t) = \frac{1}{2m_e} \sum_i \left\{ \left[\mathbf{p}_i - \frac{e}{c} \mathbf{A}(t) \right]^2 - \mathbf{p}_i^2 \right\}, \tag{E.19}$$

where $\mathbf{E}(t) = -(1/c)(d\mathbf{A}/dt)$ or $\mathbf{E}(\omega) = -(i\omega/c)\mathbf{A}(\omega)$, and the magnetic field is given by $\mathbf{B} = \nabla \times \mathbf{A}$. The desired response is the current density $\mathbf{j}$. For a transverse electromagnetic wave this is the appropriate response function.

Formulas for response function in the independent-particle approximation have the general form given in Appendix D and are given explicitly in Section 21.8. Self-consistent field expressions have exactly the same form as for the scalar dielectric function except that they involve an effective "exchange–correlation vector potential" that is the fundamental quantity in "current functional theory" [342, 343, 345, 347].

E.6 Lattice Contributions to Dielectric Response

In an ionic insulator, the motion of the ions contributes to the low-frequency dielectric response [91, 180, 1008], where the electronic contribution can be considered constant as a function of frequency ω. All quantities are properly defined *holding the macroscopic electric field* $\mathbf{E}_{mac}$ *constant*, which gives the intrinsic response. The macroscopic field is controlled by external conditions, boundary conditions, etc., and such effects should be taken into account in the specific solution. The Born effective charge tensor for each ion I is defined by

$$Z^*_{I,\alpha\beta}|e| = \frac{d\mathbf{P}_\alpha}{d\mathbf{R}_{I,\beta}}\bigg|_{\mathbf{E}_{mac}}, \tag{E.20}$$

where the macroscopic electric field is held constant. The effective charge is nonzero for some displacements in any ionic crystal, and it has been shown that in all elemental crystals with three or more atoms per cell [1009] (with the exception [1010] of two special cases out of the 230 space groups) there must also nonzero effective charges. In fact, there are large measured effective charges and infrared absorption known in elemental crystals such as trigonal Se [1009]. The polarization caused by the effective charges leads to nonanalytic terms in the force constant matrix defined in Eq. (20.9), which has the form (see eq. (4.7) of [180]):

$$C_{s,\alpha;s',\alpha'}(\mathbf{k}) = C^N_{s,\alpha;s',\alpha'}(\mathbf{k}) + \frac{4\pi e^2}{\Omega}\left[\sum_\gamma \hat{\mathbf{k}}_\gamma Z^*_{I,\gamma\alpha}\right]^\dagger \frac{1}{\epsilon(\mathbf{k})}\left[\sum_\gamma \hat{\mathbf{k}}_\gamma Z^*_{I,\gamma\beta}\right], \tag{E.21}$$

where C^N is the normal analytic part of C and $\epsilon(\mathbf{k})$ is the low-frequency electronic dielectric constant. The full dielectric function including the lattice contribution is given by Cochran and Cowley [1008] and the low-frequency limit is given in [180], Eq. (7.1). Similarly, one can define proper piezoelectric constants [807, 1011, 1012] in the absence of macroscopic fields,

$$e_{\alpha\beta\gamma} = \frac{d\mathbf{P}_\alpha}{du_{\alpha\beta}}\bigg|_{\mathbf{E}_{mac}}, \tag{E.22}$$

where $u_{\alpha\beta}$ denotes the strain tensor of Eq. (G.2). The effect can be separated into a pure strain effect and an internal displacement contribution,

$$e_{\alpha,\beta\gamma} = e^0_{\alpha,\beta\gamma} + |e|\sum_{s,\delta} Z^*_{s,\alpha\delta}\Gamma_{s,\delta,\beta\gamma}, \tag{E.23}$$

where $Z^*_{s,\alpha\delta}$ is the same effective charge tensor that governs infrared response of optic modes and Γ is defined in Eq. (G.14). This division facilitates calculations and clarifies relations of measurable physical quantities. Crystals with permanent moments present a particular problem, in that a rotation of the moment might be termed a piezoelectric effect. This is an "improper effect" and it has been shown that "proper" expressions for the polarization, such as the Berry phase form in Section 24.3, do not contain such terms [927].

SELECT FURTHER READING

Definitions of dielectric functions:

Pick, R., Cohen, M. H., and Martin, R. M., "Microscopic theory of force constants in the adiabatic approximation," *Phys. Rev. B* 1:910–920, 1970.

Pines, D., *Elementary Excitations in Solids* (Wiley, New York, 1964).

Wiser, N., "Dielectric constant with local field effects included," *Phys. Rev.* 129:62–69, 1963.

An Electrodynamics book that inludes the modern formulation of polarization:

Zangwill, A., *Modern Electrodynamics* (Cambridge University Press, Cambridge, 2013).

Exercises

E.1 Derive Eq. (E.12) for the dielectric tensor at high frequency using only the fact that electrons respond as free particles at sufficiently high frequency. It may be helpful to relate to the high-frequency limit of the harmonic oscillator response function give in Section D.4.

E.2 Show that the f sum rule, Eq. (E.13), follows from the high-frequency behavior in Eq. (E.12) and the Kramers–Kronig relations, Eq. (D.18).

E.3 Show that Eq. (E.15) results from the definition of internal and external charges in Eq. (E.4) and the definition of ϵ^{-1} in Eq. (E.14).

E.4 The expression for the macroscopic dielectric function, Eq. (E.17), can be derived by carefully applying the definition that it is the ratio of external to total internal fields given in Eq. (E.17) *for the case where there are no short wavelength external fields*, $V_{\text{ext}}(\mathbf{q} + \mathbf{G}, \omega) = 0$ for $\mathbf{G} \neq 0$, and using the definition that the inverse function is the response to external fields. Use these facts to derive Eq. (E.17).

Appendix F

Coulomb Interactions in Extended Systems

Summary

The subject of this appendix is formulations and explicit equations for the total energy that properly take into account the long-range effects of Coulomb interactions. We emphasize the Kohn–Sham independent-particle equations and expressions for total energy; however, the ideas and many of the equations also apply to many-body calculations. There are three main issues:

- Identifying various convenient expressions that each yield properly the intrinsic total energy per formula unit for an extended bulk system
- Understanding and calculating the effect on the average potential in a bulk material due to dipole terms at surfaces and interfaces
- Treating finite systems, where there is no essential difficulty, but where it is convenient to carry out the calculations in a periodic "supercell" geometry

F.1 Basic Issues

There is a simple set of guiding principles that must be followed to properly treat long-range Coulomb interactions in extended systems. If the calculations are carried out in a cell that represents an infinite system, i.e., the unit cell of a crystal, or a "supercell" constructed so that its limiting behavior represents a macroscopic system, then

- the cell must be chosen to be neutral;
- the neutral cell can be used to define a proper thermodynamic "reference state" if in addition we require that there is no average (macroscopic) electric field; and
- the average electrostatic potential is *not* an intrinsic property of condensed matter. The value is ill defined in an infinite system. In a large (but finite) sample, the value relative to vacuum depends on surface conditions.

The first condition is obvious because otherwise the Coulomb energies of the extended system diverge. The second is less obvious but is clearly required because there is no lower bound to the energy in an infinite system with an electric field. In a metal there is no problem since there can be no uniform electric field in equilibrium. However, in general, in an insulator, the total energy is the energy of this "reference state" plus changes in energy due to the presence of long-range electric fields. This is the essence of a dielectric in which the energy is a function of applied fields [480, 908], which can be described in terms of dielectric response functions derived by perturbation theory (Chapter 20).[1]

The expressions for the total energy given in Section 3.2 and in the chapters on density functional theory (see, e.g., Eqs. (7.5), (7.20), (7.22), and (7.26)) are organized into neutral groupings so that they are in the proper form to define the intrinsic, extensive properties of condensed matter in the large size (or thermodynamic) limit. These classical Coulomb contributions to the total energy given in Eq. (3.14) are determined solely by the charge density of the electrons, the nuclei, and any external charges. All effects of quantum mechanics on the electrons and correlations among the electrons can be separated into the other terms in the total energy as expressed in Eq. (3.16) and the "xc" terms in density functional theory; these are short range in nature and not subject to convergence problems.

A comment is in order regarding terminology. In density functional theory, the "external potential" has a central role. However, the external potential due to the charged nuclei diverges in an infinite system. This nomenclature should cause no difficulty so long as one maintains the principle that the long-range part of the Hartree potential is grouped with the nuclear potential in order to have a well-defined "external potential" and total energy. For example, in the bulk of a crystal the potential regarded as external may include effects of electrons at a large distance that are not part of the intrinsic bulk system.

There are three typical ways to specify the Coulomb energy and the potential of extended systems. One is to add and subtract a uniform background; then the energy can be expressed as the sum of the classical energy of nuclei (or ions) in a compensating negative background plus the total energy of the system of electrons in a compensating positive background. This has the advantage of simplicity and may be close to the real situation in materials where the electrons are nearly uniform. However, it leads to expressions for the total energy that often involve small differences between large numbers that are difficult to interpret physically. The second approach is to "smear" the ions, which allows a convenient rearrangement of terms that is especially useful in Fourier space expressions. A third method is a variation in which one finds the *difference* from isolated neutral atoms (or neutral spherical atomic-like species). Then we only deal with the *difference* between two neutral systems, which has obvious advantages since it relates to the real physical problem of the binding energy relative to atoms. However, it requires that we either specify properties of the real atom or define an arbitrary neutral reference density.

---

[1] Special care must be taken if there is a polarization in the absence of an average electric field – i.e., in pyroelectrics or polled ferroelectrics. See Chapter 24.

F.2 Point Charges in a Background: Ewald Sums

The Ewald method for summing the Coulomb interactions of point charges is based on transformation of the potential due to an infinite periodic array of charges. The result is two sums, one in reciprocal space and one in real space, each of which is absolutely convergent. The approach is intimately connected to the expressions for total energy, which must be evaluated in a consistent way to eliminate the divergent terms in both the total energy and the Kohn–Sham potential. This is the approach used in Section 13.1, in particular in Eq. (13.1). The arguments given here are the justification for the exclusion of the $\mathbf{G} = 0$ Fourier components in the expressions for the Hartree term in the Kohn–Sham potential and the $\mathbf{G} = 0$ term in the Hartree energy.

The first step is the identification of appropriate neutral groupings by adding and subtracting a uniform positive background charge density n^+, which is equivalent to adding n^+ and a uniform negative density $n^- = -n^+$. This allows us to rewrite the total energy, Eq. (3.14) (or any of the expressions in the density functional theory chapters) as the classical Coulomb energy

$$E^{CC} = E'_{\text{Hartree}}[n(\mathbf{r}) + n^+] + \int d^3r\, V'_{\text{ext}}(\mathbf{r})n(\mathbf{r}) + E'_{II}, \tag{F.1}$$

where *each term is neutral*. The effects of n^+ are incorporated in E'_{Hartree}, which is the Hartree-like energy having exactly the same form as Eq. (3.15)

$$E'_{\text{Hartree}}[n] = \frac{1}{2} \int d^3r d^3r' \frac{[n(\mathbf{r}) + n^+](n(\mathbf{r}') + n^+)}{|\mathbf{r} - \mathbf{r}'|} = \frac{1}{2} 4\pi \sum_{\mathbf{G} \neq 0}' \frac{|n(\mathbf{G})|^2}{G^2}, \tag{F.2}$$

with n replaced by the neutral density $n + n^+$. In Fourier space, the addition of n^+ simply amounts to omitting the $\mathbf{G} = 0$ term since $n + n^+$ has zero average value. In Eq. (F.1), V'_{ext} is the potential due to the nuclei (or ions) plus the negative background n^-; again, in Fourier space, one simply omits the $\mathbf{G} = 0$ term. The final term is the sum of all interactions involving the nuclei (or ions) and n^-, which is defined to be the Madelung energy and can be evaluated by the Ewald transformation.

The Ewald transformation[2] is based on the fact that expressions for lattice sums can be written in either real or reciprocal space, or a combination of the two. The explicit formulas utilize the relation ([300], p. 271),

$$\sum_{\mathbf{T}} \frac{1}{|\mathbf{r} - \mathbf{T}|} \rightarrow \frac{2}{\sqrt{\pi}} \sum_{\mathbf{T}} \int_\eta^\infty d\rho\, e^{-|\mathbf{r}-\mathbf{T}|^2\rho^2}$$

$$+ \frac{2\pi}{\Omega} \sum_{\mathbf{G} \neq 0} \int_0^\eta d\rho \frac{1}{\rho^3} e^{-|\mathbf{G}|^2/(4\rho^2)} e^{i\mathbf{G}\cdot\mathbf{r}}, \tag{F.3}$$

---

[2] The formulas were originally given by Ewald [1013], Kornfeld [1014], and Fuchs [1015] and can be found in extensive reviews – e.g., [1016, 1017].

where $\mathbf{T}$ are the lattice translation vectors and $\mathbf{G}$ are reciprocal lattice vectors. The integrals can be computed in terms of error functions, $\text{erf}(x) = \frac{2}{\sqrt{\pi}} \int_0^x du e^{-u^2}$ and $\text{erfc}(x) = 1 - \text{erf}(x)$, leading to (see Exercise F.1, [300], p. 271, and [1018]),

$$\sum_{\mathbf{T}} \frac{1}{|\mathbf{r} - \mathbf{T}|} \rightarrow \sum_{\mathbf{T}} \frac{\text{erfc}(\eta|\mathbf{r} - \mathbf{T}|)}{|\mathbf{r} - \mathbf{T}|}$$

$$+ \frac{4\pi}{\Omega} \sideset{}{'}\sum_{\mathbf{G} \neq 0} \frac{1}{|\mathbf{G}|^2} e^{\frac{-|\mathbf{G}|^2}{4\eta^2}} \cos(\mathbf{G} \cdot \mathbf{r}) - \frac{\pi}{\eta^2 \Omega}. \tag{F.4}$$

By dividing the Coulomb sum into two terms in real and reciprocal space, each term in Eqs. (F.3) and (F.4) is absolutely convergent. The value of η determines the way the sum is apportioned in real and reciprocal space: the result must be independent of η if carried to convergence, and a choice of $\eta \approx |\mathbf{G}|_{min}$ allows each sum to be computed with only a few terms.

The sum on the left-hand sides of these two equations is the electrostatic potential at a general point $\mathbf{r}$ due to a lattice of unit charges, *which is an ill-defined sum*. The arrow in each equation denotes two definitions required to specify the right-hand side. First, the sum is made finite by including a compensating background, which is accomplished by the omission of the $\mathbf{G} = 0$ term. Second, even with the compensating term, the sum is only conditionally convergent, which reflects the fact that the absolute value of the potential is not defined in an infinite system. In Eq. (F.4), the last term is chosen so that the average value of the potential is zero [1018]. Since the absolute value of the potential does not affect the total energy of a neutral system, this is sufficient for the total energy given in Eq. (F.6) below. However, the average potential is required for other properties as discussed in Section F.5; this is not specified by the conditions given so far.

The total Coulomb energy per unit cell of a periodic array of point charges plus the compensating uniform negative background (see Fig. F.1) can be expressed using the potential at each site from Eq. (F.4), omitting the self-term for each ion. Assuming that the system is neutral and has no net polarization, the expressions are absolutely convergent (with none of

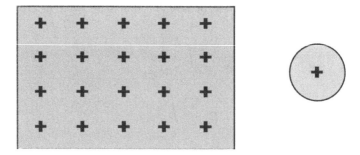

Figure F.1. A lattice of point charges in a uniform compensating background as considered in the Ewald calculation. On the right side is a nucleus in a compensating sphere, which provides a good approximation for the Coulomb energy in a close-packed lattice (see Eq. (F.10)).

the arbitrariness that occurs in the potential) and can be written for charges Z_s at positions $\tau_s, s = 1, \ldots, S$ as

$$
\gamma_E = \frac{e^2}{2} \sum_{s,s'} Z_s Z_{s'} \sum_{\mathbf{T}}' \frac{1}{|\tau_{s,s'} - \mathbf{T}|}
$$

$$
= \frac{e^2}{2} \sum_{s,s'} Z_s Z_{s'} \left[\sum_{\mathbf{T}}' \frac{\operatorname{erfc}(\eta|\tau_{s,s'} - \mathbf{T}|)}{|\tau_{s,s'} - \mathbf{T}|} + \frac{4\pi}{\Omega} \sum_{\mathbf{G} \neq 0}' \frac{1}{|\mathbf{G}|^2} e^{\frac{-|\mathbf{G}|^2}{4\eta^2}} \cos(\mathbf{G} \cdot \tau_{s,s'}) \right] \quad \text{(F.5)}
$$

$$
- \frac{e^2}{2} \left[\sum_s Z_s^2 \right] \frac{2\eta}{\sqrt{\pi}} - \frac{e^2}{2} \left[\sum_s Z_s \right]^2 \frac{\pi}{\eta^2 \Omega}, \quad \text{(F.6)}
$$

where $\tau_{s,s'} = \tau_{s'} - \tau_s$ and the primes on the sums indicate that the divergent terms are omitted. Self-terms for the ions are excluded by the omission of the $\mathbf{T} = 0$ term for $s = s'$ and by the first term in the last line that cancels a self-term included in the reciprocal space term. The $\mathbf{G} = 0$ term is omitted and the correct effects are taken into account by the second term in the last line, which is the analytic limit for $\mathbf{G} \to 0$. This term is absent in the calculation of Madelung energy for an ionic crystal where $\sum_s Z_s = 0$, but it must be included for evaluating the energy of positive ions in a background of density $n^- = -\sum_s Z_s e / \Omega$. Expression (F.6) can be used to compute the E_{II} term in the total energy in Eq. (7.5) and the needed term in Eqs. (13.1) and (13.2), and other expressions.

Finally, the reciprocal space sums in Eq. (F.6) can be written in a different form. The reciprocal space sum can be transformed to the square of a single sum over nuclei I (Exercise F.4),

$$
\sum_{s,s'}' Z_s Z_{s'} \sum_{\mathbf{G} \neq 0}' \frac{1}{|\mathbf{G}|^2} e^{\frac{-|\mathbf{G}|^2}{4\eta^2}} \cos(\mathbf{G} \cdot \tau_{s,s'}) = \sum_{\mathbf{G} \neq 0}' \frac{1}{|\mathbf{G}|^2} \left[\sum_s Z_s e^{i\mathbf{G} \cdot \tau_s} e^{\frac{-|\mathbf{G}|^2}{8\eta^2}} \right]^2, \quad \text{(F.7)}
$$

which is the Coulomb energy of a charge distribution consisting of gaussian charges at the ion sites. The real-space sum in Eq. (F.6) involving complementary error functions is a short-range sum over neighbors of the *difference* of the interaction of point charges and gaussian distributed charges.

Madelung Constant

The Madelung constant α is a dimensionless constant that characterizes the energy per cell of point charges in a lattice γ_E in terms of a typical length R

$$
\gamma_E = -\alpha \frac{(Ze)^2}{2R}. \quad \text{(F.8)}
$$

Representative values of α for are given in Table F.1, where $2R$ is the nearest-neighbor distance for ionic crystals (top line of Table F.1) and $R = R_{WS}$ for elemental crystals (bottom line of Table F.1). The neutralizing background is included in the calculation of γ_E as in Eq. (F.6) for all cases where the sum of point charges is not zero; however, it does

Table F.1. Typical values of the Madelung constant α for simple
ionic crystals and for simple elemental crystals where the
background term has been included

CsCl	NaCl	Wurtzite	Zinc-blende	
1.762,68	1.747,57	1.638,70	1.638,06	

bcc	fcc	hcp	sc	Diamond
1.791,86	1.791,75	1.791,68	1.760,12	1.670,85

not enter for ionic crystals with neutral cells of positive and negative charges (see additional comments in Exercise F.2).

For close-packed metals, the energies in Table F.1 are very close to the energy of single-point charge Ze in a sphere of uniform compensating charge, where the volume of the sphere equals that of the Wigner–Seitz cell and its radius is $R = R_{WS}$, as illustrated on the right-hand side of Fig. F.1. This can be understood simply because the cell is nearly spherical and there are no interactions between neutral spherical systems so that only internal energies need to be considered. The electrostatic potential at radius r due to the background is (Exercise F.3)

$$V(r) = Ze\left[\frac{r^2}{2R^3} - \frac{3}{2R}\right], \quad r < R, \tag{F.9}$$

where the constant is chosen to cancel the Ze/r potential from the ion at $r = R$. The total energy is the interaction of the ion with the background, plus the self-interaction of the uniform distribution (Exercise F.3)

$$E_{\text{sphere}} = (Ze)^2\left[-\frac{3}{2R} + \left(\frac{3}{2R} - \frac{9}{10R}\right)\right] = -0.90\frac{(Ze)^2}{R} = -1.80\frac{(Ze)^2}{2R}, \tag{F.10}$$

which is very close the Madelung energies for the close-packed metals in Table F.1.

Force and Stress

The part of the force on any atom due to the other nuclei or ions, treated as point charges, is easy to calculate from the analytic derivative of the Ewald energy, Eq. (F.6). The background is irrelevant in the derivative and one finds

$$-\frac{\partial\gamma_{\text{Ewald}}}{\partial\tau_s} = -\frac{e^2}{2}Z_s\sum_{s'}Z_{s'}\sum_{\mathbf{T}}{}'\left[\eta H(\eta D)\frac{\mathbf{D}}{D^2}\right]_{\mathbf{D}=\tau_{s,s'}-\mathbf{T}}$$

$$+ \frac{4\pi}{\Omega}\frac{e^2}{2}Z_s\sum_{s'}Z_{s'}\sum_{\mathbf{G}\neq 0}{}'\left[\frac{1\mathbf{G}}{|\mathbf{G}|^2}e^{\frac{-|\mathbf{G}|^2}{4\eta^2}}\sin(\mathbf{G}\cdot\tau_{s,s'})\right], \tag{F.11}$$

where $H'(x)$ is

$$H'(x) = \frac{\partial \mathrm{erfc}(x)}{\partial x} - x^{-1}\mathrm{erfc}(x). \tag{F.12}$$

The contribution of the Ewald term to the stress can be found using the forms in Appendix G. The sum in real space involves short-range two-body terms that can be expressed in the form of Eq. (G.7), and the sum in reciprocal space has the form of Eq. (G.8). The final result is (appendix of [102])

$$\frac{\partial \gamma_{\mathrm{Ewald}}}{\partial \epsilon_{\alpha\beta}} = \frac{\pi}{2\Omega\eta^2} \sum_{G \neq 0} \frac{e^{-G^2/4\eta^2}}{G^2/4\eta^2} \left| \sum_s Z_s e^{iG\cdot\tau_s} \right|^2 \left[\frac{2G_\alpha G_\beta}{G^2}(G^2/4\eta + 1) - \delta_{\alpha\beta} \right]$$

$$+ \frac{1}{2}\eta \sum_{s,s'\mathbf{T}} Z_s Z_{s'} H'(\eta D) \left. \frac{D_\alpha D_\beta}{D^2}\right|_{(D=\tau_{s'}-\tau_s+\mathbf{T}\neq 0)}$$

$$+ \frac{\pi}{2\Omega\eta^2} \left[\sum_s Z_s \right]^2 \delta_{\alpha\beta}. \tag{F.13}$$

F.3 Smeared Nuclei or Ions

The terms in the total energy can also be rearranged in a form that is readily applied in pseudopotential calculations.[3] The long-range part of the ion pseudopotential is in the local term $V_I^{\mathrm{local}}(\mathbf{r})$ defined for each ion I. If we define the charge density that would give rise to this potential as

$$n_I^{\mathrm{local}}(\mathbf{r}) \equiv -\frac{1}{4\pi}\nabla^2 V_I^{\mathrm{local}}(\mathbf{r}), \tag{F.14}$$

then the total energy for electrons in the presence of the smeared ion density can be written in terms of the total charge density

$$n^{\mathrm{total}}(\mathbf{r}) \equiv \sum_s n_I^{\mathrm{local}}(\mathbf{r}) + n(\mathbf{r}). \tag{F.15}$$

One can also define a model ion density different from Eq. (F.14); the ideas remain the same and equations given here are easily modified.

With this definition of n^{total}, the ion–ion, the Hartree, and local external terms can be combined to write the total energy, Eq. (7.5), in the form

$$E_{\mathrm{KS}} = T_s[n] + \langle \delta\hat{V}_{\mathrm{NL}} \rangle + E_{\mathrm{xc}}[n] + E'_{\mathrm{Hartree}}[n^{\mathrm{total}}] - \sum_I E_I^{\mathrm{self}} + \delta E_{II}, \tag{F.16}$$

where the non-local pseudopotential term has been added, as has also been done in Eq. (13.1). The Hartree-like term E'_{Hartree} is defined as in Eq. (F.2) with $n \to n^{\mathrm{total}}$; the

---

[3] See [646, 748] for descriptions of the ideas and practical expressions for calculations. This form is especially suited for Car–Parrinello simulations, as discussed in Section 19.4.

"self" term subtracts the ion self-interaction term included in E'_{Hartree}; and the last term δE_{II} is a short-range correction to remove spurious effects if the smeared ion densities $n_I^{\text{local}}(\mathbf{r})$ overlap.

The correspondence with the Ewald expression can be seen by choosing the densities $n_I^{\text{local}}(\mathbf{r})$ to be Gaussians, in which case this analysis is nothing but a rearrangement of the total energy using the Ewald expression, Eq. (F.6). The Fourier sum in Eq. (F.6) is included with the electron Hartree and external terms to define $\tilde{E}_{\text{Hartree}}[n^{\text{total}}]$; the real-space sum in Eq. (F.6) is simply the short-range corrections termed δE_{II} and the constants in Eq. (F.6) are the "self" terms.

Force and Stress

The force can be found by differentiating the energy, Eq. (F.16), and the force theorem, keeping in mind that n^{total} explicitly depends on the ion positions. One finds an expression analogous to Eqs. (13.3) and (F.11), with the Ewald and local terms rearranged,

$$\mathbf{F}_j^{\kappa} = -\sum_m i\mathbf{G}_m e^{i\mathbf{G}_m \cdot \tau_{\kappa,j}} V_{\text{local}}^{\kappa}(\mathbf{G}_m) n^{\text{total}}(\mathbf{G}_m) - \frac{\partial \delta E_{II}}{\partial \tau_{\kappa,j}} + \left[\mathbf{F}_j^{\kappa}\right]^{\text{NL}}, \qquad (F.17)$$

where $\left[\mathbf{F}_j^{\kappa}\right]^{\text{NL}}$ are the nonlocal final terms on the right-hand side of Eq. (13.3), and the contributions due to δE_{II} are simple short-range two-body terms. Stress is found in a form analogous to the expressions in Section F.2.

F.4 Energy Relative to Neutral Atoms

It is appealing and useful to formulate expressions for the total energy relative to atoms.[4] This can be viewed as a reformulation of the expressions in the previous section. The total energy relative to separated atoms is the difference of Eq. (F.16) from the sum of corresponding energies for the separated atoms. There is no simple expression for the difference in kinetic, nonlocal, and exchange–correlation energies, which must be calculated separately.

However, there is a simplification in the Coulomb terms that can be used to advantage. Let us define a neutral density for each atom $n_I^{\text{NA}}(\mathbf{r})$ as the sum of its electronic density $n_I(\mathbf{r})$ and the local density representing the positive ion, just as in Eq. (F.15). Then the total density can be written as

$$n^{\text{total}}(\mathbf{r}) \equiv \sum_I n_I^{\text{NA}}(\mathbf{r}) + \delta n(\mathbf{r}), \qquad (F.18)$$

where $\delta n(\mathbf{r}) = n(\mathbf{r}) - n^{\text{atom}}(\mathbf{r})$, with $n^{\text{atom}}(\mathbf{r})$ the sum of superimposed atomic densities. Substituting Eq. (F.18) into Eq. (F.2) leads directly to

---

[4] Such a form is particularly useful in local orbital methods where an atomic or atomic-like density is readily available. Informative analysis is given in [630, 646].

$$E'_{\text{Hartree}}[n^{\text{total}}] = E'_{\text{Hartree}}[n^{\text{NA}}] + \int d\mathbf{r}\, V^{\text{NA}}(\mathbf{r})\delta n(\mathbf{r}) + E'_{\text{Hartree}}[\delta n], \tag{F.19}$$

where $V^{\text{NA}}(\mathbf{r})$ is the sum of Coulomb potentials due to the neutral ion densities.

Since both n^{NA} and δn are neutral densities – i.e., having zero average value – each of the individual terms in Eq. (F.19) is well defined and can be treated individually using the Hartree-like expression, Eq. (F.2). One approach is to evaluate the first term using the fact that $n^{\text{NA}}(\mathbf{r})$ is a periodic charge density and transforming to Fourier space. However, this does not take advantage of the construction of $n^{\text{NA}}(\mathbf{r})$ as a sum of neutral, spherical densities. Using this fact, the first term can be written as a sum of intra-atom terms plus short-range interactions between the neutral atomic-like units; subtracting the unphysical self-term for the nucleus (or ion) as in Eq. (F.16), we have

$$E'_{\text{Hartree}}[n^{\text{NA}}] - \sum_I E_I^{\text{self}} = \sum_I U_I^{\text{NA}} + \sum_{I<J} U_{IJ}^{\text{NA}}(|\mathbf{R}_I - \mathbf{R}_J|), \tag{F.20}$$

where

$$U_I^{\text{NA}} = \int d\mathbf{r}\, V_I^{\text{local}}(\mathbf{r})n_I(\mathbf{r}) + \frac{1}{2}\int d\mathbf{r}\, V_I^{\text{Hartree}}(\mathbf{r})n_I(\mathbf{r}), \tag{F.21}$$

and the interaction $U_{IJ}^{\text{NA}}(|\mathbf{R}_I - \mathbf{R}_J|)$ is nonzero only for overlapping densities. If the density is strictly zero beyond a cutoff radius, then the interactions also vanish for any nonoverlapping spherical densities [630]. These expressions for the energy are used in Expression (15.14); that is particularly useful for local orbital approaches.

F.5 Surface and Interface Dipoles

Planar distributions of charge are an important special case of the effects of long-range Coulomb interactions that play a major role in surface and interface phenomena. The average electrostatic potential is shifted due to a surface or interface dipole, which gives rise to interface-dependent band offsets and surface-dependent work functions (see Sections F.5 and 13.5). The underlying cause is the long-range Coulomb interaction and the key point is that in the bulk of condensed matter *the absolute energy of a charged particle (e.g., an electron) is not an intrinsic bulk property.* One can specify energy relative to some other state (for example, the vacuum) only if the charge state of the entire system is known.

The physical problem is specified by charge density $n(\mathbf{r})$, which is nonzero only near a planar surface or interface. The density $n(\mathbf{r})$ includes both electrons and nuclei, and must be neutral for the energy per particle to be finite. If the coordinate system is fixed with $\hat{z}$ perpendicular to the plane and $\hat{x}, \hat{y}$ in the plane, then $n(\mathbf{r})$ can be divided into an average density per unit area $\sigma(z)$ plus $\delta n(\mathbf{r})$, where the latter can vary in the $\hat{x}, \hat{y}$ plane. The variations in the plane $\delta n(\mathbf{r})$ give rise to potentials that decrease exponentially [1019] as a function of $|z|$. The typical decay length is proportional to the length L_{xy} over which $\delta n(\mathbf{r})$ varies. Thus the only long-range effects are due to $\sigma(z)$.

This leaves us with the problem shown in Fig. F.2, which is equivalent to the planar capacitor shown on the left. The electrostatics is very simple and the *only* effect for z outside

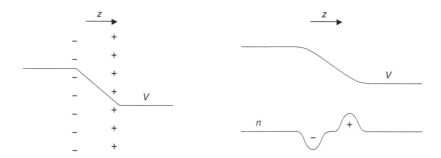

Figure F.2. Schematic for a dipole layer of charge $\sigma(z)$ and the resulting offset of the average potential. On the left is the well-known problem of a parallel-plate capacitor and on the right a schematic illustration of a realistic smooth interface density like that at a surface or interface.

the region of the surface or interface is a constant shift of the electrostatic potential that is given by integrating the electric field, which is equivalent to the dipole term

$$\Delta \bar{V}_{\text{Coulomb}} = \int dz \, z \, \sigma(z). \tag{F.22}$$

This is the dipole that must be calculated from the electronic structure in order to predict band offsets at interfaces and work functions at surfaces, as referred to in Section 13.5 and Chapter 22.

F.6 Reducing Effects of Artificial Image Charges

It is often convenient to apply periodic boundary conditions in calculations of isolated molecules, clusters, or defects in solids. The advantage is that all the machinery developed for crystals is immediately applicable. The disadvantage is unwanted effects due to the use of artificial periodic boundary conditions. There are two types of effects: artificial bands due to overlapping wavefunctions and potentials due to "image charges" from periodically repeated units. Since the bound-state wavefunctions are exponentially localized, the longest-range effects are Coulomb interactions. Thus it is very useful to identify ways of performing the calculations that minimize effects of the image potentials.

A transparent approach to the problem, with practical equations, can be found in a paper by Schultz [1020], as illustrated in Fig. F.3. The goal is to find the properties of the isolated system in part (a) using calculations with periodic cells of volume $\Omega \equiv 1/L^3$. If the density is merely repeated periodically as in (b) using the usual expressions relating the potentials and densities valid in crystals, then this artifice introduces spurious potentials due to interactions between the system and its periodic images. The effects can be understood in terms of the multipole moments of the charge density of one cell (we omit tensor indices for simplicity)

$$\langle n \rangle_M = \int d\mathbf{r} \, r^M n(r). \tag{F.23}$$

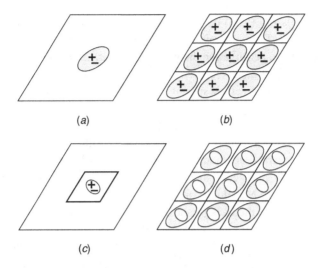

Figure F.3. Schematic illustration of the use of periodic boundary conditions to find the electrostatic potential and to solve Kohn–Sham equations for the isolated system shown in (a). The periodically repeated density shown in (b) leads to artificial interactions with the images. Subtracting a model density $n_{LM}(\mathbf{r})$ in (c) from (a) (F.24) leaves the density $n'(\mathbf{r})$, which has no moments for $M \leq M_{max}$. Calculations are done using the periodically repeated $n'(\mathbf{r})$ in (d) with Expression Eq. (F.25) for the electrostatic potential. Provided by P. Schultz; equivalent to fig. 1 of [1020].

If the cell is charged (monopole $M = 0$ moment), the sums diverge for any Ω; if there is a dipole ($M = 1$) moment, the limit as $\Omega \to \infty$ depends on the shape of the cell; quadrupole ($M = 2$) moments lead to convergent expressions for energy (with an error $\propto 1/L^5 = 1/\Omega^{5/3}$), but the potential is only conditionally convergent; the sums are convergent for higher multipoles.

A general approach to the problem [1020] is to divide the density into two parts,

$$n(\mathbf{r}) \equiv n'(\mathbf{r}) + n_{LM}(\mathbf{r}), \tag{F.24}$$

where $n_{LM}(\mathbf{r})$ is a model "local moment counter charge" density chosen to reproduce the moments, Eq. (F.23), of $n(\mathbf{r})$ for $M \leq M_{max}$. One isolated model density $n_{LM}(\mathbf{r})$ is illustrated in Fig. F.3(c) and the remaining $n'(\mathbf{r})$, which has vanishing moments for $M \leq M_{max}$, is periodically repeated in (d). The resulting Coulomb potential can be represented as the sum of two terms

$$V_{Coulomb}(\mathbf{r}) = V'_{Coulomb}(\mathbf{r}) + V_{Coulomb, LM}(\mathbf{r}), \tag{F.25}$$

where $n'(\mathbf{r})$ and $V'_{Coulomb}(\mathbf{r})$ can easily be treated in reciprocal space, whereas $V_{Coulomb, LM}(\mathbf{r})$ is determined by the model density $n_{LM}(\mathbf{r})$ with correct boundary conditions for an isolated unit, as shown in Fig. F.3(c). Note that this is *not merely a postprocessing step after a usual periodic cell calculation*; the potential calculated during the self-consistency iterations is determined from Eq. (F.25) and not from the first term alone.

There is an additional consideration in the case of a defect in a solid. Since the medium is polarizable, the change in density due to a defect is not localized and, in general, the integrals for moments Eq. (F.23) do not converge within the cell. This can be overcome by another application of the general idea of adding model densities, since the long-range terms can be found from perturbation theory for the polarization of the given material due to the slowly varying long-range electric fields.[5]

An important example deserves special mention: an atom, molecule, or defect with charge Z [1021]. Periodically repeated charged units can be treated by adding a constant neutralizing background density $n_B = -Z/\Omega$, as in the Ewald method of Section F.2. The total energy $E(\Omega)$ can then be calculated as in any other periodic system; however, it includes spurious interaction among the units and the background $\propto 1/L$. This leading term can be canceled by subtracting the energy of point charges Z in the background – i.e., $Z^2\alpha/(2L)$ – where α is the Madelung constant (Section F.2). However, there is a difference between the interaction of the background with a point charge and with the real density of the unit. This is a local effect $\propto 1/\Omega$ since the background density varies as $\propto 1/\Omega$. Correcting for this term leads to a more convergent formula for the energy valid for cubic cells [1021]

$$E(L) = E_\infty - \alpha\frac{Z^2}{2L} - \frac{2\pi Z Q}{3L^3} + O(1/L^5), \qquad (F.26)$$

where Q is the isotropic quadrupole moment $Q = \langle n \rangle_2 = \int d\mathbf{r}\, r^2 n(r)$. A different approach has been proposed by Kantorovich [1022] that applies for cells of arbitrary shape.

SELECT FURTHER READING

Kittel, C., *Introduction to Solid State Physics* (John Wiley & Sons, New York, 1996). Textbook with formulas for Ewald sums.

Coldwell-Horsfall, R. A. and Maradudin, A. A., "Zero-point energy of an electron lattice," *J. Math. Phys.* 1:395, 1960. Exposition of methods for Cloulomb sums.

Forms involving "smeared ions" and energy relative to neutral atoms:

Galli, G. and Parrinello, M., in *Computer Simulations in Material Science*, edited by M. Meyer and V. Pontikis (Kluwer, Dordrecht, 1991), pp. 283–304.

Sankey, Otto F. and Niklewski, D. J., "*Ab initio* multicenter tight-binding model for molecular dynamics simulations and other applications in covalent systems," *Phys. Rev.* B 40:3979–3995, 1989.

Soler, J. M., Artacho, E., Gale, J., Garcia, A., Junquera, J., Ordejon, P., and Sanchez-Portal, D., "The SIESTA method for *ab intio* order-N materials simulations," *J. Phys.: Condens. Matter* 14:2745–2779, 2002.

General method for reduction of effects of artificial periodic boundary conditions:

Schultz, P. A., "Local electrostatic moments and periodic boundary conditions," *Phys. Rev.* B 60:1551–1554, 1999.

Makov, G., and Payne, M. C., "Periodic boundary conditions in *ab initio* calculations," *Phys. Rev.* B 51:4014–4022, 1995.

---

[5] This approach also applies to stress and strain due to defects that obey relations analogous to those for electrostatics.

Exercises

F.1 Show that the potential in Eq. (F.4) has zero average value as claimed. As a hint in the reasoning, the final term can be considered as the limit $G \to 0$ of the middle term.

F.2 Discuss the values of the Madelung constant in Table F.1. Compare these with the result of the previous problem. Why are the values larger or smaller? Rationalize the variation of α among the structures.

F.3 The problem of a point charge at the center of a sphere with a neutralizing uniform charge density can be solved analytically. Derive the expressions given for the potential, Eq. (F.9), and energy, Eq. (F.10). Hint: use the knowledge that the potential due to the uniform distribution must vary as r^2 (why?) and that the last term in Eq. (F.9) has been chosen to make $V = 0$ at the boundary for the neutral cell (why?). (Related analysis is given for the Wigner interpolation formula for electron correlation energy by Pines [297], p. 92–94.)

F.4 Show that the two expressions for the Ewald energy, Eqs. (F.6) and (F.7), are equivalent. As a first step in the proof show that the right-hand side of Eq. (F.7) is real. Hint: expand exponentially and use the cosine addition formula $\cos(A - B) = \cos A \cos B + \sin A \sin B$.

F.5 Explain the meanings of the terms in real and reciprocal space in Eq. (F.6) in terms of the physical interactions of gaussian charge distributions, and verify the statements made in the interpretation following Eq. (F.7).

F.6 For a chosen simple crystal structure calculate the energy versus lattice constant a. Show that it varies as $1/a$. From the slope of energy versus volume, calculate the pressure. Check that this agrees with the pressure given by the stress theorem, Eq. (F.13).

F.7 Show analytically that in the simple crystal structures in Table F.1, the force on each atom vanishes. Verify this numerically using the force theorem.

F.8 Construct a crystal with two atoms per cell – e.g., diatomic molecules with spacing d placed on an fcc lattice with lattice constant a. Calculate the energy for several values of d; from the slope calculate the force on an atom and compare with the force found using the force theorem, Eq. (F.11).

F.9 Following the previous problem, calculate the stress using the stress theorem, Eq. (F.13), and compare with the slope of the energy versus lattice constant a. Give the analytic proof that the stress is given by scaling *both* d and a, and also show this numerically by direct calculation.

F.10 Consider a molecule represented by plus and minus charges so that it has a dipole moment. Place the molecules on a simple cubic lattice and evaluate the Ewald energy. Now make the cell long in one direction so that it is orthorhombic with $a = b \ll c$. (Be sure that the program sums over sufficient vectors in both real and reciprocal space for this anisotropic case.) Find the energy for dipoles along the c direction and for dipoles oriented along a. Are they different? Why? What does this have to do with Chapter 24?

F.11 Modify the program to calculate the potential at an arbitrary point. For the case in the problem above with dipoles along the c direction, show that the potential has the dipole offset given by Eq. (F.22). Vary the in-plane lattice constant $a = b$ (but still with $a = b \ll c$) and show the point stated in Section F.5 that variation of the fields in the plane decreases exponentially as a function of distance from the plane of dipoles.

Appendix G
Stress from Electronic Structure

Summary

The subject of this appendix is the macroscopic stress that enters mechanical properties of matter in the form of stress–strain relations. The stress tensor is the generalization of pressure to all the independent components of dilation and shear, and the "stress theorem" is the generalization of the virial theorem for scalar pressure to all components of the stress tensor. In condensed matter, the state of the system is specified by the forces on each atom and the stress, which is an independent variable. The conditions for equilibrium are (1) that the total force vanishes on each atom, and (2) that the macroscopic stress equals the externally applied stress.

G.1 Macroscopic Stress and Strain

Stress and strain are important concepts in characterizing the states of condensed matter [206, 806, 807, 1023]. A body is in a state of stress if it is acted upon by external forces or if one part of the body exerts forces upon another part. We consider two types of forces as illustrated in Fig. G.1: those acting interior to a volume element and those that act upon (or through) the surface of the element due to the surrounding material, which are shown as arrows in the figure. The latter forces (per unit area) are the stresses transmitted throughout the interior of the volume. Since these forces balance on any surface in equilibrium, the stress can be determined in terms of only the intrinsic internal forces; i.e., stress is an intrinsic property of a material in a given state. This brings stress into the realm of "electronic structure" as one of the properties of a body determined by the quantum state of the system of electrons and nuclei.

For condensed matter in which the stress is homogeneous, averaged over volumes of macroscopic dimensions, the state of the system is specified by the forces on each atom and the stress, which is an independent variable. The conditions for equilibrium are that the total force vanishes on each atom, whereas the macroscopic stress is fixed by externally applied forces. The equation of state is the relation of stress to the internal variables, such as the density and temperature. For example, in a homogeneous liquid, the state of the system is

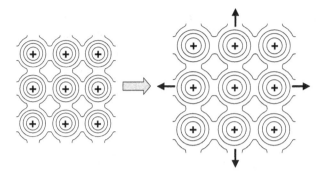

Figure G.1. Illustration of a crystal in equilibrium with no applied forces and with a strain induced by tensile forces (arrows). A uniform strain of all space including the ion cores is shown; this is the essence of the concept of "Streckung des Grundgebietes" ("stretching of the ground state") employed by Fock [270] to derive the virial theorem. Of course this is not what really happens, but it is sufficient for calculation of macroscopic stress from the generalized force theorem Eq. (G.4). An alternative approach is shown in Fig. H.1.

fully specified by the volume, pressure, and temperature, and the relation to the underlying hamiltonian is given by the virial theorem that relates pressure to the expectation value of the operators for the kinetic energy and the virial of the interaction between particles. This was first proven in quantum mechanics by Born, Heisenberg, and Jordan [1024] and later by Finkelstein [1025], Hylleraas [51], Fock [270], and Slater [1026]. In a crystal, however, there can be shear stress $\sigma_{\alpha\beta}$ in equilibrium, and the equation of state is specified in terms of stress–strain relations. The stress tensor in quantum systems was considered by Schrödinger [1027], Pauli [266], Feynman [1028], and others (e.g., [1029]), and a fundamental relation in terms of the intrinsic hamiltonian has been formulated in the form of the "stress theorem" [102, 162], which is a generalization of the elegant scaling arguments of Fock [270].

Strain is a deformation of a material that causes a displacement of a point $r_i \rightarrow r'_i$ – i.e., a displacement $\mathbf{u} = \mathbf{r}' - \mathbf{r}$. The displacement $\mathbf{u}$ as a function of the coordinate $\mathbf{r}$ specifies the deformation (see ch. 1 of [806]). Consider two nearby points joined by the vector $d\mathbf{r}$, which is deformed to $d\mathbf{r}'$. The distance between the points changes from $dl = \sqrt{(dr_1^2 + dr_2^2 + dr_3^2)}$ to the corresponding dl'. To lowest order in $\mathbf{u}$, dl' is given by

$$(dl')^2 = dl^2 + 2u_{\alpha,\beta}dr_\alpha dr_\beta, \tag{G.1}$$

where summation over repeated cartesian indices α, β is assumed, and where

$$u_{\alpha,\beta} = \frac{1}{2}\left(\frac{\partial u_\alpha}{\partial r_\beta} + \frac{\partial u_\beta}{\partial r_\alpha}\right) \tag{G.2}$$

is the strain tensor. Note that this is equivalent to a *metric tensor* that gives lengths in the deformed system in terms of undeformed coordinates [1030]

$$(dl')^2 = dr_\alpha g_{\alpha,\beta} dr_\beta; \quad g_{\alpha,\beta} = \delta_{\alpha,\beta} + 2u_{\alpha,\beta}. \tag{G.3}$$

It is also convenient to define the *unsymmetrized strain tensor* $\epsilon_{\alpha\beta}$, which is a scaling of space, $r_\alpha \rightarrow (\delta_{\alpha\beta} + \epsilon_{\alpha\beta}) r_\beta$. This is often simpler to use, but we must always remember that it is the symmetric form, Eq. (G.2), that relates to internal energies; antisymmetric terms are rotations that have no effect on relative internal coordinates.

If the strain is homogeneous over macroscopic regions,[1] then the macroscopic average stress tensor $\sigma_{\alpha\beta}$ is the derivative of the energy with respect to the strain tensor, per unit volume,

$$\sigma_{\alpha\beta} = -\frac{1}{2\Omega}\frac{\partial E_{\text{total}}}{\partial g_{\alpha\beta}} \quad \text{or} \quad \sigma_{\alpha\beta} = -\frac{1}{\Omega}\frac{\partial E_{\text{total}}}{\partial u_{\alpha\beta}}. \tag{G.4}$$

The sign of the stress is chosen as in [162, 806]: since definition (G.4) applies to the *internal* forces in the system, a negative value indicates that the internal energy decreases for positive (expansive) strain – i.e., it is under compression. For example, under hydrostatic compression, pressure is given by $P = -(1/3)\sum_\alpha \sigma_{\alpha\alpha}$.

Elastic phenomena are described by stress–strain relations – e.g., to linear order the elastic constants are given by

$$C_{\alpha\beta;\gamma\delta} = \frac{1}{\Omega}\frac{\partial^2 E_{\text{total}}}{\partial u_{\alpha\beta}\partial u_{\gamma\delta}} = -\frac{\partial\sigma_{\alpha\beta}}{\partial u_{\gamma\delta}}. \tag{G.5}$$

Symmetry [283, 285, 807] can be used to specify $C_{\alpha\beta;\gamma\delta}$ as a 6×6 array C_{ij} for a general crystal. For a cubic crystal, there are only three independent constants: $C_{11} = C_{xx,xx}$, $C_{12} = C_{xx,yy}$, and $C_{44} = C_{xy,xy}$. (See [807] or the solid state texts [280, 285, 300] for other cases.) The theory of finite strains can be treated directly from basic theory since the stress is defined by the derivative (G.4), which applies for any state with arbitrary magnitude of strain. In addition, the positions of the atoms in the unit cell are fixed by the zero-force relation (Section G.4) at any strain. Thus calculation of stress as a function of strain can be used to find linear and nonlinear stress–strain relations [102, 162]. However, care must be taken in defining stress–strain relations because *strain is not unique* since it is defined relative to a reference state.

Using the generalized force theorem, Eq. (3.25), the expression for stress, Eq. (G.4), can be evaluated using various ways of distorting the system. The example of uniform infinitesimal strain of all space (including core states) is shown in Fig. G.1; an alternative, illustrated in Fig. H.1, can be considered if the expression (G.4) is generalized to a nonuniform strain. The derivative in Eq. (G.4) can be evaluated using any of the various expressions that relate total energy E_{total} to fundamental electronic energies. The resulting expressions can appear to be very different and, indeed, even within one approach different contributions to E_{total} may be treated differently. The various types of expressions can be grouped into categories that reveal physical insight and suffice for important applications in electronic structure.

---

[1] In general, strain $u_{\alpha,\beta}$ or metric $g_{\alpha,\beta}$ is a *tensor field* that is a function of position **r**. Fields will be considered in Appendix H.

G.2 Stress from Two-Body Pair-Wise Forces

In electronic structure all fundamental forces are two-body central interactions $V_{kk'} \equiv V(|\mathbf{r}_k - \mathbf{r}_{k'}|)$, where k and k' denote any pair of particles with the relative coordinates $\mathbf{r}_{kk'} = \mathbf{r}_k - \mathbf{r}_{k'}$. In any case in which the particles are *explicitly represented* by such terms in the total energy, then the stress is given by the generalized virial (Exercise G.1)

$$\sigma_{\alpha\beta} = -\frac{1}{2\Omega} \sum_{k \neq k'} \frac{d}{d\mathbf{r}_k} V_{kk'} \frac{d\mathbf{r}_k}{d\epsilon_{\alpha\beta}} = \frac{1}{2\Omega} \sum_{k \neq k'} \mathbf{F}_{kk',\alpha} \mathbf{r}_{k,\beta}, \tag{G.6}$$

which can be written in the manifestly symmetric form

$$\sigma_{\alpha\beta} = \frac{1}{2\Omega} \sum_{k \neq k'} \frac{(\mathbf{r}_{kk'})_\alpha (\mathbf{r}_{kk'})_\beta}{r_{kk'}} \left(\frac{d}{dr_{kk'}} V \right). \tag{G.7}$$

Here the sum over k and k' is over all particles considered. Note that $\mathbf{F}_{kk',\alpha}$ is the *contribution to the force* on particle k due to particle k'; it is *not* the total force $\mathbf{F}_{k,\alpha}$ on particle k, which vanishes in equilibrium.

Equation (G.7) provides the stress due to classical particles directly in terms of the potentials and forces; it can also be viewed as a quantum mechanical operator that leads to the most general form of the potential part of the stress in a many-body system, Eq. (3.23). The formulation in Eqs. (G.7) or (G.6) also provides the needed expressions for any terms in the equation for total energy that depend on the distance between particles or parameters in the energy. This is the useful form for the real-space terms in the Ewald stress given in Eq. (F.13) and for the total energy terms in tight-binding or local orbital approaches that are expressed as a function of distances (see Eq. (14.29) and related terms in Section 15.5).

G.3 Expressions in Fourier Components

Although it might appear that Eqs. (G.7) and (3.23) are the end of the story for potential interactions, this is not the case. Even in the general many-body expression, (3.23), the long-range classical Coulomb term should be treated with special care, e.g., using expressions in Fourier space. Mean-field approaches like density functional theory do not represent particle positions directly, and the effective potential is *not* represented in terms of a potential due to specific other particles. Instead, $V_{KS}(\mathbf{r})$ is defined only by the condition that it reproduces the correct density. How does one proceed? The practical approach is simply to differentiate all the terms in E_{total}.

Expressions in Fourier space can be treated straightforwardly by using the fact that strain also scales in reciprocal space: $\mathbf{q}_\alpha \rightarrow (\delta_{\alpha\beta} - \epsilon_{\alpha\beta})\mathbf{q}_\beta$, where $\mathbf{q}$ is any vector in reciprocal space. The derivation is simplified by the fact that structure factors $S^\kappa(\mathbf{G})$, Eq. (12.17), and $\Omega n(\mathbf{G})$ are invariant. For example, the Hartree term Eq. (F.2) (which appears in the total energy expressions (3.14), term (7.16), and specific expressions in other chapters), leads to the stress contribution (Exercise G.2)

$$-\frac{1}{\Omega}\frac{\partial E_{\text{Hartree}}}{\partial \epsilon_{\alpha\beta}} = \frac{1}{2}4\pi e^2 \sum_{\mathbf{G}\neq 0}' \frac{n(\mathbf{G})^2}{G^2}\left[2\frac{G_\alpha G_\beta}{G^2} - \delta_{\alpha\beta}\right], \tag{G.8}$$

which is clearly symmetric, as it should be.

Kinetic Contributions

Scaling also applies to kinetic terms using $d/d\mathbf{r}_\alpha \rightarrow (\delta_{\alpha\beta} - \epsilon_{\alpha\beta})(d/d\mathbf{r}_\beta)$. This leads directly to a general expression, (3.23), valid in both many-body and independent-particle formulations. The expressions are particularly simple for wavefunctions expressed in Fourier space: the energy given in Eq. (13.1),

$$T_s = \frac{\hbar^2}{2m_e}\frac{1}{N_k}\sum_{\mathbf{k},i}\sum_m c_{i,m}^*(\mathbf{k})c_{i,m}(\mathbf{k})|\mathbf{k}+\mathbf{G}_m|^2, \tag{G.9}$$

leads to the kinetic contribution to the stress (Exercise G.3)

$$-\frac{1}{\Omega}\frac{\partial T_s}{\partial \epsilon_{\alpha\beta}} = \frac{\hbar^2}{m_e}\frac{1}{N_k}\sum_{\mathbf{k},i}\sum_m c_{i,m}^*(\mathbf{k})c_{i,m}(\mathbf{k})(\mathbf{k}+\mathbf{G}_m)_\alpha(\mathbf{k}+\mathbf{G}_m)_\beta. \tag{G.10}$$

In Chapter 15 use is made of the fact that tight-binding and local orbital forms of the matrix elements of the kinetic energy operator can be cast in terms of functions of distances between atoms [362, 646], so that a two-body form like Eq. (G.7) can be used instead of a generic form like Eq. (G.10).

Ewald Contribution to Stress

Using the above forms, many different expressions for stress can be found that may be more or less convenient in various methods. The application to the Ewald term is given in Section F.2. Here we reproduce the expression for the stress corresponding to the plane wave formula, Eq. (13.1), for total energy, as given in eq. (2) of [102]: the strain derivative is

$$\frac{\partial \gamma_{\text{Ewald}}}{\partial \epsilon_{\alpha\beta}} = \frac{\pi}{2\Omega\epsilon}\sum_{\mathbf{G}\neq 0}\frac{e^{-G^2/4\epsilon}}{G^2/4\epsilon}\left|\sum_\tau Z_\tau e^{i\mathbf{G}\cdot\mathbf{x}_\tau}\right|^2\left[\frac{2G_\alpha G_\beta}{G^2}(G^2/4\epsilon+1) - \delta_{\alpha\beta}\right]$$

$$+\frac{1}{2}\epsilon^{1/2}\sum_{\tau\tau'\mathbf{T}}' Z_\tau Z_{\tau'}H'(\epsilon^{1/2}D)\frac{D_\alpha D_\beta}{D^2} + \frac{\pi}{2\Omega\epsilon}\left[\sum_\tau Z_\tau\right]^2\delta_{\alpha\beta}, \tag{G.11}$$

where $\mathbf{D} = \mathbf{x}_{\tau'} - \mathbf{x}_\tau + \mathbf{T}$ and the sum is only for terms with $D \neq 0$. Note that here ϵ denotes a convergence parameter (*not* the strain $\epsilon_{\alpha\beta}$) that may be chosen for computational performance. Z_τ denotes the atomic core charge of atom τ, $\mathbf{T}$ the lattice translation vectors, and $\mathbf{x}_\tau$ the atomic positions in the unit cell. The function $H'(x)$ is

$$H'(\mathbf{x}) = \partial[\text{erfc}(\mathbf{x})]/\partial\mathbf{x} - \mathbf{x}^{-1}\text{erfc}(\mathbf{x}), \tag{G.12}$$

with erfc($\mathbf{x}$) denoting the complementary error function.

G.4 Internal Strain

The expressions for stress in the previous sections have been derived assuming a homogeneous scaling of space, including the electron wavefunctions and positions of the nuclei [102, 162]. However, this is not the whole story for the actual measured stress. The proof that this is a correct expression for the stress hinges on the requirement that the energy be minimum with respect to all internal degrees of freedom. In addition to the requirement that the electron wavefunction be at the variational minimum, one must add the requirement that each nucleus I be at the minimum energy position – i.e., that the force on each nucleus vanishes, $\mathbf{F}_I = 0$, in the presence of the strain. Only for simple crystal structures and certain symmetry strains are the positions of the nuclei fixed by symmetry. In general, one must find the positions given by condition $\mathbf{F}_I = 0$, and the displacement at which this occurs is defined to be

$$\mathbf{u}_{s,\alpha} = \sum_{\beta} \epsilon_{\alpha\beta} \tau_{s,\beta} + \mathbf{u}_{s,\alpha}^{\text{int}}, \tag{G.13}$$

where the first term represents uniform scaling of the basis and the second, the deviations or "internal strains" (see, e.g., [91, 102] and references given there). To linear order the internal strains are proportional to the external strain, defining "internal strain parameter" Γ,

$$\mathbf{u}_{s,\gamma}^{\text{int}} = \sum_{\alpha\beta} \Gamma_{s,\gamma\alpha\beta} \, \epsilon_{\alpha\beta}. \tag{G.14}$$

The effect can be understood in simple examples, such as diamond or zinc blende structures. In the unstrained crystal, planes of atoms perpendicular to the (1 1 1) direction are spaced alternately $1/4$ and $3/4$ times $\sqrt{3}a/4$; for a uniaxial strain in the (1 1 1) direction, the spacing is not determined by symmetry. The problem is equivalent to the one-dimensional chain of molecules described in Exercise G.4.

Internal strains are crucial for understanding and predicting stress–strain relations. However, internal strain parameters have been measured in only a few cases because of the difficulty of experimental measurements of atomic positions in a strained crystal. Thus this is a crucial area where theory adds information to our knowledge of elasticity even in cases where macroscopic elastic constants are well established.

SELECT FURTHER READING

Basic theory of elasticity:
Landau, L. D. and Lifshitz, E. M., *Theory of Elasticity* (Pergamon Press, Oxford, 1958).

General theory:
Nielsen, O. H. and Martin, R. M., "Quantum-mechanical theory of stress and force," *Phys. Rev. B* 32(6):3780–3791, 1985.

Applications in a plane wave basis:
Nielsen, O. H. and Martin, R. M., "Stresses in semiconductors: *Ab initio* calculations on Si, Ge, and GaAs," *Phys. Rev. B* 32(6):3792–3805, 1985.

Expressions in localized bases:

Soler, J. M., Artacho, E., Gale, J., Garcia, A., Junquera, J., Ordejon, P., and Sanchez-Portal, D., "The SIESTA method for *ab intio* order-N materials simulations," *J. Phys. : Condens. Matter* 14:2745–2779, 2002.

Feibelman, P. J., "Calculation of surface stress in a linear combination of atomic orbitals representation," *Phys. Rev. B* 50:1908–1911, 1994.

Exercises

G.1 Show that for particles interacting via two-body central potentials the contribution to the stress tensor is given by the generalized virial expression (G.6). Further, transform the expression to the symmetric form Eq. (G.7).

G.2 Derive the expression, (G.8), for the Hartree contribution to the stress tensor.

G.3 Using the argument of the scaling of reciprocal space, show that the kinetic contribution to the stress can be written in the form Eq. (G.10), which is convenient for plane wave calculations.

G.4 Find the elastic constant $C = \mathrm{d}^2 E/\mathrm{d}L^2$ and the internal strain parameter Γ defined by Eq. (G.14) for a one-dimensional chain of diatomic molecules. The atoms in a molecule are spaced a distance R_1 and are connected by a spring with constant K_1; spacing between the molecules is R_2 and they are connected by a spring with constant K_2. The cell length is $L = R_1 + R_2$. Show that the system has the expected behavior that the molecules are incompressible for $K_1 \gg K_2$.

G.5 Show that in any crystal with one atom per cell the internal strain is zero by symmetry.

G.6 As an example of the condition in the previous problem, show that for the molecular chain in Exercise G.4, internal strain vanishes for $R_1 = R_2$ and $K_1 = K_2$. For a homonuclear case, this means one atom per cell. Note that the internal strain is still zero for a diatomic ionic crystal with two different atoms so long as $R_1 = R_2$ and $K_1 = K_2$.

G.7 Show that it is *impossible* to have a chain with three inequivalent atoms per cell and still have zero internal strain.

Appendix H
Energy and Stress Densities

Summary

A *density* is a field defined at each position $\mathbf{r}$, for example the particle number density $n(\mathbf{r})$, which is a well-defined, experimentally measurable function. It would be desirable to have expressions for other densities, in particular, energy and stress densities. However, energy and stress densities are not unique on a microscopic quantum scale, even though they are the basis of the theory of elasticity on a macroscopic scale. This appendix brings out three points: (1) certain integrals of energy and stress densities are unique and very useful; (2) there are important contributions to the energy or stress density that are completely unique – these include all terms that arise from the fact that electrons are a many-body system of fermions; (3) all other terms that are non-unique can be shown to involve only the single scalar number density – there are different possible choices for these terms, each involving only derivatives of the density $n(\mathbf{r})$ or the classical Coulomb potential $V^{CC}(\mathbf{r})$ which is directly related to $n(\mathbf{r})$. It follows that all the issues of non-uniqueness are exactly the same as in a one-particle problem.

Only one density is widely used in electronic structure – the particle density $n(\mathbf{r})$. It is the fundamental measurable quantity in quantum mechanics and the fundamental density in density functional theory. Theoretical expressions for $n(\mathbf{r})$ are well defined and lead to unique results. Here we emphasize that other densities have the potential to play a useful role in electronic structure theory. In particular, energy and stress densities have the potential to be very useful in electronic structure, beyond their limited use thus far.

The difficulty in formulating energy and stress densities is their inherent nonuniqueness. The problem is that, unlike the particle density $n(\mathbf{r})$, which is defined by Eq. (3.8), there are no operators in quantum mechanics that uniquely define "energy at a point" or "stress at a point." Of course, there are expressions for the total energy and stress, but this is not sufficient to define an energy or stress density. The value at any point is always subject to "gauge transformations" that leave the total invariant.

Is there any sense in which an energy or stress density can be useful? The answer is yes, for two reasons:

1. Many important quantities can be shown to be invariant to the choice of gauge. For example, total surface energy and surface stress are defined by integrals over the surface region. Because the integral extends from the vacuum to the bulk interior of the system, it can be shown [1031–1033] that gauge-dependent terms vanish in the integrals. Similarly, the expressions for force in terms of surface integrals of the stress density in Appendix I are invariant and can be very useful (Section H.3). For such quantities, it may be convenient to choose a particular gauge, even though one must not associate any physical meaning to the gauge-dependent integrand.
2. Specific analysis can identify terms in the energy and stress densities that are well defined. As shown below, with appropriate definitions, *unique densities result from all contributions to the energy or stress that arise from the fact that electrons constitute a many-body system of fermions*

All nonunique terms in the energy or stress densities involve *only derivatives of the total density $n(\mathbf{r})$ and the classical Coulomb potential $V^{CC}(\mathbf{r})$*. It follows that *all issues of non-uniqueness are exactly the same as in a one-particle problem.*

H.1 Energy Density

The total energy of a system of electrons and nuclei can be written in the general form, Eq. (3.16), or the Kohn–Sham form, Eq. (7.5),

$$E = \langle \hat{T} \rangle + [\langle \hat{V}_{\text{int}} \rangle - E_{\text{Hartree}}] + E^{CC} = T_s + E^{CC} + E_{xc}, \tag{H.1}$$

where T_s is the independent-particle kinetic energy and the Coulomb terms are grouped to ensure they are well defined in an infinite system. An energy density $e(\mathbf{r})$ (denoted by a lowercase, italic Roman letter) or a density per particle $\epsilon(\mathbf{r}) \equiv e(\mathbf{r})/n(\mathbf{r})$ (lowercase Greek letter) is a function that when integrated over all space yields the total energy E; e.g.,

$$E = \int \mathrm{d}\mathbf{r} e(\mathbf{r}), \tag{H.2}$$

with

$$e(\mathbf{r}) = t_{\text{ip}}(\mathbf{r}) + e^{CC}(\mathbf{r}) + e_{xc}(\mathbf{r}). \tag{H.3}$$

If we separate out the ion–ion interaction E_{II}, which has no effect on the equations for the electrons except to ensure neutrality, the total energy can be written

$$E = \int \mathrm{d}\mathbf{r} n(\mathbf{r}) \epsilon(\mathbf{r}) + E_{II}, \tag{H.4}$$

with[1]

$$\epsilon(\mathbf{r}) = \tau_{ip}(\mathbf{r}) + V_{ext}(\mathbf{r}) + \frac{1}{2} V_{Hartree}(\mathbf{r}) + \epsilon_{xc}(\mathbf{r}). \tag{H.5}$$

Classical Coulomb Energy Density

The first problem in defining an energy density is the classical Coulomb term. There are two forms for the energy density in electrostatics [480, 908]:

$$E^{CC} = \frac{1}{8\pi} \int d\mathbf{r} |\mathbf{E}^{CC}(\mathbf{r})|^2 = \frac{1}{2} \int d\mathbf{r} V^{CC}(\mathbf{r})[n(\mathbf{r}) + n^+(\mathbf{r})], \tag{H.6}$$

where $\mathbf{E}^{CC} = -\nabla V^{CC}$ is the electric field due to the total charge density of the electrons and nuclei $n(\mathbf{r}) + n^+(\mathbf{r})$. Each of the integrands can be viewed as an energy density $e^{CC}(\mathbf{r})$ and each has advantages in different situations. The first expression is the Maxwell energy density assigned to the field instead of the particles. The second expression has the form of the interaction of particles with the energy assigned to the position of the particles. Even though this part of the energy density is not unique, it is purely classical and all forms can be expressed in terms of the charge density. (Note the close analogy with the "boson" part of the kinetic energy, Eq. (H.14), below.)

There is an important practical distinction between the two forms in Eq. (H.6). Only in the second case can the energy be written in the form of Eq. (H.5), with $V_{ext}(\mathbf{r})$ the Coulomb potential due to the nuclei $n^+(\mathbf{r})$ and $V_{Hartree}(\mathbf{r})$ the classical Coulomb potential due to the electrons $n(\mathbf{r})$.

Exchange–Correlation Energy Density

Chapter 7 discusses the physical reasoning for expressions for $\epsilon_{xc}(\mathbf{r})$ as a functional of the exchange–correlation hole around an electron at point $\mathbf{r}$. Even though it is not defined by the fact that its integrals must yield the total E_{xc}, $\epsilon_{xc}(\mathbf{r})$ *is uniquely specified by the definition that it is the additional energy per electron at point $\mathbf{r}$ due to exchange and correlation.* This follows from Expression (8.5) as a coupling constant integration, and it can be understood by an independent derivation [1034]: the potential part of $\epsilon_{xc}(\mathbf{r})$ is obviously unique because it is given in terms of the pair correlation functions, which are measurable functions. The kinetic energy contribution to $\epsilon_{xc}(\mathbf{r})$ is only the change in kinetic energy due to correlation; this density $\tau_c(\mathbf{r})$ is also unique [1034–1036] by extension of the arguments given below for $\tau_x(\mathbf{r})$.

Kinetic Energy Density for Independent Particles

Finally, we consider the first term in the Kohn–Sham energy, the kinetic energy of independent particles T_s. This is treated in some detail because the analysis leads to

---

[1] The factor of $1/2$ in the Hartree term might be thought of as an ad hoc assignment of $1/2$ of the energy to each particle; however, it follows from the much deeper fact that electrons are identical. Any other assignment of the energy would violate this symmetry.

expressions that are useful in construction of functionals and in analysis of actual electronic structure calculations. The kinetic energy is as a sum of terms for the two spins; here we omit the spin index for simplicity with the convention that all the expressions apply to each spin separately.

In analogy to the Coulomb energy in Eq. (H.6), the kinetic energy of N independent fermions can be expressed in different forms:

$$T_s = -\frac{1}{2} \sum_{i=1}^{N} \int d\mathbf{r} \psi_i^*(\mathbf{r}) \nabla^2 \psi_i(\mathbf{r}) = \frac{1}{2} \sum_{i=1}^{N} \int d\mathbf{r} |\nabla \psi_i(\mathbf{r})|^2. \tag{H.7}$$

The equivalence of the two forms follows from integration by parts, where boundary terms vanish for bound states since ψ_i vanish at the boundary or for period functions where boundary terms cancel. Thus either integrand,

$$t^{(1)}(\mathbf{r}) = -\frac{1}{2} \sum_{i=1}^{N} \psi_i^*(\mathbf{r}) \nabla^2 \psi_i(\mathbf{r}) \quad \text{or} \quad t^{(2)}(\mathbf{r}) = \frac{1}{2} \sum_{i=1}^{N} |\nabla \psi_i(\mathbf{r})|^2, \tag{H.8}$$

can be regarded as a "kinetic energy density" $t(\mathbf{r})$ since the integral of either density is the total kinetic energy.

How is it possible to find any part of the kinetic energy density that is unique and useful? First divide the problem into parts: the kinetic energy density of independent bosons with density $n(\mathbf{r})$ plus the excess "exchange kinetic energy density" of the fermions can be written

$$t(\mathbf{r}) = t_n(\mathbf{r}) + t_x(\mathbf{r}). \tag{H.9}$$

This can be accomplished[2] by expressing the wavefunctions as

$$\psi_i(\mathbf{r}) = u(\mathbf{r})\phi_i(\mathbf{r}); \quad u(\mathbf{r}) = n(\mathbf{r})^{1/2}. \tag{H.10}$$

Thus $\sum_{i=1}^{N} |\phi_i(\mathbf{r})|^2 = 1$ at each point $\mathbf{r}$, from which it immediately follows that (Exercise H.1)

$$\sum_{i=1}^{N} \nabla |\phi_i(\mathbf{r})|^2 = 0; \quad \sum_{i=1}^{N} \nabla^2 |\phi_i(\mathbf{r})|^2 = 0 \tag{H.11}$$

at each point $\mathbf{r}$. From the first equation in (H.11), it follows that cross terms involving $\nabla u(\mathbf{r})$ and $\nabla \phi_i(\mathbf{r})$ vanish in any expression for the kinetic energy density. Using the second equality, it is straightforward to show that $\sum_{i=1}^{N} |\nabla \phi_i(\mathbf{r})|^2 = -\sum_{i=1}^{N} \phi_i(\mathbf{r}) \nabla^2 \phi_i(\mathbf{r})$, so that

$$t_x(\mathbf{r}) = n(\mathbf{r})\tau_x(\mathbf{r}), \tag{H.12}$$

with

$$\tau_x(\mathbf{r}) = \frac{1}{2} \sum_{i=1}^{N} |\nabla \phi_i(\mathbf{r})|^2 = -\frac{1}{2} \sum_{i=1}^{N} \phi_i(\mathbf{r}) \nabla^2 \phi_i(\mathbf{r}), \tag{H.13}$$

---

[2] The approach taken here was pointed out to the author by E. Stechel.

which is manifestly invariant to the choice of form of the kinetic energy, Eq. (H.8). The "exchange kinetic energy per particle" $\tau_x(\mathbf{r})$ also has a clear physical meaning; it is the curvature of the exchange hole [1037–1039], which can be shown to be the relative kinetic energy of pairs of electrons [1040].[3] The curvature is clear from plots of the exchange hole in the Ne atom in Fig. 8.1, as well as Figs. 5.5 and 8.5, which also include correlation. Therefore, the excess exchange kinetic energy density $t_x(\mathbf{r})$ (and the density per particle $\tau_x(\mathbf{r}) = t_x(\mathbf{r})/n(\mathbf{r})$) is a unique, meaningful density.

The remaining term $t_n(\mathbf{r})$ involves only derivatives of $u(\mathbf{r}) = n(\mathbf{r})^{1/2}$. Thus the issue of non-uniqueness of the kinetic energy density has been reduced to the simplest form involving only the density. Since the density is a scalar function of one coordinate, the non-uniqueness issues are the same as for a one-particle problem, where there are two choices:

$$t_n^{(1)}(\mathbf{r}) = -\frac{1}{2}u(\mathbf{r})\nabla^2 u(\mathbf{r}) \quad \text{or} \quad t_n^{(2)}(\mathbf{r}) = \frac{1}{2}|\nabla u(\mathbf{r})|^2. \tag{H.14}$$

There is a simple physical interpretation that the kinetic energy in Eq. (H.14) is the same as that of N noninteracting bosons with density $n(\mathbf{r}) = |u(\mathbf{r})|^2$, i.e., with each boson having wavefunction $u(\mathbf{r})$.

Even though either form in Eq. (H.14) is acceptable, the second one is always positive definite and is the form chosen by Weizsacker [330] as the gradient correction to the Thomas–Fermi method as described in Section 6.2. For each spin this can be written as $t_W(\mathbf{r}) = n(\mathbf{r})\tau_W(\mathbf{r})$, where

$$\tau_W(\mathbf{r}) = \frac{1}{2}\left[\frac{\nabla u(\mathbf{r})}{u(\mathbf{r})}\right]^2 = \frac{1}{8}\left[\frac{\nabla n(\mathbf{r})}{n(\mathbf{r})}\right]^2. \tag{H.15}$$

Finally this leads to a useful way to calculate $\tau_x(\mathbf{r})$ in terms of the wavefunctions for each spin,

$$t_x(\mathbf{r}) = \frac{1}{2n(\mathbf{r})}\sum_{i=1}^{N}|\nabla\psi_i(\mathbf{r})|^2 - \tau_W(\mathbf{r}), \tag{H.16}$$

which is the difference from the Weizsacker term and which brings out its physical meaning as the excess kinetic energy due to the fermion nature of the particles. The kinetic energy density is the key ingredient in constructing meta-GGA density functionals in Section 9.4. The derivations here show the reasons why the appropriate quantity with which to construct the functionals is the difference in Eq. (H.16). Indeed, this was recognized in the construction of the SCAN functional, as discussed in [441] and Section 9.4.

Energy Density per Particle: Convenient Expressions

The expressions for the energy density per electron at each point $\mathbf{r}$ have the advantage that they are closely related to the Kohn–Sham equation, which is cast in terms of the density of

---

[3] The excess fermion kinetic energy density itself is the appropriate physically meaningful density for some properties. For example, exchange should depend on only the fermion part. In fact, exchange functionals [442] have been constructed in terms of τ_x, based on the fact that the short-range shape of the exchange hole is determined by $t_x(\mathbf{r})$.

electrons and is derived from variational equations, Eq. (7.8). Combining Expression (H.5) with (H.15) leads to (Exercise H.3)

$$\epsilon(\mathbf{r}) = \sum_i \epsilon_i |\psi_i(\mathbf{r})|^2 - \frac{1}{2} V_{\text{Hartree}}(\mathbf{r}) + [\epsilon_{\text{xc}}(\mathbf{r}) - V_{\text{xc}}(\mathbf{r})], \tag{H.17}$$

where the first term is an eigenvalue-weighted density and the other terms correct for overcounting.[4] The first term in Eq. (H.17) is essentially a projected density of states that can be used to identify local energies and bonding [1042].

H.2 Stress Density

In an inhomogeneous system, the stress field is not uniform (even for uniform strain). Is it possible to define a unique stress density field $\sigma_{\alpha\beta}(\mathbf{r})$? Forces are well-defined measurable quantities; however, the stress density related to the force density $f(\mathbf{r})$ acting on particles at point $\mathbf{r}$ is

$$\nabla_\beta \sigma_{\alpha\beta}(\mathbf{r}) = f_\alpha(\mathbf{r}). \tag{H.18}$$

For dimension $d > 1$, this relation does not uniquely determine the stress density [162, 806, 1043, 1044], since the curl of any vector field can be added to $\sigma_{\alpha\beta}(\mathbf{r})$ with no change in the forces. The stress field can also be defined as the generalization of Eq. (G.4) to an inhomogeneous metric field $g_{\alpha\beta}(\mathbf{r})$,

$$\sigma_{\alpha\beta}(\mathbf{r}) = -\frac{1}{2\Omega} \frac{\partial E_{\text{total}}}{\partial g_{\alpha\beta}(\mathbf{r})}, \tag{H.19}$$

but this still leads to a nonunique expression [1045] of the same form as given earlier by Godfrey [1043].

An illuminating point that has been clarified in [1034] is that, just as in the energy density, *all nonunique terms can be written as simple expressions in terms of derivatives of the charge density $n(\mathbf{r})$ and electrostatic potential $V^{\text{CC}}(\mathbf{r})$*. For the case of the Kohn–Sham independent-particle theory within the local density approximation for exchange and correlation ϵ_{xc}, the expressions given by Nielsen and Martin (NM) [162] (see also [1044]), Godfrey [1043], and Rogers and Rappe [1045] can all be written as

$$\sigma_{\alpha\beta}(\mathbf{r}) = -\frac{\hbar^2}{m_e} \left[n \sum_i \nabla_\alpha \phi_i \nabla_\beta \phi_i \right]_{\mathbf{r}}$$

$$- \frac{\hbar^2}{4m_e} \left[\frac{\nabla_\alpha n \nabla_\beta n}{n} + \delta_{\alpha\beta} [C - 1] \nabla^2 n - C \nabla_\alpha \nabla_\beta n \right]_{\mathbf{r}} \tag{H.20}$$

$$+ \frac{1}{4\pi} \left[E_\alpha E_\beta - \frac{1}{2} \delta_{\alpha\beta} E_\gamma E_\gamma \right]_{\mathbf{r}} + \delta_{\alpha\beta} n(\mathbf{r}) \left[\epsilon_{\text{xc}}^{\text{LDA}}(n) - V_{\text{xc}}^{\text{LDA}}(n) \right]_{\mathbf{r}},$$

---

[4] This expression is the same as that given by Cohen and Burke [1041], except that they also subtracted $\frac{1}{2} V_{\text{ext}}(\mathbf{r})$, i.e., they assigned 1/2 the interaction energy to the external potential (the nuclei). Unlike electron–electron interaction, where the factor of 1/2 follows from particle symmetry, this assignment is arbitrary. The choice made in Eqs. (H.5) and (H.17) is consistent with the definition of energy in the Kohn–Sham equations.

where all nonuniqueness in the kinetic terms is subsumed into the parameter C ($= 4\beta$ in the notation of [1045]). The other terms involving ϕ_i are unique for the same reasons as in the energy density.

H.3 Integrated Quantities

Integrals of the Energy Density: Surface Energies

The integral of the energy density over the entire system must be the unique total energy independent of any "gauge transformations." However, there can also be other cases where integral has a well-defined physical meaning. For example a surface energy can be calculated using the fact that the density must vanish in the vacuum and it must approach the bulk density well inside the material, i.e., the difference from the bulk density vanishes. This is sufficient to prove the uniqueness. The form of the energy density involving the Maxwell density $|\mathbf{E}(\mathbf{r})|^2$ and the analogous kinetic density involving $|\nabla n(\mathbf{r})|^2$ has been used [1031–1033] to calculate absolute surface energies of semiconductors that would not be possible by the usual total energy methods.

Surface Integrals of the Stress Density

The stress density provides alternative ways to calculate the macroscopic stress in a solid. The basic idea follows from the definition of stress as a force per unit area [162]. Consider a material that is in equilibrium in the presence of a macroscopic stress, i.e., all internal variables (the electron wavefunction and the positions of the nuclei) are at equilibrium. Then the macroscopic stress is the force per unit area transmitted across any surface that divides the macroscopic solid into two parts. Thus, the stress is the first derivative of the total energy for a displacement of the two half-spaces. A convenient procedure in a crystal is the nonuniform expansion illustrated in Fig. H.1, which shows a crystal with cells pulled apart on the boundaries. The linear change in energy is just the surface integral of the stress field on the boundaries. Since each of the cell boundaries can be considered to divide the crystal into two half-spaces, macroscopic stress is given by the surface integral of the stress density on the cell boundaries. The contributions to the stress calculated in this way are (1) the Coulomb forces per unit area *transmitted across the boundary* (i.e., the force on one side due to charge on the other side of the boundary), (2) the kinetic stress density at the boundary, which has the same form as a gas of particles that carry momentum across the boundary, and (3) the exchange–correlation terms that are unique but difficult to determine exactly. Finally, the result is unique for any valid form of the stress tensor, since the integral over the surface of a unit cell is invariant to gauge transformations [162].

The calculation of stress from surface integrals closely resembles the calculation of forces by the expressions given in Appendix I. In particular, the formula for the pressure in the local density and atomic sphere approximations, Eqs. (I.8) or (I.9), is a very useful special case of the more general formulation of the stress field given here. In methods that explicitly deal with core states, this approach has great advantages: the core states remain invariant and only the outer valence states evaluated at the cell boundaries are needed in the calculation of stress.

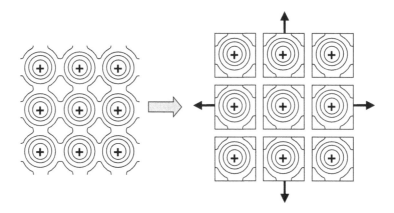

Figure H.1. A different way to expand the crystal from the uniform scaling illustrated in Fig. G.1. The regions around each nucleus are kept rigid and all the variation is the regions between the atoms, i.e., a nonuniform change of coordinates that can be expressed as the original coordinates with a nonuniform metric. Any such variation is a valid form for the generalized force theorem, and this approach has great advantages in methods that treat the cores explicitly.

This idea has been derived in several different ways[5] in the context of the atomic sphere approximation [496–499, 1046], where the pressure can be found from an integral over the sphere surface. In a monatomic close-packed solid, the atomic sphere approximation (ASA) is very good and the equations simplify because there are no Coulomb interactions between the neutral spheres, leaving only kinetic and exchange–correlation terms. Two different forms that are convenient for evaluation are given in Section I.3. These are very useful in actual calculations, and examples of results for close-packed metals are cited in Section 17.7 and Appendix I.

H.4 Electron Localization Function (ELF)

As emphasized above, the exchange kinetic energy density is well defined and is related to exchange hole curvature. This is the basis for definition of the "electron localization function" (ELF), which is a transformation of $\tau_x(\mathbf{r})$. The form proposed by Becke and Edgecomb [1047] is defined for each spin σ,

$$\mathrm{ELF}(\mathbf{r}) \equiv [1 + (\chi^\sigma(\mathbf{r}))^2]^{-1}, \qquad (\mathrm{H.21})$$

where $\chi^\sigma = t_x^\sigma / t_{\mathrm{TF}}^\sigma$, with $t_x^\sigma(\mathbf{r})$ the exchange kinetic energy density given by Eq. (H.13) for spin σ at point $\mathbf{r}$ and $t_{\mathrm{TF}}^\sigma(\mathbf{r})$ the Thomas–Fermi expression, (6.1), for spin σ in a homogeneous gas at a density equal to $n(\mathbf{r})$, which is a convenient normalization. The definition in Eq. (H.21) is chosen so that $0 \le \mathrm{ELF} \le 1$, with larger values corresponding to larger "localization," i.e., a tendency of an electron of spin σ *not* to have another same

---

[5] A good exposition of the relation of the derivations is given by Heine [499].

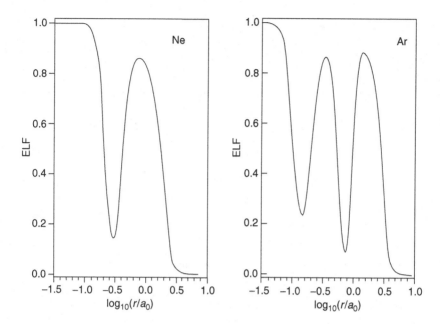

Figure H.2. The "electron localization function" (ELF): Eq. (H.21) versus radius for Ne and Ar [1047]. The minima clearly indicate the shell structure separated in space. On the other hand, the density is monotonic and has almost no structure. This indicates that the ELF (or any function of the kinetic energy density) holds the potential for improved functionals. From [1047].

spin electron in its vicinity. In any one-electron system $t_x = 0$ so that ELF $= 1$, and it is defined so that $t_x = t_{TF}$ and ELF $= 1/2$ in a homogeneous metal. Among the properties brought out by the ELF function is the shell structure, which is difficult to visualize from the density alone. For example, Fig. H.2 shows the ELF function for Ne and Ar calculated in the Hartree–Fock approximation [1047]. The shells are shown by the distinct minima, whereas the density is monotonic and has almost no structure. The ELF is unity for two electrons of opposite spin paired as in a covalent bond. The meaning of the ELF and examples of its use to identify types of bonding and nonbonding regions is brought out in the review by Savin et al. [1048]. An example of the ELF is shown in Fig. 2.6, where it is used to identify the bonding in the H_3S superconductor.

SELECT FURTHER READING

Classical theory of stress, strain, and energy fields:

Landau, L. D. and Lifshitz, E. M., *Theory of Elasticity* (Pergamon Press, Oxford, 1958).

Energy density:

Chetty, N. and Martin, R. M., "First-principles energy density and its applications to selected polar surfaces," *Phys. Rev. B* 45: 6074–6088, 1992.

Cohen, M. H., Frydel, D., Burke, K., and Engel, E., "Total energy density as an interpretative tool," *J. Chem. Phys.* 113: 2990–2994, 2000.

Electron localization function:

Becke, A. D. and Edgecombe, K. E., "A simple measure of electron localization in atomic and molecular systems," *J. Chem. Phys.* 92: 5397–5403, 1990.

Quantum theory of stress:

Nielsen, O. H. and Martin, R. M., "Quantum-mechanical theory of stress and force," *Phys. Rev. B* 32(6): 3780–3791, 1985.

Rogers, C. and Rappe, A., "Geometric formulation of quantum stress fields," *Phys. Rev. B* 65:224117, 2002.

Exercises

H.1 Show that $\sum_{i=1}^{N} |\nabla \phi_i(\mathbf{r})|^2 = -\sum_{i=1}^{N} \phi_i(\mathbf{r}) \nabla^2 \phi_i(\mathbf{r})$ follows from the requirement $\sum_{i=1}^{N} |\phi_i(\mathbf{r})|^2 = 1$ at all $(\mathbf{r})$.
Hint: use Eqs. (H.10) and (H.11).

H.2 Show that the excess fermion kinetic energy density in Eq. (H.13) follows from Eq. (H.11). The previous problem may be helpful.

H.3 Show that Eq. (H.17) follows from the definitions of the terms in the energy density given before and the Kohn–Sham equation for the eigenvalues.

H.4 Show that the formulas for the stress given by Nielsen and Martin [162] in their eqs. (33) and (34) can be written in the form of Eqs. (H.20), using definition (H.10).